Handbook
of
HEATING,
VENTILATION,
and
AIR CONDITIONING

The Mechanical Engineering Handbook Series

Series Editor
Frank Kreith
Consulting Engineer

Published Titles

Forthcoming Titles

Handbook
of
HEATING,
VENTILATION,
and
AIR CONDITIONING

Edited by
Jan F. Kreider, Ph.D., P.E.

CRC Press
Taylor & Francis Group
Boca Raton London New York

CRC Press is an imprint of the
Taylor & Francis Group, an **informa** business
A TAYLOR & FRANCIS BOOK

CRC Press
Taylor & Francis Group
6000 Broken Sound Parkway NW, Suite 300
Boca Raton, FL 33487-2742

First issued in paperback 2019

ISBN-13: 978-0-8493-9584-0 (hbk)
ISBN-13: 978-0-367-39772-2 (pbk)

Library of Congress Cataloging-in-Publication Data

Handbook of heating, ventilation, and air conditioning / edited by Jan F. Kreider.
    p. cm.
   Includes bibliographical references and index.
   ISBN 0-8493-9584-4 (alk. paper)
   1. Heating—Handbooks, manuals, etc. 2. Ventilation—Handbooks, manuals, etc. 3. Air conditioning—Handbooks, manuals, etc. I. Title.

TH7225 .K74 2000
697—dc21

00-064673
CIP

Credits: Figures 7.1.1 and 7.1.2 — With permission from ASHRAE.

Library of Congress Card Number 00-064673

Visit the Taylor & Francis Web site at
http://www.taylorandfrancis.com

and the CRC Press Web site at
http://www.crcpress.com

Dedication

To the HVAC engineers of the 21st century who will set new standards for efficient and sophisticated design of our buildings.

Preface

During the past 20 years, design and operation of the comfort systems for buildings have been transformed because of energy conservation imperatives, the use of computer-based design aids, and major advances in intelligent management systems for buildings. In the 1970s, rules of thumb were widely used by designers. Today, a strong analytical basis for the design synthesis process is standard procedure. This handbook describes the latest methods for design and operation of new and existing buildings. In addition, the principles of life cycle economics are used routinely in design selections and tradeoffs. The information in this handbook is presented in a practical way that building systems engineers will find useful.

The book is divided into eight sections:

1. Introduction to the buildings sector
2. Fundamentals
3. Economic aspects of buildings
4. HVAC equipment and systems
5. Controls
6. HVAC design calculations
7. Operation and maintenance
8. Appendices

Because of ongoing and rapid change in the HVAC industry, new material will be developed prior to the standard handbook revision cycle. By link to the CRC Web site, the author will be periodically posting new material that owners of the handbook can access.

Jan F. Kreider, Ph.D., P.E.
Boulder, Colorado

Editor

Photo by: Renée Azerbegi

Jan F. Kreider, Ph.D., P.E. is Professor of Engineering and Founding Director of the University of Colorado's (CU) Joint Center for Energy Management. He is co-founder of the Building Systems Program at CU and has written ten books on building systems, alternative energy, and other energy related topics, in addition to more than 200 technical papers. For ten years he was a technical editor of the ASME Transactions.

During the past decade Dr. Kreider has directed more than $10,000,000 in energy-related research and development. His work on thermal analysis of buildings, building performance monitoring, building diagnostics, and renewable energy-research is known all over the world. Among his major accomplishments with his colleagues are the first applications of neural networks to building control, energy management and systems identification, and of applied artificial intelligence approaches for building design and operation. He also has worked for many years to involve women in the graduate program that he founded. More than 20 women have graduated with advanced degrees in his program.

Dr. Kreider has assisted governments and universities worldwide in establishing renewable energy and energy efficiency programs and projects since the 1970s. He is a fellow of the American Society of Mechanical Engineers and a registered professional engineer and member of several honorary and professional societies. Dr. Kreider recently received ASHRAE's E.K. Campbell Award of Merit and the Distinguished Engineering Alumnus Award, the College's highest honor.

Dr. Kreider earned his B.S. degree (magna cum laude) from Case Institute of Technology, and his M.S. and Ph.D. degrees in engineering from the University of Colorado. He was employed by General Motors for several years in the design and testing of automotive heating and air conditioning systems.

Contributors

Anthony F. Armor
Electric Power Research Institute
Palo Alto, California

Peter Armstrong
Pacific Northwest National
  Laboratory
Richland, Washington

James B. Bradford
Schiller Associates, Inc.
Boulder, Colorado

Michael R. Brambley
Pacific Northwest National
  Laboratory
Richland, Washington

James Braun
Dept. of Mechanical Engineering
Purdue University
West Lafayette, Indiana

John A. Bryant
Dept. of Construction
Texas A&M University
College Station, Texas

David E. Claridge
Dept. of Mechanical Engineering
Texas A&M University
College Station, Texas

Peter S. Curtiss
Kreider & Associates, LLC
Boulder, Colorado

Ellen M. Franconi
Schiller Associates, Inc.
Boulder, Colorado

Jeffrey S. Haberl
Dept. of Architecture
Texas A&M University
College Station, Texas

Vahab Hassani
Thermal Systems Branch
National Renewable Energy
  Laboratory
Golden, Colorado

Steve Hauser
Pacific Northwest National
  Laboratory
Richland, Washington

Joe Huang
Lawrence Berkeley Laboratory
Berkeley, California

David Jump
Lawrence Berkeley Laboratory
Berkeley, California

Srinivas Katipamula
Pacific Northwest National
  Laboratory
Richland, Washington

Michael Kintner-Meyer
Pacific Northwest National
  Laboratory
Richland, Washington

Moncef Krarti
CEAE Department
University of Colorado
Boulder, Colorado

Jan F. Kreider
Kreider & Associates, LLC
Boulder, Colorado

Mingsheng Liu
College of Engineering
University of Nebraska
Lincoln, Nebraska

Paul Norton
National Renewable Energy
  Laboratory
Golden, Colorado

Dennis L. O'Neal
Dept. of Mechanical Engineering
Texas A&M University
College Station, Texas

Robert G. Pratt
Pacific Northwest National
  Laboratory
Richland, Washington

Ari Rabl
École des Mines de Paris and
University of Colorado
Boulder, Colorado

T. Agami Reddy
Civil and Architectural Engineering
Drexel University
Philadelphia, Pennsylvania

Max Sherman
Lawrence Berkeley Laboratory
Berkeley, California

Table of Contents

Section 7 Operation and Maintenance

Section 8 Appendices

Paul Norton

1

Introduction to the Buildings Sector

Jan F. Kreider
Kreider & Associates, LLC

Introduction

Buildings account for the largest sector of the U.S. economy. Construction, operation, and investment in buildings are industries to which every person is exposed daily. One of the major expenditures in the life cycle of a building is the operation of its space conditioning systems — heating, ventilation, and air conditioning (HVAC) — dwarfing the initial cost of these systems or of even the entire building itself. Therefore, it is important to use the best, most current knowledge from the design phase onward through the building life cycle to minimize cost while maintaining a productive and comfortable indoor environment.

HVAC systems are energy conversion systems — electricity is converted to cooling or natural gas is converted to heat. Because it is important to understand from the outset the nature of energy demands placed on HVAC systems, that subject is discussed immediately below. The chapter closes with a short outline of the rest of the book with its coverage of HVAC design, commissioning, operation, and problem diagnosis.

1.1 Energy Use Patterns in Buildings in the U.S.

It is instructive to examine building energy use, sector by sector, to get an idea of the numbers and to clarify the differences between large and small buildings as well as between industrial and office buildings. The next several sections discuss each.

1.1.1 Commercial Buildings

In 1997, there were 4.6 million commercial buildings, occupying 58.8 billion square feet of floor space (PNNL, 1997). These buildings consumed 126.5 thousand Btu of delivered energy use (or 252.4 thousand Btu of primary energy) per square foot of space. Figure 1.1 shows that of the four main census regions, the South contains the highest percentage of commercial buildings, 38%, and the Northeast contains the least, 16%.

Commercial Buildings Disaggregated by Floor Space

Sixty percent of U.S. commercial buildings range between 5,000 and 100,000 square feet, 82% range between 1,000 and 200,000 square feet. The size class with the largest membership is the 10,000–25,000 square foot range. Table 1.1 shows the size distribution in the U.S.

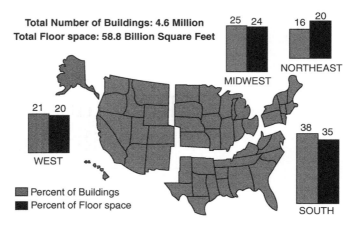

FIGURE 1.1 Commercial building geographical distribution. (From the *1995 Commercial Buildings Energy Consumption Survey.*)

TABLE 1.1   Size Distribution of U.S. Commercial Building Space

Commercial Building Size as of 1995 (percent of total floor space)	
Square Foot Range	Percent
1,001 to 5,000	10.80%
5,001 to 10,000	12.80%
10,001 to 25,000	19.80%
25,001 to 50,000	13.10%
50,001 to 100,000	13.60%
100,001 to 200,000	11.50%
200,001 to 500,000	9.40%
Over 500,000	9.00%
	100%

Commercial Energy Consumption and Intensity by Square Footage (1995)

Total consumption is fairly evenly distributed across building size categories; only the largest size category (over 500,000 square feet per building) showed a significant difference from any of the other categories. Buildings in the 10,001–25,000 square feet per building category have the lowest energy intensity of all categories.

Commercial Buildings Disaggregated by Building Type and Floor Space

The usage to which building space is put is a key influence on the type and amount of energy needed. Of the total square footage of commercial office space, 67% is used for mercantile and service, offices, warehouses and storage places, or educational facilities. The average square footage for all building types ranges between 1,001 and 25,000 square feet. The largest building types, between 20,000 and 25,000 square feet, are lodging and health care facilities. Medium sized building types, between 10,000 and 20,000 square feet, are public order and safety, offices, mercantile and service, and public assembly. Small building types, less than 10,000 square feet, include warehouse and storage facilities, education facilities, food service, and sales. Table 1.2 summarizes sector sizes and typical floor sizes.

Commercial End-Use Consumption

Mercantile and service, and office buildings consume almost 40% of total commercial energy, in terms of Btu per square foot. Education and health care facilities, lodging, and public assemblies also consume

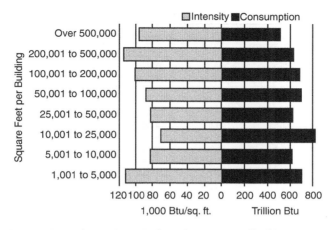

FIGURE 1.2 Energy consumption and usage intensity for eight commercial building size categories. (From the *1995 Commercial Buildings Energy Consumption Survey*.)

TABLE 1.2 Commercial Building Sector Size and Typical Floor Area

1995 Average and Percent of Commercial Building by Principal Building Type (1)

Building Type	Floor Space (%)	Average Floor Space/Building (SF)
Mercantile and Service	22%	11260
Office	18%	12870
Warehouse/Storage	14%	6670
Education	13%	1770
Public Assembly	7%	12110
Lodging	6%	22900
Health Care	4%	22220
Food Service	2%	4750
Food Sales	1%	4690
Public Order and Safety	2%	14610
Vacant (2)	9%	18480
Other	2%	—

a large amount of energy, making up another 40% of total commercial energy consumption. Table 1.3 summarizes the energy use intensities for the 12 most important categories.

End Use Consumption by Task

Finally, one must know the end use category — space heating, cooling, water heating, and lighting. Space heating and lighting are generally the largest energy loads in commercial office buildings. In 1995, energy consumed for lighting accounted for 31% of commercial energy loads. Space heating consumed 22%, and space cooling consumed 15% of commercial energy loads. On average, water heating is not high at 7%; actual load varies greatly according to building category. Health care facilities and lodging are unique in their high water heating loads; however, offices, mercantile and service facilities, and warehouses require minimal hot water. Figure 1.3 shows the distribution of energy end use by sector for 1995. Another way of considering the data in Figure 1.3 is to consider the end uses aggregated over all buildings but further disaggregated over the nine main end uses in commercial buildings. Figure 1.4 shows the data in this way.

Commercial Energy Consumption and Intensity by Principal Building Activity (1995)

Commercial buildings were distributed unevenly across the categories of most major building characteristics. For example, in 1995, 63.0 percent of all buildings and 67.1 percent of all floor space were in

TABLE 1.3   End Use Consumption Intensity by Building Category

1995 Commercial Delivered End-Use Energy Consumption Intensities by Principal Building Type[1] (1000 Btu/SF)

Building Type	Space Heating	Space Cooling	Water Heating	Lighting	Total[2]	Percent of Total Consumption
Office	24.3	9.1	8.7	28.1	97.2	19%
Mercantile and Service	30.6	5.8	5.1	23.4	76.4	18%
Education	32.8	4.8	17.4	15.8	79.3	12%
Health Care	55.2	9.9	63	39.3	240.4	11%
Lodging	22.7	8.1	51.4	23.2	127.3	9%
Public Assembly	53.6	6.3	17.5	21.9	113.7	8%
Food Service	30.9	19.5	27.5	37	245.5	6%
Warehouse and Storage	15.7	0.9	2	9.8	38.3	6%
Food Sales	27.5	13.4	9.1	33.9	213.5	3%
Vacant[3]	38	1.4	5.5	4.5	30.1	3%
Public Order and Safety	27.8	6.1	23.4	16.4	97.2	2%
Other[4]	59.6	9.3	15.3	26.7	172.2	3%
All Buildings	29	6	13.8	20.4	90.5	100%

Notes: [1] Parking garages and commercial buildings on multibuilding manufacturing facilities are excluded from CBECS 1995.
[2] Includes all end-uses.
[3] Includes vacant and religious worship.
[4] Includes mixed uses, hangars, crematoriums, laboratories, and other.
Source: EIA, Commercial Building Energy Consumption and Expenditures 1995, April 1998, Table EU-2, p. 311.

four building types: office, mercantile and service, education, and warehouse. Total energy consumption also varied by building type. Three of these — health care, food service, and food sales — had higher energy intensity than the average of 90.5 thousand Btu per square foot for all commercial buildings. Figure 1.5 shows the 13 principal building types and their total consumption and intensity.

Commercial Building Energy Consumption by Fuel Type

Five principal energy types are used in U.S. commercial buildings:

Natural gas
Fuel oil
Liquefied petroleum gas (LPG)
Other and renewables
On-site electric

Table 1.4 shows the relation between end use type in Figure 1.5 and the corresponding energy sources. Space heating, lighting, and water heating are the three largest consumers of energy. Natural gas and electricity directly competed in three of the major end uses — space heating, water heating, and cooking. In each of these three, natural gas consumption greatly exceeded electricity consumption.

Table 1.5 shows expected commercial sector energy use growth in the U.S.

1.1.2   Industrial Processes and Buildings

The industrial sector consists of more than three million establishments engaged in manufacturing, agriculture, forestry, fishing, construction, and mining. In 1997, these buildings occupied 15.5 billion square feet of floor space and 37% (34.8 quadrillion Btus) of total U.S. primary energy consumption.

After the transportation sector, the manufacturing sector consumes the most energy in the U.S. Of the 37% of primary energy consumption in the industrial sector in 1997, 33% was used for manufacturing purposes and 4% was used for nonmanufacturing purposes. Thus, manufacturing establishments consume the majority of the energy in the industrial sector even though they are far outnumbered by nonmanufacturing establishments. Because there is a lack of information regarding nonmanufacturing

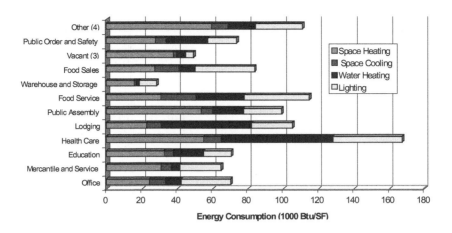

FIGURE 1.3 End use categories for commercial buildings.

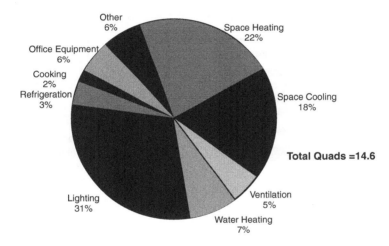

FIGURE 1.4 Commercial building energy end uses aggregated over all building types.

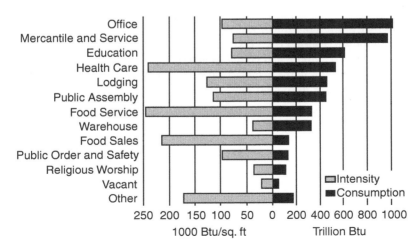

FIGURE 1.5 Energy usage and usage intensity by building type. (From the *1995 Commercial Buildings Energy Consumption Survey*.)

TABLE 1.4 Fuel Type Usage in Commercial Buildings

	Natural Gas	Fuel Oil (2)	LPG Fuel (3)	Other	Renw. En. (4)	Site Electric	Site Total	Site Percent	Primary Total
Space Heating (8)	1.58	0.37		0.11		0.16	2.22	29.10%	0.53
Space Cooling (7)	0.02					0.34	0.35	4.60%	1.08
Ventilation						0.17	0.17	2.20%	0.53
Water Heating (8)	0.75	0.07			0.02	0.09	0.93	12.10%	0.29
Lighting						1.22	1.22	15.90%	3.9
Refrigeration						0.18	0.16	2.40%	0.59
Cooking	0.23					0.02	0.25	3.30%	0.07
Office Equipment						0.4	0.4	5.30%	1.3
Other (9)	0.21	0.04	0.08	0.03	0	0.25	0.61	8.00%	0.81
Miscellaneous (10)	0.59	0.12				0.61	1.32	17.20%	1.95
Total	3.37	0.6	0.08	0.14	0.02	3.44	7.65	100%	11.03

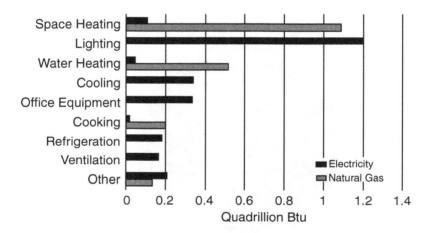

FIGURE 1.6 Gas and electric consumption in commercial buildings by task. (From the *1995 Commercial Buildings Energy Consumption Survey.*)

sectors and the majority of energy is consumed in manufacturing, the manufacturing sector is the main focus in this section.

Standard industrial classification (SIC) groups are established according to their primary economic activity. Each major industrial group is assigned a two-digit SIC code. The SIC system divides manufacturing into 20 major industry groups and nonmanufacturing into 12 major industry groups. In 1991, six of the 20 major industry groups in the manufacturing sector accounted for 88% of energy consumption for all purposes and for 40% of the output value for manufacturing:

1. Food and kindred products
2. Paper and allied products
3. Chemical and allied products
4. Petroleum and coal products
5. Stone, clay, and glass products
6. Primary metals

Table 1.6 summarizes the key characteristics of the energy using SIC categories with an overview of each. Table 1.7 shows the floor space inventory by SIC.

TABLE 1.5   Expected Future Consumption Trends for Commercial Buildings

Commercial Primary Energy Consumption by Year and Fuel Type (quads and percents of total)[3]

Year	Natural Gas		Petroleum[1]		Coal		Renewable[2]		Electricity		TOTAL	Growth Rate, 1980-Year
1980	267	25%	1.29	12%	0.09	1%	NA		6.55	62%	10.59	—
1990	27	21%	0.91	7%	0.09	1%	NA		9.12	71%	12.82	1.90%
1997	337	22%	0.73	5%	0.08	1%	0.02	0%	11.03	72%	15.24	2.20%
2000	355	22%	0.6	4%	0.09	1%	0.03	0%	11.76	73%	16.02	2.10%
2010	384	22%	0.57	3%	0.1	1%	0.03	0%	12.73	74%	17.27	1.60%
2020	4	22%	0.55	3%	0.1	1%	0.04	0%	13.4	74%	18.08	1.30%

Notes: [1] Petroleum induces distillate and residual fuels, liquid petroleum gas, kerosene, and motor gasoline.
[2] Includes site marketed and nonmarketed renewable energy.
[3] 1997 site-to-source electricity conversion = 321.
Sources: EIA, State Energy Data Report 1996, Feb. 1999, Table 13, p. 28 for 1980 and 1990; EIA, AEO 1999, Dec. 1998, Table A2, p. 113-115 for 1997-2020 and Table A18, p. 135 for nonmarketed renewable energy.

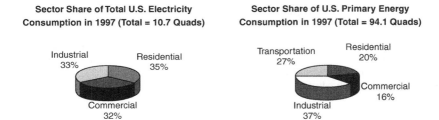

FIGURE 1.7   Primary energy and electrical consumption in the U.S. (1997).

Of a total of 15.5 billion square feet of manufacturing space, 17% is used for office space, and 83% is used for nonoffice space. Six groups account for 50% of this space: industrial machinery, food, fabricated metals, primary metals, lumber, and transportation (PNNL, 1997).

Manufacturers use energy in two major ways:

- To produce heat and power and to generate electricity
- As raw material input to the manufacturing process or for some other purpose

Three general measures of energy consumption are used by the U.S. Energy Information Administration (EIA). According to its 1991 data, the amount of total site consumption of energy for all purposes was 20.3 quadrillion Btu. About two thirds (13.9 quadrillion Btu) of this was used to produce heat and power and to generate electricity, with about one third (6.4 quadrillion Btu) consumed as raw material and feedstocks. Figure 1.8 shows the relative energy use for the energy consuming SIC sectors.

Energy Use by Standard Industrial Classification

Energy end uses for industry are similar to those for commercial buildings although the magnitudes are clearly different. Heating consumes 69% of delivered energy (45% of primary energy usage). Lighting is the second largest end use with 15% of delivered energy (27% of primary energy usage). Finally, ventilation and cooling account for 8% each.

Industrial Consumption by Fuel Type

As with commercial buildings, a variety of fuels are used in industry. Petroleum and natural gas far exceed energy consumption by any other source in the manufacturing sectors. Figure 1.9 indicates the fuel mix characteristics.

TABLE 1.6 General Characteristics of Industrial Energy Consumption SIC

Standard Industrial Code	Major Industry Group	Description
		High-Energy Consumers
20	Food and kindred products	This group converts raw materials into finished goods primarily
26	Paper and allied products	by chemical (not physical) means. Heat is essential to their
28	Chemicals and allied products	production, and steam provides much of the heat. Natural gas,
29	Petroleum and coal products	byproduct and waste fuels are the largest sources of energy for
32	Stone, clay, and glass products	this group. All, except food and kindred products, are the most
33	Primary metal industries	energy-intensive industries.
		High Value-Added Consumers
34	Fabricated metal products	This group produces high value-added transportation vehicles,
35	Industrial machinery and equipment	industrial machinery, electrical equipment, instruments, and
36	Electronic and other electric equipment	miscellaneous equipment. The primary end uses are motor-
37	Transportation equipment	driven physical conversion of materials (cutting, forming,
38	Instruments and related products	assembly) and heat treating, drying, and bonding. Natural gas is
39	Miscellaneous manufacturing industries	the principal energy source.
		Low-Energy Consumers
21	Tobacco manufactures	This group is the low energy-consuming sector and represents a
22	Textile mill products	combination of end-use requirements. Motor drive is one of the
23	Apparel and other textile products	key end uses.
24	Lumber and wood products	
25	Furniture and fixtures	
27	Printing and publishing	
30	Rubber and miscellaneous plastics	
31	Leather and leather products	

Source: Energy Information Administration, Office of Energy Markets and End Use, Manufacturing Consumption of Energy 1991, DOE/EIA-0512(91).

TABLE 1.7 Industrial Building Floor Area Distribution

		1991 Industrial Building Floor Space (10^6 square feet)		
SIC	Manufacturing Industry	Office Floor Space	Nonoffice Floor Space	Total Floor Space
20	Food	203	1207	1410
21	Tobacco	6	51	56
22	Textiles	42	581	623
23	Apparel	73	451	523
24	Lumber	53	1135	1187
25	Furniture	49	521	569
26	Paper	72	827	899
27	Printing	351	477	827
28	Chemical	185	714	899
29	Refining	20	105	125
30	Rubber	97	768	865
31	Leather	9	44	53
32	Stone, Clay	57	808	864
33	Primary Metals	81	1121	1202
34	Fabricated Metals	182	1175	1357
35	Industrial Machinery	337	1149	1485
36	Electronic Equipment	266	629	894
37	Transportation	289	776	1065
38	Instruments	225	170	395
39	Misc. Manufacturing	52	190	242
	Total	2,641	12,898	15,539

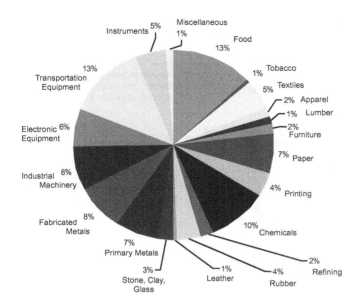

FIGURE 1.8 Energy use by SIC category.

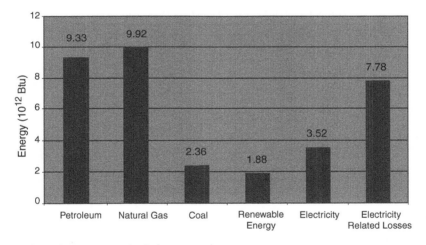

FIGURE 1.9 Industrial consumption by fuel type.

1.1.3   Residential Buildings

Although residential buildings are not often equipped with engineered HVAC systems, it is important to understand usage by this sector because it is large and many of the design and operation principles for large buildings also apply to small ones. The following data summarize residential energy use in the U.S. Figure 1.10 shows energy use by building type.

Residential Sector Overview

In 1993, there were 101.3 million households, or 76.5 million buildings with an average of 2.6 people per household. The households consisted of 69% single-family, 25% multi-family, and 6% mobile homes. These buildings consumed 107.8 million Btu of delivered energy (or 187.5 million Btu of primary energy) per household.

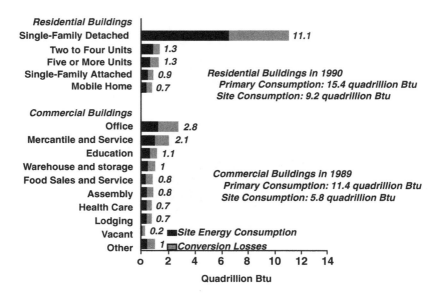

FIGURE 1.10   Comparison of commercial and residential sector energy use.

TABLE 1.8   U.S. Residential Buildings Disaggregated by Size

Household Size in Heated Floor Space as of 1995	
Square Foot Range	Percent
Fewer than 599	7.8%
600 to 999	22.6%
1,000 to 1,599	28.8%
1,600 to 1,999	12.8%
2,000 to 2,399	10.0%
2,400 to 2,999	8.5%
3,000 or more	9.6%
	100%

   More than 50% of all residences range between 600 and 1,600 square feet; 23% are between 1,600 and 2,400 square feet, and 29% are in the 1,000 to 1,600 square feet range as shown in Table 1.8.

Residential Energy Consumption Intensity

Table 1.9 and Table 1.10 summarize residential fuel utilization. Natural gas and electricity are the key residential energy sources. Table 1.11 shows expected growth through the year 2020.

1.2   What Follows

In order to cover all topics affecting the design and operation of HVAC systems in modern buildings, this book is divided into eight sections as follows:

1. Introduction to the Buildings Sector
2. Fundamentals
    2.1 Thermodynamics Heat Transfer and Fluid Mechanics Basics
    2.2 Psychrometrics and Comfort

TABLE 1.9 Energy Consumption Intensities by Ownership of Unit

	1993 Residential Delivered Energy Consumption Intensities by Ownership of Unit			
Ownership	Per Square Foot (10^3 Btu)	Per Household (10^6 Btu)	Per Household Members (10^6 Btu)	Percent of Total Consumption
Owned	52.0	118.5	44.0	75%
Rented	67.0	75.2	31.0	25%
- Public Housing	69.0	58.2	27.0	2%
- Nonpublic Housing	67.0	77.2	31.0	23%

Source: EIA, Household Energy Consumption and Expenditures 1993, Oct. 1995, Table 5.1, p. 37-38.

Table 1.10 Residential End-Use Consumption by Fuel Type and by End Use

	1997 Residential Energy End-Use Splits by Fuel Type (quads)					Site			Primary		
	Natural Gas	Fuel Oil	LPG Fuel	Other	Renw. En. (3)	Electric	Total	Percent	Electric	Total	Percent
Space Heating	3.58	0.84	0.32	0.15	0.61	0.50	6.00	54.8%	1.61	7.10	37.3%
Space Cooling	0.00					0.54	0.54	4.9%	1.72	1.72	9.1%
Water Heating	1.27	0.10	0.07		0.01	0.39	1.83	16.8%	1.24	2.69	14.2%
Lighting						0.40	0.40	3.6%	1.27	1.27	6.7%
White Goods	0.05					0.78	0.82	7.5%	2.49	2.54	13.4%
Cooking	0.16		0.03			0.23	0.42	3.9%	0.74	0.93	4.9%
Electronics						0.27	0.27	2.5%	0.86	0.86	4.5%
Motors						0.05	0.05	0.5%	0.18	0.18	0.9%
Heating Appliances						0.10	0.10	0.9%	0.31	0.31	1.6%
Other	0.09	0.00	0.01				0.10	0.9%		0.10	0.5%
Miscellaneous						0.41	0.41	3.7%	1.30	1.30	6.9%
Total	5.15	0.94	0.43	0.15	0.62	3.66	10.94	100%	11.73	19.01	100%

Table 1.11 Expected Growth in Residential Energy Use

	Residential Primary Energy Consumption by Year and Fuel Type (quads and percents of total)											Growth Rate,
Year	Natural Gas		Petroleum[1]		Coal		Renewable[2]		Electricity		TOTAL		1980-Year
1980	4.86	32%	1.75	12%	0.06	0%	NA[3]		8.41	56%	15.069	100%	—
1990	4.52	27%	1.27	8%	0.06	0%	0.63	4%	10.05	61%	16.53	100%	0.9%
1997	5.15	27%	1.47	8%	0.06	0%	0.62	3%	11.73	62%	19.01	100%	1.4%
2000	5.21	26%	1.38	7%	0.06	0%	0.62	3%	12.79	64%	20.06	100%	1.4%
2010	5.52	26%	1.23	6%	0.05	0%	0.65	3%	13.68	65%	21.13	100%	1.1%
2020	5.94	26%	1.12	5%	0.05	0%	0.70	3%	15.09	66%	22.90	100%	1.1%

Notes: [1] Petroleum includes distillate and residual fuels, liquefied petroleum gas, kerosene, and motor gasoline.
[2] Includes site marketed and non-marketed renewable energy.
[3] 1980 Renewables are estimated at 1.00 quads.
Sources: EIA, State Energy Data Report 1996, Feb. 1999, Tables 12-15, p. 22-25 for 1980 and 1990; EIA, AEO 1999, Dec. 1998, Table A2, p. 113-115 for 1997-2020 consumption and Table A18, p. 135 for nonmarketed renewable energy.

The book is indexed for all detailed topics, and adequate cross-references among the chapters have been included. The appendices include the nomenclature and selected lookup tables.

References

PNNL (1997). An Analysis of Buildings-Related Energy Use in Manufacturing, *PNNL*-11499, April.

Energy Information Administration (EIA, 1995). *1995 Commercial Buildings Energy Consumption Survey.*

2

Fundamentals

Vahab Hassani
National Renewable Energy Laboratory

Steve Hauser
Pacific Northwest National Laboratory

T. Agami Reddy
Drexel University

2.1 Thermodynamics Heat Transfer and Fluid Mechanics Basics

Vahab Hassani and Steve Hauser

Design and analysis of energy conversion systems require an in-depth understanding of basic principles of thermodynamics, heat transfer, and fluid mechanics. **Thermodynamics** is that branch of engineering science that describes the relationship and interaction between a system and its surroundings. This interaction usually occurs as a transfer of energy, mass, or momentum between a system and its surroundings. Thermodynamic laws are usually used to predict the changes that occur in a system when moving from one equilibrium state to another. The science of **heat transfer** complements the thermodynamic science by providing additional information about the energy that crosses a system's boundaries. Heat-transfer laws provide information about the mechanism of transfer of energy as heat and provide necessary correlations for calculating the rate of transfer of energy as heat. The science of **fluid mechanics**, one of the most basic engineering sciences, provides governing laws for fluid motion and conditions influencing that motion. The governing laws of fluid mechanics have been developed through a knowledge of fluid properties, thermodynamic laws, basic laws of mechanics, and experimentation.

In this chapter, we will focus on the basic principles of thermodynamics, heat transfer, and fluid mechanics that an engineer needs to know to analyze or design an HVAC system. Because of space limitations, our discussion of important physical concepts will not involve detailed mathematical derivations and proofs of concepts. However, we will provide appropriate references for those readers interested in obtaining more detail about the subjects covered in this chapter. Most of the material presented here is accompanied by examples that we hope will lead to better understanding of the concepts.

2.1.1 Thermodynamics

During a typical day, everyone deals with various engineering systems such as automobiles, refrigerators, microwaves, and dishwashers. Each engineering system consists of several components, and a system's optimal performance depends on each individual component's performance and interaction with other components. In most cases, the interaction between various components of a system occurs in the form of energy transfer or mass transfer. Thermodynamics is an engineering science that provides governing laws that describe energy transfer from one form to another in an engineering system. In this chapter, the basic laws of thermodynamics and their application for energy conversion systems are covered in the following four sections. The efficiency of the thermodynamic cycles and explanations of some advanced thermodynamic systems are presented in the succeeding two sections. Several examples have been presented to illustrate the application of concepts covered here. Because of the importance of moist air HVAC processes, these are treated in Chapter 2.2.

Energy and the First Law of Thermodynamics

In performing engineering thermodynamic analysis, we must define the *system* under consideration. After properly identifying a thermodynamic system, everything else around the system becomes that system's *environment*. Of interest to engineers and scientists is the *interaction* between the system and its environment.

In thermodynamic analysis, systems can either consist of specified matter (**controlled mass**, CM) or specified space (**control volume**, CV). In a control-mass system, energy—but not mass—can cross the system boundaries while the system is going through a thermodynamic process. Control-mass systems may be called **closed systems** because no mass can cross their boundary. On the other hand, in a control-volume system—also referred to as an **open system**—both energy and matter can cross the system boundaries. The shape and size of CVs need not necessarily be constant and fixed; however, in this chapter, we will assume that the CVs are of fixed shape and size. Another system that should be defined here is an **isolated system**, which is a system where no mass or energy crosses its boundaries.

The energy of a system consists of three components: kinetic energy, potential energy, and internal energy. The **kinetic** and **potential energy** of a system are macroscopically observable. **Internal energy** is associated with random and disorganized aspects of molecules of a system and is not directly observable. In thermodynamic analysis of systems, the energy of the whole system can be obtained by adding the individual energy components.

Conservation of Energy — The First Law of Thermodynamics

The First Law of Thermodynamics states that energy is conserved: it cannot be created or destroyed, but can change from one form to another. The energy of a closed system can be expressed as

$$E = me + \frac{mu^2}{2g_2} + \frac{mgz}{g_c},\qquad\qquad (2.1.1)$$

where E is the total energy of the system, e is its internal energy per unit mass, and the last two terms are the kinetic energy and potential energy of the system, respectively. The proportionality constant g_c is defined in the nomenclature (listed at the end of this chapter) and is discussed in the text following Eq. (2.1.73). When a system undergoes changes, the energy change within the system can be expressed by a general form of the energy-balance equation:

Energy stored   =   Energy entering   −   Energy leaving   +   Energy generated
in the system   the system   the system   in the system
                                                                                                    (e.g., chemical reactions)

For example, consider the geothermal-based heat pump shown in Figure 2.1.1. In this heat pump, a working fluid (R-22, a common refrigerant used with geothermal heat pumps, which is gaseous at room temperature and pressure) is sealed in a closed loop and is used as the transport medium for energy. Figure 2.1.2 presents a simple thermodynamic cycle for a heat pump (heating mode) and an associated pressure-enthalpy (p-h) diagram. The saturated vapor and liquid lines are shown in Figure 2.1.2, and the region between these two lines is referred to as the wet region, where vapor and liquid coexist. The relative quantities of liquid and vapor in the mixture, known as the quality of the mixture (x), is usually used to denote the state of the mixture. The **quality of a mixture** is defined as the ratio of the mass of vapor to the mass of the mixture. For example, in 1 kg of mixture with quality x, there are x kg of vapor

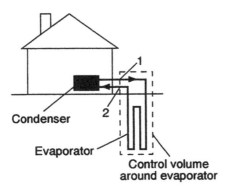

FIGURE 2.1.1 Geothermal-based (ground-source) heat pump.

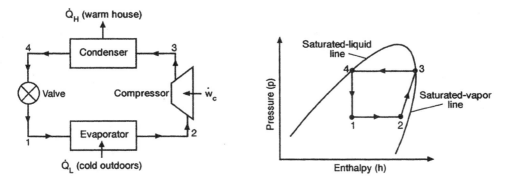

FIGURE 2.1.2 Thermodynamic cycle and *p-h* diagram for heat pump (heating mode).

and $(1 - x)$ kg of liquid. Figure 2.1.2 shows that the working fluid leaving the evaporator (point 2) has a higher quality than working fluid entering the evaporator (point 1). The working fluid in Figure 2.1.2 is circulated through the closed loop and undergoes several phase changes. Within the evaporator, the working fluid absorbs heat from the surroundings (geothermal resource) and is vaporized. The low-pressure gas (point 2) is then directed into the compressor, where its pressure and temperature are increased by compression. The hot compressed gas (point 3) is then passed through the condenser, where it loses heat to the surroundings (heating up the house). The cool working fluid exiting the condenser is a high-pressure liquid (point 4), which then passes through an expansion device or valve to reduce its pressure to that of the evaporator (underground loop).

Specifically, consider the flow of the working fluid in Figure 2.1.1 from point 1 to point 2 through the system shown within the dashed rectangle. Mass can enter and exit this control-volume system. In flowing from point 1 to 2, the working fluid goes through the evaporator (see Figure 2.1.2). Assuming no accumulation of mass or energy, the First Law of Thermodynamics can be written as

$$e_2 + \frac{u_2^2}{2g_c} + \frac{gz_2}{g_c} + p_2 v_2 = e_1 + \frac{u_1^2}{2g_c} + \frac{gz_1}{g_c} + p_1 v_1 + \frac{\dot{Q} - \dot{w}}{\dot{m}}, \qquad (2.1.2)$$

where $\dot{m}$ is the mass-flow rate of the working fluid, $\dot{Q}$ is the rate of heat absorbed by the working fluid, $\dot{w}$ is the rate of work done on the surroundings, v is the specific volume of the fluid, p is the pressure, and the subscripts 1 and 2 refer to points 1 and 2. A mass-flow energy-transport term, pv, appears in Eq. (2.1.2) as a result of our choice of control-volume system. The terms e and pv can be combined into a single term called **specific enthalpy**, $h = e + pv$, and Eq. (2.1.2) then reduces to

$$\frac{1}{2g_c}\left(u_2^2 - u_1^2\right) + \frac{g}{g_c}\Delta z_{2-1} + \Delta h_{2-1} = \frac{\dot{Q} - \dot{w}}{\dot{m}}. \tag{2.1.3}$$

For a constant-pressure process, the enthalpy change from temperature, T_1 to temperature T_2 can be expressed as

$$\Delta h_{2-1} = \int_{T_1}^{T_2} c_p \; dT = \bar{c}_p\left(T_2 - T_1\right), \tag{2.1.4}$$

where $\bar{c}_p$ is the mean specific heat at constant pressure.

Entropy and the Second Law of Thermodynamics

In many events, the state of an isolated system can change in a given direction, whereas the reverse process is impossible. For example, the reaction of oxygen and hydrogen will readily produce water, whereas the reverse reaction (electrolysis) cannot occur without some external help. Another example is that of adding milk to hot coffee. As soon as the milk is added to the coffee, the reverse action is impossible to achieve. These events are explained by the Second Law of Thermodynamics, which provides the necessary tools to rule out impossible processes by analyzing the events occurring around us with respect to time. Contrary to the First Law of Thermodynamics, the Second Law is sensitive to the *direction* of the process.

To better understand the second law of thermodynamics, we must introduce a thermodynamic property called **entropy** (symbolized by S, representing total entropy, and s, representing entropy per unit mass). The entropy of a system is simply a measure of the degree of molecular chaos or disorder at the microscopic level within a system.

The more disorganized a system is, the less energy is available to do useful work; in other words, energy is required to create order in a system. When a system goes through a thermodynamic process, the natural state of affairs dictates that entropy be produced by that process. In essence, the Second Law of Thermodynamics states that, in an isolated system, entropy can be produced, but it can never be destroyed.

$$\Delta S = S_{\text{final}} - S_{\text{initial}} \geq 0 \quad \text{for isolated system.} \tag{2.1.5}$$

Thermodynamic processes can be classified as reversible and irreversible processes. A **reversible process** is a process during which the net entropy of the system remains unchanged. A reversible process has equal chances of occurring in either a forward or backward direction because the net entropy remains unchanged. The absolute incremental entropy change for a closed system of fixed mass in a reversible process can be calculated from

$$dS = \frac{dQ}{T}, \tag{2.1.6}$$

where dS is the increase in entropy, dQ is the heat absorbed, and T is the absolute temperature. However, the net change in entropy for all the participating systems in the reversible process must equal zero; thus,

$$\Delta S = \sum_{\text{all systems}} dS = \sum_{\text{all systems}} \frac{dQ}{T} = 0. \tag{2.1.7}$$

We emphasize that most real processes are not reversible and the entropy of a real process is not usually conserved. Therefore, Eq. (2.1.6) can be written in a general form as

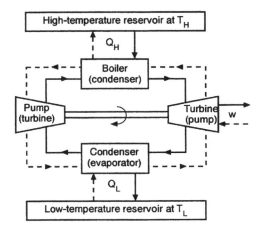

FIGURE 2.1.3 Principle of operation of a heat engine (solid lines and upper terms) and heat pump (dashed lines and lower terms in parentheses).

$$dS \geq \frac{dQ}{T}, \tag{2.1.8}$$

where the equality represents the reversible process. A reversible process in which $dQ = 0$ is called an **isentropic process**. It is obvious from Eq. (2.1.6) that for such processes, $dS = 0$, which means that no net change occurs in the entropy of the system or its surroundings.

Application of the Thermodynamic Laws to HVAC and Other Energy Conversion Systems

We can now employ these thermodynamic laws to analyze thermodynamic processes that occur in energy conversion systems. Among the most common energy conversion systems are heat engines and heat pumps. In Figure 2.1.3, the *solid* lines indicate the operating principle of a **heat engine**, where energy Q_H, is absorbed from a high-temperature thermal reservoir and is converted to work w by using a turbine, and the remainder, Q_L, is rejected to a low-temperature thermal reservoir. The **energy-conversion efficiency** of a heat engine is defined as

$$\eta_{\text{heat engine}} = \frac{\text{desired output energy}}{\text{required input energy}} = \frac{w}{Q_H}. \tag{2.1.9}$$

In the early 1800s, Nicholas Carnot showed that to achieve the maximum possible efficiency, the heat engine must be completely reversible (i.e., no entropy production, no thermal losses due to friction). Using Eq. (2.1.7), Carnot's heat engine should give

$$\Delta S = \frac{Q_H}{T_H} - \frac{Q_L}{T_L} = 0 \tag{2.1.10}$$

or

$$\frac{Q_H}{Q_L} = \frac{T_H}{T_L}. \tag{2.1.11}$$

An energy balance gives

$$w = Q_H - Q_L. \tag{2.1.12}$$

Therefore, the maximum possible efficiency is

$$\eta_{\text{rev}} = \frac{Q_H - Q_L}{Q_H} = 1 - \frac{Q_L}{Q_H} = 1 - \frac{T_L}{T_H}.$$ (2.1.13)

In real processes, however, due to entropy production, the efficiency is

$$\eta \leq 1 - \frac{T_L}{T_H}.$$ (2.1.14)

A **heat pump** is basically a heat engine with the reverse thermodynamic process. In heat pumps, work input allows for thermal energy transfer from a low-temperature reservoir to a high-temperature reservoir as shown in Figure 2.1.3 by *dashed* lines. Energy (heat), Q_L, is absorbed by a working fluid from a low-temperature reservoir (geothermal resource or solar collectors), then the energy content (temperature and pressure) of the working fluid is increased as a result of input work, *w*. The energy, Q_H, of the working fluid is then released to a high-temperature reservoir (e.g., a warm house). The efficiency of a heat pump is defined as

$$\eta_{\text{heat pump}} = \frac{\text{desired output energy}}{\text{required input energy}} = \frac{Q_H}{w} = \frac{Q_H}{Q_H - Q_L}.$$ (2.1.15)

The efficiency of a heat pump is often expressed as **coefficient of performance** (COP). The COP of a Carnot (or reversible) heat pump can be expressed as

$$COP = \frac{T_H}{T_H - T_L}.$$ (2.1.16)

Heat pumps are often used in HVAC systems to heat or cool buildings. Heat engines and heat pumps are broadly discussed by Sandord [1962], Reynolds and Perkins [1977], Wood [1982], Karlekar [1983], and Van Wylen and Sonntag [1986].

Efficiencies of Thermodynamic Cycles

To evaluate and compare various thermodynamic cycles (or systems), we further define and employ the term *efficiency*. The operating efficiency of a system reflects irreversibilities that exist in the system. To portray various deficiencies or irreversibilities of existing thermodynamic cycles, the following thermodynamic efficiency terms are most commonly considered.

$$\text{Mechanical efficiency} \quad \eta_m = \frac{w_{\text{act}}}{w_{\text{rev}}},$$ (2.1.17)

which is the ratio of the actual work produced by a system to that of the same system under reversible process. Note that the reversible process is not necessarily an adiabatic process (which would involve heat transfer across the boundaries of the system).

$$\text{Isentropic Efficiency} \quad \eta_s = \frac{w_{\text{act}}}{w_{\text{isent}}},$$ (2.1.18)

which is the ratio of actual work to the work done under an isentropic process.

$$\text{Relative efficiency} \quad \eta_r = \frac{w_{\text{rev}}}{w_{\text{isent}}},$$ (2.1.19)

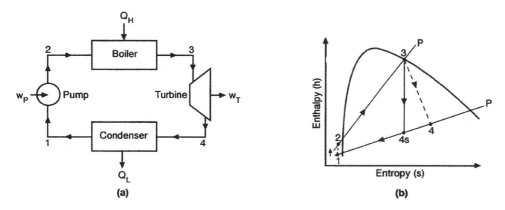

FIGURE 2.1.4 Typical Rankine cycle and its *h-s* diagram.

which is the ratio of reversible work to isentropic work.

$$\text{Thermal efficiency} \quad \eta_T = \frac{\dot{w}_{\text{out}}}{\dot{Q}_{\text{in}}}, \tag{2.1.20}$$

which is the ratio of the net power output to the input heat rate. Balmer [1990] gives a comprehensive discussion on the efficiency of thermodynamic cycles.

Some Thermodynamic Systems

The most common thermodynamic systems are those used by engineers in generating electricity for utilities and for heating or refrigeration/cooling purposes.

Modern power systems employ after Rankine cycles, and a typical **Rankine cycle** is shown in Figure 2.1.4(a). In this cycle, the working fluid is compressed by the pump and is sent to the boiler where heat Q_H is added to the working fluid, bringing it to a saturated (or superheated) vapor state. The vapor is then expanded through the turbine, generating shaft work. The mixture of vapor and liquid exiting the turbine is condensed by passing through the condenser. The fluid coming out of the condenser is then pumped to the boiler, closing the cycle. The enthalpy-entropy (h-s) diagram for the Rankine cycle is shown in Figure 2.1.4(b). The dashed line 3→4 in Figure 2.1.4(b) represents actual expansion of the steam through the turbine, whereas the solid line 3→4s represents an isentropic expansion through the turbine.

In utility power plants, the heat source for the boiler can vary depending on the type of generating plant. In geothermal power plants, for example, water at temperatures as high as 380°C is pumped from geothermal resources located several hundred meters below the earth's surface, and the water's energy is transferred to the working fluid in a boiler.

The other commonly used thermodynamic cycle is the **refrigeration cycle** (heat-pump cycle). As stated earlier, a heat engine and a heat pump both operate under the same principles except that their thermodynamic processes are reversed. Figures 2.1.2 and 2.1.3 provide detailed information about the heat-pump cycle. This cycle is sometimes called the reversed Rankine cycle.

Modified Rankine Cycles

Modifying the Rankine cycle can improve the output work considerably. One modification usually employed in large central power stations is introducing a **reheat process** into the Rankine cycle. In this modified Rankine cycle, as shown in Figure 2.1.5(a), steam is first expanded through the first stage of the turbine. The steam discharging from the first stage of the turbine is then reheated before entering the second stage of the turbine. The reheat process allows the second stage of the turbine to have a greater enthalpy change. The enthalpy-versus-entropy plot for this cycle is shown in Figure 2.1.5(b), and this figure should be compared to Figure 2.1.4(b) to further appreciate the effect of the reheat process. Note

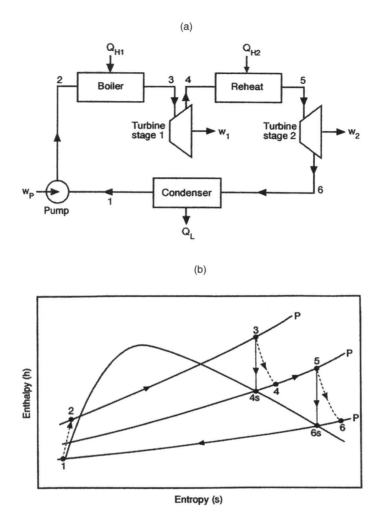

FIGURE 2.1.5 (a) Rankine cycle with a reheat process. (b) The *h-s* plot for the modified Rankine cycle of Figure 2.1.5(a).

that in the reheat process, the work output per pound of steam increases; however, the efficiency of the system may be increased or reduced depending on the reheat temperature range.

Another modification also employed at large power stations is called a **regeneration process**. The schematic representation of a Rankine cycle with a regeneration process is shown in Figure 2.1.6(a), and the enthalpy-versus-entropy plot is shown in Figure 2.1.6(b). In this process, a portion of the steam (at point 6) that has already expanded through the first stage of the turbine is extracted and is mixed in an open regenerator with the low-temperature liquid (from point 2) that is pumped from the condenser back to the boiler. The liquid coming out of the regenerator at point 3 is saturated liquid that is then pumped to the boiler.

Example 2.1.1

A geothermal heat pump, shown in Figure 2.1.7, keeps a house at 24°C during the winter. The geothermal resource temperature is –5°C. The amount of work required to operate the heat pump for a particular month is 10^6 kilojoules (kJ). What is the maximum heat input to the house during that 1-month period?

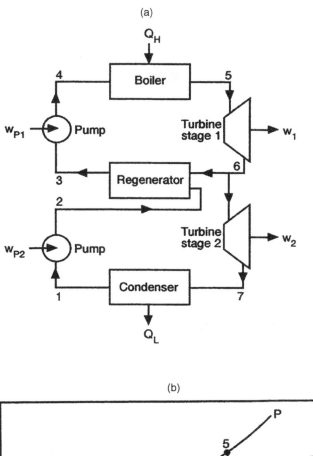

(a)

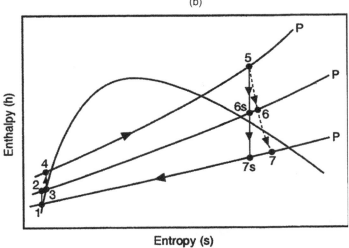

(b)

FIGURE 2.1.6 (a) Rankine cycle with regeneration process, and (b) its *h-s* diagram.

Solution:

The energy balance for the system gives

$$w_i + Q_L = Q_H.$$

(2.1.21)

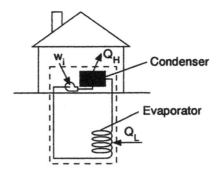

FIGURE 2.1.7   Ground-source heat pump.

Equation (2.1.7) gives the entropy change for the system:

$$\Delta S = \frac{Q_H}{T_H} - \frac{Q_L}{T_L},$$

(2.1.22)

where, from Eq. (2.1.22), we can get an expression for Q_L:

$$Q_L = \frac{Q_H T_L}{T_H} - T_L \Delta S.$$

(2.1.23)

Substituting for Q_L from Eq. (2.1.21), we get an expression for Q_H:

$$Q_H = \frac{1}{1 - \dfrac{T_L}{T_H}} \left(w_i - T_L \Delta S \right).$$

The maximum Q_H is obtained when $\Delta S = 0$; therefore,

$$Q_H \leq \frac{1}{1 - \dfrac{T_L}{T_H}} \, w_i,$$

(2.1.24)

and substituting the actual values yields

$$Q_H \leq \frac{1}{1 - \dfrac{268.15 \text{ K}}{297.15 \text{ K}}} \times 10^6 \text{ kJ}, \ or \ Q_H \leq 10^7 \text{ kJ}.$$

Example 2.1.2

Calculate the maximum COP for the heat pump of Example 2.1.1.

Solution

$$COP_{\text{heat pump}} = \frac{Q_H}{Q_H - Q_L} = \frac{10^7 \ kJ}{10^6 \ kJ} = 10.$$

Example 2.1.3

A simple heat-pump system is shown in Figure 2.1.2. The working fluid in the closed loop is R-22. The p-h diagram of Figure 2.1.2 shows the thermodynamic process for the working fluid. The following data represent a typical operating case.

$$T_1 = T_2 = -5°C \; (23°F)$$

$$T_3 = T_4 = 24°C \; (75°F)$$

$$x_1 = 0.17$$

(a) Determine the COP for this heat pump assuming isentropic compression, $s_2 = s_3$.

(b) Determine the COP by assuming a compressor isentropic efficiency of 70%.

Solution:

First, the thermodynamic properties at each station can be found using the *CRC Mechanical Engineer's Handbook.*

State Point #1:

The evaporation of working fluid R-22 occurs at a constant pressure (between points 1 and 2). This pressure can be obtained from the saturated liquid/vapor table of properties for R-22 at $T_1 = -5°C$ (23°F), which is $p_1 = 422$ kPa (61.2 psia). At point 1, the quality is $x_1 = 0.17$. Therefore, the enthalpy and entropy at this point can be obtained from:

$$h_1 = h_{f_1} + x_1 \; h_{fg_1} \quad \text{and} \quad s_1 = s_{f_1} + x \; s_{fg_1},$$

where $h_{f_1} = 39.36$ kJ/kg, $h_{fg_1} - h_{f_1} = h_{g_1} = 208.85$ kJ/kg, $s_{f_1} = 0.1563$ kJ/kg K, and $s_{fg_1} = s_{g_1} - s_{f_1} = 0.7791$ kJ/kg K.

The quantities listed are read from the table of properties for R-22. Using these properties, we obtain:

$$h_1 = 39.36 \text{ kJ/kg} + 0.17 \; (208.85 \text{ kJ/kg}) = 74.86 \text{ kJ/kg},$$

$$s_1 = 0.1563 \text{ kJ/kg K} + 0.17 \; (0.7791 \text{ kJ/kg K}) = 0.2887 \text{ kJ/kg K}$$

We then find the state properties at point 3, because they will be used to find the quality of the mixture at point 2.

State Point #3:

At point 3, the working fluid is saturated vapor at $T_3 = 24°C$ (75°F). From the table of properties, the pressure, enthalpy, and entropy at this point are $p_3 = 1,014$ kPa (147 psia), $h_3 = 257.73$ kJ/kg, and $s_3 = 0.8957$ kJ/kg K.

State Point #2:

The temperature at this point is $T_2 = -5°C$ (23°F), and because we are assuming an isentropic compression, the entropy is $s_2 = s_3 = 0.8957$ kJ/kg K. The quality of the mixture at point 2 can be calculated from

$$x_2 = \frac{s_2 - s_{f_2}}{s_{fg_2}},$$

where $s_{fg_2} = s_{g_2} - s_{f_2}$, and the quantities s_{g_2} and s_{f_2} can be obtained from the table of properties at $T_2 = -5°C$ (23°F).

Note that the saturation properties for points 1 and 2 are the same because both points have the same pressure and temperature. Therefore,

$$s_{f_2} = s_{f_1} = 0.1563 \text{ kJ/kg K},$$

$$h_{f_2} = h_{f_1} = 39.36 \text{ kJ/kg},$$

$$s_{fg_2} = s_{fg_1} = 0.7791 \text{ kJ/kg K},$$

$$h_{fg_2} = h_{fg_1} = 208.85 \text{ kJ/kg},$$

$$s_2 = \frac{s_2 - s_{f_2}}{s_{fg_2}} = \frac{0.8957 \text{ kJ/kg K} - 0.1563 \text{ kJ/kg K}}{0.7791 \text{ kJ/kg K}} = 0.95.$$

Knowing the quality at point 2, the enthalpy at point 2 can be calculated:

$$h_2 = h_{f_2} + x_2 \ h_{fg_2} = 39.36 \text{ kJ/kg} + 0.95 \ (208.85 \text{ kJ/kg}) = 237.77 \text{ kJ/kg}.$$

State Point #4

At point 4, we have saturated liquid at $T_4 = 24°C$ (75°F). Therefore, from the table of properties, $s_4 = 0.2778$ kJ/kg K and $h_4 = 74.16$ kJ/kg.

(a) The coefficient of performance for a heat pump is

$$\text{COP} = \frac{\text{rate of energy transfer to house}}{\text{compressor shaft power}}$$

$$= \frac{h_3 - h_4}{h_3 - h_2} = \frac{183.57 \text{ kJ/kg}}{19.96 \text{ kJ/kg}} = 9.2.$$

(b) If the isentropic efficiency is 70%, the *p-h* diagram is as shown in Figure 2.1.8. The isentropic efficiency for the compressor is defined as

$$h_s = \frac{h_{3s} - h_2}{h_3 - h_2}.$$

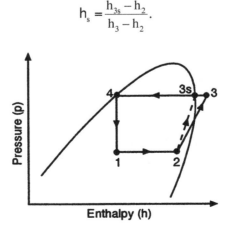

FIGURE 2.1.8   The *p-h* diagram for a heat-pump cycle with a 70% isentropic efficiency for the compressor.

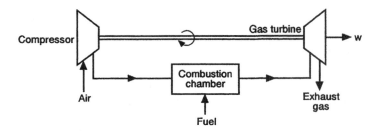

FIGURE 2.1.9   A basic gas-turbine or Brayton-cycle representation.

Using this relationship, h_3 can be calculated as follows:

$$h_3 = \frac{h_{3s} \pm h_2}{\eta_s} + h_2 = \frac{257.73 \text{ kJ/kg} \pm 237.77 \text{ kJ/kg}}{0.70} + 237.77 \text{ kJ/kg} = 266.28 \text{ kJ/kg}.$$

Therefore, the COP is

$$\text{COP} = \frac{h_3 - h_4}{h_3 - h_2} = \frac{266.28 \text{ kJ/kg} - 74.16 \text{ kJ/kg}}{266.28 \text{ kJ/kg} - 237.77 \text{ kJ/kg}} = 6.7.$$

Advanced Thermodynamic Power Cycles

Over the past 50 years, many technological advances have improved the performance of power plant components. Recent developments in exotic materials have allowed the design of turbines that can operate more efficiently at higher inlet temperatures and pressures. Simultaneously, innovative thermodynamic technologies (processes) have been proposed and implemented that take advantage of improved turbine isentropic and mechanical efficiencies and allow actual operating thermal efficiencies of a power station to approach 50%. These improved technologies include (1) modification of existing cycles (reheat and regeneration) and (2) use of combined cycles. In the previous section, we discussed reheat and regeneration techniques. In the following paragraphs, we give a short overview of the combined-cycle technologies and discuss their operation.

The basic gas-turbine or **Brayton cycle** is shown in Figure 2.1.9. In this cycle, ambient air is pressurized in a compressor and the compressed air is then forwarded to a combustion chamber, where fuel is continuously supplied and burned to heat the air. The combustion gases are then expanded through a turbine to generate mechanical work. The turbine output runs the air compressor and a generator that produces electricity.

The exhaust gas from such a turbine is very hot and can be used in a bottoming cycle added to the basic gas-turbine cycle to form a **combined cycle**. Figure 2.1.10 depicts such a combined cycle where a **heat-recovery steam generator** (HRSG) is used to generate steam required for the bottoming (Rankine)

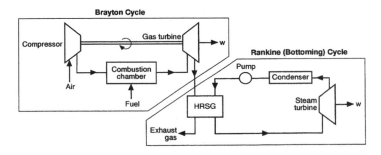

FIGURE 2.1.10   A combined cycle known as the steam-and-gas-turbine cycle.

cycle. The high-temperature exhaust gases from the gas-turbine (Brayton) cycle generate steam in the HRSG. The steam is then expanded through the steam turbine and condensed in the condenser. Finally, the condensed liquid is pumped to the HRSG for heating. This combined cycle is referred to as a **steam and gas turbine cycle**.

Another type of bottoming cycle proposed by Kalina [1984] uses a mixture of ammonia and water as a working fluid. The multicomponent mixture provides a boiling process that does not occur at a constant temperature; as a result, the available heat is used more efficiently. In addition, Kalina employs a distillation process or working-fluid preparation subsystem that uses the low-temperature heat available from the mixed-fluid turbine outlet. The working-fluid mixture is enriched by the high-boiling-point component; consequently, condensation occurs at a relatively constant temperature and provides a greater pressure drop across the turbine. The use of multicomponent working fluids in Rankine cycles provides variable-temperature boiling; however, the condensation process will have a variable temperature as well, resulting in system inefficiencies. According to Kalina, this type of bottoming cycle increases the overall system efficiency by up to 20% above the efficiency of the combined-cycle system using a Rankine bottoming cycle. The combination of the cycle proposed by Kalina and a conventional gas turbine is estimated to yield thermal efficiencies in the 50 to 52% range.

2.1.2 Fundamentals of Heat Transfer

In Section 2.1, we discussed thermodynamic laws and through some examples we showed that these laws are concerned with interaction between a system and its environment. Thermodynamic laws are always concerned with the **equilibrium** state of a system and are used to determine the amount of energy required for a system to change from one equilibrium state to another. These laws do not quantify the mode of the energy transfer or its rate. Heat transfer relations, however, complement thermodynamic laws by providing **rate equations** that relate the heat transfer rate between a system and its environment.

Heat transfer is an important process that is an integral part of our environment and daily life. The heat-transfer or heat-exchange process between two media occurs as a result of a temperature difference between them. Heat can be transferred by three distinct modes: conduction, convection, and radiation. Each one of these heat transfer modes can be defined by an appropriate rate equation presented below:

Fourier's Law of Heat Conduction—represented here by Eq. (2.1.25) for the one-dimensional steady-state case:

$$\dot{Q}_{cond} = -kA\frac{dT}{dx}. \tag{2.1.25}$$

Newton's Law of Cooling—which gives the rate of heat transfer between a surface and a fluid:

$$\dot{Q}_{conv} = hA\Delta T, \tag{2.1.26}$$

where h is the average heat-transfer coefficient over the surface with area A.

Stefan–Boltzmann's Law of Radiation—which is expressed by the equation:

$$\dot{Q}_r = A_1 F_{1-2} \sigma \left(T_1^4 - T_2^4\right). \tag{2.1.27}$$

Conduction Heat Transfer

Conduction is the heat-transfer process that occurs in solids, liquids, and gases through molecular interaction as a result of a temperature gradient. The energy transfer between adjacent molecules occurs without significant physical displacement of the molecules. The rate of heat transfer by conduction can be predicted by using Fourier's law, where the effect of molecular interaction in the heat-transfer medium

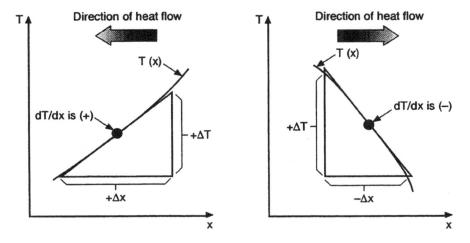

FIGURE 2.1.11 The sign convention for conduction heat flow.

is expressed as a property of that medium and is called the **thermal conductivity**. The study of conduction heat transfer is a well-developed field where sophisticated analytical and numerical techniques are used to solve many problems in buildings including heating and cooling load calculation.

In this section, we discuss basics of steady-state one-dimensional conduction heat transfer through homogeneous media in cartesian and cylindrical coordinates. Some examples are provided to show the application of the fundamentals presented, and we also discuss fins or extended surfaces.

One-Dimensional Steady-State Heat Conduction

Fourier's law, as represented by Eq. (2.1.25), states that the rate of heat transferred by conduction is directly proportional to the temperature gradient and the surface area through which the heat is flowing.

The proportionality constant k is the thermal conductivity of the heat-transfer medium. Thermal conductivity is a thermophysical property and has units of W/m K in the SI system, or Btu/h ft °F in the English system of units. Thermal conductivity can vary with temperature, but for most materials it can be approximated as a constant over a limited temperature range. A graphical representation of Fourier's law is shown in Figure 2.1.11.

Equation (2.1.25) is only used to calculate the rate of heat conduction through a one-dimensional homogenous medium (uniform k throughout the medium). Figure 2.1.12 shows a section of a plane wall with thickness L, where we assume the other two dimensions of the wall are very large compared to L. One side of the wall is at temperature T_1, and the other side is kept at temperature T_2, where $T_1 > T_2$. Integrating Fourier's law with constant k and A, the rate of heat transfer through this wall is

$$\dot{Q} = kA\frac{T_1 - T_2}{L}, \tag{2.1.28}$$

where k is the thermal conductivity of the wall.

The Concept of Thermal Resistance

Figure 2.1.12 also shows the analogy between electrical and thermal circuits. Consider an electric current I flowing through a resistance R_e, as shown in Figure 2.1.12. The voltage difference $\Delta V = V_1 - V_2$ is the driving force for the flow of electricity. The electric current can then be calculated from

$$I = \frac{\Delta V}{R_e}. \tag{2.1.29}$$

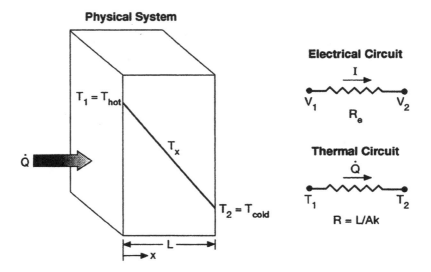

FIGURE 2.1.12   Analogy between thermal and electrical circuits for steady-state conduction through a plane wall.

Like electric current flow, heat flow is governed by the temperature difference, and it can be calculated from

$$\dot{Q} = \frac{\Delta T}{R},$$

(2.1.30)

where, from Eq. (2.1.28), $R = L/Ak$ and is called **thermal resistance.** Following this definition, the thermal resistance for convection heat transfer given by Newton's Law of Cooling becomes $R = 1/(hA)$. Thermal resistance of composite walls (plane and cylindrical) has been discussed by Kakac and Yener [1988], Kreith and Bohn [1993], and Bejan [1993]. The following example shows how we can use the concept of thermal resistance in solving heat-transfer problems in buildings.

Example 2.1.4

One wall of an uninsulated house, shown in Figure 2.1.13, has a thickness of 0.30 m and a surface area of 11 m². The wall is constructed from a material (brick) that has a thermal conductivity of 0.55 W/m K. The outside temperature is −10°C, while the house temperature is kept at 22°C. The convection heat-transfer coefficient is estimated to be h_o = 21 W/m² K in the outside and h_i = 7 W/m² K in the inside. Calculate the rate of heat transfer through the wall, as well as the surface temperature at either side of the wall.

Solution:

The conduction thermal resistance is

$$R_{t,cond} = \frac{L}{Ak} = \frac{0.3 \ \text{m}}{11 \ \text{m}^2 \times 0.55 \ \text{W/m K}} = 0.0496 \frac{\text{K}}{\text{W}}.$$

Note that the heat-transfer rate per unit area is called **heat flux** and is given by

$$q'' = \frac{\dot{Q}}{A} = \frac{T_1 - T_2}{L/k}.$$

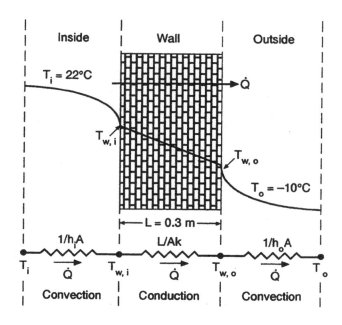

FIGURE 2.1.13   Heat loss through a plane wall.

In this case, the resistance to heat transfer over a 1–m by 1–m area of the wall is

$$R_{t,\text{cond}} = L/k = \frac{0.3 \text{ m}}{0.55 \text{ W/m K}} = 0.5455 \frac{\text{m}^2 \text{ K}}{\text{W}}.$$

The convection resistances for inside and outside, shown in Figure 2.1.13, are

$$R_{i,\text{conv}} = \frac{1}{h_i A} = \frac{1}{7 \text{ W/m}^2 \text{ K} \times 11 \text{ m}^2} = 0.0130 \frac{\text{K}}{\text{W}}$$

$$R_{o,\text{conv}} = \frac{1}{h_o A} = \frac{1}{21 \text{ W/m}^2 \text{ K} \times 11 \text{ m}^2} = 0.0043 \frac{\text{K}}{\text{W}}.$$

Note that the highest resistance is provided by conduction through the wall. The total heat flow can be calculated from

$$\dot{Q} = \frac{\Delta T}{\Sigma R} = \frac{T_i - T_o}{\dfrac{1}{h_i A} + \dfrac{L}{kA} + \dfrac{1}{h_o A}} = \frac{295.15 \text{ K} - 263.15 \text{ K}}{0.013 \dfrac{\text{K}}{\text{W}} + 0.0496 \dfrac{\text{K}}{\text{W}} + 0.0043 \dfrac{\text{K}}{\text{W}}} = 478.3 \text{ W}.$$

The surface temperatures can then be calculated by using the electric analogy depicted in Figure 2.1.13. For the inside surface temperature,

$$\dot{Q} = \frac{T_i - T_{w,i}}{\dfrac{1}{h_i A}},$$

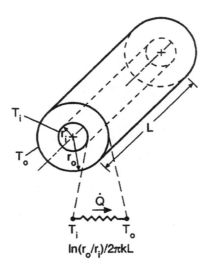

FIGURE 2.1.14 Conduction through hollow cylinders.

or

$$T_{w,i} = T_i - \frac{\dot{Q}}{h_i A} = 295.15 \text{ K} - \frac{478.3 \text{ W}}{77 \text{ W/K}} = 288.94 \text{ K} = 15.79°C.$$

Similarly, for the outside surface temperature:

$$T_{w,o} = \frac{\dot{Q}}{h_o A} + T_o = \frac{478.3 \text{ W}}{231 \text{ W/K}} + 263.15 \text{ K} = 265.22 \text{ K} = -7.93°C.$$

Conduction Through Hollow Cylinders

A cross section of a long hollow cylinder such as pipe insulation with internal radius r_i and external radius r_o is shown in Figure 2.1.14. The internal surface of the cylinder is at temperature T_i and the external surface is at T_o, where $T_i > T_o$. The rate of heat conduction in a radial direction is calculated by

$$\dot{Q} = \frac{2\pi L k (T_i - T_o)}{\ln(r_o/r_i)}, \tag{2.1.31}$$

where L is the length of the cylinder that is assumed to be long enough so that the end effects may be ignored. From Eq. (2.1.31) the resistance to heat flow in this case is

$$R = \frac{\ln(r_o/r_i)}{2\pi k L}. \tag{2.1.32}$$

Equation (2.1.31) can be used to calculate the heat loss through insulated pipes, as presented in the following example.

Example 2.1.5

The refrigerant of the heat pump discussed in Example 2.1.3 is circulating through a thin-walled copper tube of radius $r_i = 6$ mm, as shown in Figure 2.1.15. The refrigerant temperature is T_i, ambient temperature is T_∞, and $T_i < T_\infty$. The outside convection heat-transfer coefficient is $h_o = 7$ W/m² K.

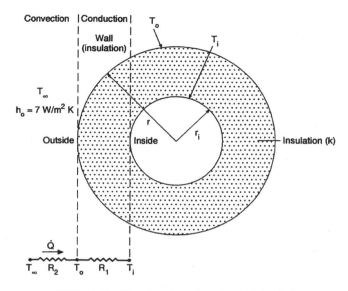

FIGURE 2.1.15   Heat loss through a pipe with insulation.

(a) If we decide to insulate this tube, what would be the optimum thickness of the insulation?

(b) Show the behavior of heat flow through the tube at different insulation thicknesses such as 0, 3, 6, 10, 15, and 20 mm, and plot the results for $\dot{Q}/L$ versus radius r. Assume an insulation material with thermal conductivity $k = 0.06$ W/m K.

Solution:

(a) In thermal analysis of radial systems, we must keep in mind that there are competing effects associated with changing the thickness of insulation. Increasing the insulation thickness increases the conduction resistance; however, the area available for convection heat transfer increases as well, resulting in reduced convection resistance. To find the optimum radius for insulation, we first identify the major resistances in the path of heat flow. Our assumptions are that (1) the tube wall thickness is small enough that conduction resistance can be ignored, (2) heat transfer occurs at steady state, (3) insulation has uniform properties, and (4) radial heat transfer is one-dimensional

The resistances per unit length are

$$R_1 = \frac{\ln(r/r_i)}{2\pi k} \quad \text{and} \quad R_2 = \frac{1}{2\pi r h_o},$$

where r, the outer radius of insulation, is unknown. The total resistance is

$$R_t = R_1 + R_2 = \frac{\ln(r/r_i)}{2\pi k} + \frac{1}{2\pi r h_o}$$

and the rate of heat flow per unit length is

$$\dot{Q} = \frac{T_\infty - T_i}{R_t}.$$

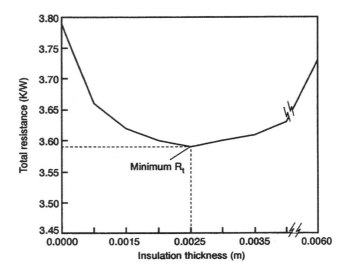

FIGURE 2.1.16   Total resistance versus insulation thickness for an insulated tube.

The optimum thickness of the insulation is obtained when the total heat flow is minimized or when the total resistance is maximized. By differentiating R_t with respect to r, we obtain the condition under which R_t is maximum (or minimum). Therefore,

$$\frac{dR_t}{dr} = \frac{1}{2\pi rk} - \frac{1}{2\pi r^2 h_o} = 0,$$

from which we obtain $r = k/h_o$. To determine if R_t is maximum or minimum at $r = k/h_o$, we take the second derivative and find its quantity at $r = k/h_o$.

$$\frac{d^2 R_t}{dr^2} = -\frac{1}{2\pi k(k/h_o)^2} + \frac{1}{\pi h_o(k/h_o)^3} = \frac{1}{2\pi k^3/h_o^2} > 0.$$

Therefore, R_t is a *minimum* at $r = k/h_o$, which means that the heat flow is maximum at this insulation radius. An optimum radius of insulation does not exist; however, the radius obtained in this analysis is referred to as the **critical radius**, r_c, and this radius should be avoided when selecting insulation for pipes. The economic optimum insulation can be found using techniques in Chapter 3.2.

(b) For this example the critical radius is $r_c = k/h_o = 0.06$ W/m K/7 W/m² K = 0.0086 m = 8.6 mm, and $r_i = 6$ mm, so $r_c > r_i$. This means that by adding insulation, we will increase the heat loss from the tube. Using the expression for R_t, we can plot the total resistance versus the insulation thickness as shown in Figure 2.1.16. Note that the minimum total resistance occurs at an insulation thickness of about 0.025 m (corresponding to the r_c calculated earlier). Also note that as the insulation thickness is increased, the conduction resistance increases; however, the convection resistance decreases as listed in Table 2.1.1.

Convection Heat Transfer

Energy transport (heat transfer) in fluids usually occurs by the motion of fluid particles. In many engineering problems, fluids come into contact with solid surfaces that are at different temperatures than the fluid. The temperature difference and random/bulk motion of the fluid particles result in an energy transport process known as convection heat transfer. Convection heat transfer is more complicated than conduction because the motion of the fluid, as well as the process of energy transport, must be studied

TABLE 2.1.1   Effect of Insulation Thickness on Various Thermal Resistances

Insulation Thickness (m)	Outer Radius, r (m)	Convection Resistance, R_2, (K/W)	Conduction Resistance, R_1, (K/W)	Total Resistance, R_t, (K/W)
0	0.0060	3.79	0	3.79
0.0010	0.0070	3.25	0.41	3.66
0.0015	0.0075	3.03	0.59	3.62
0.0020	0.0080	2.84	0.76	3.60
0.0025	0.0085	2.67	0.92	3.59
0.0030	0.0090	2.53	1.07	3.60
0.0035	0.0095	2.39	1.22	3.61
0.0040	0.0100	2.27	1.36	3.63
0.0060	0.0120	1.89	1.84	3.73

simultaneously. Convection heat transfer can be created by external forces such as pumps and fans in a process referred to as **forced convection**. In the absence of external forces, the convection process may result from temperature or density gradients inside the fluid; in this case, the convection heat-transfer process is referred to as **natural convection**. We will discuss this type of convection in more detail in the next section. There are other instances where a heat-transfer process consists of both forced and natural convection modes and they are simply called **mixed-convection** processes.

The main unknown in the convection heat-transfer process is the heat-transfer coefficient (see Eq. 2.1.26). Figure 2.1.17 serves to explain the convection heat-transfer process by showing the temperature and velocity profiles for a fluid at temperature T_∞ and bulk velocity U_∞ flowing over a heated surface. As a result of viscous forces interacting between the fluid and the solid surface, a region known as **velocity boundary layer** is developed in the fluid next to the solid surface. In this region the fluid velocity is zero at the surface and increases to the bulk fluid velocity u_∞. Because of the temperature difference between the fluid and the surface, a region known as **temperature boundary layer** also develops next to the surface, where the temperature at the fluid varies from T_w (surface temperature) to T_∞ (bulk fluid temperature). The velocity-boundary-layer thickness δ and temperature-boundary-layer thickness δ_t and their variation along the surface are shown in Figure 2.1.17.

Depending on the thermal diffusivity and kinematic viscosity of the fluid, the velocity and temperature boundary layers may be equal or may vary in size. Because of the no-slip condition at the solid surface, the fluid next to the surface is stationary; therefore, the heat transfer at the interface occurs only by conduction.

If the temperature gradient were known at the interface, the heat exchange between the fluid and the solid surface could be calculated from Eq. (2.1.25), where k in this case is the thermal conductivity of the fluid and dT/dx (or dT/dy in reference to Figure 2.1.17) is the temperature gradient at the interface. However, the temperature gradient at the interface depends on the macroscopic and microscopic motion of fluid particles. In other words, the heat transferred from or to the surface depends on the nature of the flow field.

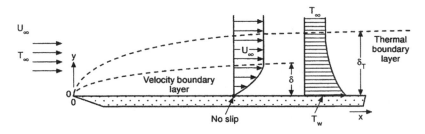

FIGURE 2.1.17   Temperature and velocity profiles for convection heat-transfer process over a heated surface.

Therefore, in solving convection problems, engineers need to determine the relationship between the heat transfer through the solid-body boundaries and the temperature difference between the solid-body wall and the bulk fluid. This relationship is given by Eq. (2.1.26), where h is the convection coefficient averaged over the solid surface area. Note that h depends on the surface geometry and the fluid velocity, as well as on the fluid's physical properties. Therefore, depending on the variation of the above quantities, the heat-transfer coefficient may change from one point to another on the surface of the solid body. As a result, the *local* heat-transfer coefficient may be different than the *average* heat-transfer coefficient. However, for most practical applications, engineers are mainly concerned with the *average* heat-transfer coefficient, and in this section we will use only average heat-transfer coefficients unless otherwise stated.

Natural-Convection Heat Transfer

Natural-convection heat transfer results from density differences within a fluid. These differences may result from temperature gradients that exist within a fluid. When a heated (or cooled) body is placed in a cooled (or heated) fluid, the temperature difference between the fluid and the body causes heat flow between them, resulting in a density gradient inside the fluid. As a result of this density gradient, the low-density fluid moves up and the high-density fluid moves down. The heat-transfer coefficients (and consequently, the rate of heat transfer in natural convection) are generally less than that in forced convection because the driving force for mixing of the fluid is less in natural convection.

Natural-convection problems can be divided into two categories: **external natural convection** and **internal natural convection**. Natural-convection heat transfer from the external surfaces of bodies of various shapes has been studied by many researchers. Experimental results for natural-convection heat transfer are usually correlated by an equation of the type

$$\text{Nu} = \frac{hL}{k} = f\left(\text{Ra}\right), \tag{2.1.33}$$

where the **Nusselt number**, Nu, provides a measure of the convection heat transfer occurring between the solid surface and the fluid. Knowing Nu, the convection heat-transfer coefficient, h, can be calculated. Note that in Eq. (2.1.33), k is the fluid conductivity and Ra is the **Rayleigh number**, which represents the ratio of buoyancy force to the rate of change of momentum. The Rayleigh number is given by

$$\text{Ra} = \frac{g\beta\left(T \pm T_\infty\right)L^3}{\nu\alpha}, \tag{2.1.34}$$

where β is the coefficient of thermal expansion equal to $1/T$ (T is the absolute temperature expressed in Kelvin) for an ideal gas, L is a characteristic length, ν is the kinematic viscosity of the fluid, and α is its thermal diffusivity. A comprehensive review of the fundamentals of natural-convection heat transfer is provided by Raithby and Hollands [1985]. Table 2.1.2 gives correlations for calculating heat transfer from the external surfaces of some common geometries.

Experiments conducted by Hassani and Hollands [1989], Sparrow and Stretton [1985], Yovanovich and Jafarpur [1993], and others have shown that the external natural-convection heat transfer from bodies of arbitrary shape exhibit Nu-Ra relationships similar to regular geometries such as spheres and short cylinders. An extensive correlation for predicting natural-convection heat transfer from bodies of arbitrary shape was developed by Hassani and Hollands [1989]; it is useful for most situations on the surfaces of buildings.

Internal natural-convection heat transfer occurs in many engineering problems such as heat loss from building walls, electronic equipment, double-glazed windows, and flat-plate solar collectors. Some of the geometries and their corresponding Nusselt numbers are listed in Table 2.1.3. Anderson and Kreith [1987] provide a comprehensive summary of natural-convection processes that occur in various solar thermal systems.

TABLE 2.1.2 Natural-Convection Correlations for External Flows

Configuration	Correlation	Restrictions	Source
Vertical plate with constant T_w, T_∞	$\bar{Nu}_L = \left\{ 0.825 + \dfrac{0.387 \, Ra_L^{1/6}}{[1 + (0.492/Pr)^{9/16}]^{8/27}} \right\}^2$	$10^{-1} < Ra_L < 10^{12}$ All Pr	Churchill and Chu [1975]
	$\bar{Nu}_L = 0.56 \, (Ra_L \cos\theta)^{1/4}$	$10^5 < Ra_L \cos\theta < 10^{11}$ $0 \le \theta \le 89$	Fujii and Imura [1972]
Horizontal plate with hot surface facing upward or cold surface facing downward	$\bar{Nu}_L = 0.54 \, Ra_L^{1/4}$ $\bar{Nu}_L = 0.15 \, Ra_L^{1/3}$ $\left(L = \dfrac{\text{surface area}}{\text{perimeter}}\right)$	$10^5 < Ra_L < 10^7$ $10^7 < Ra_L < 10^{10}$ Pr > 0.5	McAdams [1954] and Incropera and DeWitt [1990]
Horizontal plate with hot surface facing downward or cold surface facing upward	$\bar{Nu}_L = 0.27 \, Ra_L^{1/4}$ $\left(L = \dfrac{\text{surface area}}{\text{perimeter}}\right)$	$10^5 < Ra_L < 10^{10}$ Pr > 0.5	McAdams [1954] and Incropera and DeWitt [1990]

TABLE 2.1.2 (continued) Natural-Convection Correlations for External Flows

Configuration	Correlation	Restrictions	Source
Horizontal cylinders	$\bar{Nu}_D = \left\{ 0.6 + \dfrac{0.387\, Ra_D^{1/6}}{\left[1 + (0.559/Pr)^{9/16}\right]^{8/27}} \right\}^2$	$10^{-5} < Ra_D < 10^{12}$ All Pr	Churchill and Chu [1975]
Vertical cylinders of height L	$\bar{Nu}_L = 0.68 + \dfrac{0.67\, Ra_L^{1/4}}{\left[1 + (0.492/Pr)^{9/16}\right]^{4/9}}$	$1 < Ra_L < 10^9$	Churchill and Chu [1975]
	$\bar{Nu}_L = 0.13\, (Ra_L)^{1/3}$	$\dfrac{D}{L} \geq 35 \left(\dfrac{Pr}{Ra_L}\right)^{1/4}$ $Ra_L > Pr \times 10^9$	McAdams [1954]
Sphere	$\bar{Nu}_D = 2 + \dfrac{0.589\, Ra_D^{1/4}}{\left[1 + (0.469/Pr)^{9/16}\right]^{4/9}}$	$Pr \geq 0.7$ $Ra_D < 10^{11}$	Churchill [1983]
Other immersed bodies such as cubes, bispheres, spheroids			Hassani and Hollands [1989]

Note: Ra is Rayleigh number, Pr is Prandtl number, and Nu is Nusselt number.

TABLE 2.1.3 Natural-Convection Correlations for Internal Flows

Configuration	Correlation	Restrictions	Source
Space enclosed between two horizontal plates heated from below	$$\overline{Nu}_L = 1 + 1.44\left[1 - \frac{1708}{Ra_L}\right]^* + \left[\left(\frac{Ra_L}{5830}\right)^{1/3} - 1\right]^*$$	Air, $1700 < Ra_L < 10^8$	Hollands et al. [1975]
	$$\overline{Nu}_L = 1 + 1.44\left[1 - \frac{1708}{Ra_L}\right]^* + \left[\left(\frac{Ra_L}{5830}\right)^{1/3} - 1\right]^* + 2.0\frac{Ra_L^{1/3}}{140}\left[1 - \ln(Ra_L^{1/3}/140)\right]$$	Water, $1700 < Ra_L < 3.5 \times 10^9$	Hollands et al. [1975]
	The quantities contained between the parenthesis with asterisk, $( )^*$, must be set equal to zero if they become negative.		
Space enclosed between two vertical plates heated from one side	$$\overline{Nu}_L(90°) = 0.22\left(\frac{H}{L}\right)^{-1/4}\left(\frac{Pr}{0.2 + Pr}Ra_L\right)^{0.28}$$	$2 < \dfrac{H}{L} < 10, \quad Pr < 10, \quad Ra_L < 10^{10}$	Catton [1978]
	$$\overline{Nu}_L(90°) = 0.18\left(\frac{Pr}{0.2 + Pr}Ra_L\right)^{0.29}$$	$\left\{\begin{array}{l}1 < \dfrac{H}{L} < 2, \quad 10^{-3} < Pr < 10^5 \\[2mm] 10^3 < \dfrac{Ra_L\,Pr}{0.2 + Pr}\end{array}\right.$	
	$$\overline{Nu}_L = 0.42\,Ra_L^{0.25}\,Pr^{0.012}\left(\frac{H}{L}\right)^{-0.3}$$	$\left\{\begin{array}{l}10 < \dfrac{H}{L} < 40, \quad 1 < Pr < 2 \times 10^4 \\[2mm] 10^4 < Ra_L < 10^7\end{array}\right.$	MacGregor and Emery [1969]

TABLE 2.1.3 (continued)   Natural-Convection Correlations for Internal Flows

Configuration	Correlation	Restrictions	Source
	$\overline{Nu}_L(\theta) = 1 + \left[\overline{Nu}_L(90°) - 1\right]\sin\theta$	$90° < \theta < 180°$, air	Arnold et al. [1976], Catton (1978), Arnold et al. [1974], and Ayyaswamy and Catton [1973]. Hollands et al. [1976]
	$\overline{Nu}_L(\theta) = \overline{Nu}_L(90°)(\sin\theta)^{1/4}$	$\theta^* < \theta < 90°$, air	
	$\overline{Nu}_L(\theta) = \left[\dfrac{\overline{Nu}_L(90°)}{\overline{Nu}_L(0°)}(\sin\theta^*)^{1/4}\right]^{\theta/\theta^*}$	$0° < \theta < \theta^*$, $\dfrac{H}{L} < 10$, air	
	$\overline{Nu}_L(\theta) = 1 + 1.44\left[1 - \dfrac{1708}{Ra_L\cos\theta}\right]^*\left[1 - \dfrac{(\sin 1.8\theta)^{1.6}\times 1708}{Ra_L\cos\theta}\right] + \left[\left(\dfrac{Ra_L\cos\theta}{5830}\right)^{1/3} - 1\right]^*$	$0° < \theta < \theta^*$, $\dfrac{H}{L} > 10$, air	
	where $\begin{array}{l\|ccccc} \text{H/L} & 1 & 3 & 6 & 12 & >12 \\ \hline \theta^* & 25° & 53° & 60° & 67° & 70° \end{array}$		
Natural convection in inclined enclosures.	The quantities contained between the parentheses with asterisk, ( )*, must be set equal to zero if they become negative.		
	$\overline{Nu}_D = C(Gr_D Pr)^n$	$\begin{array}{c\|cc} Gr_D Pr & C & n \\ \hline 10^4\text{-}10^9 & 0.59 & 1/4 \\ 10^9\text{-}10^{12} & 0.13 & 1/3 \end{array}$	Kreith [1970]
Spherical cavity interior			

Configuration	Correlation	Restrictions	Source
Long concentric cylinders 	$$\frac{k_{eff}}{k} = 0.386 \left[\frac{\ln(D_o/D_i)}{b^{3/4}(1/D_i^{3/5} + 1/D_o^{3/5})^{5/4}}\right] \times \left(\frac{Pr}{0.861 + Pr}\right)^{1/4} Ra_b^{1/4}$$ where $b = (D_o - D_i)/2$	$0.70 \le Pr \le 6000$ $$10 \le \left[\frac{\ln(D_o/D_i)}{b^{3/4}(1/D_i^{3/5} + 1/D_o^{3/5})^{5/4}}\right]^4 \qquad Ra_b \le 10^7$$	Raithby and Hollands [1974]
Concentric spheres 	$$\frac{k_{eff}}{k} = 0.74 \left[\frac{b^{1/4}}{D_o D_i (D_i^{-7/5} + D_o^{-7/5})^{5/4}}\right] \times \left(\frac{Pr}{0.861 + Pr}\right)^{1/4} Ra_b^{1/4}$$ where $b = (D_o - D_i)/2$	$0.70 \le Pr \le 4200$ $$10 \le \left[\frac{b}{(D_o D_i)^4 (D_i^{-7/5} + D_o^{-7/5})^5}\right] \qquad Ra_b \le 10^7$$	Raithby and Hollands [1974]

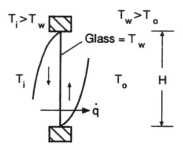

FIGURE 2.1.18 Heat loss through a single-pane window.

The natural-convection heat transfer for long concentric horizontal cylinders and concentric spheres has been studied by Raithby and Hollands [1985]. Their proposed correlations are listed in Table 2.1.3, where D_o and D_i represent the diameters of outer cylinder (or sphere) and inner cylinder (or sphere), respectively. The Rayleigh number is based on the temperature difference across the gap and a characteristic length defined as $b = (D_o - D_i)/2$. The **effective thermal conductivity** k_{eff} in their correlation is the thermal conductivity that a stationary fluid in the gap must have to transfer the same amount of heat as the moving fluid. Raithby and Hollands also provide correlations for natural-convection heat transfer between long eccentric horizontal cylinders and eccentric spheres.

Example 2.1.6

One component of the total heat loss from a room is the heat loss through a single-pane window in the room, as shown in Figure 2.1.18. The inside temperature is kept at $T_i = 22°C$, and the outside temperature is $T_o = -5°C$. The window height H is 0.5 m, and its width is 2 m. The weather is calm, and there is no wind blowing. Assuming uniform glass temperature T_w, calculate the heat loss through the window.

Solution:

The air flow pattern next to the window is shown in Figure 2.1.18. When warm room air approaches or contacts the window, it loses heat and its temperature drops. Because this cooled air next to the window is denser and heavier than the room air at that height, it starts moving down and is replaced by warmer room air at the top of the window. A similar but opposite air movement occurs at the outside of the window. The total heat loss can be calculated from

$$\dot{Q} = h_i A(T_i - T_w) = h_o A(T_w - T_o),$$ (2.1.35)

where h_i and h_o are the average natural-convection heat-transfer coefficients for inside and outside, respectively. Using the correlation recommended by Fujii and Imura [1972] for a vertical plate with constant temperature T_w (from Table 2.1.2) and substituting for angle of inclination $\theta = 0$, we get

$$\overline{Nu_H} = 0.56 \ Ra_H^{1/4},$$

where

$$Ra_H = \frac{g\beta \ \Delta T \ H^3}{\nu\alpha},$$

and H is the height of the window pane. Note that the temperature difference in the expression for Rayleigh number depends on the medium for which the heat-transfer coefficient is sought. For example, for calculating the interior surface coefficient h_i, we write

$$\overline{\mathrm{Nu}}_{H,i} = \frac{h_i H}{k} = 0.56 \left[\frac{g\beta(T_i - T_w)H^3}{\nu\alpha} \right]^{1/4}, \tag{2.1.36}$$

and for calculating h_o, we write

$$\overline{\mathrm{Nu}}_{H,o} = \frac{h_o H}{k} = 0.56 \left[\frac{g\beta(T_w - T_o)H^3}{\nu\alpha} \right]^{1/4}. \tag{2.1.37}$$

Note that all the properties in Eqs. (2.1.36) or (2.1.37) should be calculated *at film temperature* $T_f = (T_i + T_w)/2$ or $T_f = (T_o + T_w)/2$. To calculate $\dot{Q}$ and air properties, we need to know T_w. To estimate T_w, we assume that air properties over the temperature range of interest to this problem do not change significantly (refer to air property tables to verify this assumption). Using this assumption, we find the ratio between Eqs. (2.1.36) and (2.1.37) as

$$\frac{h_o}{h_i} = \left(\frac{T_w - T_o}{T_i - T_w} \right)^{1/4}, \tag{2.1.38}$$

which provides a relationship between h_o, h_i, and T_w. Another equation of this kind can be obtained from Eq. (2.1.35):

$$\frac{h_o}{h_i} = \frac{T_i - T_w}{T_w - T_o}. \tag{2.1.39}$$

Solving Eqs. (2.1.38) and (2.1.39), we can show that

$$T_w = \frac{T_o + T_i}{2} = 8.5°C.$$

Now, by calculating h_i or h_o and substituting into Eq. (2.1.35), the total heat transfer can be calculated. In this case, we choose to solve for h_o. Therefore, the air properties should be calculated at

$$T_f = \frac{T_o + T_w}{2} = 1.75°C.$$

Air properties at $T_f = 1.75°C$ are $k = 0.0238$ W/m K, $\nu = 14.08 \times 10^{-6}$ m²/s, $\alpha = 19.48 \times 10^{-6}$ m²/s, and $\beta = 1/T_f = 0.00364$ K⁻¹. Using these properties, the Rayleigh number is

$$\mathrm{Ra}_{H,o} = \frac{g\beta(T_w - T_o)H^3}{\nu\alpha} = 220 \times 10^6.$$

From Eq. (2.1.37), we obtain

$$h_o = \frac{k}{H}(0.56)\,\mathrm{Ra}_{H,o}^{1/4} = 3.25 \ \mathrm{W/m^2 \ K},$$

and

$$\dot{Q} = h_o A (T_w - T_o)$$

$$= 3.25 \ \mathrm{W/m^2 \ K} \times 1 \ \mathrm{m^2} \times (281.65 \ \mathrm{K} - 268.15 \ \mathrm{K}) = 43.87 \ \mathrm{W}.$$

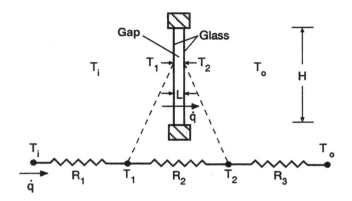

FIGURE 2.1.19   Heat loss through a double-pane window.

Example 2.1.7

The single-pane window of the previous example is replaced by a double-pane window as shown in Figure 2.1.19. The outside and inside temperatures are the same as in Example 2.1.6 ($T_i = 22°C$, $T_o = -5°C$). The glass-to-glass spacing is L = 20 mm, the window height $H = 0.5$ m, and the width is 2 m. Find the heat loss through this window and compare it to the heat loss through the single-pane window. Ignore conduction resistance through the glass.

Solution:

The thermal circuit for the system is shown in Figure 2.1.19. Temperatures T_1 and T_2 are unknown and represent the average glass temperature (i.e., we assume that the glass temperature is uniform over the entire surface because of the low thermal resistance of glass). As in Example 2.1.6, we first estimate temperatures T_1 and T_2. The rate of heat transfer is

$$\dot{Q} = h_1 A(T_i - T_1) = h_2 A(T_1 - T_2) = h_3 A(T_2 - T_o). \tag{2.1.40}$$

The heat-transfer coefficients h_1 and h_3 for natural-convection heat transfer between the glass surface and interior/exterior can be calculated using Eqs. (2.1.36) and (2.1.37), and the ratio between h_1 and h_3 is

$$\frac{h_1}{h_3} = \left(\frac{T_i - T_1}{T_2 - T_o} \right)^{1/4}. \tag{2.1.41}$$

Another relationship between h_1, h_3, T_1, and T_2, is obtained from Eq. (2.1.40):

$$\frac{h_1}{h_3} = \left(\frac{T_2 - T_o}{T_i - T_1} \right). \tag{2.1.42}$$

Solving Eqs. (2.1.41) and (2.1.42), we get:

$$T_1 + T_2 = T_i + T_o. \tag{2.1.43}$$

We need an additional equation that provides a relationship between T_1 and T_2, and we obtain this equation from the correlation that expresses the natural-convection heat transfer in the enclosed area of the double-pane window. We choose the correlation recommended by MacGregor and Emery [1969] from Table 2.1.3:

$$h_2 = \frac{k}{L} \ 0.42 \ \mathrm{Ra}_L^{0.25} \left(\frac{H}{L}\right)^{-0.3} \quad \text{for Pr} = 0.72, \tag{2.1.44}$$

where

$$\mathrm{Ra}_L = \frac{g\beta(T_1 - T_2)L^3}{\nu\alpha}.$$

Using some mathematical manipulations Eq. (2.1.44) can be written as

$$h_2 = \frac{k}{L} 0.16 \left(\frac{T_1 - T_2}{T_i - T_o}\right)^{1/4} \left(\frac{L}{H}\right)^{3/4} \left(\frac{g\beta(T_i - T_o)H^3}{\nu\alpha}\right)^{1/4}, \tag{2.1.45}$$

where $L/H = 0.04$. The heat-transfer coefficient h_1 can be calculated from Eq. (2.1.36) (used to calculate h_i) and can be written as

$$h_1 = \frac{k}{L} 0.56 \left(\frac{T_i - T_2}{T_i - T_o}\right)^{1/4} \left(\frac{g\beta(T_i - T_o)H^3}{\nu\alpha}\right)^{1/4}, \tag{2.1.46}$$

where the Rayleigh number has been written in terms of $(T_i - T_o)$ instead of $(T_i - T_1)$. Substituting for L/H, the ratio between h_2 and h_1 is

$$\frac{h_2}{h_1} = \frac{0.63 \ (T_1 - T_2)^{1/4}}{(T_i - T_1)^{1/4}}. \tag{2.1.47}$$

Note that in finding h_2/h_1, we have assumed that the properties do not change much in the temperature range of interest. From Eq. (2.1.40), we have

$$\frac{h_2}{h_1} = \frac{T_i - T_1}{T_1 - T_2}. \tag{2.1.48}$$

Therefore, solving Eqs. (2.1.47) and (2.1.48), we obtain T_1 in terms of T_i and T_o:

$$T_1 = 0.71 \ T_i + 0.29 \ T_o. \tag{2.1.49}$$

Substituting for T_i and T_o, we obtain $T_1 = 14.2°C$, and substituting for T_1, T_i, and T_o in Eq. (2.1.43), we get $T_2 = 2.8°C$. Knowing T_1 and T_2, we can calculate Ra_L. To calculate Ra_L, we should obtain air properties at $T_f = (T_1 + T_2)/2 = 8.5°C$, which are $k = 0.0244$ W/m K, $\nu = 14.8 \times 10^{-6}$ m²/s, $\alpha = 20.6 \times 10^{-6}$ m²/s, and $\beta = 1/T_f = 0.00355$ K⁻¹. Therefore,

$$\mathrm{Ra}_L = \frac{9.81 \ \mathrm{m}/\mathrm{s}^2 \times 0.00355 \ \mathrm{K}^{-1} \times (287.35 \ \mathrm{K} - 275.95 \ \mathrm{K}) \times (0.02)^3 \ \mathrm{m}^3}{14.8 \times 10^{-6} \ \mathrm{m}^2/\mathrm{s} \times 20.6 \times 10^{-6} \ \mathrm{m}^2/\mathrm{s}} = 10,420$$

h_2 from Eq. (2.1.44) is

$$h_2 = 1.97 \ \mathrm{W}/\mathrm{m}^2 \ \mathrm{K},$$

and

$$\dot{Q} = h_2 \, A \left(T_1 - T_2 \right) = 22.46 \ \text{W}.$$

Comparing the heat loss to that of Example 2.1.6 for a single-pane window, we note that the heat loss through a single-pane window is almost twice as much as through a double-pane window for the same inside and outside conditions.

Forced-Convection Heat Transfer

Forced-convection heat transfer is created by auxiliary means such as pumps and fans or natural phenomena such as wind. This type of process occurs in many engineering applications such as flow of hot or cold fluids inside ducts and various thermodynamic cycles used for refrigeration, power generation, and heating or cooling of buildings. As with natural convection, the main challenge in solving forced-convection problems is to determine the heat-transfer coefficient.

The forced-convection heat transfer processes can be divided into two categories: **external-flow forced convection** and **internal-flow forced convection**. External forced-convection problems are important because they occur in various engineering applications such as heat loss from external walls of buildings on a windy day, from steam radiators, from aircraft wings, or from a hot wire anemometer. To solve these problems, researchers have conducted many experiments to develop correlations for predicting the heat transfer. The experimental results obtained for external forced-convection problems are usually expressed or correlated by an equation of the form

$$\text{Nu} = f(\text{Re}) \, g(\text{Pr}),$$

where f and g represent the functional dependance of the Nusselt number on the Reynolds and Prandtl numbers. The Reynolds number is a nondimensional number representing the ratio of inertia to viscous forces, and the **Prandtl number** is equal to ν/α, which is the ratio of momentum diffusivity to thermal diffusivity.

Table 2.1.4a lists some of the important correlations for calculating forced-convection heat transfer from external surfaces of common geometries. Listed in Table 2.1.4a is the correlation for the forced-convection heat-transfer to or from a fluid flowing over a bundle of tubes, which is relevant to many industrial applications such as the design of commercial heat exchangers. Figures 2.1.20 and 2.1.21 show different configurations of tube bundles in cross-flow whose forced-convection correlations are presented in Table 2.1.4a.

Forced-convection heat transfer in confined spaces is also of interest and has many engineering applications. Flow of cold or hot fluids through conduits and heat transfer associated with that process is important in many HVAC engineering processes. The heat transfer associated with internal forced convection can be expressed by an equation of the form

$$\text{Nu} = f(\text{Re}) \, g(\text{Pr}) \, e(x/D_H),$$

where $f(\text{Re})$, $g(\text{Pr})$, and $e(x/D_H)$ represent the functional dependance on Reynolds number, Prandtl number, and x/D_H, respectively. The functional dependance on x/D_H becomes important for short ducts in laminar flow. The quantity D_H is called the **hydraulic diameter** of the conduit and is defined as

$$D_H = 4 \times \frac{\text{flow cross-sectional area}}{\text{wetted perimeter}} \tag{2.1.50}$$

and is used as the characteristic length for Nusselt and Reynolds numbers.

Fully developed laminar flow through ducts of various cross-section has been studied by Shah and London [1978], and they present analytical solutions for calculating heat transfer and friction coefficients.

TABLE 2.1.4(a) Forced-Convection Heat-Transfer Correlations for External Flows*

Configuration	Correlation	Restrictions	Source
Flat plate in parallel flow	$\overline{Nu}_x = 0.664\, Re_x^{1/2}\, Pr^{1/3}$	Laminar $Pr \geq 0.6$	Incropera and DeWitt [1990]
Flat plate in parallel flow	$Nu_x = 0.0296\, Re_x^{4/5}\, Pr^{1/3}$	Turbulent, local, $0.6 < Pr < 60$, $Re_x < 10^8$	Incropera and DeWitt [1990]
Circular cylinder in cross flow	$\overline{Nu}_D = C\, Re_D^m\, Pr^n \left(\dfrac{Pr_\infty}{Pr_s}\right)^{1/4}$ $n=0.36$ for $Pr > 10$ $n=0.37$ for $Pr \leq 10$ $\begin{array}{ccc} \underline{Re_D} & \underline{C} & \underline{m} \\ 1\text{--}40 & 0.75 & 0.4 \\ 40\text{--}1000 & 0.51 & 0.5 \\ 10^3\text{--}2\times10^5 & 0.26 & 0.6 \\ 2\times10^5\text{--}10^6 & 0.076 & 0.7 \end{array}$	$0.7 < Pr < 500$ $1 < Re_D < 10^6$ Properties at T_∞	Zukauskas [1972]
Non-circular cylinder in cross flow in a gas	$\overline{Nu}_D = C\, Re_D^m\, Pr^{1/3}$ For C and m, see Table 4b.	$0.4 < Re_D < 4 \times 10^5$ $Pr \geq 0.7$	Jakob [1949]
Short cylinder in a gas	$\overline{Nu}_D = 0.123\, Re_D^{0.651} + 0.00416 \left(\dfrac{D}{L}\right)^{0.85} Re_D^{0.792}$	$7 \times 10^4 < Re_D < 2.2 \times 10^5$ $L/D < 4$	Quarmby and Al-Fakhri [1980]

TABLE 2.1.4(a) (continued)　Forced-Convection Heat-Transfer Correlations for External Flows*

Configuration	Correlation	Restrictions	Source
Sphere in a gas or liquid	$\bar{Nu}_D = 2 + \left(0.4\,Re_D^{1/2} + 0.06\,Re_D^{2/3}\right) Pr^{0.4} \left(\dfrac{\mu_\infty}{\mu_s}\right)^{1/4}$	$3.5 < Re_D < 7.6 \times 10^4$ $0.7 < Pr < 380$ $1.0 < \dfrac{\mu_\infty}{\mu_s} < 3.2$ Properties at T_∞	Whitaker [1972]
Tube bundle in cross-flow	$\bar{Nu}_D = Pr^{+0.36}\left(\dfrac{Pr}{Pr_s}\right)^{1/4} C \left(\dfrac{S_T}{S_L}\right)^n Re_D^{\,m}$	See Figures 20 and 21 Properties at T_∞	Zukauskas [1972]

C	m	n	
0.8	0.4	0	$10 < Re_D < 100$, in-line
0.9	0.4	0	$10 < Re_D < 100$, staggered
0.27	0.63	0	$1000 < Re_D < 2 \times 10^5$, in-line $S_T/S_L \geq 0.7$
0.35	0.60	0.2	$1000 < Re_D < 2 \times 10^5$, staggered $S_T/S_L < 2$
0.40	0.60	0	$1000 < Re_D < 2 \times 10^5$, staggered $S_T/S_L \geq 2$
0.021	0.84	0	$Re_D > 2 \times 10^5$, in-line
0.022	0.84	0	$Re_D > 2 \times 10^5$, staggered $Pr > 1$

$\bar{Nu}_D = 0.019\,Re_D^{0.84}$　　$Re_D > 2 \times 10^5$, staggered $Pr = 0.7$

Configuration	Correlation	Restrictions	Source
Flow over staggered tube bundle, gas and liquid	$\bar{Nu}_D = 0.0131\,Re_D^{0.883}\,Pr^{0.36}$	$4.5 \times 10^5 < Re_D < 7 \times 10^6$ $Pr > 0.5$ $S_T/D = 2$,　$S_L/D = 1.4$	Achenbach [1989]

* All properties calculated at $(T_\infty + T_s)/2$ unless otherwise stated under the column "condition." Properties with the subscript "s" are calculated at T_s (surface temperature).

TABLE 2.1.4(b) Constants for Noncircular Cylinders in Cross Flow of a Gas

Configuration	Re_D	C	m
Square			
Flow direction ◇ D	$5 \times 10^3 - 10^5$	0.246	0.588
Flow direction ▢ D	$5 \times 10^3 - 10^5$	0.102	0.675
Hexagon			
Flow direction ⬡ D	$5 \times 10^3 - 1.95 \times 10^4$ $1.95 \times 10^4 - 10^5$	0.160 0.0385	0.638 0.782
Flow direction ⬡ D	$5 \times 10^3 - 10^5$	0.153	0.638
Vertical plate			
Flow direction ▯ D	$4 \times 10^3 - 1.5 \times 10^4$	0.228	0.731

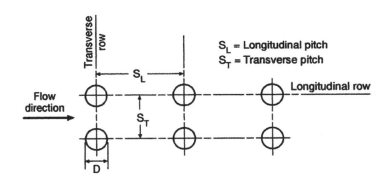

FIGURE 2.1.20 In-line tube arrangement for tube bundle in cross-flow forced convection.

Solving internal tube-flow problems requires knowledge of the nature of the tube-surface thermal conditions. Two special cases of tube-surface conditions cover most engineering applications: constant tube-surface heat flux and constant tube-surface temperature. The axial temperature variations for the fluid flowing inside a tube are shown in Figure 2.1.22. Figure 2.1.22(a) shows the mean fluid-temperature variations inside a tube with constant surface heat flux. Note that the mean fluid temperature, $T_m(x)$, varies linearly along the tube. Figure 2.1.22(b) shows the mean fluid-temperature variations inside a tube with constant surface temperature. Some of the recommended correlations for forced convection of incompressible flow inside tubes and ducts are listed in Table 2.1.5.

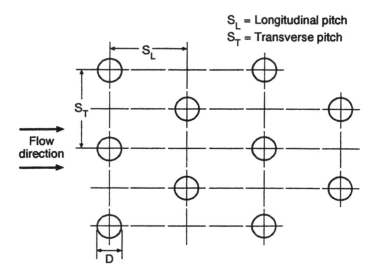

FIGURE 2.1.21   Staggered tube arrangement for tube bundle in cross-flow forced convection.

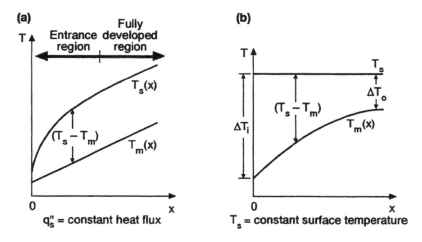

FIGURE 2.1.22   Axial fluid temperature variations for heat transfer in a tube for (a) constant surface heat flux, and (b) constant surface temperature.

Now that we have reviewed both natural- and forced-convection heat-transfer processes, it is useful to compare the order of magnitude of the heat-transfer coefficient for both cases. Table 2.1.6 provides some approximate values of convection heat-transfer coefficients.

Example 2.1.8

A solar-thermal power plant is depicted in Figure 2.1.23. In this system, solar radiation is reflected from tracking mirrors onto a stationary receiver. The receiver consists of a collection of tubes that are radiatively heated, and a working fluid (coolant) flows through them; the heat absorbed by the working fluid is then used to generate electricity. Consider a central-receiver system that consists of several horizontal circular tubes each with an inside diameter of 0.015 m. The working fluid is molten salt that enters the tube at 400°C at a rate of 0.015 kg/s. Assume that the average solar flux approaching the tube is about 10,000 W/m².

TABLE 2.1.5 Forced-Convection Correlations for Incompressible Flow Inside Tubes and Ducts[*,†]

Configuration	Correlation	Restrictions	Source
Fully developed laminar flow in long tubes:			
a. With uniform wall temperature	$\overline{Nu}_D = 3.66$	Pr > 0.6	Kays and Perkins [1985]
b. With uniform heat flux	$\overline{Nu}_D = 4.36$	Pr > 0.6	Incropera and DeWitt [1990]
c. Friction factor (liquids)	$f = \left(\dfrac{64}{Re_D}\right)\left(\dfrac{\mu_s}{\mu_b}\right)^{0.14}$		
d. Friction factor (gas)	$f = \left(\dfrac{64}{Re_D}\right)\left(\dfrac{T_s}{T_b}\right)^{0.14}$		
Laminar flow in short tubes and ducts with uniform wall temperature	$\overline{Nu}_{D_H} = 3.66 + \dfrac{0.0668\ Re_{D_H}\ Pr\ \dfrac{D_H}{L}\left(\dfrac{\mu_b}{\mu_s}\right)^{0.14}}{1+0.045\left(Re_{D_H}\ Pr\ \dfrac{D_H}{L}\right)^{0.66}}$	$100 < Re_{D_H}\ Pr\ \dfrac{D_H}{L} < 1500$ Pr > 0.7	Hausen [1983]
Fully developed turbulent flow through smooth, long tubes and ducts:			
a. Nusselt number	$\overline{Nu}_{D_H} = 0.027\ Re_{D_H}^{0.8}\ Pr^{0.33}\left(\dfrac{\mu_b}{\mu_s}\right)^{0.14}$	$6\times10^3 < Re_{D_H} < 10^7$ $0.7 < Pr < 10^4$ $60 < L/D_H$	Sieder and Tate [1936]
b. Friction factor	$f = \dfrac{0.184}{Re_{D_H}^{0.2}}$	$10^4 < Re_{D_H} < 10^6$	Kays and London [1984]

[*] All physical properties are evaluated at the bulk temperature T_b except μ_s, which is evaluated at the surface temperature T_s.

[†] Incompressible flow correlations apply to gases and vapors when average velocity is less than half the speed of sound (Mach number < 0.5).

TABLE 2.1.6 Order of Magnitude of Convective Heat-Transfer Coefficients h_c

	W/m² K	Btu/h ft²°F
Air, free convection	6–30	1–5
Superheated steam or air, forced convection	3–300	5–50
Oil, forced convection	60–1800	10–300
Water, forced convection	300–18,000	50–3000
Water, boiling	3000–60,000	500–10,000
Steam, condensing	6000–120,000	1000–20,000

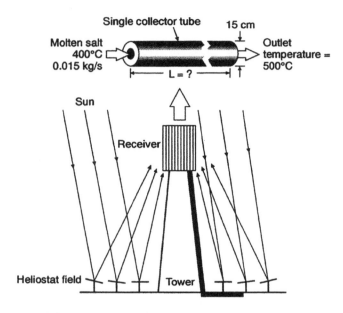

FIGURE 2.1.23 A solar-thermal central-receiver system.

(a) Find the necessary length of the tube to raise the working-fluid temperature to 500°C at the exit.

(b) Determine the tube-surface temperature at the exit.

Solution:

We will assume steady-state conditions, fully developed flow, and incompressible flow with constant properties. The axial temperature variations for heat transfer in a tube for constant heat flux is shown in Figure 2.1.22(a).

(a) The heat capacity of molten salt at $T_m = (T_i + T_o)/2 = 450$°C is $c_p = 1,520$ J/kg K. The total heat transferred to the working fluid is $q''A_t = \dot{m}c_p (T_o - T_i)$, where q'' is the solar flux and $A_t = \pi DL$ is the surface area of the tube (assuming that the solar flux is incident over the entire perimeter of the tube). Therefore,

$$L = \frac{\dot{m}c_p (T_o - T_i)}{q'' \pi D} = \frac{0.015 \text{ kg/s} \times 1520 \text{ J/kg K} \times (773 \text{ K} - 673 \text{ K})}{10^4 \text{ W/m}^2 \times \pi \times 0.015 \text{ m}} = 4.8 \text{ m}$$

(b) Molten salt properties at $T_o = 500$°C are $\mu = 1.31 \times 10^{-3}$ Ns/m², $k = 0.538$ W/m K, and Pr = 3.723. The peak tube-surface temperature can be obtained from $q'' = h(T_s - T_o)$, where h is the local convection coefficient at the exit. To find h, the nature of the flow must first be established by calculating the Reynolds number:

$$Re = \frac{uD}{v} = \frac{4\dot{m}}{\pi\mu D} = \frac{4 \times 0.015 \text{ kg/s}}{\pi \times 1.31 \times 10^{-3} \text{ Ns/m}^2 \times 0.015 \text{ m}} = 972.$$

Because Re < 2,300, the flow inside the tube is laminar. Therefore, from Table 2.1.5, $Nu_D = hD/k = 4.36$, and

$$h = \frac{Nu_D k}{D} = \frac{4.36 \times 0.538 \text{ W/m K}}{0.015 \text{ m}} = 156.4 \text{ W/m}^2 \text{ K}.$$

The surface temperature at the exit is

$$T_s = \frac{q''}{h} + T_o = \frac{10^4 \text{ W/m}^2}{156.4 \text{ W/m}^2 \text{ K}} + 773 \text{ K} = 836.9 \text{ K}.$$

Extended Surfaces or Fins

According to Eq. (2.1.25), the rate of heat transfer by conduction is directly proportional to the heat flow area. To enhance the rate of heat transfer, we can increase the effective heat-transfer surface area. Based on this concept, extended surfaces or fins are widely used in industry to increase the rate of heat transfer for heating or cooling purposes. Various types of extended surfaces are shown in Figure 2.1.24. The simplest type of extended surface is the fin with a uniform cross-section, as shown in Figure 2.1.24(d). The temperature distribution and fin heat-transfer rate can be found by solving a differential equation that expresses energy balance on an infinitesimal element in the fin as given by

$$\frac{d^2T(x)}{dx^2} - \frac{hP}{kA}\left[T(x) - T_\infty\right] = 0, \tag{2.1.51}$$

where P is the cross-sectional perimeter of the fin, k is the thermal conductivity of the fin, A is the cross-sectional area of the fin, and h is the mean convection heat-transfer coefficient between the fin and its surroundings. To solve Eq. (2.1.51), we need two boundary conditions: one at $x = 0$ (base of the fin) and the other at $x = L$ (tip of the fin). The boundary condition used at the base of the fin is usually $T(x = 0) = T_b$, the temperature of the main body to which the fin is attached. The second boundary condition at the tip of the fin ($x = L$) may take several forms:

1. The fin temperature approaches the environment temperature:

$$T \approx T_\infty \quad at \quad x = L.$$

2. There is no heat loss from the end surface of the fin (insulated end):

$$\frac{dT}{dx} = 0 \quad at \quad x = L.$$

3. The fin-end surface temperature is fixed:

$$T = T_L \quad at \quad x = L.$$

4. There is convection heat loss from the end surface of the fin:

$$-k \frac{dT}{dx}\bigg|_{x=L} = h_L \left(T_L - T_\infty\right).$$

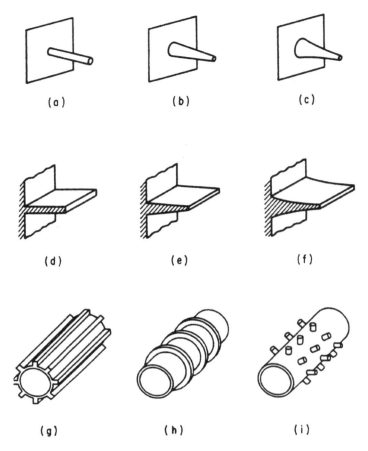

FIGURE 2.1.24   Various types of extended surfaces. Designs (d) – (f) are often used in HVAC heating or cooling coils.

Using the boundary condition at $x = 0$ along with one of the four boundary conditions for $x = L$, we can solve Eq. (2.1.51) and obtain the temperature distribution for a fin with a uniform cross section. Knowing the temperature distribution of the fin, the fin heat-transfer rate q_{fin} can be obtained by applying Fourier's law at the base of the fin:

$$-kA \left.\frac{dT}{dx}\right|_{x=0} = -kA \left.\frac{d\theta}{dx}\right|_{x=0}, \qquad\qquad (2.1.52)$$

where A is the cross-sectional surface area of the fin and $\theta(x) = T(x) - T_\infty$. Figure 2.1.25 is a schematic representation of the temperature distribution in a fin with boundary condition 4. Table 2.1.7 lists equations of temperature distribution and rate of heat transfer for fins of uniform cross section with all four different tip boundary conditions.

Fins or extended surfaces are used to increase the heat-transfer rate from a surface. However, the presence of fins introduces an additional conduction resistance in the path of heat dissipating from the base surface. If a fin is made of highly conductive material, its resistance to heat conduction is small, creating a small temperature gradient from the base to the tip of the fin. However, fins show a temperature distribution similar to that shown in Figure 2.1.25. Therefore, the thermal performance of fins is usually assessed by calculating **fin efficiency**.

The efficiency of a fin is defined as the ratio of the actual heat loss to the maximum heat loss that would have occurred if the total surface of the fin were at the base temperature, that is,

TABLE 2.1.7 Equations for Temperature Distribution and Rate of Heat Transfer for Fins of Uniform Cross Section*

Case	Tip Condition $(x = L)$	Temperature Distribution (θ/θ_b)	Fin Heat-Transfer Rate (q_{fin})
1	Infinite fin $(L \to \infty)$: $\theta(L) = 0$	$e^{-\mu x}$	M
2	Adiabatic: $\left.\dfrac{d\theta}{dx}\right\|_{x=L} = 0$	$\dfrac{\cosh m(L-x)}{\cosh mL}$	$M \tanh mL$
3	Fixed temperature: $\theta(L) = \theta_L$	$\dfrac{(\theta_L/\theta_b)\sinh mx + \sinh m(L-x)}{\sinh mL}$	$M \dfrac{\cosh mL - (\theta_L/\theta_b)}{\sinh mL}$
4	Convection heat transfer: $h\theta(L) = k\left.\dfrac{d\theta}{dx}\right\|_{x=L}$	$\dfrac{\cosh m(L-x) + (h/mk)\sinh m(L-x)}{\cosh mL + (h/mk)\sinh mL}$	$M \dfrac{\sinh mL + (h/mk)\cosh mL}{\cosh mL + (h/mk)\sinh mL}$

$^* \theta \equiv T - T_\infty; \ \theta_b \equiv \theta(0) = T_b - T_\infty; \ m^2 \equiv \dfrac{hP}{kA}; \ M \equiv \sqrt{hPkA}\ \theta_b.$

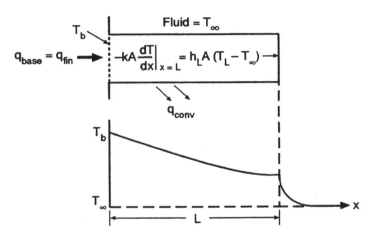

FIGURE 2.1.25 Schematic representation of temperature distribution in a fin with boundary condition 4 at its tip.

$$\eta_{fin} = \frac{q_{fin}}{q_{max}} = \frac{q_{fin}}{hA_f(T_b - T_\infty)}, \qquad (2.1.53)$$

where A_f is the total surface area of the fin, and q_{fin} for fins with uniform cross section is obtained from Table 2.1.7.

Radiation Heat Transfer

Thermal radiation is a heat-transfer process that occurs between any two objects that are at different temperatures. All objects emit thermal radiation by virtue of their temperature. Scientists believe that the thermal radiation energy emitted by a surface is propagated through the surrounding medium either by electromagnetic waves or is transported by photons. In a vacuum, radiation travels at the speed of light C_0 (3×10^8 m/s in a vacuum); however, the **speed of propagation** c in a medium is less than C_0 and is given in terms of index of refraction of the medium, as in Eq. (2.1.54). The radiation wavelength depends on the source frequency and refractive index of the medium through which the radiation travels, according to the equation

$$c = \lambda v = \frac{C_0}{n},$$ (2.1.54)

where n = index of refraction of the medium
$C_0 = 3 \times 10^8$ m/s (9.84×10^8 ft/s)
λ = wavelength, m (ft)
v = frequency, s^{-1}

Thermal radiation can occur over a wide spectrum of wavelengths, namely between 0.1 and 100 μm. The spectral distribution and the magnitude of the emitted radiation from an object depends strongly on its absolute temperature and the nature of its surface. For example, at the surface temperature of the sun, 5,800 K, most energy is emitted at wavelengths near 0.3 μm. However, thermal processes within buildings occur at 10 μm. This particular radiation-process property has caused environmental concerns such as global warming (or the greenhouse effect) in recent years. Global warming is a result of the increased amount of carbon dioxide in the atmosphere. This gas absorbs radiation from the sun at shorter wavelengths but is opaque to emitted radiation from the earth at longer wavelengths, thereby trapping the thermal energy and causing a gradual warming of the atmosphere, as in a greenhouse.

A perfect radiator—called a **blackbody**—emits and absorbs the maximum amount of radiation at any wavelength. The amount of heat radiated by a blackbody is

$$\dot{Q}_r = \sigma A T_b^4,$$ (2.1.55)

where σ = the Stefan–Boltzmann constant = 5.676×10^{-8} W/m^2 K^4 (or 0.1714×10^{-8} Btu/h ft^2 °R^4)
T_b = absolute temperature of the blackbody, K (°R)
A = surface area, m^2 (ft^2)

The spectral (or monochromatic) **blackbody emissive power** according to Planck's Law is

$$E_{b\lambda}(T) = \frac{C_1 \lambda^{-5}}{e^{C_2/\lambda T} - 1},$$ (2.1.56)

where $E_{b\lambda}(T)$ = spectral emissive power of a blackbody at absolute temperature T, $\dfrac{W}{m^3}\left(\dfrac{Btu}{h\,ft^2\mu}\right)$

λ = wavelength, m (μ)

T = absolute temperature of blackbody, K (°R)

C_1 = constant, 3.7415×10^{-16} W m^2 $\left(1.187 \times 10^8 \dfrac{Btu\ \mu^4}{h\,ft^2}\right)$

C_2 = constant, 1.4388×10^{-2} m K (2.5896×10^4 μ °R)

The spectral blackbody emissive power for different temperatures is plotted in Figure 2.1.26, which shows that as the temperature increases, the emissive power and the wavelength range increase as well. However, as temperature increases, the wavelength at which maximum emissive power occurs decreases. Wien's Displacement Law provides a relationship between the maximum power wavelength λ_{max} and the absolute temperature at which $E_{b\lambda}$ is maximum:

$$\lambda_{max}\ T = 2.898 \times 10^{-3}\ mK = 5216.4\ \mu\ °R.$$

To obtain the total emissive power of a blackbody, we integrate the spectral emissive power over all wavelengths:

$$E_b = \int_0^\infty E_{b\lambda}\ d\lambda = \sigma T_b^4.$$ (2.1.57)

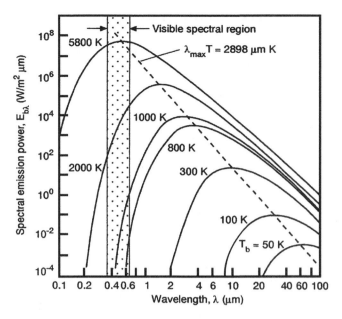

FIGURE 2.1.26 Spectral blackbody emissive power for different temperatures.

Equation (2.1.57) is the same as Eq. (2.1.55) except that it is expressed per unit area. At a given temperature T_b, the quantity E_b of Eq. (2.1.57) is the area under the curve corresponding to T_b in Figure 2.1.26.

Engineers sometimes encounter problems where it is necessary to find the fraction of the total energy radiated from a blackbody in a finite interval between two specific wavelengths λ_1 and λ_2. This fraction for an interval from 0 to λ_1 can be determined from:

$$B(0 \to \lambda_1) = \frac{\displaystyle\int_0^{\lambda_1} E_{b\lambda}\ d\lambda}{\displaystyle\int_0^{\infty} E_{b\lambda}\ d\lambda} = \frac{\displaystyle\int_0^{\lambda_1} E_{b\lambda}\ d\lambda}{\sigma T_b^4}.$$

This integral has been calculated for various λT quantities, and the results are presented in Table 2.1.8. The fraction of total radiation from a blackbody in a finite wavelength interval from λ_1 to λ_2 can then be obtained from

$$B(\lambda_1 \to \lambda_2) = \frac{\displaystyle\int_0^{\lambda_2} E_{b\lambda}\ d\lambda - \int_0^{\lambda_1} E_{b\lambda}\ d\lambda}{\sigma T_b^4} = B(0 \to \lambda_2) - B(0 \to \lambda_1)$$

where quantities $B(0 \to \lambda_2)$ and $B(0 \to \lambda_1)$ can be read from Table 2.1.8.

Radiation Properties of Objects

When radiation strikes the surface of an object, a portion of the total incident radiation is reflected, a portion is absorbed, and if the object is transparent, a portion is transmitted through the object, as depicted in Figure 2.1.27.

The fraction of incident radiation which is reflected is called the **reflectance (or reflectivity)** ρ, the fraction transmitted is called the **transmittance (or transmissivity)** τ, and the fraction absorbed is called the

TABLE 2.1.8   Blackbody Radiation Functions

λT $(mK \times 10^3)$	$B(0 \to \lambda)$	λT $(mK \times 10^3)$	$B(0 \to \lambda)$
0.2	0.341796×10^{-26}	6.2	0.754187
0.4	0.186468×10^{-11}	6.4	0.769234
0.6	0.929299×10^{-7}	6.6	0.783248
0.8	0.164351×10^{-4}	6.8	0.796180
1.0	0.320780×10^{-3}	7.0	0.808160
1.2	0.213431×10^{-2}	7.2	0.819270
1.4	0.779084×10^{-2}	7.4	0.829580
1.6	0.197204×10^{-1}	7.6	0.839157
1.8	0.393499×10^{-1}	7.8	0.848060
2.0	0.667347×10^{-1}	8.0	0.856344
2.2	0.100897	8.5	0.856344
2.4	0.140268	9.0	0.890090
2.6	0.183135	9.5	0.903147
2.8	0.227908	10.0	0.914263
3.0	0.273252	10.5	0.923775
3.2	0.318124	11.0	0.931956
3.4	0.361760	11.5	0.939027
3.6	0.403633	12	0.945167
3.8	0.443411	13	0.955210
4.0	0.480907	14	0.962970
4.2	0.516046	15	0.969056
4.4	0.548830	16	0.973890
4.6	0.579316	18	0.980939
4.8	0.607597	20	0.985683
5.0	0.633786	25	0.992299
5.2	0.658011	30	0.995427
5.4	0.680402	40	0.998057
5.6	0.701090	50	0.999045
5.8	0.720203	75	0.999807
6.0	0.737864	100	1.000000

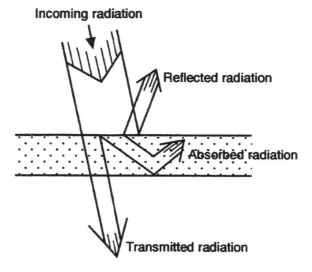

FIGURE 2.1.27   Schematic of reflected, transmitted, and absorbed radiation.

absorptance (**or absorptivity**) α. There are two types of radiation reflections: specular and diffuse. A **specular reflection** is one in which the angle of incidence is equal to the angle of reflection, whereas a **diffuse reflection** is one in which the incident radiation is reflected uniformly in all directions. Highly polished surfaces such as mirrors approach the specular reflection characteristics, but most industrial surfaces (rough surfaces) have diffuse reflection characteristics. By applying an energy balance to the surface of the object as shown in Figure 2.1.27, the relationship between these properties can be expressed as

$$\alpha + \rho + \tau = 1. \tag{2.1.58}$$

The relative magnitude of each one of these components depends on the characteristics of the surface, its temperature, and the spectral distribution of the incident radiation. If an object is opaque ($\tau = 0$), it will not transmit any radiation. Therefore

$$\alpha + \rho = 1 \text{ and } \tau = 0 \text{ for an opaque object.} \tag{2.1.59}$$

If an object has a perfectly reflecting surface (a good mirror), then it will reflect all the incident radiation, and

$$\rho = 1, \ \alpha = 0, \text{ and } \tau = 0 \text{ for a perfectly reflective surface.} \tag{2.1.60}$$

The **emissivity**, ε, of a surface at temperature T is defined as the ratio of total energy emitted to the energy that would be emitted by a blackbody at the same temperature T:

$$\varepsilon = \frac{E(T)}{\sigma T^4}, \tag{2.1.61}$$

where $E(T)$ represents the radiation energy emitted from the surface. For a blackbody, Eq. (2.1.61) gives $\varepsilon_b = 1$. The absorptivity for a blackbody is also equal to unity; therefore, $\varepsilon_b = \alpha_b = 1$.

A special type of surface called a gray surface or **graybody** is a surface with spectral emissivity and absorptivity that are both independent of the wavelength. Therefore, for a graybody, $\bar{\alpha} = \alpha_\lambda = \bar{\varepsilon} = \varepsilon_\lambda$ where $\bar{\varepsilon}$ and $\bar{\alpha}$ are the average values of emissivity and absorptivity, respectively. In many engineering problems, surfaces are not gray surfaces. However, one can employ graybody assumptions by using suitable $\bar{\alpha}$ and $\bar{\varepsilon}$ values.

Table 2.1.9 provides emissivities of various surfaces at several wavelengths and temperatures. A more extensive list of experimentally measured radiation properties of various surfaces has been provided by Gubareff et al. [1960] and Kreith and Bohn [1993]; note that the listed quantities in Table 2.1.9 are hemispherical emissivities. Detailed directional and spectral measurements of radiation properties of surfaces are limited in the literature. Because of the difficulties in performing these detailed measurements, most of the tabulated properties are averaged quantities, such as those presented in Table 2.1.9. Properties averaged with respect to wavelength are termed *total* quantities, and properties averaged with respect to direction are termed *hemispherical* quantities. Hemispherical spectral emissivity of a surface is the ratio of (1) the spectral radiation emitted by a unit surface area of an object into all directions of a hemisphere surrounding that area to (2) the spectral radiation emitted by a unit surface area of a blackbody (at the same temperature) into all directions of that hemisphere.

The Radiation Shape Factor (View Factor)
In this section, we will only deal with surfaces that have diffuse reflection characteristics, because most real surfaces used in different industries can be assumed to have diffuse reflection characteristics. In solving radiation problems, we must find out how much of the radiation leaving one surface is being intercepted by another surface.

TABLE 2.1.9   Hemispherical Emissivities of Various Surfaces[a]

Material	Wavelength and Average Temperature				
	9.3 mm 310 K	5.4 mm 530 K	3.6 mm 800 K	1.8 mm 1700 K	0.6 mm Solar ~6,000 K
			Metals		
Aluminum					
polished	~0.04	0.05	0.08	~0.19	~0.30
oxidized	0.11	~0.12	0.18		
24-ST weathered	0.40	0.32	0.27		
surface roofing	0.22				
anodized (at 1,000°F)	0.94	0.42	0.60	0.34	
Brass					
polished	0.10	0.10			
oxidized	0.61				
Chromium					
polished	~0.08	~0.17	0.26	~0.40	0.49
Copper					
polished	0.04	0.05	~0.18	~0.17	
oxidized	0.87	0.83	0.77		
Iron					
polished	0.06	0.08	0.13	0.25	0.45
cast, oxidized	0.63	0.66	0.76		
galvanized, new	0.23			0.42	0.66
galvanized, dirty	0.28			0.90	0.89
steel plate, rough	0.94	0.97	0.98		
oxide	0.96		0.85		0.74
molten				0.3–0.4	
Magnesium	0.07	0.13	0.18	0.24	0.30
Molybdenum filament			~0.09	~0.15	~0.20[b]
Silver					
polished	0.01	0.02	0.03		0.11
Stainless steel					
18-8, polished	0.15	0.18	0.22		
18-8, weathered	0.85	0.85	0.85		
Steel tube					
oxidized		0.94			
Tungsten filament	0.03			~0.18	0.35[c]
Zinc					
polished	0.02	0.03	0.04	0.06	0.46
galvanized sheet	~0.25				
		Building and Insulating Materials			
Asbestos paper	0.93	0.93			
Asphalt	0.93		0.90		0.93
Brick					
red	0.93				0.70
fire clay	0.90		~0.70	~0.75	
silica	0.90		0.75	0.84	
magnesite refractory	0.90			~0.40	
Enamel, white	0.90				
Marble, white	0.95		0.93		0.47
Paper, white	0.95		0.82	0.25	0.28
Plaster	0.91				
Roofing board	0.93				
Enameled steel, white				0.65	0.47
Asbestos cement, red				0.67	0.66

TABLE 2.1.9 (continued)   Hemispherical Emissivities of Various Surfaces[a]

Material	Wavelength and Average Temperature				
	9.3 mm 310 K	5.4 mm 530 K	3.6 mm 800 K	1.8 mm 1700 K	0.6 mm Solar ~6,000 K
	Paints				
Aluminized lacquer	0.65	0.65			
Cream paints	0.95	0.88	0.70	0.42	0.35
Lacquer, black	0.96	0.98			
Lampblack paint	0.96	0.97		0.97	0.97
Red paint	0.96				0.74
Yellow paint	0.95		0.50		0.30
Oil paints (all colors)	~0.94	~0.90			
White (ZnO)	0.95		0.91		0.18
	Miscellaneous				
Ice	~0.97[d]				
Water	~0.96				
Carbon					
T-carbon, 0.9% ash	0.82	0.80	0.79		
filament	~0.72			0.53	
Wood	~0.93				
Glass	0.90				(Low)

[a] Since the emissivity at a given wavelength equals the absorptivity at that wavelength, the values in this table can be used to approximate the absorptivity to radiation from a source at the temperature listed. For example, polished aluminum will absorb 30% of incident solar radiation.

[b] At 3,000 K.

[c] At 3,600 K.

[d] At 273 K.

Sources: Fischenden and Saunders [1932]; Hamilton and Morgan [1962]; Kreith and Black [1980]; Schmidt and Furthman [1928]; McAdams [1954]; Gubareff et al. [1960].

The **radiation shape** factor F_{1-2} is defined as

$$F_{1-2} = \frac{\text{diffuse radiation leaving surface } A_1 \text{ and being intercepted by surface } A_2}{\text{total diffuse radiation leaving surface } A_1}.$$

For example, consider two black surfaces A_1 and A_2 at temperatures T_1 and T_2, as shown in Figure 2.1.28. The radiation leaving surface A_1 and reaching A_2 is

$$\dot{Q}_{1\to2} = A_1 F_{1-2} E_{b1}, \tag{2.1.62}$$

and the radiation leaving surface A_2 and reaching surface A_1 is

$$\dot{Q}_{2\to1} = A_2 F_{2-1} E_{b2}, \tag{2.1.63}$$

From Eqs. (2.1.62) and (2.1.63), we can calculate the net radiation heat exchange between these two black surfaces:

$$\Delta\dot{Q}_{1\to2} = A_1 F_{1-2} E_{b1} - A_2 F_{2-1} E_{b2}.$$

Shape factors for some geometries that have engineering applications are presented in Table 2.1.10. For more information and an extensive list of shape factors, refer to Siegel and Howell [1972].

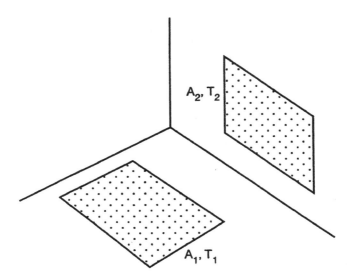

FIGURE 2.1.28 Sketch illustrating the nomenclature for shape factor between the two surfaces A_1 and A_2.

Example 2.1.9

A flat-plate solar collector with a single glass cover to be used for building water heating is shown in Figure 2.1.29. The following quantities are known:

The solar irradiation, $G_s = 750$ W/m^2
The absorptivity of the cover plate to solar radiation, $\alpha_{cp,s} = 0.16$
The transmissivity of the cover plate to solar radiation, $\tau_{cp} = 0.84$
The emissivity of the cover plate to longwave radiation, $\varepsilon_{cp} = 0.9$
The absorptivity of the absorber plate to solar radiation, $\alpha_{ap,s} = 1.0$
The emissivity of the absorber plate to longwave radiation, $\varepsilon_{ap} = 0.1$
The convection coefficient between the absorber plate and the cover plate, $h_i = 2$ W/m^2 K
The convection coefficient between the cover plate and ambient, $h_o = 5$ W/m^2 K
The absorber-plate temperature, $T_{ap} = 120$°C
The ambient air temperature $T_\infty = 30$°C
The effective sky temperature, $T_{sky} = -10$°C

Using this information, calculate the useful heat absorbed by the absorber plate.

Solution:

We will assume the following:

- Steady-state conditions
- Uniform surface heat-flux and temperature for the cover plate and the absorber plate
- Opaque, diffuse-gray surface behavior for longwave radiation
- Well-insulated absorber plate

To find the useful heat absorbed by the absorber plate, perform an energy balance on a unit area of the absorber plate, as in Figure 2.1.30:

$$\alpha_{ap,s}\ \tau_{cp,s}\ G_s = q_{conv,i} + q_{rad,ap-cp} + q_u, \tag{2.1.64}$$

TABLE 2.1.10 Minicatalog of Geometric View Factors

Configuration	Geometric View Factor
	Two infinitely long plates of width L, joined along one of the long edges: $$F_{1-2} = F_{2-1} = 1 - \sin\frac{\alpha}{2}$$
	Two infinitely long plates of different widths (H, L), joined along one of the long edges and with a 90° angle between them: $$F_{1-2} = \frac{1}{2}[1 + x - (1 + x^2)^{1/2}]$$ where x = H/L
	Triangular cross-section enclosure formed by three infinitely long plates of different widths (L_1, L_2, L_3): $$F_{1-2} = \frac{L_1 + L_2 - L_3}{2L_1}$$
	Circular disk and plane element positioned on the disc centerline: $$F_{1-2} = \frac{R^2}{H^2 + R^2}$$
	Parallel discs positioned on the same centerline: $$F_{1-2} = \frac{1}{2}\left\{ X - \left[X^2 - 4\left(\frac{x_2}{x_1}\right)^2 \right]^{1/2} \right\}$$ where $$x_1 = \frac{R_1}{H}, \ x_2 = \frac{R_2}{H}, \text{ and } X = 1 + \frac{1 + x_2^2}{x_1^2}$$

TABLE 2.1.10 (continued)   Minicatalog of Geometric View Factors

Configuration	Geometric View Factor
	Infinite cylinder parallel to an infinite plate of finite width (L_1 - L_2): $$F_{1-2} = \frac{R}{L_1 - L_2}\left(\tan^{-1}\frac{L_1}{H} - \tan^{-1}\frac{L_2}{H}\right)$$
	Two parallel and infinite cylinders: $$F_{1-2} = F_{2-1} = \frac{1}{\pi}\left[\left(X^2 - 1\right)^{1/2} + \sin^{-1}\left(\frac{1}{X}\right) - X\right]$$ where $X = 1 + \dfrac{L}{2R}$
	Concentric cylinders of infinite length: $$F_{1-2} = 1$$ $$F_{2-1} = \frac{R_1}{R_2}$$ $$F_{2-2} = 1 - \frac{R_1}{R_2}$$
	Row of equidistant infinite cylinders parallel to an infinite plate: $$F_{1-2} = 1 - (1 - x^2)^{1/2} + x\tan^{-1}\left(\frac{1 - x^2}{x^2}\right)^{1/2}$$ where $x = D/L$

where $q_{conv,i} = h_i\,(T_{ap} - T_{cp})$ is the convection heat exchange between the absorber plate and the cover plate and $q_{rad,ap\text{-}cp} = \sigma\left(T_{ap}^4 - T_{cp}^4\right)\big/\left(1/\varepsilon_{ap} + 1/\varepsilon_{cp} - 1\right)$ is the heat exchange by radiation between them. Note that the shape factor between two parallel plates is equal to one. The left-hand side of Eq. (2.1.64) represents the solar irradiation transmitted through the cover plate and absorbed by the absorber plate. Substituting for $q_{conv,i}$ and $q_{rad,ap\text{-}cp}$ in Eq. (2.1.64), we obtain (for $\alpha_{ap,s} = 1$)

$$\tau_{cp,s}\ G_s = h_i\left(T_{ap} - T_{cp}\right) + \frac{\sigma\left(T_{ap}^4 - T_{cp}^4\right)}{1/\varepsilon_{ap} + 1/\varepsilon_{cp} - 1} + q_u. \tag{2.1.65}$$

TABLE 2.1.10 (continued) Minicatalog of Geometric View Factors

Configuration	Geometric View Factor
	Sphere and disc positioned on the same centerline: $$F_{1-2} = \frac{1}{2}\left[1 - \frac{1}{\sqrt{1+x^2}}\right]$$ where $x = \dfrac{R_2}{H}$
	Sphere and a sector of disk positioned on the same centerline: $$F_{1-2} = \frac{\alpha}{4\pi}\left[1 - \frac{1}{\sqrt{1+x^2}}\right]$$ where $x = \dfrac{R_2}{H}$
	Concentric spheres: $$F_{1-2} = 1$$ $$F_{2-1} = \left(\frac{R_1}{R_2}\right)^2$$ $$F_{2-2} = 1 - \left(\frac{R_1}{R_2}\right)^2$$

To find q_u from Eq. (2.1.65), T_{cp} should be known, which is obtained from an energy balance on the cover plate, as in Figure 2.1.31:

$$\alpha_{cp,s}\, G_s + q_{conv,i} + q_{rad,ap-cp} = q_{conv,o} + q_{rad,cp-sky}, \tag{2.1.66}$$

where $q_{conv,o} = h_o\,(T_{cp} - T_\infty)$ is the heat loss by convection and $q_{rad,cp-sky} = \varepsilon_{cp}\sigma\left(T_{cp}^4 - T_{sky}^4\right)$ is the heat exchange by radiation between the cover plate and sky. Equation (2.1.66) can be written as

$$\alpha_{cp,s}\, G_s + h_i\left(T_{ap} - T_{cp}\right) + \frac{\sigma\left(T_{ap}^4 - T_{cp}^4\right)}{1/\varepsilon_{ap} + 1/\varepsilon_{cp} - 1} = h_o\left(T_{cp} - T_\infty\right) + \varepsilon_{cp}\,\sigma\left(T_{cp}^4 - T_{sky}^4\right). \tag{2.1.67}$$

Substituting for known quantities in Eq. (2.1.67) T_{cp} is calculated to be $T_{cp} = 44.6°C$. Substituting for T_{cp} and other known quantities in Eq. (2.1.65), q_u is 402.5 W/m².

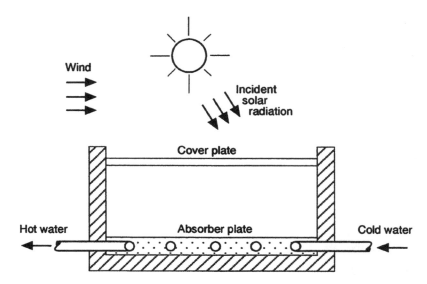

FIGURE 2.1.29   Flat-plate solar collector with a single glass cover.

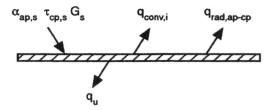

FIGURE 2.1.30   Energy balance on a unit area of the absorber plate.

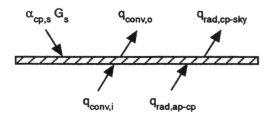

FIGURE 2.1.31   Energy balance on a unit area of the cover plate.

2.1.3   Fundamentals of Fluid Mechanics

The distribution of heated and cooled fluids by pipes and ducts, is an essential part of all HVAC processes and systems. The fluids encountered in these processes are gases, vapors, liquids, or mixtures of liquid and vapor (2–phase flow). This section briefly reviews certain basic concepts of fluid mechanics that are often encountered in analyzing and designing HVAC systems.

Fluid flowing through a conduit will encounter shear forces that result from viscosity of the fluid. The fluid undergoes continuous deformation when subjected to these shear forces. Furthermore, as a result of shear forces, the fluid will experience pressure losses as it travels through the conduit.

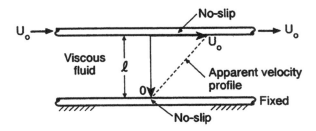

FIGURE 2.1.32   A fluid sheared between two parallel plates.

Viscosity, μ, is a property of fluid best defined by Newton's Law of Viscosity:

$$\tau = \mu \frac{du}{dy},\tag{2.1.68}$$

where τ is the frictional shear stress, and du/dy represents the measure of the motion of one layer of fluid relative to an adjacent layer. The following observation will help to explain the relationship between viscosity and shear forces. Consider two very long parallel plates with a fluid between them, as shown in Figure 2.1.32. Assume a uniform pressure throughout the fluid. The upper plate is moving with a constant velocity u_0, and the lower plate is stationary. Experiments show that the fluid adjacent to the moving plate will adhere to that plate and move along with the plate at a velocity equal to u_0, whereas the fluid adjacent to the stationary plate will have zero velocity. The experimentally verified velocity distribution in the fluid is linear and can be expressed as

$$u = \frac{y}{\ell}\,u_0,\tag{2.1.69}$$

where ℓ is the distance between the two parallel plates. The force necessary to keep the upper plate moving at a constant velocity if u_0 should be large enough to overcome (or balance) the frictional forces in the fluid. Again, experimental observations indicate that this force is proportional to the ratio u_0/ℓ. One can conclude from Eq. (2.1.69) that u_0/ℓ is equal to the rate of change of velocity, du/dy. Therefore, the frictional force per unit area (shear stress), τ, is proportional to du/dy, and the proportionality constant is μ, which is a property of the fluid known as viscosity. Therefore, we obtain Eq. (2.1.68), which is known as Newton's Law of Viscosity (or friction). The quantity μ is a measure of the viscosity of the fluid and depends on the temperature and pressure of the fluid. Equation (2.1.68) is analogous to Fourier's Law of Heat Conduction given by Eq. (2.1.25). Fluids that do not obey Newton's Law of Viscosity are called non-Newtonian fluids. Fluids with zero viscosity are known as inviscid or ideal fluids. Molasses and tar are examples of highly viscous liquids; water and air on the other hand, have low viscosities. The viscosity of a gas increases with temperature, but the viscosity of a liquid decreases with temperature. Reid, Sherwood, and Prausnitz [1977] provide a thorough discussion on viscosity.

Flow Characteristics

The flow of a fluid may be characterized by one or a combination of the following descriptor pairs: laminar/turbulent, steady/unsteady, uniform/nonuniform, reversible/irreversible, rotational/irrotational. In this section, however, we will focus our attention only on laminar and turbulent flows.

In **laminar flow**, fluid particles move along smooth paths in layers, with one layer sliding smoothly over an adjacent layer without significant macroscopic mixing. Laminar flow is governed by Newton's Law of Viscosity. Turbulent flow is more prevalent than laminar flow in engineering processes. In

turbulent flow, the fluid particles move in irregular paths, causing an exchange of momentum between various portions of the fluid; adjacent fluid layers mix and this mixing mechanism is called eddy motion. In this type of flow, the velocity at any given point under steady-state conditions fluctuates in all directions about some time-mean value. Turbulent flow causes greater shear stresses throughout the fluid, producing more irreversibilities and losses. An equation similar to Newton's Law of Viscosity may be written for turbulent flows:

$$\tau = (\mu + \eta)\, \frac{du}{dy}, \tag{2.1.70}$$

where the factor η is the eddy viscosity, which depends on the fluid motion and density. Unlike the fluid viscosity, μ, the eddy viscosity is not a fluid property and is determined through experiments.

The type of flow is primarily determined by the value of a nondimensional number known as a Reynolds number, which is the ratio of inertia forces to viscous forces given by

$$\mathrm{Re} = \frac{\rho\, u_{\mathrm{avg}}\, D_H}{\mu}, \tag{2.1.71}$$

where u_{avg} is the average velocity and D_H is the hydraulic diameter defined by Eq. (2.1.50). The value of the Reynolds number can be used as the criterion to determine whether the flow is laminar or turbulent. In general, laminar flow occurs in closed conduits when $\mathrm{Re} < 2{,}100$; the flow goes through transition when $2{,}100 < \mathrm{Re} < 6{,}000$ and becomes turbulent when $\mathrm{Re} > 6{,}000$.

For fluid flow over flat plates, laminar flow is generally accepted to occur at $\mathrm{Re}_x = \rho u x/\mu < 3 \times 10^5$, where x is the distance from the leading edge of the plate and u is the free-stream velocity. Note that if the flow approaching the flat plate is turbulent, it will remain turbulent from the leading edge of the plate forward.

When a fluid is flowing over a solid surface, the velocity of the fluid layer in the immediate neighborhood of the surface is influenced by viscous shear; this region of the fluid is called the **boundary layer**. Boundary layers can be laminar or turbulent depending on their length, the fluid viscosity, the velocity of the bulk fluid, and the surface roughness of the solid body.

Analysis of Flow Systems

Most engineering problems require some degree of system analysis. Regardless of the nature of the flow, all fluid-flow situations are subject to the following relations:

1. Newton's Law of Motion, $\Sigma F = \dfrac{1}{g_c}\dfrac{d(mu)}{dt}$
2. Conservation of mass
3. The First and Second Laws of Thermodynamics
4. Boundary conditions such as zero velocity at a solid surface.

In an earlier section, the First Law of Thermodynamics was applied to a system shown in Figure 2.1.1. With some modifications, the same energy balance can be applied to any fluid-flow system. For example, a term representing the frictional pressure losses should be added to the left-hand side of Eq. (2.1.2), as expressed by the following equation:

$$e_2 + \frac{u_2^2}{2g_c} + \frac{gz_2}{g_c} + \frac{p_2}{\rho_2} + \frac{\dot{w}_f}{\dot{m}} = e_1 + \frac{u_1^2}{2g_c} + \frac{gz_1}{g_c} + \frac{p_1}{\rho_1} + \frac{\dot{Q} + \dot{w}}{\dot{m}}, \tag{2.1.72}$$

where $\dot{w}_f$ represents the frictional pressure losses and is the rate of work done on the fluid (note the sign change from $-\dot{w}$ to $+\dot{w}$ in Eq. (2.1.72), because the work is done on the fluid). In the remainder of this section, we will focus on obtaining an expression for $\dot{w}_f$ and analyzing different sources of frictional pressure losses.

Using Newton's Law of Motion, the weight of a body, w, can be defined as the force exerted on the body as a result of the acceleration of gravity, g,

$$w = \frac{g}{g_c} m. \tag{2.1.73}$$

In the English system of units, 1 lbm weighs 1 lbf at sea level because the proportionality constant g_c is numerically equal to the gravitational acceleration (32.2 ft/s²). However, in the SI system, 1 kg of mass weighs 9.81 N at sea level because $g_c = 1$ kg m/N s² (or $g_c = 10^3$ kg m³/kJ s²) and $g = 9.81$ m/s².

Equation (2.1.73) can be used to determine the static pressure of a column of fluid. For example, a column of fluid at height z that experiences an environment or atmospheric pressure of p_0 over its upper surface will exert a pressure of p at the base of the fluid column given by

$$p = p_0 + \rho z \frac{g}{g_c}, \tag{2.1.74}$$

where ρ is the density of the fluid. The base pressure as expressed by Eq. (2.1.74) is a function of fluid height or fluid head and does not depend on the shape of the container. Knowing the fluid head is very important, especially in specifying a pump, as it is common practice to specify the performance of the pump in terms of fluid head. Therefore, we can calculate the required mechanical power from

$$\dot{w}_{pump} = zg \, \dot{m}. \tag{2.1.75}$$

Equation (2.1.75) expresses the pump power at 100% efficiency; in reality, however, mechanical pumps have efficiencies of less than 100%. Therefore, the required mechanical power $\dot{w}$ is

$$\dot{w} = \frac{\dot{w}_{pump}}{\eta_{pump}}. \tag{2.1.76}$$

A pump used in a system is expected to overcome various types of pressure losses such as frictional pressure losses in the piping; pressure losses due to fittings, bends, and valves; and pressure losses due to sudden enlargements and contractions. All these pressure losses should be calculated for a system and summed up to obtain the total pressure drop through a system.

The frictional pressure losses in the piping are caused by the shearing force at the fluid-solid interface. Through a force balance, we can obtain the frictional pressure loss of an incompressible fluid in a pipe between two points as

$$p_1 - p_2 = 4f \frac{L}{D} \frac{\rho u^2}{2g_c}, \tag{2.1.77}$$

where L is the length of the pipe between points 1 and 2, D is the pipe diameter, u is the average fluid velocity in the pipe, and f is the dimensionless friction factor. For laminar flow inside a pipe, the friction factor is

$$f = \frac{16}{\mathrm{Re}_{D_H}}, \tag{2.1.78}$$

where the Reynolds number is based on the hydraulic diameter D_H. The friction factor for turbulent flow depends on the surface roughness of the pipe and on the Reynolds number. The friction factor for various surface roughnesses and Reynolds numbers is presented in Figure 2.1.33, which is called the **Moody diagram**. The relative roughnesses of the various commercial pipes are given in Figure 2.1.34.

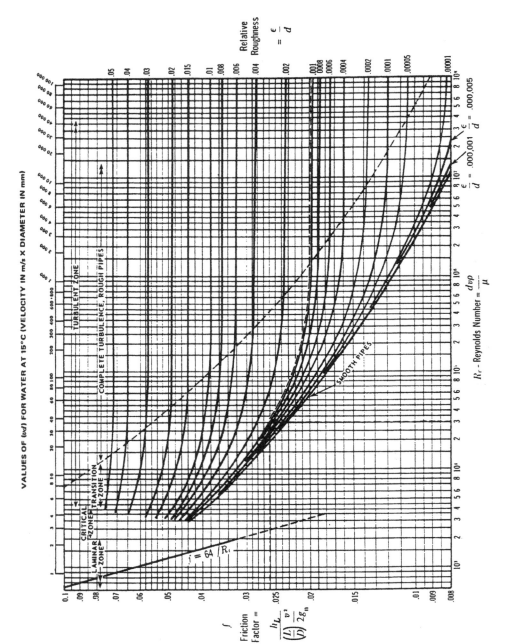

FIGURE 2.1.33 Friction factors for various surface roughness and Reynolds numbers. Data extracted from *Friction Factor for Pipe Flow* by L.F. Moody (1944), with permission of the publisher, The American Society of Mechanical Engineers.

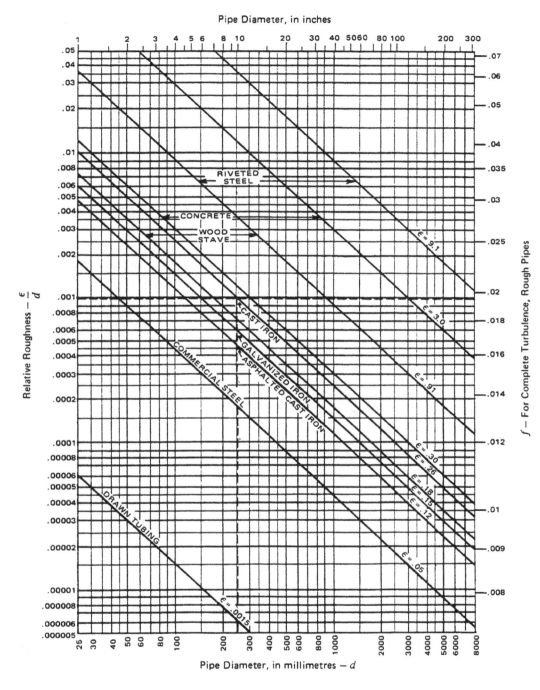

FIGURE 2.1.34 Relative roughness of commercial pipe. Data extracted from *Friction Factor for Pipe Flow* by L.F. Moody (1944), with permission of the publisher, The American Society of Mechanical Engineers.

Pressure losses due to fittings, bends, and valves are generally determined through experiments. This type of pressure loss can be correlated to the average fluid velocity in the pipe by

$$\Delta p_b = k_b \frac{\rho u^2}{2g_c}, \tag{2.1.79}$$

where k_b is a pressure-loss coefficient obtained from a handbook or from the manufacturer, and u is the average fluid velocity in the pipe upstream of the fitting, bend, or valve. For typical values of k_b, refer to Perry, Perry, Chilton, and Kirkpatrick [1963], Freeman [1941], and the *Standards of Hydraulic Institute* [1948].

Pressure losses due to sudden enlargement of the cross section of the pipe can be calculated using

$$\Delta p = \alpha \left(1 - \frac{A_s}{A_L}\right) \frac{\rho u^2}{2g_c} = k_e \frac{\rho u^2}{2g_c}, \tag{2.1.80}$$

where A_s/A_L is the ratio of the cross-sectional area of the smaller pipe to that of the larger pipe, α is the nondimensional pressure-loss coefficient ($\alpha = 1$ for turbulent flow and 2 for laminar flow), and u is the average fluid velocity in the smaller pipe. Note that a gradual increase in pipe cross section will have little effect on pressure losses. In case of sudden contraction of pipe size, the pressure drop can be calculated from

$$\Delta p = 0.55 \, \alpha \left(1 - \frac{A_s}{A_L}\right) \frac{\rho u_c^2}{2g_c} = k_c \frac{\rho u_c^2}{2g_c}, \tag{2.1.81}$$

where A_s/A_L and α are as defined for Eq. (2.1.80), and u_c is the average fluid velocity in the smaller pipe (contraction). Adding the various pressure losses, the total pressure loss in a system can be calculated from

$$\frac{\dot{w}_f}{\dot{m}} = \Sigma \frac{\Delta p}{\rho} = 4f \frac{L}{D} \frac{u^2}{2g_c} + \Sigma k_b \frac{u^2}{2g_c} + k_e \frac{u^2}{2g_c} + k_c \frac{u_c^2}{2g_c}. \tag{2.1.82}$$

For a system under consideration, a pump must be chosen that can produce sufficient pressure head to overcome all the losses presented in Eq. (2.1.82). For system engineering applications, Eq. (2.1.82) can be simplified to

$$\frac{\dot{w}_f}{\dot{m}} = \Sigma \frac{2fLu^2}{g_c D}, \tag{2.1.83}$$

where f is as defined for Eq. (2.1.77), u is the average velocity inside the conduit, and D is the appropriate diameter for the section of the system under consideration. The summation accounts for the effect of changes in pipe length, diameter, and relative roughness. The length L represents not only the length of the straight pipe of the system, but also, equivalent lengths of straight pipe that would have the same effects as the fittings, bends, valves, and sudden enlargements or contractions. Figure 2.1.35 provides a nomogram to determine such equivalent lengths.

Example 2.1.10

Figure 2.1.36 shows a system layout for a small solar collector where water at 35°C (95°F) is pumped from a tank (surface-area heat exchanger) through three parallel solar collectors and back to the tank. The water flow rate is 0.9 m³/min (23.8 gal/min). All the piping is 1–in. Sch 40 steel pipe (cross-sectional area = 0.006 ft² = 5.57×10^{-4} m², with inside diameter = 1.049 in. = 0.0266 m). The pressure drop through each solar collector is estimated to be 1.04 kPa (0.15 psi) for a flow rate of 0.03 m³/min

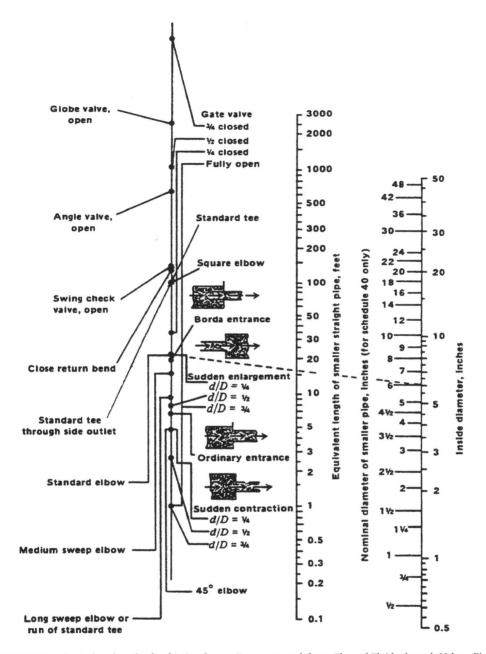

FIGURE 2.1.35 Equivalent lengths for friction losses. Data extracted from *Flow of Fluids through Valves, Fittings and Pipe*, Publication 410M (1988), with permission of the publisher, Crane Company.

(7.9 gal/min). Find the appropriate pump size for this system using the lengths and fittings specified in Figure 2.1.36. Assume a pump efficiency of about 75% and that the heat gain through the collectors is equal to the change in the internal energy of the water from point 1 to point 2.

Solution:

To find the pump 1 size, we apply an energy-balance similar to Eq. (2.1.72) between points 1 and 2 shown in Figure 2.1.36. Point 1 represents the water free-surface in the tank, whereas point 2 represents

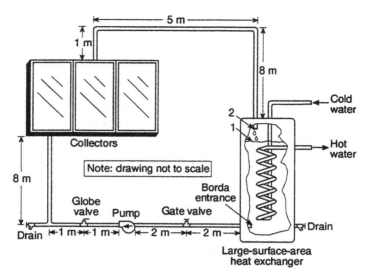

FIGURE 2.1.36   Layout of a small solar-collector system.

the water inlet to the tank after the water has circulated through the collectors. Because both points (1 and 2) have the same pressure, we have

$$\frac{\Delta p_{2-1}}{\rho} = \frac{p_2}{\rho_2} - \frac{p_1}{\rho_1} = 0.$$

Similarly, because there is no significant height difference between these two points, we have

$$\frac{g}{g_c}\,\Delta z_{2-1} = \frac{gz_2}{g_c} - \frac{gz_1}{g_c} = 0.$$

The velocity at the water free-surface (point 1) is $u_1 = 0$. For point 2, the velocity is

$$u_2 = \frac{q_2}{A_2} = \frac{0.09 \text{ m}^3/\text{min}}{5.57 \times 10^{-4} \text{ m}^2} = 161.6 \text{ m/min} = 2.7 \text{ m/s}.$$

Therefore,

$$\frac{\Delta u_{2-1}^2}{2g_c} = \frac{u_2^2}{2g_c} - \frac{u_1^2}{2g_c} = \frac{(2.7 \text{ m/sec})^2}{2\left(10^3 \dfrac{\text{kg}}{\text{kJ}}\dfrac{\text{m}^2}{\text{sec}^2}\right)} = 3.7 \times 10^{-3} \text{ kJ/kg}.$$

Also note that

$$\Delta e_{2-1} = e_2 - e_1 = \frac{\dot{Q}}{\dot{m}}.$$

Therefore, Eq. (2.1.72) reduces to

$$\frac{\Delta u_{2-1}^2}{2g_c} + \frac{\dot{w}_f}{\dot{m}} = \frac{\dot{w}}{\dot{m}}. \tag{2.1.84}$$

The frictional pressure losses $\dot{w}_f$ should be determined for the whole system between points 1 and 2. Equation (2.1.83) can be used to determine $\dot{w}_f$; however, the total equivalent length should be determined first. The tot al straight piping in the system is

$$L_s = 2\ m + 2\ m + 1\ m + 1\ m + 8\ m + 1\ m + 5\ m + 8\ m = 28\ \iota$$

Using Figure 2.1.35, the equivalent lengths for bends and valves are obtained as follows:

Borda entrance: 0.79 m (2.6 ft)
Open gate valve: 0.18 m (0.6 ft)
Open globe valve: 7.90 m (26.0 ft)
Standard tee: 1.80 m (5.9 ft)
Standard elbow: 0.81 m (2.7 ft)

Therefore, with two standard elbows in this system, the equivalent length for bends, elbows, and valves becomes L_b = 0.79 m + 0.18 m + 7.9 m + 1.8 m + 2 (0.82 m) = 12.31 m, and the total equivalent of 1–in. Sch 40 pipe is L = L_b + L_s = 40.31 m. To calculate the friction factor, we must calculate the Reynolds number. Assuming an average fluid density of ρ = 988 kg/m³ and an absolute viscosity of μ = 555×10^{-6} Ns/m², the Reynolds number is

$$\mathrm{Re} = \frac{\rho u D_H}{\mu} = \frac{988\ \mathrm{kg/m^3} \times 2.7\ \mathrm{m/s} \times 0.0266\ \mathrm{m}}{555 \times 10^{-6}\ N\mathrm{s/m^2}} = 128 \times 10^3.$$

From Figure 2.1.34, the relative roughness of the pipe obtained is e/D = 0.0018, and by using Figure 2.1.33 (the Moody diagram), the friction factor obtained is $f \approx 0.006$. Substituting in Eq. (2.1.83), the work required to overcome the frictional losses is obtained from

$$\frac{\dot{w}_f}{\dot{m}} = \frac{2 f L u^2}{g_c D} + Fr_c,$$

where Fr_c is the required work to overcome pressure loss through the collectors. Since the collectors are in parallel, the total pressure loss is equal to the pressure drop through each collector. Therefore,

$$Fr_c = 1.04\ kN/m^2 \times \frac{1}{988\ \mathrm{kg/m^3}} = 0.0011\ \mathrm{kJ/kg},$$

and

$$\frac{\dot{w}_f}{\dot{m}} = \frac{2 \times 0.006 \times 40.31\ \mathrm{m} \times (2.7\ \mathrm{m/s})^2}{10^3\ \dfrac{\mathrm{kg}}{\mathrm{kJ}} \dfrac{\mathrm{m^2}}{\mathrm{s^2}} \times 0.0266\ \mathrm{m}} + 0.0011\ \mathrm{kJ/kg} = 0.134\ \mathrm{kJ/kg}.$$

Substituting for $\frac{\Delta u_{2-1}^2}{2 g_c}$ and $\dot{w}_f$ in Eq. (2.1.84), we can calculate the input power to the pump (the mass flow rate of the fluid is 0.494 kg/s):

$$\frac{\dot{w}}{\dot{m}} = \frac{\Delta u_{2-1}^2}{2 g_c} + \frac{\dot{w}_f}{\dot{m}} = 3.7 \times 10^{-3}\ \mathrm{kJ/kg} + 0.134\ \mathrm{kJ/kg} = 0.138\ \mathrm{kJ/kg}.$$

With a 75% efficiency, the actual mechanical energy required will be

$$w_{act} = \frac{0.138 \text{ kJ/kg}}{0.75} = 0.184 \text{ kJ/kg}.$$

The appropriate pump size is

$$\dot{w}_{act} = 0.184 \text{ kJ/kg} \times 0.494 \text{ kg/s} = 0.091 \text{ kJ/s} = 0.12 \text{ hp}.$$

2.1.4 Heat Exchangers

A **heat exchanger** is a device designed to transfer energy between two fluids. Heat exchangers are often used to transfer thermal energy from a source (e.g., a boiler or chiller) to a point of use (e.g., a cooling or heating coil). They are particularly important for improving overall process efficiency of energy-efficient systems. Heat exchangers can be expensive and must be designed carefully to maximize effectiveness and minimize cost. Depending on their application, heat exchangers can have different shapes, designs, and sizes. The major types of heat exchangers include boilers, condensers, radiators, evaporators, cooling towers, regenerators, and recuperators. All heat exchangers are identified by their geometric shape and the direction of flow of the heat-transfer fluids inside them. Figure 2.1.37 depicts some common heat exchangers. In the following paragraphs, we describe the operating principles of some of the more common heat exchangers.

A **direct-contact heat exchanger** is designed so that two fluids are physically brought into contact, with no solid surface separating them. In this type of heat exchanger, fluid streams form a mutual interface through which the heat transfer takes place between the two fluids. Direct-contact (DC) heat exchangers are best used when the temperature difference between the hot and cold fluids is small. An example of a direct-contact heat exchanger is a cooling tower, where water and air are brought together by letting water fall from the top of the tower and having it contact a stream of air flowing upward. Evaporative coolers are another common DC device.

Regenerators are heat exchangers in which the hot and cold fluids flow alternately through the same space. As a result of alternating flow, the hot fluid heats the core of the heat exchanger, where the stored heat is then transferred to the cold fluid. Regenerators are used most often with gas streams, where some mixing of the two streams is not a problem and where the cost of another type of heat exchanger would be prohibitive. For example, heat recovery in very energy efficient homes is often done with "air-to-air" regenerators to maintain an acceptable quality of air inside the homes.

The **recuperator** is the heat exchanger encountered most often. It is designed so that the hot and cold fluids do not come into contact with each other. Energy is exchanged from one fluid to a solid surface by convection, through the solid by conduction, and from the other side of the solid surface to the second fluid by convection. The evaporator tube bundle in a chiller is such as device.

In Section 2.1.2, we described these heat-transfer processes and developed some simple equations that are applied here to determine basic equipment performance. Designing a heat exchanger also requires estimating the pressure flow losses that can be carried out, based on the information provided in Section 2.1.3. Finally, appropriate materials must be selected and a structural analysis done; Frass and Ozisik [1965] provide a good discussion of these topics.

Heat-Exchanger Performance

The performance of a heat exchanger is based on the exchanger's ability to transfer heat from one fluid to another. Calculating the heat transfer in heat exchangers is rather involved because the temperature of one or both of the fluids is changing continuously as they flow through the exchanger. There are three main flow configurations in heat exchangers: parallel flow, counter flow, and cross flow. In **parallel-flow** heat exchangers, both fluids enter from one end of the heat exchanger flowing in the same direction and they both exit from the other end. In **counter-flow** heat exchangers, hot fluid enters from one end and flows in an opposite direction to cold fluid entering from the other end. In **cross-flow** heat exchangers, baffles are used to force the fluids to move perpendicular to each other, to take advantage of higher heat-transfer coefficients encountered in a cross-flow configuration. Figure 2.1.38 shows the

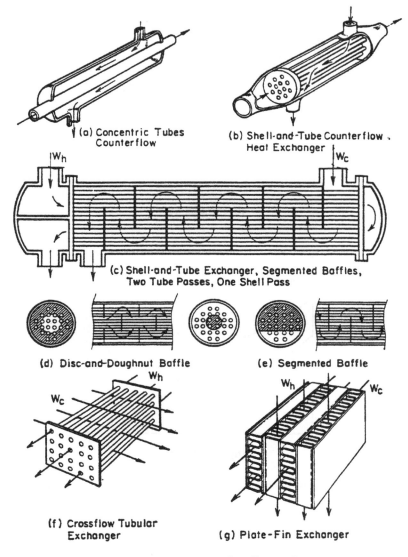

(a) Concentric Tubes
Counterflow

(b) Shell-and-Tube Counterflow
Heat Exchanger

(c) Shell-and-Tube Exchanger, Segmented Baffles,
Two Tube Passes, One Shell Pass

(d) Disc-and-Doughnut Baffle

(e) Segmented Baffle

(f) Crossflow Tubular
Exchanger

(g) Plate-Fin Exchanger

FIGURE 2.1.37   Some examples of heat exchangers.

temperature variation of the fluids inside the heat exchanger for a parallel-flow and a counter-flow heat exchanger. In parallel-flow heat exchangers, the temperature difference ΔT_i between the two fluids at the inlet of the heat exchanger is much greater than ΔT_o, the temperature difference at the outlet of the heat exchanger. In counter-flow heat exchangers, however, the temperature difference between the fluids shows only a slight variation along the length of the heat exchanger. Assuming that the heat loss from the heat exchanger is negligible, usually the case in a practical design, the heat loss of the hot fluid should be equal to the heat gain by the cold fluid. Therefore, we can write

$$\dot{Q} = \dot{m}_c c_{p,c} \left(T_{co} - T_{ci} \right) = \dot{m}_h \, c_{p,h} \left(T_{hi} - T_{ho} \right), \qquad (2.1.85)$$

where the subscripts c and h refer to cold and hot fluids, respectively. Note that in heat-exchanger analysis, the terms $\dot{m}_c c_{p,c}$ and $\dot{m}_h c_{p,h}$ are called the **capacity rates** of the cold and hot fluids, respectively, and are usually represented by C_c and C_h.

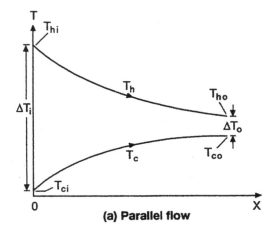

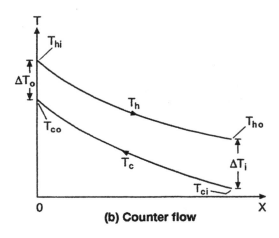

FIGURE 2.1.38 Fluid temperature variation for (a) parallel-flow configuration, and (b) counter-flow configuration.

The heat-transfer rate between a hot and a cold fluid can be written as

$$\dot{Q} = UA\ \Delta T_m,$$

(2.1.86)

where U is the overall heat-transfer coefficient (and is assumed to be constant over the whole surface area of the heat exchanger) and ΔT_m is an appropriate mean temperature difference to be defined later. In fact, the overall heat-transfer coefficient is not the same for all locations in the heat exchanger, and its local value depends on the local fluid temperatures as was shown in Example 2.1.8. For most engineering applications, designers of heat exchangers are usually interested in the *overall* average heat-transfer coefficient. Common practice is to calculate the overall heat-transfer coefficient based on some kind of mean fluid temperatures.

Expanding on the definition of thermal resistance described earlier, the heat transfer in a heat exchanger can be expressed as

$$\dot{Q} = UA\ \Delta T_m = \frac{\Delta T_m}{\displaystyle\sum_{i=1}^{n} R_i},$$

(2.1.87)

TABLE 2.1.11 Approximate Overall Heat-Transfer Coefficients
for Preliminary Estimates*

Heat-Transfer Duty	Overall Heat-Transfer Coefficient (W/m² K)
Steam to water	
Instantaneous heater	2,200–3,300
Storage-tank heater	960–1,650
Steam to oil	
Heavy fuel	55–165
Light fuel	165–330
Light petroleum distillate	275–1,100
Steam to aqueous solutions	550–3,300
Steam to gases	25–275
Water to compressed air	55–165
Water to water, jacket water coolers	825–1,510
Water to lubricating oil	110–330
Water to condensing oil vapors	220–550
Water to condensing alcohol	250–660
Water to condensing R22	440–830
Water to condensing ammonia	830–1,380
Water to organic solvents, alcohol	275–830
Water to boiling R22	275–830
Water to gasoline	330–500
Water to gas, oil, or distillate	200–330
Water to brine	550–1,100
Light organics to light organics	220–420
Medium organics to medium organics	110–330
Heavy organics to heavy organics	55–220
Heavy organics to light organics	55–330
Crude oil to gas oil	170–300

* Reproduced from *Principles of Heat Transfer* by F. Kreith, International
Textbook Co., Scranton, PA, 1958, p. 463. With permission.

where ΣR_i represents the total thermal resistance to heat transfer between fluid streams in the heat exchanger. For example, consider the simple case of heat transfer inside a shell-and-tube heat exchanger, where a hot fluid at T_h is flowing inside a steel tube with inside radius r_i and outside radius r_o as shown in Figure 2.1.14. The cold fluid at T_c is flowing in the shell side over the steel tube, where the convection heat-transfer coefficient between the cold fluid and the exterior of steel tube is h_o. For this case, the total resistance to heat transfer can be written as

$$\frac{1}{UA} = \sum_{i=1}^{5} R_i = \frac{1}{2\pi r_i L h_i} + R_{f,i} + \frac{\ln\left(r_o/r_i\right)}{2\pi k_p L} + R_{f,o} + \frac{1}{2\pi r_o L h_o}, \qquad (2.1.88)$$

where L is the length of the heat exchanger, k_p is the thermal conductivity of steel, and h_i and h_o are the convection heat-transfer coefficients of hot and cold fluid sides, respectively. Terms $R_{f,i}$ and $R_{f,o}$ represent the fouling resistances on the cold and hot heat-transfer surfaces. The overall heat-transfer coefficient can be based on either the hot surface area (in this case, $A_i = 2\pi r_i L$) or on the cold surface area ($A_o = 2\pi r_o L$). Therefore, the numerical value of U will depend on the area selected; however, it is always true that $UA \equiv U_i A_i \equiv U_o A_o$.

Table 2.1.11 gives some typical values of overall heat-transfer coefficients that are useful in preliminary system analysis and design. For all but the simplest heat exchangers, designing the best heat exchanger for a given application involves using a model that accurately sums the temperature difference and the resistance over the entire surface of the heat exchanger. Most engineers use sophisticated computer models for designing heat exchangers. These computer models incorporate the most accurate algorithms for a myriad of applications. A reasonably good estimate of heat-exchanger performance can be calculated by hand by using one of various readily available handbooks (e.g., *Handbook of Heat Exchanger Design* [1983]).

The other important term in Eq. (2.1.86) for calculating the heat-transfer rate is the mean temperature difference ΔT_m. The mean temperature difference for a heat exchanger depends on its flow configuration and the degree of fluid mixing in each flow stream.

For a simple single-pass heat exchanger with various temperature profiles (e.g., parallel flow, counter flow, and constant surface temperature), the mean temperature of Eq. (2.1.87) can be calculated from

$$\Delta T_m = \frac{\Delta T_i - \Delta T_o}{\ln\left(\Delta T_i / \Delta T_o\right)}, \qquad (2.1.89)$$

where ΔT_i represents the greatest temperature difference between the fluids and ΔT_o represents the least temperature difference, and only if the following assumptions hold:

1. U is constant over the entire heat exchanger.
2. The flow of fluids inside the heat exchanger is in steady-state mode.
3. The specific heat of each fluid is constant over the entire length of the heat exchanger.
4. Heat losses from the heat exchanger are minimal.

The mean temperature difference ΔT_m given by Eq. (2.1.89) is known as the **logarithmic mean temperature difference** (LMTD).

Example 2.1.11

Lubricating oil from a building standby generator at initial temperature of 115°C and flow rate of 2 kg/s is to be cooled to 70°C in a shell-and-tube heat exchanger. Cold water at a flow rate of 2 kg/s and initial temperature of 20°C is used as the cooling fluid in the heat exchanger. Calculate the heat-exchanger area required by employing first a counter-flow and then a parallel-flow heat-exchanger arrangement. The overall heat-transfer coefficient is $U = 900$ W/m² K, and the specific heat of the oil is $c_{p,h} = 2.5$ kJ/kg K.

Solution:

First, we use Eq. (2.1.85) to calculate the water outlet temperature. The specific heat of water can be assumed to be constant over the temperature range of interest, and it is $c_{p,c} = 4.182$ kJ/kg K.

$$\dot{m}_c c_{p,c}\left(T_{co} - T_{ci}\right) = \dot{m}_h c_{p,h}\left(T_{hi} - T_{ho}\right) \quad \text{(first law)}$$

or

$$T_{co} = \frac{\dot{m}_h c_{p,h}}{\dot{m}_c c_{p,c}}\left(T_{hi} - T_{ho}\right) + T_{ci}.$$

$$T_{co} = \frac{2 \text{ kg/s } \left(2.5 \text{ kJ/kg K}\right)}{2 \text{ kg/s } \left(4.182 \text{ kJ/kg K}\right)}\left(115°C - 70°C\right) + 20°C.$$

$$T_{co} = 46.9°C.$$

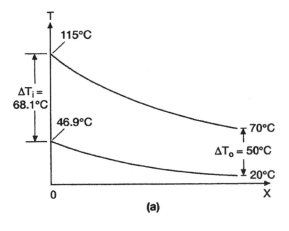

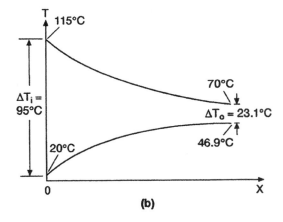

FIGURE 2.1.39 Temperature differences of Example 2.1.11 for (a) counter-flow arrangement, and (b) parallel-flow arrangement.

The total heat transferred from hot fluid to the cold fluid is

$$\dot{Q} = \dot{m}_c \; c_{p,c} \; \left(T_{co} - T_{ci}\right) = 2 \; \text{kg/s} \; \left(4.182 \; \text{kJ/kg K}\right) \left(46.9°C - 20C\right) = 225 \; \text{kJ/s}.$$

For a *counter-flow* arrangement, the temperature differences are shown in Figure 2.1.39(a). The greatest temperature difference is $\Delta T_i = 68.1°C$, and the least temperature difference is $\Delta T_o = 50°C$. Using Eq. (2.1.89), the mean temperature can be calculated as

$$\Delta T_m = \frac{\Delta T_i - \Delta T_o}{\ln\left(\Delta T_i / \Delta T_o\right)} = \frac{68.1°C - 50°C}{\ln\left(68.1/50\right)} = 58.58°C \; .$$

The heat-exchanger surface area for counter-flow arrangement can be obtained from Eq. (2.1.86):

$$\dot{Q} = UA \; \Delta T_m,$$

or

$$A = \frac{\dot{Q}}{U \, \Delta T_m} = \frac{225 \times 10^3 \text{ J/s}}{900 \text{ W/m}^2 \text{ K} \times 58.58°\text{C}} = 4.27 \text{ m}^2.$$

A similar procedure can be followed for the *parallel-flow* arrangement. The water outlet temperature T_{co} and the total heat transfer calculated earlier still hold for this arrangement. However, the temperature differences are as shown in Figure 2.1.39(b). The mean temperature for this arrangement is

$$\Delta T_m = \frac{(\Delta T_i - \Delta T_o)}{\ln (\Delta T_i / \Delta T_o)} = \frac{95°\text{C} - 23.1°\text{C}}{\ln (95/23.1)} = 50.84°\text{C},$$

and the area required is

$$A = \frac{\dot{Q}}{U \, \Delta T_m} = \frac{225 \times 10^3 \text{ J/s}}{900 \text{ W/m}^2 \text{ K} \times 50.84°\text{C}} = 4.92 \text{ m}^2.$$

Therefore, the heat-exchanger surface area required for parallel flow is more than that required for the counter-flow arrangement if all the other conditions are assumed to be the same. Consequently, whenever possible, it is advantageous to use the counter-flow arrangement because it will require less heat-exchanger surface area to accomplish the same job. In addition, as seen from Figure 2.1.39, with the counter-flow arrangement, the outlet temperature of the cooling fluid may be raised much closer to the inlet temperature of the hot fluid.

The LMTD expression presented by Eq. (2.1.89) does not hold for more complex flow configurations such as cross flow or multipass flows. To extend the LMTD definition to such configurations, a correction factor is defined as

$$F = \frac{\Delta T_m}{\Delta T_{m,cF}}, \tag{2.1.90}$$

where $\Delta T_{m,cF}$ is calculated from Eq. (2.1.89) for a counter-flow configuration. Bowman et al. [1940] provide charts for calculating the correction factor F for various flow configurations of heat exchangers. A sample of their charts is shown in Figure 2.1.40 for one fluid mixed and the other fluid unmixed. The term *unmixed* means that a fluid stream passes through the heat exchanger in separated flow channels or passages with no fluid mixing between adjacent flow passages. Note that the correction factor F in Figure 2.1.40 is a function of two dimensionless parameters Z and P defined as

$$Z = \frac{T_{hi} - T_{ho}}{T_{co} - T_{ci}} \equiv \frac{C_c}{C_h} \tag{2.1.91}$$

and

$$P = \frac{T_{co} - T_{ci}}{T_{hi} - T_{ci}}, \tag{2.1.92}$$

where the term Z is the ratio of the capacity rates of the cold and hot streams, and the term P is referred to as the temperature effectiveness of the cold stream. Kays and London [1984] provide a comprehensive representation of F charts.

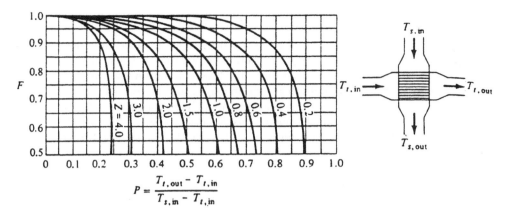

FIGURE 2.1.40 Correction factor for counter-flow LMTD for cross-flow heat exchangers with the fluid on the shell side mixed, the other fluid unmixed, and one pass through tube. Data extracted from "Mean Temperature Difference in Design" by Bowman et al. (1940) in *ASME Proceedings*, with permission of the publisher, The American Society of American Engineers.

Heat-Exchanger Design Methods

Heat-exchanger designers usually use two well-known methods for calculating the heat-transfer rate between fluid streams—the UA-LMTD and the effectiveness-NTU (number of heat-transfer units) methods.

UA-LMTD Method

In this method, the relationship between the total heat-transfer rate, the heat-transfer area, and the inlet and outlet temperatures of the two streams is obtained from Eqs. (2.1.87) and (2.1.90). Substituting Eq. (2.1.90) in Eq. (2.1.87) gives

$$\dot{Q} = F \times T_{m,cF} \times UA. \tag{2.1.93}$$

For a given heat-exchanger configuration, one can calculate UA by identifying heat-transfer resistances and summing them as in Eq. (2.1.88), calculating $\Delta T_{m,cF}$ from Eq. (2.1.89), reading the F value from an appropriate chart, and substituting them in Eq. (2.1.93) to find the heat-transfer rate. The UA-LMTD method is most suitable when the fluid inlet and outlet temperatures are known or can be determined readily from an energy-balance expression similar to Eq. (2.1.85). There may be situations where only inlet temperatures are known. In these cases, using the LMTD method will require an iterative procedure. However, an alternative is to use the effectiveness-NTU method described in the following section.

Effectiveness-NTU Method

In this method, the capacity rates of both hot and cold fluids are used to analyze the heat-exchanger performance. We will first define two dimensionless groups that are used in this method—the number of heat-transfer units (NTU) and the heat exchanger effectiveness ε

The NTU is defined as

$$\text{NTU} = \frac{UA}{C_{min}}, \tag{2.1.94}$$

where C_{min} represents the smaller of the two capacity rates C_c and C_h. NTU is the ratio of the heat-transfer rate per degree of mean temperature-difference between the fluids, Eq. (2.1.86), to the heat-transfer rate per degree of temperature change for the fluid of minimum heat-capacity rate. NTU is a measure of the physical size of the heat exchanger: the larger the value of the NTU, the closer the heat exchanger approaches its thermodynamic limit.

The heat-exchanger effectiveness is defined as the ratio between the actual heat-transfer rate $\dot{Q}$ and the maximum possible rate of heat that thermodynamically can be exchanged between the two fluid streams. The actual heat-transfer rate can be obtained from Eq. (2.1.85). To obtain the maximum heat-transfer rate, one can assume a counter-flow heat exchanger with infinite surface area, where one fluid undergoes a temperature change equal to the maximum temperature-difference available, $\Delta T_{max} = T_{hi} - T_{ci}$. The calculation of $\dot{Q}_{max}$ is based on the fluid having the smaller capacity rate C_{min}, because of the limitations imposed by the Second Law of Thermodynamics (see Bejan [1993] for more detail). Therefore,

$$\dot{Q}_{max} = C_{min}\left(T_{hi} - T_{ci}\right), \tag{2.1.95}$$

where Eq. (2.1.95) is not limited to counter-flow heat exchangers and can be applied equally to other configurations. Therefore, the effectiveness can be expressed as

$$\varepsilon = \frac{\dot{Q}}{\dot{Q}_{max}} = \frac{C_c\left(T_{co} - T_{ci}\right)}{C_{min}\left(T_{hi} - T_{ci}\right)}, \tag{2.1.96}$$

or

$$\varepsilon = \frac{C_h\left(T_{hi} - T_{ho}\right)}{C_{min}\left(T_{hi} - T_{ci}\right)}. \tag{2.1.97}$$

Knowing the effectiveness of a heat exchanger, one can calculate the actual rate of heat transfer by using Eq. (2.1.96) or from

$$\dot{Q} = \varepsilon C_{min}\left(T_{hi} - T_{ci}\right). \tag{2.1.98}$$

Expressions for the effectiveness of heat exchangers with various flow configurations have been developed and are given in heat-transfer texts (e.g., Bejan [1993], Kreith and Bohn [1993]). For example, the effectiveness of a counter-flow heat exchanger is given by

$$\varepsilon = \frac{1 - \exp\left[-\mathrm{NTU}\left(1 - R\right)\right]}{1 - R\exp\left[-\mathrm{NTU}\left(1 - R\right)\right]}, \tag{2.1.99}$$

where $R = C_{min}/C_{max}$. The effectiveness of a parallel-flow heat exchanger is given by

$$\varepsilon = \frac{1 - \exp\left[-\mathrm{NTU}\left(1 + R\right)\right]}{1 + R}. \tag{2.1.100}$$

The effectiveness for heat exchangers of various flow configurations has been evaluated by Kays and London [1984] and is presented in a graph format similar to the one shown in Figure 2.1.41. In this figure, the heat-exchanger effectiveness has been plotted in terms of NTU and R. Note that for an evaporator and a condenser, $R = 0$ because the fluid remains at a constant temperature during the phase change.

The two design-and-analysis methods just described are equivalent, and both can be equally employed for designing heat exchangers. However, the NTU method is preferred for rating problems where at least one exit temperature is unknown. If all inlet and outlet temperatures are known, the UA-LMTD method does not require an iterative procedure and is the preferred method.

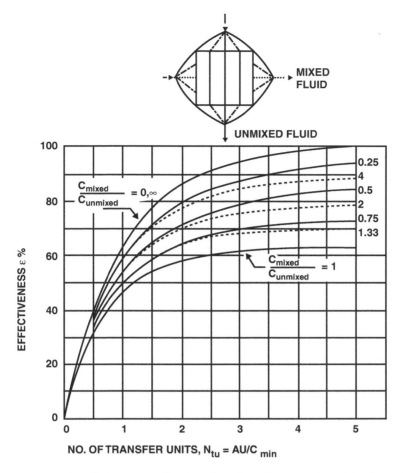

FIGURE 2.1.41 Heat-transfer effectiveness as a function of number of heat-transfer units and capacity-rate ratio; cross-flow exchanger with one fluid mixed. Data extracted from *Compact Heat Exchangers, 3rd ed.*, by W.M. Kays and A.L. London (1984), with permission of the author.

Nomenclature

A	heat-transfer surface area, m² (ft²)
c	speed of propagation of radiation energy, m/s (ft/s)
C_c	capacity rate of cold fluid equal to $\dot{m}_c (c_p)_c$
C_h	capacity rate of hot fluid equal to $\dot{m}_h (c_p)_h$
C_0	speed of light in a vacuum, 3×10^8 m/s (9.84×10^8 ft/s)
c_p	specific heat, kJ/kg K (Btu/lbm R)
D_H	hydraulic diameter, defined by Eq. (2.1.50)
E	total energy of a thermodynamic system, kJ (Btu)
e	internal energy of a thermodynamic system, kJ/kg (Btu/lbm)
E_b	total emissive power of a blackbody, W/m² (Btu/h ft²)
$E_{b\lambda}$	spectral blackbody emissive power, given by Eq. (2.1.56), W/m³ (Btu/h ft²μ)
f	dimensionless friction factor, see Eq. (2.1.77)
F_{1-2}	radiation shape factor between surfaces 1 and 2
g	gravitational acceleration, 9.81 m/s² (32.2 ft/s²)
g_c	Newton constant, equal to 32.2 ft × lbm/(lbf × s²), or 10^3 kg × m²/(kJ × s²)
h	enthalpy, kJ/kg (Btu/lbm), or convection heat-transfer coefficient, W/m² K (Btu/h ft² °F)

h_f	enthalpy of saturated liquid, kJ/kg (Btu/lbm)
h_g	enthalpy of saturated vapor, kJ/kg (Btu/lbm)
k	thermal conductivity, W/m K (Btu/h ft °F)
k_b	pressure-loss coefficient due to bends and fittings, see Eq. (2.1.79)
k_c	pressure-loss coefficient due to sudden contraction of pipe, see Eq. (2.1.81)
k_e	pressure-loss coefficient due to sudden enlargement of pipe, see Eq. (2.1.80)
k_p	conductivity of steel plate, W/m K (Btu/h ft °F)
L	length of the circular cylinder or heat exchanger, m (ft)
m	mass of a system, kg (lbm)
$\dot{m}$	mass-flow rate of working fluid, kg/s (lbm/s)
n	index of refraction of a medium
Nu	Nusselt number, defined by Eq. (2.1.33)
p	pressure, N/m² (lbf/in²)
Pr	Prandtl number, equal to ν/α
Q_H	heat absorbed from a high-temperature reservoir, kJ (Btu)
Q_L	heat rejected to a low-temperature reservoir, kJ (Btu)
$\dot{Q}$	rate of energy (heat) transfer, W (Btu/h)
q''	heat flux
$\dot{Q}_r$	heat-transfer rate by radiation, W (Btu/h)
R	thermal resistance, defined by Eq. (2.1.30), m² K/W (h ft²°F/Btu)
Ra	Rayleigh number, defined by Eq. (2.1.34)
Re	Reynolds number, equal to uD_H/ν
$R_{f,i}$	fouling resistance of cold (inner) heat transfer surface, m² K/W (h ft²°F/Btu)
$R_{f,o}$	fouling resistance of hot (outter) heat transfer surface, m² K/W (h ft²°F/Btu)
r_i	inner radius of cylinder, m (ft)
r_o	outer radius of cylinder, m (ft)
S	entropy, kJ/K (Btu/R)
s_f	entropy of saturated liquid, kJ/kg K (Btu/lbm R)
s_g	entropy of saturated vapor, kJ/kg K (Btu/lbm R)
T	temperature, °C (°F)
T_H	temperature of higher-temperature reservoir, K (R)
T_L	temperature of low-temperature reservoir, K (R)
U	overall heat transfer coefficient, given by Eq. (2.1.88), W/m²K (Btu/h ft²°F)
u	velocity, m/s (ft/s)
v	specific volume (volume per unit mass), m³/kg (ft³/lbm)
w	work or energy, kJ (Btu)
$\dot{w}$	rate of work (done on the surroundings), kJ/s (Btu/h)
w_{act}	actual work of a system, kJ (Btu)
$\dot{w}_f$	frictional pressure losses, kJ/s (Btu/h)
w_{isent}	work done under an isentropic process, kJ (Btu)
w_{rev}	work of a system under reversible process, kJ (Btu)
x	distance along x axis, m (ft)
y	distance along y axis, m (ft)
z	distance along z axis, or elevation of a thermodynamic system, m (ft)

Greek Symbols

α	thermal diffusivity, m²/s (ft²/h), or absorptivity of a surface
$\bar{\alpha}$	average absorptivity of a surface
β	thermal expansion coefficient, K⁻¹ (R⁻¹)
δ	velocity boundary-layer thickness, mm (in.)
δ_T	temperature boundary-layer thickness, mm (in.)

ΔT_i equal $T_{ho} - T_{ci}$ for counter-flow, and $T_{hi} - T_{ci}$ for parallel-flow heat exchanger, see Figure 2.1.38

ΔT_o equal to $T_{hi} - T_{co}$ for counter-flow, and $T_{ho} - T_{co}$ for parallel-flow heat exchanger, see Figure 2.1.38

ε emissivity of a surface

$\bar{\varepsilon}$ average absorptivity of a surface

η_m mechanical efficiency, defined by Eq. (2.1.17)

η_s isentropic efficiency, defined by Eq. (2.1.18)

η_r relative efficiency, defined by Eq. (2.1.19)

η_{rev} energy conversion efficiency for a reversible system

η_T thermal efficiency, defined by Eq. (2.1.20)

λ wavelength of radiation energy, m (ft)

μ viscosity of fluid, Ns/m²

ν kinematic viscosity, m²/s (ft²/h), or frequency, s⁻¹

ρ reflectivity of a surface, or density (mass per unit volume) kg/m³ (lbm/ft³)

σ Stefan-Boltzmann constant, Wm⁻²K⁻⁴ (Btu/h ft²R⁴)

τ transmissivity of a surface, or frictional shear stress, N/m² (lbf/ft²)

υ thermal radiation source frequency, s⁻¹

References

Achenbach, E. 1989. Heat transfer from a staggered tube bundle in cross-flow at high Reynolds numbers, *Int. J. Heat Mass Transfer,* 32:271–280.

Anderson, R.S. and Kreitn, F. 1987. Natural convection in active and passive solar thermal systems, *Advances in Heat Transfer,* Vol. 18, pp. 1–86. Academic Press, New York.

Arnold, J.N., Bonaparte, P.N., Catton, I., and Edwards, D.K. 1974. *Proceedings 1974 Heat Transfer Fluid Mech. Inst.,* Stanford University Press, Stanford, CA.

Arnold, J.N., Catton, I., and Edwards, D.K. 1976. Experimental investigation of natural convection in inclined rectangular regions of differing aspect ratios, *J. Heat Transfer,* 98:67–71.

ASHRAE Handbook of Fundamentals, 1993. ed., I-P, American Society of Heating, Ventilating, and Air Conditioning Engineers, Atlanta, GA.

Ayyaswamy, P.S. and Catton, I. 1973. The boundary layer regime for natural convection in a differentially heated tilted rectangular cavity, *J. Heat Transfer,* 95:543–545.

Balmer, R.T. 1990. *Thermodynamics,* West Publishing Company, St. Paul, MN.

Bejan, A. 1993. *Heat Transfer,* John Wiley & Sons, Inc., New York.

Bohn, M.S., Williams, T.A., and Price, H.W. 1995. "Combined Cycle Power Tower," presented at *Int. solar Energy Conference,* ASME/JSME/JSES joint meeting, Maui, HI.

Bowman, R.A., Mueller, A.C., and Nagle, W.M. 1940. "Mean Temperature Difference in Design," *Trans. ASME,* Vol. 62, pp. 283–294.

Carnot, S. 1960. *Reflections on the Motive Power of Fire (and Other Papers on the Second law of Thermodynamics by E. Clapeyron and R. Clausius),* ed. E. Mendoza. Dover, New York.

Catton, I. 1978. Natural convection in enclosures. In *Proceedings Sixth International Heat Transfer Conference, Toronto,* Vol. 6, pp. 13–31. Hemisphere, Washington, D.C.

Churchill, S.W. 1983. Free convection around immersed bodies. In *Heat Exchanger Design Handbook,* ed. E.U. Schlünder, Section 2.5.7. Hemisphere, New York.

Churchill, S.W. and Chu, H.H.S. 1975. Correlating equations for laminar and turbulent free convection from a vertical plate, *Int. J. Heat Mass Transfer,* 18:1323–1329.

Fischenden, M. and Saunders, O.A. 1932. *The Calculation of Heat Transmission.* His Majesty's Stationary Office, London.

Fraas, A.P. and Ozisik, MN. 1965. *Heat Exchanger Design,* John Wiley & Sons, Inc., New York.

Freeman, A. 1941. *Experiments upon the Flow of Water in Pipes and Pipe Fittings,* American Society of Mechanical Engineers, New York.

Fujii, T. and Imura, H. 1972. Natural convection heat transfer from a plate with arbitrary inclination, *Int. J. Heat Mass Transfer,* 15:755–767.

Gubareff, G.G., Janssen, E.J., and Torborg, R.H. 1960. *Thermal Radiation Properties Survey,* Honeywell Research Center, Minneapolis, MN.

Hamilton, D.C. and Morgan, W.R. 1962. Radiant Interchange Configuration Factors, NACA TN2836, Washington, D.C.

Hassani, A.V. and Hollands, K.G.T. 1989. On natural convection heat transfer from three-dimensional bodies of arbitrary shape, *J. Heat Transfer,* 111:363–371.

Hausen, H. 1983. *Heat Transfer in Counter Flow, Parallel Flow and Cross Flow,* McGraw-Hill, New York.

Hollands, K.G.T., Raithby, G.D., and Konicek, L.J. 1976. Correlation equations for free convection heat transfer in horizontal layers of air and water, *Int. J. Heat Mass Transfer,* 18:879–884.

Hollands, K.G.T., Unny, T.E., Raithby, G.D., and Konicek, L.J. 1976. Free convection heat transfer across inclined air layers, *J. Heat Transfer,* 98:189–193.

Howell, J.R. 1982. *A Catalog of Radiation Configuration Factors,* McGraw-Hill, New York.

Incropera, F.P. and DeWitt, D.P. 1990. *Introduction to Heat Transfer,* 2nd ed., Wiley, New York.

Jakob, M. 1949. *Heat Transfer,* Vol. 1, Wiley, New York.

Kakac, S. and Yener, Y. 1988. *Heat Conduction,* 2nd ed. Hemisphere, Washington, D.C.

Kalina, A.I. 1984. Combined-cycle system with novel bottoming cycle, *J. of Eng. for Gas Turbines and Power,* 106:737–742.

Karlekar, B.V. 1983. *Thermodynamics for Engineers,* Prentice-Hall, Englewood Cliffs, NJ.

Kays, W.M. and London, A.L. 1984. *Compact Heat Exchangers,* 3rd ed., McGraw-Hill, New York.

Kays, W.M. and Perkins, K.R. 1985. Forced convection, internal flow in ducts. In *Handbook of Heat Transfer Applications,* eds. W.R. Rosenow, J.P. Hartnett, and E.N. Ganic, Vol. 1, Chap. 7. McGraw-Hill, New York.

Kreith, F. 1970. Thermal design of high altitude balloons and instrument packages. *J. Heat Transfer,* 92:307–332.

Kreith, F. and Black, W.Z. 1980. *Basic Heat Transfer,* Harper & Row, New York.

Kreith, F. and Boehm, eds. 1987. *Direct Condenser Heat Transfer,* Hemisphere, New York.

Kreith, F. and Bohn, M.S. 1993. *Principles of Heat Transfer,* 5th ed., West Publishing Company, St. Paul, MN.

McAdams, W.H. 1954. *Heat Transmission,* 3rd ed., McGraw-Hill, New York.

MacGregor, R.K. and Emery, A.P. 1969. Free convection through vertical plane layers: moderate and high Prandtl number fluid, *J. Heat Transfer,* 91:391.

Moody, L.F. 1944. Friction factors for pipe flow, *Trans. ASME,* 66:671.

Perry, J.H., Perry, R.H., Chilton, C.H., and Kirkpatrick, S.D. 1963. *Chemical Engineer's Handbook,* 4th ed. McGraw-Hill, New York.

Planck, M. 1959. *The Theory of Heat Radiation,* Dover, New York.

Quarmby, A. and Al-Fakhri, A.A.M. 1980. Effect of finite length on forced convection heat transfer from cylinders, *Int. J. Heat Mass Transfer,* 23:463–469.

Raithby, G.D. and Hollands, K.G.T. 1974. A general method of obtaining approximate solutions to laminar and turbulent free convection problems. In *Advances in Heat Transfer,* Academic Press, New York.

Raithby, G.D. and Hollands, K.G.T. 1985. Natural convection. In *Handbook of Heat Transfer Fundamentals,* 2nd ed., eds. W.M. Rosenow, J.P. Hartnett, and E.N. Ganic, McGraw-Hill, New York.

Reid, R.C., Sherwood, T.K., and Prausnitz, J.M. 1977. *The Properties of Gases and Liquids,* 3rd ed., McGraw-Hill, New York.

Reynolds, W.C. and Perkins, H.C. 1977. *Engineering Thermodynamics,* 2nd ed., McGraw Hill, New York.

Sanford, J.F. 1962. *Heat Engines,* Doubleday, Garden City, NY.

Schlünder, E.U., ed. 1983. *Handbook of Heat Exchanger Design,* Vol. 1. Hemisphere, Washington, D.C.

Schmidt, H. and Furthman, E. 1928. Ueber die Gesamtstrahlung fäster Körper. Mitt. Kaiser-Wilhelm-Inst. Eisenforsch., Abh. 109, Dusseldorf.

Shah, R.K. and London, A.L. 1978. *Laminar Flow Forced Convection in Ducts,* Academic Press, New York.

Sieder, E.N. and Tate, C.E. 1936. Heat transfer and pressure drop of liquids in tubes, *Ind. Eng. Chem.,* 28:1429.

Siegel, R. and Howell, J.R. 1972. *Thermal Radiation Heat Transfer,* 3rd ed., Hemisphere, New York.

Sparrow, E.M. and Stretton, A.J. 1985. Natural convection from variously oriented cubes and from other bodies of unity aspect ratio, *Int. J. Heat Mass Transfer,* 28(4):741–752.

Standards of Hydraulic Institute—Tentative Standards, Pipe Friction, 1948. Hydraulic Institute, New York.

Van Wylen, G.J. and Sonntag, R.E. 1986. *Fundamentals of Classical Thermodynamics,* 3rd ed., Wiley, New York.

Whitaker, S. 1972. Forced convection heat transfer correlations for flow in pipes, past flat plates, single cylinders, single spheres, and for flow in packed beds and tube bundles, *AIChE J.,* 18:361–371.

Wood, B.D. 1982. *Applications of Thermodynamics,* 2nd ed., Addison-Wesley, Reading, MA.

Yovanovich, M.M. and Jafarpur, K. 1993. Models of laminar natural convection from vertical and horizontal isothermal cuboids for all Prandtl numbers and all Rayleigh numbers below 10^{11}, *Fundamentals of Natural Convection,* ASME, HTD, Vol. 264, pp. 111–126.

Zukauskas, A.A. 1972. Heat transfer from tubes in cross flow, *Advances in Heat Transfer,* Vol. 8, pp. 93–106. Academic Press, New York.

2.2   Psychrometrics and Comfort

T. Agami Reddy

A large fraction of the energy used in buildings goes toward maintaining indoor thermal comfort conditions for the occupants. This section presents the basic thermodynamic relations of the important air conditioning processes and illustrates their application to the analysis of energy flows in one-zone spaces. A discussion of factors affecting human thermal comfort is also provided.

2.2.1   Atmospheric Composition and Pressure

Atmospheric air is not only a mixture of several gases, water vapor, and numerous pollutants; it also varies considerably from location to location. The composition of dry air is relatively constant and varies slightly with time, location, and altitude. The standard composition of dry air has been specified by the International Joint Committee on Psychrometric Data in 1949 as shown in Table 2.2.1.

TABLE 2.2.1   Composition of Dry Air

Constituent	Molecular Mass	Volume Fraction
Oxygen	32.000	0.2095
Nitrogen	28.016	0.7809
Argon	39.944	0.0093
Carbon Dioxide	44.010	0.0003

The *ASHRAE Fundamentals Handbook* (1997) gives the following definition of the U.S. standard atmosphere:

(a) Acceleration due to gravity is constant at 32.174 ft/sec^2 (9.807 m/s^2).

(b) Temperature at sea level is 59.0°F (15°C or 288.1 K).

(c) Pressure at sea level is 29.921 inches of mercury (101.039 kPa).

(d) The atmosphere consists of dry air, which behaves as a perfect gas.

The total atmospheric pressure at different altitudes is given in standard gas tables. For altitudes up to 60,000 ft (18,291 m) the following equation can be used:

$$P = a + b \cdot H \tag{2.2.1}$$

where the constants a and b are given in the following table, H is the elevation above sea level, and pressure P is in inches of Hg or in kPa.

Constant	H ≤ 4000 ft (1220 m)		H > 4000 ft (1220 m)	
	IP	SI	IP	SI
a	29.92	101.325	29.42	99.436
b	−0.001025	−0.01153	−0.0009	−0.010

In HVAC applications, the mixture of various constituents that compose dry air is considered to be a single gas. The molecular mass of dry air can be assumed to be 28.965.

2.2.2 Thermodynamic Properties of Moist Air

A *property* is any attribute or characteristic of matter that can be observed or evaluated quantitatively. Thermodynamic properties, i.e., those concerned with energy and its transformation, of primary interest to HVAC are described below.

A. *Temperature* t of a substance indicates its thermal state and its ability to exchange energy with a substance in contact with it. Reference points are the freezing point of water (0°C in the Celsius scale and 32°F in the Fahrenheit scale) and the boiling point of water (100°C in the Celsius scale and 212°F in the Fahrenheit scale). Often the absolute temperature scale T is more relevant:

$$\text{SI units : Kelvin scale:} \quad T = t + 273.15° \text{ (K) with t in °C}$$
$$\text{IP units : Rankine scale:} \quad T = t + 459.67° \text{ (°R) with t in °F.}$$

B. *Pressure* p is the normal or perpendicular force exerted by a fluid per unit area against which the force is exerted. Absolute pressure is the measure of pressure above zero; gauge pressure is measured above existing atmospheric pressure. The unit of pressure is pound per square inch (psi) or Pascal (Pa). Standard atmospheric pressure is 101.325 kPa or 14.696 psi.

C. *Density* ρ of a fluid is its mass per unit volume. It is more common to use its reciprocal, the *specific volume* v, the volume occupied per unit mass. The density of air at standard atmospheric pressure and 77°F (25°C) is approximately 0.075 lbm/ft³ (1.2 kg/m³). Density and specific volume of a vapor or gas are affected by both pressure and temperature. Tables are used to determine both; in some cases, they can be calculated from basic thermodynamic property relations. The density of liquids is usually assumed to be a function of temperature only.

D. *Internal energy* u refers to the energy possessed by a substance due to the motion and/or position of the molecules. This form of energy consists of two parts: the internal kinetic energy due to the velocity of the molecules, and the internal potential energy due to the attractive forces between molecules. Changes in the average velocity of molecules are indicated by temperature changes of a substance.

E. *Enthalpy* h is another important property defined as (u+pv) where u is the internal energy (itself a property) and (pv) is the *flow work*, i.e., the work done on the fluid to force it into a control volume. Enthalpy values are based on a specified datum temperature value.

F. *Specific heat* c of a substance is the quantity of energy required to raise the temperature of a unit mass by 1°R or 1 K. For gases, one distinguishes between two cases:

specific heat at constant volume $c_v = \left(\dfrac{\delta u}{\delta T}\right)_v$ and specific heat at constant pressure $c_p = \left(\dfrac{\delta h}{\delta T}\right)_p$.

The former is appropriate in air conditioning because the processes occur at constant pressure. Approximately, $c_p = 0.24$ Btu/lbm °F (1.00 kJ/kg K) for dry air, 1.0 Btu/lbm °F (4.19 kJ/kg K) for liquid water, and 0.444 Btu/lbm °F (1.86 kJ/kg K) for water vapor.

One needs to distinguish between *extensive* properties and *intensive* properties. While extensive properties are those, such as volume V (ft^3 or m^3) or enthalpy H (Btu or kJ), that depend on the mass of the substance, intensive properties, such as temperature and pressure, do not. Our notation is to use lowercase symbols for intensive properties, i.e., properties per unit mass. Thus, v would denote the specific volume, while u and h denote the internal energy and enthalpy per unit mass, respectively. The use of the term *specific* (for example, specific enthalpy) is recommended in order to avoid ambiguity.

G. *Phase* is a quantity of matter homogeneous throughout in chemical composition and physical structure. A pure substance is one that is uniform and invariable in chemical composition. Thus, a pure substance may exist in more than one phase, such as a mixture of liquid water and water vapor (steam). On the other hand, a mixture of gases (such as air) is not a pure substance. However, if no change of phase is involved (as in most HVAC processes), air can be assumed to be a pure substance.

H. *Ideal gas law.* The ideal gas law is a relationship between the pressure, specific volume, and absolute temperature of the substance:

$$pv = RT \qquad\qquad (2.2.2a)$$

where   p = absolute pressure, lbf/ft^3 (Pa)
v = specific volume, ft^3/lbm (m^3/kg)
T = absolute temperature, °R (K), and
R = gas constant = 53.352 ft lbf/lbm °R (287 J/kg K) for air and 85.78 ft lbf/lbm °R (462 J/kg K) for water vapor.

Alternatively, the ideal gas law can also be written as

$$pv = \left(\frac{R^*}{MW}\right)T \qquad\qquad (2.2.2.b)$$

where MW is the molecular weight of the substance, and R* is the universal gas constant = 1545.32 ft lbf/lbmol °R (8.3144 kJ/kg-mol K), and whose value is independent of the substance.

I. *Liquid-vapor properties.* Understanding the behavior of substances such as steam or refrigerants during their transition between liquid and vapor phases is very important in air conditioning and refrigeration systems. The ideal gas law cannot be used for such cases and one must use property tables or charts specific to the substance. Properties of liquid and saturated water vapor are given in Table 2.2.2 based on temperature and pressure. The second column gives the corresponding saturation pressure. Specific volume and specific enthalpy at the saturated liquid condition and at the saturated vapor condition are listed in the columns. Separate tables are also available to determine properties of superheated vapor.

J. A thermodynamic *process* is one where the state of a system under study undergoes a change. For example, the dehumidification of air in a cooling coil is a thermodynamic process. A process is described in part by the series of states passed through by the system. Often some interaction between the system and the surroundings occurs during the process. Thermodynamic analysis basically involves studying the interaction of work, heat, and the properties of the substance contained in the system.

TABLE 2.2.2 Properties of Saturated Steam and Saturated Water

		Temperature Table Specific vol, cu ft/lb		Enthalpy, Btu/lb					Pressure Table Specific vol, cu ft/lb		Enthalpy, Btu/lb		
Temp F	Abs press, psi	Sat liquid v_f	Sat vapor v_g	Sat liquid h_f	Evap h_{fg}	Sat vapor h_g	Abs press, psi	Temp, F	Sat liquid v_f	Sat vapor v_g	Sat liquid h_f	Evap h_{fg}	Sat vapor h_g
32	0.08854	0.01602	33306	0.00	1075.8	1075.8	0.5	79.58	0.01608	641.4	47.6	1048.8	1096.4
35	0.09995	0.01602	2947	3.02	1074.1	1077.1	1.0	101.74	0.01614	333.6	69.7	1036.3	1106.0
40	0.12170	0.01602	2444	8.05	1071.3	1079.3	2.0	126.08	0.01623	173.73	94.0	1022.2	1116.2
45	0.14752	0.01602	2036.4	13.06	1068.4	1081.5	3.0	141.48	0.01630	118.71	109.4	1013.2	1122.6
50	0.17811	0.01603	1703.2	18.07	1065.6	1083.7	4.0	152.97	0.01636	90.63	120.9	1006.4	1127.3
55	0.2141	0.01603	1430.7	23.07	1062.7	1085.8	5.0	162.24	0.01640	73.53	130.1	1001.0	1131.1
60	0.2563	0.01604	1206.7	28.06	1059.9	1088.0	6.0	170.06	0.01645	61.98	138.0	996.2	1134.2
65	0.3056	0.01605	1021.4	33.05	1057.1	1090.2	7.0	176.85	0.01649	53.64	144.8	992.1	1136.9
70	0.3631	0.01606	867.9	38.04	1054.3	1092.3	8.0	182.86	0.01653	47.34	150.8	988.5	1139.3
75	0.4298	0.01607	740.0	43.03	1051.5	1094.5	9.0	188.28	0.01656	42.40	156.2	985.2	1141.4
80	0.5069	0.01608	633.1	48.02	1048.6	1096.6	10	193.21	0.01659	38.42	161.2	982.1	1143.3
85	0.5959	0.01609	543.5	53.00	1045.8	1098.8	14.7	212.00	0.01672	26.80	180.0	970.4	1150.4
90	0.6982	0.01610	468.0	57.99	1042.9	1100.9	20	227.96	0.01683	20.089	196.2	960.1	1156.3
95	0.8153	0.01612	404.3	62.98	1040.1	1103.1	25	240.07	0.01692	16.303	208.5	952.1	1160.6
100	0.9492	0.01613	350.4	67.97	1037.2	1105.2	30	250.33	0.01701	13.746	218.8	945.3	1164.1
105	1.1016	0.01615	304.5	72.95	1034.3	1107.3	40	267.25	0.01715	10.498	236.0	933.7	1169.7
110	1.2748	0.01617	265.4	77.94	1031.6	1109.5	50	281.01	0.01727	8.515	250.1	924.0	1174.1
115	1.4709	0.01618	231.9	82.93	1028.7	1111.6	60	292.71	0.01738	7.175	262.1	915.5	1177.6
120	1.6924	0.01620	203.27	87.92	1025.8	1113.7	70	302.92	0.01748	6.206	272.6	907.9	1180.6
125	1.9420	0.01622	178.61	92.91	1022.9	1115.8	80	312.03	0.01757	5.472	282.0	901.1	1183.1
130	2.2225	0.01625	157.34	97.9	1020.0	1117.9	90	320.27	0.01766	4.896	290.6	894.7	1185.3
135	2.5370	0.01627	138.95	102.9	1017.0	1119.9	100	327.81	0.01774	4.432	298.4	888.8	1187.2
140	2.8886	0.01629	123.01	107.9	1014.1	1122.0	110	334.77	0.01782	4.049	305.7	883.2	1188.9
145	3.281	0.01632	109.15	112.9	1011.2	1124.1	120	341.25	0.01789	3.728	312.4	877.9	1190.4
150	3.718	0.01634	97.07	117.9	1008.2	1126.1	130	347.32	0.01796	3.455	318.8	872.9	1191.7
155	4.203	0.01637	86.52	122.9	1005.2	1128.1	140	353.02	0.01802	3.220	324.8	868.2	1193.0
160	4.741	0.01639	77.29	127.9	1002.3	1130.2	150	358.42	0.01809	3.015	330.5	863.6	1194.1
165	5.335	0.01642	69.19	132.9	999.3	1132.2	160	363.53	0.01815	2.834	335.9	859.2	1195.1
170	5.992	0.01645	62.06	137.9	996.3	1134.2	170	368.41	0.01822	2.675	341.1	854.9	1196.0
175	6.715	0.01648	55.78	142.9	993.3	1136.2	180	373.06	0.01827	2.532	346.1	850.8	1196.9
180	7.510	0.01651	50.23	147.9	990.2	1138.1	190	377.51	0.01833	2.404	350.8	846.8	1197.6
185	8.383	0.01654	45.31	152.9	987.2	1140.1	200	381.79	0.01839	2.288	355.4	843.0	1198.4
190	9.339	0.01657	40.96	157.9	984.1	1142.0	250	400.95	0.01865	1.8438	376.0	825.1	1201.1
200	11.526	0.01663	33.64	168.0	977.9	1145.9	300	417.33	0.01890	1.5433	393.8	809.0	1202.8
212	14.696	0.01672	26.80	180.0	970.4	1150.4	350	431.72	0.01913	1.3260	409.7	794.2	1203.9
220	17.186	0.01677	23.15	188.1	965.2	1153.4	400	444.59	0.0193	1.1613	424.0	780.5	1204.5
240	24.969	0.01692	16.323	208.3	952.2	1160.5	450	456.28	0.0195	1.0320	437.2	767.4	1204.6
260	35.429	0.01709	11.763	228.6	938.7	1167.3	500	467.01	0.0197	0.9278	449.4	755.0	1204.4
280	49.203	0.01726	8.645	249.1	924.7	1173.8	600	486.21	0.0201	0.7698	471.6	731.6	1203.2
300	67.013	0.01745	6.466	269.6	910.1	1179.7	700	503.10	0.0205	0.6554	491.5	709.7	1201.2
350	134.63	0.01799	3.342	321.6	870.7	1192.3	800	518.23	0.0209	0.5687	509.7	688.9	1198.6
400	247.31	0.01864	1.8633	375.0	826.0	1201.0	900	531.98	0.0212	0.5006	526.6	668.8	1195.4
450	422.6	0.0194	1.0993	430.1	774.5	1204.6	1000	544.61	0.0216	0.4456	542.4	649.4	1191.8
500	680.8	0.0204	0.6749	487.8	713.9	1201.7	1200	567.22	0.0223	0.3619	571.7	611.7	1183.4
550	1045.2	0.0218	0.4240	549.3	640.8	1190.0	1500	596.23	0.0235	0.2760	611.6	556.3	1167.9

FIGURE 2.2.1   A simple flow system.

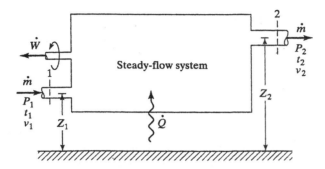

FIGURE 2.2.2   Schematic thermodynamic system for steady flow.

K. *Equilibrium* is a condition of balance maintained by an equality of opposing forces. There are different types of equilibrium: thermal, mechanical, chemical. The study of thermodynamics has to do with determining end states, and not with the dynamics of the process (e.g., how fast the process change occurs). The concept of equilibrium is important as it is only in an equilibrium state that the thermodynamic properties have meaning. We implicitly assume that the system moves from one state of equilibrium to another very slowly as it undergoes a process, a condition called quasi-equilibrium or a quasi-static process.

L. *Energy* is the capacity for producing an effect. It can be stored within the system as potential energy, kinetic energy, internal energy, etc. It can also be transferred to or from the system by work or heat transfer. Heat is transferred across the boundary of a system to another system or surroundings by virtue of a temperature difference between the two systems.

M. *Conservation of mass* simply states that mass of a substance can be neither created nor destroyed in the processes analyzed. Consider a simple flow system (shown in Figure 2.2.1) when a fluid stream flows into and out of a control volume. If the mass in the system at time θ is m(θ), then the mass at time $\theta + \delta\theta$ is m($\theta + \delta\theta$). Assuming that during the time increment dθ, an increment δm_{in} enters the system and δm_{out} leaves the system, conservation of mass relation results in

$$m(\theta) + \delta m_{in} = m(\theta + \delta\theta) + \delta m_{out}$$

which can be simplified as

$$\frac{dm}{d\theta} = \dot{m}_{in} - \dot{m}_{out} \tag{2.2.3}$$

where $\dot{m} = \dfrac{\delta m}{\delta \theta}$

For steady flow, $\dfrac{dm}{d\theta} = 0$ and $\dot{m}_{in} = \dot{m}_{out}$ $\qquad$ (2.2.4)

N. *Conservation of energy* or the first law of thermodynamics is the basis of most of the analysis done in HVAC. It formulates a relationship between the various quantities shown in Figure 2.2.2. It is described in Chapter 2.1.

2.2.3 Psychrometric Properties of Moist Air

Psychrometry is the study of the properties of moist air, i.e., a mixture of air and water vapor. A thorough understanding of psychrometry is essential since it is fundamental to understanding the various processes related to air conditioning. Atmospheric air is never totally dry; it always contains varying degrees of water vapor. Just like relatively small amounts of trace materials drastically impact the physical properties of steel alloys, small amounts of moisture have a large influence on human comfort.

The amount of water vapor contained in air may vary from near zero (totally dry) to a maximum determined by the temperature and pressure of the mixture. Properties of moist air can be determined from tables such as Table 2.2.3, from equations and steam tables as shown below, or from the psychrometric chart (described later). Moist air up to about three atmospheres pressure can be assumed to obey the perfect gas law. Assuming dry air to consist of one gas only, the total pressure p_t of moist air, given by the Gibbs-Dalton Law for a mixture of perfect gases, is equal to the individual contributions of dry air and water vapor.

$$p_t = p_a + p_v \tag{2.2.5}$$

where p_a is the partial pressure of dry air, and p_v is the partial pressure of water vapor. It is because $p_v \ll p_a$ that we can implicitly assume water vapor also follows the perfect gas law for atmospheric air.

The thermodynamic state of an air-vapor mixture is fully determined if three independent intensive properties are specified. Since one can assume for most of the HVAC processes being studied that the total atmospheric pressure does not change, a chart known as the psychrometric chart, applicable to a specific value of total pressure (commonly the standard atmospheric pressure), is used. The psychrometric chart not only provides a quick means for determining values of moist air properties, it is also very useful in solving numerous process problems with moist air and allows quick visualization of how the process occurs. Hence, for better comprehension, we describe the manner in which it is generated along with the description of the pertinent moist air properties.

The primary moist air properties shown on a psychrometric chart are described below:

A. *Dry-bulb temperature* T_{db} or t is the temperature of air one would measure with an ordinary thermometer. This property is the x-axis of the psychrometric chart (Figure 2.2.3).

B. *Saturation pressure* of water vapor: p_s or $p_{v,sat}$ can be determined or obtained from steam tables (see Table 2.2.2). For example, the saturation pressure p_s of water vapor from Table 2.2.2 at a temperature $t = 80°F$ is $p_s = 0.5069$ psia, which is about 30 times less than the corresponding partial pressure of dry air p_a.

C. *Humidity ratio* or specific humidity or absolute humidity W is defined as the ratio of the mass of water vapor to that of dry air, i.e.,

$$W = \frac{\text{mass of water vapor}}{\text{mass of dry air}}$$

Using the ideal gas law under saturated air conditions,

$$W = \frac{m_v}{m_a} = \frac{p_s V M W_v / R^* T}{p_a V M W_a / R^* T} = \frac{p_s M W_v}{(p_t - p_s) M W_a}$$

where V is an arbitrary volume of the air and water vapor mixture, R^* is the universal gas constant, MW_a is the molecular weight of dry air (28.965), and MW_v is the molecular weight of water (18.015). The above formula then reduces to

$$W = 0.622 \frac{p_s}{p_t - p_s} \tag{2.2.6}$$

BLE 2.2.3 Moist Air, Standard Atmospheric Pressure, 14.696 psi

Temp.	Humidity Ratio lb$_w$/lb$_a$ W_s	Volume ft³/lb dry air v_a	v_{as}	v_s	Enthalpy But/lb dry air h_a	h_{as}	h_s	Entropy But/(lb dry air·F) s_a	s_{as}	s_s	Condensed Water Enthalpy But/lb h_w	Entropy Btu/ (lb·F) s_w	Vapor Press. in Hg p_s	Temp. F
0	0.0007875	11.579	0.015	11.594	0.0	0.835	0.835	0.00000	0.00192	0.00192	−158.89	−0.3243	0.037671	0
5	0.0010207	11.706	0.019	11.725	1.201	1.085	2.286	0.00260	0.00247	0.00506	−156.52	−0.3192	0.048814	5
0	0.0013158	11.832	0.025	11.857	2.402	1.402	3.804	0.00517	0.00315	0.00832	−154.13	−0.3141	0.062901	10
5	0.0016874	11.959	0.032	11.991	3.603	1.801	5.404	0.00771	0.00400	0.01171	−151.71	−0.3089	0.080623	15
0	0.0021531	12.085	0.042	12.127	4.804	2.303	7.107	0.01023	0.00505	0.01528	−149.27	−0.3038	0.102798	20
5	0.0027339	12.212	0.054	12.265	6.005	2.930	8.935	0.01272	0.00636	0.01908	−146.80	−0.2987	0.130413	25
0	0.0034552	12.338	0.068	12.406	7.206	3.711	10.917	0.01519	0.00796	0.02315	−144.31	−0.2936	0.164631	30
2	0.0037895	12.389	0.075	12.464	7.687	4.073	11.760	0.01617	0.00870	0.02487	−143.30	−0.2915	0.180479	32
6	0.004452	12.490	0.089	12.579	8.648	4.793	13.441	0.01811	0.01016	0.02827	4.05	0.0081	0.21181	36
0	0.005216	12.591	0.105	12.696	9.609	5.624	15.233	0.02004	0.01183	0.03187	8.07	0.0162	0.24784	40
4	0.006094	12.692	0.124	12.816	10.570	6.582	17.152	0.02196	0.01374	0.03570	12.09	0.0242	0.28918	44
8	0.007103	12.793	0.146	12.939	11.531	7.684	19.215	0.02386	0.01592	0.03978	16.10	0.0321	0.33651	48
2	0.008259	12.894	0.171	13.065	12.492	8.949	21.441	0.02575	0.01840	0.04415	20.11	0.0400	0.39054	52
6	0.009580	12.995	0.200	13.195	13.453	10.397	23.850	0.02762	0.02122	0.04884	24.11	0.0478	0.45205	56
0	0.011087	13.096	0.233	13.329	14.415	12.052	26.467	0.02947	0.02442	0.05389	28.11	0.0555	0.52193	60
4	0.012805	13.198	0.271	13.468	15.376	13.942	29.318	0.03132	0.02804	0.05936	32.11	0.0632	0.60113	64
8	0.014758	13.299	0.315	13.613	16.337	16.094	32.431	0.03315	0.03214	0.06529	36.11	0.0708	0.69065	68
2	0.016976	13.400	0.365	13.764	17.299	18.543	35.841	0.03496	0.03677	0.07173	40.11	0.0783	0.79167	72
6	0.019491	13.501	0.422	13.923	18.260	21.323	39.583	0.03676	0.04199	0.07875	44.10	0.0858	0.90533	76
0	0.022340	13.602	0.487	14.089	19.222	24.479	43.701	0.03855	0.04787	0.08642	48.10	0.0933	1.03302	80
4	0.025563	13.703	0.561	14.264	20.183	28.055	48.238	0.04033	0.05448	0.09481	52.09	0.1006	1.17608	84
8	0.029208	13.804	0.646	14.450	21.145	32.105	53.250	0.04209	0.06192	0.10401	56.09	0.1080	1.33613	88
2	0.033323	13.905	0.742	14.647	22.107	36.687	58.794	0.04384	0.07028	0.11412	60.08	0.1152	1.51471	92
6	0.037972	14.006	0.852	14.858	23.069	41.871	64.940	0.04558	0.07968	0.12525	64.07	0.1224	1.71372	96
0	0.043219	14.107	0.976	15.084	24.031	47.730	71.761	0.04730	0.09022	0.13752	68.07	0.1296	1.93492	100
4	0.049140	14.208	1.118	15.326	24.993	54.354	79.346	0.04901	0.10206	0.15108	72.06	0.1367	2.18037	104
8	0.055826	14.309	1.279	15.588	25.955	61.844	87.799	0.05071	0.11537	0.16608	76.05	0.1438	2.45232	108
2	0.063378	14.411	1.462	5.872	26.917	70.319	97.237	0.05240	0.13032	0.18272	80.05	0.1508	2.75310	112
6	0.071908	14.512	1.670	16.181	27.879	79.906	107.786	0.05408	0.14713	0.20121	84.04	0.1577	3.08488	116
0	0.081560	14.613	1.906	16.519	28.842	90.770	119.612	0.05575	0.16605	0.22180	88.04	0.1647	3.45052	120
4	0.092500	14.714	2.176	16.890	29.805	103.102	132.907	0.05740	0.18739	0.24480	92.03	0.1715	3.85298	124
8	0.104910	14.815	2.485	17.299	30.767	117.111	147.878	0.05905	0.21149	0.27054	96.03	0.1783	4.29477	128
2	0.119023	14.916	2.837	17.753	31.730	133.066	164.796	0.06068	0.23876	0.29944	100.02	0.1851	4.77919	132
6	0.135124	15.017	3.242	18.259	32.693	151.294	183.987	0.06230	0.26973	0.33203	104.02	0.1919	5.30973	136
0	0.153538	15.118	3.708	18.825	33.656	172.168	205.824	0.06391	0.30498	0.36890	108.02	0.1985	5.88945	140
4	0.174694	15.219	4.245	19.464	34.620	196.183	230.802	0.06551	0.34530	0.41081	112.02	0.2052	6.52241	144
8	0.199110	15.320	4.869	20.189	35.583	223.932	259.514	0.06710	0.39160	0.45871	116.02	0.2118	7.21239	148
2	0.227429	15.421	5.596	21.017	36.546	256.158	292.705	0.06868	0.44507	0.51375	120.02	0.2184	7.96306	152
6	0.260512	15.522	6.450	21.972	37.510	293.849	331.359	0.07025	0.50723	0.57749	124.02	0.2249	8.77915	156
0	0.29945	15.623	7.459	23.082	38.474	338.263	376.737	0.07181	0.58007	0.65188	128.02	0.2314	9.6648	160
4	0.34572	15.724	8.664	24.388	39.438	391.095	430.533	0.07337	0.66622	0.73959	132.03	0.2378	10.6250	164
8	0.40131	15.825	10.117	25.942	40.402	454.630	495.032	0.07491	0.76925	0.84415	136.03	0.2442	11.6641	168
2	0.46905	15.926	11.894	27.820	41.366	532.138	573.504	0.07644	0.89423	0.97067	140.04	0.2506	12.7880	172
6	0.55294	16.027	14.103	30.130	42.331	628.197	670.528	0.07796	1.04828	1.12624	144.05	0.2569	14.0010	176
0	0.65911	16.128	16.909	33.037	43.295	749.871	793.166	0.07947	1.24236	1.32183	148.06	0.2632	15.3097	180
4	0.79703	16.229	20.564	36.793	44.260	908.061	952.321	0.08098	1.49332	1.57430	152.07	0.2694	16.7190	184
8	0.98272	16.330	25.498	41.828	45.225	1121.174	1166.399	0.08247	1.82963	1.91210	156.08	0.2756	18.2357	188
2	1.24471	16.431	32.477	48.908	46.190	1422.047	1468.238	0.08396	2.30193	2.38589	160.10	0.2818	19.8652	192
6	1.64070	16.532	43.046	59.578	47.155	1877.032	1924.188	0.08543	3.01244	3.09797	164.12	0.2880	21.6152	196
0	2.30454	16.633	60.793	77.426	48.121	2640.084	2688.205	0.08690	4.19787	4.28477	168.13	0.2941	23.4906	200

Source: Abridged by permission from *ASHRAE Handbook, Fundamentals Volume*, 1985.

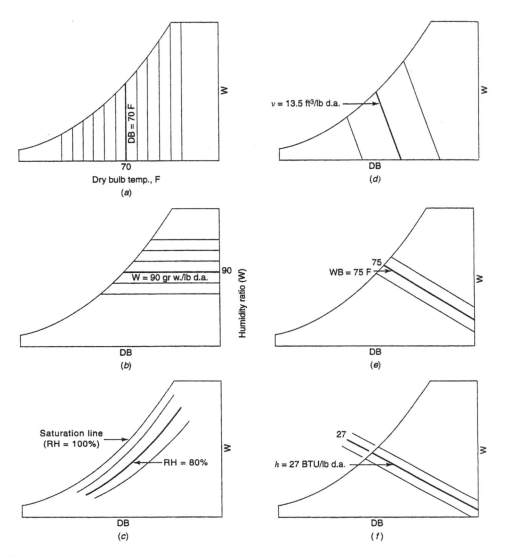

FIGURE 2.2.3 Construction of psychrometric chart, showing lines of constant property values. (*a*) Lines of constant dry-bulb temperature (DB) on the psychrometric chart. (*b*) Lines of constant humidity ratio (W) on the psychrometric chart. (*c*) Lines of constant relative humidity (RH) on the psychrometric chart. (*d*) Lines of constant specific volume (*v*) on the psychrometric chart. (*e*) Lines of constant wet-bulb temperature (WB) on the psychrometric chart. (*f*) Lines of constant enthalpy (*h*) on the psychrometric chart.

Note that because $p_s \ll p_t$, the relation between W and p_s is close to being linear in p_s. As shown in Figure 2.2.3, the y-axis of the psychrometric chart is allocated to the humidity ratio.

D. *Relative humidity* RH (or ϕ) is defined as the ratio of the partial pressure of water vapor p_v divided by the saturated pressure of water vapor at the same dry-bulb temperature, i.e.,

$$\phi = \frac{p_v}{p_s} \tag{2.2.7}$$

As shown in Figure 2.2.3(c), one can now add lines of constant RH by marking vertical distances between the saturation line and the base of the chart.

E. *Specific enthalpy* h of moist air is equal to the sum of dry air enthalpy and that of water vapor. Though the air-vapor mixture is likely to be superheated, there is not much error in assuming the enthalpy of the water vapor to be equal to the saturated value at the same temperature. Thus, it is convenient to estimate moist air enthalpy as

$$h = c_p t + W \cdot h_g \qquad (2.2.8a)$$

where h_g is the enthalpy of saturated steam at temperature t which can be determined from Table 2.2.2. Enthalpy values are always based on a datum value, usually 32°F or 0°C.
Alternative expressions for moist air enthalpy are:

SI units: $h = t + W(2501.3 + 1.86 \cdot t)$ in kJ/kg of dry air $\qquad (2.2.8b)$
IP units: $h = 0.24 \cdot t + W(1061.2 + 0.444 \cdot t)$ in Btu/lb.

As shown in Figure 2.2.3(f), lines of constant enthalpy can now be drawn since h is a function of dry-bulb temperature and specific humidity.

F. *Specific volume* v is the volume of the mixture (say in m³ or in ft³) per unit mass of dry air. The perfect gas law can be used to estimate it:

$$v = \frac{R_a T}{P_t - P_s} \qquad (2.2.9)$$

Lines of constant specific volume are shown in Figure 2.2.3(d) by selecting a value of v and solving for p_s (and hence W using Equation 2.2.6) for different values of T. Recall that a specific psychrometric chart can be used for only a specific pressure p_t.

G. *Adiabatic saturation temperature* is the temperature reached by air when it passes through a spray of water such that there is thermal and vapor pressure equilibrium between the air and the water, as shown in Figure 2.2.4. The process is assumed to be adiabatic in that no heat is either added or lost from the chamber. Since the air is not fully saturated, a certain amount of water evaporates into the air whose latent heat is supplied by the air. Hence the moist air dry-bulb temperature decreases while its specific humidity increases. A small amount of make-up water to compensate for the evaporated water is supplied at a temperature equal to that of the sump.

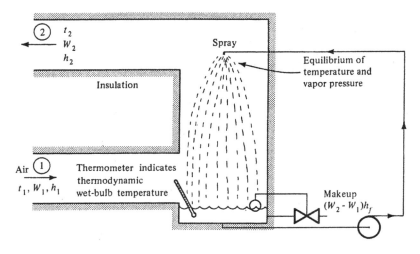

FIGURE 2.2.4 Adiabatic saturation.

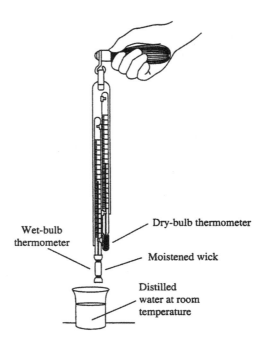

Wet-bulb
thermometer

Dry-bulb thermometer

Moistened wick

Distilled
water at room
temperature

FIGURE 2.2.5   Sling psychrometer device for conveniently measuring wet- and dry-bulb temperatures.

H. The *adiabatic wet-bulb temperature* is the temperature of sump water in an adiabatic saturator. An energy balance on the adiabatic saturator shown in Figure 2.2.4 yields

$$h_1 = h_2 - (W_2 - W_1) \cdot h_f \tag{2.2.10}$$

where h_f is the enthalpy of saturated liquid at the thermodynamic wet-bulb temperature. Because of the latent heat term, lines of constant wet-bulb temperature are not quite identical to lines of constant enthalpy, though close. Lines of constant wet-bulb temperature are shown in Figure 2.2.3(e).

The adiabatic saturator is not a practical device for measuring the adiabatic saturation temperature. Instead, the hand-held sling psychrometer is often used (Figure 2.2.5). The apparatus consists of two thermometers, one measuring the dry-bulb temperature, and the other, which has a wetted wick covering the bulb, measures the wet-bulb temperature. The instrument has a handle which allows the thermometers to be spun so as to induce air movement over the bulbs that is adequate for proper heat transfer between the bulb and the ambient air. Though the wet-bulb temperature is not the same as the adiabatic saturation temperature, the difference is small. A detailed discussion of these differences can be found in Kuehn et al. (1998). Electronic devices are also commonly used to measure humidity levels in air.

I. *Dew point temperature* T_{dp} of a given mixture is the temperature of *saturated* moist air at the same pressure, temperature, and humidity ratio as the given mixture. When a surface reaches the dew point temperature, the moisture will start condensing from the surrounding air.

Note that specific volume, specific enthalpy, and specific humidity are all defined per mass of *dry* air and not per mass of moist air. This convention dispels the confusion created when mass transfer takes place (i.e., when water vapor is either added to or removed from the air).

A complete psychrometric chart corresponding to standard atmospheric pressure is shown in Figure 2.2.6. Although computer programs allow for more accurate and faster determination of the moist air properties (as well as the conversion of different systems of units and arbitrary atmospheric pressures), the psychrometric chart is still used extensively by HVAC professionals for several aspects of design and analysis.

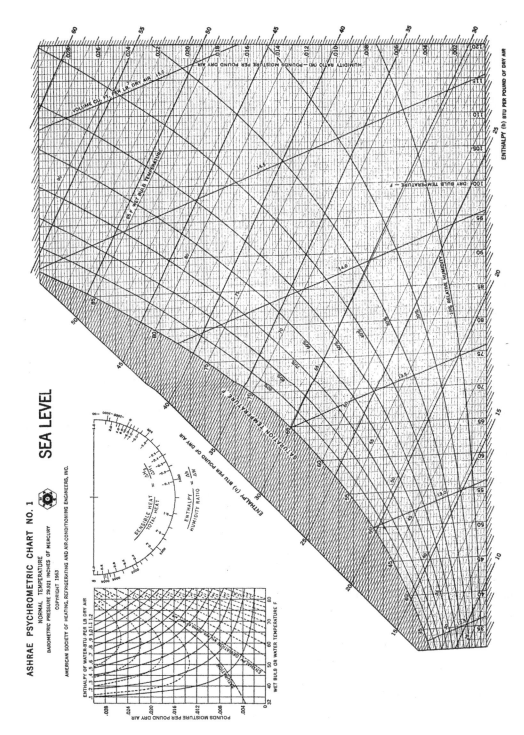

FIGURE 2.2.6 ASHRAE Psychrometric Chart No. 1. With permission.

Example 1: Determination of moist air properties.

Calculate values of humidity ratio, specific volume, and enthalpy for air at 60°F and RH = 80% using the equations presented above. Assume standard atmospheric pressure.

From the steam tables, saturated vapor pressure at 60°F: p_s = 0.2563 psia
The vapor pressure of the moist air: $p_v = p_s \times RH = 0.2563 \times 0.8 = 0.2050$ psia

Humidity ratio: $W = 0.622 \dfrac{p_v}{P - p_v} = 0.622 \dfrac{0.205}{14.696 - 0.205} = 0.0088$ lbw/lba

Density: $\dfrac{1}{v} = \rho = \dfrac{P_a}{R_a T} = \dfrac{(14.696 - 0.205)}{53.352 \cdot (60 + 459.67)} \times 144 = 0.07526$ lba/ft^3

(the multiplier 144 converts ft^2 into in^2).

Thus specific volume: v = (1/ 0.07526) = 13.287 ft^3/lba.

Finally, enthalpy is determined from

$$h = 0.240 \cdot t + W (1061.2 + 0.444\, t)$$

$$= 0.240 \times 60 + 0.0088 (1061.2 + 0.444 \times 60) = 110.13 \text{ Btu/lba.}$$

2.2.4   Psychrometric Processes

Analysis of moist air processes with various HVAC devices essentially involves a few fundamental processes, discussed below. Consider a duct containing a device through which moist air is flowing. The device could be a cooling or heating coil and/or a humidifier. The analysis of moist air processes flowing through such a device is based on the laws of conservation of mass and energy. Although in actual practice the properties of the moist air may not be uniform across the duct cross section (especially downstream of the device), such phenomena are neglected, and the focus is on bulk or fully mixed conditions. Further, assuming (1) steady state conditions and (2) a perfectly insulated duct, the following equations apply:

Mass balance of dry air: $\qquad m_{a1} = m_{a2}$ $\qquad\qquad\qquad\qquad\qquad\qquad$ (2.2.11a)

Mass balance on water vapor: $\qquad m_{a1} W_1 + m_w = m_{a2} W_2$ $\qquad\qquad\qquad$ (2.2.11b)

Heat balance: $\qquad m_{a1} h_1 + m_w h_w + Q = m_{a2} h_2$ $\qquad\qquad\qquad$ (2.2.11c)

where   Q is the rate of heat added to the air stream (in W or Btu/hr),
$\qquad$ m_a is the mass flow rate of dry air (in kga/s or lba/hr), and
$\qquad$ m_w is the rate of water added to the air stream.

A. Sensible Heating and Cooling

A process is called *sensible* (either heating or cooling) when it involves a change in dry-bulb temperature only (i.e., the moisture content specified by the specific humidity is unchanged in a sensible heating or cooling process). This could apply to either heating (an increase in T_{db}) or to cooling (a decrease in T_{db}). In such a case, $m_w = 0$, and $W_1 = W_2$. The above equations reduce to

$$Q = m_a (h_2 - h_1)$$
$\qquad\qquad\qquad\qquad\qquad\qquad\qquad\qquad\qquad\qquad\qquad\qquad\qquad\qquad$ 2.2.12

where $m_a = m_{a1} = m_{a2}$

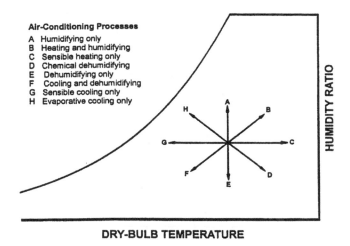

Air-Conditioning Processes
A Humidifying only
B Heating and humidifying
C Sensible heating only
D Chemical dehumidifying
E Dehumidifying only
F Cooling and dehumidifying
G Sensible cooling only
H Evaporative cooling only

HUMIDITY RATIO

DRY-BULB TEMPERATURE

FIGURE 2.2.7   Different psychrometric processes.

The process of sensible heating or cooling is represented as a straight line on the psychrometric chart as shown in Figure 2.2.7. Such a process occurs when moist air flows across a cooling coil when condensation does not occur.

Example 2: Sensible heating
Moist air enters a steam-heating coil at 40°F dry-bulb temperature and 36°F wet-bulb temperature at a rate of 2000 ft³/min. The air leaves the coil at a dry-bulb temperature of 140°F. Determine the heat transfer rate that occurs at the coil.

Assumptions: steady state, standard atmospheric conditions

$$\text{Heat transfer rate } Q = m_a (h_2 - h_1)$$

where states 1 and 2 represent the entering and exiting air stream conditions, respectively.

From the psychrometric chart, we find: v_1 = 12.66 ft³/lba, h_1 = 13.47 Btu/lba, h_2 = 37.70 Btu/lba

The mass flow rate of moist air: $m_a = \dfrac{Q_a}{v_1} = \dfrac{2000}{12.66}$ = 158 lba/min = 9479 lba/hr

Finally, Q = 9479 · (37.70 − 13.47) = 229,670 Btu/hr.

B. Cooling and Dehumidification

This process occurs when conditioning outdoor air in summer or in internal spaces where heat and moisture are removed by cooling coils in a conditioned space. For this process to occur, moist air is cooled to a temperature below its dew point. Some of the water vapor condenses out of the air stream. Although the actual process path varies depending on the type of surface, surface temperature, and flow conditions, the heat and mass transfer can be expressed in terms of the initial and final states.

As shown in Figure 2.2.8, a certain amount of moisture condenses out of the air stream. Although this condensation occurs at various temperatures ranging from the initial dew point to its final saturation temperature, it is assumed that condensed water is cooled to the final air temperature t_2 before it drains out. The above equations reduce to

Rate of water condensation: $\qquad m_w = m_a(W_1 - W_2)$ (2.2.13a)

Rate of total heat transfer: $\qquad Q = m_a[(h_1 - h_2) - (W_1 - W_2)h_{w2}]$ (2.2.13b)

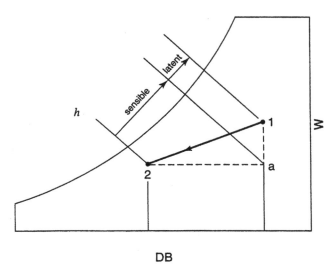

FIGURE 2.2.8   Cooling and dehumidification process.

The above equation gives the total rate of heat transfer from the moist air. The last term is usually small compared to the other term and is often neglected (as illustrated in Example 3). Cooling and dehumidification processes involve both sensible and latent heat transfer. The sensible heat transfer q_s results in a decrease in dry-bulb temperature, while the latent heat transfer q_l is associated with the decrease in specific humidity. These quantities can be estimated as follows. Let point "a" be the intersection point between the constant dry-bulb temperature line from point 1 and the constant specific humidity line at point 2 (see Figure 2.2.8). Then

Rate of sensible heat transfer: $Q_s = m_a(h_a - h_2)$ (2.2.14a)

Rate of latent heat transfer: $Q_l = m_a(h_1 - h_a)$ (2.2.14b)

It is customary to characterize the relative contributions of sensible and latent heat transfer rates by the sensible heat ratio (SHR) where SHR is defined as follows:

$$\text{SHR} = \frac{q_s}{q_s + q_l}$$ (2.2.15)

Example 3: Cooling and dehumidification.
Moist air enters a cooling coil at 80°F dry-bulb temperature and 67°F wet-bulb. It is cooled to 58°F and 80% RH. The volume flow rate is 2000 cfm and the condensate leaves at 60°F. Find the total coil heat transfer rate, as well as the individual sensible and latent heat transfer rates.

Assumptions: steady state, standard atmospheric conditions

Heat transfer rate

$$Q = m_a(h_2 - h_1) + m_w \cdot h_w$$

Note that Q will be negative, since cooling is being done.
From the psychrometric chart:

h_1 = 31.6 Btu/lba,
h_2 = 22.9 Btu/lba,
v_1 = 13.85 ft³/lba,
W_1 = 0.0112 lbw/lba,
W_2 = 0.0082 lbw/lba,

and enthalpy of liquid water at 60°F from the steam tables (Table 2.2.2),

$$h_w = 28.06 \text{ Btu/lbw}$$

The mass flow rate of dry air:

$$m_a = \frac{Q_a}{v_1} = \frac{2000}{13.85} = 144.4 \text{ lba/min} = 8664 \text{ lba/hr}$$

Rate at which water is condensed:

$$m_w = m_a \cdot (W_1 - W_2) = 8664 \cdot (0.0112 - 0.0082) = 25.99 \text{ lbw/hr}$$

Finally,

$$Q = 8664 \cdot (31.6 - 22.9) - 25.99 \times 28.06 = 75379.06 - 729.3 = 74{,}650 \text{ Btu/hr}$$

$$Q = 6.22 \text{ tons}$$

(As stated in the text above, the contribution of the heat contained in the condensing water is very small; here it is about 1% only). Often this term is ignored in psychrometric calculations.

In order to determine the contributions of the sensible and latent heat transfer rates, one determines point "a" on the psychrometric chart. This can be done by drawing a horizontal line from point 2 (constant humidity ratio) and a vertical line from point 1 (constant dry-bulb temperature). The enthalpy at this point h_a = 27 Btu/lba.

Sensible heat transfer rate:

$$q_s = m_a \cdot (h_a - h_2) = 8664 \cdot (27 - 22.9) = 35{,}522 \text{ Btu/hr} = 2.96 \text{ tons}$$

Latent heat transfer rate:

$$q_l = m_a \cdot (h_1 - h_a) = 8664 \cdot (31.6 - 27) = 39{,}850 \text{ Btu/hr} = 3.32 \text{ tons}$$

The sensible heat ratio is

$$SHR = 35{,}522 / (35{,}522 + 39{,}854) = 0.47$$

C. Heating and Humidification

In winter when the outdoor air is cool and dry, the building air supply stream must be heated and humidified to meet comfort criteria. In this case, we have

Rate of water evaporation: $\quad m_w = m_a(W_2 - W_1)$ (2.2.16a)

Rate of heat transfer: $\quad Q = m_a[(h_2 - h_1) - (W_2 - W_1)h_{w2}]$ (2.2.16b)

Alternatively the direct addition of moisture to an air stream without any heat addition can be adopted to humidify an air stream (i.e., there is no heating coil). The heat input term Q is zero in Equation 2.2.16b.

The water is typically in the form of spray or vapor (steam). Practical examples of this process include humidifiers, cooling towers, and evaporative coolers.

The direction of the conditioning line between states 1 and 2 depends on the enthalpy of the moisture added. Two unique cases will be discussed. The cases depend on the state of the humidifying spray. First, if the spray is saturated vapor ($h_w = h_g$) at the same dry-bulb temperature of the incoming air stream, the process line will proceed at a constant dry-bulb temperature (line A of Figure 2.2.7). The enthalpy and wet-bulb temperature of the air will increase, but the dry-bulb temperature of the air will remain constant.

Another case occurs if the humidifying spray is saturated water at the wet-bulb temperature of the air. The exiting air stream will have the same wet-bulb temperature as the entering air stream. The dry-bulb temperature of the air will decrease. The leaving enthalpy of the air will be close to the entering enthalpy because the constant enthalpy lines on the psychrometric chart are approximately parallel to the constant wet-bulb lines. This process is shown by line H in Figure 2.2.7.

If the enthalpy of the spray is larger than the enthalpy of saturated water vapor at the dry-bulb temperature, then the air stream will be sensibly heated during humidification. In contrast, if the spray enthalpy is less than the enthalpy of saturated water vapor at the entering dry-bulb temperature of the air, the air will be sensibly cooled during the humidification process.

Example 4: Heating and humidification.

Outdoor winter air enters a heating and humidification system at a rate of 900 cfm and at 32°F dry-bulb and 28°F wet-bulb conditions. The air absorbs 75,000 Btu/hr of energy and 25 lb/hr of saturated steam at 212°F. Determine the dry-bulb and wet-bulb temperature of the leaving air.

Determine relevant properties:

$$h_1 = 10.2 \text{ Btu/lb}, v_1 = 12.42 \text{ ft}^3\text{/lba}, W_1 = 0.0023 \text{ lbw/lba}$$

The mass flow rate of dry air:

$$m_a = \frac{Q}{v_1} = \frac{900}{12.42} = 72.45 \text{ lba/min}$$

Equation 2.2.16a can be used to solve for the final humidity ratio:

$$W_2 = \frac{m_a W_1 + m_w}{m_a} = W_1 + \frac{m_w}{m_a} = 0.0023 + \frac{25}{72.45 \times 60} = 0.008051 \text{ lbw/lba}$$

Equation 2.2.16b is used to determine the exit air enthalpy:

$$h_2 = h_1 + \frac{q}{m_a} + \frac{m_w}{m_a} h_w = 10.2 + \frac{75000}{72.45 \times 60} + \frac{25}{72.45 \times 60} \cdot 1150.5 = 34.07 \text{ Btu/lba}$$

With the exit humidity ratio known, the final state is established and other properties can be read from the psychrometric chart.

D. Adiabatic Mixing of Air Streams

The mixing of two air streams is very common in HVAC systems. For example, we assume adiabatic conditions (i.e., no heat transfer across the duct walls). Then we have

Mass balance of dry air:	$m_{a1} + m_{a2} = m_{a3}$	(2.2.17a)
Mass balance on water vapor:	$m_{a1}W_1 + m_{a2}W_2 = m_{a3}W_3$	(2.2.17b)
Heat balance:	$m_{a1}h_1 + m_{a2}h_2 = m_{a3}h_3$	(2.2.17c)

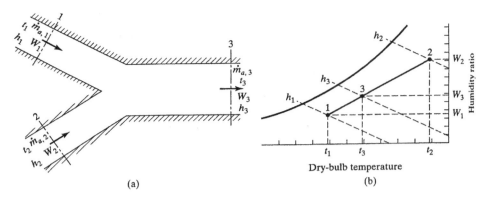

FIGURE 2.2.9 Schematic adiabatic mixing of two streams of moist air.

Combining the above yields

$$\frac{h_2 - h_3}{h_3 - h_1} = \frac{W_2 - W_3}{W_3 - W_1} = \frac{m_{a1}}{m_{a2}} \qquad (2.2.18)$$

The state of the mixed streams lies on a straight line between states 1 and 2 with the point 3 determined in inverse proportion to the masses of the two incoming air streams (see Figure 2.2.9):

$$\frac{m_{a1}}{m_{a2}} = \frac{\text{segment } 3 - 2}{\text{segment } 1 - 3}$$

Example 5: Adiabatic mixing of air streams.

A stream of 5000 cfm outdoor air at 40°F dry-bulb temperature and 35°F wet-bulb temperature is adiabatically mixed with 15,000 cfm of recirculated air at 75°F dry-bulb and 50% RH. Find the dry-bulb temperature and wet-bulb temperature of the resulting mixture.

The following property data are needed:

$$W_1 = 0.00315 \text{ lbw/lba}, \ v_1 = 12.65 \text{ ft}^3/\text{lba}, \ h_1 = 13.2 \text{ Btu/lba}$$

$$W_2 = 0.00915 \text{ lbw/lba}, \ v_2 = 13.68 \text{ ft}^3/\text{lba}, \ h_2 = 28.32 \text{ Btu/lba}$$

Mass flow rate of stream 1:

$$m_{a1} = \frac{Q_1}{v_1} = \frac{500}{12.65} = 395.3 \text{ lba/min}$$

Mass flow rate of stream 2:

$$m_{a2} = \frac{Q_2}{v_2} = \frac{15,000}{13.68} = 1096.5 \text{ lba/min}$$

Humidity ratio of mixed air:

$$W_3 = \frac{1}{(m_{a1} + m_{a2})}(m_{a1} \cdot W_1 + m_{a2} \cdot W_2) = \frac{395.3 \times 0.00315 + 1096.5 \times 0.00915}{1491.8} = 0.00756 \text{ lbw/lba}$$

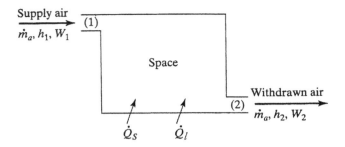

FIGURE 2.2.10 Schematic flow processes for an air conditioned space.

The dry-bulb temperature and the enthalpy can be similarly determined. It is left to the reader to verify that:

$$T_{3db} = 66°F \text{ and } h_3 = 24.3 \text{ Btu/lba}$$

2.2.5 Psychrometric Analysis of Basic HVAC Systems

Buildings generally consist of a number of rooms which may have different energy and moisture gains or losses — the loads. Loads exhibit both seasonal and diurnal variation. Adjacent rooms with similar loads are usually lumped together into one *zone* which is controlled by one thermostat. Air handlers in an HVAC system can be designed to condition one zone (called single zone systems) or multiple zones (called multiple zone systems). Residences and small commercial buildings are usually designed and operated as single zone spaces.

Condition Line for a Space

A space is air conditioned to offset the heating and/or cooling loads of the space as a result of envelope heat transmission, ventilation air requirements, and internal loads due to occupants, lights, and equipment. The calculations involved in air conditioning design reduce to the determination of the mass of dry air to be circulated, its dry-bulb temperature, and its humidity level that will result in comfortable indoor conditions for the occupants.

Let Q_s and Q_l be the sensible and latent loads on a space to be air conditioned (see Figure 2.2.10). The latent load is due to the sum of all rates of moisture gain designated by m_w. Assuming steady conditions,

$$Q_s + Q_l = m_a(h_2 - h_1) \tag{2.2.19a}$$

$$m_w = m_a(W_2 - W_1) \tag{2.2.19b}$$

Thus, the enthalpy-moisture ratio q' is

$$q' = \frac{Q_s + Q_l}{m_w} = \frac{h_2 - h_1}{W_2 - W_1} \tag{2.2.20}$$

The above equation suggests that for the supply air to satisfy simultaneously both the sensible and latent loads, its condition must lie on a straight line called the *condition line* (or load line). In case the space needs to be cooled, supply temperature $t_1 < t_2$ and the condition line will appear as shown in Figure 2.2.11. Point 1 must lie on this condition line. How far point 1 is from point 2 is determined by practical considerations, for example, on a prespecified air flow rate or on a prespecified temperature difference between states 1 and 2. During summer conditions, an approximate range of variation of $(t_1 - t_2)$ is 15–25°F (8–15°C).

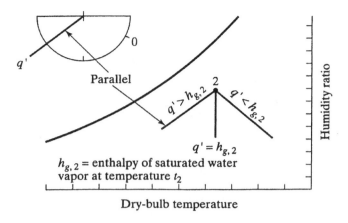

FIGURE 2.2.11 The condition line.

As shown in Figure 2.2.11, there is a simple graphical construction for drawing the load line through a specified point on a psychrometric chart. The semicircular protractor in the upper left-hand corner of the chart has a series of marks corresponding to different SHR values (see Figure 2.2.6 and Equation 2.2.15). One determines the SHR for the known design conditions and draws a line through point 2 parallel to the line in the protractor circle corresponding to the calculated SHR.

Example 6: The air in a space is to be maintained at comfort conditions of 75°F and 50% RH. The sensible and latent load gains for the space are 89,000 Btu/hr and 52,000 Btu/hr respectively. The supply air to the room is to be at 60°F. Determine the dew point temperature and the required volume flow rate of the supply air.

First, calculate the sensible heat ratio:

$$\text{SHR} = \frac{89{,}000}{89{,}000 + 52{,}000} = 0.631$$

Next, locate the return air condition (the same as room conditions) as state 2 on the psychrometric chart. The condition line is drawn from this point with the aid of the chart protractor as a line parallel to the SHR = 0.631 line. Where this line intersects the vertical line representing 60°F dry-bulb temperature is the required supply state 1, i.e., the supply air condition. The dew point temperature is read as 49°F for this condition using Figure 2.2.6.

From the psychrometric chart: $h_1 = 22.45$ Btu/lba, $v_1 = 13.25$ ft³/lba, and $h_2 = 28.15$ Btu/lba. A simple heat balance yields the required air mass flow rate:

$$m_a = \frac{89{,}000 + 52{,}000}{28.15 - 22.45} = 24{,}700 \text{ lba/hr}$$

Finally the supply volumetric flow rate is:

$$= \frac{m_a \cdot v_1}{60} = \frac{24{,}700 \times 13.25}{60} = 5{,}465 \text{ ft}^3/\text{min}$$

Cooling Coil Performance Calculations

Section 2.2.4B discussed the cooling and dehumidification process. The discussion assumed the inlet and outlet conditions of the air stream to be specified, and the objective was to determine the

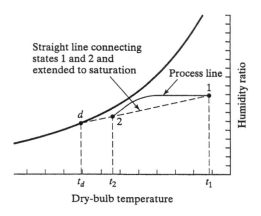

FIGURE 2.2.12   Schematic psychrometric chart of a real cooling/dehumidifying process.

associated rates of sensible, latent, and total heat transfer. Here we briefly describe the physical phenomenon that occurs as the air stream flows over the cooling coil, and how this process can be represented on the psychrometric chart. Next, we present a simplified model of characterizing the cooling coil performance.

We need to distinguish between two cases: an ideal process and an actual process. The ideal process corresponds to when all the air molecules of the air stream come in perfect contact with the cooling coil surface and are cooled to the dew point temperature, after which dehumidification occurs along the saturated line to the final state, shown as point *d* on Figure 2.2.12. In the real process, all the air molecules do not come into intimate contact with the cooling coil surface. The cooling coil is basically a heat exchanger, consisting of a series of flat, parallel cooled metal surfaces, which form passages for the air stream to flow through. Often there are several rows of coils in parallel. The temperature of the air flowing through is nonuniform both at a given cross-section and along the depth of the coil. The air particles near the surface of the heat exchanger surface follow the idealized process, while those near the center line of the passage are not fully cooled. This results in dehumidification even though the average air temperature in the passage is above the dew point. Since processes on the psychrometric chart represent average or bulk conditions, the real cooling process as the air stream flows through the cooling coil resembles the curved process line shown in Figure 2.2.12.

The phenomenon described above serves as the basis of a simple means of characterizing the cooling coil performance when subjected to varying inlet conditions. The *bypass factor method* assumes that the passage of the air stream through a cooling coil of given geometric design consists of two streams: one stream (m_a') that comes into intimate contact with the heat exchanger surface, and another that totally bypasses the cooling coil and hence remains at the condition of the entering air. The former portion of the air stream is cooled along the saturation line to the apparatus dew point temperature t_d as shown in Figure 2.2.12. The final state of the air leaving the cooling coil can be determined if the relative flow rates of both air streams are known. The coil is characterized by a *coil bypass factor b* defined as

$$b \equiv \frac{m_a - m_a'}{m_a} = \frac{t_2 - t_d}{t_1 - t_d} \tag{2.2.21}$$

where, as shown in Figure 2.2.12, t_1 and t_2 are the dry-bulb temperatures of the air stream entering and leaving the cooling coil.

The usefulness of this approach lies in the fact that the bypass factor of a given coil is a coil characteristic which remains constant under a wide range of operating temperature conditions; however, it is likely to change with varying air flow rate.

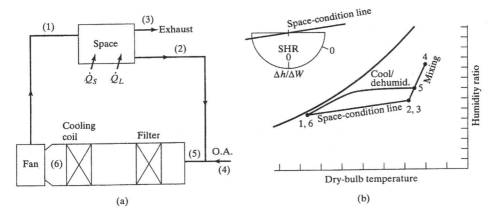

FIGURE 2.2.13 Schematic elementary summer air conditioning system.

Psychrometrics of Single Zone Systems

This section discusses and illustrates general principles of how the fundamental processes presented in the previous section are applied to buildings. We limit the discussion to three common HVAC systems used to condition one-zone spaces. Treatment of multiple zones and the analysis of more complex HVAC systems can be found in Kreider and Rabl (2001). In the discussion that follows, we assume that the temperature and humidity conditions required for maintaining comfort in the space are specified in advance.

Figure 2.2.13a shows the most elementary HVAC system suitable for summer air conditioning, while Figure 2.2.13b shows the corresponding state points on the psychrometric chart. A certain amount of outdoor air is required for ventilation purposes (to meet indoor comfort criteria). Hence, outdoor air is introduced into the system at (4) while the same amount of air is exhausted at (3). This air is shown as being exhausted directly from the space; it is also common to exhaust it from a convenient location in the ducting downstream of the room. A fan is used to move air through the system. Although only a supply air fan is shown, larger systems might use a supply fan and a return fan. Fans add thermal energy to the air stream due to their inefficiencies and may result in air stream heating of 1–3°F. In simplified psychrometric analyses of HVAC systems, this load is either neglected or combined with that of the conditioned space. Hence states (6) and (1) are shown as identical in Figure 2.2.13b. Filters, which are essential for control of particulate matter carried by the air, do not affect the state of the moist air. A cooling coil is the only processing device; it reduces both the dry-bulb temperature and the humidity of zone supply air.

The simple system shown above has several limitations. First, since the cooling coil is the only processing device, only one property of the moist air can be controlled. In comfort air conditioning systems, this is usually the space dry-bulb temperature. Second, though the system can be designed to meet the peak summer loads, such a system is not suitable for part load operation which occurs most of the year. A reheat coil, as shown in Figure 2.2.14a, is commonly used. Here, the cooling coil reduces the supply air specific humidity level at state (6) to that required at state (1). By doing so, the dry-bulb temperature is lower than that required at state (1) to meet the sensible loads of the space. The heating coil now heats the air stream to state (1) as illustrated in Figure 2.2.14b. Hence, as the outdoor conditions vary over the course of the year, the reheat coil acts as a final regulation device, adding only the required amount of heat to condition the air to the level desired at the supply of the room. However, cooling of the supply air stream and subsequent reheat are wasteful of energy.

One approach to reducing the energy waste is to vary the rate of air supply to the room so the condition of the supply air stream at state (1) is equal to that leaving the cooling coil (state 6). Variable air volume

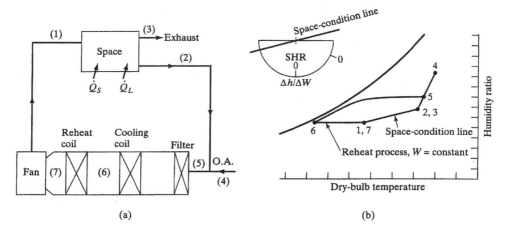

FIGURE 2.2.14   Schematic summer air conditioning system with reheat.

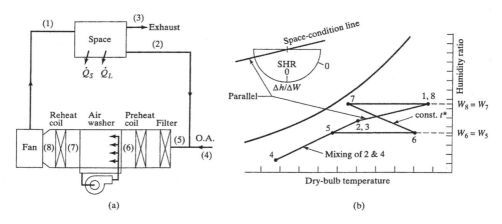

FIGURE 2.2.15   Schematic winter air conditioning system.

(VAV) systems, in contrast to constant volume (CV) systems, have been widely used for the past 20 years. Another means of reducing energy use is to use a heat recovery device (such as a rotary heat wheel or a run-around coil) between the exhaust air stream and the supply air stream (Kuehn et al., 1998).

During winter operation, heating must be supplied to a space to meet comfort conditions. Outdoor air in winter is much drier than that required for indoor human comfort. Hence, the basic system required for winter air conditioning, shown in Figure 2.2.15a, includes an air washer that allows the specific humidity of the supply air to be controlled. A preheat coil, also shown, prevents water in the air washer from possibly freezing when $T_5 < 32°F$. The various state points for this system operating in winter are shown on the psychrometric chart (Figure 2.2.15b). The process in the air washer (state points 6 to 7) follows the constant wet-bulb line, shown as t^*.

Outdoor conditions and internal loads require that the HVAC system supply heating in winter and cooling in summer. Such an HVAC system would be similar to that shown in Figure 2.2.15a with the added feature that a cooling coil would be placed downstream of the preheat coil. The preheat coil now has the primary responsibility of preventing the mixed air from becoming so cold in winter that the cooling coil is likely to suffer physical damage due to coil frosting.

Example 7:   A given space is to be maintained at 78°F dry-bulb and 65°F wet-bulb. The total heat gain to a zone Q_t is 60,000 Btu/hr of which 42,000 Btu/hr is sensible heat transfer (i.e., SHR = 0.70). The outdoor air requirement of the occupants totals 500 cfm, and the outdoor air condition is 90°F dry-bulb and 55% RH.

Determine:

    a. the quantity and state of the supply air to the space
    b. required capacity of the cooling and dehumidification equipment

  Assumptions: Neglect the heat added by the supply fan.
                  Use the notation followed in Figure 2.2.13.

The problem is under-specified since neither the supply air flow rate nor the supply air temperature are specified. Usually one of these needs to be known in order to solve the problem as stated above. A typical rule of thumb is that the supply air dry-bulb temperature entering the space is to be 20°F cooler than the room temperature. Using this rule, T_{1db} = 58°F.

The following exercise shows the steps that an HVAC design engineer would undertake to size HVAC equipment.

Start with determining relevant properties of states that are fully specified.

Outdoor air: T_{4db} = 90°F, RH_4 = 55%. Then v_4 = 14.23 ft³/lba, W_4 = 0.0170 lbw/lba, and h_4 = 40.8 Btu/lba

Return or room air: T_{2db} = 78°F, T_{2wb} = 65°F. Then W_2 = 0.0102 lbw/lba and h_2 = 30 Btu/lba

Also, $T_6 = T_1$ because we ignore the fan temperature rise.

The supply air mass flow is determined from a sensible heat balance. (Note that an enthalpy or total heat balance cannot be used since the humidity of the supply air is unknown.)

Then m_{a1} = $\dfrac{Q_s}{c_p(T_2 - T_1)}$ = $\dfrac{42{,}000}{0.245(78 - 58)}$ = 8571 lba/hr

First, we use the total load to determine the enthalpy of the room supply air:

$$h_1 = -\frac{Q_t}{m_{a1}} + h_2 = -\frac{60{,}000}{8571} + 30 = 23 \text{ Btu/lba}$$

Second, we analyze the adiabatic mixing process at the outdoor air intake. The mass flow rate of outdoor air intake m_{a4} = $\dfrac{cfm}{v_4}$ = $\dfrac{500 \times 60 \text{ min/hr}}{14.23}$ = 2110 lba/hr

Hence, the amount of recirculated air:

$$m_{a2} = 8571 - 2110 = 6461 \text{ lba/hr}$$

An energy balance at the mixing point yields

$$h_5 = \frac{m_{a4}h_4 + m_{a2}h_2}{m_{a1}} = \frac{2110 \times 40.8 + 6461 \times 30}{8571} = 32.6 \text{ Btu/lba}$$

Similarly, humidity and temperature balances yield

$$W_5 = 0.01191 \text{ lbw/lba, and}$$

$$T_{5db} = 82°F.$$

Finally, the total cooling capacity of the cooling coil can be determined:

$$Q_c = m_{a1}(h_1 - h_5) = 8571(23 - 32.6) = -82,282 \text{ Btu/hr} = 6.86 \text{ cooling tons}$$

How much of this cooling load is sensible can also be deduced as follows:

$$Q_{sc} = m_{a1} \cdot c_p \cdot (T_5 - T_1) = 8571 \times 0.245(58 - 32) = -50,397 \text{ Btu/hr}$$

Similar to the room SHR, a cooling coil SHR can also be defined. For this system:

$$SHR_c \equiv \frac{Q_{sc}}{Q_c} = \frac{50,397}{82,282} = 0.612 \,.$$

Note that this is different from the room SHR = 0.7.

At this point, coils and fans can be specified by the designer. If heating is required, similar heating load and sizing calculations are needed.

2.2.6 Human Comfort

Proper HVAC system design and operation requires simple objective criteria that ensure comfort by the majority of occupants.

Human indoor comfort can be characterized by to the occupants' feeling of well-being in the indoor environment. It depends on several interrelated and complex phenomena involving subjective as well as objective criteria. Categories of factors include personal (such as metabolism and clothing), measurable (environmental parameters), and psychological (such as color, sound, light). There is an increasing realization that work productivity suffers if indoor comfort is compromised. Hence, providing for proper indoor comfort is acquiring greater importance than in the past. Because of individual differences, it is impossible to specify a thermal environment that will satisfy everyone. ASHRAE Standard 55-1992 specifies criteria for the thermal environment that is acceptable to at least 80% of the occupants. These criteria are described below.

Thermal Balance of the Human Body

The body is a heat engine. It converts chemical energy of the food consumed into both heat to sustain metabolism and work. The harder the body exercises or works, the greater the need to reject heat in order for the body to maintain thermal balance. The human body primarily rejects heat to the environment from the body surface by convection, radiation, or evaporation.

The total energy production rate of the body can be written as

$$Q + W = M A_{sk} \tag{2.2.22}$$

where Q = heat production rate
  W = rate of work
  M = metabolic rate
  A_{sk} = total surface area of skin.

The *metabolic rate* is customarily expressed in units of *mets* (or M) where

$$1 \text{ M} = 1 \text{ met} = 18.4 \text{ Btu/h ft}^2 = 58.2 \text{ W/m}^2$$

Since the area A_{sk} for adults is of the order of 16 to 22 ft^2 (1.5 to 2 m^2), heat production rates by adults are about 340 Btu/h (110W) for typical indoor activities. Metabolic rates in units of mets for various activities are shown in Table 2.2.4.

Analysis of the rate of heat produced in the body is based on a steady state energy balance. Because the body maintains the interior body temperature at a fairly constant value (about 98.2°F or 36.8°C) by controlling the perspiration rate and blood flow, a steady state model suggested by Fanger (1970) is often

TABLE 2.2.4 Typical Metabolic Heat Generation for Various Activities

	Btu/hr · ft^2	W/m^2	met[a]
Resting			
Sleeping	13	40	0.7
Reclining	15	45	0.8
Seated, quiet	18	60	1.0
Standing, relaxed	22	70	1.2
Walking (on level surface)			
2.9 ft/s (0.88 m/s)	37	115	2.0
4.4 ft/s (1.3 m/s)	48	150	2.6
5.9 ft/s (1.8 m/s)	70	220	3.8
Office activities			
Reading, seated	18	55	1.0
Writing	18	60	1.0
Typing	20	65	1.1
Filing, seated	22	70	1.2
Filing, standing	26	80	1.4
Walking about	31	100	1.7
Lifting, packing	39	120	2.1
Driving/flying			
Car driving	18–37	60–115	1.0–2.0
Aircraft, routine	22	70	1.2
Aircraft, instrument landing	33	105	1.8
Aircraft, combat	44	140	2.4
Heavy vehicle	59	185	3.2
Miscellaneous occupational activities			
Cooking	29–37	95–115	1.6–2.0
Housecleaning	37–63	115–200	2.0–3.4
Seated, heavy limb movement	41	130	2.2
Machine work			
sawing (table saw)	33	105	1.8
light (electrical industry)	37–44	115–140	2.0–2.4
heavy	74	235	4.0
Handling 110-lb (50-kg) bags	74	235	4.0
Pick-and-shovel work	74–88	235–280	4.0–4.8
Miscellaneous leisure activities			
Social dancing	44–81	140–255	2.4–4.4
Calisthenics/exercise	55–74	175–235	3.0–4.0
Tennis, singles	66–74	210–270	3.6–4.0
Basketball	90–140	290–440	5.0–7.6
Wrestling, competitive	130–160	410–505	7.0–8.7

Source: Adapted from *ASHRAE Fundamentals 1993*, p. 8.7.
[a] 1 met = 18.43 Btu/hr · ft^2 (58.2 W/m^2).

used. Its basis is that the heat produced in the body is equal to the total amount of heat rejected by the body to its environment.

$$\dot{Q} = \dot{Q}_{con} + \dot{Q}_{rad} + \dot{Q}_{evap} + \dot{Q}_{res.sens} + \dot{Q}_{res,lat} \qquad (2.2.23)$$

where the first three terms refer to the convection, radiation, and evaporation from the skin, and the last two terms to sensible and latent heat of respiration.

Clothing adds thermal resistance to heat flowing from the skin and must be included in the model. The *insulating value of clothing* is measured in units of *clo*, defined as

$$1 \text{ clo} = 0.88 \text{ ft}^2 \text{ h } °F/Btu \text{ (0.155 m}^2 \text{ K/W)} \qquad (2.2.24)$$

TABLE 2.2.5   Typical Insulation and Permeability Values for Clothing Ensembles

Ensemble Description	I_{cl} (clo)	I_T (clo)	A_{cl}/A_{sk}
Walking shorts, short-sleeve shirt	0.36	1.02	1.10
Trousers, short-sleeve shirt	0.57	1.20	1.15
Trousers, long-sleeve shirt	0.61	1.21	1.20
Same as above plus suit jacket	0.96	1.54	1.23
Same as above plus vest and t-shirt	1.14	1.69	1.32
Trousers, long-sleeve shirt, long-sleeve sweater, t-shirt	1.01	1.56	1.28
Same as above plus suit jacket and long underwear bottoms	1.30	1.83	1.33
Sweat pants, sweat shirt	0.74	1.35	1.19
Long-sleeve pajama top, long pajama trousers, short 3/4 sleeve robe, slippers, no socks	0.96	1.50	1.32
Knee-length skirt, short-sleeve shirt, pantyhose, sandals	0.54	1.10	1.26
Knee-length skirt, long-sleeve shirt, full slip, pantyhose	0.67	1.22	1.29
Knee-length skirt, long-sleeve shirt, half slip, pantyhose, long-sleeve sweater	1.10	1.59	1.46
Same as above, replace sweater with suit jacket	1.04	1.60	1.30
Ankle-length skirt, long-sleeve shirt, suit jacket, pantyhose	1.10	1.59	1.46
Long-sleeve coveralls, t-shirt	0.72	1.30	1.23
Overalls, long-sleeve shirt, t-shirt	0.89	1.46	1.27
Insulated coveralls, long-sleeve thermal underwear, long underwear bottoms	1.37	1.94	1.26

Source: Adapted from *ASHRAE Fundamentals.* With permission.
Note: clo = 0.88 ft^2 · °F/Btu (0.155 m^2 K/w)

The unit of clo is based on the insulating value of the typical American man's business suit in 1941. Table 2.2.5 gives values of the thermal resistance of various clothing ensembles. The values for the clothing alone are given the symbol I_{cl}, and those for the total thermal resistance between the skin and the environment are given the symbol I_T. The outside surface area of the clothing should be used rather than skin area when computing heat losses. The surface area ratio between the total surface area of the clothing and the skin area A_{cl}/A_{sk} is also given in Table 2.2.5.

The convective transfer can be written as

$$\dot{Q}_{con} = A_{cl} h_{con}(T_{cl} - T_a) \qquad (2.2.25)$$

where

A_{cl} is the surface area of the clothing and skin in contact with the air,
T_{cl} is the mean temperature of the clothing and skin in contact with the air,
h_{con} is the average convective heat transfer coefficient from clothing, and
T_a is the dry-bulb temperature of the surrounding air.

The following correlations can be used to determine h_{con} in units of Btu/hr ft^2 °F:

(a) for a seated person:

$$h_{con} = 0.55 \qquad\qquad\qquad 0 \leq V \leq 40 \text{ ft/min}$$

$$= 0.061 \cdot V^{0.6} \qquad\qquad 40 \leq V \leq 800 \text{ ft/min} \qquad (2.2.26a)$$

(b) for an active person in still air

$$h_{con} = (M - 0.85)^{0.39} \qquad\qquad 1.1 \le M \le 30 \qquad\qquad (2.2.26b)$$

where M is the metabolic rate in units of met.

The radiative process is more complicated because different surfaces of the environment (or room) may have different temperatures. For example, during winter, a window exposed to the outside may be at a much lower temperature than that of other surfaces. In summer, a sunlit interior wall may be several degrees warmer than an unlit one. Because these differences are usually small, a linearized radiative heat loss can be assumed without much inaccuracy. The emissivities of various indoor surfaces are close to 0.9 and we assume that the surfaces are black. A *mean radiant temperature* T_r of the environment is defined such that

$$\sigma \sum_n F_{cl-n}(T_{cl}^4 - T_n^4) = \sigma(T_{cl}^4 - T_r^4) \qquad\qquad (2.2.27)$$

where
the sum includes all surfaces with which the body can exchange direct radiation,
F_{cl-n} is the radiation view factor from the body to the nth surrounding surface, and
σ is the Stefan–Boltzman constant.

In other words, T_r is the temperature of an imaginary isothermal enclosure with which a human body would exchange the same amount of radiation as with the actual environment.

The radiative heat loss from the body can now be written as

$$\dot{Q}_{rad} = A_{cl}h_{rad} \; (T_{cl} - T_r) \qquad\qquad (2.2.28)$$

where h_{rad} is the linearized radiative heat transfer coefficient. A numerical value of 0.83 Btu/hr ft^2 °F (4.7 W/m^2 K) is advocated for this coefficient for normal nonmetallic clothing.

We can combine the convective and radiative coefficients into one single total heat transfer coefficient:

$$h_{c+r} \equiv h_{con} + h_{rad} \qquad\qquad (2.2.29)$$

Then, Equations 2.2.25 and 2.2.28 can be written together as:

$$\dot{Q}_{con} + \dot{Q}_{rad} = A_{cl} \cdot h_{c+r} \cdot (T_{cl} - T_{op}) \qquad\qquad (2.2.30)$$

where T_{op} is called the *operative temperature* defined as

$$T_{op} \equiv \frac{h_{con}T_a + h_{rad}T_r}{h_{c+r}} \qquad\qquad (2.2.31)$$

Often T_{op} is close to the simple arithmetic average of T_a and T_r. The comfort indices above do not include humidity effects. One can include the evaporative skin loss in the above treatment by defining the *adiabatic equivalent temperature* which is a linear combination of T_{op} and the vapor pressure of the air, which depends on skin wetness and on clothing permeability. This quantity would then allow combining environmental dry-bulb temperature and humidity as well as the surrounding surface temperatures into a single temperature index that completely determines the total heat loss from the skin. It is used as the basis of the ASHRAE comfort chart described below.

Another temperature index, the *effective temperature* ET*, is also used in the analysis of thermal comfort. Like the adiabatic equivalent temperature, it is a linear combination of T_{op} and the vapor pressure. More strictly, it is the temperature of an isothermal black enclosure with 50% RH where the body surface would experience the same heat loss as in the actual space. The 50% level is chosen as the reference value because it is the most common and widely accepted level of indoor humidity.

Example 8: Determine the operative temperature of an enclosure whose dry-bulb temperature is 75°F and mean radiant temperature is 80°F, and where the occupants walk around at 4.4 ft/s.

From Table 2.2.4, the activity level corresponds to a met level of 2.6. The convective heat transfer coefficient h_{con} is determined from Equation 2.2.26b:

$$h_{con} = (2.6 - 0.85)^{0.39} = 1.244 \text{ Btu/hr} \cdot \text{ft}^2 \cdot °F$$

We use the standard value for the radiative heat-transfer coefficient of 0.83 Btu/hr ft² °F. Thus from Equation 2.2.31,

$$T_{op} = \frac{1.244 \times 75 + 0.83 \times 80}{1.244 + 0.83} = 77.0°F$$

Conditions for Thermal Comfort

The environmental parameters discussed above are the primary factors used to characterize human comfort. There are also secondary effects such as nonuniformity of the environment, visual stimuli, age, and outdoor climate (ASHRAE, 2001). Studies on 1600 college-age students revealed correlations between comfort level, temperature, humidity, sex, and length of exposure. Several trends are described below.

The thermal sensation scale adopted for voting by the students is called the ASHRAE thermal sensation scale, and is represented as follows:

+3   hot
+2   warm
+1   slightly warm
 0   neutral
−1   slightly cool
−2   cool
−3   cold

The relevant index of acceptability of the indoor environment is the predicted mean vote (PMV) for which an empirical correlation has been developed:

$$PMV = a^*t + b^*P_v + c^* \tag{2.2.32}$$

where the numerical values of the coefficients a*, b*, and c* are given in Table 2.2.6.

In general, the distribution of votes will always show considerable scatter. A useful index of acceptability of an environment is the percentage of people dissatisfied (PPD), defined as people voting outside the range of −1 to +1. When the PPD is plotted versus the mean vote of a large group characterized by the PMV, one typically finds a distribution such as shown in Figure 2.2.16 This graph shows that even under optimal conditions (i.e., a mean vote of zero), at least 5% of occupants are dissatisfied with the thermal comfort.

ASHRAE Standard 55 specifies boundaries of the comfort zones where 80% of occupants are thermally comfortable. The specification of these zones is based on the concept of ET* as depicted in Figure 2.2.17. This chart is similar to the psychrometric chart, but the abscissa is the operative temperature T_{op} rather than the dry-bulb temperature. The acceptable values of temperature and humidity are indicated by the shaded zones in Figure 2.2.17. Numerical values for the comfort zone are are provided in Table 2.2.7. In winter, an effective temperature of 70 to 72°F (21 to 22°C) is optimum for normally clothed people. In summer, about 76°C would be optimum. The further one strays from these values, the greater the percentage of people who are likely to be dissatisfied. The chart shows different zones for winter and for summer because comfort depends on the insulation value of the clothing, as discussed earlier.

TABLE 2.2.6 Coefficients a^*, b^*, and c^* Used to Calculate Predicted Mean Vote

Exposure Period, hr	Sex	Coefficients					
		English Units, t (°F), P_v (psia)			SI Units, t (°C), P_v (kPa)		
		a^*	b^*	c^*	a^*	b^*	c^*
1.0	Male	0.122	1.61	−9.584	0.220	0.233	−5.673
1.0	Female	0.151	1.71	−12.080	0.272	0.248	−7.245
1.0	Combined	0.136	1.71	−10.880	0.245	0.248	−6.475
3.0	Male	0.118	2.02	−9.718	0.212	0.293	−5.949
3.0	Female	0.153	1.76	−13.511	0.275	0.255	−8.622
3.0	Combined	0.135	1.92	−11.122	0.243	0.278	−6.802

Source: Adapted from *ASHRAE Fundamentals 1993*, p. 8.16.
Note: For young adult subjects with sedentary activity and wearing clothing with a thermal resistance of approximately 0.5 clo, $t_r = t$, air velocities ≤ 40 ft/min (0.2 m/s).

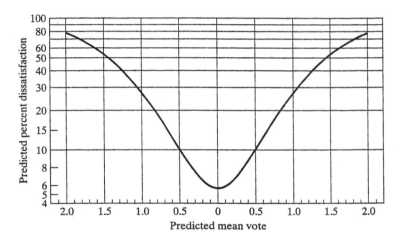

FIGURE 2.2.16 Predicted percentage of dissatisfied (PPD) as a function of predicted mean vote (PMV). [Reprinted with permission from *ASHRAE Fundamentals 1993* (IP & SI), p. 8.18.]

TABLE 2.2.7 Optimum and Acceptable Ranges of Operative Temperature for People during Light, Primarily Sedentary Activity (≤ 1.2 met) at 50% Relative Humidity and Mean Air Speed ≤ 0.15 m/s (30 fpm)

Season	Description of typical clothing	I_d (clo)	Optimum operative temperature	Operative temperature range (10% dissatisfaction criterion)
Winter	heavy slacks, long-sleeve shirt and sweater	0.9	22°C 71°F	20–23.5°C 68–75°F
Summer	light slacks and short-sleeve shirt	0.5	24.5°C 76°F	23–26°C 73–79°F
	minimal	0.05	27°C 81°F	26–29°C 79–84°F

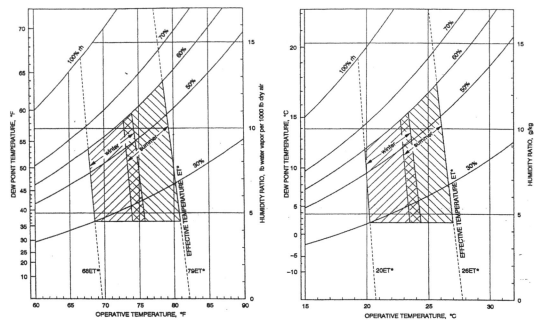

FIGURE 2.2.17 Acceptable ranges of operative temperature and humidity for people in typical summer and winter clothing during light, primarily sedentary activity (≤ 1.2 met). The ranges are based on a 10% dissatisfaction criterion.

The left and right boundaries of the shaded zones are lines of constant adiabatic equivalent temperature, hence lines of constant heat loss. They are sloped from upper left to lower right because the evaporative heat loss from the body decreases as the humidity ratio of the air increases.

The comfort chart is applicable to reasonably still air (velocities less than 30 ft/min or 0.15 m/s), with occupants seated at rest or doing light work (met = 1.2) and to spaces where enclosing surfaces are at the mean temperature equal to the dry-bulb temperature. When the activity level is between 1.2 and 3 mets, Standard 55-1992 recommends that the operative temperature be determined as:

$$T_{op} = T_{op,std} - 5.4 \ (1.0 + clo)(M - 1.2) \ °F$$

$$T_{op} = T_{op,std} - 3.0 \ (1.0 + clo)(M - 1.2) \ K \qquad (2.2.33)$$

where clo=clothing insulation, and the subscript std designates the value below 1.2 met.

The ASHRAE Comfort chart applies to clothing levels of clo=0.9 during winter and clo=0.5 during summer. The relationship between clothing insulation and operative temperature so as to be within the ASHRAE 80% acceptability limits is shown in Figure 2.2.18. As the operative temperature increases, one can still maintain comfort by reducing the clothing insulation value, i.e., by dressing lightly.

Within the thermally acceptable temperature ranges stated earlier, there is no minimum air movement that is necessary for thermal comfort. The temperature may be increased above these levels provided one can increase the air speed. An estimate is provided by Figure 2.2.19 which corresponds to clothing and activities pertinent to summer comfort zone. However, temperature should not be increased more than 5.4°F (3°C) above the comfort zone and air speeds should be kept less than 160 ft/min (0.8 m/s). Studies have shown that women of all age groups prefer an effective temperature about one degree higher than that preferred by men, while both men and women over 40 years prefer an effective temperature about one degree higher than that desired by younger people.

Finally, thermal stratification in the room also affects comfort. Vertical air temperature difference should not exceed 5°F (3 K) between 4 in (0.1 m) and 67 in (1.7 m) above the floor. Other specific recommendations are provided in ASHRAE Standard 55-1992.

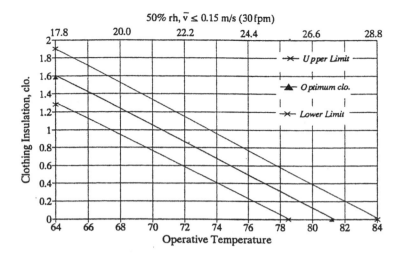

FIGURE 2.2.18 The recommended range of clothing insulation providing acceptable thermal conditions at a given operative temperature for people during light, primarily sedentary activity (≤1.2 met). The limits are based on a 10% dissatisfaction criterion.

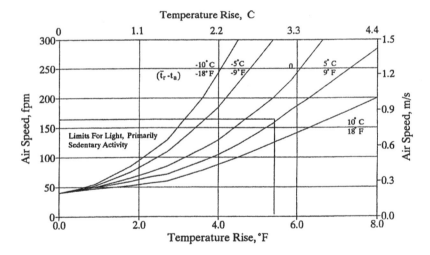

FIGURE 2.2.19 Air speed required to offset increased temperature. The air speed increases in the amount necessary to maintain the same total heat transfer from the skin. This figure applies to increases in temperature above those allowed in the summer comfort zone with both $\bar{t}_r$ and t_a increasing equally. The starting point of the curves at 0.2 m/s (40 fpm) corresponds to the recommended air speed limit for the summer comfort zone at 26°C (79°F) and typical ventilation (i.e., turbulence intensity between 30% and 60%). Acceptance of the increased air speed requires occupant control of the local air speed.

References

ANSI/ASHRAE Standard 55-92, 1992. *Thermal Environmental Conditions for Human Occupancy,* American Society of Heating, Refrigerating and Air Conditioning Engineers, Atlanta, GA.

ASHRAE Handbook, 1997. *Fundamentals,* American Society of Heating, Refrigerating and Air Conditioning Engineers, Atlanta, GA.

Fanger, P.O., 1970. *Thermal Comfort Analysis and Applications in Environmental Engineering,* McGraw-Hill, New York, NY.

Kreider, J.F. and Rabl, A., 2001. *Heating and Cooling of Buildings*, McGraw-Hill, New York, NY.

Kuehn, T.H., Ramsey, J.W., and Threlkeld, J.L., 1998. *Thermal Environmental Engineering*, 3rd ed., Prentice-Hall, Upper Saddle River, NJ.

McQuiston, F.C. and Parker, J.D., 1994. *Heating, Ventilating and Air Conditioning*, 4th ed., John Wiley and Sons, New York, NY.

Pita, E.G., 1998. *Air Conditioning Principles and Systems*, 3rd ed., Prentice-Hall, Upper Saddle River, NJ.

Stoecker, W.F. and Jones, J.W., 1982. *Refrigeration and Air Conditioning*, 2nd ed., McGraw-Hill, New York, NY.

3

Economic Aspects of Buildings

Anthony F. Armor
Electric Power Research Institute

Jan F. Kreider
Kreider & Associates, LLC

Ari Rabl
École des mines de Paris and University of Colorado

3.1 Central and Distributed Utilities

Anthony F. Armor and Jan F. Kreider

Electricity is the primary energy source for operating much of the HVAC equipment in buildings including fans, chillers, pumps, and electrical auxiliaries. It is also the primary energy source for lighting in the U.S. Given the importance of electricity and the current rapid evolution of the electrical industry in the U.S., it is important for the HVAC designer to understand both present and future options for electricity generation.

This chapter discusses the expected near-term situation in traditional generation and distribution industries and a new paradigm, the generation of electricity using many, small distributed generators. Distributed generation (sometimes called distributed resources — DR) is expected to grow rapidly in the first decade of the 21st century to the extent that by 2010, up to 20% of new generation capacity in the U.S. will be from DR.

This section reviews the status and likely future application of competing electrical generation technologies, particularly those with near-term potential. Capacity additions in the United States in the next 10 years will be based on gas, coal, and to some extent on renewables. Repowering of older plants will likely be increasingly attractive.

Gas turbine-based plants will dominate in the immediate future. The most advanced combustion turbines achieve more than 40% efficiency in simple cycle mode and greater than 50% lower heating value (LHV) efficiency in combined cycle mode (gas turbine plus steam turbine). In addition, combustion turbine/combined cycle (CT/CC) plants offer siting flexibility, phased construction, and capital costs between $400/kW and $800/kW. These advantages, coupled with adequate natural gas supplies and the assurance of coal gasification backup, make this technology a prime choice for green field and repowered plants.

There is also good reason why the pulverized coal plant may still be a primary choice for many generation companies. Scrubbers have proved to be more reliable and effective than early plants indicated. Up to 99% SO_2 removal efficiency is possible. About 60 GW of U.S. coal-fired generation is currently equipped with flue gas desulfurization (FGD) systems. Also, the pulverized-coal (PC) plant has the capability for much improved heat rate (about 8500 Btu/kWh) even with full flue gas desulfurization.

Atmospheric and pressurized fluidized bed combustion (FBC) offer reductions in both SO_2 and NO_x and also permit the efficient combustion of low rank fuels. In the U.S., there are now over 150 operating units for power generation and ten vendors of FBC boilers, four of which offer units up to 250 MW in size.

Gasification power plants exist at the 100 MW and 160 MW levels and are planned up to 450 MW. Much of the impetus is now coming from the DOE clean coal program, where three gasification projects are in progress.

In small unit sizes, often suitable for distributed generation, technical progress will be made (although large-scale applications still remain modest) in renewables (solar, wind, biomass), microturbines and fuel cells. They promise high efficiencies, low emissions, and compact plants.

Capital cost and efficiency will remain the determining issues in the application of all these possible generation options.

Overall Industry Needs

The U.S. electric power industry consists of a network of small and large companies, both private and public — more than 3000 in all. They generate more than 700 GW of electric power — by far the greatest concentration of electric power production in the world, about equal to the next five countries combined. However, although advanced generation technologies are beginning to find their way into the power industry, most installed capacity is 20 or more years old and equipment efficiency reflects this vintage. Likewise, the transmission and distribution system is aging.

The National Energy Strategy of 1991 set general U.S. policy for the future, and one specific directive was to enhance the efficiency of generation, transmission, and use of electricity. There are two key drivers for this directive: one is to reduce emissions of undesired air, water, and ground pollutants, and the other is to conserve our fossil fuels. The desire to move to a more energy-conscious mode of operation will clearly give impetus to the renewables as significant (although probably not major) future power sources and will encourage efficiency advances in both fossil and nuclear plants.

The CRC *Handbook of Energy Efficiency* (Chapter 7) summarizes the fuel supply outlook through 2010.

3.1.1   Management of Existing Fossil Plant Assets

Generation companies are now looking at power plants in a more profit-focused manner, treating them as company assets, to be invested in a way that maximizes the company bottom line or the profit for the company. As the average age of fossil unit inches upward, executives are often asking "If I invest a dollar into this plant to improve heat rate, availability, or some other plant performance measure, will this produce more in base profit to my company than investing, say, in some other plant, or building new capacity, or buying power from outside?" One important aspect of this new business strategy concerns the "use" that is being made of any particular plant, since increased plant usage implies more company value for that plant.

Here are some measures of plant utilization:

Heat rate is the quantifiable measure of how efficiently we can convert fuel into MW. It is inherently limited by cycle and equipment design and by how we operate the plant. In a simple condensing cycle the heat rate of a fossil fuel plant cannot fall much below 8500 Btu/kWh, even with super-critical cycles and double reheat of the feedwater.

Capacity factor (CF) is a measure that indicates how a plant is loaded over the year. Few single cycle fossil units achieve 90% capacity factor these days, and this has an impact on the measured heat rate of the unit. Under ideal conditions for effective asset management, and apart from downtime for maintenance, CF should be close to 100% for the purposes of getting the most out of the plant asset. Market conditions and the reserve margins of the utility often dictate otherwise.

Price of electricity is a determining factor in how units are dispatched. Electricity cost is largely dictated by fuel cost, which typically makes up 70% of the cost of operating a power plant. It is interesting that none of the top ten U.S. units in heat rate makes the top ten in electricity cost.

Finally, some have suggested a term called overall *energy efficiency*, which describes how well a plant utilizes the basic feedstock (coal, oil, or gas). If we can produce other products from a fossil plant besides electricity, the value of that asset goes up, and, of course, the "effective" heat rate drops significantly.

Using these measures we are seeing many, perhaps most, of the major U.S. generation companies take a close look at their plant assets to judge their bottom-line value to the company. In the growing competitive generation business, an upgrade or maintenance investment in a power plant will be determined largely by the return on investment the company can expect at the corporate level. In order to achieve these corporate goals, it is necessary to have a good handle on equipment life and the probability of failures. Also needed are options for improving heat rate, for increasing output (by repowering), and for improving plant productivity to make the assets competitive.

3.1.2   Clean Coal Technology Development

At an increasing rate in the last few years, innovations have been developed and tested aimed at reducing emissions through improved combustion and environmental control in the near term, and in the longer term by fundamental changes in how coal is preprocessed before converting its chemical energy to electricity. Such technologies are referred to as clean coal technologies — described by a family of precombustion, combustion/conversion, and postcombustion technologies. They are designed to provide the coal user with added technical capabilities and flexibility and at lower net cost than current environmental control options. They can be categorized as follows:

- **Precombustion**, in which sulfur and other impurities are removed from the fuel before it is burned.
- **Combustion**, in which techniques to prevent pollutant emissions are applied in the boiler while the coal burns.
- **Postcombustion**, in which the flue gas released from the boiler is treated to reduce its content of pollutants.
- **Conversion**, in which coal, rather than being burned, is changed into a gas or liquid that can be cleaned and used as a fuel.

Coal Cleaning

Cleaning of coal to remove sulfur and ash is well established in the United States with more than 400 operating plants, mostly at the mine. Coal cleaning removes primarily pyritic sulfur (up to 70% SO_2 reduction is possible) and in the process increases the heating value of the coal, typically by about 10% but occasionally by 30% or higher. Additionally, if slagging is a limiting item, increased MW may be possible, as at one station which increased generation from 730 MW to 779 MW. The removal of organic sulfur, chemically part of the coal matrix, is more difficult but may be possible using microorganisms or through chemical methods, and research is underway. Finally, heavy metal trace elements can be removed also, conventional cleaning removing (typically) 30 to 80% of arsenic, mercury, lead, nickel, and antimony.

Pulverized-Coal-Fired Plants

Built in 1959, Eddystone 1 at PECO Energy was, and still is, the supercritical power plant with the highest steam conditions in the world. Main steam pressure was 5000 psi when built, and steam temperature 1200°F for this double reheat machine. PECO Energy will continue to operate Eddystone I to the year 2010, an impressive achievement for a prototype unit.

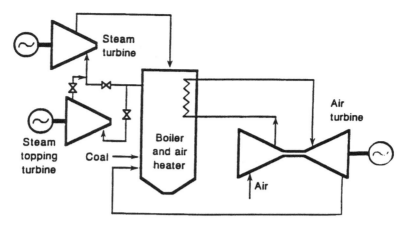

FIGURE 3.1.1 A pulverized-coal combined cycle with topping steam turbine has a projected heat rate of 7200 Btu/kWh. The air turbine uses 1800°F air, or 2300°F air with supplemental firing. The topping turbine uses steam at 1300°F.

But the most efficient pulverized-coal-fired plant of the future is likely to be a combined cycle plant, perhaps with a topping steam turbine, as shown in Figure 3.1.1. With an 1800°F air turbine and 1300°F topping steam turbine, the heat rate of this cycle is about 7200 Btu/kWh — very competitive with any other proposed advanced cycles in the near term.

3.1.3 Emissions Control

Worldwide about 40% of electricity is generated from coal. Installed coal-fired generating capacity, more than 1000 GW, is largely made up of 340 GW in North America, 220 GW in Western Europe, Japan, and Australia, 250 GW in Eastern Europe and the former USSR, and 200 GW in China and India. In the decade to the year 2000, about 190 GW of new coal-fired capacity was added. The control of particulates, sulfur dioxides, and nitrogen oxides from those plants is one of the most pressing needs of today and of the future. This is accentuated when the impact of carbon dioxide emissions, with its contribution to global warming, is considered. To combat these concerns, a worldwide move toward environmental retrofitting of older fossil-fired power plants is underway, focused largely on sulfur dioxide scrubbers and combustion or postcombustion optimization for nitrogen oxides.

Sulfur Dioxide Removal

When it is a matter of retrofitting an existing power plant, no two situations are identical: fuels, boiler configurations, and even space available for new pollution control equipment all play a role in the decision on how a utility will meet new emission reduction requirements. For example, a decision to install a sorbent injection technology rather than flue gas desulfurization (FGD) for SO_2 reduction may depend not only on the percentage reduction required but also on the space constraints of the site and on the capacity factor of the plant (with a lower capacity factor, the lower capital cost of sorbent injection is advantageous compared to FGD). Most utilities, though, have been selecting either wet or dry scrubbing systems for desulfurization. Generically these can be described as in the following sections.

Conventional Lime/Limestone Wet Scrubber. By 1994 the United States already had more than 280 flue gas desulfurization (FGD) systems in operation on 95,000 MW at utility stations; now the U.S. experience is approaching 1000 unit years.

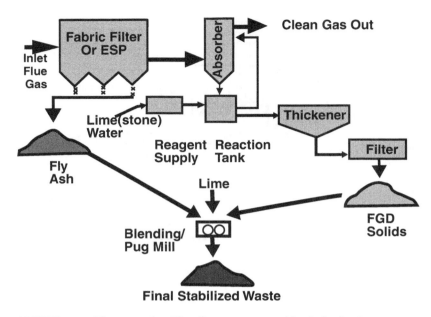

FIGURE 3.1.2   The conventional lime/limestone wet scrubber is the dominant system.

The dominant system is the wet limestone design, limestone being one-quarter the cost of lime as a reagent. In this system (Figure 3.1.2) the limestone is ground and mixed with water in a reagent preparation area. It is then conveyed to a spray tower called an absorber, as a slurry of 90% water and 10% solids, and sprayed into the flue gas stream. The SO_2 in the flue gas is absorbed in the slurry and collected in a reaction tank, where it combines with the limestone to produce water and calcium sulfate or calcium sulfate crystals. A portion of the slurry is then pumped to a thickener where these solids/crystals settle out before going to a filter for final dewatering. Mist eliminators installed in the system ductwork at the spray tower outlet collect slurry/moisture entrained in the flue gas stream. Calcium sulfate is typically mixed with fly ash (1:1) and lime (5%) and disposed of in a landfill.

Various improvements can be made to this basic process, including the use of additives for performance enhancement and the use of a hydrocyclone for dewatering, replacing the thickener, and leading to a salable gypsum byproduct. The Chiyoda-121 process (Figure 3.1.3) reverses the classical spray scrubber and bubbles the gas through the liquid. This eliminates the need for spray pumps, nozzle headers, separate oxidation towers, and thickeners. Bechtel has licensed this process in the U.S.; the first commercial installation is at the University of Illinois on a heating boiler, and a DOE clean coal demonstration is underway. The waste can be sold as gypsum or disposed of in a gypsum stack.

Spray Drying.   Spray drying (Figure 3.1.4) is the most advanced form of dry SO_2 control technology. Such systems tend to be less expensive than wet FGD but typically remove a smaller percentage of the sulfur (90% compared with 98%). They are used when burning low-sulfur coals and utilize fabric filters for particle collection, although recent tests have shown applicability to high-sulfur coals also.

Spray driers use a calcium oxide reagent (quicklime), which, when mixed with water, produces a calcium hydroxide slurry. This slurry is injected into the spray drier, where it is dried by the hot flue gas. As the drying occurs, the slurry reacts to collect SO_2. The dry product is collected at the bottom of the spray tower and in the downstream particulate removal device, where further SO_2 removal may take place. It may then be recycled to the spray drier to improve SO_2 removal and alkali utilization.

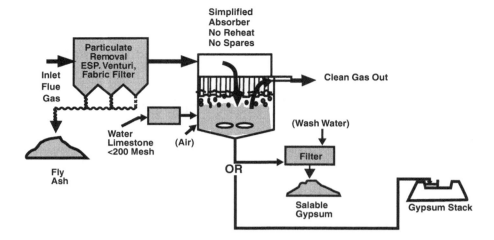

FIGURE 3.1.3 The Chiyoda-121 scrubber simplifies the process by bubbling the flue gas through the liquid, eliminating some equipment needs.

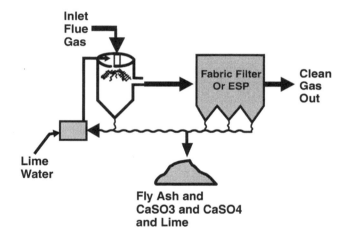

FIGURE 3.1.4 Spray driers use a calcium oxide reagent mixed with water, which is dried by the flue gas. A dry product is collected in a fabric filter.

For small, older power plants with existing electrostatic precipitators (ESPs), the most cost-effective retrofit spray dry configuration locates the spray dryer and fabric filter downstream of the ESP, separating in this manner the spray dryer and fly ash waste streams. The fly ash can then be sold commercially.

Control of Nitrogen Oxides

Nitrogen oxides can be removed either during or after coal combustion. The least expensive option and the one generating the most attention in the U.S. is combustion control, first through adjustment of the fuel/air mixture, and second through combustion hardware modifications. Postcombustion processes seek to convert NO_x to nitrogen and water vapor through reactions with amines such as ammonia and urea. Selective catalytic reduction (SCR) injects ammonia in the presence of a catalyst for greater effectiveness. So the options (Figure 3.1.5) can be summarized as

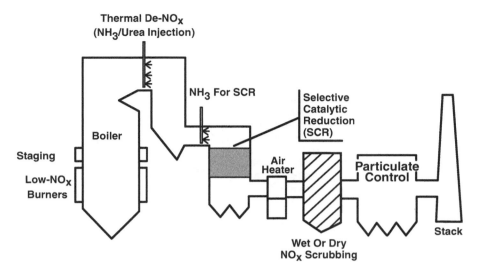

FIGURE 3.1.5   Control options for NO_x include operational, hardware, and postcombustion modifications.

Operational changes. Reduced excess air, and biased firing, including taking burners out of service.
Hardware combustion modifications. Low NO_x burners, air staging, and fuel staging (reburning).
Postcombustion modifications. Injection of ammonia or urea into the convection pass, selective
     catalytic reduction (SCR), and wet or dry NO_x scrubbing (usually together with SO_2 scrubbing).

Low NO_x burners can reduce NO_x by 50%, and SCR by 80%, but the low NO_x burner option is much
more cost-effective in terms of cost per ton of NO_x removed. Reburning is intermediate in cost per
removed ton and can reduce NO_x by 50%, or 75% in conjunction with low NO_x burners.

Fluidized-Bed Plants.   Atmospheric fluidized-bed boilers (see Figure 3.1.6) offer reductions in both
SO_2 and NO_x and also permit the efficient combustion of low-rank fuels. In the U.S. there are now over
150 operating units generating over 5000 MW, and ten vendors of FBC boilers, four of which offer units
up to 250 MW in size. The rapid growth in number of FBC units and capacity is shown in Figure 3.1.7.

The future for fluidized-bed utility boilers is evident in the move by several countries (Sweden, Japan,
Spain, Germany, United States) toward the pressurized fluidized-bed combined cycle design. The modular
aspect of the PFBC unit is a particularly attractive feature leading to short construction cycles and low-
cost power. This was particularly evident in the construction of the Tidd plant, which first generated
power from this combined cycle (Figure 3.1.8) on November 29, 1990.

3.1.4   Combustion Turbine Plants

Combustion turbine-based plants comprise the fastest growing technology in power generation. Between
now and 2005, natural gas-fired combustion turbines and combined cycles burning gas will account for
50 to 70% of new generation to be ordered worldwide. GE forecasts that combustion turbines and
combined cycles will account for 45% of new orders globally, and for 66% of new U.S. orders. Almost
all of these CT and CC plants will be gas-fired, leading a major expansion of gas for electricity generation.

Estimates suggested 23 GW of new gas-fired CT capacity in the United States between 1990 and 2000
(Figure 3.1.9). Worldwide there are even more striking examples. In the United Kingdom, 100% of all
new generation ordered or under construction is gas-fired combined cycle.

Aircraft Technology

In the 1960s, gas turbines derived from military jet engines formed a major source of utility peaking
generation capacity. Fan-jets, though, which replaced straight turbojets, were much more difficult and

FLUID BED COMBUSTION

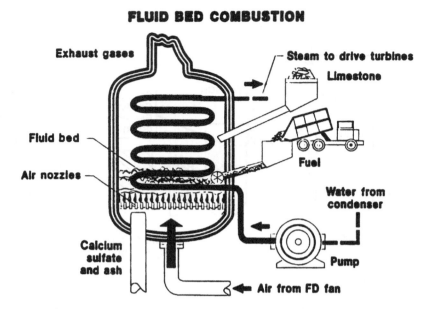

FIGURE 3.1.6   The addition of limestone or Dolomite to the combustion chamber allows the coal limestone mixture to be burned in a suspended bed, fluidized by an underbed air supply. The sulfur in the coal reacts with the calcium to produce a solid waste of calcium sulfate. The combustion temperature is low (1500°F), reducing NO_x emissions.

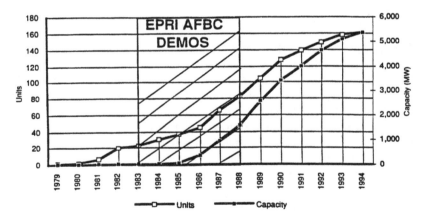

FIGURE 3.1.7   The growth in number and size of AFBC plants has been significant since the demonstration of 100+ MW sizes in U.S. utility plants.

expensive to convert to utility use, and the resulting aeroderivative turbines have been little used. The main reason for the high cost was the need to replace the fan and add a separate power turbine.

As bypass ratios, and hence fan power, have increased, the most recent airline fan-jets can be converted to utility service without adding a separate power turbine. Furthermore, modifications of these engines, with intercooling and possibly reheat, appear to be useful for advanced power cycles such as chemical recuperation and the humid air turbine.

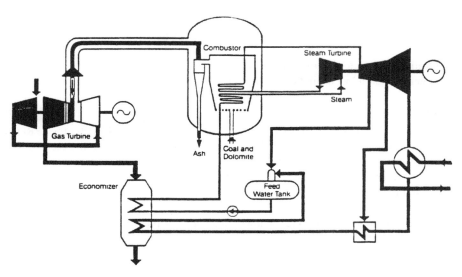

FIGURE 3.1.8 Pressurized, fluidized-bed combustor with combined cycle. This 70 MW system operated at the Tidd plant of American Electric Power.

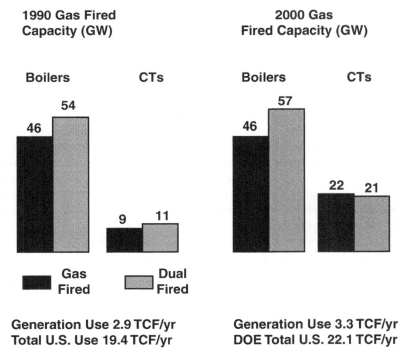

FIGURE 3.1.9 Combustion turbine additions in the United States in the decade before the year 2000. (From GRI/DOE [1991], NERC [1990].)

Humidified Air Power Plants

A new class of combustion turbine-based approaches are termed humidified air power plants. In these combustion turbine cycles the compressor exit air is highly humidified prior to combustion. This reduces the mass of dry air needed and the energy required to compress it.

The continuous plant cycle for this concept is termed the Humid Air Turbine (HAT). This cycle using, for example, extensive modification of the TPM FT4000, has been calculated to have a heat rate on natural gas about 5% better than the latest high-technology combined cycle. The HAT cycle is adaptable to coal gasification leading to the low emissions and high efficiency characteristics of gasification combined cycle plants but at a low capital cost, since the steam turbine bottoming cycle is eliminated.

Gasification Plants

One option of particular importance is that of coal gasification (Figure 3.1.10). After the EPRI Coolwater demonstration in 1984 at the 100 MW level, the technology has moved ahead in the U.S. largely through demonstrations under the CCT program. Overseas, the 250 MW Buggenham plant in Holland was operational in 1994, and the PSI/Destec 265 MW and TECO 260 MW clean coal demos both operated in 1996. Beyond this, there is a 300 MW plant for Endesa, Spain and a 330 MW unit for RWE in Germany (Figure 3.1.11).

Gasification-based plants have among the lowest emissions of pollutants of any central station fossil technology. With the efficiency advantages of combined cycles, CO_2 emissions are also lower. Fuel flexibility is an additional benefit, since the gasifier can accommodate a wide range of coals, plus petroleum coke. Integrated gasification combined cycle (IGCC) plants permit a hedge against long-term increases in natural gas prices. Natural gas-fired combustion turbines can be installed initially, and gasifiers at a later time when a switch to coal becomes prudent.

Concurrent with the advances in gasification are efficiency improvements in combustion turbines. The new F-type CTs operate at 2300°F, and 2500°F machines are likely soon. This makes the IGCC a very competitive future option.

A Look at the Future of Combustion Turbines

Combustion turbines and combined cycles grew steadily more important in all generation regimes; peaking, mid-range, and base load. They account for the majority of new generation ordered and installed. If the present 2300°F firing temperature machines operate reliably and durably, CT and CC plants will begin to retire older steam plants and uneconomic nuclear plants. With no clear rival other than fuel cells, which are only now emerging, CT technology may dominate fossil generation, and new advanced CT cycles — with intercooling, reheat, possibly chemical recuperation, and most likely humidification — should result in higher efficiencies and lower capital costs. Integrated gasification, which guarantees a secure bridge to coal in the near term, will come into its own as gas prices rise under demand pressure. By 2015, coal through gasification is expected to be the economic fuel for a significant fraction of new base-load CT/CC generation. The rate at which these trends develop depends in a large measure on the relative costs by burning gas or coal.

3.1.5 Distributed Electrical Generation Basics

A confluence of events in the U.S. electrical generation and transmission industry has produced a new paradigm for distributed electrical generation and distribution in the U.S. Electrical deregulation, reluctance of traditional utilities to commit capital to large central plants and transmission lines, and a suite of new, efficient generation hardware have all combined to bring this about. Persistent environmental concerns have further stimulated several new approaches. This section describes the term *distributed generation technologies* and their differentiating characteristics along with their readiness for the U.S. market. In order to decide which approaches are well suited to a specific project, an assessment methodology is needed. A technically sound approach is therefore described and example results are given in the final section of this chapter.

Distributed resource generation (DR) is any small scale electrical power generation technology that provides electric power at a site closer to customers than central station generation, and that is usually interconnected to the distribution system or directly to the customer's facilities. According to the Distributed Power Coalition of America (DPCA), research indicates that distributed power has the

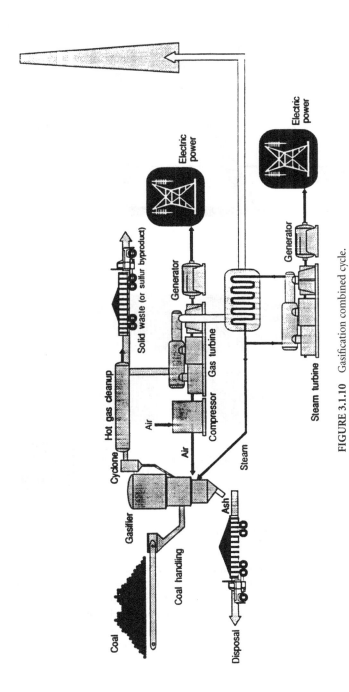

FIGURE 3.1.10 Gasification combined cycle.

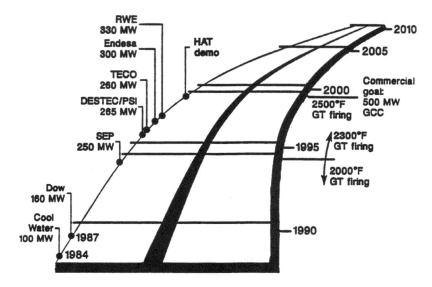

FIGURE 3.1.11   Gasification power plant time line.

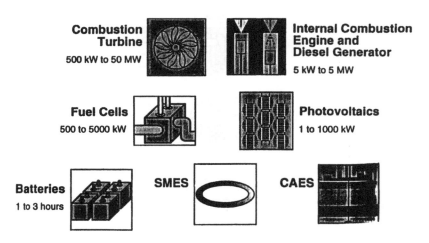

FIGURE 3.1.12   Distributed generation and energy storage technologies.

potential to capture about 20 percent of additions to generating capacity, or 35 gigawatts (GW), over the next two decades. The Electric Power Research Institute estimates that the DR market could amount to 2.5–5 GW/year by 2010. DR technologies include small combustion turbine generators, internal combustion reciprocating engines and generators, photovoltaic panels, and fuel cells.* Other technologies including solar thermal conversion, stirling engines, and biomass conversion are considered as DR. We arbitrarily limit the term DR, in this chapter, to refer to plants with capacities less than a few MW.

---

* Wind power is not included since, in the U.S., it has been primarily deployed as a central power generation technology.

TABLE 3.1.1 Cost Projections for Representative Generation Technologies for a Plant in the Northeast United States

Plants	Pulverized Coal Plants		Atmospheric FBC	
	Subcritical Wet Scrubber 300 MW	Supercritical Wet Scrubber 400 MW	Circulating No Scrubber 200 MW	Pressurized FBC Combined Cycle 340 MW
Capital cost (1993)[a] ($/kW)	1607	1600	1805	1318
Nonfuel O&M costs				
Variable (mills/kWh)	3.0	2.8	5.4	1.8
Fixed ($/kW yr)	46.6	43.1	37.0	37.6
Efficiency (%)	36.0	39.0	35.0	41.0

Plants	Coal Gasification Combined Cycle 500 MW	Coal Gasification Humid Air Turbine 500 MW	Coal Gasification Molten Carbonate Fuel Cell 400 MW	Gas-Fired Combined Cycle 225 MW
Capital cost (1993) ($/kW)	1648	1447	2082	595
Nonfuel O&M costs				
Variable (mills/kWh)	0.5	1.3	1.1	0.4
Fixed ($/kW yr)	49.9	40.4	57.2	26.5
Efficiency (%)	42.0	42.0	50.0	46.0

[a] Costs of new plants are likely to reduce, in real terms, over the next 10 years due to technology developments and increased worldwide competition for markets in the developing countries. New technologies (PFBC, IGCC, fuel cells) will lower capital costs as production capacity grows.

Source: Technical Assessment Guide, EPRI TR-102275-V1R7, June 1993. Electrical Power Research Institute, Palo Alto, CA.

Distributed generation can provide a multitude of services to both utilities and consumers, including standby generation, peak shaving capability, baseload generation, or cogeneration. Less well understood benefits including ancillary services — VAR support, voltage support, and network stability among others — may ultimately be of more economic benefit than simple energy-related benefits.

DR technologies can have environmental benefits ranging from truly green power (photovoltaics) to significant mitigation of one or more pollutants often associated with coal-fired generation. Natural gas-fired DR turbine generators, for example, release less than a quarter of the emissions of SO_2, less than 1/100th NO_X, and 40 percent less CO_2 than many new coal-boiler power plants; these DR units are clean enough to be situated among residential and commercial establishments (DCPA, 1998).

Electric restructuring has spurred the consideration of DR power because the buyers and sellers of electricity will have to be more responsive to market forces. Central utilities suffer from the burden of significant stranded costs. DR avoids this cost. DR is a priority in parts of the country where the spinning reserve margins are shrinking, where growing industrial and commercial uses as well as transmission and distribution (T&D) constraints are limiting power flows (DCPA, 1998).

Additional impetus was added to DR efforts during the summers of 1998, 1999, and 2000 by the heat waves that staggered the U.S. and caused power cuts in the west and across the Rust Belt. The shortages and outages were the result of a combination of factors — climbing electricity demand, permanent or temporary shutdown of some of the region's nuclear facilities, unusually hot weather, and 1998 summer tornadoes that downed a transmission line (McGinley, 1998).

Forces Propelling DR Today

The DR era appears to have arrived based on evidence from many and diverse sources. U.S. utility deregulation is an "enabler" for widespread DR adoption but is not a required condition of rapid growth in small generation sales. Experts list the following reasons for expected DR applications in the next 20 years:

- Utilities are seeking ways to avoid large capital investments in new generating capacities. Incremental investments in smaller plants are preferred.
- Distribution system loading is near the limit with the result that power quality is suffering and power outages are becoming more prevalent. DR of power bypasses most of the distribution system.
- Several small and efficient DR technologies are nearing maturity.
- Telecommunication and computational systems compatible with the widespread deployment of DR will exist in the very near future.
- U.S. utilities are being restructured as a result of deregulation of electrical utilities.
- DR owned by local building operators or energy service companies avoid the electricity price volatility seen in the past two years during peak load periods.

In spite of these notable reasons for DR growth, it must be recognized that DR is a "disruptive technology," and, as was the case with past technologies, it may offer worse economic or technical performance than traditional approaches. However, as commercialization continues, these new technologies will be characterized by rapid performance improvements and larger market share. Because the small scale technologies described next tend to be simpler and smaller than older ones, they may well be less expensive to own and operate.

TABLE 3.1.2 Summary of Distributed Generation Technologies

Criterion	IC Engine	Microturbine	PVs	Fuel Cells
Dispatchability	Yes	Yes	No	Yes
Capacity range	50 kW–5MW	25 kW–25 MW	1 kW–1 MW	200 kW–2 MW
Efficiency[a]	35%	29–42%	6–19%	40–57%
Capital cost ($/kW)	200–350	750–1000	6600	3750–5700
O&M cost[b] ($/kWh)	0.01	0.005–0.0065	0.001–0.004	0.0017
NO_X (lb/Btu) — Nat. gas	0.3	0.10	n/a	0.003–0.02
NO_X (lb/Btu) — Oil	3.7	0.17	n/a	n/a
Technology status	Commercial	Commercial in larger sizes	Commercial	Commercial scale demos

[a] Efficiencies of fossil and renewable DR technologies are not directly comparable. The method described later in this book includes all effects needed to assess energy production.
[b] O&M costs do not include fuel. Capital costs have been adjusted based on quotes.

Source: From DCPA, 1998.

The Technologies

Table 3.1.2 provides an overview of feasible, present or near term DR technologies. Each listed technology is summarized below in alphabetical order.

Fuel Cells

A fuel cell is a device in which hydrogen and oxygen combine without combustion to produce electricity in the presence of a catalyst. One design is shown in Figure 3.1.13. Several competing technologies have been demonstrated and are listed below with their nominal operating temperatures.

- Phosphoric acid (PA) — 300°F
- Proton exchange membrane (PEM) — 200°F
- Molten carbonate (MC) — 1200°F
- Solid oxide (SO) — 1300°F

As indicated in Table 3.1.2, the costs of fuel cells are too high to be competitive now, but industry experts have indicated that prices should fall because of mass production. Where environmental regulations are strict, fuel cells offer the only truly clean solution to electricity production outside of the renewables sector.

The key barriers to fuel cell usage include

- Cost — predicted cost reductions have not materialized; in fact, one large firm recently announced a 60% price increase.
- Hydrogen fuel — widespread adoption will require a new fuel distribution infrastructure in the U.S. or on-site reforming of natural gas (methane).
- Maintenance costs are uncertain.
- Transient response to building load variations is unacceptable for load following for some technologies.

Contrasting these barriers are some very attractive FC features:

- The only byproduct is water — NO_x emissions are very low (<1PPM).
- Efficiency is good — 50–60% (LHV basis).
- Thermal or electrical cogeneration is possible in buildings.
- Modularity is excellent — nearly any building-related load can be matched well (kW to MW range).

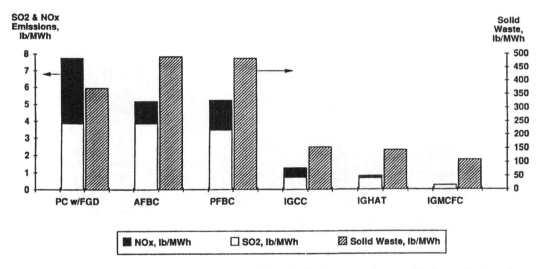

FIGURE 3.1.13 Emissions and solid waste from coal-based technologies are lowest with the gasification plants. IGCC (Integrated Gasification Combined Cycle), IGHAT (Integrated Gasification Humid Air Turbine), and IGMCFC (Integrated Gasification Molten Carbonate Fuel Cell).

Internal Combustion Engines

Reciprocating internal combustion engines (ICEs) are the traditional technology for emergency power around the world. Operating experience with Diesel and Otto cycle units is extensive. The cost of units is the least of any DR technology, but maintenance costs are among the greatest. Furthermore, diesel and gasoline engines produce unacceptable emission levels in air quality nonattainment areas of the U.S.

Natural gas (NG) ICE generators offer a partial solution to the emissions problem but do not solve it entirely. However, the NG-fired ICE is the key competition to all DR technologies considered here.

The key barriers to ICE usage include

- Maintenance cost — the highest among the DR technologies due to the large number of moving parts.
- NO_x emissions are highest among the DR technologies (15–20 PPM even for lean burn designs).
- Noise is low frequency and more difficult to control than for other technologies; adequate attenuation is possible.

Attractive ICE features include

- Capital cost is lowest of the DR approaches.
- Efficiency is good — 32–36% (LHV basis).
- Thermal or electrical cogeneration is possible in buildings.
- Modularity is excellent — nearly any building-related load can be matched well (kW to MW range).
- Part load efficiency is good (the need for this is described later).

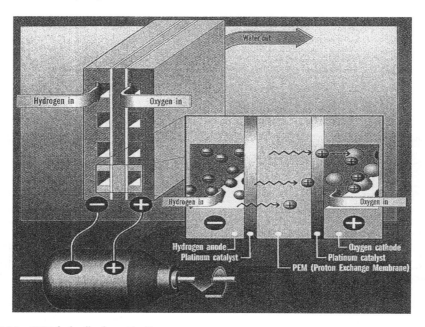

FIGURE 3.1.14   PEM fuel cell schematic diagram.

Microturbines

A microturbine (MT) is a Brayton cycle engine using atmospheric air and natural gas fuel to produce shaft power. Figure 3.1.15 shows the essential components of this device. Although a dual shaft approach is shown in the figure, a single-shaft design is also used in which the power produced in the expander is supplied to both the compressor and the load by a single shaft. The dual shaft design offers better control but at the cost of another rotating part and two more high speed bearings. Electrical power is produced by a permanent magnet generator attached to the output shaft or to a gear reducer driven by the output shaft.

Figure 3.1.16 is a photograph of a small microturbine showing most of the key components except for the recuperator. The recuperator is used in most units because about half of the heat supplied to the working fluid can be transferred from the exhaust gas to the combustion air. Without a recuperator, the overall efficiency of an MT is 15–17% whereas with an 85% effective recuperator the efficiency can be as high as 33%. MTs without recuperators are basically burners that produce a small amount of electricity with thermal output to be used for cogeneration.

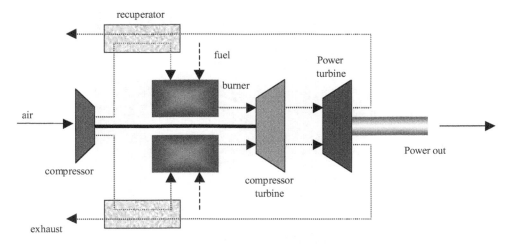

FIGURE 3.1.15   Schematic diagram of dual shaft microturbine design.

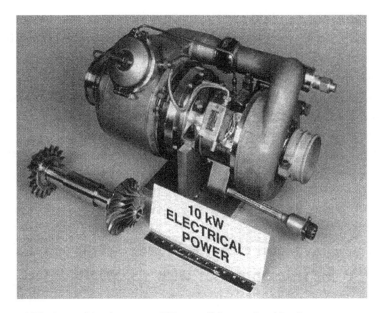

FIGURE 3.1.16   10-kW microturbine (courtesy of Honeywell International Inc.).

A handful of MT manufacturers have announced products in the U.S. Sizes range from 25 kW to 150 kW with double digit power ratings being the most common. As of December 2000, fewer than 1000 MTs had been shipped to U.S. locations.

Attractive MT features include

- Capital cost is low.
- Efficiency is good — 30–33% (LHV basis).
- Emissions are modest (<10 PPM NO_x is quoted by manufacturers).
- Thermal or electrical* cogeneration is possible in buildings.

_____

*Electrical cogeneration refers to the use of exhaust gases to power a bottoming turbine cycle, often an organic Rankine cycle to produce additional electric power.

FIGURE 3.1.17   Solar PV panel (courtesy of NREL; photo by W. Gretz).

- Modularity is excellent — nearly any building-related load can be matched well by multiple units of small capacity.

The key barriers to MT usage include

- Maintenance cost — unknown but expected to be lower than ICEs because of many fewer moving parts.
- Part load efficiency is questionable — manufacturer's data vary.
- Limited field experience.
- Use of air bearings is desirable to reduce maintenance, but air filtration requirements are stringent.
- High frequency noise is produced but is relatively easy to control.

Solar Photovoltaics

Photovoltaic (PV — see Figure 3.1.17) cells directly convert sufficiently energetic photons in sunlight to electricity. Because sunlight is a diffuse resource, large array areas are needed to produce significant power. However, offsetting this is the zero cost of the fuel itself. In 2000 there was a PV market worldwide of the order of 200 MW* per year.

Prices for PV arrays have dropped by at least two orders of magnitude in the past three decades, but they still appear to be too high for many applications in the U.S. where the present utility grid offers an alternative. However, in mountainous areas where the grid does not exist or in developing countries where electricity infrastructure investments may never be made, PVs can produce power more cheaply than the common ICE alternative.

Attractive features of PV systems include

- Emissions are zero.
- Fossil fuel consumption is zero.
- Low temperature thermal cogeneration (using building-integrated modules) is possible for space heating in buildings.

---

*The rating of solar systems on a kW or MW basis is misleading because the rating conditions are such that quoted outputs are nearly optimal. To compare solar and nonsolar energy costs, one must determine the cost of electricity for each technology as described later in this book.

- Modularity is excellent — nearly any building-related load can be matched well by multiple units of small capacity.
- Maintenance is negligible except where batteries are involved.
- Part load efficiency is excellent.

The key barriers to PV usage include

- The price of delivered power exceeds other DR resources; subsidies exist in some states that make PV-produced power competitive.
- Temporal match of power produced to load is imperfect; batteries or other systems are often needed.

Other Approaches
Although not treated in this chapter, several additional DR techniques hold promise for the future. They can be assessed using the approaches described later in this chapter. Included in this future list are stirling engines, solar stirling engines, and solar thermal conversion.

New Characteristics Common to All DR Systems

The previous section described four DR systems that appear to have significant near term feasibility. The next section describes exactly how one rationally selects the best, or best mix of, DR approaches. However, we first need to enumerate the key features that distinguish DR approaches from traditional central generation methods.

Need for New Controls Methods
Present utility power dispatch approaches cannot accommodate thousands of small, distributed generators of power and other ancillary services. Generally speaking, DR owners or their agents will operate their systems as much as possible when their marginal cost of producing power is less than the marginal cost of competing power. Legal, operating, and marketing costs for individual, small generators may also be prohibitive.

The financial dispatch of these generators requires knowledge of the following:

- Real time building loads and accurate predictions of near future loads.
- Real time cost of power controlled by the independent system operator (ISO) or its equivalent; dynamic pricing is the wave of the future.
- Dynamic power production characteristics of DR units.
- Real time cost of producing power from the available DR resources.

It appears that what will evolve to handle the dispatch problem is a new virtual power plant independent system operator which appears to the (ISO*) as a single dispatchable entity. It is the job of the VPP to determine which DR resources should be used to sell power to the grid and when. Real time knowledge of both local loads and grid-wide demand for power are required. New methods of trading and transaction processing seem to be inevitable as DR grows rapidly. By treating many small generators in the aggregate, the per unit cost of transactions and information may be reduced. To gain an appreciation of this, consider that the California ISO will not purchase ancillary services from generators less than 10 MW in size — equivalent to 200 50-kW microturbines.

Need for New Interconnection Codes and Standards
Each DR technology requires different electrical safety protocols and standards. It appears that the PV industry had made the most significant progress of the new DR systems. Numerous IEEE, ASME, and other standards bodies are involved as of the writing of this handbook.

---

*Also included would be the power exchange (PX) if it exists for electricity futures marketing.

3.1.6 Distributed Generation Economic Assessment

The technical and economic assessment of DR resources requires that a uniform method of assessment be used for all technologies. The methodology must include defensible estimates of

- Thermal and electrical loads on the DR system
- DR electrical production including part load effects
- Economic analysis including all reasonable and known* costs and benefits of DR.

This section summarizes a first principles approach for assessing the economic feasibility of DR systems, whether renewable or fossil-based. The same approach can be used to compare central and distributed generation sources competing in the same utility region. The context here is the U.S. energy economy. However, DR has much larger potential in the developing world where central T&D and generation do not exist. The methods described next can also be used to determine the most suitable DR technology for these markets.

Loads

It is often not possible to obtain the exact load shape for a specific building. Long-term monitoring can be costly and is usually not seen as a justifiable expense. For this reason, it is helpful to work with libraries of "standard" load shapes that can be adjusted to match the total and peak consumption of a specific building.

Once a "generic" load shape for a specified building class has been determined, the hourly values are adjusted to match the (1) actual total consumption, (2) peak demand, and (3) load factor for the selected billing period.

Economic Parameters

The proper calculation of the life cycle cost or savings of distributed generation requires a number of different economic parameters. An investment in distributed generation must have an acceptable rate of return. There are two components to the economic valuation: the first is a standard calculation based on initial costs and fuel use, and the second incorporates transmission, distribution, and other credits that vary by location and utility.

The life cycle analysis of DR is similar to other energy systems. The discount, inflation, and fuel cost escalation rates are used along with the estimated system lifetime to estimate the life cycle savings of the system. The negative cash flows (i.e., out of pocket) include the installed cost of the system, the incremental cost of gas consumed by the DR system (for IC engine, turbine, and fuel cell technologies), annual and intermittent maintenance costs, and any replacement costs. The offsetting positive cash flows include the avoided cost of electricity, the avoided cost of gas if heat recovery is used, and any tax credits and depreciation.

Utility Rates

The use of distributed generation makes sense only if

- DR power can be produced at a lower cost than the utility-supplied power
- The need for guaranteed power overrides mere power cost considerations
- The sale of power produced by a DR system results in a positive cash flow to the DR owner.

Determining the total energy bill is usually not a trivial calculation. It is easy to calculate if a real-time pricing rate is used. However, most commercial and industrial rates incorporate time-of-use components, block components, or both. These make it difficult to predict the incremental cost of the "next" kWh or therm used by the building at any arbitrary point within the billing cycle. To properly estimate the utility

---

*In the present work, environmental benefits are not included in the economic analysis. Once economic values can be established and agreed to, the methodology presented can accommodate them readily.

bill for a given building, the energy consumption and demand for each hour of the billing period must be known. Stranded cost recovery charges must also be included.

Ancillary benefits from installing distributed resources have traditionally been targeted at the T&D system level. For electric utilities, installing small-scale distributed generation systems in the range of 200- to 5000-kW capacity can reduce costs by avoiding transmission and distribution upgrades. In addition, they can provide more reliable service to their customers. Some of the benefits from a utility perspective include

- Reduced T&D line losses
- T&D system upgrade deferrals
- Modular generation equipment investments
- Reactive power and voltage support; other power quality improvements
- Environmental compliance benefits
- Improved system reliability
- Reduced equipment maintenance intervals

These costs and benefits are described in detail in Curtiss, Kreider, and Cohen (1999).

Environmental Considerations

Emission reductions from clean DR sources can save utilities money and provide for public benefits of cleaner air. The Clean Air Act Amendment of 1990 required lower ozone emissions in virtually all metropolitan areas by the year 2000. Just as clean technologies can benefit from environmental compliance considerations, DR technologies which exceed the limits of the environmental requirements will have to comply and therefore be in competition with other cleaner DR technologies. In some cases, the environmental permitting and regulations will add to the cost of DR technologies.

Performance Modeling

To assess accurately the economic feasibility of distributed generation at a site, a reasonably complete model should be developed that incorporates the following algorithms:

- A *weather simulation* module should provide hourly site temperatures, solar radiation, wind speed, etc.
- A *building simulation* algorithm should generate hourly building loads.
- A *DR equipment simulation* should use any site weather data from the weather simulation and the building loads to determine the portion of building electrical and thermal loads that are offset by on-site production.
- Some form of *DR control optimization* must be included in the algorithms. This can be as simple as selecting between peak shaving control or baseline support control. A smarter control approach than either of these will result in greater economic benefit to the DR hardware investor.
- A *rate analysis and calculation* algorithm will be necessary to accurately determine the utility bill based on any block and time-of-use specifications.

An Example Microturbine Assessment

The HVAC engineer is often called early in the design phase to assess the viability of various energy supplies for a project. For traditional utility suppliers, the techniques are well known and use standard building simulation tools. However, for DR assessments there are additional steps. This section, using a concrete example, illustrates the approach by summarizing results from a simple analysis using a DR screening tool. In this case study, a hospital was chosen with the load shapes as shown in Figure 3.1.18 — 450 kW of electrical capacity were added to the building as a package of 10 microturbines. The turbines were assumed to have a lifetime of 7 years and an installed cost of $750/kW. No heat recovery (i.e., cogeneration) was used in the analysis. This example illustrates the importance of control strategy and utility rate. The conclusion from the study is that local utility rates (both gas and electric) must be used

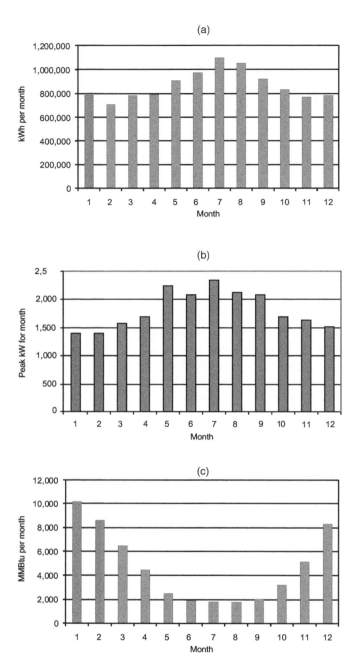

FIGURE 3.1.18 Monthly energy values used in example. (a) Building electrical load in kWh, (b) building demand in kW, and (c) thermal load in MMBtu.

for each study. It is not possible to generalize DR feasibility results for one building in a given location to another building in another location.

The first control mode was a peak-shaving technique where the turbines were operated only when the building electrical load was greater than 1000 kW. At this point, the turbines would come on one by one as the load increased above 1000 kW until all ten were operating. The second control method assumed that the turbines would run whenever possible in an attempt to maximize the annual run time. Both of these techniques were analyzed using two different rates. The rates were based on actual commercial rates from a major metropolitan utility; one rate is a general service rate with no demand component, the other is a time-of-use rate with time-of-use periods for both consumption and demand. The natural gas rate was not changed throughout these studies.

1. Results using general service (GS) rate

The monthly electricity costs for each of the control methods using the general service rate are shown in Figure 3.1.19. No ancillary credits or environmental costs are included in these results; the values shown are based strictly on the utility rate. The spread between the effective electrical and natural gas costs is such that the use of DR is advantageous in this building. It is cheaper to use natural gas to generate electricity on site than it is to purchase the electricity directly from the utility.

2. Results for time-of-use (TOU) rate

A similar analysis was performed using the time-of-use rate for the same building in the same location as above. The monthly electricity cost components are shown in Figure 3.1.20 and are directly comparable to the previous figure.

TABLE 3.1.3   Cost Comparisons

(a) Summary of microturbine performance using general service rate

| | Annual Energy Costs ($1000) | | Change in Annual Cost | | | |
	Elec.	Gas	Elec.	Gas	Total	Int. rate of return
a. Without DG	$1283	$237	—	—	—	—
b. Peak limiting	$1071	$328	−17%	+39%	−8%	12%
c. DG always on	$805	$451	−37%	+91%	−17%	45%

(b) Summary of microturbine performance using time-of-use rate

| | Annual Energy Costs ($1000) | | Change in Annual Cost | | | |
	Elec.	Gas	Elec.	Gas	Total	Int. rate of return
a. Without DG	$951	$237	—	—	—	—
b. Peak limiting	$773	$328	−19%	+39%	−7%	2%
c. DG always on	$638	$451	−33%	+91%	−8%	4%

The summary of the annual costs for the TOU rate is given in Table 3.1.3. These results for the same building, equipment, and control methods do not look promising at all for the GS rate results as shown in part (a) of Table 3.1.3. This is because the effective, aggregated cost of electricity using the time-of-use rate is not as great as with the general service rate. This example serves to illustrate the importance of knowing both detailed load shape and rate schedule information. *Rules of thumb or using results from one DR study for a different situation will result in incorrect economic assessments.* This analysis involved the same building, location, DR equipment, and control methods, yet the economic advantage of the installation of the microturbines depends solely on the utility rates that apply. Finally, we observe that the rate of return is very strongly affected by the control methodology.

Conclusions

This overview of DR in the U.S. has illustrated several key points:

1. Several viable DR technologies now exist.
2. A uniform assessment approach is needed to select the appropriate DR option(s) for a given application; the method must be able to include nonenergy aspects of DR such as ancillary services and emergency power benefits. A separate assessment is needed for each project; rules of thumb do not suffice.
3. New approaches are needed for control and financial dispatch of DR-produced power.

(a)

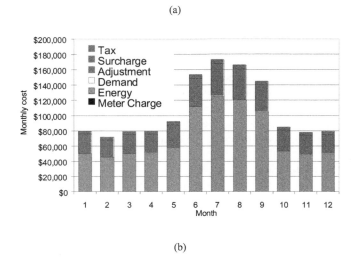

(b)

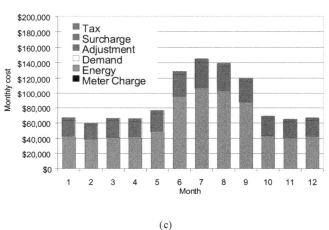

(c)

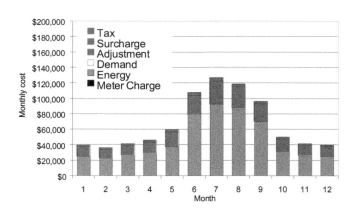

FIGURE 3.1.19 Monthly electricity cost for general service rate. (a) Without microturbines. (b) Turbines run when building load is > 1000 kW. (c) Turbines run at all times.

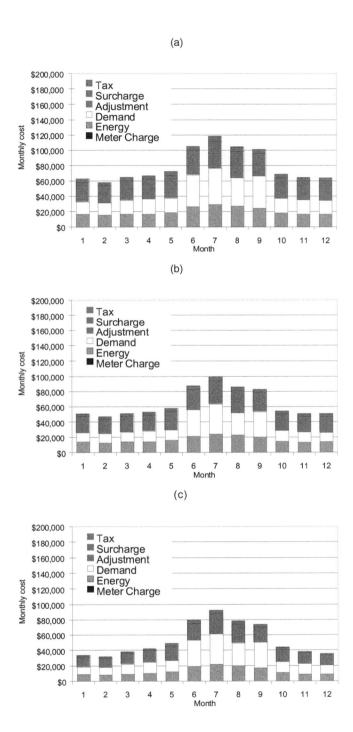

FIGURE 3.1.20 Monthly electricity costs for time-of-use rate. (a) without turbines. (b) turbines run when building load is > 1000 kW. (c) turbines run at all times.

References

Distributed Power Coalition of America (DCPA) (1998). URL: www.dpc.org.

McGinley, S. (1998). Statement: The Power Shortage Answer, URL: www.dpc.org.

Joint Center for Energy Management (JCEM) and Vesica Architecture and Planning (1995). *Community Energy Assessment and Decision Support (CEADS)*. Available from Kreider and Associates, Boulder, CO.

FCCG (1997). *FCCG Update, Newsletter of the Fuel Cell Commercialization Group*, Volume 7, Number 1.

GRI (1998). *Distributed Generation Guidebook for Municipal Utilities*, Gas Research Institute.

Hoff, T. E., Wenger, H. J., and Farmer, B. K., The Value of Deferring Electric Utility Capacity Investments with Distributed Generation. *Energy Policy*, March, 1996.

Hirst, E. and Kirby, B., Creating Competitive Markets for Ancillary Services, *ORNL/CON*-448, October, 1997.

Curtiss, P., Kreider, J., and Cohen D. (1999). A Methodology for Technical and Financial Assessment of Distributed Generation in the U.S., *Proceedings of the ASME Solar Energy Division*, Maui, HI.

3.2   Economics and Costing of HVAC Systems

Ari Rabl

3.2.1   Economic Analysis and Optimization

We do not live in paradise and our resources are limited. Therefore it behooves us to try to reduce the cost of heating and cooling to a minimum, subject, of course, to the constraint of providing the desired indoor environment and services. But while capital costs and a operating costs are readily stated in financial terms, other factors such as comfort, convenience, and aesthetics may be difficult or impossible to quantify. Furthermore, there is uncertainty: future energy prices, future rental values, future equipment performance, and future uses of a building are uncertain.

As a way around the difficulties, it is best to approach the design optimization in the following manner. First, one evaluates the total cost for each proposed design or design variation by properly combining all capital and operating costs. Then, knowing the cost of each design, one can select the "best," much like selecting the best product in a store where each product carries a price tag. Proceeding in this way, one separates the factors that can be quantified unambiguously (i.e., cost, according to the price tag), from those that are less tangible (e.g., aesthetics). The calculation of the price is the essence of engineering economics and forms the main part of this section. Optimization and some effects of uncertainty are addressed at the end.

3.2.2   Comparing Present and Future Costs

The Effect of Time on the Value of Money

Before one can compare first costs (i.e., capital costs) and operating costs, one must apply a correction because a dollar (or any other currency unit) to be paid in the future does not have the same value as a dollar available today. This time dependence of money is due to two, quite different, causes. The first is inflation, the well known and ever present erosion of the value of our currency. The second reflects the fact that a dollar today can buy goods to be enjoyed immediately or it can be invested to increase its value by profit or interest. Thus a dollar that becomes available in the future is less desirable than a dollar today; its value must be discounted. This is true even if there is no inflation. Both inflation and discounting are characterized in terms of annual rates.

Let us begin with inflation. To avoid confusion, subscripts are added to the currency signs, indicating the year in which the currency is specified. For example, during the middle of the 1980s the inflation rate r_{inf} in western industrial countries was around $r_{inf} = 4\%$. Thus, a dollar in 1986 is worth only $1/(1+0.04)$ as much as the same dollar one year before:

$$1.00 \, \$_{1986} = \frac{1}{1 + r_{inf}} \, \$_{1985} = \frac{1}{1 + 0.04} = 0.96 \, \$_{1985} \, .$$

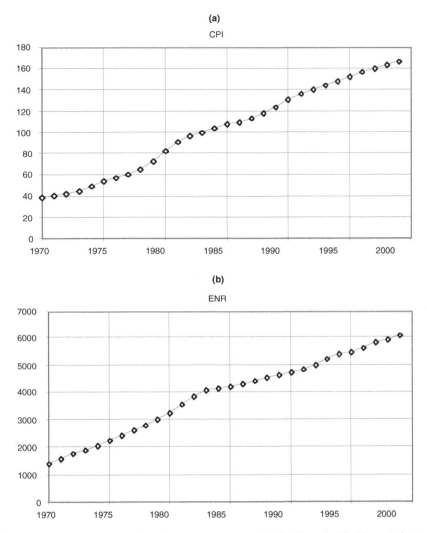

FIGURE 3.2.1 History of various cost indices. (a) CPI = consumer price index (from ftp://ftp.bls.gov/pub/special.requests/cpi/cpiai.txt), (b) ENR = *Engineering News Record* construction cost index.

The definition and measure of the inflation rate are actually not without ambiguities since different prices escalate at different rates and the inflation rate depends on the mix of goods assumed. The most common measure is probably the consumer price index (CPI), an index that was arbitrarily set at 100 in 1983. Its evolution is shown in Figure 3.2.1, along with another index of interest to the HVAC designer — the *Engineering News Record* construction cost index. In terms of the CPI, the average inflation rate from year ref to year ref+n is given by[*]

$$(1 + r_{inf})^n = \frac{CPI_{ref+n}}{CPI_{ref}}.$$
(3.2.1)

---

[*]For simplicity, we write the equations as if all growth rates were constant. Otherwise the factor $(1 + r)^n$ would have to be replaced by the product of factors for each year $(1 + r_1)(1 + r_2) \dots (1 + r_n)$. Such a generalization is straightforward, but tedious and of dubious value in practice as it is already risky to predict averages trends without trying to guess a detailed scenario.

Suppose $\$_{1985}$ 1.00 was invested at an interest rate $r_{int} = 10\%$, the *nominal* or *market* rate, as usually quoted by financial institutions. Then, after one year this dollar had grown to $\$_{1986}$ 1.10, but it is worth only $\$_{1985}$ 1.10/1.04 = $\$_{1985}$ 1.06. To show the increase in the real value, it is convenient to define the real interest rate r_{int0} by the relation

$$1 + r_{int0} = \frac{1 + r_{int}}{1 + r_{int}} \tag{3.2.2}$$

or

$$r_{int0} = \frac{r_{int} - r_{inf}}{1 + r_{inf}}.$$

The simplest way of dealing with inflation is to eliminate it from the analysis right at the start by using *constant currency* and expressing all growth rates (interest, energy price escalation, etc.) as real rates, relative to constant currency. After all, one is concerned about the real value of cash flows, not about their nominal values in a currency eroded by inflation. Constant currency is obtained by expressing the *current* or *inflating* currency of each year (i.e., the nominal value of the currency) in terms of equivalent currency of an arbitrarily chosen reference year *ref*. Thus, the current dollar of year *ref+n* has a constant dollar value of

$$\$_{ref} = \frac{\$_{ref + n}}{(1 + r_{inf})^n} \tag{3.2.3}$$

A *real growth rate* r_0 is related to the *nominal growth rate* r analogous to Equation 3.2.2:

$$r_0 = \frac{r - r_{inf}}{1 + r_{inf}}. \tag{3.2.4}$$

For low inflation rates one can use the approximation

$$r_0 \approx r - r_{inf} \quad \text{if} \quad r_{inf} \quad \text{is small.} \tag{3.2.5}$$

Later, as proved in the section entitled "Constant Currency Versus Inflating Currency", an analysis in terms of constant currency and real rates is exactly equivalent to one with inflating currency and nominal rates, if the investment is paid out of equity (i.e., without a loan) and without a tax deduction for depreciation or interest. Slight real differences between the two approaches can arise from the formulas for depreciation and for loan payments (in the U.S., loan payments are usually arranged to have fixed amounts in current currency, and the real value of annual loan payments differs between the two approaches). Therefore, the inflating dollar approach is commonly chosen in the U.S. business world.

However, when the constant dollar approach is correct, it offers several advantages. Having one variable less, it is simpler and clearer. What is more important, the long term trends of real growth rates are fairly well known, even if the inflation rate turns out to be erratic. For example, from 1955 to 1980, the real interest rate on high quality corporate bonds consistently hovered around 2.2% despite large fluctuations of inflation (Jones, 1982), while the high real interest rates of the 1980s were probably a short term anomaly. Riskier investments, such as the stock market, might promise higher returns, but they, too, tend to be more constant in constant currency.

Likewise, prices tend to be more constant when stated in terms of real currency. This is illustrated in Figure 3.2.2 by comparing some energy prices in real and in inflating dollars. For example, the market price (price in inflating currency) of crude oil reached a peak of $36 in 1981, ten times higher than the market price during the 1960s, while in terms of constant currency the price increase over the same period was only a factor of four. Crude oil during oil crises is, of course, an example of extreme price fluctuations. For other goods, the price in constant currency is far more stable (it would be exactly constant in the absence of relative price shifts among different goods). Therefore, it is instructive to think in terms of real rates and real currency.

(a)

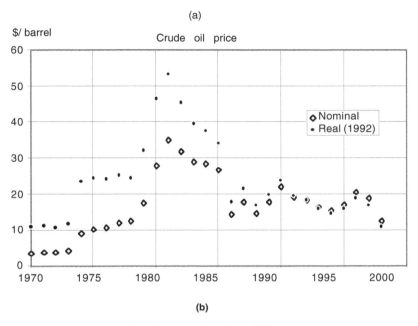

(b)

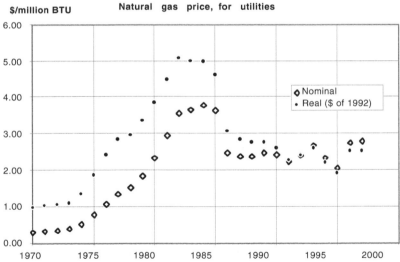

FIGURE 3.2.2 Energy/prices, in constant dollars (solid symbols) and in inflating dollars (hollow symbols). (a) crude oil; (b) natural gas (for utility companies); (c) electricity (average retail price). (From http://www.eia.doe.gov.)

Example 1

Find the nominal and real escalation rates for residential electricity prices between 1970 and 1995.

Given: data in Figure 3.2.2(c); real prices are in $_{1992}$.

Find: real growth rate r_0 and nominal growth rate r.

Solution

In 1970 the price was $p_{1970} = 1.7¢_{1970}/kWh = 5.6¢_{1992}/kWh$.

In 1995 the price was $p_{1995} = 6.9¢_{1995}/kWh = 6.4¢_{1992}/kWh$.

The number of years is n = 1995 − 1970 = 25.

Hence, the real growth rate is given by

(c)

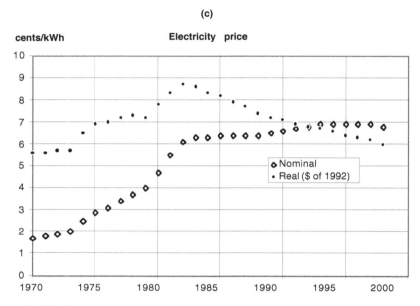

FIGURE 3.2.2 (*continued*)

$$r_0 = -1 + \sqrt[n]{\frac{p_{1995}}{p_{1970}}}$$

$$= -1 + \sqrt[25]{\frac{6.4}{5.6}} = -1 + 1.005 = 0.005 .$$

The nominal growth rate is

$$r = -1 + \sqrt[25]{\frac{6.9}{1.7}} = -1 + 1.058 = 0.058 .$$

Comments: This example highlights the importance of distinguishing between real and nominal growth rates. While the apparent price has grown by almost 6% per year, the real price increased by only 0.5% per year; the inflation rate averaged about 5.3% during this period.

Discounting of Future Cash Flows

As mentioned above, even if there were no inflation, a future cash amount F is not equal to its *present value* P; it must be discounted. The relation between P and its future value F_n n years from now is given by the *discount rate* r_d, defined such that

$$P = \frac{F_n}{(1 + r_d)^n}. \qquad (3.2.6)$$

The higher the discount rate, the lower the present value of future transactions.

To determine the appropriate value of the discount rate one has to ask at what value of r_d one is indifferent between an amount P today and an amount $F_n = P/(1 + r_d)^n$ n years from now. That indifference depends on circumstances and individual preferences. Consider a consumer who would put his money in a savings account with 5% interest. His discount rate is 5% because by putting the $1000 into this account he in fact accepts the alternative of $(1 + 5\%) \times \$1000$ a year from now. If instead he would use it to pay off a car loan at 10%, then his discount rate would be 10%; paying off the loan is like putting the money into a savings account which pays at the loan interest rate. If the money would allow him to avoid an emergency loan at 20%, then his discount rate would be 20%. At the other extreme, if he would hide the money in his mattress, his discount rate would be zero.

The situation becomes more complex when there are several different investment possibilities offering different returns at different risks, such as savings accounts, stocks, real estate, or a business venture. By and large, if one wants the prospect of a higher rate of return one has to accept a higher risk. Thus, as a more general rule, we can say that the appropriate discount rate for the analysis of an investment is the rate of return on alternative investments of comparable risk. In practice that is sometimes quite difficult to determine, and it may be desirable to have an evaluation criterion that bypasses the need to choose a discount rate. Such a criterion is obtained by calculating the profitability of an investment in terms of an unspecified discount rate and then solving for the value of the rate at which the profitability goes to zero. That method, called *internal rate of return*, is explained later in the section entitled "Internal Rate of Return".

As with other growth rates, one can specify the discount rate with or without inflation. If F_n is given in terms of constant currency, designated as F_{n0}, then it must be discounted with the real discount rate r_{d0}. The latter is, of course, related to the market discount rate r_d by

$$r_{d0} = \frac{r_d - r_{inf}}{1 + r_{inf}}. \qquad (3.2.7)$$

According to Equation 3.2.4. Present values can be calculated with real rates and real currency or with market rates and inflating currency; the result is readily seen to be the same because multiplying the numerator and denominator of Equation 3.2.6 by $(1 + r_{inf})^n$ one obtains

$$P = \frac{F_n}{(1 + r_d)^n} = \frac{F_n(1 + r_{inf})^n}{(1 + r_{inf})^n(1 + r_d)^n}$$

which is equal to

$$P = \frac{F_{n0}}{(1 + r_{d0})^n}$$

since

$$F_{n0} = \frac{F_n}{(1 + r_{inf})^n} \qquad (3.2.8)$$

by Equation 3.2.3.

The ratio P/F_n of present and future value is called *present worth factor*. We designate it with the mnemonic notation

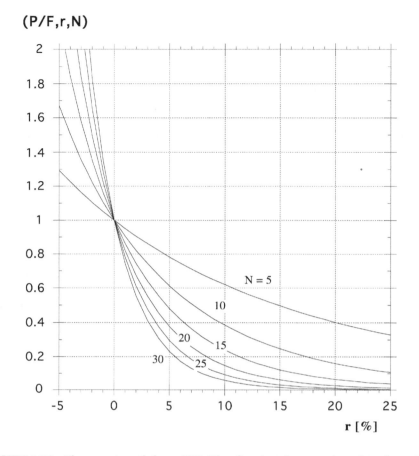

(P/F,r,N)

FIGURE 3.2.3 The present worth factor (P/F,r,N) as function of rate r and number of years N.

$$(P/F,r,n) = P/F_n = (1 + r)^{-n} \tag{3.2.9}$$

It is plotted in Figure 3.2.3. Its inverse

$$(F/P,r,n) = \frac{1}{(P/F,r,n)} \tag{3.2.10}$$

is called *compound amount factor*. These factors are the basic tool for comparing cash flows at different times. Note that we have chosen the end-of-year convention by designating F_n as the value at the end of the nth year. Also, we have assumed annual intervals, which is generally an adequate time step for engineering economic analysis; accountants, by contrast, tend to work with monthly intervals, corresponding to the way most regular bills are paid. The basic formulas are the same, but the numerical results differ because of differences in the compounding of interest; this point is explained more fully later in the section called "Discrete and Continuous Cash Flows" in which we pass to the continuous limit by letting the time step approach zero.

Example 2

What might be an appropriate discount rate for analyzing the energy savings from a proposed new cogeneration plant for a university campus? Consider the fact that from 1970 to 1988, the endowment of the university grew by a factor of 8 (current dollars) due to profits from investments.

Given:

growth factor in current dollars = 8.0,

increase in CPI = 118.3/38.9 = 3.04, from Figure 3.2.1a,

over N = 18 years.

Find: real discount rate r_{do}

Solution

There are two equivalent ways of solving for r_{do}.

First method: take the real growth factor, 8.0/3.04, and set it equal to

$(1 + r_{do})^N$.

The result is $r_{do} = 5.52\%$.

Second method: calculate market rate r_d by setting the market growth in current dollars equal to $(1 + r_d)^N$ and calculate inflation by setting the CPI increase equal to $(1 + r_{inf})^N$.

We find $r_d = 12.246\%$ and $r_{inf} = 6.371\%$. Then solve Equation 3.2.7 for r_{do}, with the result $r_{do} = \dfrac{0.12246 - 0.06371}{1 + 0.06371} = 5.52\%$, the same as with the first method.

Comments: Choosing a discount rate is not without pitfalls. For the present example, the comparison with the real growth of other long-term investments seems appropriate; of course, there is no guarantee that the endowment will continue growing at the same real rate in the future.

Equivalent Cash Flows and Levelizing

It is convenient to express irregular or variable payments as equivalent uniform payments in regular intervals; in other words, one replaces a nonuniform series by an equivalent uniform series. We refer to this technique as *levelizing*. It is useful because regularity facilitates understanding and planning. To develop the formulas, calculate the present value P of a series of N equal annual payments A. If the first payment occurs at the end of the first year, its present value is $A/(1 + r_d)$. For the second year it is $A/(1 + r_d)^2$, etc. Adding all the present values from year 1 to N, we find the total present value

$$P = \frac{A}{1 + r_d} + \frac{A}{(1 + r_d)^2} + \dots + \frac{A}{(1 + r_n)^N} \tag{3.2.11}$$

This is a simple geometric series, and the result is readily summed to

$$P = A\frac{1 - (1 + r_d)^{-N}}{r_d} \quad \text{for} \quad r_d \neq 0 \tag{3.2.12}$$

For zero discount rate, this equation is indeterminate, but its limit $r_d \to 0$ is A N, reflecting the fact that the N present values all become equal to A in that case. Analogous to the notation for the present worth factor, we designate the ratio of A and P by

$$(A/P, r_d, N) = \begin{cases} \dfrac{r_d}{1 - (1 + r_d)^N} & \text{for} \quad r_d \; 0 \\[2mm] \dfrac{1}{N} & \text{for} \quad r_d = 0 \end{cases} \tag{3.2.13}$$

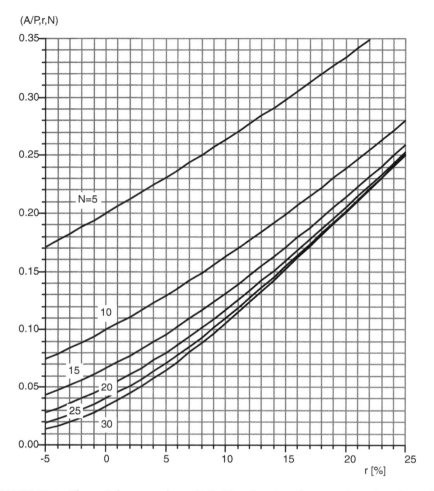

FIGURE 3.2.4   The capital recovery factor (A/P,r,N) as function of rate r and number of years N.

It is called *capital recovery factor* and is plotted in Figure 3.2.4. For the limit of long life, N →, it is worth noting that $(A/P,r_d,N) \to r_d$ if $r_d > 0$. The inverse is known as *series present worth factor* since P is the present value of a series of equal payments A.

With the help of present worth factor and capital recovery factor, any single expense C_n that occurs in year n, for instance a major repair, can be expressed as an equivalent annual expense A that is constant during each of the N years of the life of the system. The present value of C_n is $P = (P/F,r_d,n) C_n$ and the corresponding annual cost is

$$A = (A/P,r_d,N)(P/F,r_d,N)C_n \tag{3.2.14}$$

$$= \frac{r_d}{1-(1+r_d)^{-N}}(1+r_d)^{-n}C_n.$$

Example 3

A system has a salvage value of $1000 at the end of its useful life of N = 20 years. What is the equivalent levelized annual value if the discount rate is 8%?

Given: $C_{20} = \$1000$, $N = n = 20$ and $r_d = 0.08$.

Find: A.

Lookup values:

$(A/P,r_d,N) = 0.1019$ for Figure 3.2.4 or Equation 3.2.13,

and $(P/F,r_d,n) = 0.2145$ from Figure 3.2.3 or Equation 3.2.9.

Solution

Insert into Equation 3.2.14

$A = (A/P,r_d,N)\ (P/F,r_d,n)\ C_{20} = 0.1019 \times 0.2145 \times 1000\ \$/yr = 21.86\ \$/yr.$

A very important application of the capital recovery factor is the calculation of loan payments. In principle, a loan could be repaid according to any arbitrary schedule, but, in practice, the most common arrangement is based on constant payments in regular intervals. The portion of A due to interest varies, in a way calculated later in the section on "Principal and Interest," but to find the relation between A and the loan amount L we need not worry about that. Let us first consider a loan of amount L_n that is to be repaid with a single payment F_n at the end of n years. With n years of interest, at loan interest rate r_l, the payment must be

$$F_n = L_n(1 + r_l)^n.$$

Comparison with the present worth factor shows that the loan amount is the present value of the future payment F_n, discounted at the loan interest rate.

A loan that is to be repaid in N equal installments can be considered as the sum of N loans, the nth loan to be repaid in a single installment A at the end of the nth year. Discounting each of these payments at the loan interest rate and adding them we find the total present value; it is equal to the total loan amount

$$L = P = \frac{A}{1 + r_l} + \frac{A}{(1 + r_l)^2} + \ldots + \frac{A}{(1 + r_l)^N} \tag{3.2.15}$$

This is just the series of the capital recovery factor. Hence, the relation between annual loan payment A and loan amount L is

$$A = (A/P,r_l,N)\ L. \tag{3.2.16}$$

Now the reason for the name *capital recovery factor* becomes clear; it is the rate at which a bank recovers its investment in a loan.

Example 4

A home buyer obtains a mortgage of $100,000 at interest rate 8% over 20 years. What are the annual payments?

Given: $L = \$100,000$, $r_l = 8\%$, $N = 20$ yr.

Find: A.

Solution

From Figure 3.2.4 the capital recovery factor is 0.1019, and the annual payments are $10,190, approximately one tenth of the loan amount.

Some payments increase or decrease at a constant annual rate. It is convenient to replace a growing or diminishing cost by an equivalent constant or *levelized* cost. Suppose the price of energy is p_e at the start of the first year, escalating at an annual rate r_e while the discount rate is r_d. If the annual energy consumption Q is constant, then the present value of all the energy bills during the N years of system life is

$$P_e = Qp_e \left\{ \left(\frac{1 + r_e}{1 + r_d} \right)^1 + \left(\frac{1 + r_e}{1 + r_d} \right)^2 + \dots + \left(\frac{1 + r_e}{1 + r_d} \right)^N \right\}$$ (3.2.17)

assuming the end-of-year convention described above. As in Equation 3.2.3 we introduce a new variable $r_{d,e}$ defined by

$$1 + r_{d,e} = \frac{1 + r_d}{1 + r_e}$$ (3.2.18)

or

$$r_{d,e} = \frac{r_d - r_e}{1 + r_e} (\approx r_d - r_e \quad \text{if} \quad r_e \ll 1),$$ (3.2.19)

which allows us to write P_e as

$$P_e = (P/A, r_{d,e}, N) \, Q \, p_e.$$ (3.2.20)

Since $(A/P, r_d, N)$ is the inverse of $(A/P, r_d, N)$, we can write this as

$$P_e = (P/A, r_d, N) \, Q \left[\frac{(A/P, r_d, N)}{(A/P, r_{d,e}, N)} p_e \right]$$

If the quantity in brackets were the price, this would only be the formula without escalation. Let us call this quantity the *levelized* energy price $\bar{p}_e$

$$\bar{p}_e = \left[\frac{(A/P, r_d, N)}{(A/P, r_{d,e}, N)} p_e \right]$$ (3.2.21)

It allows us to calculate the costs as if there were no escalation. Levelized quantities can fill a gap in our intuition which is ill prepared to gauge the effects of exponential growth over an extended period. The levelizing factor

$$\text{levelizing factor} = \frac{(A/P, r_d, N)}{(A/P, r_{d,e}, N)}$$ (3.2.22)

tells us, in effect, the average of a quantity that changes exponentially at a rate r_e while being discounted at a rate r_d over a lifetime of N years. It is plotted in Figure 3.2.5 for a wide range of the parameters.

Example 5

The price of fuel is $p_e = 5$ \$/GJ at the start of the first year, growing at a rate $r_e = 4\%$ while the discount rate is $r_d = 6\%$. What is the equivalent levelized price over $N = 20$ years?

Given: $p_e = 5$ \$/GJ, $r_e = 4\%$, $r_d = 6\%$, $N = 20$ yr.

Find: $\bar{p}_e$

Solution

From Figure 3.2.5 the levelizing factor is 1.44. Hence the levelized fuel price is $\bar{p}_e = 1.44 \times 5$ \$/GJ = 7.20 \$/GJ.

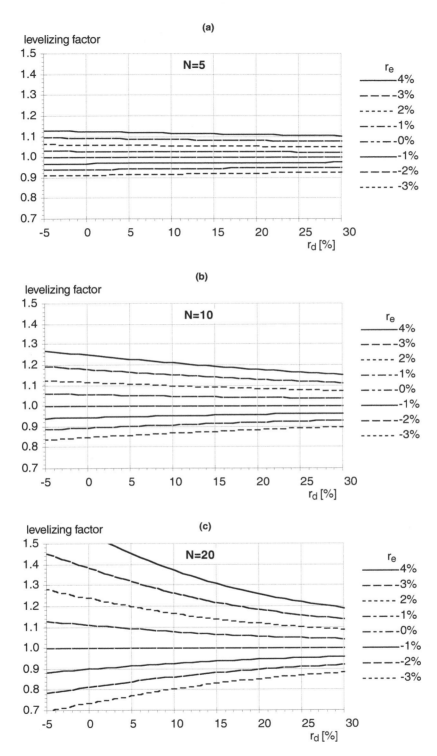

FIGURE 3.2.5   Levelizing factor $(A/P,r_d,N)/(A/P,r_{d,e},N)$ as a function of r_d and $r_{d,e}$. (a) N = 5 yr; (b) N = 10 yr; (c) N = 20 yr.

Several features may be noted in Figure 3.2.5. First, the levelizing factor increases with cost escalation r_e, being unity if $r_e = 0$. Second, for a given escalation rate, the levelizing factor decreases as the discount rate increases, reflecting the fact that a high discount rate de-emphasizes the influence of high costs in the future.

Discrete and Continuous Cash Flows

The above formulas suppose that all costs and revenues occur in discrete intervals. That is common engineering practice, in accord with the fact that bills are paid in discrete installments. Thus, growth rates are quoted as annual changes even if growth is continuous. It is instructive, however, to consider the continuous case.

TABLE 3.2.1 Discrete and Continuous Formulas for Economic Analysis, with Growth Rate r and Time Horizon N

Quantity known	Quantity to be found	Factor	Expression for discrete analysis	Expression for continuous analysis
P	F	(F/P,r,N)	$(1 + r)^N$	$\exp(rN)$
F	P	(P/F,r,N)	$(1 + r)^{-N}$	$\exp(-rN)$
P	A	(A/P,r,N)	$\dfrac{r}{1 - (1 + r)^{-N}}$	$\dfrac{\exp(r) - 1}{1 - \exp(-rN)}$
A	P	(P/A,r,N)	$\dfrac{1 - (1 + r)^{-N}}{r}$	$\dfrac{1 - \exp(-rN)}{\exp(r) - 1}$

Note: The rates for the discrete and continuous formulas are related by Equation 3.2.24.

Let us establish the connection between continuous and discrete growth by way of an apocryphal story about the discovery of *e*, the basis of natural logarithms. Before the days of compound interest, a mathematician who was an inveterate penny pincher thought about possibilities of increasing the interest he earned on his money. He realized that if the bank gives interest at a rate of r per year, he could get even more by taking the money out after half a year and reinvesting it to earn interest on the interest as well. With *m* compounding intervals per year the money would grow by a factor

$$(1 + r/m)^m$$

and the larger m, the larger this factor. Of course, he looked at the limit m → and found the result

$$\lim_{m \varnothing}(1 + r/m)^m = e^r \quad \text{with} \quad e = 2.71828\ldots \tag{3.2.23}$$

At the end of one year the growth factor is $(1 + r_{ann})$ with annual compounding at a rate r_{ann}, while with continuous compounding at a rate r_{cont} the growth factor is $\exp(r_{cont})$. If the two growth factors are to be the same, the growth rates must be related by

$$1 + r_{ann} = \exp(r_{cont}). \tag{3.2.24}$$

With this replacement of rates, the continuous formulas in Table 3.2.1 yield the same results as the discrete ones. Similarly, with m compounding intervals at rate r_m is equivalent to annual compounding if one takes

$$1 + r_{ann} = (1 + r_m/m)^m. \tag{3.2.25}$$

Example 6

A bank quotes a nominal interest rate of 10% (i.e., annual growth without compounding). What is the equivalent annual growth rate with monthly, daily, and continuous compounding?

Solution

For monthly compounding take m = 12 in Equation 3.2.25, with the result

$r_{12} = (1 + 0.1/12)^{12} - 1 = 0.104713$

With daily compounding we have

$r_{365} = (1 + 0.1/365)^{365} - 1 = 0.105155$

and with continuous compounding

$r_{cont} = \exp(0.1) - 1 = 0.105171.$

Comment: Beyond monthly compounding the differences are very small.

For small rates the first three terms in the series expansion of the exponential give an approximation

$$r_{ann} \approx r_{cont} (1 + r_{cont}/2)$$

which is convenient if one does not have a calculator at hand.

The Rule of Seventy for Doubling Times

Most of us do not have a good intuition for exponential growth. As a helpful tool we present therefore the rule of seventy for doubling times. The doubling time N_2 is related to the continuous growth rate r_{cont} by

$$2 = \exp(N_2 r_{cont}) \tag{3.2.26}$$

Solving the exponential relation for N_2 we obtain

$$N_2 = \frac{\ln(2)}{r_{cont}} = \frac{0.693...}{r_{cont}}.$$

The product of doubling time and growth rate in units of percent is very close to 70 years

$$N_2 r_{cont} \times 100 = 69.3 \ ... \ yr \approx 70 \ yr. \tag{3.2.27}$$

In terms of annual rates, the relation would be

$$N_2 = \frac{\ln(2)}{\ln(1 + r_{ann})},$$

numerically close to Equation 3.2.27 for small rates, but less convenient.

Example 7

Population growth rates average 2% for the world as a whole and reach 4% in certain countries. What are the corresponding doubling times?

Solution

70/2 = 35 yr for the world,

and 70/4 = 17.5 yr for countries with 4% growth.

Example 8

A consultant presents an economic analysis of an energy investment with N = 20 yr, assuming a 10% escalation rate for energy prices without stating the inflation rate. Is that reasonable?

Solution

A growth rate of 10% implies a doubling time of 7 years. There would be almost three doublings in 20 years, with a final energy price almost eight times the original. In constant dollars that would clearly be an absurd hypothesis. A totally different conclusion emerges if the inflation were 6%: then the real growth rate would be only $10 - 6 = 4\%$, and the doubling time 17.5 years in constant currency; although extreme, that is not inconceivable given our experience since the 1973 oil shock.

Comment: There are two lessons:

1. Never state growth rate or discount rate without indicating the corresponding inflation.
2. Be careful about assuming large growth rates over long time horizons. Use the rule of 70 to check whether the implications for the end of the time horizon make sense.

3.2.3 The Life Cycle Cost

Cost Components

A rational decision is based on the true total cost, which is the sum of the present values of all cost components and is called *life cycle cost*. The cost components relevant for the HVAC engineer are

- Capital cost (total initial investment)
- Energy costs
- Costs for maintenance, including major repairs
- Resale value
- Insurance
- Taxes

There is some arbitrariness in this assignment of categories. One could make a separate category for repairs, or one could include energy among O&M (operation and maintenance) cost, as is done in some industries. There is, however, a good reason for keeping energy apart. In buildings, energy costs dominate the other O&M costs, and they can grow at a different rate. Electric rates usually contain charges for peak demand in addition to charges for energy (see the section on "Demand Charge" discussed later). As a general rule, if an item is important it merits separate treatment.

Rental income needs to be included if one wants to evaluate the profitability of the building, or if one wants to compare design options that would affect the rent. It can be left out of the picture if one is concerned only with comparing design options that do not differ in their effect on rental income. The same is true for cleaning, security, and fire protection. *Usually, when comparing two options, there is no need to include terms that would be the same for each.* For example, when choosing between two chillers, one can restrict one's attention to the costs associated directly with the chillers (capital cost, energy, maintenance), without worrying about the heating system if it is not affected. In some cases it becomes necessary to account for the effects of taxes, because of tax deductions for interest payments and depreciation; these items are discussed below followed by the equation for the complete system cost.

Principal and Interest

In the U.S., interest payments are deductible from the income tax, while payments for the reimbursement of the loan are not. A tax paying investor, therefore, needs to know what fraction of a loan payment is due to interest. As explained earlier in the section on "Equivalent Cash Flows and Levelizing," it is assumed that a loan of duration N_l is repaid in N_l regular and equal payments A. (In this section we take N and N_l in years, but the formulas are valid for any choice of units. For billing purposes, the month is frequently chosen as the payment period. Slight numerical differences in the payments are due to compound interest.) Consider the nth payment, and let I_n = interest and P_n = principal (loan reimbursement); their sum A is constant

$$I_n + P_n = A. \tag{3.2.28}$$

Up to this point, n–1 payments have been made, so the debt remaining (on a loan of amount L) is

$$\text{remaining debt} = L - P_1 - P_2 - \ldots P_{n-1}. \tag{3.2.29}$$

At a loan interest rate r_l the interest for the nth period is

$$I_n = r_l(L - P_1 - P_2 - \ldots - P_{n-1}). \tag{3.2.30}$$

Comparing I_{n+1} with I_n one finds

$$I_{n+1} = I_n - r_l P_n \tag{3.2.31}$$

By means of Equation 3.2.28, one can eliminate P_n with the result

$$I_{n+1} = (1 + r_l)I_n - r_l A. \tag{3.2.32}$$

This recursion relation has the solution

$$I_n = (1 + r_l)^{n-1} r_l L + [1 - (1 + r_l)^{n-1}] A \tag{3.2.33}$$

as can readily be proved by mathematical induction. Since A and L are related by $A = (A/P,r_l,N_l) L$, where N_l is the duration of the loan and $(A/P,r_l,N_l)$ the capital recovery factor of Equation 3.2.13, this can be rewritten in the form

$$I_n/A = 1 - (1 + r_l)^{n-1-N_l}. \tag{3.2.34}$$

It is worth noting the period n enters only in the combination $(n - N_l)$, implying the fractional allocation to principal and interest depends only on the number of periods $(n - N_l)$ left in the loan, not on the original life of the loan. A loan has no memory, so to speak.

In general, the loan interest rate r_l differs from the discount rate r used for the economic analysis, and the loan life N_l may be different from the system life N. Inserting $A = (A/P,r_l,N_l)$ into Equation 3.2.34, we find that the interest payment I is related to loan amount L by

$$I_n = [1 - (1 + r_l)^{n-1-N_l}] (A/P,r_l,N_l) L. \tag{3.2.35}$$

The present value P_{int} of the total interest payments is found by discounting each I with the discount rate r and summing over n

$$P_{int} = \sum_{n=1}^{N_l} \frac{1 - (1 + r_l)^{n-1-N_l}}{(1 + r_d)^n} (A/P,r_l,N_l)L . \tag{3.2.36}$$

Using the formula for geometric series, this can be transformed to

$$P_{int} = \left\{ \frac{(A/P,r_l,N_l)}{(A/P,r_d,N_l)} - \frac{(A/P,r_l,N_l) - r_l}{(1 + r_l)(A/P,r_{d1},N_l)} \right\} L \tag{3.2.37}$$

with

$$r_{d1} = (r_d - r_l)/(1 + r_l).$$

If the incremental tax rate is τ, the total tax payments are reduced by τP_{int} (assuming a constant tax rate; otherwise, the tax rate would have to be included in the summation).

Example 9

A solar water heating system costing \$2000 is financed with a 5-year loan at $r_l = 8\%$. The tax rate is $\tau = 40\%$. How much is the tax deduction for interest worth if the discount rate is $r_d = 8\%$?

Given:

$\quad$ L = \$2000, $r_l = 8\%$, $r_d = 8\%$, $\tau = 40\%$.

Find:

$\quad P_{int}$

Lookup values:

$\quad (A/P,r_{dl},N_l) = 0.20$ and $(A/P,r_l,N_l) = 0.2505$ from Equation 3.2.13

Solution

$\quad$ We have $r_{dl} = 0$, since $r_l = r_d$.

$\quad$ Also $(A/P,r_{dl},N_l) = 0.20$,

$\quad$ and $(A/P,r_l,N_l) = 0.2505 = (A/P,r_d,N_l)$.

$\quad$ Thus the present value of the interest payments is, from Equation 3.2.37,

$$P_{int} = \left\{ 1 - \frac{0.2505 - 0.08}{1.08 \times 0.20} \right\} \times \$2000 = \$421.$$

$\quad$ At the stated tax rate that is worth $0.40 \times \$421 = \$168$.

Depreciation and Tax Credit

U.S. tax law allows business property to be depreciated. This means that for tax purposes the value of the property is assumed to decrease by a certain amount each year, and this decrease is treated as a tax deductible loss. For the economic analysis, one needs to express the depreciation as an equivalent present value. The details of the depreciation schedule have been changing with the tax reform of the 1980s. Instead of trying to present the full details, which can be found in Internal Revenue Service publications, we merely note the general features. In any year n, a certain fraction $f_{dep,n}$ of the capital cost (minus salvage value) can be depreciated. For example, in the simple case of straight line depreciation over N_{dep} years

$$f_{dep,n} = 1/N_{dep} \text{ for straight line depreciation.} \tag{3.2.38}$$

To obtain the total present value, one multiplies by the present worth factor and sums over all years from 1 to N

$$f_{dep} = \sum_{n=1}^{N_{dep}} f_{dep,n}(P/F,r_d,n) . \tag{3.2.39}$$

For straight line depreciation the sum is

$$f_{dep} = \frac{(P/A,r_d,N_{dep})}{N_{dep}} \text{ for straight line depreciation.} \tag{3.2.40}$$

A further feature of some tax laws is the tax credit. For instance, in the U.S. for several years around 1980, tax credits were granted for certain renewable energy systems. If the tax credit rate is τ_{cred} for an investment C_{cap}, the tax liability is reduced by $\tau_{cred} C_{cap}$.

Example 10

A machine costs \$10,000 and is depreciated with straight line depreciation over 5 years, and the salvage value after 5 years is \$1000. Find the present value of the tax deduction for depreciation if the incremental tax rate is $\tau = 40\%$, and the discount rate $r_d = 15\%$.

Given:

$$C_{cap} = 10 \text{ k\$},$$

$$C_{salv} = 1 \text{ k\$},$$

$$N = 5 \text{ yr},$$

$$\tau = 0.4,$$

$$r_d = 0.15.$$

Find: $\tau \times f_{dep} (C_{cap} - C_{salv})$

Lookup values:

$$f_{dep} = \frac{(P/A,0.15,5 \text{ yr})}{5 \text{ yr}} = \frac{(1/0.2983)}{5} = 3.3523/5 = 0.6705, \text{ from Equation 3.2.40.}$$

Solution

For tax purposes, the net amount to be depreciated is the difference

$$C_{cap} - C_{salv} = (10 - 1) \text{ k\$} = 9 \text{ k\$},$$

and with straight line depreciation $1/N_{dep} = 1/5$ of this can be deducted from the tax each year.

Thus, the annual tax is reduced by $\tau \times (1/5) \times 9 \text{ k\$} = 0.40 \times 1.8 \text{ k\$} = 0.72 \text{ k\$}$ for each of the five years.

The present value of this tax reduction is

$$\tau \times f_{dep} (C_{cap} - C_{salv}) = 0.40 \times 0.6705 \times 9 \text{ k\$} = 2.41 \text{ k\$}.$$

Comment: The present value of the reduction would be equal to $5 \times 0.72 \text{ k\$} = 3.6 \text{ k\$}$ if the r_d were zero. The discount rate of 15% reduces the present value by almost a third to $0.6705 = f_{dep}$.

Demand Charge

The cost of producing electricity has two major components: fuel and capital (for power plant and distribution system). As a consequence, the cost of electricity varies with the total load on the grid. To the extent that it is practical, utility companies try to base the rate schedule on their production cost. Thus, the rates for large customers contain two items: one part of the bill is proportional to the energy, and the other is proportional to the peak demand. (For most individual houses, the bill contains only an energy charge because the cost of separate meters was once considered to be too high.) If the monthly demand charge is $p_{dem} = 10$ \$/kW and the energy charge $p_e = 0.07$ \$/kWh, a customer with monthly energy consumption Q_m and peak demand P_{max} will receive a total bill of

$$\text{monthly bill} = Q_m p_e + P_{max} p_{dem} \qquad (3.2.41)$$

There are many small variations from one utility company to another. In most cases, p_e and p_{dem} depend on time of·day and time of year, being higher during the system peak than off·peak. In regions with extensive air conditioning, the system peak occurs in the afternoon of the hottest days. In regions with much electric heating, the peak is correlated with outdoor temperature. Some companies use what is called a "ratcheted" demand charge; it has the effect of basing the demand charge on the annual rather than the monthly peak.

Example 11

A 100 ton electric chiller with COP = 3 is used for 8 months of the year (running at 100% capacity at least once per month during 4 months and at 50% capacity at least once per month during 4 months), and the total load is equivalent to 1000 hours at peak capacity (a typical value around the belt from New York to Denver). What is the annual electricity bill, if p_e = 0.10 \$/kWh$_e$ and p_{dem} = 10 \$/kW$_e$ per month?

Given:

$P_{max,t}$ = 100 ton × 3.516 kW$_t$/ton, with COP = 3 kW$_t$/kW$_e$,

annual energy = P_{max} × 1000 h,

demand P_{max} for 4 months and 0.5 × P_{max} for 4 months,

p_e = 0.10 \$/kWh$_e$,

p_{dem} = 10 \$/kW$_e$ per month.

Find: annual bill.

Solution

Peak demand P_{max} = (100 ton × 3.516 kW$_t$/ton)/(3 kW$_t$/kW$_e$) = 117.2 kW$_e$.

annual energy Q = $P_{max,e}$ × 1000 h = 117,200 kWh$_e$.

annual bill = Q p_e + P_{max} p_{dem} × (4 × 1 + 4 × 0.5)

= 117,200 kWh$_e$ × 0.10 \$/kWh$_e$ + 117.2 kW$_e$ × 10 \$/kW$_e$ × 6

= \$11,720 + \$7032

= \$18,752.

Comments: In a real building, the precise value of the peak demand may be difficult to predict because it depends on the coincidence of the demands of individual pieces of equipment.

The total cost per kWh depends on the load profile. The more uneven the profile, the higher the cost. To take an extreme example, suppose the chiller were used only one hour per year, at full capacity. Then, with the rate structure of this example, the demand charge would be \$1172 while the energy charge would be only \$11.72, all that for consuming 1 kWh of energy. The total cost per kWh would be \$1172 + 11.72 = \$1183.72 for 117.2 kWh, an effective electricity price of 10.10 \$/kWh. This illustrates the interest of load leveling devices, such as cool storage for electric chillers.

The Complete Formula

The equations for a business investment can be stated in terms of before-tax cost or after-tax cost. Consider the purchase of fuel, with a market price of 5 \$/GJ, by a business that is subject to an income tax rate τ = 40%. Fuel, like all business expenditures, is tax deductible. And it is ultimately paid by profits. To purchase 1 GJ, one takes \$5 of profits before taxes; this reduces the tax liability by \$5 × 40% = \$2, resulting in a net cost of only \$3 after taxes.

We could do the accounting before or after taxes; the former counts the cash payments, the latter the net (after-tax) values. The two modes differ by a factor $(1 - \tau)$ where τ is the income tax rate. For example, if the market price of fuel is 5 \$/GJ and the tax rate τ = 40%, then the before-tax cost of fuel is 5 \$/GJ and the after-tax cost $(1 - \tau) \times 5$ \$/GJ $= 3$ \$/GJ. Stated in terms of *after-tax cost*, the complete equation for the life cycle cost of an energy investment can be written in the form

$$C_{life} \tag{3.2.42}$$

$$= C_{cap} \{(1 - f_l) \qquad \qquad \text{down payment}$$

$$+ f_l \frac{(A/P,r_1,N_1)}{(A/P,r_d,N_1)} \qquad \qquad \text{cost of loan}$$

$$- \tau f_l \left[\frac{(A/P,r_1,N_1)}{(A/P,r_d,N_1)} - \frac{(A/P,r_1,N_1) - r_1}{(1 + r_1)(A/P,r_{d,1},N_1)} \right] \qquad \text{tax deduction for interest}$$

$$- \tau_{cred} \qquad \qquad \text{tax credit}$$

$$- \tau f_{dep} \} \qquad \qquad \text{depreciation}$$

$$- C_{salv} \left(\frac{1 + r_{inf}}{1 + r_d} \right)^N (1 - \tau) \qquad \qquad \text{salvage}$$

$$+ Q \, P_e \frac{1 - \tau}{(A/P,r_{d,e},N)} \qquad \qquad \text{cost of energy}$$

$$+ P_{max} \, P_{dem} \frac{1 - \tau}{(A/P,r_{d,dem},N)} \qquad \qquad \text{cost of demand}$$

$$+ A_m \frac{1 - \tau}{(A/P,r_{d,M},N)} \} \qquad \qquad \text{cost of maintenance}$$

where

A_M = annual cost for maintenance [in first year \$]
C_{cap} = capital cost [in first year \$]
C_{salv} = salvage value [in first year \$]
f_{dep} = present value of depreciation, as fraction of C_{cap}
f_l = fraction of investment paid by loan
N = system life [yr]
N_l = loan period [yr]
P_e = energy price [in first year \$/GJ]
Q = annual energy consumption [GJ]
r_d = market discount rate
r_e = market energy price escalation rate
$r_{d,e}$ = $(r_d - r_e)/(1 + r_e)$
r_{dem} = market demand charge escalation rate
$r_{d,dem}$ = $(r_d - r_{dem})/(1 + r_{dem})$
r_{inf} = general inflation rate
r_l = market loan interest rate
$r_{d,l}$ = $(r_d - r_1)/(1 + r_1)$
r_M = market escalation rate for maintenance costs
$r_{d,M}$ = $(r_d - r_M)/(1 + r_M)$
τ = incremental tax rate
τ_{cred} = tax credit

If there are several forms of energy, e.g., gas and electricity, the term $Q \, p_e$ is replaced by a sum over the individual energy terms. Many other variations and complications are possible; for example, the salvage tax rate could be different from τ.

Example 12

Find the life cycle cost of the chiller of Example 11 under the following conditions:

Given:

    system life $N = 20$ yr,

    loan life $N_l = 10$ yr,

    depreciation period $N_{dep} = 10$ yr, straight line depreciation

    discount rate $r_d = 0.15$,

    loan interest rate $r_l = 0.15$,

    energy escalation rate $r_e = 0.01$,

    demand charge escalation rate $r_{dem} = 0.01$,

    maintenance cost escalation rate $r_M = 0.01$,

    inflation $r_{inf} = 0.04$,

    loan fraction $f_l = 0.7$,

    tax rate $\tau = 0.5$,

    tax credit rate $\tau_{cred} = 0$,

    capital cost (at 400 \$/ton) $C_{cap} = 40$ k\$,

    salvage value $C_{salv} = 0$,

    annual cost of maintenance $A_M = 0.8$ k\$/yr (= 2% of C_{cap}),

    capacity 100 ton $= 351.6$ kW$_t$,

    peak electric demand 351.6 kW$_t$/COP $= 117.2$ kW$_e$,

    annual energy consumption $Q = 100$ kton·h $= 351.6$ MWh$_t$,

    electric energy price $p_e = 10$ cents/kWh$_e = 100$ \$/MWh$_e$,

    demand charge $p_{dem} = 10$ \$/kW$_e$·month, effective during 6 months of the year.

    The rates are market rates.

Find: C_{life}.

Lookup values:

$r_{d,l} = 0.0000$	$(A/P,r_l,N_l) = 0.1993$
$r_{d,e} = 0.1386$	$(A/P,r_d,N_l) = 0.1993$
$r_{d,dem} = 0.1386$	$(A/P,r_{d,l},N_l) = 0.1000$
$r_{d,M} = 0.1386$	$(A/P,r_d,N) = 0.1598$
	$(A/P,r_{d,e},N) = 0.1498$
$(1+r_{inf})/(1+r_d) = 0.9043$	$(A/P,r_{d,dem},N) = 0.1498$

$f_{dep} = 0.502$ from Equation 3.2.40 $(A/P,r_{d,M},N) = 0.1498$

Solution

Components of C_{life} [all in k$] as per Equation 3.2.42

down payment	12.0
cost of loan	28.0
tax deduction for interest	−8.0
tax credit	0.0
depreciation	−10.0
salvage value	0.0
cost of energy	39.1
cost of demand charge	23.5
cost of maintenance	2.7
Total = C_{life}	87.3

Comments:

a. A spreadsheet is recommended for this kind of calculation. Standard business calculators contain most needed functions.
b. The cost of energy and demand is higher than the capital cost.

Cost per Unit of Delivered Service

Sometimes it is necessary to know the cost per unit of delivered service (for example, cost per ton-hour of cooling), analogous to the cost per driven mile for cars. This can be calculated as a ratio of levelized annual cost and annual delivered service. The levelized annual cost is obtained by multiplying the life cycle cost by the capital recovery factor for discount rate and system life. There appear two possibilities: the real discount rate r_{d0} and the market discount rate r_d. The quantity $(A/P,r_{d0},N)\,C_{life}$ is the annual cost in constant dollars (of the initial year), whereas $(A/P,r_d,N)\,C_{life}$ is the annual cost in inflating dollars. The latter is difficult to interpret because it is an average over dollars of different real value. Therefore, we levelize with the real discount rate because it expresses everything in first year dollars, consistent with the currency of C_{life}. Thus, we write the annual cost in initial dollars as

$$A_{life} = (A/P,r_{d0},N)C_{life}. \tag{3.2.43}$$

The effective total cost per delivered service is therefore

$$\text{effective cost per energy} = A_{life}/Q \tag{3.2.44}$$

where Q = annual delivered service (assumed constant, for simplicity).

We do not simply divide C_{life} by the service $N\,Q$ delivered by the system over its life time because that would not be consistent; C_{life} is the present value, while $N\,Q$ contains service flows (and thus monetary values) that are associated with future times. One must allocate service flows and costs within the same time frame which is accomplished by dividing the levelized annual cost by the levelized annual service; the latter is equal to Q because we have assumed that the consumption is constant from year to year.

Example 13

What is the cost per ton-hour for the chiller of Example 12?

Given: C_{life} = 87.3 k\$ and Q = 100 kton·h.

Find: A_{life}/Q.

Lookup values:

r_{do} = 0.1058 from Equation 3.214

$(A/P,r_{do},N)$ = 0.1221 from Equation 3.2.13.

Solution

Levelized annual cost in first year dollars

$A_{life} = (A/P,r_{do},N)\ C_{life} = 0.1221 \times 87.3$ k\$ = 10,659 \$/yr.

Cost per ton-hour = A_{life}/Q = 0.107 \$/ton·h.

Constant Currency Versus Inflating Currency

In the life cycle cost equation, all cost components have been converted to equivalent present values (i.e., first year costs). Let us see to what extent the result is the same whether one uses constant currency and real rates or inflating currency and market rates. In the term for energy cost only the variable $r_{d,e} = (r_d - r_e)/(1 + r_e)$ depends on this choice. Inserting real rates according to

$$(1 + r_{d0}) = \frac{1 + r_d}{1 + r_{inf}} \tag{3.2.45}$$

and

$$(1 + r_{e0}) = \frac{1 + r_e}{1 + r_{inf}}, \tag{3.2.46}$$

one finds that

$$r_{d,e} = \frac{(1 + r_{d0})(1 + r_{inf}) - (1 + r_{e0})(1 + r_{inf})}{(1 + r_{e0})(1 + r_{inf})}, \tag{3.2.47}$$

and after cancelling the factor $(1 + r_{inf})$, one sees that this is equal to

$$r_{d,e} = \frac{(1 + r_{d0}) - (1 + r_{e0})}{(1 + r_{e0})}. \tag{3.2.48}$$

The energy cost is the same, whether one uses real rates or market rates. The same holds for the maintenance cost term. The salvage term is also independent of this choice because

$$(1 + r_{inf})/(1 + r_d) = 1/(1 + r_{d0}).$$

By contrast, the ratio of capital recovery factors in the loan terms is not invariant, as one can see by inserting numerical values. For example, with r_{d0} = 0.08, r_{l0} = 0.12, and r_{inf} = 0.05 one finds, with N = 20 yr,

$$\frac{(A/P, r_{10}, N)}{(A/P, r_{d0}, N)} = 1.31 .$$

The corresponding market rates are $r_d = 0.134$ and $r_l = 0.176$, and the ratio becomes

$$\frac{(A/P, r_1, N)}{(A/P, r_d, N)} = 1.26 .$$

The difference arises from the fact that the cash flows are different. For a loan that is based on real rates, the annual payments are constant in constant currency, whereas for one based on market rates the payments are constant in inflating currency. Similarly, the depreciation terms can depend on inflation.

It follows that the two approaches, constant currency and inflating currency, yield identical results for equity investments ($f_l = 0$) without depreciation. But, if f_l or f_{dep} are not zero, there can be differences. Numerically the effect is not large, at most on the order of ten percent for inflation rates below ten percent [Dickinson and Brown, 1979]. The effect has opposite signs for the loan term and the depreciation term, leading to partial cancellation of the error.

3.2.4   Economic Evaluation Criteria

Life Cycle Savings

Having determined the life cycle cost of each relevant design alternative, one can select the "best," i.e., the one that offers all desirable features at the lowest life cycle cost. Frequently, one takes one design as reference and considers the difference between it and each alternative design. The difference is called *life cycle savings* relative to the reference case

$$S = -\Delta C_{life} \text{ with } \Delta C_{life} = C_{life} - C_{life,ref}. \tag{3.2.49}$$

Often the comparison can be quite simple because only those terms that are different between the designs need to be considered. For simplicity, we write the equations of this section only for an equity investment without tax. Then, the loan fraction f_l in Equation 3.2.42 is zero and most of the complications of that equation drop out. Of course, the concepts of life cycle savings, internal rate of return, and payback time are perfectly general, and tax and loan can readily be included.

A particularly important case is the comparison of two designs that differ only in capital cost and operating cost; the one that saves operating costs has higher initial cost (otherwise the choice would be obvious, without any need for an economic analysis). Setting f_l, τ_{cred}, C_{salv}, P_{max}, A_m, and $\tau = 0$ in Equation 3.2.42 and taking the difference between the two designs, one obtains the life cycle savings as

$$S = \frac{-\Delta Q p_e}{(A/P, r_{d,e}, N)} - \Delta C_{cap} \tag{3.2.50}$$

where
$\Delta Q = Q - Q_{ref} = $ difference in annual energy consumption,
$\Delta C_{cap} = C_{cap} - C_{cap,ref} = $ difference in capital cost, and
$r_{d,e} = (r_d - r_e)/(1 + r_e)$.

(If the reference design has higher consumption and lower capital cost, ΔQ is negative, and ΔC_{cap} is positive with this choice of signs.)

Example 14

Compared to a one-stage model, a two-stage absorption chiller is more efficient, but its first cost is higher. Find the life cycle savings of a two-stage model for the followings situation.

Given:

Required chiller capacity 1000 kW$_t$,

operating at 1000 hours per year full load equivalent.

A single stage absorption chiller has COP = 0.7 and costs 100 $/kW$_t$ (reference system),

while a two-stage absorption chiller has COP = 1.1 and costs 130 $/kW$_t$.

gas price p$_e$ = 4 $/GJ at the start,

escalating at r$_e$ = 0% (real),

discount rate r$_d$ = 8% (real).

Find:

Life cycle savings for the two-stage chiller.

Lookup value:

$(A/P,r_d,N) = 0.1019$.

Solution

The annual energy consumption is

1000 kW$_t$ × 1000 hr/COP = 1.0 MWh$_t$/COP. This equals

5.143 × 10^3 GJ$_{gas}$ for COP = 0.7 and

3.273 × 10^3 GJ$_{gas}$ for COP = 1.1;

thus the difference in energy cost is

$\Delta Q\ p_e = (3.273 - 5.143) \times 10^3\ GJ \times 4\ \$/GJ = -\$7481$ per year;

the difference in capital cost is

$\Delta C_{cap} = (130 - 100) \times 1000 = \$30,000$

From Equation 3.2.50 we find the life cycle savings

$S = -\Delta Q\ p_e/(A/P,r_{d,e},N) - \Delta C_{cap}$

$= \$7481/0.1019 - 30,000 = \$73,415 - 30,000 = \$43,415.$

Comment:

Even though the discount rate in this example is rather high (5% might be more appropriate), the life cycle savings are large. The investment certainly pays off.

Internal Rate of Return

The life cycle savings are the true savings if all the input is known correctly and without doubt. But future energy prices or system performance are uncertain, and the choice of the discount rate is not clear cut. An investment in a building or its equipment is uncertain, and it must be compared with competing investments that have their own uncertainties. The limitation of the life cycle savings approach can be circumvented if one evaluates the profitability of an investment by itself, expressed as a dimensionless rate. Then one can rank different investments in terms of profitability and in terms of risk. General business experience can serve as a guide for expected profitability as a function of risk level. Among investments of comparable risk the choice can then be based on profitability.

More precisely, the profitability is measured as *internal rate of return* r_r defined as that value of the discount rate r_d at which the life cycle savings S are zero:

$$S(r_d) = 0 \text{ at } r_d = r_r. \tag{3.2.51}$$

For an illustration, take the case of Equation 3.2.50 with energy escalation rate $r_e = 0$ (so that $r_{d,e} = r_d$), and suppose an extra investment ΔC_{cap} is made to provide annual energy savings $(-\Delta Q)$. The initial investment ΔC_{cap} provides an annual income from energy savings

$$\text{annual income} = (-\Delta Q)p_e. \tag{3.2.52}$$

If ΔC_{cap} were placed in a savings account instead, bearing interest at a rate r_r the annual income would be

$$\text{annual income} = (A/P, r_r, N)\Delta C_{cap}. \tag{3.2.53}$$

The investment behaves like a savings account whose interest rate r_r is determined by the equation

$$(A/P, r_r, N)\Delta C_{cap} = (-\Delta Q)p_e. \tag{3.2.54}$$

Dividing by $(A/P, r_r, N)$, we see that the right and left sides correspond to the two terms in Equation 3.2.50 for the life cycle savings

$$S = \frac{-\Delta Q p_e}{(A/P, r_d, N)} - \Delta C_{cap}, \tag{3.2.55}$$

and that r_r is indeed the discount rate r_d for which the life cycle savings are zero; it is the internal rate of return. Now the reason for the name is clear — it is the profitability of the project by itself, without reference to an externally imposed discount rate. When the explicit form of the capital recovery factor is inserted, one obtains an equation of the Nth degree, generally not solvable in closed form. Instead, one resorts to an iterative or graphical solution. (There could be up to N different real solutions, and multiple solutions can indeed occur if there are sign changes in the stream of annual cash flows. However, not to worry, the solution is unique for the case of interest here; an initial investment that brings a stream of annual savings.)

Example 15

What is the rate of return for Example 14?

Given:

$$S = \frac{-\Delta Q p_e}{(A/P, r_{d,e}, N)} - \Delta C_{cap},$$

with $r_{d,e} = r_d$ (because $r_e = 0$),

$(-\Delta Q) p_e = \$7481$, and

$\Delta C = \$30,000.$

Find: r_r

Solution

$S = 0$ for

$$(A/P,r_p,N) = \frac{-\Delta Q p_e}{\Delta C_{cap}}$$

$$= \frac{7481}{30,000} = 0.2494 \text{ with } N = 20.$$

By iteration one finds $r_r = 0.246 = 24.6\%$.

Payback Time

The *payback time* N_p is defined as the ratio of extra capital cost ΔC_{cap} to first year savings

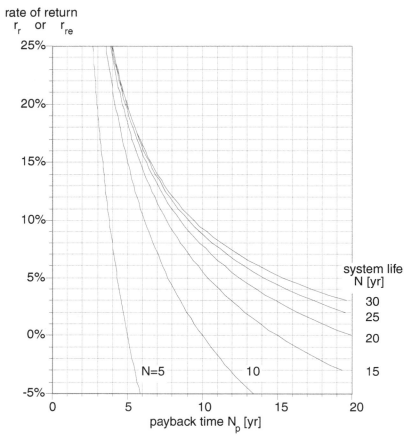

FIGURE 3.2.6 Relation between rate of return r_r, system life N, and payback time N_p. If r_e = escalation rate of annual savings, 0, the vertical axis is the variable $r_{r,e}$ from which r_r is obtained as $r_r = r_{r,e}(1 + r_e) + r_e$.

$$N_p = \frac{\Delta C_{cap}}{\text{first year savings}}. \qquad (3.2.56)$$

(The inverse of N_p is sometimes called *return on investment*.) If one neglects discounting, one can say that after N_p years the investment has paid for itself, and any revenue thereafter is pure gain. The shorter N_p, the higher the profitability. As selection criterion, the payback time is simple, intuitive, and obviously wrong because it neglects some of the relevant variables. Attempts have been made to correct for that by constructing variants such as a discounted payback time (by contrast to which Equation 3.2.56 is sometimes called simple payback time), but the resulting expressions become so complicated that one might as well work directly with life cycle savings or internal rate of return.

The simplicity of the simple payback time is, however, irresistible. When investments are comparable to each other in terms of duration and function, the payback time can give an approximate ranking that is sometimes clear enough to discard certain alternatives right from the start, thus avoiding the effort of detailed evaluation.

To justify the use of the payback time, recall Equation 3.2.54 for the internal rate of return and note that it can be written in the form

$$(A/P,r_p,N) = 1/N_p, \quad \text{or} \quad (P/A,r_p,N) = N_p. \tag{3.2.57}$$

The rate of return is uniquely determined by the payback time N_p and the system life N. This equation implies a simple graphical solution for finding the rate of return if one plots $(P/A,r_p,N)$ on the x-axis versus r_r on the y-axis as in Figure 3.2.6. Given N and N_p, one simply looks for the intersection of the line $x = N_p$ (i.e., the vertical line through $x = N_p$) with the curve labeled by N; the ordinate (y-axis) of the intersection is the rate of return r_r.

This graphical method can be generalized to the case where the annual savings change at a constant rate r_e. In that case, Equation 3.2.20 implies that the rate of return is replaced by $r_{r,e} = (r_r - r_e)/(1 + r_e)$, and Figure 3.2.6 yields $r_{r,e}$ rather than r_r. In other words Equation 3.2.57 becomes

$$(P/A,r_{r,e},N) = N_p, \text{ with } r_{r,e} = (r_r - r_e)/(1 + r_e). \tag{3.2.58}$$

The graph yields $r_{r,e}$ which is readily solved for

$$r_r = r_{r,e} (1 + r_e) + r_e. \tag{3.2.59}$$

In particular, if r_e is equal to the general inflation rate r_{inf}, then $r_{r,e}$ is the real rate of return $r_{r,0}$.

Example 16

Find payback time for Example 15 and check rate of return graphically.

Given:

   first year savings $(-\Delta Q)$ p_e = $7481,

   extra investment ΔC_{cap} = $30,000.

Find:

   N_p and r_p for r_e = 0 and 2%.

Solution

N_p = 30,000/7481 = 4.01 yr;

   it is independent of r_e.

Then r_r = 0.246 for r_e = 0, from Figure 3.2.6,

and r_r = 0.271 for r_e = 2%, from Equation 3.2.58.

Generally a real (i.e., corrected for inflation) rate of return above 10% can be considered excellent if there is low risk — a look at savings accounts, bonds, and stocks shows that it is difficult to find better. From the graph we see immediately that $r_{r,e}$ is above 10% if the payback time is shorter than 8.5 yr (6 yr), for a system life of 20 yr (10 yr). And $r_{r,e}$ is close to the real rate of return if the annual savings growth is close to the general inflation rate.

3.2.5   Complications of the Decision Process

In practice, the decision process is likely to bump into some obstacles. Suppose, for example, that the annual operating cost of a proposed office building can be reduced by $1000 if one installs daylight sensors and dimmers for the lights, at an extra cost of $2000. The payback time is only 2 years. It looks like an irresistible investment opportunity, with a rate of return well above 25%, as shown by Figure 3.2.6 (the exact value depends somewhat on lifetime and taxes, but that is beside the point). However, quite a few hurdles stand in the way.

First, to find out about this opportunity, the design engineer has to obtain the necessary information. Requesting catalogs, reading technical reviews of the equipment, and carrying out the calculations of cost and performance all take time and effort. Under the pressures of the job, the engineer may not be willing to spend the extra time or neglect other items that compete for his attention.

Suppose our engineer has done a good analysis and tries to convince the builder to spend the extra money. In the case of a speculative office building, the builder is likely to say "why should I pay a penny more, if only the future tenant will reap the benefit?" So, the design engineer is forced to aim for lowest first cost.

Even if the builder is willing to spend a bit more for efficiency, with hopes that the prospect of reduced energy bills will make it easier to find tenants, the decision is not obvious. Can the builder trust the claims of the sales brochure or the calculations of the engineer? Daylight controls are relatively new, and perhaps the builder has heard that some of the first models did not live up to expectations. If malfunctions reduce the productivity of the workers, the hassle and the costs could nullify the expected savings. So the builder may refuse to take what he or she perceives as an excessive risk. The threat of a liability suit is a potent inhibitor; that is why the building industry has a reputation for extreme conservatism.

This example illustrates the basic mechanisms that frequently prevent the adoption of efficient technologies:

- Lack of information or excessive cost of obtaining the information
- Purchase decision made by someone who does not have to pay the operating costs
- Uncertainty (about future costs, reliability, etc.)

Any one of these hurdles can be sufficient to reject an investment. In the above example and decision to reject the lighting controller, it looks as if the discount rate was higher than 25%. Quite generally, these mechanisms have the effect of raising the apparent discount rate or foreshortening the time horizon. The resulting decisions appear irrational: people do not spend as much for energy efficiency as would be optimal according to a life cycle cost analysis with the correct discount rate. In reality, this irrationality is but a reflection of other problems.

In the world of business, risk and uncertainty are pervasive — so much so that most decision makers insist on very short payback times, almost always less than five years and frequently less than two. However, this decision depends on the business and circumstances. There are industries like electric power plants, where profits are sure (albeit moderate); once a power plant has been built it is expected to run smoothly for at least thirty years. Here the discounts rates are low and payback times are longer than ten years. Governments, charged with the long-term welfare of its citizens, also tend to have a long time horizon.

What does all this mean for the HVAC engineer? The more a design choice involves unproven technology or is dependent on occupant behavior for proper functioning, the more risky it is. For example, a daylighting strategy that relies on manual control of shading devices by the occupants may not bring the intended savings because the occupants may not follow the intentions of the designer. Likewise, when considering a new design or a new piece of equipment without a track record, it is not irrational to demand short payback times.

By contrast, paying extra for an efficient boiler or chiller is a safe investment (assuming the equipment has a good reputation) because the occupant does not care how the heating or cooling is produced as long as the environment is comfortable, pleasant, and healthy. Also, the building will certainly be heated and cooled over its entire life. Here a life cycle cost analysis with the correct discount rate is certainly in order, and it would be shortsighted to insist on payback times below two years.

Finally, what about the problem of the builder or landlord who refuses to pay for measures that would only reduce the energy bill of the tenant or of a future owner? This difficulty is serious, indeed. In an ideal market, the information about reduced energy cost would translate itself as higher rent or resale value, but, in practice, this process is slow and inefficient (there is a *market failure*, in the language of the economists). This situation justifies energy-efficiency standards such as the ASHRAE Standard 90.1 and their enforcement by government regulations.

3.2.6 Cost Estimation

Capital Costs

For mass produced consumer products, such as cars or cameras, the capital cost (i.e., the purchase price) is easy to determine by looking at catalogs or newspaper ads or by calling the store. Even then there may be uncertainties — when you actually go the store, a discount may be offered on the spot to beat a competitor. Different prices can be found in different stores for identical products, not only because of differences in service or transportation but also because of the sheer difficulty of obtaining the price information.

And, of course, price is not the only criterion. Even more important, and more difficult to ascertain and compare, are the various characteristics of a good: the features it offers, the quality, the operating costs, among others. Economists have even coined a special term, *cost of information*, which demonstrates how universal is the difficulty of finding the pertinent information.

For HVAC equipment, the problems tend to be more complicated than for consumer goods. Transport and installation are important items in addition to the cost of the equipment at the factory. The determination of the cost can become a major undertaking, especially for complex or custom made systems. The capital cost of a system or component is known with certainty only when one has a firm contract from a vendor. Asking for bids on each design variation, however, is simply not feasible — the cost of information would become prohibitive.

The more a design engineer wants to be sure of coming close to the optimal design, the more he or she needs to learn about the details of the cost calculation. Information on costs is available from a number of sources, for example Boehm (1987). An important feature is the variation of the cost with size. Because of fixed costs and economies of scale, simple proportionality between cost and size is not the rule. But usually one can assume the following functional form over a limited range of sizes

$$C = C_r\left(\frac{S}{S_r}\right)^m \quad \text{for} \quad S_{min} < S < S_{max} \tag{3.2.60}$$

where
  C = cost at size S,
  C_r = cost at a reference size S_r, and
  m = exponent.

Typically, m is in the range of 0.5 to 1.0; exponents less than unity are a reflection of economies of scale. On a logarithmic plot, m is the slope of $\ln(C)$ versus $\ln(S)$. If m is not known, a value of 0.6 can be recommended as default. Table 3.2.A1 of the Appendix summarizes cost data for HVAC equipment in this form.

When interpreting such cost figures, one has to be careful about what is included and what is not. Is it the cost at the factory FOB (free on board, i.e., excluding transportation), the cost delivered to the site, or the cost installed? For items such as cool storage, the space requirements may impose additional costs. And finally, what are the specific features and how is the quality?

Costs change not only with general inflation but with the evolution of technology. The first models of a novel product tend to be expensive. Gradually, mass production, technological advance, and competition combine to drive the prices down. General inflation or increases in the cost of some input, for example energy, will push in the opposite direction. The resulting evolution of the price of the product may be difficult to predict. Cost reductions due to technological advance are more likely with products of high technology (e.g., energy management systems) than with mature products that cannot be miniaturized (e.g., fans and motors). In some cases, there is an improvement in a product rather than a reduction of its cost; variable speed motors, for instance, are more expensive than constant speed motors but allow better control or higher system efficiency.

Cost tabulations are based on sales or projects of the past, and they must be updated to the present by means of correction factors. For that purpose one could use general inflation (i.e., the CPI discussed in the earlier section on "The Effect of Time on the Value of Money"), but that is less reliable than specific cost indices for that class of equipment or that sector of the economy. The following two indices are particularly pertinent for buildings and HVAC equipment. One is the Marshall and Swift Equipment Cost Index, values of which are published regularly in *Chemical Engineering*. Another one is the construction cost index published by *Engineering News Record*, plotted in Figure 3.2.1(b).

It is important during the design process to have a realistic understanding of all the relevant costs, yet the effort of obtaining these costs should not be prohibitive. Konkel (1987) describes a method that seems to be a good compromise between these conflicting requirements. The basic idea is to group certain portions of a project into what is called *unit operations*. The components of the unit operations, called *unit assemblies*, are itemized, priced, and plotted by size of the unit operation. A boiler is an example of a unit operation; its unit assemblies include burner, air intake, flue, shutoff valves, piping, fuel supply, expansion tank, water makeup valves, and deaerator. Their sizes and costs vary with the size of the boiler. Once the size-price relations have been found for each component, the size-price relation for the boiler as a whole is readily derived. Knowing the size-price relation for the unit operations, the designer can estimate the total cost of a project and its design variations without too much effort.

Maintenance and Energy

Maintenance cost and energy prices may evolve differently from general inflation and from each other. It is instructive to correct energy prices for general inflation, as in Figure 3.2.2 where the prices of oil, gas, and electricity are shown in both current and constant dollars. One can see that some adjustments have occurred since the oil shocks of the 1970s.

What should we assume for the future? Projections of energy prices are published periodically by several organizations, for instance the American Gas Association and the National Institute of Standards and Technology (Lippiat and Ruegg, 1990). Most analysts predict real escalation rates in the range of 0 to 3%, averaged over the next two decades. This is based on the gradual exhaustion of cheap oil and gas reserves, and the fact that alternatives, i.e., coal, nuclear, and solar, are more expensive to utilize. Who knows? Further turmoil in the Middle East? What progress will be made in fusion and how will public acceptance of nuclear power evolve? How much can be saved by improved efficiency, and at what cost? What constraints will be imposed by environmental concerns?

Data on maintenance costs can be obtained, for example, from the *BOMA Experience Exchange Report* published annually by the Building Owners and Managers Association International (BOMA, 1987). Specifically for maintenance costs of HVAC equipment in office buildings, a succinct equation can be found in ASHRAE (1991). It states the annual cost A_M for maintenance, in dollars per floor area A_{floor}, in the form

$$\frac{A_M}{A_{floor}} = C_{base} + a\,n + h + c + d \qquad (3.2.61)$$

where

C_{base} = value for the base system (fire-tube boilers for heating, centrifugal chillers for cooling, and VAV for distribution, during first year),

n = age of equipment in years,

a = coefficient for age of equipment,

and the coefficients h, c, and d allow the adjustment to other systems.

TABLE 3.2.2   HVAC Maintenance Costs of Equation 3.2.61

	$/ft^2	$/m^2
C_{base}	0.3335	3.590
Coefficient **a** for **age**/yr	0.0018	0.019
Heating Equipment, coefficient **h**		
Water-tube boiler	0.0077	0.083
Cast iron boiler	0.0094	0.101
Electric boiler	–0.0267	–0.287
Heat pump	–0.0969	–1.043
Electric resistance	–0.1330	–1.432
Cooling Equipment, coefficient **c**		
Reciprocating chiller	–0.0400	–0.431
Absorption chiller (single stage)	0.1925	2.072
Water source heat pump	–0.0472	–0.508
Distribution System, coefficient **d**		
Single zone	0.0829	0.892
Multizone	–0.0466	–0.502
Dual duct	–0.0029	–0.031
Constant volume	0.0881	0.948
Tow-pipe fan coil	–0.0277	–0.298
Four-pipe fan coil	0.0580	0.624
Induction	0.0682	0.734

Note: Cost C_{base} of base system and coefficients for adjustment. Units of dollars per floor area, 1983 U.S. dollars.
Source: Adapted from ASHRAE (1991).

Numerical values for C_{base}, a, h, c, and d are listed in Table 3.2.2. These values are 1983 dollars. They still need to be adjusted to the year of interest by multiplication by the corresponding ratio of CPI (consumer price index) values, as explained in Section 3.2.2. In using this equation one should keep in mind that it is based on a survey of office buildings originally published in 1986. Extrapolation to other building types or newer technologies may introduce large and unknown uncertainties.

Example 17

Estimate the annual HVAC maintenance cost for an office building that has floor area 1000 m^2 and is n = 10 yr old in the year 2003. The system consists of an electric boiler, a reciprocating chiller, and a constant volume distribution system. Suppose the CPI is 180 in 2003.

Given:

A_{floor} = 1000 m^2,

n = 10 yr,

CPC_{2003}/CPI_{1983} = 180/100 = 1.80.

Lookup values:

From Table 3.2.2

C_{base} = 3.59, a = 0.019, h = –0.287, c = –0.431, d = 0.948 [in $\$_{1983}$/m^2].

Find: A_M

Solution

Using Equation 3.2.61

$$A_M = 1000 \text{ m}^2 \times (3.59 + 0.019 \times 10 - 0.287 - 0.431 + 0.948) \ \$_{1983}/\text{m}^2 = 4010 \ \$_{1983}/\text{m}^2$$

To convert to $\$_{2003}$ multiply by the CPI ratio

$$A_M = 4010 \ \$_{1983}/\text{m}^2 \times 1.80 = 7218 \ \$_{2003}/\text{m}^2.$$

3.2.7 Optimization

In principle, the process of optimizing the design of a building is simple — evaluate all possible design variations and select the one with the lowest life cycle cost. Who would not want to choose the optimum? In practice, it would be a daunting task to find the true optimum among all conceivable designs. The difficulties, some of which have already been discussed, are

- The enormous number of possible design variations (building configuration and materials, HVAC systems, types and models of equipment, control modes)
- Uncertainties (costs, energy prices, reliability, occupant behavior, future uses of building)
- Imponderables (comfort, convenience, aesthetics)

Fortunately, there is a certain tolerance for moderate errors, as we show below, which facilitates the job greatly because one can reduce the number of steps in the search for the optimum. Also, within narrow ranges, some variables can be suboptimized without worrying about their effect on others.

Some quantities are easier to optimize than others. Optimizing the heating and cooling equipment, for a given building envelope, is less problematic than trying to optimize the envelope — the latter touches on the imponderables of aesthetics and image.

It is instructive to illustrate the optimization process with a very simple example: the thickness of insulation on a wall. The annual heat flow Q across the insulation is

$$Q = A \, k \, D/t \tag{3.2.62}$$

where
A = area [m²],
k = conductivity [W/m·K],
D = annual degree-seconds [K·s], and
t = thickness of insulation [m].
The capital cost of the insulation is

$$C_{cap} = A \, t \, p_{ins} \tag{3.2.63}$$

with p_{ins} = price of insulation [\$/m³].
The life cycle cost is

$$C_{life} = C_{cap} + Q \frac{p_e}{(A/P, r_{d,e}, N)} \tag{3.2.64}$$

where p_e = first year energy price, and $r_{d,e}$ is related to discount rate and energy escalation rate as in Equation 3.2.19. We want to vary the thickness t to minimize the life cycle cost, keeping all the other quantities constant. (This model is a simplification that neglects fixed cost of insulation as well as possible feedback of t on D.) Eliminating t in favor of C_{cap}, one can rewrite Q as

$$Q = K/C_{cap} \tag{3.2.65}$$

with a constant

$$K = A^2 \, k \, D \, p_{ins}. \tag{3.2.66}$$

Then the life cycle cost can be written in the form

$$C_{life} = C_{cap} + P \, K/C_{cap} \tag{3.2.67}$$

where the variable

$$P = \frac{P_e}{(A/P, r_{d,e}, N)} \tag{3.2.68}$$

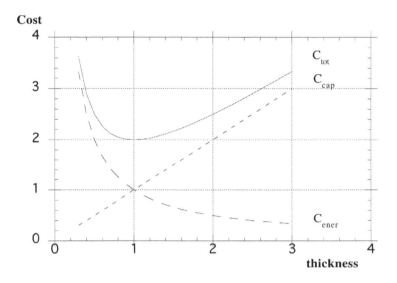

FIGURE 3.2.7   Optimization of insulation thickness. Insulation cost = C_{cap}, energy cost = C_{ener}, life cycle cost = C_{tot}.

contains all the information about energy price and discount rate. K is fixed, and the insulation investment C_{cap} is to be varied to find the optimum. C_{life} and its components are plotted in Figure 3.2.7. As t is increased, capital cost increases, energy cost decreases; C_{life} has a minimum at some intermediate value. Setting the derivative of C_{life} with respect to C_{cap} equal to zero yields the optimal value C_{cap0}

$$C_{cap0} = \sqrt{K \, P} \,. \tag{3.2.69}$$

Now an interesting question: what is the penalty for not optimizing correctly? In general, the following causes could prevent correct optimization:

- Insufficient accuracy of the algorithm or program for calculating the performance
- Incorrect information on economic data (e.g., the factor P in Equation 3.2.69)
- Incorrect information on technical data (e.g., the factor K in Equation 3.2.69)
- Unanticipated changes in the use of the building

Misoptimization would produce a design at a value C_{cap} different from the true optimum C_{cap0}. For the example of insulation thickness, the effect on the life cycle cost can be seen directly with the solid curve in Figure 3.2.7. For example, a ±10% error in C_{cap0} would increase C_{life} by only +1%. Thus, the penalty is not excessive for small errors.

This relatively large insensitivity to misoptimization is a feature much more general than the insulation model. As shown by Rabl (1985), the greatest sensitivity likely to be encountered in practice corresponds to the curve

$$\frac{C_{life,true}(C_{cap0,guess})}{C_{life,true}(C_{cap0,true})} = \frac{x}{1 + \log(x)}, \quad \text{(``upper bound'')} \tag{3.2.70}$$

Cost Penalty

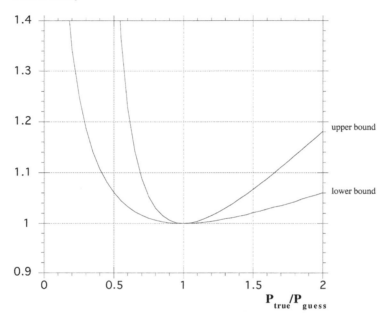

FIGURE 3.2.8   Life cycle cost penalty versus energy price ratio.

also shown in Figure 3.2.8, with the label "upper bound." Even here the minimum is broad; if the true energy price differs by ±10% from the guessed price, the life cycle cost increases only 0.4% (0.6%) over the minimum. Even when the difference in prices is 30%, the life cycle cost penalty is less than 8%.

Errors in the factor K (due to wrong information about price or conductivity of the insulation material) can be treated the same way because K and P play an entirely symmetric role in the above equations. Therefore, curves in Figure 3.2.8 also apply to uncertainties in other input variables.

The basic phenomenon is universal: any smooth function is flat at an extremum. The only question is how flat. For energy investments, that question has been answered with the curves of Figures 3.2.7 and 3.2.8. We can conclude that misoptimization penalties are definitely less then 1% (10%), when the uncertainties of the input variables are less than 10% (30%).

Nomenclature

A	annual payment
A_{life}	levelized annual cost
A_M	annual cost for maintenance [in first year $]
(A/P,r,N)	capital recovery factor
C	cost at size S
C_{cap}	capital cost [in first year $]
C_{life}	life cycle cost

$\mathbf{CPI}$	consumer price index
$\mathbf{C_r}$	cost at a reference size S_r
$\mathbf{C_{salv}}$	salvage value [in first year $]
$\mathbf{f_{dep}}$	present value of total depreciation, as fraction of C_{cap}
$\mathbf{f_{dep,n}}$	depreciation during year n, as fraction of C_{cap}
$\mathbf{f_l}$	fraction of investment paid by loan
$\mathbf{I_n}$	interest payment during nth year
$\mathbf{L}$	loan amount
$\mathbf{m}$	exponent of relation between cost and size of equipment
$\mathbf{N}$	system life [yr]
$\mathbf{n}$	year
$\mathbf{N_2}$	doubling time
$\mathbf{N_{dep}}$	depreciation period [yr]
$\mathbf{N_p}$	payback time [yr]
$\mathbf{N_l}$	loan period [yr]
$\mathbf{\underline{p}_e}$	energy price
$\mathbf{\bar{p}_e}$	levelized energy price
$\mathbf{(P/F,r,n)}$	present worth factor
$\mathbf{p_{dem}}$	demand charge [$/kW·month]
$\mathbf{p_{ins}}$	price of insulation [$/m^3]
$\mathbf{P_{int}}$	present value of interest payments
$\mathbf{P_{max}}$	peak demand [kW]
$\mathbf{P_n}$	principal during nth payment period n
$\mathbf{r_0}$	$(r - r_{inf})/(1 + r_{inf})$
$\mathbf{r_d}$	market discount rate
$\mathbf{r_{d,e}}$	$(r_d - r_e)/(1 + r_e)$
$\mathbf{r_{d,l}}$	$(r_d - r_l)/(1 + r_l)$
$\mathbf{r_{d,M}}$	$(r_d - r_M)/(1 + r_M)$
$\mathbf{r_{dif}(\omega_{ss},\omega)}$	I_{dif}/H_{dif}
$\mathbf{r_e}$	market energy price escalation rate
$\mathbf{r_{inf}}$	general inflation rate
$\mathbf{r_l}$	market loan interest rate
$\mathbf{r_M}$	market escalation rate for maintenance costs
$\mathbf{r_r}$	internal rate of return
$\mathbf{S}$	life cycle savings $(= -C_{life} + C_{life,ref})$
$\mathbf{S}$	size of equipment
$\mathbf{s}$	annual savings
$\mathbf{t}$	thickness of insulation [m]
$\mathbf{\tau}$	incremental tax rate
$\mathbf{\tau_{cred}}$	tax credit

The subscript $_0$ designates real growth rates r_0, related to the corresponding nominal (or market) rates r by $1 + r_{ann} = \exp(r_{cont})$. The subscript $_{ann}$ designates annual growth rates, related to the corresponding continuous rates (with subscript $_{cont}$).

References

ASHRAE TC1.8, 1985, Analysis of survey data on HVAC maintenance costs, ASHRAE Technical Committee 1.8 Research Project 382.

Boehm, R. F., 1987, *Design Analysis of Thermal Systems.* John Wiley & Sons, New York, NY.

BOMA, 1987, *BOMA Experience Exchange Report: Income/Expense Analysis for Office Buildings.* Building Owners and Managers Association International. Washington, DC.

Chemical Engineering, 2001, The M&S equipment cost index is published periodically in the journal *Chemical Engineering*.

Dickinson, W. C. and Brown, K. C., 1979, Economic analysis for solar industrial process heat, Report UCRL-52814, Lawrence Livermore Laboratory, Livermore, CA.

Economic Indicators, U.S. Government Printing Office, Washington, DC.

EIA, 1983, *Energy Review*, Energy Information Administration, U.S. Department of Energy, Washington, DC.

EN-R Construction Cost Index, 2000, *Engineering News Record*.

Jones, B. W., 1982, *Inflation in Engineering Economic Analysis*, Wiley Interscience, New York.

Konkel, J. H., 1987, *Rule of Thumb Cost Estimating for Building Mechanical Systems: Accurate Estimating and Budgeting Using Unit Assembly Costs*, McGraw-Hill, New York, NY.

Lippiat, B. C. and Ruegg, R. T., 1990, Energy prices and discount factors for life-cycle cost analysis 1990, Report NISTIR 85-3273-4, Annual Supplement to *NBS Handbook* 135 and *NBS Special Publication* 709, National Institute of Standards and Technology, Applied Economics Group, Gaithersburg, MD.

Rabl, A., 1985, Optimizing investment levels for energy conservation: individual versus social perspective and the role of uncertainty, *Energy Economics*, 259, 1985.

Riggs, J. L., 1996, *Engineering Economics*, McGraw-Hill, New York, NY.

Ross, M. and Williams, R. H., 1980, *Our Energy: Regaining Control*, McGraw-Hill, New York, NY.

USDOC, 2001, *Statistical Abstract of the United States*, U.S. Department of Commerce, Washington, DC.

Appendix

TABLE 3.2.A1 Typical Equipment Cost C as Function of Size S, in the Form $C = C_r (S/S_r)^m$

Component or System Description	m	C_r[k\$]	S_r	S Range, Units
Pumps, Fans, Blowers, and Compressors				
Pump, centrifugal, horizontal, 50 ft head,	0.26	2	10	$0.2\leftrightarrow16$ kW
ci, radial flow, no motor. FOB.	0.43	5.3	100	$16\leftrightarrow400$ kW
	0.34	3.2	0.5	$0.05\leftrightarrow30$ m³/min
Pump, same as above but with motor.	0.39	2.5	10	$1\leftrightarrow23$ kW
	0.58	7.5	100	$23\leftrightarrow250$ kW
	0.59	4.3	1	$0.04\leftrightarrow30$ m³/min
Pump, positive displacement, P_{in} = 150 psi,				
P_{out} <1000 psi, no motor, gears. ci. FOB.				
(P_{out} =5000 $\leftrightarrow$ 10000 psi, ×2.5).	0.52	4	10	$1\leftrightarrow70$ kW
Fan, centrifugal, radial bladed, 2.5 kPa,				
del., no motor.	0.78	5.3	10	$2\leftrightarrow100$ m³/s
del., with motor.	0.93	9.3	10	$2\leftrightarrow50$ m³/s
Fan, propeller with motor. FOB.	0.58	0.45	1	$0.5\leftrightarrow6$ m³/s
	0.36	1.5	10	$6\leftrightarrow50$ m³/s
Blower, centrifugal, 28 kPa, del., no motor.	0.61	160	30	$12\leftrightarrow70$ m³/s
With motor, drive: 1.6 × (no motor cost).				
Blower, rotary sliding vane, 275 kPa,				
del., no drive.	0.4	9.9	0.1	$0.01\leftrightarrow0.4$ m³/s
Compressor, centrifugal, <7000 kPa,				
del., with electric motor.	0.9	450	10^3	$2\leftrightarrow4000$ kW
Compressor, same as above, but no				
motor. FOB.	0.53	290	10^3	$(5\leftrightarrow40) \times 10^2$ kW
exit pressure (MPa) factors: 1.7, × 8; 6.9, ×1;				
14, ×1.15; 34, ×1.4; 48, ×1.5				
Electrical motors, AC, enclosed, fan cooled.	0.68	0.67	10	$1\leftrightarrow10$ HP
Other types	0.87	0.67	10	$10\leftrightarrow1000$ HP
Heat Exchangers				
(Costs can vary tremendously with material and flow design)				
Shell and tube, 150 psi., floating head, cs, 16 ft long, del.	0.71	21	100	$2\leftrightarrow2000$ m²
Factors: 400 psi, ×1.25; 1000 psi, ×1.55; 3000 psi, ×2.5;				
5000 psi, ×3.1.				
Plate and frame, CS frame, 304ss plates.	0.78	0.1	1	$100\leftrightarrow5000$ ft²
Air cooler, finned tube, cs, 150 psi, includes motor and fan. FOB.	0.8	70	280	$20\leftrightarrow2000$ m²
Heat recovery unit, for engine/generator.	0.45	0.95	1	$200\leftrightarrow1500$ kW
O&M costs=\$0.67/kWh.				
Heat recovery unit, water and firetube boilers (flue gas flow, scf/h).	0.75	110	200	$30\leftrightarrow2000$ scf/hr
Immersion heater, electric, FOB.	0.87	1.9	50	$10\leftrightarrow200$ kW
Cooling tower, induced or forced draft, approach temp.=5.5°C,	1.0	70	10	$4\leftrightarrow60$ m³/min
wet-bulb temp.=23.8°C, range=5.5°C, directly installed,				
all costs except foundations, water pumps, and distribution pipes.				
	0.64	560	100	$60\leftrightarrow700$ m³/min
In terms of cooling capacity	1.0	72	3.6×10^3	$10^3\leftrightarrow10^4$ kW

Factors to correct to other conditions: approach T, °C	Wet-bulb T, °C	Cooling range, °C
2.75, ×1.50	10, ×1.92	3, ×0.78
4, ×1.22	15, ×1.43	5, ×0.92
7, ×0.85	20, ×1.14	10, ×1.3
11, ×0.49	25, ×0.95	15, ×1.62
14, ×0.39	32, ×0.66	22, ×1.93

incl. foundations and basin × 1.7 to 3.0.

	m	C_r	S_r	S Range
*Water distribution to and from cooling tower, installed.	0.7	160	1	$0.1\leftrightarrow2$ m³/s
*Water treatment. Demineralizing, ion exchange, input 1330 ppm,	1.0	3200	0.1	$0.0004\leftrightarrow0.8$ m³/s
output 30–40 ppm solids, installed. (FOB, ×0.7)				
Factors to correct to other conditions: inlet feed: 1000 ppm, ×0.5;				
500 ppm, ×0.25; 200 ppm, ×0.18;100 ppm, ×0.13				
*Steam deaerator, cs, FOB.	0.78	67	1	$(0.05\leftrightarrow40) \times 10^5$ kg/hr

TABLE 3.2.A1 (*continued*)   Typical Equipment Cost C as Function of Size S, in the Form $C = C_r (S/S_r)^m$

Component or System Description	m	C_r[k$]	S_r	S Range, Units
Furnaces, Boilers, Heaters				
Firetube package boiler, FOB.	0.59	40	200	40↔800 HP
Stoker, economizer, dust collectors.	0.37	170	5000	$(2↔10) \times 10^3$ lb/hr
	0.56	500	25000	$(1↔5) \times 10^4$ lb/hr
Gas fired, del., 50–200 psi sat. steam.	0.64	16	10^3	$(0.2↔10) \times 10^3$ kg/hr
Water tube, FOB.	0.67	340	12	$(4↔40) ↔ 10^4$ lb/hr
Water heaters				
Gas-fired tank, FOB.	1.1	0.26	40	30 ↔ 100 gal
Electric heated tank, FOB.	1.0	0.26	50	30 ↔ 100 gal
Electric immersion, without tank, FOB.	0.87	1.3	50	10 ↔ 200 kW
Electric resistance heaters				
For household heating, cost = $550 + $40/kW.				
Waste heat steam boiler, unfired, 150 psi, del.	0.81	160	10¢	$(0.1↔10) \times 10^4$ kg/hr
Box-type furnace, 500 psi, cs, del.	0.75	144	12	10 ↔ 400 kW
Refrigerating Systems, Heat Pumps				
Air conditioners, Room, FOB.	0.8	1.2	2	0.33 ↔15 tons
Room, totally installed.	0.83	2.2	2	0.5 ↔15 tons
Maintenance costs, $/year.	0.38	0.2	2	0.33 ↔10 tons
Central chillers, vapor compression				
Reciprocating package, FOB.	0.5	13.6	50	10↔185 tons
Roof reciprocating, air-cooled condensor, FOB.	0.71	19.5	50	20↔85 tons
Centrifugal or screw compressor, FOB.	0.66	92	500	80↔2000 tons
O&M annual costs, reciprocating.	0.77	2	50	10↔185 tons
O&M annual costs, centrifugal or screw.	0.42	8	500	105↔2000 tons
Central chillers, LiBr absorption				
Single effect, installed.	0.66	160	500	100↔1400 tons
Double effect, installed.	0.7	230	500	400↔1200 tons
O&M, single or double effect, per year.	0.56	5.8	500	100↔1400 tons
Air-to-air heat pumps				
Equipment only.	0.86	2.4	3	1↔50 tons
Installed.	0.9	4.9	3	1↔50 tons
O&M per year.	0.5	0.3	3	1↔50 tons
Water-to-air heat pumps				
Equipment only.	0.64	1.65	3	1↔25 tons
Installed.	0.69	3.4	3	1↔25 tons
Maintenance, years 2–5.	0.5	0.3	3	1↔25 tons
Miscellaneous				
Storage tanks				
Vertical steel field erected tanks.	0.68	0.017	1	$10^3↔10^5$ gal
Carbon steel.	0.56	1.4	100	$100↔10^5$ gal
Large volume cs, floating roof.	0.78	385	2×10^6	$(2↔10) \times 10^6$ gal
Generally:				
Concrete $0.75-0.90/gal.				
Fiberglass $1.50/gal for 2000 gal size.				
Pipe type $1.00/gal.				
Pipe insulation (contractor price)				
Elastomer 3/4-in. thickness, $0.52/ft.				
Phenolic foam 1-in. thickness; $1.10/ft.				
Fiberglass 1-in. thickness, $0.70/ ft.				
Urethane $1.00/ft.				
Rule of thumb 10% of total mechanical costs.				

Notes: All costs adjusted for M&S index = 800. For fans and blowers, flow is in normal m3/s (at 0°C and 1.0 bar). Abbreviations used: ci=cast iron; cs=carbon steel; ss=stainless steel; m³/min denotes cubic meters per minute of feed flow; del.=delivered; sat.=saturated; S_r = Reference size, in same units as *S Range* values.

Source: Extracted from Appendix D of Boehm (1987) to which the reader is referred for further detail and references.

* Additional option.

4

HVAC Equipment and Systems

James B. Bradford
Schiller Associates, Inc.

John A. Bryant
Texas A&M University

Ellen M. Franconi
Schiller Associates, Inc.

Moncef Krarti
University of Colorado

Jan F. Kreider
Kreider & Associates, LLC

Dennis L. O'Neal
Texas A&M University

4.1 Heating Systems

Jan F. Kreider

This chapter discusses equipment used for producing heat from fossil fuels, electricity, or solar power. The emphasis is on design-oriented information, including system characteristics, operating efficiency, the significance of part load characteristics, and criteria for selecting from among the vast array of heat producing equipment available.

The heating plants discussed in this chapter are often called the *primary systems*. Systems intended to distribute heat produced by the primary systems are called *secondary systems* and include ducts and pipes, fans and pumps, terminal devices, and auxiliary components. Such secondary systems for heating and cooling are described in Chapter 4.3. The terms *primary* and *secondary* are equivalent to the terms *plant* and *system* used by some building analysts and HVAC system modelers.

The goal of this chapter is to have the reader understand the operation of various heat generation or transfer systems and their performance:

- Furnaces
- Boilers
- Heat pumps
- Heat exchangers
- Part load performance and energy calculations for each

The primary sources of heat for building heating systems are fossil fuels — natural gas, fuel oil, and coal. Under certain circumstances electricity is used for heat in commercial buildings although the economic penalties for so doing are significant. Solar radiation power can be converted to heat for commercial building applications, including perimeter zone heating and service water heating.

4.1.1 Natural Gas and Fuel Oil-Fired Equipment

This section describes fossil fuel-fired *furnaces and boilers* — devices which convert the chemical energy in fuels to heat. Furnaces are used to heat air streams that are in turn used for heating the interior of buildings. Forced air heating systems supplied with heat by furnaces are the most common type of residential heating system in the U.S. Boilers are pressure vessels used to transfer heat, produced by burning a fuel, to a fluid. The most common heat transfer fluid used for this purpose in buildings is water, in the form of either liquid or vapor. The key distinction between furnaces and boilers is that air is heated in the former and water is heated in the latter.

The fuels used for producing heat in boilers and furnaces include natural gas (i.e., methane), propane, fuel oil (at various grades numbered from 1 through 6), wood, coal, and other fuels including refuse-derived fuels. It is beyond the scope of this handbook to describe the design of boilers and furnaces or how they convert chemical energy to heat in detail. Rather we provide the information needed by HVAC designers for these two classes of equipment. Since boilers and furnaces operate at elevated temperatures (and pressures for boilers), they are hazardous devices. As a result, a body of standards has been developed to assure the safe operation of this equipment.

Furnaces

Modern furnaces use forced convection to remove heat produced within a furnace's firebox. There are many designs to achieve this; four residential classifications based on airflow type are shown in Figure 4.1.1. The *upflow* furnace shown in Figure 4.1.1a has a blower located below the firebox heat exchanger with heated air exiting the unit at the top. Return air from the heated space enters this furnace type at the bottom. The upflow design is used in full-sized mechanical rooms where sufficient floor-to-ceiling space exists for the connecting ductwork. This is the most common form of residential furnace.

Downflow furnaces (Figure 4.1.1b) are the reverse with air flowing downward as it is heated by passing over the heat exchanger. This design is used in residences without basements or in upstairs mechanical spaces in two-story buildings. *Horizontal* furnaces of the type shown in Figure 4.1.1c use a horizontal air flow path with the air mover located beside the heat exchanger. This design is especially useful in applications where vertical space is limited, such as in attics or crawl spaces of residences.

A combination of upflow and horizontal furnaces is available and is named the *basement* or low-boy furnace (Figure 4.1.1d). With the blower located beside the firebox, air enters the top of the furnace, is heated, and exits from the top. This design is useful in applications where head room is restricted.

The combustion side of the heat exchanger in gas furnaces can be at either atmospheric pressure (the most common design for small furnaces) or at super-atmospheric pressures produced by combustion air blowers. The latter are of two kinds, forced draft (blower upstream of combustion chamber) or induced draft (blower downstream of combustion chamber); furnaces with blowers have better control of parasitic heat losses through the stack. As a result, efficiencies are higher for such *power combustion furnaces*.

In addition to natural gas, liquefied propane gas (LPG) and fuel oil can be used as energy sources for furnaces. LPG furnaces are very similar to natural gas furnaces. The only differences between the two are energy content (1000 Btu/ft^3 for natural gas and 2500 Btu/ft^3 for propane) and supply pressure to the burner. Gas furnaces can be adapted for LPG use and vice versa in many cases. Fuel oil burner systems differ from gas burner systems owing to the need to atomize oil before combustion. The remainder of the furnace is not much different from a gas furnace, except that heavier construction is often used.

Other furnaces for special applications are also available. These include (1) unducted space heaters located within the space to be heated and relying on natural convection for heat transfer to the space;

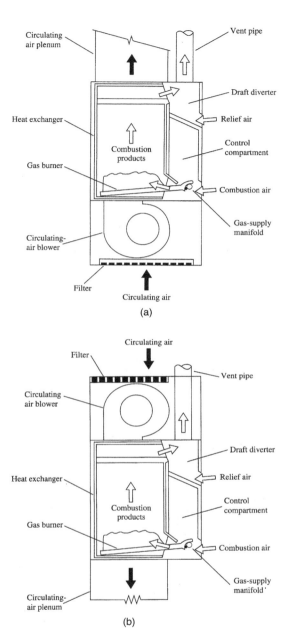

FIGURE 4.1.1 Examples of furnaces for residential space heating: (a) vertical, (b) downflow, (c) horizontal, and (d) low boy. (From Rabl, A. and Kreider J.F., *Heating and Cooling of Buildings*, McGraw-Hill, New York, NY, 1994. With permission.)

(2) wall furnaces attached to walls and requiring very little space; and (3) direct fired unit heaters used for direct space heating in commercial and industrial applications. Unit heaters are available in sizes between 25,000 and 320,000 Btu/hr (7 to 94 kW).

On commercial buildings, one often finds furnaces incorporated into *package units* (or "rooftop units") consisting of air conditioners and gas furnaces (or electric resistance coils). Typical sizes of these units range from 5 to 50 tons of cooling (18 to 175 kW) with a matched to 50% over sized furnace. Smaller

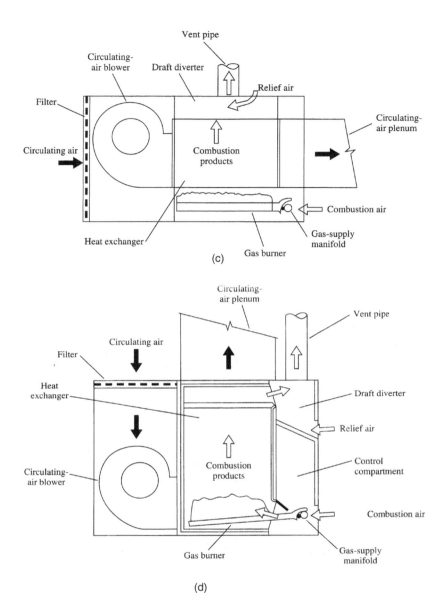

FIGURE 4.1.1 (*continued*)

units are designed to be used for a single zone in either the heating only or cooling only mode. Larger units above 15 tons (53 kW) can operate simultaneously in heating and cooling modes to condition several zones. In the heating mode, these commercial-sized package units operate with an air temperature rise of about 85°F (47 K).

Furnace Design and Selection for HVAC Applications

Selection of a furnace is straightforward once the fuel source and heat load (see Chapter 6.1) are known. The following factors must be accounted for in furnace sizing and type selection:

- *Design heat loss* of area to be heated — Btu/hr or kW
- Morning recovery capacity from night setback
- Constant *internal gains or waste heat recovery* that reduce the needed heat rating of a furnace

- *Humidification load* (see Chapters 4 and 7)
- *Fan and housing size* sufficient to accommodate air conditioning system
- *Duct heat losses* if heat so lost is external to the heated space
- *Available space* for furnace location

Residential furnaces are available in sizes ranging from 35,000 to 175,000 Btu/hr (10 to 51 kW). Commercial sizes range upward to 1,000,000 Btu/hr (300 kW).

Economic criteria including initial cost and life cycle operating cost must be considered using the techniques of Chapter 3.2 to make the final selection. Although high efficiency may cost more initially, it is often worthwhile to make the investment when the overall economic picture is considered. However, in many cases first cost is the primary determinant of selection. In these cases, the HVAC engineer must point out to the building owner or architect that the building lifetime penalties of using inexpensive but inefficient heating equipment are considerable, many times the initial cost difference.

The designer is advised to avoid the customary tendency to oversize furnaces. An oversized furnace operates at less efficiency than a properly sized one due to the penalties of part load operation. If a proper heat load calculation is done (with proper attention to the recognized uncertainties in infiltration losses and warm-up transients), only a small safety factor should be needed, for example, 10%. The safety factor is applied to account for heat load calculation uncertainties and possible future, modest changes in building load due to usage changes. Oversizing of furnaces also has other penalties, including excessive duct size and cost, along with poorer control of comfort due to larger temperature swings in the heated space.

Furnace Efficiency and Energy Calculations

The steady state efficiency η_{furn} is defined as the ratio of fuel supplied less flue losses, all divided by the fuel supplied:

$$\eta_{\text{furn}} = \frac{\dot{m}_{\text{fuel}}h_{\text{fuel}} - \dot{m}_{\text{flue}}h_{\text{flue}}}{\dot{m}_{\text{fuel}}h_{\text{fuel}}} \tag{4.1.1}$$

in which the subscripts identify the fuel input and flue gas exhaust mass flow rates and enthalpies h. Gas flows are usually expressed in ft^3/hr (l/s). To find the mass flow rate, one must know the density which in turn depends on the gas main pressure. The ideal gas law can be used for such calculations. Efficiency values are specified by the manufacturer at a single value of fuel input rate.

This instantaneous efficiency is of limited value in selecting furnaces owing to the fact that furnaces often operate in a cyclic, part load mode where instantaneous efficiency may be lower than that at peak operating conditions. Part load efficiency is low since cycling causes inefficient combustion, cyclic heating and cooling of furnace heat exchanger mass, and thermal cycling of distribution ductwork. A more useful performance index is the *Annual Fuel Utilization Efficiency* (AFUE) which accounts for other loss mechanisms over a season. These include stack losses (sensible and latent), cycling losses, infiltration, and pilot losses (ASHRAE Equipment, 1996). An ASHRAE standard (103-1982R) is used for finding the AFUE for residential furnaces.

Table 4.1.1 shows typical values of AFUE for residential furnaces. The table shows that efficiency improvements can be achieved by eliminating standing pilots, by using a forced draft design, or by condensing the products of combustion to recover latent heat normally lost to the flue gases. Efficiency can also be improved by using a vent damper to reduce stack losses during furnace off periods. Although this table is prepared using residential furnace data, it can be used for commercial-sized furnaces as well. Few data have been published for commercial systems because it has not been mandated by law as it has been for residential furnaces. The AFUE has the shortcoming that a specific usage pattern and equipment characteristics are assumed. The next section discusses a more accurate method for finding annual performance of heat-producing primary systems.

The AFUE can be used to find annual energy consumption directly from its definition below. The fuel consumption during an average year Q_{yr} is given by

TABLE 4.1.1   Typical Values of AFUE for Furnaces

Type of gas furnace	AFUE, %
1. Atmospheric with standing pilot	64.5
2. Atmospheric with intermittent ignition	69.0
3. Atmospheric with intermittent ignition and automatic vent damper	78.0
4. Same basic furnace as type 2, except with power vent	78.0
5. Same as type 4 but with improved heat transfer	81.5
6. Direct vent with standing pilot, preheat	66.0
7. Direct vent, power vent, and intermittent ignition	78.0
8. Power burner (forced-draft)	75.0
9. Condensing	92.5

Type of oil furnace	AFUE, %
1. Standard	71.0
2. Same as type 1 with improved heat transfer	76.0
3. Same as type 2 with automatic vent damper	83.0
4. Condensing	91.0

Source: From ASHRAE. With permission.

$$Q_{fuel, yr} = \frac{Q_{yr}}{AFUE} \qquad \text{MMBtu/yr (GJ/yr)} \qquad (4.1.2)$$

Where Q_{yr} is the annual heat load. Using this approach, it is a simple matter to find the savings one might expect, on the average, by investing in a more efficient furnace.

Example 1   *Energy saving using a condensing furnace*

A small commercial building is heated by an old atmospheric type, gas furnace. The owner proposes to install a new pulse type (condensing) furnace. If the annual heat load Q_{yr} on the warehouse is 200 GJ, what energy saving will the new furnace produce?

Assumptions: AFUE is an adequate measure of seasonal performance and furnace efficiency does not degrade with time.

Find: $\Delta Q_{fuel} = Q_{fuel,old} - Q_{fuel,new}$

Lookup values: AFUEs from Table 4.1.1

$$AFUE_{old} = 0.645 \qquad AFUE_{new} = 0.925$$

Solution

Equation 4.1.1 is used to find the solution. The energy saving is given by

$$\Delta Q_{fuel} = Q_{yr}\left(\frac{1}{AFUE_{old}} - \frac{1}{AFUE_{new}}\right)$$

Substituting the tabulated values for AFUE we have

$$\Delta Q_{fuel} = 200\left(\frac{1}{0.645} - \frac{1}{0.925}\right) = 93.9 \text{ GJ/yr}$$

The saving of energy using the modern furnace is substantial, almost equivalent to 50% of the annual heating load.

In addition to energy consumption, the designer must also be concerned with a myriad of other factors in furnace selection. These include:

- Air side temperature rise — affects duct design and air flow rate
- Air flow rate — affects duct design
- Control operation — for example, will night or unoccupied day/night setback be used or not? Is fan control by thermal switch or time delay relay?
- Safety issues — combustion gas control, fire hazards, high temperature limit switch

4.1.2 Boilers

A boiler is a device made from copper, steel, or cast iron to transfer heat from a combustion chamber (or electric resistance coil) to water in either the liquid phase, vapor phase, or both. Boilers are classified both by the fuel used and by the operating pressure. Fuels include gas, fuel oils, wood, coal, refuse-derived fuels, or electricity. This section focuses on fossil fuel fired boilers.

Boilers produce either hot water or steam at various pressures. Although water does not literally boil in hot water "boilers," they are called boilers, nevertheless. Steam is an exceptionally effective heat transport fluid due to its very large heat of vaporization and coefficient of heat transfer, as noted in Chapter 2.1.

Pressure classifications for boilers for buildings are

- *Low Pressure:* Steam boilers with operating pressures below 15 psig (100 kPa). Hot water boilers with pressures below 150 psig (1000 kPa); temperatures are limited to 250°F (120°C).
- *High Pressure:* Steam boilers with operating pressures above 15 psig (100 kPa). Hot water boilers with pressures above 150 psig (1000 kPa); temperatures are above 250°F (120°C).

Heat rates for steam boilers are often expressed in lb_m of steam produced per hour (or kW). The heating value of steam for these purposes is rounded off to 1000 Btu/lb_m. Steam boilers are available at heat rates of 50 to 50,000 lb_m of steam per hour (15 to 15,000 kW). This overlaps the upper range of furnace sizes noted in the previous section. Steam produced by boilers is used in buildings for space heating, water heating, and absorption cooling. Water boilers are available in the same range of sizes as are steam boilers: 50 to 50,000 MMBtu/hr (15 to 15,000 kW). Hot water is used in buildings for space and water heating.

Since the energy contained in steam and hot water within and flowing through boilers is very large, an extensive codification of regulations has evolved to assure safe operation. In the U.S. the ASME Boiler and Pressure Vessel Code governs construction of boilers. For example, the Code sets the limits of temperature and pressure on low pressure water and steam boilers listed above.

Large boilers are constructed from steel or cast iron. Cast iron boilers are modular and consist of several identical heat transfer sections bolted and gasketed together to meet the required output rating. Steel boilers are not modular but are constructed by welding various components together into one assembly. Heat transfer occurs across tubes containing either the fire or the water to be heated. The former are called *fire-tube boilers* and the latter *water-tube boilers*. Either material of construction can result in equally efficient designs. Small, light boilers of moderate capacity are sometimes needed for use in buildings. For these applications, the designer should consider the use of copper boilers.

Figure 4.1.2 shows a cross-section of a steam boiler of the type used in buildings.

Boiler Design and Selection for Buildings

The HVAC engineer must specify boilers based on a few key criteria. This section lists these but does not discuss the internal design of boilers and their construction. Boiler selection is based on the following criteria:

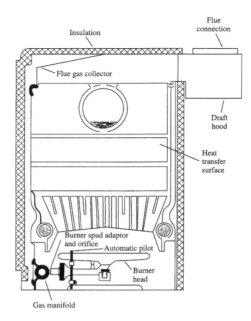

FIGURE 4.1.2   Boiler cross-sectional drawing showing burner, heat exchanger, and flue connection. (From Rabl, A. and Kreider J.F., *Heating and Cooling of Buildings*, McGraw-Hill, New York, NY, 1994. With permission.)

- *Boiler fuel* — type, energy content, heating value including altitude effects if gas fired (no effect for coal or fuel oil boilers).
- *Required heat output* — net output rating in MMBtu/hr (kW)
- Operating pressure and working fluid
- Efficiency and part load characteristics
- *Other* — space needs, control system, combustion air requirements, safety requirements, ASME code applicability

The boiler heat output required for a building is determined by summing the *maximum heating requirement* of all zones or loads serviced by the boiler during peak demand for steam or hot water and adding to that (1) parasitic losses including piping losses and (2) initial loop fluid warm-up. Simply adding all of the *peak heating unit capacities* of all the zones in a building can result in an oversized boiler since the zones do not all require peak heating simultaneously. The ratio of the total of all zone loads under peak conditions to the total heating capacity installed in a building is called the *diversity*.

Additional boiler capacity may be needed to recover from night setback in massive buildings. This transient load is called the *pickup* load and must be accounted for in both boiler and terminal heating unit sizing.

Boilers are often sized by their *sea-level input* fuel ratings. Of course, this rating must be multiplied by the applicable efficiency to determine the gross output of the boiler. In addition, if a gas boiler is not to be located at sea level, the effect of altitude must be accounted for in the rating. Some boiler designs use a forced draft burner to force additional combustion air into the firebox to offset part of the effect of altitude. Also, enriched or pressurized gas may be provided at high altitude so that the heating value per unit volume is the same as at sea level. If no accommodation to altitude is made, the output of a gas boiler drops by approximately 4% per 1000 ft (13% per km) of altitude above sea level. For example, a gas boiler located in Denver, Colorado (5000 ft, 1500 m) will have a capacity of only 80% of its sea level rating.

Table 4.1.2 shows the type of data provided by manufacturers for the selection of boilers for a specific project. Reading across the table, the fuel input needs are first tabulated for the 13 boiler models listed. The fifth column is the sea level boiler output at the maximum design heat rate. The next four columns

TABLE 4.1.2 Example of Manufacturer's Boiler Capacity Table

Boiler unit number, steam, or water (1)	IBR burner capacity			Gross IBR output, Btu/h (5)	Net IBR ratings			Net heat transfer area, ft² H₂O (9)	Boiler hp (10)	Net firebox volume, ft³ (11)	Stack gas volume, ft³/min (12)	Positive pressure in firebox in WG (13)	IBR chimney size vent dia., in (14)
	Light oil, gal/h (2)	Gas, kBtu/h (3)	Min. gas press. req'd., in WG (4)		Steam, ft²/h (6)	Steam, Btu/h (7)	Water, Btu/h (8)						
▲486°F●	6.30	882	5.5	720,000	2,250	540,100	626,100	4,175	21.5	11.02	395	0.34	10
▲586°F●	8.25	1,155	7.0	940,000	2,940	705,200	817,400	5,450	28.1	14.45	517	0.35	10
▲686°F●	10.20	1,428	5.5	1,160,000	3,625	870,200	1,008,700	6,725	34.6	18.08	640	0.35	10
▲786°F●	12.15	1,701	6.0	1,380,000	4,355	1,044,700	1,200,000	8,000	41.2	21.61	762	0.36	12
▲886°F●	14.10	1,974	5.0	1,600,000	5,115	1,227,900	1,391,300	9,275	49.6	25.14	884	0.37	12
▲986°F●	16.05	2,247	6.0	1,820,000	5,875	1,409,800	1,582,600	10,550	54.3	28.67	1,006	0.38	14
▲1086°F●	18.00	2,520	6.5	2,040,000	6,600	1,583,900	1,773,900	11,825	60.9	32.20	1,128	0.39	14
▲1186°F●	19.95	2,793	7.0	2,260,000	7,310	1,754,700	1,965,200	13,100	67.5	35.73	1,251	0.40	14
▲1286°F●	21.95	3,073	7.0	2,480,000	8,025	1,925,500	2,156,500	14,375	74.1	39.26	1,376	0.41	14
▲1386°F●	23.90	3,346	6.5	2,700,000	8,735	2,096,300	2,347,800	15,650	80.6	42.79	1,498	0.42	14
▲1486°F●	25.90	3,626	7.5	2,920,000	9,445	2,267,100	2,539,100	16,925	87.2	46.32	1,623	0.43	16
▲1586°F●	27.85	3,899	7.5	3,140,000	10,160	2,437,900	2,730,400	18,200	93.8	49.85	1,746	0.44	16
▲1686°F●	29.75	4,165	8.5	3,350,000	10,835	2,600,900	2,913,000	19,420	100.1	53.38	1,865	0.45	16

Note: 1 bhp = 33,475 Btu/h = 9.8 kW.
Source: From Rabl, A. and Kreider, J.F., *Heating and Coling of Buildings*, McGraw-Hill, New York, 1994. With permission.

convert the heat rate to steam and hot water production rates. The following column expresses heat rate in still a different way, using units of boiler horsepower (= 33,475 Btu/hr or 9.81 kW). The final four columns provide information needed for designing the combustion air supply system and the chimney.

A rule of thumb to check boiler selection in heating climates in the U.S. is that the input rating (columns two and three of the table, for example) in Btu/hr expressed on the basis of *per heated square foot* of building is usually in the range of one third to one fifth of the design temperature difference (difference between indoor and outdoor, winter design temperature). For example, if the design temperature difference for a 100,000 square foot building is 80°F, the boiler input would be expected to be in the range between 1.6 MMBtu/hr ([80/5] × 100,000 ft²) and 2.7 MMBtu/hr ([80/3] × 100,000 ft²). The difference between the two depends on the energy efficiency of the building envelope and its infiltration controls. Boiler efficiency also has an effect on this design check.

Proper control of boilers in response to varying outdoor conditions can improve efficiency and occupant comfort. A standard feature of boiler controls is the *boiler reset* system. Since full boiler capacity is needed only at peak heating conditions, better comfort control results if capacity is reduced with increasing outdoor temperature. Capacity reduction of zone hot water heating is easy to accomplish by simply reducing the water temperature supplied by the boiler. An example reset schedule might specify boiler water at 210°F at an outdoor temperature of –20°F and at 70°F a water temperature of 140°F. This schedule is called a 1:1 schedule since for every degree rise in outdoor temperature the boiler output drops by 1.0°F.

Auxiliary Steam Equipment

Steam systems have additional components needed to provide safety or adequate control in building thermal systems. This section provides an overview of the most important of these components including steam traps and relief valves.

Steam traps are used to separate both steam condensate and noncondensable gases from live steam in steam piping systems and at steam equipment. Steam traps "trap" or confine steam in heating coils, for example, while releasing condensate to be revaporized again in the boiler. The challenge in trap selection is to assure that the condensate and gases are removed promptly and with little to no loss of live steam. For example, if condensate is not removed from a heating coil, it will become waterlogged and have

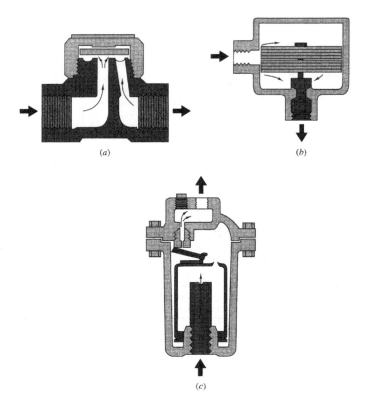

FIGURE 4.1.3 Steam traps. (a) Disc trap; (b) thermostatic trap; (c) mechanical, inverted bucket trap. (From Rabl, A. and Kreider J.F., *Heating and Cooling of Buildings*, McGraw-Hill, New York, NY, 1994. With permission.)

much reduced heating capacity. A brief description of the most common types of traps which should be used in HVAC applications follows.

Thermodynamic traps are simple and inexpensive. The most common type, the *disc trap*, is shown in Figure 4.1.3a. This type of trap operates on kinetic energy changes as condensate flows through and flashes into steam within the trap. Steam flashed (i.e., converted from hot liquid to vapor) from hot condensate above the disc holds the trap closed until the disc is cooled by cooler condensate. Steam line pressure then pushes the disc open. It remains open until all cool condensate has been expelled and hot condensate is again present and flashes again to close the valve. The disc action is made more rapid by the flow of condensate beneath the disc; the high velocities produce a low pressure area there in accordance with Bernoulli's equation and the disc slams shut. These traps are rugged and make a characteristic clicking sound, making operational checking easy. They can stick open if a particle lodges in the seat. This design has relatively high operating cost due to its live steam loss.

Thermostatic traps use the temperature difference between steam and condensate to control condensate flow. One type of thermostatic trap is shown in Figure 4.1.3b. The bimetal unit within the housing opens the valve as condensate cools, thereby allowing condensate to exit the trap. Significant subcooling of the condensate is needed to open the valve and operation can be slow. Other more complex designs have more rapid response and reduced need for subcooling. The bimetal element can be replaced with a bellows filled with an alcohol/water mixture permitting closer tracking of release setting as steam temperature changes. A trap that has a temperature/pressure characteristic greater than the temperature/pressure of saturated steam will lose live steam, whereas a trap with a T/p characteristic lying below the steam curve will build up condensate. The ideal trap has an opening T/p characteristic identical with the T/p curve of saturated steam.

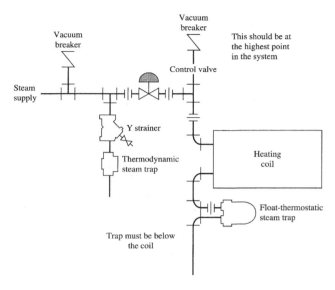

FIGURE 4.1.4 Piping arrangement for heating coil steam trap application. (From Rabl, A. and Kreider J.F., *Heating and Cooling of Buildings*, McGraw-Hill, New York, NY, 1994. With permission.)

Mechanical traps operate on the density difference between condensate and live steam to displace a float. Figure 4.1.3c shows one type of mechanical trap — the *inverted-bucket trap* — that uses an open, upside down bucket with a small orifice. Steam flowing with the condensate (that fills the housing outside of the bucket) fills the inverted bucket and causes it to float since the confined steam is less dense than the liquid water surrounding the bucket. Steam bleeds through the small hole in the bucket and condenses within the trap housing. As the bucket fills with condensate it becomes heavier and eventually sinks and opens the valve. Steam pressure forces condensate from the trap. The design of this trap continuously vents noncondensable gases, although the capacity for noncondensable gas flow (mostly air) rejection is limited by the size of the small hole in the top of the bucket. This hole is limited in size by the need to control parasitic steam loss through the same hole. Dirt can block the hole causing the trap to malfunction. The trap must be mounted vertically. An inverted bucket trap has significantly smaller parasitic live steam losses than the thermodynamic disc trap.

Steam traps are used to drain condensate from steam headers and from equipment where condensing steam releases its heat to another fluid. Steam piping is sloped so that condensate flows to a collecting point where it is relieved by the trap. At equipment condensate collection points, the trap is placed below the equipment where the condensate drains by gravity. Figure 4.1.4 shows a typical trap application for both purposes. The left trap drains the header, and the right trap drains the condensate produced in the heating coil.

Selection of traps requires knowledge of the condensate rejection rate (lb_m/hr, kg/s) and the suitability of various trap designs to the application. Table 4.1.3 summarizes the applications of the three types of traps discussed above (Haas, 1990).

The operating penalties for malfunctioning steam traps (clogged, dirty, or corroded) can dwarf the cost of a trap because expensive heat energy is lost if live steam is lost from malfunctioning traps. One of the first things to inspect in an energy audit of a new or existing steam system is the condition of the traps. For example, if steam is produced in a gas fired boiler of typical efficiency, a 0.25 in (0.64 cm) orifice in a steam trap will lose about $2000 worth of steam in a year. (The cost of gas in this example is $3.00/thousand ft^3 [$0.11/m^3], the usual units used by utilities; this converts to approximately $3.00/million Btu or $2.84/GJ).

A pressure *relief valve* is needed to control possible overpressure in boilers for safety reasons. Valves are specified by their ability to pass a given amount of steam or hot water at the boiler outlet condition. This

TABLE 4.1.3   Operating Characteristics of Steam Traps

System needs	Thermodynamics	Float-thermostatic	Inverted bucket
Maximum pressure (psig)	1740	465	2755
Maximum capacity (lb/h)	5250	100,000	20,500
Discharge	Hot	Hot	Hot
temperature, °F		(Close to saturated-steam temperature)	
Discharge	On/off	Continuous	On/off
Air venting	Good	Excellent	Fair
Dirt handling	Fair	Good	Good
Freeze resistance	Good	Poor	Poor
Superheat	Excellent	Poor	Fair
Waterhammer	Excellent	Fair	Excellent
Varying load	Good	Excellent	Good
Change in psi	Good	Excellent	Fair
Backpressure	Maximum 80%	Good	Good
Usual failure	Open	Closed/air vent open	Open

Source: From Rabl, A. and Kreider, J.F., *Heating and Coling of Buildings*, McGraw-Hill, New York, 1994. With permission.

dump rate can either be specified in units of mass per time or in units of energy flow per time. Pressure relief valves must be used wherever heat can be added to a confined volume of water. Water could become confined in the piping of an HVAC system, for example, if automatic control valves failed closed or if isolation valves were improperly closed by a system operator. Not only boilers must be protected, but also heat exchangers and water pipe lengths that are heated externally by steam tracing or solar heat. The volume expansion characteristics of water can produce tremendous pressures if heat is added to confined water. For example, water warmed by only 30°F (17°C) will increase in pressure by 1100 psi (7600 kPa). The method for sizing boiler relief valves is outlined in Wong (1989). The discharge from boiler relief valves must be piped to a drain or other location where injury from live steam will be impossible.

Combustion Calculations — Flue Gas Analysis

The combustion of fuel in a boiler is a chemical reaction and as such is governed by the principles of stoichiometry. This section discusses the combustion of natural gas (for our purposes assumed to be 100% methane) in boilers as an example of fuel burning for heat production. It also outlines how the flue gas from a boiler can be analyzed to ascertain the efficiency of the combustion process. Continuous monitoring of flue gases by a building's energy management system can result in early identification of boiler combustion problems. In a new building, one should test a boiler to determine its efficiency as installed and to compare output to that specified by the designer.

Combustion analysis involves using the basic chemical reaction equation and the known composition of air to determine the composition of flue gases. The inverse problem, finding the precombustion composition, is also of importance when analyzing flue gases. The chemical reaction for stoichiometric combustion of methane is

$$CH_4 + 2O_2 \rightarrow CO_2 + 2H_2O \tag{4.1.3}$$

Recalling that the molecular weights are

Hydrogen (H_2): 2
Methane (CH_4): 16
Oxygen (O_2): 32
Carbon Dioxide (CO_2): 44
Water (H_2O): 18

we can easily determine that 4.0 lb of oxygen per lb of methane are required for complete combustion. Since air is 23% oxygen by weight, 17.4 lb_m (or kg) of air per lb_m (or kg) of fuel are required, theoretically.

It is easy to show that on a volumetric basis (recall Avogadro's law which states that one mole of any gas at the same temperature and pressure occupies the same volume) the equivalent requirements are 2.0 ft³ of oxygen per ft³ of methane for complete combustion. This oxygen requirement is equivalent to 8.7 ft³ of air per ft³ of methane. A rule of thumb to check the preceding calculation is that 0.9 ft³ of air are required for 100 Btu of fuel heating value (about 0.25 m³ of air per MJ of heating value). For example, the heating value of natural gas is about 1000 Btu/ft³ requiring 9 ft³ of air according to the above rule. This compares well with the value of 8.7 ft³ previously calculated.

Combustion air is often provided in excess of this amount to guarantee complete combustion. Incomplete combustion yields toxic carbon monoxide (CO) in the flue gas. This incomplete combustion is to be avoided not only as energy waste but as air pollution. The amount of excess air involved in combustion is usually expressed as the *excess air fraction* $f_{exc\ air}$:

$$f_{exc\ air} = \frac{\text{air supplied} - \text{stoichiometric air}}{\text{stoichiometric air}} \tag{4.1.4}$$

In combustion calculations for gaseous fuels, the air amounts in Equation 4.1.4 are usually expressed on a volumetric basis, whereas for all other fuels a mass basis is used.

The amount of excess air provided is critical to the efficiency of a combustion process. Excessive air both reduces combustion temperature (reducing heat transfer rate to the working fluid) and results in excessive heat loss through the flue gases. Insufficient excess air results in incomplete combustion and loss of chemical energy in the flue gases. The amount of excess air provided varies with the fuel and with the design of the boiler (or furnace). Recommendations of the manufacturer should be followed. The optimum excess air fraction is usually between 10 and 50%.

Flue gas analysis is a method of determining the amount of excess air in a combustion process. This information can be used to find an approximate value of boiler efficiency. Periodic, regular analysis can provide a trend of boiler efficiency with time, indicating possible problems with the burner or combustion equipment in a boiler or furnace. Flue gas analysis is often expressed as the volumetric fraction of flue gases — oxygen, nitrogen, and carbon monoxide. If these three values are known, the excess air (%) can be found from (ASHRAE, 1997)

$$f_{exc\ air} = \frac{O_2 - 0.5CO}{0.264N_2 - (O_2 - 0.5CO)} \tag{4.1.5}$$

in which the chemical symbols represent the volume fractions in the flue gas analysis expressed in %. The following example indicates how this expression is used.

Example 2   Flue gas analysis

The volumetric analysis of flue gas from combustion of methane in a gas boiler is measured to be

    10.5% carbon dioxide
    3.2% oxygen
    86.3% nitrogen
    0% carbon monoxide

Find the amount of excess air. Is it within the recommended range suggested above?
Equation 4.1.5 will be used as follows:

$$f_{exc\ air} = \frac{3.2\% - (0.5 \times 0\%)}{0.264(86.3\%) - [3.2\% - (0.5 \times 0\%)]} = 0.163$$

The excess air is 16.3%, within the 10 - 50% range above.

TABLE 4.1.4 Stoichiometric and Excess Air Values of CO_2 for Combustion of Common Fossil Fuels

Type of fuel	Theoretical or maximum CO_2, %	CO_2 at given excess-air values		
		20%	40%	60%
Gaseous fuels				
Natural gas	12.1	9.9	8.4	7.3
Propane gas (commercial)	13.9	11.4	9.6	8.4
Butane gas (commercial)	14.1	11.6	9.8	8.5
Mixed gas (natural and carbureted water gas)	11.2	12.5	10.5	9.1
Carbureted water gas	17.2	14.2	12.1	10.6
Coke oven gas	11.2	9.2	7.8	6.8
Liquid fuels				
No. 1 and No. 2 fuel oil	15.0	12.3	10.5	9.1
No. 6 fuel oil	16.5	13.6	11.6	10.1
Solid fuels				
Bituminous coal	18.2	15.1	12.9	11.3
Anthracite	20.2	16.8	14.4	12.6
Coke	21.0	17.5	15.0	13.0

Source: From Rabl, A. and Kreider, J.F., *Heating and Coling of Buildings*, McGraw-Hill, New York, 1994. With permission.

The efficiency of a steam boiler can be found from field measurements by

$$\eta_{boil} = \frac{\dot{Q}_{steam}}{\dot{m}_{fuel}(HHV)} \tag{4.1.6}$$

where

$\dot{Q}_{steam}$ is the steam output rate, Btu/hr (kW)
$\dot{m}_{fuel}$ is the fuel supply rate, lbm/hr (kg/s)
HHV is the higher heating value of the fuel, Btu/lb (kJ/kg)

The previous discussion described the combustion of methane and at what rate air is to be supplied for proper combustion. Of course, many other fuels are used to fire boilers. Table 4.1.4 contains data which can be used to quickly estimate the excess air from a flue gas analysis for other fuels.

Coal and fuel oil contain carbon and hydrogen along with sulfur, the combustion of all of which produce heat. However, sulfur oxide formed during combustion is a corrosive acid if dissolved in liquid water. In order to avoid corrosion of boilers and stacks, liquid water must be avoided anywhere in a boiler by maintaining sufficiently high stack temperatures to avoid condensation. (Stainless steel stacks and fireboxes provide an alternative solution since they are not subject to corrosion, but they are very costly.) In addition, sulfur oxides are one of the sources of acid rain. Therefore, their emissions must be carefully controlled.

Boiler Efficiency and Energy Calculations

A simpler, overall efficiency equation can be used for boiler energy estimates if the steam rate required for using Equation 4.1.6 is not measurable:

$$\eta_{boil} = \frac{HHV - losses}{HHV} \tag{4.1.7}$$

The loss term includes five parts:

1. Sensible heat loss in flue gases
2. Latent heat loss in flue gases due to combustion of hydrogen
3. Heat loss in water in combustion air
4. Heat loss due to incomplete combustion of carbon
5. Heat loss from unburned carbon in ash (coal and fuel oil)

Boilers can be tested for efficiency in laboratories and rated in accordance with standards issued by the Hydronics Institute (formerly IBR, the Institute of Boiler and Radiator Manufacturers, and the SBI, the Steel Boiler Institute), the American Gas Association (AGA), and other industry groups. In addition to cast iron boiler ratings, IBR ratings are industry standards for baseboard heaters and finned-tube radiation. The ratings of the Hydronics Institute apply to steel boilers (IBR and SBI are trademarks of the Institute). SBI and IBR ratings apply to oil- and coal-fired boilers while gas boilers are rated by the AGA.

As noted earlier, efficiency under specific test conditions has very limited usefulness in calculating the annual energy consumption of a boiler due to significant drop off of efficiency under part load conditions. For small boilers (up to 300 MBtu/hr [90 kW]), the U.S. Department of Energy has set a method for finding the AFUE (defined in the earlier section on furnaces). Annual energy consumption must be known in order to perform economic analyses for optimal boiler selection.

For larger boilers, data specific to a manufacturer and an application must be used in order determine annual consumption. Efficiencies of fossil fuel boilers vary with heat rate depending on their internal design. If the boiler has only one or two firing rates, the continuous range of heat inputs needed to meet a varying heating load is achieved by cycling the boiler on and off. However, as load decreases, efficiency decreases since the boiler spends progressively more and more time in transient warm-up and cool-down modes during which relatively little heat is delivered to the load. At maximum load, the boiler cycles very little and efficiency can be expected to be near the rated efficiency of the boiler. Part load effects can reduce average efficiency to less than half of the peak efficiency. Of course, for an oversized boiler, average efficiency is well below the peak efficiency since it operates at part load for the entire heating season. This operating cost penalty persists for the life of a building, long after the designer who oversized the system has forgotten the error.

To quantify part load effects we define the *part-load ratio*, PLR (a quantity between 0 and 1), as

$$\text{PLR} \equiv \frac{\dot{Q}_o}{\dot{Q}_{o,\text{full}}} \tag{4.1.8}$$

where

$\dot{Q}_o$ is the boiler heat output at part load; Btu/hr, kW
$\dot{Q}_{o,\text{full}}$ is the rated heat output at full load; Btu/hr, kW

It is not practical to calculate from basic principles how boiler *input* depends on the value of PLR since the processes to be modeled are very complex and nonlinear. The approach used for boilers (and other heat producing equipment in this chapter) involves using test data to calculate the boiler input needed to produce an output $\dot{Q}_o$. If efficiency were constant and if there were no standby losses, the function relating input to output would merely be a constant, the efficiency. For real equipment the relationship is more complex. A common function used to relate input to output (i.e., to PLR) is a simple polynomial (at least for a boiler) such as

$$\frac{\dot{Q}_i}{\dot{Q}_{i,\text{full}}} = A + B(\text{PLR}) + C(\text{PLR})^2 + \dots \tag{4.1.9}$$

where

$\dot{Q}_i$ is the heat input required to meet the part load level quantified by PLR
$\dot{Q}_{i,\text{full}}$ is the heat input at rated full load on the boiler

The first term of Equation 4.1.9 represents standby losses, for example, those resulting from a standing pilot light in a gas boiler. Since the part load characteristic is not far from linear for most boilers, a quadratic or cubic expression is sufficient for annual energy calculations. Part load data are not as readily available as are standard peak ratings. If available, the data may often be in tabular form. The designer will need to make a quick regression of the data to find A, B, and C, using commonly available spreadsheet or statistical software in order to be able to use the tabular data for annual energy calculations as described in the following.

The remainder of this section examines a particularly simple application of a boiler — building space heating — to see how important part load effects can be. The *annual* energy input $Q_{i,\text{yr}}$ of a space heating boiler can be calculated from the following basic equation:

$$Q_{i,\text{yr}} = \int_{yr} \frac{\dot{Q}_o(t)}{\eta_{\text{boil}}(t)} dt \tag{4.1.10}$$

where

η_{boil} is the boiler efficiency — a function of time since the load on the boiler varies with time
$\dot{Q}_o(t)$ is the heat load on the boiler which varies with time as well, Btu/hr (kW)

The argument of the integral is just the instantaneous, time-varying energy input to the boiler. However, since the needed output — not the input — is usually known as a result of building load calculations, the form in Equation 4.1.10 is that practically used by designers. The time dependence in this expression is determined in turn by the temporal variation of load on the boiler as imposed by the HVAC system in response to climatic, occupant, and other time-varying loads.

A simple case is a boiler used solely for space heating. As described in Chapter 6.1, the heating load is determined first to order by the difference between indoor and outdoor temperature; all characteristics of the building's load and use remaining fixed. Therefore, the heat rate in Equation 4.1.10 is determined by outdoor temperature if the interior temperature remains constant. In this very simple case one could replace the integral in Equation 4.1.10 with a sum utilizing the bin approach as follows:

$$Q_{i,\text{yr}} = \sum_{j=0}^{N} = \frac{\dot{Q}_o(T_j) n_j(T_j)}{\eta_{\text{boil}}(T_j)} \tag{4.1.11}$$

where

$\eta_{\text{boil}}(T_j)$ is the efficiency of the boiler in a given ambient temperature bin j; the efficiency depends strongly but indirectly on ambient temperature T_j since the load, which determines PLR, depends on temperature.
$\dot{Q}_o(T_j)$ is the boiler load (i.e., building heat load) which depends on ambient temperature as described above.
$n_j(T_j)$ is the number of hours in the temperature bin j for which the value of efficiency and heat input apply.

This expression assumes that the sequence of hours during the heating season is of no consequence. The following example illustrates how bin weather data can be used to take proper account of part load efficiency of a boiler used for space heating.

TABLE 4.1.5 Summary of Solution for Example 3 — Boiler Energy Analysis

		Calculating Annual Boiler Energy Use					
Bin range, °F	Bin size, h	Heating load, kBtu/h	PLR	$\dot{Q}_i$, kBtu/h	Boiler effic.	Fuel used, MBtu	Net output, MBtu
55 to 60	762	0	0.00	875	0.000	667	0
50 to 55	783	500	0.07	1844	0.271	1444	391
45 to 50	716	1000	0.14	2750	0.364	1969	716
40 to 45	665	1500	0.21	3594	0.417	2390	997
35 to 40	758	2000	0.29	4375	0.457	3316	1516
30 to 35	713	2500	0.36	5094	0.491	3632	1782
25 to 30	565	3000	0.43	5750	0.522	3249	1695
20 to 25	399	3500	0.50	6344	0.552	2531	1396
15 to 20	164	4000	0.57	6875	0.582	1127	656
10 to 15	106	4500	0.64	7344	0.613	778	477
5 to 10	65	5000	0.71	7750	0.645	504	325
0 to 5	80	5500	0.79	8094	0.680	647	440
−5 to 0	22	6000	0.86	8375	0.716	184	132

Example 3   Annual energy consumption of a gas boiler accounting for part load effects

A gas boiler is used to supply space heat to a building. The load varies linearly with ambient temperature as shown in Table 4.1.5 below. If the efficiency of the boiler is 80% at peak, rated conditions, find the seasonal average efficiency, annual energy input, and annual energy output using the data in the table. The boiler input at rated conditions is 8750 MBtu/hr corresponding to −12.5°F temperature bin at which the load is 7000 MBtu/hr.

This boiler is turned off in temperature bins higher than 57.5°F roughly corresponding to the limit of the heating season; therefore, the standby losses above this temperature are zero.

The values of the coefficients in the part load characteristic Equation 4.1.9 are

$$A = 0.1$$

$$B = 1.6$$

$$C = -0.7$$

Load data shown in the third column of Table 4.1.5 exhibit the linearity of load with ambient temperature.

The part load characteristic equation is

$$\frac{\dot{Q}_i}{\dot{Q}_{i,full}} = 0.1 + 1.6(\text{PLR}) - 0.7(\text{PLR})^2 \tag{4.1.12}$$

The key equation for the solution is Equation 4.1.11. Since we are given the part load energy input equation instead of the efficiency at part load, this expression takes a somewhat simpler form

$$Q_{i,yr} = \sum_{j=0}^{j=N} \dot{Q}_i(T_j)n_j(T_j)$$

To find the total energy used by the boiler, one sums the "fuel used" column of Table 4.1.5 to find that 22,439 MMBtu are used to meet the annual load of 10,525 MMBtu. The ratio of these two numbers is the overall annual boiler efficiency, 47%. This value is 41% less than the peak efficiency of 80%. Clearly, one must take part load effects into account in annual energy calculations.

One method of avoiding the poor efficiency of this system would be to use two (or more) smaller boilers, the combined capacity of which would total the needed 7000 kBtu/h. Properly chosen, the smaller boilers would have operated more nearly at full load more of the time resulting in higher seasonal efficiency. However, smaller boilers cost more than one large boiler with the capacity equal to the total of the smaller boilers. Multiple boiler systems also offer standby security; if one boiler should fail, the other could carry at least part of the load. A single boiler system would entirely fail to meet the load.

The final decision must be made based on economics, giving proper account to the increased reliability of a system composed of several smaller boilers. Constraints are imposed on such decisions by initial budget, fuel type, owner and architect decisions, and available space.

4.1.3 Service Hot Water

Heated water is used in buildings for various purposes, including basins, sinks for custodial service, showers, and specialty services including kitchens in restaurants and the like. This section overviews service (or domestic) water heating methods for buildings. For details refer to ASHRAE (1999).

Water is heated by equipment that is either part of the space heating system, i.e., the boiler, or by a standalone water heater. The standalone equipment is similar to a small boiler except that water chemistry must be accounted for by use of anodic protection for the tank and by water softening in geographic areas where hardness can cause scale (lime) deposits in the water heater tank.

Two types of systems are used for water heating — *instantaneous* or *storage*. The former heats water on demand as it passes through the heater which uses either steam or hot water. Output temperatures can vary with this system unless a control valve is used on the heated water (not the heat supply) side of the water heater (usually a heat exchanger). Instantaneous water heaters are best suited to relatively uniform loads. They avoid the cost and heat losses of the storage tank but require larger and more expensive heating elements.

Storage type systems are used to accommodate varying loads or loads where large peak demands make it impractical to use instantaneous systems. Water in the storage tank is heated by an immersion steam coil, by direct firing, or by an external heat exchanger. In sizing this system, the designer must account for standby losses from the tank jacket and connected hot water piping. For any steam-based system cold supply water can be preheated using the steam condensate.

In order to size the equipment two items must be known:

- Hourly peak demand for the year — gal/hr, l/hr
- Daily consumption — gal/day, l/d

Of course, the volumetric usage rates must be converted to energy terms by multiplying by the specific heat and water temperature rise.

$$\dot{Q}_{water} = \dot{m}_{water} c_{water} (T_{set} - T_{source}) \qquad (4.1.13)$$

where

$\dot{Q}_{water}$ is the water heat rate, either on a daily or hourly basis; Btu/d or Btu/hr, kJ/d or W

$\dot{m}_{water}$ is the water mass flow rate, either on a daily or hourly basis, calculated from the volumetric flow listed above

$(T_{set} - T_{source})$ are the required hot water supply temperature and water source temperatures, respectively.

c_{water} is the specific heat of water

Table 4.1.6 summarizes water demands for various types of buildings, and Table 4.1.7 lists nominal set points of water heaters for several end uses. When using the lower settings in the table, the designer must be aware of the potential for *Legionella pneumophila* (Legionnaire's Disease). This microbe has been

TABLE 4.1.6 Hot Water Demands and Use for Various Types of Buildings

Type of building[a]	Maximum hour	Maximum day	Average day
Men's dormitories	3.8 gal (14.4 L)/student	22.0 gal (83.4 L)/student	13.1 gal (49.7 L)/student
Women's dormitories	5.0 gal (19 L)/student	26.5 gal (100.4 L)/student	12.3 gal (46.6 L)/student
Motels: No. of units[a]			
20 or less	6.0 gal (22.7 L)/unit	35.0 gal (132.6 L)/unit	20.0 gal (75.8 L)/unit
60	5.0 gal (19.7 L)/unit	25.0 gal (94.8 L)/unit	14.0 gal (53.1 L)/unit
100 or more	4.0 gal (15.2 L)/unit	15.0 gal (56.8 L)/unit	10.0 gal (37.9 L)/unit
Nursing homes	4.5 gal (17.1 L)/bed	30.0 gal (113.7 L)/bed	18.4 gal (69.7 L)/bed
Office buildings	0.4 gal (1.5 L)/person	2.0 gal (7.6 L)/person	1.0 gal (3.8 L)/person
Food service establishments:			
Type A: full-meal restaurants and cafeterias	1.5 gal (5.7 L)/max meals/h	11.0 gal (41.7 L)/max meals/h	2.4 gal (9.1 L)/average meals/h[c]
Type B: drive-ins, grilles, luncheonettes, sandwich and snack shops	0.7 gal (2.6 L)/max meals/h	6.0 gal (22.7 L)/max meals/h	0.7 gal (2.6 L)/average meals/day[c]
Apartment houses: No. of apartments			
20 or less	12.0 gal (45.5 L)/apt.	80.0 gal (303.2 L)/apt.	42.0 gal (159.2 L)/apt.
50	10.0 gal (37.9 L)/apt.	73.0 gal (276.7 L)/apt.	40.0 gal (151.6 L)/apt.
75	8.5 gal (32.2 L)/apt.	66.0 gal (250 L)/apt.	38.0 gal (144 L)/apt.
100	7.0 gal (26.5 L)/apt.	60.0 gal (227.4 L)/apt.	37.0 gal (140.2 L)/apt.
200 or more	5.0 gal (19 L)/apt.	50.0 gal (195 L)/apt.	35.0 gal (132.7 L)/apt.
Elementary schools	0.6 gal (2.3 L)/student	1.5 gal (5.7 L)/student	0.6 gal (2.3 L)/student[b]
Junior and senior high schools	1.0 gal (3.8 L)/student	3.6 gal (13.6 L)/student	1.8 gal (6.8 L)/student[b]

[a] The average usage of a U.S. residence is 60 gal/day (227 L/h) with a peak usage of 6 gal/h (22.7 L/h) (ASHRAE, 1987).
[b] Interpolate for intermediate values.
[c] Per day of operation. Temperature basis: 140°F.
Source: From ASHRAE. With permission.

TABLE 4.1.7 Representative Hot Water Use Temperatures

Use	Temperature °F	Temperature °C
Lavatory		
Handwashing	105	40
Shaving	115	45
Showers and tubs	110	43
Therapeutic baths	95	35
Commercial and institutional laundry	180	82
Residential dishwashing and laundry	140	60
Surgical scrubbing	110	43
Commercial spray-type dishwashing		
Single or multiple tank hood(s) or rack(s)		
Wash	150 min	65 min
Final rinse	180–195	82–90
Single tank conveyor		
Wash	160 min	71 min
Final rinse	180–195	82–90

Note: Table values are water use temperatures, not necessarily water heater set points.
Source: From ASHRAE. With permission.

traced to infestations of shower heads; it is able to grow in water maintained at 115°F (46°C). This problem can be limited by using domestic water temperatures in the 140°F (60°C) range.

Hot water can be supplied from a storage type system at the maximum rate

$$\dot{V}_{water} = \dot{V}_r + \frac{f_{useful}V_{tank}}{\Delta t} \qquad (4.1.14)$$

where

$\dot{V}_{water}$ is the volumetric hot water supply rate; gal/hr, l/s

$\dot{V}_r$ is the water heater recovery rate; gal/hr, l/s

f_{useful} is the useful fraction of the hot water in the tank before dilution lowers temperature excessively; 0.60–0.80

V_{tank} is the tank volume; gal,, (L)

Δt is the duration of peak demand, h, (s)

Jacket losses are assumed to be small.

4.1.4 Electric Resistance Heating

Electricity can be used as the heat source in both furnaces and boilers. Electric units are available in the full range of sizes from small residential furnaces (5 to 15 kW) to large boilers for commercial buildings (200 kW to 20 MW). Electric units have four attractive features:

- Relatively lower initial cost
- Efficiency near 100%
- Near zero part load penalty
- Flue gas vents are not needed

The cost of electricity (both energy and demand charges, see Chapters 3.1 and 3.2) diminishes the apparent advantage of electric boilers and furnaces, however. Nevertheless, they continue to be installed where first cost is a prime concern. However, the prudent designer should consider the overwhelming life cycle costs of electric systems. Electric boiler and furnace sizing follows the methods outlined above for fuel-fired systems. In many cases, the thermodynamic and economic penalties of pure resistance heating can be reduced by using electric heat pumps, the subject of the next section.

Environmental concerns must also be considered when considering electric heating. Low conversion and transmission efficiencies (relative to direct combustion of fuels for water heating) result in relatively higher CO_2 emissions. SO_2 emissions from coal power plants are also an environmental concern.

4.1.5 Electric Heat Pumps

A *heat pump* extracts heat from environmental or other medium temperature sources (such as the ground, groundwater, or building heat recovery systems), raises its temperature sufficiently to be of value in meeting space heating or other loads, and delivers it to the load. This chapter emphasizes heat pumps used for space heating with outdoor air or groundwater as the heat source.

Figure 4.1.5 shows a heat pump cycle on the *T-s* diagram; Figure 4.1.6 shows it on the more frequently used p-h diagram. It is exactly the refrigeration cycle discussed in Chapter 2. Vapor is compressed in step 3-4 and heat is extracted from the condenser in step 4–1. This heat is used for space heating in the systems discussed in this section. In step 1-2, isenthalpic throttling takes place to the low side pressure. Finally, heat extracted from the environment, or other low temperature heat source, is used to boil the refrigerant in the evaporator in step 2-3.

An ideal Carnot heat pump would appear as a rectangle in the *T-s* diagram. The coefficient of performance (COP) of a Carnot heat pump is given in Chapter 2.1 which shows there to be inversely proportional to the difference between the high and low temperature reservoirs. The same result applies generally to heat pumps using real fluids. Although the high side temperature (T_1) remains essentially fixed (ignoring for now the effect of night thermostat setback), the low side temperature closely tracks the widely varying outdoor temperature. As a result, the *capacity and COP of air source heat pumps are strong functions of outdoor temperature*. This feature of heat pumps must be accounted for by the designer since heat pump capacity diminishes as the space heating load on it increases. Heat pumps can be

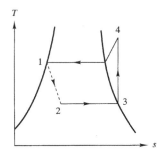

FIGURE 4.1.5   Heat pump T-s diagram showing four steps of the simple heat pump process. (From Rabl, A. and Kreider J.F., *Heating and Cooling of Buildings*, McGraw-Hill, New York, NY, 1994. With permission.)

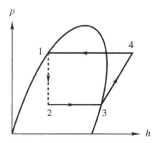

FIGURE 4.1.6   Heat pump p-h diagram showing four steps of the simple heat pump process. (From Rabl, A. and Kreider J.F., *Heating and Cooling of Buildings*, McGraw-Hill, New York, NY, 1994. With permission.)

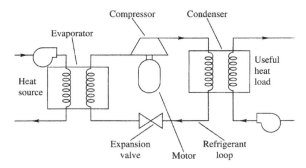

FIGURE 4.1.7   Liquid source heat pump mechanical equipment schematic diagram showing motor driven centrifugal compressor, condenser, and evaporator. (From Rabl, A. and Kreider J.F., *Heating and Cooling of Buildings*, McGraw-Hill, New York, NY, 1994. With permission.)

supplemented by fuel heat or electric resistance heating depending on the cost of each. Figure 4.1.7 shows a water source heat pump system that is not subject to outdoor temperature variations if groundwater or a heat recovery loop is used as the heat source.

The attraction of heat pumps is that they can deliver more thermal power than they consume electrically during an appreciable part of the heating season. In moderate climates requiring both heating and cooling, the heat pump can also be operated as an air conditioner, thereby avoiding the additional cost of a separate air conditioning system. Figure 4.1.8 shows one way to use a heat pump system for both heating and cooling by reversing flow through the system.

Typical Equipment Configurations

Heat pumps are available in sizes ranging from small residential units (10 kW) to large central systems (up to 15 MW) for commercial buildings. Large systems produce heated water at temperatures up to

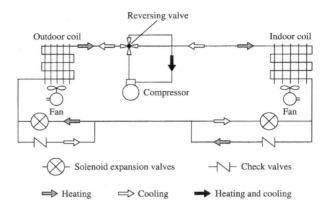

FIGURE 4.1.8   Air-to-air heat pump diagram. A reciprocating compressor is used. This design allows operation as a heat pump or an air conditioner by reversing the refrigerant flow. (From Rabl, A. and Kreider J.F., *Heating and Cooling of Buildings*, McGraw-Hill, New York, NY, 1994. With permission.)

220°F (105°C). Central systems can use both environmental and internal building heat sources. In many practical circumstances the heat gains in the core zones of a commercial building could satisfy the perimeter heat losses in winter. A heat pump could be used to efficiently condition both types of zones simultaneously.

Heat pumps require a compressor and two heat exchangers. In the energy bookkeeping that one does for heat pumps, the power input to the compressor is added to the heat removed from the low temperature heat source to find the heat delivered to the space to be heated. Increased heating capacity at low air source temperatures can be achieved by oversizing the compressor. To avoid part load penalties in moderate weather, a variable speed compressor drive can be used.

The outdoor and indoor heat exchangers use forced convection on the air side to produce adequate heat transfer coefficients. In the outdoor exchanger, the temperature difference between the boiling refrigerant and the air is between 10 and 25°F (6–14°C). If the heat source is internal building heat, water is used to transport heat to the heat pump evaporator and smaller temperature differences can be used.

A persistent problem with air source heat pumps is the accumulation of frost on the outdoor coil at coil surface temperatures just above the freezing point. The problem is most severe in humid climates; little defrosting is needed for temperatures below 20°F (−7°C) where humidities are below 60%. *Reverse cycle defrosting* can be accomplished by briefly operating the heat pump as an air conditioner (by reversing the flow of refrigerant) and turning the outdoor fan off. Hot refrigerant flowing through the outside melts the accumulated frost. This energy penalty must be accounted for in calculating the COP of heat pumps. Defrost control can be initiated either by time clock or, better, by a sensor measuring either the refrigerant condition (temperature or pressure) or, ideally, by the air pressure drop across the coil.

The realities of heat pump performance, as discussed above, reduce the capacity of real systems from the Carnot ideal. Figure 4.1.9 shows ideal Carnot COP values as a function of source temperature for a high side temperature of 70°F (21°C). The intermediate curve shows performance for a Carnot heat pump with real (i.e., finite temperature difference) heat exchangers. Finally, the performance of a real heat pump is shown in the lower curve. Included in the lower curve are the effects of heat exchanger losses, use of real fluids, compressor inefficiencies, and pressure drops. The COP of real machines is much lower (about 50%) than that for an ideal Carnot cycle with heat exchanger penalties.

The *energy efficiency ratio* EER is the ratio of heating capacity (Btu per hour) to the electric input rate (watts). EER thus has the units of Btu per watt-hour. The dimensionless COP is found from the EER by dividing it by the conversion factor 3.413 Btu/W · h.

Heat Pump Selection

The strong dependence of heat pump output on ambient temperature must be accounted for when selecting central plant equipment. If outdoor air is used as the heat source, peak heating requirements

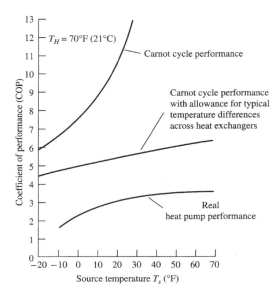

FIGURE 4.1.9 COP of ideal Carnot, Carnot (with heat exchanger penalty), and real heat pumps. (From Rabl, A. and Kreider J.F., *Heating and Cooling of Buildings*, McGraw-Hill, New York, NY, 1994. With permission.)

will invariably exceed the capacity of any economically feasible unit. Therefore, auxiliary heating is needed for such systems. Supplemental heat should always be added downstream of the heat pump condenser to ensure that the condenser operates at as low a temperature as possible, thereby improving the COP.

The amount of auxiliary heat needed and the type (electricity, natural gas, oil, or other) must be determined by an economic analysis and fuel availability (heat pumps are often used when fossil fuels are unavailable). The key feature of such analysis is the combined effect of part load performance and ambient source temperature on system output and efficiency. The following section shows how the temperature bin approach can be used for such an assessment. Figure 4.1.10 shows the conflicting characteristics of heat pumps and buildings in the heating season. As ambient temperature drops, loads increase, but heating capacity drops. The point at which the two curves intersect is called the *heat pump balance point*. To the left, auxiliary heat is needed; to the right, the heat pump must be modulated since excess capacity exists.

Recovery from night thermostat setback must be carefully thought out by the designer if an air source heat pump is used. A step change up in the thermostat setpoint on a cold winter morning will inevitably cause the auxiliary heat source to come on. If this heat source is electricity, high electrical demand charges may result, and the possible economic advantage of the heat pump will be reduced. One approach to avoid activation of the electric resistance heat elements uses a longer setup period with gradually increasing thermostat setpoint. A smart controller could control the start-up time based on known heat pump performance characteristics and outdoor temperature. Alternatively, fuel could be used as the auxiliary heat source. During building warmup, all outside air dampers remain closed as is common practice for any commercial building heating system.

Heat pump efficiency is greater if lower high side temperatures can be used. In order to produce adequate space heat in such conditions, a larger coil may be needed in the air stream. However, if the coil is sized for the cooling load, it will nearly always have adequate capacity for heating. In such a case, adequate space heat can be provided at relatively low air temperatures, 95–110°F (35–43°C). Table 4.1.8 summarizes advantages and disadvantages of air and water source heat pumps.

Controls for heat pumps are more complex than for fuel-fired systems since outdoor conditions, coil frosting, and heat load must all be considered. In addition, to avoid excessive demand charges, the controller must avoid *coincident* operation of resistance heat and the compressor at full capacity (attempting to meet a large load on a cold day).

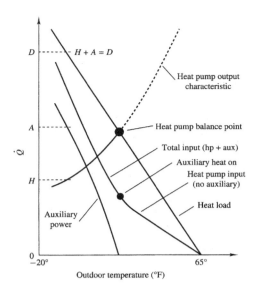

FIGURE 4.1.10 Graph of building heat load, heat pump capacity, and auxiliary heat quantity as a function of outdoor air temperature for a typical, air source residential heat pump. (From Rabl, A. and Kreider J.F., *Heating and Cooling of Buildings*, McGraw-Hill, New York, NY, 1994. With permission.)

Table 4.1.8   Advantages and Disadvantages of Air and Water Source Heat Pumps

Type	Advantages	Disadvantages
Air source	Indoor distribution permits air conditioning and humidity control Outdoor air source readily available Simple installation Least expensive Established commercial technology	Defrost required Low capacity at cold outdoor temperature Lower efficiency because of large evaporator $\Delta T \approx 30°F$ Indoor air distribution temperature must be high for comfort reasons Reliability at low temperature is only fair, due to frosting effects Must keep evaporator clear of leaves, dirt, etc.
Water source	Multiple family and commercial installation as central system In commercial installations, good coupling to cooling towers No refrigerant reversal needed; reverse water flow instead	Needs water source at useful temperature Efficiency penalty due to space heat exchanger ΔT

Part Load Performance

As discussed in detail above, air source heat pumps are particularly sensitive to the environment. Earlier we examined how the performance of boilers changed when the heating load changed with outdoor temperature. The COP of an air source heat pump has an even greater dependence on environmental conditions. This section provides an example to illustrate the magnitude of the effect. Since air source heat pumps are often used on residences, it provides a residential scale example.

Example 4   Seasonal Heat Pump Performance Using the Bin Method

A residence in a heating climate has a total heat transmission coefficient K_{tot} = 650 Btu/(h · °F) (343 W/K). An air source heat pump with a capacity of 39,900 Btu/h (11.7 kW) at 47°F (8.3°C) (standard rating point in the U.S.) is to be evaluated. Find the heating season electrical energy usage, seasonal COP (often called the *seasonal performance factor*, SPF), and energy savings relative to electric

TABLE 4.1.9 Heat Pump and Building Load Data – Example 4

Bin temp. °F	Heating load Btu/h	Heat pump COP	Heat pump output Btu/h	Heat pump input Btu/h	Auxiliary power Btu/h	Heating system COP
62	1,950	2.64	1,950	739	0	2.64
57	5,200	2.68	5,200	1,940	0	2.68
52	8,450	2.64	8,450	3,201	0	2.64
47	11,700	2.63	11,700	4,449	0	2.63
42	14,950	2.50	14,950	5,980	0	2.50
37	18,200	2.39	18,200	7,615	0	2.39
32	21,450	2.23	21,450	9,619	0	2.23
27[a]	24,700	2.07	24,700	11,932	0	2.07
22	27,950	1.97	25,100	12,741	2,850	1.79
17	31,200	1.80	22,400	12,444	8,800	1.47
12	34,450	1.70	19,900	11,706	14,550	1.31
7	37,700	1.54	17,600	11,429	20,100	1.20
2	40,950	1.39	15,400	11,079	25,550	1.12
−3	44,200	1.30	13,500	10,385	30,700	1.08
−8	47,450	1.17	11,700	10,000	35,750	1.04

[a] Heat pump balance point.

resistance heating. Use the bin data and heat pump performance data given in Table 4.1.9. The house heating base temperature is 65°F (18.3°C) accounting for internal gains. Figure 4.1.11 shows the energy flows as a function of outdoor temperature.

The preceding table includes these data in order:

1. Center point of temperature bin, T_{bin}
2. Heating demand, $\dot{Q} = K_{tot}(65°F - T_{bin})$
3. COP from manufacturer's data, a function of temperature, including defrost
4. Heat pump output; above the balance point, $\dot{Q}$; below the heat pump balance point, manufacturer's data
5. Heat pump input, the heat pump output divided by COP
6. Auxiliary power; the positive difference, if any, between $\dot{Q}$ and heat pump output
7. *Heating system* COP given by $\dot{Q}$ divided by the sum of auxiliary power and heat pump input

The energy calculations are summarized in Table 4.1.10. The bottom line in the table contains energy totals. With the heat pump, the total electricity requirement is 48.7 MBtu/yr (51.4 GJ/yr). If pure resistance heating were used, the total electricity requirement would be 98.36 MBtu/yr (103.8 GJ/yr).

The SPF for the heat pump is the seasonal output divided by the seasonal input to the heat pump:

$$SPF_{hp} = \frac{Q_{o,yr}}{Q_{i,yr}} = \frac{90.33 \text{ MBtu}}{40.66 \text{ MBtu}} = 2.22$$

The SPF for the heating system is the seasonal heat load divided by the seasonal input to the heat pump and the auxiliary heater:

$$SPF_{sys} = \frac{Q_{o,yr}}{Q_{i,yr} + Q_{i,aux,yr}} = \frac{98.36 \text{ MBtu}}{(40.66 + 8.03) \text{ MBtu}} = 2.02$$

The advantage of a constant temperature heat source is apparent from this example. If ground water or building exhaust air (both essentially at constant temperature) were used as the heat source rather than outdoor air, there would not be a drop off in capacity as with the outdoor air source device just when heat is most needed.

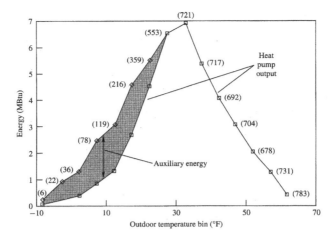

FIGURE 4.1.11   Heat pump energy use for the bin method example. The numbers at each bin temperature indicate the number of hours of occurrence in each bin. (From Rabl, A. and Kreider J.F., *Heating and Cooling of Buildings*, McGraw-Hill, New York, NY, 1994. With permission.)

TABLE 4.1.10   Heat Pump Energy Calculations – Example 4

Bin temp. °F	Bin hours h	Heating energy MBtu	Heat pump output MBtu	Heat pump input MBtu	Aux. heat input MBtu	Total input MBtu
62	783	1.53	1.53	0.58	0.00	0.58
57	731	3.80	3.80	1.42	0.00	1.42
52	678	5.73	5.73	2.17	0.00	2.17
47	704	8.24	8.24	3.13	0.00	3.13
42	692	10.35	10.35	4.14	0.00	4.14
37	717	13.05	13.05	5.46	0.00	5.46
32	721	15.47	15.47	6.94	0.00	6.94
27[a]	553	13.66	13.66	6.60	0.00	6.60
22	359	10.03	9.01	4.57	1.02	5.60
17	216	6.74	4.84	2.69	1.90	4.59
12	119	4.10	2.37	1.39	1.73	3.12
7	78	2.94	1.37	0.89	1.57	2.46
2	36	1.47	0.55	0.40	0.92	1.32
-3	22	0.97	0.30	0.23	0.68	0.90
-8	6	0.28	0.07	0.06	0.21	0.27
Total		98.36	90.33	40.66	8.03	48.70

[a] Heat pump balance point.

4.1.6   Low Temperature Radiant Heating

Heating systems in many parts of the world use warmed floors and/or ceilings for space heating in buildings. Although this system is unusual in the U.S., the good comfort and quiet operation provided by this approach make it worth considering for some applications. In Europe it is far more common. Radiant systems are well suited to operation with heat pump, solar, and other low temperature systems. This section discusses the principles of low temperature space heating. This form of heating is distinct from high temperature radiant heating using either electricity or natural gas to provide a high temperature source from which radiation can be directed for localized heating.

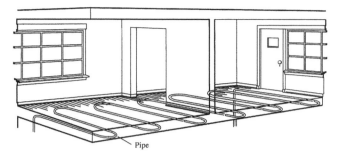

FIGURE 4.1.12 Residential radiant floor heating system. (From Rabl, A. and Kreider J.F., *Heating and Cooling of Buildings*, McGraw-Hill, New York, NY, 1994. With permission.)

Figure 4.1.12 shows how a radiant floor might be configured in a residence. The same concept can also be used in the ceiling in both residential and commercial buildings. The term *radiant* is a misnomer since between 30% (ceilings) and 50% (floors) of heat transferred from radiant panels is actually by convection. However, we will use the industry's nomenclature for this heating system.

The radiation heat output of radiant panels is given by the Stefan-Boltzmann equation as discussed in Chapter 2.1.

$$\dot{Q} = \epsilon_{eff} F_{h,u} \sigma (T_h^4 - T_u^4) \tag{4.1.15}$$

where

ϵ_{eff} is $1/[(1/\epsilon_h + 1/\epsilon_u) - 1]$, the effective emittance of the space, and the subscripts h and u refer to the unheated and heated (by radiant panels) surfaces of the space; the effective emittance is approximately 0.8.

$F_{h,u}$ is the view factor between the heating surface and the unheated surfaces; its value is 1.0 in the present case.

T_h is the heating surface temperature.

T_u is the mean of the unheated surface temperatures.

σ is the Stefan-Boltzmann constant (see Chapter 2).

Convection from the heating surface can be found using the standard free convection expressions in Chapter 2.1.

The designer's job is to determine the panel area needed, its operating temperature, the heating liquid flow rate, and construction details. The panel size is determined based on standard heat load calculations (Chapter 6.1). Proper account should be made of any losses from the back of the radiant panels to unheated spaces. Panel temperatures should not exceed 85°F (29.5°C) for floors and 115°F (46°C) for ceilings.

Water temperatures are typically 120°F (49°C) for floors and up to 155°F (69°C) for ceilings. Panels can be piped in a series configuration if pipe runs are not excessively long (the final panels in a long series run will not perform up to specifications due to low fluid temperatures). Long series loops also have excessively high pressure drops. If large areas are to be heated, a combination of series and parallel connections can be used. Manufacturers can advise regarding the number of panels that can be connected in series without performance penalties.

If radiant floors are to be built during building construction rather than using prefabricated panels in ceilings, the following guidelines can be used. Tubing spacing for a system of the type shown above should be between 6 and 12 in (15 and 30 cm). The tubing diameter ranges between 0.5 and 1.0 in (0.6 and 2.5 cm). Flow rates are determined by the rate of heat loss from the panel, which in turn depends on the surface temperature and hence the fluid temperature. This step in the design is iterative. Panel design follows this process:

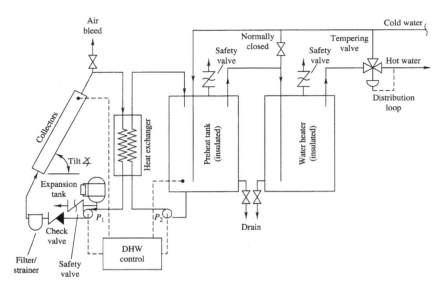

FIGURE 4.1.13 Solar water heating system including collectors, pumps, heat exchanger, and storage tanks along with piping and ancillary fittings. Collectors are tilted up from the horizontal at a fixed angle roughly equal to the local latitude. (From Rabl, A. and Kreider J.F., *Heating and Cooling of Buildings*, McGraw-Hill, New York, NY, 1994. With permission.)

- Determine room heat load.
- Decide on location of panels (roof or floor).
- Find panel heat flux including both radiation and convection contributions at 80°F (27°C) for floor panels and 110°F (43°C) for ceiling panels.
- Divide heat load by heat flux to find needed panel area.
- If panel area exceeds available floor or ceiling area, raise panel temperature (not exceeding temperatures noted earlier) and repeat steps 3 and 4.
- If the panel area is still insufficient, consider both floor and ceiling panels.

Control of radiant heating systems has proven to be a challenge in the past due to the large time constant of these systems. Both under- and overheating are problems. If the outdoor temperature drops rapidly, this system will have difficulty responding quickly. On the other hand, after a morning warmup followed by high solar gains on a sunny winter day, the radiant system may overshoot. The current generation of "smart" controls should help improve the comfort control of these systems.

4.1.7 Solar Heating

Solar energy is a source of low temperature heat that has selected applications to buildings. Solar water heating is a particularly effective method of using this renewable resource since low to moderate temperature water (up to 140°F, 60°C) can be produced by readily available, flat plate collectors (Goswami, Kreider, and Kreith, 2000).

Figure 4.1.13 shows one system for heating service water for residential or commercial needs using solar collectors. The system consists of three loops; it is instructive to describe the system's operation based on these three.

First, the collector loop (filled with a nonfreezing solution if needed) operates whenever the DHW controller determines that the collector is warmer, by a few degrees, than the storage tank. Heat is transferred from the solar-heated fluid by a counterflow or plate heat exchanger to the storage tank in the second loop of the system. Storage is needed since the availability of solar heat rarely matches the water heating load. The check valve in the collector loop is needed to prevent reverse flow at night in systems where the collectors (which are cold at night) are mounted above the storage tank.

The third fluid loop is the hot water delivery loop. Hot water drawn off to the load is replaced by cold water supplied to the solar preheat tank, where it is heated as much as possible by solar heat. If solar energy is insufficient to heat the water to its setpoint, conventional fuels can finish the heating in the water heater tank, as shown on the right of Figure 4.1.13. The tempering valve in the distribution loop is used to limit the temperature of water dispatched to the building if the solar tank should be above the water heater setpoint in summer.

The energy delivery of DHW systems can be found using the f-chart method described in Duffie and Beckman (1992). As a rough rule of thumb, one square foot of collector can provide one gallon of hot water per day (45 L/m^2) on the average in sunny climates. Design pump flows are to be 0.02 gal/min per square foot of collector [$0.01 \text{ L/(s} \cdot \text{m}^2)$], and heat exchanger effectivenesses of at least 0.75 can be justified economically. Tanks should be insulated so that no more than 2% of the stored heat is lost overnight.

Solar heating should be assessed on an economic basis. If the cost of delivered solar heat, including the amortized cost of the delivery system and its operation, is less than that of competing energy sources, an incentive exists for using the solar resource. The collector area needed on commercial buildings can be large; if possible, otherwise unused roof space can be used to hold the collector arrays. See Chapter 6.4 for a complete and detailed description of solar system analysis and design.

References

ASHRAE, 1999, *HVAC Applications*, ASHRAE, Atlanta, GA.

ASHRAE, 2000, *Equipment*, ASHRAE, Atlanta, GA.

ASHRAE, 2001, *Fundamentals*, ASHRAE, Atlanta, GA.

Haas, J. H., 1990, Steam traps — key to process heating, *Chemical Engineering*, vol. 97, pp. 151–156, January.

Goswami, Y., Kreider, J. F., and Kreith, F., 2000, *Principles of Solar Engineering*, Taylor and Francis, New York, pp. 694.

Rabl, A. and Kreider, J.F., 1994, *Heating and Cooling of Buildings*, McGraw-Hill, New York, NY.

Wong, W. Y., 1989, Safer relief valve sizing, *Chemical Engineering*, vol. 96, pp. 137–140, May.

4.2 Air Conditioning Systems

Dennis L. O'Neal and John A. Bryant

Air conditioning has rapidly grown over the past 50 years, from a luxury to a standard system included in most residential and commercial buildings. In 1970, 36% of residences in the U.S. were either fully air conditioned or utilized a room air conditioner for cooling (Blue, et al., 1979). By 1997, this number had more than doubled to 77%, and that year also marked the first time that over half (50.9%) of residences in the U.S. had central air conditioners (Census Bureau, 1999). An estimated 83% of all new homes constructed in 1998 had central air conditioners (Census Bureau, 1999). Air conditioning has also grown rapidly in commercial buildings. From 1970 to 1995, the percentage of commercial buildings with air conditioning increased from 54 to 73% (Jackson and Johnson, 1978, and DOE, 1998).

Air conditioning in buildings is usually accomplished with the use of mechanical or heat-activated equipment. In most applications, the air conditioner must provide both cooling and dehumidification to maintain comfort in the building. Air conditioning systems are also used in other applications, such as automobiles, trucks, aircraft, ships, and industrial facilities. However, the description of equipment in this chapter is limited to those commonly used in commercial and residential buildings.

Commercial buildings range from large high-rise office buildings to the corner convenience store. Because of the range in size and types of buildings in the commercial sector, there is a wide variety of equipment applied in these buildings. For larger buildings, the air conditioning equipment is part of a total system design that includes items such as a piping system, air distribution system, and cooling tower.

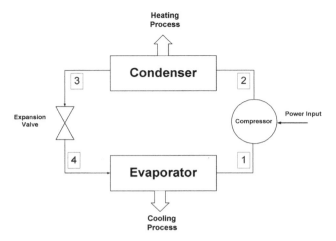

FIGURE 4.2.1 Simplified schematic and pressure/enthalpy diagram of the vapor compression cycle used in many air conditioners.

Proper design of these systems requires a qualified engineer. The residential building sector is dominated by single family homes and low-rise apartments/condominiums. The cooling equipment applied in these buildings comes in standard "packages" that are often both sized and installed by the air conditioning contractor.

The chapter starts with a general discussion of the vapor compression refrigeration cycle then moves to refrigerants and their selection. Chillers and their auxiliary systems are then covered, followed by packaged air conditioning equipment.

4.2.1 Vapor Compression Cycle

Even though there is a large range in sizes and variety of air conditioning systems used in buildings, most systems utilize the vapor compression cycle to produce the desired cooling and dehumidification. This cycle is also used for refrigerating and freezing foods and for automotive air conditioning. The first patent on a mechanically driven refrigeration system was issued to Jacob Perkins in 1834 in London, and the first viable commercial system was produced in 1857 by James Harrison and D.E. Siebe (Thevenot 1979).

Besides vapor compression, there are two less common methods used to produce cooling in buildings: the absorption cycle and evaporative cooling. These are described later in the chapter. With the vapor compression cycle, a working fluid, which is called the refrigerant, evaporates and condenses at suitable pressures for practical equipment designs.

The four basic components (Figure 4.2.1) in every vapor compression refrigeration system are the compressor, condenser, expansion device, and evaporator. The compressor raises the pressure of the refrigerant vapor so that the refrigerant saturation temperature is slightly above the temperature of the cooling medium used in the condenser. The type of compressor used depends on the application of the system. Large electric chillers typically use a centrifugal compressor while small residential equipment uses a reciprocating or scroll compressor.

The condenser is a heat exchanger used to reject heat from the refrigerant to a cooling medium. The refrigerant enters the condenser and usually leaves as a subcooled liquid. Typical cooling mediums used in condensers are air and water. Most residential-sized equipment uses air as the cooling medium in the condenser, while many larger chillers use water.

After leaving the condenser, the liquid refrigerant expands to a lower pressure in the expansion valve. The expansion valve can be a passive device, such as a capillary tube or short tube orifice, or an active device, such as a thermal expansion valve or electronic expansion valve. The purpose of the valve is to regulate the flow of refrigerant to the evaporator so that the refrigerant is superheated when it reaches the suction of the compressor.

At the exit of the expansion valve, the refrigerant is at a temperature below that of the medium (air or water) to be cooled. The refrigerant travels through a heat exchanger called the evaporator. It absorbs energy from the air or water circulated through the evaporator. If air is circulated through the evaporator, the system is called a *direct expansion system*. If water is circulated through the evaporator, it is called a *chiller*. In either case, the refrigerant does not make direct contact with the air or water in the evaporator.

The refrigerant is converted from a low quality, two-phase fluid to a superheated vapor under normal operating conditions in the evaporator. The vapor formed must be removed by the compressor at a sufficient rate to maintain the low pressure in the evaporator and keep the cycle operating.

All mechanical cooling results in the production of heat energy that must be rejected through the condenser. In many instances, this heat energy is rejected to the environment directly to the air in the condenser or indirectly to water where it is rejected in a cooling tower. With some applications, it is possible to utilize this waste heat energy to provide simultaneous heating to the building. Recovery of this waste heat at temperatures up to 65°C (150°F) can be used to reduce costs for space heating.

Capacities of air conditioning are often expressed in either tons or kilowatts (kW) of cooling. The ton is a unit of measure related to the ability of an ice plant to freeze one short ton (907 kg) of ice in 24 hr. Its value is 3.51 kW (12,000 Btu/hr). The kW of thermal cooling capacity produced by the air conditioner must not be confused with the amount of electrical power (also expressed in kW) required to produce the cooling effect.

4.2.2 Refrigerants Use and Selection

Up until the mid-1980s, refrigerant selection was not an issue in most building air conditioning applications because there were no regulations on the use of refrigerants. Many of the refrigerants historically used for building air conditioning applications have been chlorofluorocarbons (CFCs) and hydrochlorofluorocarbons (HCFCs). Most of these refrigerants are nontoxic and nonflammable. However, recent U.S. federal regulations (EPA 1993a; EPA 1993b) and international agreements (UNEP, 1987) have placed restrictions on the production and use of CFCs and HCFCs. Hydrofluorocarbons (HFCs) are now being used in some applications where CFCs and HCFCs were used. Having an understanding of refrigerants can help a building owner or engineer make a more informed decision about the best choice of refrigerants for specific applications. This section discusses the different refrigerants used in or proposed for building air conditioning applications and the regulations affecting their use.

The American Society of Heating, Refrigerating and Air Conditioning Engineers (ASHRAE) has a standard numbering system (Table 4.2.1) for identifying refrigerants (ASHRAE, 1992). Many popular CFC, HCFC, and HFC refrigerants are in the methane and ethane series of refrigerants. They are called halocarbons, or halogenated hydrocarbons, because of the presence of halogen elements such as fluorine or chlorine (King, 1986).

Zeotropes and azeotropes are mixtures of two or more different refrigerants. A zeotropic mixture changes saturation temperatures as it evaporates (or condenses) at constant pressure. The phenomena is called temperature glide. At atmospheric pressure, R-407C has a boiling (bubble) point of –44°C (–47°F) and a condensation (dew) point of –37°C (–35°F), which gives it a temperature glide of 7°C (12°F). An azeotropic mixture behaves like a single component refrigerant in that the saturation temperature does not change appreciably as it evaporates or condenses at constant pressure. R-410A has a small enough temperature glide (less than 5.5°C, 10°F) that it is considered a near-azeotropic refrigerant mixture.

ASHRAE groups refrigerants (Table 4.2.2) by their toxicity and flammability (ASHRAE, 1994). Group A1 is nonflammable and least toxic, while Group B3 is flammable and most toxic. Toxicity is based on the upper safety limit for airborne exposure to the refrigerant. If the refrigerant is nontoxic in quantities less than 400 parts per million, it is a Class A refrigerant. If exposure to less than 400 parts per million is toxic, then the substance is given the B designation. The numerical designations refer

TABLE 4.2.1   Common Refrigerants with Their Applications and Characteristics

Refrigerant	Typical or Proposed Applications	Normal boiling point, °C	Safety Group
	Methane Series		
11	Low pressure centrifugal chillers	24	A1
12	Refrigeration, medium pressure chillers, auto A/C	−30	A1
22	Package A/C, heat pumps	−41	A1
32	Component of R-407C and R-410A	−52	A2
	Ethane Series		
123	Low pressure chillers	27	B1
125	Component of R-407C and R-410A	−49	A1
134a	Chillers, refrigeration, auto A/C	−26	A1
	Propane Series		
290	Proposed replacement for R-22	−42	A3
	Zeotropes		
407C	Package A/C, heat pumps	−44	A1
410A	Package A/C, heat pumps	−53	A1
	Azeotropes		
500	Medium pressure centrifugal chillers	−33	A1
502	Refrigeration, low temperature heat pumps	−45	A1
	Hydrocarbons		
600	Refrigeration	0	A3
600a	Refrigeration	−12	A3
	Inorganic Compounds		
717	Industrial Refrigeration	−33	B2
744	Proposed automotive A/C	−78	A1

Source: From ASHRAE (1997); Smit et al. (1996).

TABLE 4.2.2   Toxicity and Flammability Rating System

Flammability	Group	Group
High	A3	B3
Moderate	A2	B2
Non	A1	B1
Threshold Limit Value (parts per million)	<400	>400

Source: From ASHRAE (1994).

to the flammability of the refrigerant. The last column of Table 4.2.1 shows the toxicity and flammability rating of common refrigerants.

Refrigerants in the A1 group usually fulfill the basic requirements for an ideal refrigerant for comfort air conditioning because they are nontoxic and nonflammable. Common refrigerants in the A1 group used in building air conditioning applications include R-11, R-12, R-22, R-134a, and R-410A.

R-11, R-12, R-123, and R-134a are refrigerants commonly used in centrifugal chiller applications. Both R-11, a CFC, and R-123, an HCFC, have low-pressure high-volume characteristics ideally suited for use in centrifugal compressors. Before the ban on production of CFCs, R-11 and R-12 were the refrigerants

TABLE 4.2.3 Comparative Performance of Commonly Used Refrigerants at +2°C Evaporating Temperature and +40°C Condensing Temperature[a]

Refrigerant Number	Evaporator Pressure (kPa)	Condenser Pressure (kPa)	Net Cooling Effect (kJ/kg)	Refrigerant Circulated (kg/min)	Compressor Displacement (L/min)	Coefficient of Performance
11	44	173	155	3.9	144	6.5
12	329	960	114	5.3	27	6.0
22	531	1533	157	3.8	17	5.9
123	36	155	142	4.2	173	6.3
134a	315	1017	143	4.2	27	5.9
290	508	1368	270	2.2	20	5.8
407C	488	1738	151	4.0	20	4.8
410A	848	2417	157	3.8	12	5.5
717	464	1559	1072	0.6	13	7.1

[a] Refrigerant circulation rate is based on 10 kW cooling capacity.

of choice for chiller applications. The use of these two refrigerants is currently limited to maintenance of existing systems. Both R-123 and R-134a are now being used extensively in new chillers. R-123 provides an efficiency advantage (Table 4.2.3) over R-134a. However, R-123 has a B1 safety classification, which means it has a lower toxicity threshold than R-134a. If an R-123 chiller is used in a building, ASHRAE Standard 15 (ASHRAE, 1992) provides guidelines for safety precautions when using this or any other refrigerant that is toxic or flammable.

Refrigerant 22 is an HCFC, is used in many of the same applications, and is still the refrigerant of choice in many reciprocating and screw chillers as well as small commercial and residential packaged equipment. It operates at a much higher pressure than either R-11 or R-12. Restrictions on the production of HCFCs will start in 2004. In 2010, R-22 cannot be used in new air conditioning equipment. R-22 cannot be produced after 2020 (EPA, 1993b).

R-407C and R-410A are both mixtures of HFCs. Both are considered replacements for R-22. R-407C is expected to be a drop-in replacement refrigerant for R-22. Its evaporating and condensing pressures for air conditioning applications are close to those of R-22 (Table 4.2.3). However, replacement of R-22 with R-407C should be done only after consulting with the equipment manufacturer. At a minimum, the lubricant and expansion device will need to be replaced. The first residential-sized air conditioning equipment using R-410A was introduced in the U.S. in 1998. Systems using R-410A operate at approximately 50% higher pressure than R-22 (Table 4.2.3); thus, R-410A cannot be used as a drop-in refrigerant for R-22. R-410A systems utilize compressors, expansion valves, and heat exchangers designed specifically for use with that refrigerant.

Ammonia is widely used in industrial refrigeration applications and in ammonia water absorption chillers. It is moderately flammable and has a class B toxicity rating but has had limited applications in commercial buildings unless the chiller plant can be isolated from the building being cooled (Toth, 1994, Stoecker, 1994). As a refrigerant, ammonia has many desirable qualities. It has a high specific heat and high thermal conductivity. Its enthalpy of vaporization is typically 6 to 8 times higher than that of the commonly used halocarbons, and it provides higher heat transfer compared to halocarbons. It can be used in both reciprocating and centrifugal compressors.

Research is underway to investigate the use of natural refrigerants, such as carbon dioxide (R-744) and hydrocarbons in air conditioning and refrigeration systems (Bullock, 1997, and Kramer, 1991). Carbon dioxide operates at much higher pressures than conventional HCFCs or HFCs and requires operation above the critical point in typical air conditioning applications. Hydrocarbon refrigerants, often thought of as too hazardous because of flammability, can be used in conventional compressors and have been used in industrial applications. R-290, propane, has operating pressures close to R-22 and has been proposed as a replacement for R-22 (Kramer, 1991). Currently, there are no commercial systems sold in the U.S. for building operations that use either carbon dioxide or flammable refrigerants.

Table 4.2.3 shows a comparative performance of refrigerants at evaporating and condensing temperatures typical of cooling applications. The data show the relatively large cooling effect produced with ammonia. For the specific conditions, R-11 and R-123 have the lowest evaporating pressures. The coefficient of performance (COP) listed in the far right column of Table 4.2.3 is a measure of the thermodynamic efficiency of an air conditioner (or chiller) with that particular refrigerant. It is defined as the cooling output (kW) divided by the power input (kW) to the compressor. The actual COP in a system will not only depend on the refrigerant, but on the design of the compressor, heat exchangers, and expansion device. At cooling conditions, there is substantial drop off in efficiency between R-123 and R-134a. Manufacturers who have replaced their R-11 chillers with R-134a chillers have had to make substantial modifications in the chiller designs to maintain comparable efficiencies. Two R-22 replacement refrigerants, R-407C and R-410A, both have lower COPs than R-22.

Rowland and Molina (1974) hypothesized that CFCs were responsible for destroying ozone in the stratosphere. By the late 1970s, the U.S. and Canada had banned the use of CFCs in aerosols. In the mid 1980s, a 40% depletion in the ozone layer was measured (Salas and Salas, 1992). In September 1987, forty-three countries signed an agreement called the Montreal Protocol in which the participants agreed to freeze CFC production levels by 1990 and then to decrease production by 20% by 1994 and 50% by 1999. The protocol was ratified by the U.S. in 1988 and subjected the air conditioning industry to major CFC restrictions. Title IV of the Clean Air Act of November 1990 required elimination of the production of CFCs by 2000 (EPA, 1993a) and placed a schedule on the phasing out of the production of HCFCs by 2030.

Two ratings were developed to classify the harmful effects of a refrigerant on the environment (EPA, 1993b). The first, the ozone depletion potential (ODP), quantifies the potential damage that the refrigerant molecule has in destroying ozone in the stratosphere. The estimated atmospheric life of a given CFC or HCFC is an important factor in determining the value of the ODP.

The second rating is known as the halocarbon global warming potential (HGWP). It relates the potential for a refrigerant in the atmosphere to contribute to the greenhouse effect. Like CO_2, refrigerants such as CFCs, HCFCs, and HFCs can block energy from the earth from radiating back into space. One molecule of R-12 can absorb as much energy as almost 5000 molecules of CO_2. Both the ODP and HGWP are normalized to the value of R-11.

Table 4.2.4 shows the ODP and HGWP for a variety of refrigerants. As a class of refrigerants, the CFCs have the highest ODP and HGWP. Because HCFCs tend to be more unstable compounds and therefore have much shorter atmospheric lifetimes, their ODP and HGWP values are much smaller than those of the CFCs. All HFCs and their mixtures have zero ODP because fluorine does not react with ozone. However, some of the HFCs, such as R-125 and R-134a, do have HGWP values that are as large as or larger than some of the HCFCs. Hydrocarbons provide zero ODP and HGWP.

In recent years, attempts have been made to develop an alternate criteria, called the total equivalent warming impacts (TEWI) for evaluating the global warming impact of different refrigerants (Sand, Fischer, and Baxter, 1999). TEWI includes the total energy use of the equipment over its expected lifetime as well as the global warming caused by release of the refrigerant charge in the system. The TEWI depends on assumptions about the usage and efficiency of the system. Sand, Fischer, and Baxter (1999) estimated that energy usage in a system accounts for over 90% of the global warming potential of the system. A high efficiency system using R-22 could have a lower TEWI than a lower efficiency system using a zero HGWP refrigerant. The TEWI represents a more systems approach to global warming impact than does HGWP by itself.

4.2.3 Chilled Water Systems

Chilled water systems were used in less than 4% of commercial buildings in the U.S. in 1995. However, because chillers are usually installed in larger buildings, chillers cooled over 28% of the U.S. commercial building floor space that same year (DOE, 1998). Five types of chillers are commonly applied to commercial buildings: reciprocating, screw, scroll, centrifugal, and absorption. The first four utilize the vapor

TABLE 4.2.4   Ozone Depletion Potential and Halocarbon
Global Warming Potential of Popular Refrigerants and Mixtures

Refrigerant Number	Ozone Depletion Potential (ODP)	Halogen Global Warming Potential (HGWP)
Chlorofluorocarbons		
11	1.0	1.0
12	1.0	3.05
22	0.051	0.37
123	0.016	0.019
Hydrofluorocarbons		
32	0	0.13
125	0	0.58
134a	0	0.285
Hydrocarbons		
50	0	0
290	0	0
Zeotropes		
407C	0	0.22
410A	0	0.44
Azeotropes		
500	0.74	2.4
502	0.23	5.1

Source: Compiled from Salas and Salas (1992),
NR (1995), and Didion (1996).

compression cycle to produce chilled water. They differ primarily in the type of compressor used. Absorption chillers utilize thermal energy (typically steam or combustion source) in an absorption cycle with either an ammonia-water or water-lithium bromide solution to produce chilled water.

Overall System

Figure 4.2.2 shows a simple representation of a dual chiller application with all the major auxiliary equipment. An estimated 86% of chillers are applied in multiple chiller arrangements like that shown in the figure (Bitondo and Tozzi, 1999). In chilled water systems, return water from the building is circulated through each chiller evaporator where it is cooled to an acceptable temperature (typically 4 to 7°C) (39 to 45°F). The chilled water is then distributed to water-to-air heat exchangers spread throughout the facility. In these heat exchangers, air is cooled and dehumidified by the cold water. During the process, the chilled water increases in temperature and must be returned to the chiller(s).

The chillers shown in Figure 4.2.2 are water-cooled chillers. Water is circulated through the condenser of each chiller where it absorbs heat energy rejected from the high pressure refrigerant. The water is then pumped to a cooling tower where the water is cooled through an evaporation process. Cooling towers are described in a later section. Chillers can also be air cooled. In this configuration, the condenser would be a refrigerant-to-air heat exchanger with air absorbing the heat energy rejected by the high pressure refrigerant.

Chillers nominally range in capacities from 30 to 18,000 kW (8 to 5100 ton). Most chillers sold in the U.S. are electric and utilize vapor compression refrigeration to produce chilled water. Compressors for these systems are either reciprocating, screw, scroll, or centrifugal in design. A small number of centrifugal chillers are sold that use either an internal combustion engine or steam drive instead of an electric motor to drive the compressor.

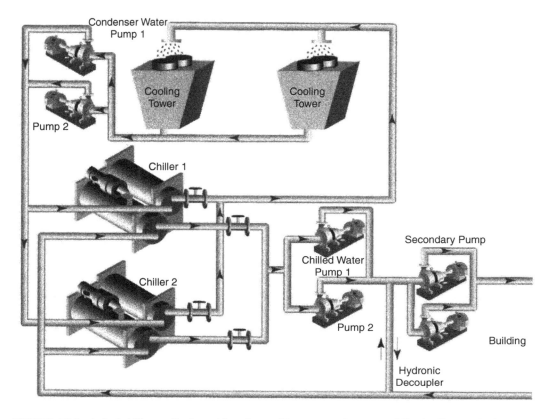

FIGURE 4.2.2　A dual chiller application with major auxiliary systems (courtesy of Carrier Corporation).

　　The type of chiller used in a building depends on the application. For large office buildings or in chiller plants serving multiple buildings, centrifugal compressors are often used. In applications under 1000 kW (280 tons) cooling capacities, reciprocating or screw chillers may be more appropriate. In smaller applications, below 100 kW (30 tons), reciprocating or scroll chillers are typically used.

Vapor Compression Chillers

Table 4.2.5 shows the nominal capacity ranges for the four types of electrically driven vapor compression chillers. Each chiller derives its name from the type of compressor used in the chiller. The systems range in capacities from the smallest scroll (30 kW; 8 tons) to the largest centrifugal (18,000 kW; 5000 tons). Chillers can utilize either an HCFC (R-22 and R-123) or HFC (R-134a) refrigerant. The steady state efficiency of chillers is often stated as a ratio of the power input (in kW) to the chilling capacity (in tons). A capacity rating of one ton is equal to 3.52 kW or 12,000 btu/h. With this measure of efficiency, the smaller number is better. As seen in Table 4.2.5, centrifugal chillers are the most efficient; whereas, reciprocating chillers have the worst efficiency of the four types. The efficiency numbers provided in the table are the steady state full-load efficiency determined in accordance to ASHRAE Standard 30 (ASHRAE, 1995). These efficiency numbers do not include the auxiliary equipment, such as pumps and cooling tower fans that can add from 0.06 to 0.31 kW/ton to the numbers shown (Smit et al., 1996).

　　Chillers run at part load capacity most of the time. Only during the highest thermal loads in the building will a chiller operate near its rated capacity. As a consequence, it is important to know how the efficiency of the chiller varies with part load capacity. Figure 4.2.3 shows a representative data for the efficiency (in kW/ton) as a function of percentage full load capacity for a reciprocating, screw, and scroll chiller plus a centrifugal chiller with inlet vane control and one with variable frequency drive (VFD) for the compressor. The reciprocating chiller increases in efficiency as it operates at a smaller percentage of

TABLE 4.2.5 Capacity Ranges and Efficiencies of Vapor Compression Chillers Used for Commercial Building Air Conditioning

Type of Chiller	Nominal Capacity Range (kW)	Refrigerants Used in New Systems	Range in Full Load Efficiency (kW/ton)
Reciprocating	50 to 1750	R-22	0.80 to 1.00
Screw	160 to 2350	R-134a, R-22	0.60 to 0.75
Scroll	30 to 200	R-22	0.81 to 0.92
Centrifugal	500 to 18,000	R-134a, R-123	0.50 to 0.70

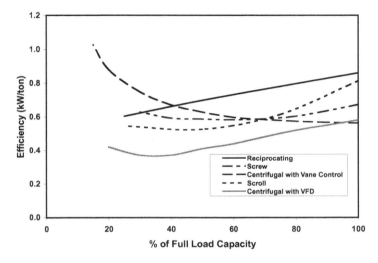

FIGURE 4.2.3 Chiller efficiency as a function of percentage of full load capacity.

full load. In contrast, the efficiency of a centrifugal with inlet vane control is relatively constant until the load falls to about 60% of its rated capacity and its kW/ton increases to almost twice its fully loaded value.

In 1998, the Air Conditioning and Refrigeration Institute (ARI) developed a new standard that incorporates into their ratings part load performance of chillers (ARI 1998c). Part load efficiency is expressed by a single number called the integrated part load value (IPLV). The IPLV takes data similar to that in Figure 4.2.3 and weights it at the 25%, 50%, 75%, and 100% loads to produce a single integrated efficiency number. The weighting factors at these loads are 0.12, 0.45, 0.42, and 0.01, respectively. The equation to determine IPLV is:

$$IPLV = \frac{1}{\dfrac{0.01}{A} + \dfrac{0.42}{B} + \dfrac{0.45}{C} + \dfrac{0.12}{D}}$$

where,

A = efficiency at 100% load
B = efficiency at 75% load
C = efficiency at 50% load
D = efficiency at 25% load

Most of the IPLV is determined by the efficiency at the 50% and 75% part load values. Manufacturers will provide, on request, IPLVs as well as part load efficiencies such as those shown in Figure 4.2.3.

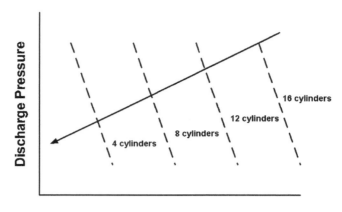

FIGURE 4.2.4   Volume-pressure relationships for a reciprocating compressor.

The four compressors used in vapor compression chillers are each briefly described below. While centrifugal and screw compressors are primarily used in chiller applications, reciprocating and scroll compressors are also used in smaller unitary packaged air conditioners and heat pumps.

Reciprocating Compressors

The reciprocating compressor is a positive displacement compressor. On the intake stroke of the piston, a fixed amount of gas is pulled into the cylinder. On the compression stroke, the gas is compressed until the discharge valve opens. The quantity of gas compressed on each stroke is equal to the displacement of the cylinder. Compressors used in chillers have multiple cylinders, depending on the capacity of the compressor. Reciprocating compressors use refrigerants with low specific volumes and relatively high pressures. Most reciprocating chillers used in building applications currently employ R-22.

Modern high-speed reciprocating compressors are generally limited to a pressure ratio of approximately nine. The reciprocating compressor is basically a constant-volume variable-head machine. It handles various discharge pressures with relatively small changes in inlet-volume flow rate as shown by the heavy line (labeled 16 cylinders) in Figure 4.2.4. Condenser operation in many chillers is related to ambient conditions, for example, through cooling towers, so that on cooler days the condenser pressure can be reduced. When the air conditioning load is lowered, less refrigerant circulation is required. The resulting load characteristic is represented by the solid line that runs from the upper right to lower left of Figure 4.2.4.

The compressor must be capable of matching the pressure and flow requirements imposed by the system. The reciprocating compressor matches the imposed discharge pressure at any level up to its limiting pressure ratio. Varying capacity requirements can be met by providing devices that unload individual or multiple cylinders. This unloading is accomplished by blocking the suction or discharge valves that open either manually or automatically. Capacity can also be controlled through the use of variable speed or multi-speed motors. When capacity control is implemented on a compressor, other factors at part-load conditions need to considered, such as (a) effect on compressor vibration and sound when unloaders are used, (b) the need for good oil return because of lower refrigerant velocities, and (c) proper functioning of expansion devices at the lower capacities.

With most reciprocating compressors, oil is pumped into the refrigeration system from the compressor during normal operation. Systems must be designed carefully to return oil to the compressor crankcase to provide for continuous lubrication and also to avoid contaminating heat-exchanger surfaces.

Reciprocating compressors usually are arranged to start unloaded so that normal torque motors are adequate for starting. When gas engines are used for reciprocating compressor drives, careful matching of the torque requirements of the compressor and engine must be considered.

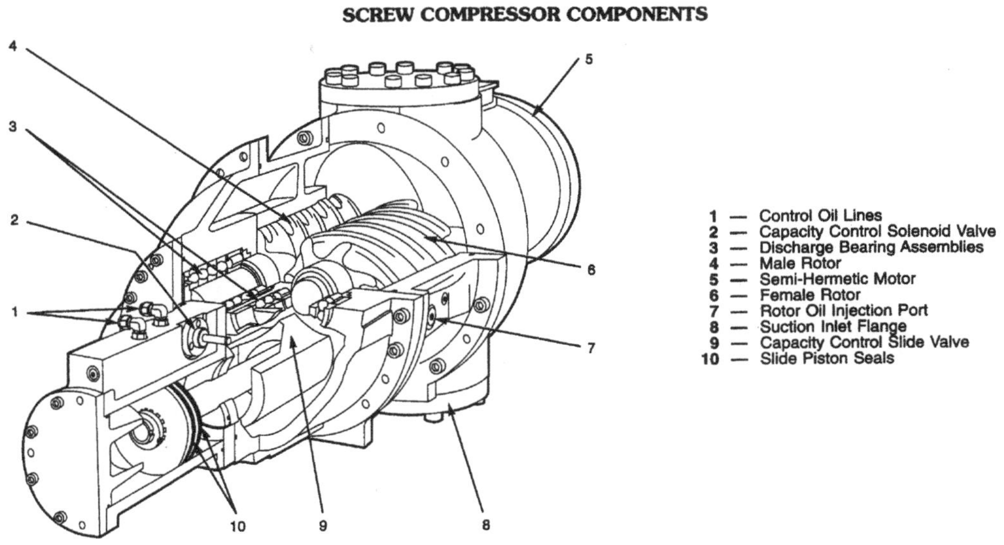

SCREW COMPRESSOR COMPONENTS

1 — Control Oil Lines
2 — Capacity Control Solenoid Valve
3 — Discharge Bearing Assemblies
4 — Male Rotor
5 — Semi-Hermetic Motor
6 — Female Rotor
7 — Rotor Oil Injection Port
8 — Suction Inlet Flange
9 — Capacity Control Slide Valve
10 — Slide Piston Seals

FIGURE 4.2.5 Illustration of a twin-screw compressor design (courtesy of Carrier Corporation).

Screw Compressors

Screw compressors, first introduced in 1958 (Thevenot, 1979), are positive displacement compressors. They are available in the capacity ranges that overlap with reciprocating compressors and small centrifugal compressors. Both twin-screw and single-screw compressors are used in chillers. The twin-screw compressor is also called the helical rotary compressor. Figure 4.2.5 shows a cutaway of a twin-screw compressor design. There are two main rotors (screws). One is designated male (4 in the figure) and the other female (6 in the figure).

The compression process is accomplished by reducing the volume of the refrigerant with the rotary motion of screws. At the low pressure side of the compressor, a void is created when the rotors begin to unmesh. Low pressure gas is drawn into the void between the rotors. As the rotors continue to turn, the gas is progressively compressed as it moves toward the discharge port. Once reaching a predetermined volume ratio, the discharge port is uncovered and the gas is discharged into the high pressure side of the system. At a rotation speed of 3600 rpm, a screw compressor has over 14,000 discharges per minute (ASHRAE, 1996).

Fixed suction and discharge ports are used with screw compressors instead of valves, as used in reciprocating compressors. These set the *built-in volume ratio* — the ratio of the volume of fluid space in the meshing rotors at the beginning of the compression process to the volume in the rotors as the discharge port is first exposed. Associated with the built-in volume ratio is a pressure ratio that depends on the properties of the refrigerant being compressed. Screw compressors have the capability to operate at pressure ratios of above 20:1 (ASHRAE, 1996). Peak efficiency is obtained if the discharge pressure imposed by the system matches the pressure developed by the rotors when the discharge port is exposed. If the interlobe pressure in the screws is greater or less than discharge pressure, energy losses occur but no harm is done to the compressor.

Capacity modulation is accomplished by slide valves that provide a variable suction bypass or delayed suction port closing, reducing the volume of refrigerant compressed. Continuously variable capacity control is most common, but stepped capacity control is offered in some manufacturers' machines. Variable discharge porting is available on some machines to allow control of the built-in volume ratio during operation.

Oil is used in screw compressors to seal the extensive clearance spaces between the rotors, to cool the machines, to provide lubrication, and to serve as hydraulic fluid for the capacity controls. An oil separator is required for the compressor discharge flow to remove the oil from the high-pressure refrigerant so that performance of system heat exchangers will not be penalized and the oil can be returned for reinjection in the compressor.

Screw compressors can be direct driven at two-pole motor speeds (50 or 60 Hz). Their rotary motion makes these machines smooth running and quiet. Reliability is high when the machines are applied properly. Screw compressors are compact so they can be changed out readily for replacement or maintenance. The efficiency of the best screw compressors matches or exceeds that of the best reciprocating compressors at full load. High isentropic and volumetric efficiencies can be achieved with screw compressors because there are no suction or discharge valves and small clearance volumes. Screw compressors for building applications generally use either R-134a or R-22.

Scroll Compressors

The principle of the scroll compressor was first patented in 1905 (Matsubara el al., 1987). However, the first commercial units were not built until the early 1980s and were sold in Japan in residential heat pump systems (Senshu et al., 1985). Of the different electric driven chillers discussed in this section, scroll chillers have the smallest range in capacity. Only one U.S. manufacturer currently offers a scroll chiller, and these are limited to capacities below 200 kW (57 tons). Scroll compressors are built in sizes as small as 3 kW (.05 ton). Scroll compressors are primarily used in direct expansion air conditioners, heat pumps, and some refrigeration applications. Chillers using scroll compressors currently only use R-22. However, direct expansion air conditioners with scroll compressors that use R-410A have recently been introduced into the market, and it would be reasonable to expect a switch to an HFC in chillers in the near future.

Scroll compressors have two spiral-shaped scroll members that are assembled 180° out of phase (Figure 4.2.6). One scroll is fixed while the other "orbits" the first. Vapor is compressed by sealing it off at the edge of the scrolls and reducing the volume of the gas as it moves invward toward the discharge port. Figure 4.2.6a shows the two scrolls at the instance that vapor has entered the compressor and compression begins. The orbiting motion of the second scroll forces the pocket of vapor toward the discharge port while decreasing its volume (Figures 4.2.6b–h). In Figures 4.2.6c and f, the two scrolls open at the ends and allow new pockets of vapor to be admitted for compression. Compression is a nearly continuous process in a scroll compressor.

Scroll compressors offer several advantages over reciprocating compressors. First, relatively large suction and discharge ports can be used to reduce pressure losses. Second, the separation of the suction and discharge processes reduces the heat transfer between those processes. Third, with no valves and re-expansion losses, they have higher volumetric efficiencies. The plots of part load efficiencies in Figure 4.2.3 show that scroll compressors have better efficiencies down to 25% part load than do reciprocating compressors. Capacities of systems with scroll compressors can be varied by using variable speed motors or by using multiple suction ports at different locations within the two spiral members.

Centrifugal Compressors

The reciprocating, screw, and scroll compressors are all positive displacement compressors. They each work by taking a fixed volume of low pressure refrigerant and reducing it to achieve compression. In contrast, the centrifugal compressor uses dynamic compression. The primary operating component of the compressor is the impeller. The center of the impeller has vanes that draw the low pressure refrigerant vapor into radial passages internal to the impeller. The impeller rotates and accelerates the gas and increases its kinetic energy. When the gas leaves the impeller, it flows to a circular diffuser passage, the *volute*, where the gas is decelerated and the pressure is increased (Trane, 1980). Centrifugal chillers have operational speeds from 3600 to 35,000 rpm.

Figure 4.2.7 shows a cutaway of a centrifugal compressor that has three impellers. Refrigerant flows from the bottom left to the right through the compressor. This compressor has three stages of compression. Centrifugal compressors with multiple stages can generate a pressure ratio up to 18:1, but their high discharge temperatures limit the efficiency of the simple cycle at these high pressure ratios. As a result, they operate with evaporator temperatures in the same range as reciprocating compressors. Multistage centrifugal compressors are built for direct connection to high-speed drives.

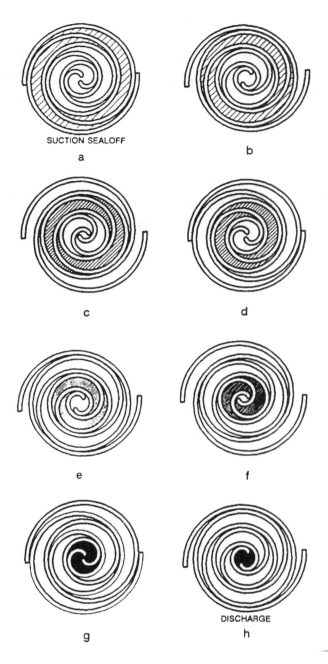

FIGURE 4.2.6   Operation of the scrolls in a scroll compressor. (From ASHRAE, 1996).

Figure 4.2.7 also shows inlet vanes at the entrance to the first impeller. These vanes control the capacity of the compressor by adjusting the angle at which the low pressure refrigerant enters the impeller. Inlet vanes also help to stabilize compressor performance over a wide range of load conditions and to prevent surge (Trane, 1980).

Under certain low load conditions, flow can reverse through the impeller. This phenomena is called surge and is unique to centrifugal compressors. Surge increases inefficiency, static pressure fluctuations, vibration, and noise in the compressor (Trane, 1980).

FIGURE 4.2.7 Cutaway of a three stage centrifugal compressor with guide vanes at the inlet (courtesy of the Trane Company).

The centrifugal compressor has a more complex pressure-volume characteristic than reciprocating machines, as shown by the system characteristic curve in Figure 4.2.8a. Changing discharge pressure may cause relatively large changes in inlet volume. Adjustment of variable inlet vanes allows the compressor to operate anywhere below the system line to conditions imposed by the system. A variable-speed controller offers an alternative way to match the compressor's characteristics to the system load, as shown in the lower half of Figure 4.2.8b. The maximum head capability is fixed by the operating speed of the compressor. Both methods have advantages: generally, variable inlet vanes provide a wider range of capacity reduction; variable speed usually is more efficient. Maximum efficiency and control can be obtained by combining both methods of control.

The centrifugal compressor has a surge point — a minimum-volume flow below which stable operation cannot be maintained. The percentage of load at which the surge point occurs depends on the number of impellers, design-pressure ratio, operating speed, and variable inlet-vane setting. The system design and controls must keep the inlet volume above this point.

Provision for minimum load operation is strongly recommended for all installations because there will be fluctuations in plant load. The difference between the operating characteristics of the positive displacement compressor and the centrifugal compressor are important considerations in chiller plant design to achieve satisfactory performance. Unlike positive displacement compressors, the centrifugal compressor will not rebalance abnormally high system heads. The drive arrangement for the centrifugal compressor must be selected with sufficient speed to meet the maximum head anticipated. The relatively flat head characteristics of the centrifugal compressor necessitates different control approaches than for positive displacement machines, particularly when parallel compressors are utilized. These differences, which account for most of the troubles experienced in centrifugal-compressor systems, cannot be over-looked in the design of a chiller system.

Absorption Chillers

The first absorption machine was patented in 1859 by Ferdinand Carr (Thevenot, 1979) and used an ammonia/water solution. The design was produced in Europe and the U.S., and by 1876 over 600 absorption systems had been sold in the U.S. These systems were primarily used for producing ice. During the late 1800s and early 1900s, different combinations of fluids were tested in absorption machines. Lithium bromide and water were not used until 1940 (Thevenot, 1979). Through the 1960s, both absorption and centrifugal chillers competed for large-building air conditioning. However, with the rising prices of oil and gas in the 1970s, absorption chillers became more costly to operate than centrifugal

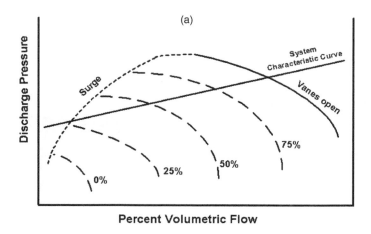

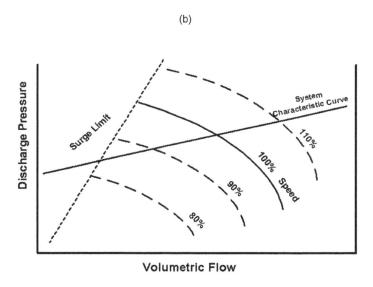

FIGURE 4.2.8 Volume-pressure relationships in a centrifugal compressor. (a) With variable inlet-van control at constant rotational speed. (b) With variable speed control at a constant inlet vane opening.

chillers (Wang, 1993). With the introduction of the more efficient two stage absorption systems in the 1980s and reduction in oil and gas costs by the mid 1980s, absorption systems again became a competitive option for cooling in buildings.

Absorption systems offer at least three advantages over conventional electric vapor compression systems. First, they do not use CFC or HCFC refrigerants. The solutions used in absorption systems are not refrigerants that could someday be eliminated because of ozone depletion or global warming concerns. Second, absorption systems can utilize a variety of heat sources, including natural gas, steam, solar-heated water, and waste heat from a turbine or industrial process. If the source of heat is from waste heat, such as from a co-generation system, absorption systems may provide the lowest cost alternative for providing chilled water for air conditioning. Because sources of energy besides electricity are used, installation of an absorption system can be used to reduce peak electrical demand in situations where electrical demand charges are high. Third, because of the absence of heavy rotating parts, absorption systems produce much less vibration and noise compared to large centrifugal systems (Carrier, 1964).

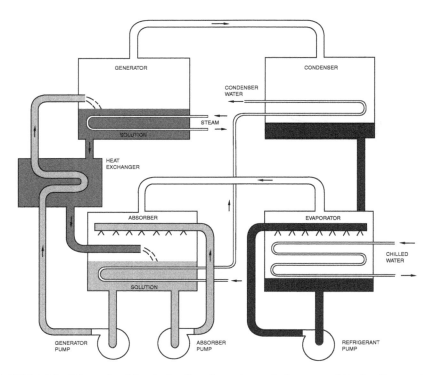

FIGURE 4.2.9 Schematic of the single-effect absorption cycle (courtesy of Carrier Corporation).

Two different absorption systems are currently used for air conditioning applications: (1) a water-lithium bromide system where water is the refrigerant and lithium bromide is the absorbent and (2) an ammonia-water system where ammonia is the refrigerant and water is the absorbent. Absorption systems sold in the U.S. for commercial building applications are almost exclusively water-lithium bromide chillers. One manufacturer sells an ammonia-water system for residential applications in the U.S. Most refrigeration applications of absorption systems are with ammonia and water. Because of the larger applications of water-lithium bromide systems in commercial buildings, the focus of the discussion in this section is on those systems. A description of ammonia-water absorption technology can be found in ASHRAE (1996).

Water-lithium bromide absorption machines can be classified by the method of heat input and whether the cycle is singe or multiple effect (ASHRAE, 1996). Indirect fired chillers use steam or hot liquids as a heat source. Direct fired chillers use the heat energy from the firing of fossil fuels. Heat-recovery chillers use waste gases as the heat source. Single effect and double effect chillers are described below.

The basic components for a single-effect water-lithium bromide absorption system are shown in Figure 4.2.9. The major components of the system are listed below (Carrier 1964):

- **Evaporator** — The evaporator is the section where chilled water is cooled by evaporating the refrigerant (water) over chilled water tubes. Operating pressures in the evaporator must be near a vacuum (less than 1 kPa; 0.1 psi) for evaporation to occur at a low enough temperature to produce chilled water in the tubes for air conditioning purposes.

- **Absorber** — Water vapor from the evaporator is absorbed by the lithium bromide into a liquid solution in this section. Condenser water is circulated through pipes into the liquid solution in the absorber to remove heat energy released during the absorption process. The absorber operates at the same pressure as the evaporator.

- **Generator** — The liquid solution from the absorber is pumped through a heat exchanger to the generator which is a part of the high pressure side of the system. Typical pressures in this section range from 5 to 7 kPA (0.7 to 1.0 psi), which is considerably below atmospheric pressure. In the generator, the water-rich liquid solution is heated where the water boils off from the solution and is transported to the condenser section. The solution can be heated with steam or other hot fluids through piping or it can be heated by a burner. Once water has evaporated from the water-lithium bromide solution, the dilute water-lithium bromide solution is then returned to the absorber.

- **Condenser** — The relatively high-pressure water vapor from the generator is condensed to liquid in the condenser. This is accomplished by circulating water in piping through the condenser that can absorb the latent heat energy of the water vapor. Once condensed into liquid, the water then flows through an expansion valve and into the low pressure evaporator, which completes the cycle. The condenser water that flows through the pipes is typically sent to a cooling tower where it can be cooled and then recirculated back to the system.
- **Pumps** — At least three pumps are required. The absorber and refrigerant (evaporator) pumps are primarily used to recirculate liquids in their respective sections. The generator pump moves the concentrated water-lithium bromide solution from the absorber to the generator.
- **Heat Exchanger** — The dilute solution of water-lithium bromide is much hotter than the concentrated solution being pumped from the absorber. The heat exchanger reduces energy use by heating the concentrated liquid flowing to the generator as it cools the hot dilute solution flowing from the generator to the absorber. If the dilute solution passing through the heat exchanger does not contain enough refrigerant (water) and is cooled too much, crystallization of the lithium bromide can occur. Leaks or process upsets that cause the generator to over-concentrate the solution are indicated when this occurs. The slushy mixture formed does not harm the machine but does interfere with operation. External heat and added water may be required to redissolve the mixture
- **Purge Unit** — The pressures throughout the system are far below atmospheric pressure. The purge unit is required to remove air and other noncondensable gases that can leak into the system and to maintain the required system pressures.

Absorption systems also include components not shown in Figure 4.2.9 or listed above (ASHRAE, 1994). Palladium cells are used to continuously remove small amounts of hydrogen generated by corrosion. Corrosion inhibitors protect internal parts from the corrosive absorbent solution in the presence of air. Performance additives enhance the heat and mass transfer coefficients of the water-lithium bromide solutions. Flow control from the generator to the absorber is typically achieved with a control valve.

The COP of an absorption system is the cooling achieved in the evaporator divided by the heat input to the generator. The COP of a single-effect water-lithium bromide chiller generally is from 0.65 to 0.70. The heat rejected by the cooling tower from both the condenser and the absorber is the sum of the waste heat supplied plus the cooling effect produced. Thus, absorption systems require larger cooling towers and cooling water flows than do vapor compression systems.

Single-effect water-lithium bromide chillers can still be found operating in older buildings and chiller plants. However, most new systems being sold by major manufacturers are double-effect chillers because of their improved efficiency over single-effect technology (EPRI, 1996). Double-effect absorption systems have a two-stage generator (Figure 4.2.10) with heat input temperatures greater than 150°C (300°F). The basic operation of the double-effect machine is the same as the single-effect machine except that an additional generator, condenser, and heat exchanger are used. Energy from an external heat source is used to boil the dilute lithium bromide (absorbent) solution. The vapor from the primary generator flows in tubes to the second-effect generator. It is hot enough to boil and concentrate absorbent, which creates more refrigerant vapor without any extra energy input. Dual-effect machines typically use steam or hot liquids as input. Coefficients of performance above 1.0 can be obtained with these machines.

Figure 4.2.11 illustrates the part load performance of a dual effect chiller. The steam consumption ratio (SCR) is plotted as a function of capacity at several condenser water temperatures. For the 29.4°C (85°F) curve, at 50% capacity, the system uses only about 5% less steam than at full load. Information such as that shown in Figure 4.2.11 is available from manufacturers. With these data, performance of

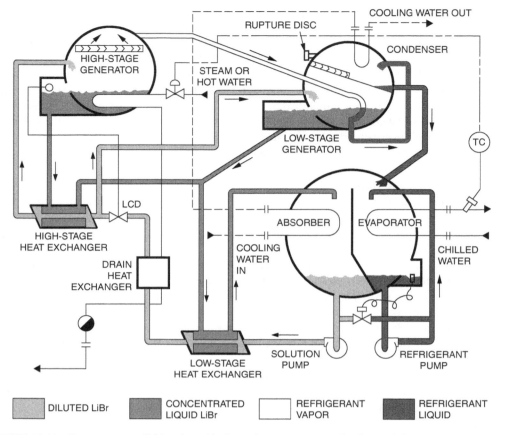

FIGURE 4.2.10 Two-stage water-lithium bromide absorption system. LCD = level control device, TC = temperature controller (capacity control) [Courtesy of Carrier Corporation].

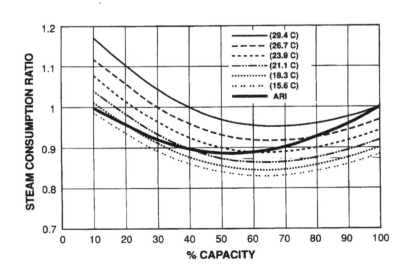

FIGURE 4.2.11 Part load performance curve for a dual-effect absorption chiller.

the system can be estimated over a wide range of conditions. The ARI load line is based on a 1.4°C (2.5°F) reduction in condenser water temperature for every 10% reduction in load. Absorption chillers are tested in accordance with ARI Standard 560 (ARI, 1992).

Chiller Controls

New chillers are equipped with electronic control systems which provide for safe operation and capacity control to meet the cooling load. The controls allow for setpoint temperature of the chilled water leaving the evaporator. This section describes some of the controls found on many chillers. For a specific chiller, the manufacturer's literature should be consulted.

Expansion Valves

The primary purpose of an expansion device is to control the amount of refrigerant entering the evaporator and to ensure that only superheated vapor reaches the compressor. In the process, the refrigerant entering the valve expands from a relatively high-pressure subcooled liquid to a saturated low-pressure liquid/vapor mixture. Other types of flow control devices, such as pressure regulators and float valves, can also be found in some refrigeration systems. Discussion of these can be found in Wang (1993). The most common expansion valves found in modern chillers are thermal expansion valves and electronic expansion valves, discussed below. Three other expansion devices are found in refrigeration systems or smaller air conditioning equipment: constant pressure expansion valves, short tube restrictors, and capillary tubes.

Thermostatic Expansion Valve — The thermostatic expansion valve (TXV) uses the superheat of the gas leaving the evaporator to control the refrigerant flow into the evaporator. Its primary function is to provide superheated vapor to the suction of the compressor. A TXV is mounted near the entrance to the evaporator and has a capillary tube extending from its top that is connected to a small bulb (Figure 4.2.12). The bulb is mounted on the refrigerant tubing near the evaporator outlet. The capillary tube and bulb are filled with the thermostatic charge (ASHRAE, 1998). This charge often consists of a vapor or liquid that is the same substance as the refrigerant used in the system. The response of the TXV and the superheat setting can be adjusted by varying the type of charge in the capillary tube and bulb.

The operation of a TXV is straightforward. Liquid enters the TXV and expands to a mixture of liquid and vapor at pressure P_2 (Figure 4.2.12). The refrigerant evaporates as it travels through the evaporator and reaches the outlet where it is superheated. If the load on the evaporator is increased, the superheat leaving the evaporator will increase. This increase in flow will increase the temperature and pressure (P_1) of the charge within the bulb and capillary tube. Within the top of the TXV is a diaphragm. With an increase in pressure of the thermostatic charge, a greater force is exerted on the diaphragm, which forces the valve port to open and allow more refrigerant into the evaporator. The larger refrigerant flow reduces the evaporator superheat back to the desired level.

The capacity of TXVs is determined on the basis of opening superheat values. TXV capacities are published for a range in evaporator temperatures and valve pressure drops. TXV ratings are based on liquid only entering the valve. The presence of flash (two-phase) gas will reduce the capacity substantially.

Electronic Expansion Valve — The electronic expansion valve (EEV) has become popular in recent years on larger or more expensive systems where its cost can be justified. EEVs can be heat motor activated, magnetically modulated, pulse width modulated, and step motor driven (ASHRAE, 1998). These are an integral part of the refrigeration system in a chiller. The EEV size and type are chosen by the manufacturer. EEVs can be used with digital control systems to provide control of the refrigeration system based on input variables from throughout the system. They offer more precise control of the refrigerant system than do TXVs. Also, some manufacturers make EEVs that are capable of flow in either direction through the EEV. This allows one EEV to replace two TXVs in heat pump applications. The selection of the valve size is similar to that for TXVs. The system refrigerant, evaporator load, liquid temperature, desired capacity, and pressure drop across the valve must be known.

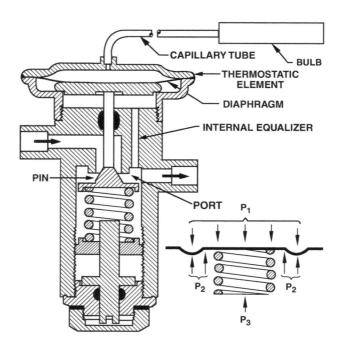

FIGURE 4.2.12 Cross section of a thermal expansion valve. P_1 = thermostatic element pressure, P_2 = evaporator pressure, and P_3 = pressure equivalent of the superheat spring force (reprinted with permission of ASHRAE).

Head Pressure Controls

Air cooled chillers operating during low outdoor temperatures require some means of maintaining an adequate condensing pressure to ensure proper system performance. The two primary reasons for needing adequate condensing pressure include (1) higher condensing pressure helps keep the refrigerant upstream of the expansion valve subcooled to prevent premature flashing in the liquid line, and (2) higher pressures ensure that the pressure differential between the condenser and evaporator are high enough so the expansion valve can provide proper control of the system.

The two most common ways of controlling head (condenser) pressure include a three-way pressure regulating valve and airflow control through the condenser. The three-way pressure regulating valve has two refrigerant inlet ports and one outlet port. One inlet comes from the outlet of the condenser and the other inlet is connected to a refrigerant line that bypasses the condenser. Under normal, high outdoor temperature operation, only refrigerant from the condenser passes through the valve. However, at low ambient temperatures, the valve begins to close off flow through the condenser and increases its pressure. This action forces refrigerant through the bypass around the condenser. This type of valve is usually preset to hold condenser pressure above a specific value.

As the outdoor temperature drops, the air conditioning load typically decreases and the capacity of the condenser increases. To reduce condenser capacity and maintain an acceptable minimum condenser pressure, the air flow through the condenser can be reduced at lower outdoor temperatures. Larger condensers have multiple fans. Individual fans can be sequenced to maintain condenser pressure. Another approach is to use variable speed control on at least one fan.

Head pressure controls are needed to maintain proper operation of the refrigerant side of the system. However, the artificially high condenser pressures created by head pressure controls decrease the efficiency of the chiller and increase energy use at low outdoor temperatures.

Capacity Controls

The type of capacity control used in a chiller depends on the type of chiller. With the growing affordability of variable speed drive technology, many chillers are equipped with variable speed electric drives. With variable speed drives, capacity is controlled directly by the speed of the compressor(s). Variable speed drives offer excellent energy saving opportunities compared to some other capacity control technologies.

Reciprocating chillers with multiple cylinders often use cylinder unloading to reduce capacity as the thermal load on the building drops. Unloading is accomplished by either bypassing gas to the suction chamber, blocking the suction or discharge valve, or closing the suction valve (Wang, 1993).

Capacity modulation for both screw and centrifugal compressors is discussed in earlier sections describing these two compressors.

A capacity control technique found on some older systems is hot-gas bypass control. With this technique, some of the hot discharge gas from the compressor bypasses the condenser and expansion valve and is introduced between the expansion valve and the evaporator. Hot-gas bypass provides a wide range of control of cooling capacity in the evaporator. However, the technique does not provide any energy saving at low loads, is discouraged by current building standards, and is prohibited in federal buildings (Wang, 1993).

Safety Controls

Chiller safety controls are provided to shut the chiller down in case of a malfunction. A short summary of some of the more important safety controls is provided below. A more complete discussion can be found in Wang (1993) and ASHRAE (1998).

Low pressure controls ensure that the compressor operates only if the suction pressure is above a set value. At low suction pressures, the refrigerant flow rate can drop below the rate needed to cool electric motors in hermetic systems. Low pressures can occur if the chilled water flow drops too low or if the chiller has lost refrigerant.

High pressure controls shut the compressor down if the discharge pressure reaches a high enough value to possibly cause damage to the compressor. High discharge temperatures are usually associated with high discharge pressures. The lubricant in the refrigeration system can begin to break down at high discharge temperatures.

Low temperature control in chillers keeps the chilled water from freezing in the evaporator. If the water freezes, the evaporator can be damaged.

Oil pressure failure control protects the compressor. Insufficient lubrication of the compressor can result from low oil pressure. Thus, the compressor would be shut down if this condition is indicated.

Motor overload controls shut down the motor to keep it from overheating caused by overloading. Thermal sensors inside the motor sense temperature in the motor windings. The electric motor current can also be measured to prevent it from exceeding a preset fixed value.

Centrifugal Chiller Controls

Controls unique to centrifugal chillers include *surge protection, air purge,* and *demand limit* controls. Surge occurs in a centrifugal compressor when the refrigerant flow is reversed and refrigerant flows from the discharge to the suction in the compressor. If surge is detected, the condenser water temperature is lowered to reduce the condenser refrigerant pressure and to eliminate the surge.

Gases, such as air and water vapor, can leak into the chiller. Purging is normally done automatically at fixed intervals to eliminate these gases. In older chillers, for every kilogram of air purged from the system, (1.5–9 kg) (3–20 lb) of refrigerant could be exhausted (Carrier, 1999). Purge systems in new chillers reduce refrigerant losses by a factor of 10 to 15 (Carrier, 1999).

Centrifugal chillers are normally applied in large buildings where electrical demand charges may be high during certain parts of the day. The demand limit controls can be used to limit current draw to 40–100% of full load. Limiting the power consumption also limits the capacity of the system.

Absorption Chiller Controls

Absorption chillers require a variety of limit and safety controls (ASHRAE, 1994). Low-temperature chilled water control allows the user to set the exiting chilled water temperature. The low-temperature refrigerant limit control reduces the loading on the chiller as the refrigerant temperature drops. If the refrigerant temperature drops enough, this control will shut off the machine. The absorbent concentration limit

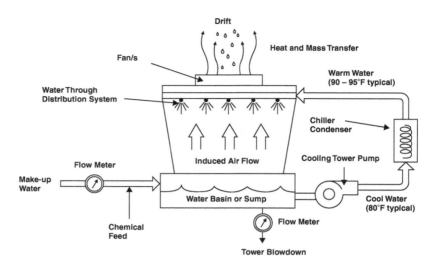

FIGURE 4.2.13 Mechanical draft cooling tower.

control prevents lithium bromide from crystallizing by reducing the loading of the chiller based on temperature and pressure measurements in the water-lithium bromide solutions. High pressure and temperature limit controls limit the pressure of the generator and the maximum operating temperature of the solution near the burners on direct fired machines. Absorption systems also include flow control switches for chilled water, condenser water, and pump motor coolant to shut down the system if flow is stopped in these circuits. Some systems provide either a modulating valve or variable speed pump to control the flow of concentrated solution from the absorber to the generator.

The cooling water entering the absorber tubes is usually limited to between 7 and 43°C (45–109°F) (ASHRAE, 1994). If temperatures drop below 7°C (45°F) or there is a sudden decrease in the cooling water temperature, crystallization of the absorbent solution can occur in the heat exchanger. Most systems have a control that limits the heat input in the generator to the entering cooling water temperature in the absorber.

Cooling Towers

If a chiller is used to provide chilled water for building air conditioning, then the heat energy that is absorbed through that process must be rejected. The two most common ways to reject thermal energy from the vapor compression process are either directly to the air or through a cooling tower. In a cooling tower, water is recirculated and evaporatively cooled through direct contact heat transfer with the ambient air. This cooled water can then be used to absorb and reject the thermal energy from the condenser of the chiller. The most common cooling tower used for HVAC applications is the mechanical draft cooling tower (Figure 4.2.13). The mechanical draft tower uses one or more fans to force air through the tower, a heat transfer media or fill that brings the recirculated water into contact with the air, a water basin (sump) to collect the recirculated water, and a water distribution system to ensure even dispersal of the water into the tower fill.

Figure 4.2.14 shows the relationship between the recirculating water and air as they interact in a counterflow cooling tower. The evaporative cooling process involves simultaneous heat and mass transfer as the water comes into contact with the atmospheric air. Ideally, the water distribution system causes the water to splash or atomize into smaller droplets, increasing the surface area of water available for heat transfer. The approach to the wet-bulb is a commonly used indicator of tower size and performance. It is defined as the temperature difference between the cooling water leaving the tower and the wet-bulb of the air entering the tower. Theoretically, the water being recirculated in a tower could reach the wet-bulb temperature, but this does not occur in actual tower operations.

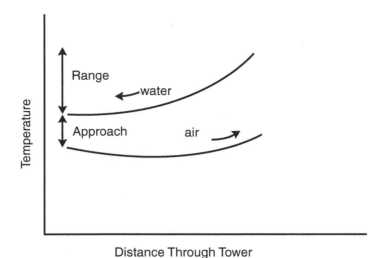

FIGURE 4.2.14 Air/water temperature relationship in a counterflow cooling tower.

The range for a chiller/tower combination is determined by the condenser thermal load and the cooling water flow rate, not by the capacity of the cooling tower. The range is defined as the temperature difference between the water entering the cooling tower and that leaving. The driver of tower performance is the ambient wet-bulb temperature. The lower the average wet-bulb temperature, the "easier" it is for the tower to attain the desired range, typically 6°C (10°F) for HVAC applications. Thus, in a hot, dry climate towers can be sized smaller than those in a hot and humid area for a given heat load.

Cooling towers are widely used because they allow designers to avoid some common problems with rejection of heat from different processes. The primary advantage of the mechanical draft cooling tower is its ability to cool water to within 3–6°C (5–10°F) of the ambient wet-bulb temperature. This means more efficient operation of the connected chilling equipment because of improved (lower) head pressure operation which is a result of the lower condensing water temperatures supplied from the tower.

Cooling Tower Designs

The ASHRAE *Systems and Equipment Handbook* (1996) describes over 10 types of cooling tower designs. Three basic cooling tower designs are used for most common HVAC applications. Based upon air and water flow direction and location of the fans, these towers can be classified as counterflow induced draft, crossflow induced draft, and counterflow forced draft.

One component common to all cooling towers is the heat transfer packing material, or fill, installed below the water distribution system and in the air path. The two most common fills are splash and film. Splash fill tends to maximize the surface area of water available for heat transfer by forcing water to break apart into smaller droplets and remain entrained in the air stream for a longer time. Successive layers of staggered splash bars are arranged through which the water is directed. Film fill achieves this effect by forcing water to flow in thin layers over densely packed fill sheets that are arranged for vertical flow. Towers using film type fill are usually more compact for a given thermal load, an advantage if space for the tower site is limited. Splash fill is not as sensitive to air or water distribution problems and performs better where water quality is so poor that excessive deposits in the fill material are a problem.

Counterflow Induced Draft — Air in a counterflow induced draft cooling tower is drawn through the tower by a fan or fans located at the top of the tower. The air enters the tower at louvers in the base and then comes into contact with water that is distributed from basins at the top of the tower. Thus, the relative directions are counter (down for the water, up for the air) in this configuration. This arrangement

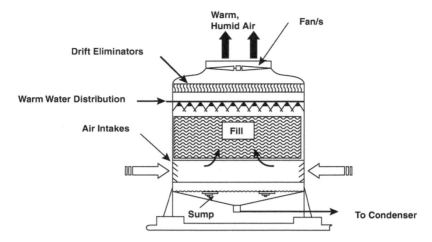

FIGURE 4.2.15   Counterflow induced draft cooling tower.

is shown in Figure 4.2.15. In this configuration, the temperature of the water decreases as it falls down through the counterflowing air, and the air is heated and humidified. Droplets of water that might have been entrained in the air stream are caught at the drift eliminators and returned to the sump. Air and some carryover droplets are ejected through the fans and out the top of the tower. The water that has been cooled collects in the sump and is pumped back to the condenser.

Counterflow towers generally have better performance than crossflow types because of the even air distribution through the tower fill material. These towers also eject air at higher velocities which reduces problems with exhaust air recirculation into the tower. However, these towers are also somewhat taller than crossflow types and thus require more condenser pump head.

Crossflow Induced Draft — As in the counterflow cooling tower, the fan in the crossflow tower is located at the top of the unit (Figure 4.2.16). Air enters the tower at side or end louvers and moves horizontally through the tower fill. Water is distributed from the top of the tower where it is directed into the fill and is cooled by direct contact heat transfer with the air in crossflow (air horizontal and water down). Water collected in the sump is pumped back to the chiller condenser. The increased airflow possible with the crossflow tower allows these towers to have a much lower overall height. This results in lower pump head required on the condenser water pump compared to the counterflow tower. The reduced height also increases the possibility of recirculating the exhaust air from the top of the tower back into the side or end air intakes which can reduce the tower's effectiveness.

Counterflow Forced Draft — Counterflow forced draft cooling towers have the fan mounted at or near the bottom of the unit near the air intakes (Figure 4.2.17). As in the other towers, water is distributed down through the tower and its fill, and through direct contact with atmospheric air it is cooled. Thermal operation of this tower is similar to the counterflow induced draft cooling tower. Fan vibration is not as severe for this arrangement compared to induced draft towers. There is also some additional evaporative cooling benefit because the fan discharges air directly across the sump which further cools the water. There are some disadvantages to this tower. First, the air distribution through the fill is uneven, which reduces tower effectiveness. Second, there is risk of exhaust air recirculation because of the high suction velocity at the fan inlets, which can reduce tower effectiveness. These towers find applications in small- and medium-sized systems.

Materials

Cooling towers operate in a continuously wet condition that requires construction materials to meet challenging criteria. Besides the wet conditions, recirculating water could have a high concentration of mineral salts due to the evaporation process. Cooling tower manufacturers build their units from a

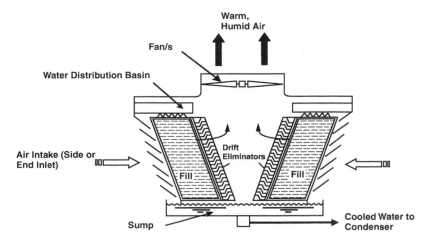

FIGURE 4.2.16 Crossflow induced draft cooling tower.

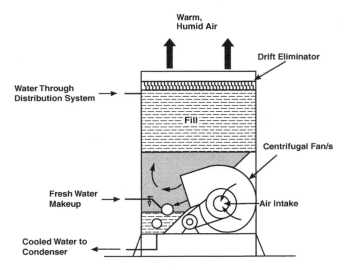

FIGURE 4.2.17 Counterflow forced draft cooling tower.

combination of materials that provide the best combination of corrosion resistance and cost. Wood is a traditional material used in cooling tower construction. Redwood or fir are often used and are usually pressure treated with preservative chemicals. Chemicals such as chromated copper arsenate or acid copper chromate help prevent decay due to fungi or destruction by termites.

Galvanized steel is commonly used for small- to mid-sized cooling tower structures. Hardware is usually made of brass or bronze. Critical components, such as drive shafts, hardware mounting points, etc., may be made from 302 or 304 stainless steel. Cast iron can be found in base castings, motor housings, and fan hubs. Metals coated with plastics are finding application for special components.

Many manufacturers make extensive use of fiberglass-reinforced plastic (FRP) in their structure, pipe, fan blades, casing, inlet louvers, and connection components. Polyvinyl chloride (PVC) is used for fill media, drift eliminators, and louvers. Fill bars and flow orifices are commonly injection molded from polypropylene and acrylonitrile butadiene styrene (ABS).

Concrete is normally used for the water basin or sump of field erected towers. Tiles or masonry are used in specialty towers when aesthetics are important.

TABLE 4.2.6 Cooling Tower Design Parameters

Value	Typical Range or Description
Heat load — kW (Btu/hr)	Determined for the specific application
Condenser water flow rate — L/s (gpm)	0.06 L/s/kW (3 gpm/ton) of rejected load is commonly used to size the cooling tower water recirculation rate
Entering condenser water temperature	32°C (90°F) to 46°C (115°F) is a common range for HVAC and refrigeration applications. A nominal value would be 35°C (95°F).
Leaving condenser water temperature	27°C (80°F) to 32°C (90°F) is a common range for HVAC and refrigeration applications. This value depends on ambient wet-bulb temperature. A nominal value would be about 27°C (80°F).
Outdoor wet-bulb temperature	Depends upon geographical location. The designer should consult local weather archives (use 1 or 2.5% summer value). A typical conservative design value is 25°C (78°F).
Range	Depends on water recirculation rate and load to be rejected. Range can be as high as 8°C (15°F). A typical value is 5.5°C (10°F).
Approach	Varies from 3°C (5°F) to 7°C (12°F) for HVAC applications. Approach less than 3°C (5°F) is not economical (extremely large tower required).

Performance

Rejection of the heat load produced at the chilling equipment is the primary goal of a cooling tower system. This heat rejection can be accomplished with an optimized system that minimizes the total compressor power requirements of the chiller and the tower loads such as the fans and condenser pumps. Several criteria must be determined before the designer can complete a thorough cooling tower analysis, including selection of tower range, water-to-air ratio, approach, fill type and configuration, and water distribution system. Table 4.2.6 lists some of the common design criteria and normally accepted ranges for cooling towers.

Most common HVAC applications requiring a cooling tower will use an "off the shelf" unit from a cooling tower manufacturer. Manufacturer representatives are usually well informed about their products and their proper application. After the project design process has produced the information called for in Table 4.2.6, it is time to contact one or more cooling tower representatives and seek their input on correct tower selection.

Control Scheme with Chillers — Most cooling towers are subject to large changes in load and ambient wet-bulb temperature during normal operations. For a typical cooling tower, the tower fan energy consumption is approximately 10% of the electric power used by the chiller compressor. The condenser pumps are about 2–5% of the compressor power. Controlling the capacity of a tower to supply adequately cooled water to the condenser while minimizing energy use is a desirable operational scheme. Probably the most common control scheme employed for towers serving an HVAC load is to maintain a fixed leaving water temperature, usually 27°C (80°F). Fan cycling is a common method to achieve this cooling tower control strategy and is applicable to multiunit and multicell tower installations. However, this control method does not minimize total energy consumed by the chiller/cooling tower system components.

Lowering the condensing water temperature increases a chiller's efficiency. As long as the evaporator temperature is constant, a reduced condenser temperature will yield a lower pressure difference between the evaporator and condenser and reduce the load on the compressor. However, it is important to recognize that the efficiency improvements initially gained through lower condenser temperatures are limited. Improved chiller efficiency may be offset by increased tower fan and pumping costs. Maintaining a constant approach at some minimum temperature is desirable as long as the condensing temperature does not fall below the chiller manufacturer's recommendations.

Since most modern towers use two- or three-speed fans, a near optimal control scheme can be developed as follows (Braun and Diderrich, 1990):

- Tower fans should be sequenced to maintain a constant approach during part load operation to minimize chiller/cooling tower energy use.
- The product of range and condensing water flow rate, or the heat energy rejected, should be used to determine the sequencing of the tower fans.
- Develop a simple relationship between tower capacity and tower fan sequencing.

De Saulles and Pearson (1997) found that savings for a setpoint control versus the near optimal control for a cooling tower were very similar. Their control scheme called for the tower to produce water at the lowest setpoint possible, but not less than the chiller manufacturer would allow, and to compare that operation to the savings obtained using near optimal control as described above. They found that the level of savings that could be achieved was dependent on the load profile and the method of optimization. Their simulations showed 2.5 to 6.5% energy savings for the single setpoint method while the near optimal control yielded savings of 3 to 8%. Use of variable speed fans would increase the savings only in most tower installations. It is more economical to operate multiple cooling tower fans at the same speed than to operate one at maximum before starting the next fan. Variable speed fans should be used when possible in cooling towers.

The system designer should ensure that any newly installed cooling tower is tested according to ASME Standard PTC 23 (ASME 1986) or CTI Standard ATC-105. These field tests ensure that the tower is performing as designed and can meet the heat rejection requirements for the connected chiller or refrigeration load.

Selection Criteria

The criteria listed in Table 4.2.6 are usually known *a priori* by the designer. If not known explicitly, then commonly accepted values can be used. These criteria are used to determine the tower capacity needed to reject the heat load at design conditions. Other considerations besides the tower's capacity include economics, servicing, environmental considerations, and aesthetics. Many of these factors are interrelated, but, if possible, they should all be evaluated when selecting a particular tower design.

Because economics is an important part of the selection process, two methods are commonly used — life-cycle costing and payback analysis. These procedures compare equipment on the basis of owning, operation, and maintenance costs. Other criteria can also affect final selection of a cooling tower design: building codes, structural considerations, serviceability, availability of qualified service personnel, and operational flexibility for changing loads. In addition, noise from towers can become a sensitive environmental issue. If local building code sound limits are an issue, sound attenuators at the air intakes and the tower fan exit should be considered. Aesthetics can be a problem with modern architectural buildings or on sites with limited land space. Several tower manufacturers can erect custom units that can completely mask the cooling tower and its operation.

Applications

Unlike chillers, pumps, and air handlers, the cooling tower must be installed in an open space with careful consideration of factors that might cause recirculation (recapture of a portion of warm and humid exhaust air by the same tower) or restrict air flow. A poor tower siting situation might lead to recirculation, a problem not restricted to wet cooling towers. Similar recirculation can occur with air-cooled condensing equipment as well. With cooling tower recirculation, performance is adversely affected by the increase in entering wet-bulb temperature. The primary causes of recirculation are poor siting of the tower adjacent to structures, inadequate exhaust air velocity, or insufficient separation between the exhaust and intake of the tower.

Multiple tower installations are susceptible to interference — when the exhaust air from one tower is drawn into a tower located downwind. Symptoms similar to the recirculation phenomenon then plague the downwind tower. For recirculation, interference, or physically blocking air-flow to the tower the result is larger approach and range which contribute to higher condensing pressure at the chiller. Both recirculation and interference can be avoided through careful planning and layout.

Another important consideration when siting a cooling tower installation is the effect of fogging, or plume, and carryover. Fogging occurs during cooler weather when moist warm air ejected from the tower comes into contact with the cold ambient air, condenses, and forms fog. Fog from cooling towers can limit visibility and can be an architectural nuisance. Carryover is when small droplets of entrained water in the air stream are not caught by the drift eliminators and are ejected in the exhaust air stream. These droplets then precipitate out from the exhaust air and fall to the ground like a light mist or rain (in extreme cases). Carryover or drift contains minerals and chemicals from the water treatment in the tower and can cause staining or discoloration of the surfaces it settles upon. To mitigate problems with fog or carryover, as with recirculation, the designer should consider nearby traffic patterns, parking areas, prevailing wind direction, large glass areas, or other architectural considerations.

Operation and Maintenance

Winter Operation — If chillers or refrigeration equipment are being used in cold weather, freeze protection should be considered to avoid formation of ice on or in the cooling tower. Capacity control is one method that can be used to control water temperature in the tower and its components. Electric immersion heaters are usually installed in the tower sump to provide additional freeze protection. Since icing of the air intakes can be especially detrimental to tower performance, the fans can be reversed to de-ice these areas. If the fans are operating in extremely cold weather, ice can accumulate on the leading edges of the fan blades, which can cause serious imbalance in the fan system. Instrumentation to detect out-of-limits vibration or eccentricity in rotational loads should be installed. As with any operational equipment, frequent visual inspections during extreme weather are recommended.

Water Treatment — The water circulating in a cooling tower must be at an adequate quality level to help maintain tower effectiveness and prevent maintenance problems from occurring. Impurities and dissolved solids are concentrated in tower water because of the continuous evaporation process as the water is circulated through the tower. Dirt, dust, and gases can also find their way into the tower water and either become entrained in the circulating water or settle into the tower sump. To reduce the concentration of these contaminants, a percentage of the circulating water is drained or blown-down. In smaller evaporatively cooled systems, this process is called a bleed-off and is continuous. Blow-down is usually 0.8 to 1.2% of the total water circulation rate and helps to maintain reduced impurity concentrations and to control scale formation. If the tower is served with very poor water quality, additional chemical treatments might be needed to inhibit corrosion, control biological growth, and limit the collection of silt. If the tower installation presents continuing water quality problems, a water treatment specialist should be consulted.

Legionellosis — *Legionellosis* has been connected with evaporative condensers, cooling towers, and other building hydronic components. Researchers have found that well-maintained towers with good water quality control were not usually associated with contamination by *Legionella pneumophila* bacteria. In a position paper concerning *Legionellosis*, the Cooling Tower Institute (CTI, 1996) stated that cooling towers are prone to colonization by *Legionella* and have the potential to create and distribute aerosol droplets. Optimum growth of the bacteria was found to be at about 37°C (99°F) which is an easily attained temperature in a cooling tower.

The CTI proposed recommendations regarding cooling tower design and operation to minimize the presence of *Legionella*. They do not recommend frequent or routine testing for *Legionella pneumophila* bacteria because there is difficulty interpreting test results. A clean tower can quickly be reinfected, and a contaminated tower does not mean an outbreak of the disease will occur.

Maintenance — The cooling tower manufacturer usually provides operating and maintenance (O&M) manuals with a new tower installation. These manuals should include a complete list of all parts used and replaceable in the tower and also details on the routine maintenance required for the cooling tower. At a minimum, the following should also be included as part of the maintenance program for a cooling tower installation.

- Periodic inspection of the entire unit to ensure it is in good repair.
- Complete periodic draining and cleaning of all wetted surfaces in the tower. This gives the opportunity to remove accumulations of dirt, slime, scale, and areas where algae or bacteria might develop.
- Periodic water treatment for biological and corrosion control.
- Continuous documentation on operation and maintenance of the tower. This develops the baseline for future O&M decisions and is very important for a proper maintenance policy.

TABLE 4.2.7 Packaged Equipment Advantages and Disadvantages

Disadvantages	Advantages
Limited performance choices because of fixed component sizing.	Individual room control is allowed.
Unitary systems generally not good for close space humidity control.	Cooling and heating are available at any time and are independent of operation in other spaces.
Space temperature control is usually two-position which causes temperature swing.	Individual ventilation, when included with the unit, is available whenever the unit is operated.
Packaged system life is relatively short.	Unit capacities are certified by the manufacturer.
Energy usage will be higher than a central system because of fixed capacity increments and tendency to oversize equipment.	Equipment in unoccupied spaces can easily be turned off which is an easy energy conservation opportunity.
Full use of economizer cycle is usually not possible.	Unitary equipment operation is usually very simple.
Air distribution control is restricted on individual room units.	Packaged equipment requires less floor space than central systems.
Sound levels of equipment can be objectionable.	First costs are low.
Outside air for ventilation is usually limited or set at a fixed quantity.	Equipment can be located such that shorter duct runs or reduced duct space is allowed.
Aesthetics of units can be unappealing.	Installation is relatively simple; no factory-trained personnel are required.
Filtering option for air flow through units can be limited.	
Condensate from units can be a nuisance.	
Maintenance can be an issue because of the number of units, location, or difficult access.	

4.2.4 Packaged Equipment

Central HVAC systems are not always the best application for a particular cooling or heating load. Initial costs for central systems are usually much higher than unitary or packaged systems. There may also be physical constraints on the size of the mechanical components that can be installed in the building. Unitary or packaged systems come factory assembled and provide only cooling or combined heating and cooling. These systems are manufactured in a variety of configurations that allow the designer to meet almost any application. Cabinet or skid-mounted for easy installation, typical units generally consist of an evaporator, blower, compressor, condenser, and, if a combined system, a heating section. The capacities of the units ranges from approximately 5 kW to 460 kW (1.5 to 130 tons). Typical unitary systems are single-packaged units (window units, rooftop units), split-system packaged units, heat pump systems, and water source heat pump systems. Unitary systems do not last as long (only 8 to 15 years) as central HVAC equipment and are often less efficient.

Unitary systems find application in buildings up to eight stories in height, but they are more generally used in one-, two-, or three-story buildings that have smaller cooling loads. They are most often used for retail spaces, small office buildings, and classrooms. Unitary equipment is available only in pre-established capacity increments with set performance characteristics, such as total L/s (cfm) delivered by the unit's air handler. Some designers combine central HVAC systems with packaged equipment used on perimeter building zones. This composite can solve humidity and space temperature requirements better than packaged units alone. This also works well in buildings where it is impractical for packaged units to serve interior spaces.

Table 4.2.7 lists some of the advantages and disadvantages of packaged and unitary HVAC equipment. Table 4.2.8 lists energy efficiency ratings (EERs) for typical electric air- and water-cooled split and single package units with capacity greater than 19 kW (65,000 Btuh).

Typically, commercial buildings use unitary systems with cooling capacities greater than 18 kW (5 tons). In some cases, however, due to space requirements, physical limitations, or small additions, residential-sized unitary systems are used. If a unitary system is 10 years or older, energy savings can be achieved by replacing unitary systems with properly sized, energy-efficient models.

TABLE 4.2.8   Unitary Package System Rating

Product Type[a] and Size	Recommended EER[b]	Best Available EER (1998)
Air source 19–40 kW (65–135 MBtuh)	10.3 or more	13.5
Air source 40–73 kW (135–240 MBtuh)	9.7 or more	11.5
Air source >73 kW (240 MBtuh)	10.0 or more	11.7
Water source 19–40 kW (65–135 MBtuh)	11.5 or more	12.5
Water source >40 kW (135 MBtuh)	11.0 or more	11.0
Product Type	Recommended SEER[c,d]	Best Available SEER (1998)
Residential Air Conditioner[e]	12.0 or more	18.0

[a] Electric air- and water-cooled split system and single package units with capacity over 19 kW (65,000 Btu) are covered here.

[b] EER, or energy efficiency ratio, is the cooling capacity in kW (Btu/h) of the unit divided by its electrical input (in watts) at standard (ARI) conditions of 35°C (95°F) for air-cooled equipment, and 29°C (85°F) entering water for water-cooled models.

[c] Based on ARI 210/240 test procedure.

[d] SEER (seasonal energy efficiency ratio) is the total cooling output kW (Btu) provided by the unit during its normal annual usage period for cooling divided by the total energy input (in Wh) during the same period.

[e] Split system and single package units with total capacity under 19 kW (65,000 Btu) are covered here. This analysis excludes window units and packaged terminal units.

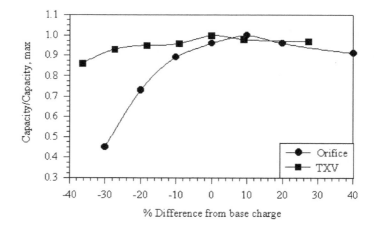

FIGURE 4.2.18   Comparison between TXV and short-tube orifice systems capacity for a range of charging conditions and 95°F (35°C) outdoor temperature. (From Rodriquez et al., 1996).

As with any HVAC equipment, proper maintenance and operation will ensure optimum performance and life for a system. Split-system air conditioners and heat pumps are the most common units applied in residential and small commercial applications. These units are typically shipped to the construction site as separate components; after the condenser (outdoor unit) and the evaporator (indoor unit) are mounted, the refrigerant piping is connected between them. The air conditioning technician must ensure that the unit is properly charged with refrigerant and check for proper operation. If the system is under- or over-charged, performance can be adversely affected. Rodriquez et al. (1996) found that performance of an air conditioning system equipped with a short tube orifice was affected by improper charge (Figure 4.2.18).

The plot in Figure 4.2.18 clearly shows that for a 20% under-charge in refrigerant, a unit with a short tube orifice suffers a 30% decrease in cooling capacity. This same study also investigated the effects of return-air leakage. A common problem with new installations is improper sealing of duct connections

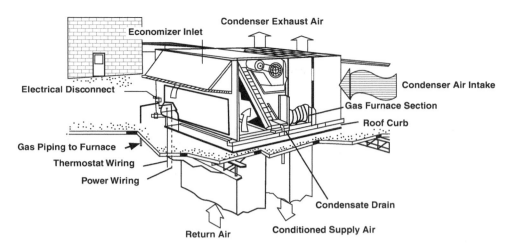

FIGURE 4.2.19 Rooftop packaged heating and air conditioning unit. (Adapted from Carrier Corporation).

at the diffusers and grills as well as around the return-air plenum. Leakage amounts as low as 5% in the return air ducts resulted in capacity and efficiency reductions of almost 20% for high humidity climates. These reductions dropped to about 7% for low humidity climates. The results of the charging and leakage studies suggest the need for the installation contractor, maintenance contractor, and system owner to ensure the proper installation of the air conditioning system.

Packaged Units

Packaged units are complete HVAC units that are usually mounted on the exterior of a structure (roof or wall) freeing up valuable indoor floor space (Figure 4.2.19). They can also be installed on a concrete housekeeping pad at ground level. Because they are self-contained, complete manufactured units, installation costs are usually lower than for a site-built HVAC system.

Single-package units consist of a blower section, filter bank, evaporator coil, at least one compressor (larger units may have more than one), and an air-cooled condensing section. Units may also come equipped with a heating section. Heating is accomplished using either natural gas or electricity. Heat pump systems can be used in situations where electricity is the only source of energy. Unitary heat pumps are restricted in size to no more than 70 kW (20 tons).

As packaged units age and deteriorate, their efficiency often decreases while the need for maintenance increases. Upgrading existing packaged units to high-efficiency models will result in substantial long-term energy savings. In the last 10 to 15 years, manufacturers have made significant improvements in the efficiency of packaged units. The efficiency of energy transfer at both the evaporator and condenser coils has been improved, high-efficiency motors are now standard, and blower and compressor designs have improved in high-efficiency packaged units. Scroll compressors are now commonplace on medium-sized (70 to 210 kW; 20 to 60 ton) rooftop units. Energy efficiencies of newer units have a SEER in the range of 9.50 to 13.0. It is not uncommon to find older units operating at efficiencies as low as 6.0, and most operate at less than 9.0. Gas-fired heating sections typically have an annual fuel utilization efficiency (AFUE) of about 80%. All newer packaged rooftop units are equipped with factory-installed microprocessor controls. These controls make maintaining equipment easier and improve energy efficiency of both the unit and the overall HVAC system. Control features include temperature setback and on/off scheduling. Larger systems can be delivered with variable air volume capability. Also, most units have an optional communication interface for connection to an energy management control system.

Vertical Packaged Units

Vertical packaged units are typically designed for indoor or through-the-wall installation. These units are applied in hotels and apartments. Some designs have a water-cooled condenser, which can be fed

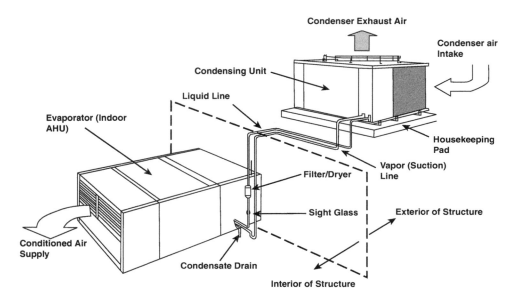

FIGURE 4.2.20   Split system diagram (courtesy of the Trane Co.).

from a cooling tower and/or city water. Many others use standard air-cooled condensers. Both style units have all other components mounted inside the package. Ductwork, if needed, can be connected to the unit to distribute the air.

Split-System Packaged Units

Split-system packaged units can have the condenser mounted on an outdoor housekeeping pad or on a rooftop. Refrigerant piping connects the compressor section to an indoor air handling unit and evaporator coil. Unless they are heat pump type units, they cannot provide heat to the space. Heating coils can be installed in the air handling section, particularly if there is a central source of heat such as hot water or steam from a boiler. Alternatively, the indoor unit can be coupled to a gas-fired furnace section to provide heating.

Air Source Heat Pumps

Air source heat pump (ASHP) systems are typically rooftop units, either packaged complete or as split systems. Split-package heat pumps are designed with an air handling unit located inside the conditioned space, while the condenser and compressor are packaged in units for outdoor installation on a house-keeping pad or on the roof. During cooling mode, the heat pump operates an air conditioner. During heating mode, the system is reversed and extracts energy from the outside air and provides it to the space. Each of these cycles is shown schematically in Figures 4.2.21 and 4.2.22, respectively. The size of unitary heat pump systems ranges from approximately 5 to 70 kW (1½ to 20 tons). In some cases, existing packaged cooling units with electric resistance heat can be upgraded to heat pumps for improved energy efficiency.

Heat pump applications are best suited to mild climates, such as the southeastern portion of the U.S., and to areas where natural gas for heating is less available. Space heating needs may exceed the capacity of the heat pump during extremely cold weather. This is because the units are most often sized to satisfy the cooling load requirements. As the outdoor temperature drops, the coefficient of performance (COP) of the heat pump decreases. A 26 kW (7½ ton) rooftop heat pump unit that has a high temperature (8.3°C) COP of 3.0 can have a low temperature (−8.3°C) COP of 2.0 or less. Because the capacity also drops with outdoor temperature, heat pumps require supplemental electric resistance heat to maintain temperature in the building. Figure 4.2.23 shows typical trends in capacity and COP for an air source heat pump. Chapter 4.2 discusses the characteristics of heat pumps.

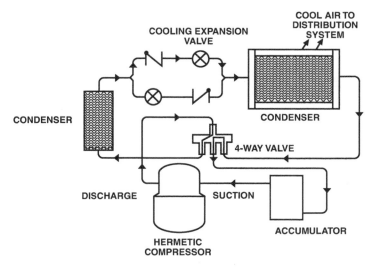

FIGURE 4.2.21 Air or water source heat pump in cooling mode (courtesy of the Trane Co.).

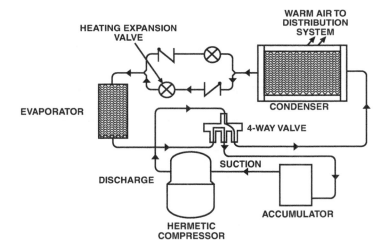

FIGURE 4.2.22 Heat pump schematic showing heating cycle (courtesy of the Trane Co.).

When the ambient air temperature approaches 0°C (32°F), heat pumps operating in the heating mode will begin to build a layer of frost on the outdoor heat exchanger. After a sufficient run-time under these conditions, the unit must go into a defrost cycle. This short (<10 min) cycle melts the frost from the heat exchanger and at the end of the cycle returns the unit to normal heating operation. During the defrost cycle, supplemental heating must be used to supply comfort heating indoors. The electrical energy penalty can become significant under extreme ambient frosting conditions (consistently cold and moist) which coincide with high space heating requirements. Various methods have been used to engage the start of the defrost cycle. A timed cycle can be set to start defrosting at a determined interval, typically about 1.5 hours. The defrost cycle can be terminated either by a control element sensing the coil pressure or a thermostat measuring the temperature of the liquid refrigerant in the outdoor coil. When this temperature reaches about 26°C (80°F) the cycle ends and the unit returns to normal heating operation. Another method utilizes two temperature sensors. One measures outdoor air, and the other responds to refrigerant temperature in the outdoor coil. As frost builds, the temperature difference between these sensors increases, and at a predetermined setpoint, the defrost cycle is started.

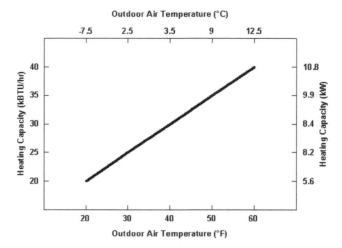

FIGURE 4.2.23　System heating capacity as a function of outdoor air temperature.

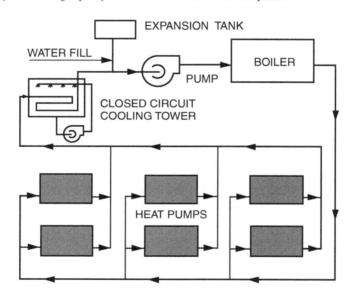

FIGURE 4.2.24　Water loop heat pump system (courtesy of SMACNA).

Water Source Heat Pumps

Water loop, or water source heat pumps use water instead of air to transfer energy between the building and outside. In an air-to-air heat pump system, energy is removed from the indoor air and rejected to the outside air during the cooling cycle. The reverse happens during the heating cycle. However, in a water loop heat pump, water replaces the outdoor air as the source or sink for energy, depending on the cycle in use. In hot weather, a cooling tower removes heat from the water loop; in cooler weather, a central boiler heats the water. As shown in Figure 4.2.24, water loop heat pump systems allow for simultaneous heating and cooling by multiple separate and distinct units and thus increase individual comfort. Recovering heat from cooled areas and recycling it into other areas adds to the system's efficiency. Size of water source heat pumps ranges from approximately 2 to 88 kW ($\frac{1}{2}$ to 25 tons). Efficiencies of water source units are generally higher than their air-to-air counterparts, with an EER of 11.0 and COP of 3.8 to 4.0 not uncommon. High-efficiency water-source heat pumps have an EER as high as 14.0 to 15.0 and a COP as high as 4.4.

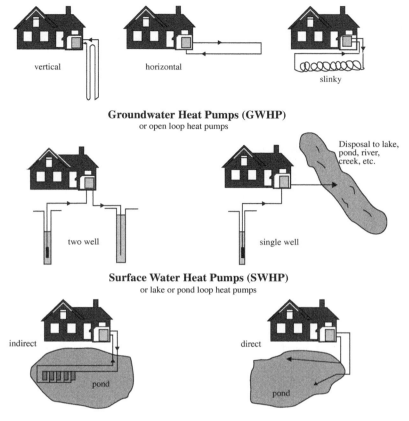

FIGURE 4.2.25 Different ground source heat pump configurations.

Geothermal Heat Pumps

Geothermal heat pumps (Figure 4.2.25) are heat pumps that draw energy from or deposit energy to the ground or groundwater. In the winter, a geothermal heat pump (GHP) transfers thermal energy from the ground or groundwater to provide space heating. In the summer, the energy transfer process is reversed; the ground or groundwater absorbs thermal energy from the conditioned space and cools the air. A GHP benefits from the nearly constant year round ground and groundwater. These temperatures are higher on average than winter air temperatures and lower on average than summer temperatures. The heat pump does not have to work as hard to extract thermal energy from or transfer energy to the ground or groundwater at a moderate temperature as from the cold air in winter or hot air in summer. The energy efficiency of a GHP is thus higher than that of a conventional ASHP. Many GHPs are also more efficient than fossil fuel furnaces in the heating mode.

Each system may also have a desuperheater to supplement the building's water heater, or a full-demand water heater to meet all of the building's hot water needs. The desuperheater transfers excess thermal energy from the GHP's compressor to a hot water tank. In summer, hot water is provided free; in the winter, water heating costs can be reduced by up to 50%. Although residential GHPs are generally more expensive to install than ASHPs, they operate more efficiently than ASHPs. GHPs can also be installed without a backup heat source over a very wide range of climates (EPA, 1993). For commercial buildings, GHPs are very competitive with boilers, chillers, and cooling towers.

The primary difference between an ASHP and a GHP is the investment in a ground loop for energy collection and rejection that is required for the GHP system. Whether a GHP is cost effective relative to a conventional ASHP depends upon generating annual energy cost savings that are high enough to pay for the ground loop in a relatively short time.

Performance Ratings

Most heating and cooling performance ratings are useful for comparing units of the same type (i.e., ASHP to ASHP, or GHP to GHP). The ratings used for different types of equipment (furnaces, ASHP, GHP), however, are not generally comparable. As a result, it is useful to know what the ratings values include. All heat pumps are rated by the ARI. Results are published in the *Directory of Certified Applied Air Conditioning Products* (for GHPs) and the *Directory of Certified Unitary Products* (for ASHPs). For water source heat pumps (the type of heat pump used in all GHP systems), cooling performance is defined by the EER. Electrical input includes compressor, fans, and "pumping" allowance (for the groundwater or ground loop). Heating performance is defined by the the COP. This is the heating effect produced by the unit divided by the energy equivalent of the electrical input resulting in a dimensionless (no units) value.

Both the COP and EER values for groundwater heat pumps are single point (valid only at the specific test conditions used in the rating) steady state values. In contrast, HSPF and SEER values published for air source equipment are seasonal values that depend on both steady state and transient tests. Ratings for GHPs are published under two different headings: ARI Standard 325 (ARI 1998b) and ARI Standard 330 (ARI 1998a). These ratings are intended for specific applications and cannot be used interchangeably. Standard 325 is intended for groundwater heat pump systems. Performance (EER and COP) is published at two water temperatures: 21°C and 10°C (70° and 50°F). The pumping penalty used in Standard 325 (ARI 1998b) is higher than the pumping allowance for Standard 330. Standard 330 is intended for closed loop or ground-coupled GHPs and is based upon entering water temperature of 25°C (77°F) in the cooling mode and 0°C (32°F) in the heating mode. One of the limitations of this rating is that the temperatures used are reflective of a northern climate. Southern installations would see higher temperatures entering the heat pump and, thus, have better winter and poorer summer performance than indicated.

The major difference between ratings for ASHPs and GHPs is that the air source values are seasonal. They are intended to reflect the total heating or cooling output for the season divided by the total electrical input for the season. These ratings (HSPF — heating, SEER — cooling) cannot be directly compared to the GHP EER and COP numbers. ASHPs are rated under Standard 210/240 (ARI 1994). To simplify the process, a number of assumptions are made regarding operation of the heat pump. The rating is based on a moderate U.S. climate and, as a result, is not reflective of either very cold or very warm areas of the country.

4.2.5 Evaporative Cooling

Evaporative air conditioning is an effective method of cooling hot, dry air. Evaporative air conditioning uses no refrigerant gases or mechanical vapor compression in producing the cooling effect. The decrease in electrical consumption and zero use of CFCs possible with evaporative air conditioning equipment means they help reduce greenhouse gas emissions and ozone depletion problems (Foster, 1991). Evaporative air conditioning is the cooling effect provided by the adiabatic evaporation of water in air. Air is drawn through wetted pads or sprays and its sensible heat energy goes towards evaporating some water which reduces the air dry-bulb temperature. In the ideal evaporative process (applies to cooling towers as well) the temperature approaches the ambient air wet-bulb temperature. A typical evaporative air conditioner or "swamp cooler" is shown in Figure 4.2.26.

These coolers contain evaporative media and a water circulating pump to lift the sump water to a distributing system which directs water down through the media and back to the sump. The fan pulls air through the evaporative media where it is cooled by direct contact with the wetted surface area and the water, and then it is delivered to the space to be cooled. Residential-sized units are either side- or down-draft depending on the evaporative media configuration. Evaporative air conditioning units are currently rated by total air delivery, and common sizes range from 5600 to 113,300 l/s (2000 to 40,000 cfm).

Two primary methods of evaporative cooling are used. *Direct cooling*, in which the water evaporates directly into the air-stream, reduces the temperature and humidifies the air. With *indirect cooling*, primary

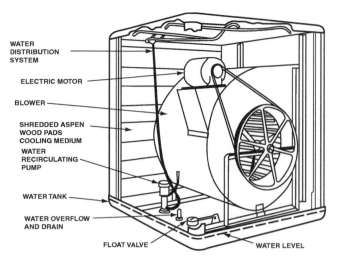

FIGURE 4.2.26 Side-draft evaporative air conditioning unit (courtesy of the Trane Co.).

air is sensibly cooled with a heat exchanger, while the secondary air operates as in the traditional direct cooling mode on the "wet" side of the heat exchanger. Clean, commercial, wetted media evaporative air conditioners typically operate at an evaporation efficiency of approximately 80%. Water use in evaporative coolers depends on air flow, the effectiveness of the wetted media, and the wet-bulb temperature of the incoming air. Fans are usually centrifugal, forward-curved types complete with motor and drive.

Evaporative air conditioning consumes significantly less energy than vapor compression refrigeration equipment of similar cooling capacity. These units operate with a fan and a small water pump. Direct systems in low humidity zones can show energy savings of 60 to 80% over mechanical cooling systems. System selection is usually based on air quantity required to properly cool a space and the system static pressure required for the duct system. To provide comfort cooling in most applications of evaporative air conditioners, 60–120 l/s/m^2 (2–4 cfm/ft^2) is adequate.

Evaporative air conditioning is useful in many commercial and industrial applications, such as schools, commercial greenhouses, laundries, warehouses, factories, kitchens (make-up air), and poultry houses, among others. The largest application area for evaporative air conditioners is in the southwestern U.S. This area experiences warm, dry weather during the cooling season and presents a good opportunity for evaporative air conditioners. They can be employed in all types of buildings that require cooling during times when ambient wet-bulb temperatures are below 18°C (65°F) and where cooling loads cannot be met with outside air only (economizer cycle). Direct evaporative air conditioners are not suitable for areas with strict humidity control requirements. See Kreider et al. (2001) for details.

Regular inspection and maintenance of evaporative air conditioners is required to ensure proper service and effectiveness of the unit. Water treatment (or bleed) is required to help prevent excessive scaling on the interior wetted surfaces of the unit.

REFERENCES

ARI 1992, *Absorption Water Chilling and Water Heating Packages*, ARI Standard 560-92, Air Conditioning and Refrigeration Institute, Arlington, VA.

ARI 1994, *Unitary Air-Conditioning and Air-Source Heat Pump Equipment*, ARI Standard 210/240, Air Conditioning and Refrigeration Institute, Arlington, VA.

ARI 1998a, *Ground Source Closed-Loop Heat Pumps*, ARI Standard 330-98, Air Conditioning and Refrigeration Institute, Arlington, VA.

ARI 1998b, *Ground Water-Source Heat Pumps*, ARI Standard 325-98, Air Conditioning and Refrigeration Institute, Arlington, VA.

ARI 1998, *Standard for Water Chilling Packages Using the Vapor Compression Cycle*, ARI Standard 550, Air Conditioning and Refrigeration Institute, Arlington, VA.

ASHRAE 1992, *Number Designation and Safety Classification of Refrigerants*, ANSI/ASHRAE Standard 34-1992, American Society of Heating, Refrigeration, and Air Conditioning Engineers, Atlanta, GA.

ASHRAE, 1994, *Safety Code for Mechanical Refrigeration*, ANSI/ASHRAE Standard 15-1994. American Society of Heating, Refrigeration, and Air Conditioning Engineers, Atlanta, GA.

ASHRAE 1995, *Methods of Testing Liquid Chilling Packages*, ASHRAE Standard 30, American Society of Heating, Refrigerating, and Air Conditioning Engineers, Atlanta, GA.

ASHRAE, 1996, *ASHRAE Handbook of HVAC Systems and Equipment*, American Society of Heating, Refrigerating, and Air Conditioning Engineers, Atlanta, GA.

ASHRAE, 1997, *ASHRAE Handbook of Fundamentals*, American Society of Heating, Refrigerating, and Air Conditioning Engineers, Atlanta, GA.

ASHRAE, 1998, *ASHRAE Handbook of Refrigeration*, American Society of Heating, Refrigerating, and Air Conditioning Engineers, Atlanta, GA.

ASME, 1986, Atmospheric Water Cooling Equipment, *Performance Test Code* PTC 23-86, American Society of Mechanical Engineers, New York, NY.

Bitondo, M.J. and Tozzi, M.J., 1999, *Chiller Plant Control Multiple Chiller Controls*, Carrier Corporation White Paper, Syracuse, NY, August.

Blue, J.L. et al., 1979, *Building Energy Use Data Book*, 2nd Edition, ORNL-5552, Oak Ridge National Laboratory, Oak Ridge, TN, December.

Braun, J.E. and Diderrich, G.T., 1990, Near-optimal Control of Cooling Towers for Chilled Water Systems, *ASHRAE Transactions*, 96(2), pp. 806–813.

Bullock, C.E., 1997, Theoretical Performance of Carbon Dioxide in Subcritical and Transcritical Cycles, *Refrigerants for the 21st Century*, Proc. ASHRAE/NIST Refrigerants Conf., Gaithersburg, MD, pp. 20–26.

Carrier Corporation 1999, The Money Leaking from the Mechanical Room: A Practical Guide to Addressing Chiller Leaks, *HVAC Analysis*, Vol. 2, No. 2, Syracuse, NY.

Carrier Corporation 1964, *System Design Manual, Part 7, Refrigeration Equipment*, Syracuse, NY.

CTI, 1990, Acceptance Test Code for Water-Cooling Towers, *Standard* ATC 105-90, Cooling Tower Institute, Houston, TX.

CTI, 1996, Legionellosis Position Statement, Cooling Tower Institute, Houston, TX.

De Saulles, T. and Pearson, C.C., 1997, Energy Performance Evaluation of Set Point Control and Near Optimal Control for Cooling Towers Used in Water Chilling Systems, http://www.virtual-conference.com/cibse97/conference/papers/56-content.htm.

Didion, D. 1996, Pratical Considerations in the Use of Refrigerant Mixtures, Presented at the ASHRAE Winter Meeting, Atlanta, Georgia.

EPA, 1993, *Space Conditioning: The Next Frontier*, Report 430-R-93-004, Environmental Protection Agency, Washington, D.C., April.

Foster, R.E., 1991, *Evaporative Air-Conditioning Technologies and Contributions to Reducing Greenhouse Gases*, Asia-Pacific Conf. CFC Issues and Greenhouse Effects, Singapore, May.

Holihan, P., 1998, Analysis of Geothermal Heat Pump Manufacturer's Survey Data, *Renewable Energy 1998: Issues and Trends*, Energy Information Administration.

Jackson, J. and Johnson, W., 1998, *Commercial Energy Use: A Disaggregation by Fuel, Building Type, and End Use*, ORNL/CON-14, Oak Ridge National Laboratory, Oak Ridge, TN, February.

King, G., 1986, *Basic Refrigeration*, Business News Publishing Company, Troy, Michigan.

Kramer, D., 1991, Why Not Propane?, *ASHRAE Journal*, 33, (6), pp. 52–55.

Kreider, J.F., Rabl, A., and Curtiss, P., 2001, *Heating and Cooling of Buildings*, McGraw-Hill, New York, NY.

Matsubara, K., Suefuji, K., and Kuno, H., 1987, The Latest Compressor Technologies for Heat Pumps in Japan, in *Heat Pumps*, K. Zimmerman and R.H. Powell, Jr., Eds., Lewis Chelsea, MI.

Molina, M.J. and Rowland, F.S., 1974, Stratospheric Sink for Chlorofluoromethanes: Chlorine Atoms Catalyzed Destruction of Ozone, *Nature*, 249, pp. 810–812.

National Refrigerants, 1992, *Refrigerant Reference Guide*, Philadelphia, PA.

Rodriquez, A.G., O'Neal, D.L., Bain, J.A, and Davis, M.A., 1996, *The Effect of Refrigerant Charge, Duct Leakage, and Evaporator Air Flow on the High Temperature Performance of Air Conditioners and Heat Pumps, Final Report*, EPRI TR-106542, July.

Salas, C. E. and Salas, M., 1992, *Guide to Refrigeration CFCs*, Fairmont Press, Liburn, GA.

Sand, J.R., Fischer, S.K., and Baxter, V.D., 1999, Comparison of TEWI for Fluorocarbon Alternative Refrigerants and Technologies in Residential Heat Pumps and Air-Conditioners, *ASHRAE Transactions*, Vol. 105, Pt. 1, pp. 1209–1218.

Senshu, T. Araik, A., Oguni, K., and Harada, F., 1985, Annual Energy-Saving Effect of Capacity Modulated Air Conditioner Equipped with Inverter-Driven Scroll Compressor, *ASHRAE Transactions*, Vol. 91, Part 2.

Smit, K., Keder, J., and Tidball, R., 1996, *Electric Chiller Handbook*, TR-105951, Electric Power Research Institute, Palo Alto, CA, February.

Stoecker, W.F., 1994, Comparison of Ammonia with Other Refrigerants for District Cooling Plant Chillers, *ASHRAE Transactions*, Vol. 100, Pt. 1, pp. 1126–1135.

Thevenot, R. 1979, *A History of Refrigeration Throughout the World*, International Institute of Refrigeration, Paris, France, pp. 39–46.

Trane, 1980, *Centrifugal Water Chillers*, The Trane Company, La Crosse, WI.

Troth, S. J., 1994, Air Conditioning with Ammonia for District Cooling, *ASHRAE Journal*, Vol. 36, No. 7, pp. 28–36.

United Nations Environmental Program (UNEP), 1987, *Montreal Protocol on Substances that Deplete the Ozone Layer — Final Act.*

U.S. Census Bureau, 1999, *Statistical Abstract of the United States*, Washington, D.C.

U.S. Department of Energy, 1998, *A Look at Commercial Buildings in 1995: Characteristics, Energy Consumption, and Energy Expenditures*, DOE/EIA-0625(95), Washington, D.C., October.

U.S. Environmental Protection Agency, 1993a, Class I Nonessential Products Ban, Section 610 of the Clean Air Act Amendments of 1990, *Federal Register*, 4768, January 15, 1993.

U.S. Environmental Protection Agency, 1993b, The Accelerated Phaseout of Ozone-Depleting Substances, *Federal Register*, December 10, 1993.

Wang, S. K., 1994, *Handbook of Air Conditioning and Refrigeration*, McGraw-Hill, Inc., New York, NY.

4.3 Ventilation and Air Handling Systems

Ellen M. Franconi and James B. Bradford

Secondary systems transfer heating and cooling energy between central plants and building spaces. This chapter introduces all popular air handling systems and discusses common configurations. Descriptions and design considerations are presented for the following air system components: air filters, humidifiers, coils, fans, ducts, terminal units, and diffusers. The chapter concludes with a discussion of air system controls and an overview of system design procedures.

Air-handling systems encompass the components and function of mechanical ventilation systems and provide air conditioning as well. Thus, air handling systems include a central cooling coil in addition to fans, heating, humidification, heat reclamation, and cooling through use of outdoor air.

Most modern, large commercial buildings require air conditioning to maintain occupant comfort. Historically, commercial buildings had shallow floor plans to maximize natural lighting. Their space-conditioning loads were shell-dominant. With the advent of fluorescent fixtures, air conditioning, and larger building designs, commercial buildings have become internal-gain dominant — requiring year-round air conditioning regardless of climate.

All-air systems meet the entire cooling load with cold air supplied to the conditioned space. Heating may also be supplied through the air system, at the zone, or both. Instead of air systems, water systems may be used to meet air conditioning loads. Because air has a much lower heat capacity than water, a

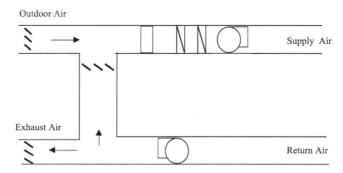

FIGURE 4.3.1   Simple air-handling unit.

much larger volume of air must be moved to meet the same load. Air systems are convenient because they incorporate ventilation within the system. Also, they are well-suited for utilizing an air-side econ-omizer and controlling humidity. Air systems require no piping in the occupied building space. However, air systems do require more building space to accommodate the ductwork. Leaking ducts may not damage building interiors or structure, but leaks are not usually detected and they can decrease system efficiency.

4.3.1   Anatomy of Air Handling Systems

The function of an air handling system is to supply conditioned air to one or several building zones. The term *zone* refers to a thermal space that has comfort conditions controlled by a single thermostat. The air handling system supplies air at a specific flow rate and temperature to the zone in order to meet its heating or cooling load.

A building's HVAC air system may be distributed or centralized. For example, distributed systems can use local direct-expansion, packaged systems that have window, wall, or exterior mounted installations located close to the zones served. Centralized systems are installed in building mechanical rooms and provide heating and cooling to the zones through extensive ductwork. Central air systems distribute cooling and heating provided by the building plant. Generally, plant equipment for centralized systems includes chillers, cooling towers, and boilers.

While this distinction between air system types exists, the basic anatomy of all-air systems is similar. Figure 4.3.1 presents the layout for a typical system. The fundamental equipment components found in most air handling units (AHU) include dampers, air filtration devices, coils, and fans. The system may also include humidifiers and heat recovery devices.

As shown in Figure 4.3.1, during operation of a typical AHU, the outdoor air mixes with return air and passes through an air filter and a variety of air conditioning devices. The cooling coils cool, or the heating coils heat, the mixed air to maintain a supply, return, or zone air setpoint temperature. In systems that control both temperature and humidity, a heating or reheat coil may be present downstream of the cooling coil. The reheat coil raises the temperature of the cooled supply air if the cooling coil over-cools the air in order to remove humidity. Preheat coils may be present upstream of the cooling coil if it is necessary to heat the outside air stream to prevent freezing of the cooling coil during cold outdoor periods. If installed, the humidifier adds moisture to the air stream during winter months.

The supply fan draws and/or blows the air through the AHU equipment components, ductwork, terminal units, and diffusers to supply the required air flow rate to the zone. Return-air fans may be required in central systems to overcome the return system pressure drop and move the air from the building to the AHU or exhaust the air from the building.

Central System Advantages and Disadvantages

There are several advantages that central, all-air system designs have over other types of systems, including water systems and distributed (single zone) air systems. Some advantages include

- The location of equipment is in a centralized, unoccupied location that consolidates and facilitates maintenance.
- Piping is not within the conditioned space, reducing the possibility of damage in occupied areas.
- Air systems make it possible to cool the building with outdoor air.
- Air systems provide flexibility in zoning and comfort control.

Some of the disadvantages associated with all-air central distribution systems are

- Additional space is required for duct work.
- The central distribution fan may frequently need to operate during unoccupied hours in cold climates.
- Proper operation and zone comfort rely on a thorough air-balancing of the system.
- Extensive cooling and reheating of supply air may be required for systems serving zones with diverse loads.

System Configurations

While air handlers share a basic form and set of components, the components can be arranged and controlled in different ways. In general, the configuration categories describe the number of zones, duct air-path, and fan type. Specifically, the system categories include

- single or multiple zone
- single or dual air paths
- constant or variable air volume

Single or Multiple Zone

Many air handling systems provide conditioned air to a single zone. For large multistory buildings, however, it is not practical to use many AHUs that each serve only one zone. Instead, air handling systems designed to serve several zones, each with is own thermostat control, are used. The multiple zone air system design presents challenges to the engineer to accommodate diverse loads while maintaining system efficiency.

The system diagram shown in Figure 4.3.1 is complete in its representation for the simplest of all air systems — a single zone system. In multiple zone systems in large buildings, one AHU may supply space conditioning to all zones on 20 or more floors. Figure 4.3.2 shows a system schematic that represents the layout of a multiple zone, centralized system.

In a single zone system, the supply air flow rate is constant when the fan is on. To satisfy variations in the zone load in off-design conditions, the system fan may cycle on and off, or the supply-air temperature may vary. For multiple zone systems, a means for varying the amount of cooling and heating supplied to each zone must be included as part of the system design. The means for meeting diverse zone loads include varying the temperature of the air introduced to the zone or varying its volumetric flow rate. More details of how different configurations and controls are used in multizone systems are provided in the system category descriptions below.

Single or Dual Duct Systems

The distribution system air path may be either single duct or dual duct. Systems are defined as single duct if they have a single air path for supplying both heating and cooling by the system. All single zone systems are the single duct type, while multiple zones may have a single duct or a dual duct arrangement. In multiple zone, single duct systems, there is a primary air stream that serves each zone and a terminal unit that can reduce air flow and/or add heat to the supply air stream to meet the zone setpoint.

A dual-duct system supplies heating and cooling in separate ducts, each referred to as a *deck*. Typically, the hot deck is maintained at 90–95°F. The cold deck is maintained at 50–55°F. The heated and cooled air streams are blended by thermostatically controlled mixing boxes to provide the proper temperature and flow of air to each zone. In these systems, the warm and cool air streams may be mixed near the

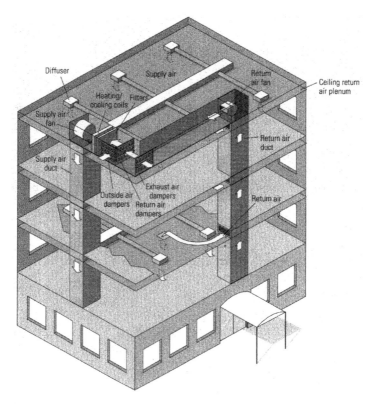

FIGURE 4.3.2 Multiple zone system schematic (courtesy of E-source, Boulder, CO).

central supply fan or near the zone. Systems that mix supply air near the zone are referred to as *dual duct systems* while systems that mix supply air near the central air handler are called *multizone systems.* While the convention is to name such systems as described above, the system names are poorly chosen since both types have dual ducts and serve multizones. Figures 4.3.3 and 4.3.4 present system schematics for the two types of systems, a dual duct and a multizone system, respectively. As shown, the two systems are functionally the same, but the location of the air blending is different. These systems are by nature inefficient since hot and cold energy streams are used simultaneously to meet zone loads. This approach goes against second law of thermodynamics design guidelines. In addition, the multizone system violates low-pressure, fan-power-saving design principles since the proximity of the mixing box to the fan results in high pressure drops and box damper leakage due to high flow velocities.

Variable or Constant Air Volume Systems

As mentioned previously, single duct, single zone systems are constant air volume systems. Single duct multiple zone systems, dual duct systems, and multizone* systems may be either *constant air volume systems* or *variable air volume systems.* Constant air volume (CAV) systems supply a constant flow rate of air to the building whenever the fans are on. CAV systems use the simplest type of AHU. In variable air volume (VAV) systems, flow modulation is achieved through fan dampering or motor speed adjustment.

Both CAV and VAV systems are engineered to meet the same peak building zone loads. However, the means by which the systems meet off-design conditions vary. Frequently in multiple zone CAV systems, supply air is cooled to a constant temperature sufficient for meeting design loads. In off-design conditions,

---

*The terms "multiple zone" and "multizone" are distinct. "Multiple zone" describes buildings with more than one zone; "multizone" is a type of air handler that creates conditioned air streams for several zones by mixing hot and cold streams responding to thermostat signals from each zone.

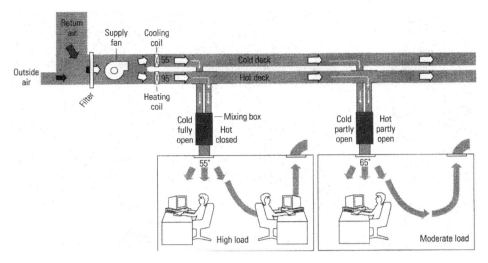

FIGURE 4.3.3 Dual duct system schematic (courtesy of E-source, Boulder, CO).

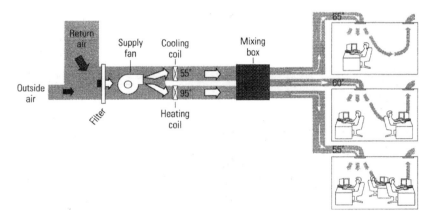

FIGURE 4.3.4 Multizone system schematic (courtesy of E-source, Boulder, CO).

thermostatically controlled zone-terminal reheat occurs to raise the temperature of the cooled air before it enters the zone. Figure 4.3.5 presents an example of a CAV system with a constant supply-air temperature serving several zones. In the figure, the zone with lower internal gains and smaller cooling load requires reheat at the zone terminal box.

VAV systems respond to changing cooling loads by modulating the zone air flow rate instead of the zone air temperature. The flow control of VAV systems is based on maintaining a constant pressure at some point in the main supply air duct. As zone terminal box dampers open and close, the duct pressure changes. The fan flow modulates to maintain the pressure setpoint. Flow variation is achieved by adjusting fan inlet dampers, fan outlet dampers, or the fan motor rpm. Direct fan motor control results in the lowest fan energy use in VAV systems. Figure 4.3.6 presents an example of a variable-air volume system responding to changes in cooling load. In the schematic, the two zones receive air at the same temperature but one has a reduced flow compared to design.

By reducing or eliminating the need for reheat, VAV systems use significantly less energy than CAV systems in meeting the same building loads. VAV systems also tend to be more expensive and more complicated to operate and maintain. Nevertheless, conversion of CAV systems to VAV is a popular energy conservation measure. Energy savings are achieved not only through reductions in fan power but also from reductions in cooling coil and reheat coil loads.

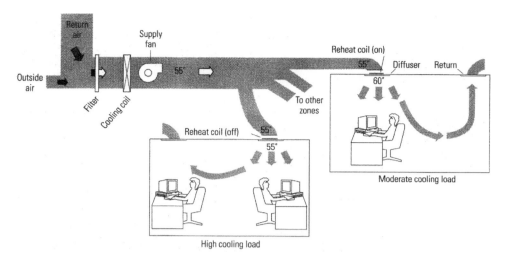

FIGURE 4.3.5 Single duct, multiple zone, constant air volume (CAV) system with reheat (courtesy of E-source, Boulder, CO).

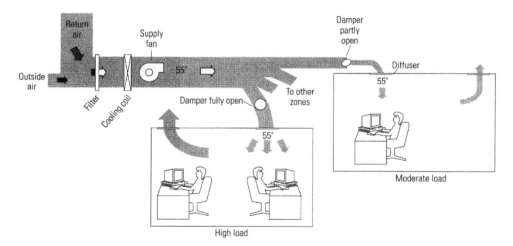

FIGURE 4.3.6 Single duct, multiple zone, variable air volume (VAV) system (courtesy of E-source, Boulder, CO).

System Design Considerations

A properly designed system meets zone loads under both design and off-design conditions. This requires calculation of maximum zone loads and an understanding of the flexibility that various components offer to meet loads in off-design conditions. Methods outlining the calculation of design zone loads are outlined in Chapter 6.1. In addition to zone loads, there are system dependent factors that influence coil loads and equipment size, including

- Supply fan heat gain
- Duct heat transfer
- Duct air leakage
- Component leakage

Details about these topics are described in other sections of this chapter.

Considering the second law of thermodynamics can provide insight for improving distribution system performance through design and operational changes. Second law guidance aids in the detection and avoidance of unnecessary depletion of useful work in a process. Guidelines particularly relevant to HVAC distribution systems include

- Minimize the mixing of streams with different temperatures and pressures.
- Do not discard heat at high temperatures to ambient or to cooling water.
- Do not heat refrigerated streams with hot streams.
- Heat (or refrigeration) is more valuable, the further the temperature is from ambient.
- The larger the mass flow, the larger the opportunity to save (or waste) energy.

Most of these considerations are common sense. Yet, as noted in the description of the CAV system with reheat, they are not adhered to. While there are many criteria other than thermodynamic performance for developing an acceptable design, any opportunities for reducing the depletion of useful work should be recognized when making design decisions.

Air Filtration

Air filtration is used in air handling systems to remove unwanted particulates, smoke, or gases from the air stream. Methods for smoke and gas removal from an air stream are normally reserved for industrial and special processes and are not usually necessary for typical HVAC applications.

There are two main types of particulate air filters used in modern HVAC air systems:

- *Viscous impingement filters* use filter media coated with a viscous substance, such as oil, which acts as an adhesive that catches particles in the flow stream.
- *Dry-type extended surface filters* consist of fibrous bats or blankets of varying thicknesses and density. The bats may be made of various materials such as bonded glass fiber, cellulose, wool felt, or synthetic materials.

Filters are often arranged in a pleated or v-configuration which extends the filter surface and reduces the pressure drop across the media. A common practice is to use inexpensive prefilters upstream of more effective filters to reduce premature loading. This practice extends the filter life and reduces the replacement frequency of the more effective, more expensive final filters.

Filter Testing and Rating

ASHRAE Standard 52.1 (ASHRAE 1992b) provides test procedures for testing and rating filtration devices. Of the many methods used for the testing and rating of air filtration devices, three of the most commonly cited methods include the *arrestance test*, the *dust-spot efficiency test*, and the *DOP penetration test*.

The arrestance test is conducted in a controlled laboratory setting. It consists of releasing a known quantity of material known as *ASHRAE test dust* into the filter to be rated. ASHRAE test dust is comprised of 72% standardized air cleaner test dust, 23% fine powdered carbon, and 5% cotton linters. The percent arrestance, which is the measure of the filter effectiveness, is calculated as follows:

$$\text{Arrestance} = 100 \cdot \left[1 - \frac{\text{weight gain of filter}}{\text{weight of dust fed}} \right]$$

Airborne particulates can result in soiled interior building surfaces. The discoloration rate of white filter paper simulates this effect. The dust spot efficiency test uses this approach to measure the effectiveness of the filter in reducing soiling of surfaces. The test measures the changes in light transmitted across the filter to evaluate its effectiveness. Efficiency is calculated from the following equation:

$$DS = 100 \cdot \left[1 - \frac{Q_1 \cdot (T_{20} - T_{21}) \cdot T_{10}}{Q_2 \cdot (T_{10} - T_{11}) \cdot T_{20}} \right]$$

where,

DS = percentage dust spot efficiency

Q_1 = total quantity of air drawn through an upstream target

Q_2 = total quantity of air drawn through a downstream target

T_{10} = initial light transmission of upstream target

T_{11} = final light transmission of upstream target

T_{20} = initial light transmission of downstream target

T_{21} = final light transmission of downstream target

The di-octyl phthalate (DOP) test is reserved for very high efficiency filters, typical of those used in clean rooms. In the test, a smoke cloud of DOP, an oily, high boiling point liquid, is fed into the filter. To determine efficiency, the concentrations of the DOP upstream and downstream of the filter are compared. The percent efficiency, DP, is calculated from the following equation:

$$DP = 100 \cdot \left[1 - \left[\frac{downstream\ concentration}{upstream\ concentration} \right] \right]$$

Application and Performance of Filtration Systems

The required level of filtration varies with the application. The amount of filtration for use in an industrial application may require only the removal of large particles, while the filtration requirement in applications such as clean rooms may be extremely rigorous. Table 4.3.1 outlines the type and effectiveness of filters that should be used for various applications. Note in the table that *A* designates arrestance, *DS*, dust spot efficiency, and *DP*, DOP efficiency percent.

Several types of filters are shown in Figure 4.3.7. From left to right, they include two pleated, disposable filters, a HEPA filter, and a bag filter. The pleated filters are least expensive and least efficient. The HEPA filter is most expensive and most efficient. Pleated filters are appropriate for general HVAC applications. Bag filters are appropriate for most hospital spaces. HEPA filters are appropriate for clean rooms and other aseptic applications.

Filters create air pressure drop in a system. Therefore, they affect air flow and/or fan power draw (kW). Manufacturers provide data regarding the pressure drop associated with both clean filters and dirty filters. When selecting a fan for a particular air handling unit, it is necessary to account for the pressure drop associated with a *fully loaded* (dirty) filter that is nearing the end of its service life or cleaning cycle. Table 4.3.2 gives typical pressure drop values for various filters.

Humidification and Adiabatic Cooling

In some applications, it is desirable or useful to supply humidification to a conditioned space. Some reasons for adding humidification capabilities to the air handling system include

- Space humidity control, where the primary goal is to maintain a humidity setpoint
- Air sensible cooling through adiabatic cooling
- Air cleaning

Air Humidification

Control of space humidity can be important for maintenance of high indoor air quality. Both high and low relative humidity levels can promote the growth of fungus and microbiological organisms. Special process rooms or industrial spaces sometimes require control of humidity to some particular setpoint, which can require humidification control.

Devices used to introduce humidity into a supply air stream include

- *Direct steam injection*, where steam is injected into the supply air stream
- *Pan humidifiers*, where water is evaporated into the air via a heated pan installed in the supply ducting
- *Wetted elements*, where water is applied to an open-textured media in the supply air stream
- *Atomizing devices*, where water is broken into a fine mist via atomizers. One of the more common types of atomizers are nozzles.

TABLE 4.3.1 Performance of Filtration Systems

Application	Prefilter	Filter	Final Filter	Remarks
Warehouses, mechanical rooms		50–80% A 25–30% DS	None	Large particles only; provides coil protection
General offices and laboratories	None	75–90% A 35–60% DS	None	Average housecleaning; provides pollen and some smudge reduction
Conference rooms, cleaner office spaces, specialty rooms	75–80% A 25–40% DS	>98% A 80–85% DS	None	Good housecleaning; no dust settling, significant smudge reduction
Hospitals, R&D, "gray" clean rooms	75–85% A 25–40% DS	>98% A 80–85% DS	95% DP or elect.	High bacteria reduction, effective smudge reduction
Aseptic areas, clean rooms	75–85% A 25–40% DS	>98% A 80–85% DS	99.97% DP (HEPA)	Protection against bacteria, radioactive dust; very clean room

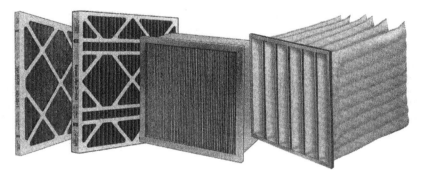

FIGURE 4.3.7 Pleated, HEPA, and bag air filters (courtesy of Airguard Industries).

TABLE 4.3.2 Filter Pressure Drop

Filter Type	Average Efficiency	Rated Face Velocity (fpm)	Clean Filter Pressure Drop (inches water)	Dirty Filter Pressure Drop (inches water)
Flat	85% A	500	0.10–0.20	1.00
Pleated	90% A	500	0.15–0.40	1.00
Bag	90% A	625	0.25–0.40	1.00
HEPA	99.97% DP	250–500	0.65–1.35	2.00

Humidifiers increase the moisture content (thus, latent heat) in the air stream while the stream enthalpy remains essentially constant. The humidification may have a sensible cooling or heating effect on the supply air stream. The change in state of the supply air is dependent on the supply air temperature and humidity, and on the state of the water absorbed by the air.

Evaporative Cooling

A common, but underutilized, application of air humidification is sensible air cooling through evaporative cooling. In an evaporative cooler, water is introduced into the supply air via wetted elements or atomizers (also called "air washers"). The water that is not evaporated by the air is captured in a sump and recirculated through the media or atomizers. Figure 4.3.8 shows a very common evaporative cooler arrangement.

The evaporative, or *adiabatic*, cooling that occurs in an evaporative cooler is the process of evaporating water into the air, thereby causing a sensible heat loss (cooling) of an air stream equal to the air's latent heat gain. Since the sensible loss is equal to the latent gain of the air stream, the process is adiabatic, thus the term *adiabatic cooling*. Since there is very little heat loss or gain in the supply air stream, it maintains a constant enthalpy as the air moves through the cooler.

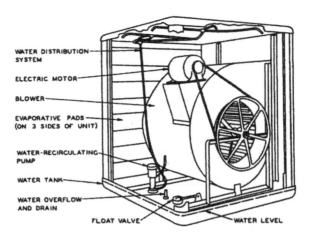

FIGURE 4.3.8   Typical evaporative cooler.

4.3.2   Coils

Coils are a special class of heat exchangers designed to transfer heat to or from an air stream. Coils are used in HVAC systems to provide air heating, preheating, reheating, cooling, and dehumidification. Coils come in various configurations to accommodate the needs of a design engineer in meeting the requirements of particular applications.

Types and Configuration

All HVAC air conditioning coils share the common feature that air is the working fluid on one side of the heat exchanger. The working fluid inside the coil, the primary fluid, may be one of the following.

Liquid — For coils with liquid as the working fluid, heat is transferred from or to the liquid when heating or cooling the air. The liquid working fluid may be water or a water/glycol mixture. Liquid coils are used for heating and cooling applications.

Refrigerant or DX — Direct expansion (DX) coils use refrigerant as the working fluid. They are applicable only for cooling applications. They are named DX systems because the refrigerant is directly expanded as it changes phase from liquid to gas and cools the air stream.

Steam — Steam coils are heating coils that use steam as the working fluid. In a phase change process that is the reverse of a DX coil, steam is condensed in the coil to heat the air stream. The latent heat of vaporization is released as the steam changes phases from gas to liquid.

Combustion gas — In heating coils, the working fluid may be high temperature, fossil fuel combustion gases. Devices that use combustion gases as the working fluid are commonly known as furnaces.

HVAC coils that use liquid, refrigerant, or steam as the working fluid share many components. A cutaway view of a typical coil is shown in Figure 4.3.9. Coil components, as shown in the figure, include fins, tubes, and headers.

The coil *fins* are extended surfaces that increase heat transfer area and improve heat transfer characteristics on the air side of the coil. Fins are usually constructed of aluminum; stainless steel is used for corrosion protection. Other fin materials may be used for special heat transfer properties. The fins are usually press-fitted or brazed onto the primary coil heat transfer surface to assure good contact and high heat transfer rates. Coils are generally specified with a particular number of fins per inch. The number of fins per inch in HVAC applications usually ranges between 6 and 12.

Coil *tubes* carry the working fluid (refrigerant, steam, water, etc.) to or from which heat is transferred to provide the desired air conditioning effect. Tubes are often constructed of copper but may be made of other materials depending on the application. The tubes can be manufactured in various configurations that affect coil heat transfer and pressure drop characteristics. Tube coils can be configured in (1) one

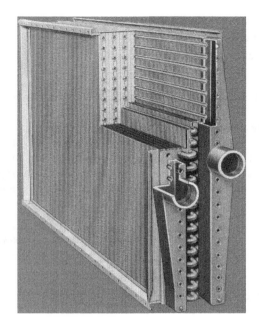

FIGURE 4.3.9   Typical coil configuration (courtesy of the Trane Co.).

or multiple circuits and (2) one or multiple rows. Circuits refer to the number of parallel paths in the coil through which the primary fluid passes. Rows run perpendicular to the depth of the coil. The number of rows equals the number of times a circuit crosses the air stream. In Figure 4.3.9, the primary fluid flows into 22 tubes and each crosses the coil 4 times. Thus, this coil has 22 circuits and 4 rows. However, determining a coil's configuration from external observations is not always reliable.

Coil *headers* distribute the primary fluid into the tube circuits, and they also serve to collect the fluid at the end of the circuits. In Figure 4.3.9, the inlet and outlet coil headers are apparent in the front of the image. Coils are sometimes arranged with face and bypass dampers, which modulate to direct the air across or around the heat transfer surfaces. The dampers control the amount of conditioning the coil imparts on the air stream. Miscellaneous appurtenances for coils also include drain pans for condensate collection and removal, valves for shutoff and control, air removal devices, and flexible piping connectors.

Coil Performance and Selection

Coil heat transfer characteristics are a function of several variables, including number of rows, coil fins per inch, entering air conditions and flow rate, and working fluid entering conditions and flow rate. An air system designer's task is to select a coil that will provide the necessary heat transfer to maintain the supply air setpoint temperature. Evaluating coil performance is complex, involving both heat and mass transfer. Computerized sizing programs are available from coil manufacturers, and simplified sizing methods are also presented in manufacturers' catalogues. Chapter 2.1 summarizes the technical basis for coil performance calculations.

To start the sizing process, the maximum or design load of the zones served by the unit must be determined. Generally for packaged or central systems, design air flow rate is based on the system *cooling* load. This maximum load can be estimated from the following parameters:

- the peak latent and sensible zone loads
- the desired quantity of outside air
- the return air temperature and humidity
- the outside air temperature and humidity

Knowing the above information, the coil selection process can proceed according to the following steps:

1. Specify supply air conditions and determine flow rate.

The design air flow rate is the amount of supply air required to offset the peak zone loads. This flow rate can be determined from the supply air temperature, the return air temperature, and the sum of the zone peak cooling loads. A value of 55°F is often used as the design supply air temperature. The design air flow rate is determined using a sensible heat energy balance as shown below:

$$\dot{m}_{air} = \frac{Q_{sensible,zones}}{Cp_{air} \cdot (T_{return\,air} - T_{supply\,air})}$$

where

Q = peak cooling load (Btu/hr)
T = temperature (°F)
Cp_{air} = heat capacity (for air 0.24 Btu/°F lb)
m = mass flow rate (lb/hr)

In situations where space humidity control is desired, the necessary air humidity leaving the coil can be calculated from the return air humidity ratio and the space latent heat gains.

$$W_{supply\,air} = W_{return\,air} - \frac{Q_{latent,zones}}{\dot{m}_{dry\,air} \cdot h_{fg,water}}$$

where

$$\dot{m}_{dry\,air} = \frac{\dot{m}_{air}}{1 + W_{return\,air}}$$

and

W = humidity ratio (lbs. of water/lbs of dry air)
Q = peak cooling load (Btu/hr)
h_{fg} = heat of fusion (Btu/lb)
m = mass flow rate (lb/hr)

A good estimate of the supply air humidity ratio can also be calculated by using the total air flow rate (not dry air flow rate) in the first equation above as noted in Chapter 2.2.

The design air flow rate and air-exiting-the-coil conditions are now defined. The supply air temperature and humidity ratio specify the state of the air and pinpoint its location on a psychrometric chart.

2. Calculate the coil entering conditions and loads.

The coil load is generally not equal to the sum of the zone loads. The coil load is a function of several design parameters, including the supply air temperature, air flow rate, percentage of outside air, and outside air temperature and humidity at design conditions. To calculate the coil load, the mixed air conditions (the temperature and humidity of the air entering the coil) can be easily calculated using mixing equations:

$$T_{mixed\ air} = F_{outside\ air} \cdot T_{outside\ air} + (1 - F_{outside\ air}) \cdot T_{return\ air}$$

$$W_{mixed\ air} = F_{outside\ air} \cdot W_{outside\ air} + (1 - W_{outside\ air}) \cdot T_{return\ air}$$

where

F = mass flow rate/supply air mass flow rate

Finally, the design cooling sensible and latent coil loads are calculated as follows:

$$Q_{sensible} = \dot{m}_{air} Cp_{air} (T_{mixed\ air} - T_{supply\ air})$$

$$Q_{latent} = \dot{m}_{dry\ air} h_{fg} (W_{mixed\ air} - W_{supply\ air})$$

A very close approximation of latent load can be determined by using the total air flow rate (not the dry air flow rate) in the equation above.

3. Select coil to meet the coil loads at the design conditions.

A large number of parameters affect the ability of a coil to provide the desired supply air conditions given the air flow rate and coil inlet conditions. Some of these parameters include

- Coil face area — The coil face area affects the heat transfer surface area, the air velocity in the coil, and the size of the air handling unit.
- The sensible heat ratio (SHR) — The SHR is the ratio of the coil sensible load to the total load (see Chapter 2.2).
- Coil depth and fin spacing — These parameters affect the coil heat transfer surface area and air velocity in the coil.
- Primary fluid entering temperature and flow rate — The primary fluid conditions and flow are important because they affect the heat capacity of the primary fluid, as well as the heat transfer characteristics inside the coil tubes.
- Thermodynamic limits — The coil effectiveness is the fraction of the theoretical maximum heat that may be transferred to the air stream. This value can not exceed a value of one. For HVAC applications, a typical cooling coil effectiveness value is 0.5.

The goal of the selection process is to select a coil that has adequate capacity to add or remove the necessary heat from the air stream. It also must have a design SHR that matches the desired SHR as closely as possible. Another important consideration is the minimization of life-cycle costs. Energy costs for air-side fans and water-side pumps decrease as pressure drop across the coil decreases. Coil air-side pressure drop is affected by flow rate, face area, and fin spacing. Water-side pressure drop is affected by flow rate, tube size, and number of rows. All of these design parameters impact coil first costs. Thus, the objective is to balance the first cost of the coil with annual energy operating costs.

Computerized selection procedures predict coil performance accurately and easily. Most software permits the performance of the coil to be evaluated in all operating conditions. The same software also facilitates performing life-cycle cost analyses since many different coils that meet all thermal criteria must be evaluated quickly to find the optimal coil. Manual selection procedures are outlined in manufacturers' catalogs. These procedures rely on design performance data presented in tables and charts.

For either computerized or manual methods, the following procedure is generally followed in selecting a water cooling coil.

1. Determine volumetric supply air flow rate from design mass air flow rate (CFM = lbs/hr/4.5).
2. Specify maximum face velocity (typically 300 to 800 fpm).
3. Calculate minimum coil face area.
4. Select an available coil size.
5. Recalculate face velocity based on actual size.
6. Determine the enthalpy of the air entering and exiting the coil under design conditions.
7. Determine total coil load from enthalpies and air flow rate.
8. Specify a water temperature entering the coil (typically 45°F).
9. Specify a water temperature exiting the coil (typically 10–20°F higher than the entering temperature).
10. Calculate required flow rate of water.
11. Assume number of circuits to give 2–5 GPM per feed.
12. From coil performance data, select the lowest number of rows and fin spacing that will meet or exceed total design coil load.
13. Calculate SHR.
14. Depending on SHR value, use appropriate air friction charts to determine air-side pressure drop.
15. Determine total water-side pressure drop across the header and circuit. Circuit pressure drop is dependent on tube size, water velocity, and number of passes.
16. Repeat sizing procedure to select other coils and evaluate each based on minimizing life-cycle costs.

4.3.3 Fans

Fans move air through ducts and system equipment to provide heating, cooling, and ventilation to the building zones. A fan utilizes a power-driven, rotating impeller that creates a pressure differential causing air flow.

Fan Types

Fans are classified by the direction of air flow through the impeller. Two main categories of fans are centrifugal and axial. In centrifugal fans, air flows in a direction radially outward from the shaft. In axial fans, air flows in a direction parallel to the shaft. Within the two fan classifications, there are various subclasses. For either fan type, the energy imparted to the air is mostly in the form of static pressure and partially as velocity pressure. Figure 4.3.10 shows the configuration and fundamental components of both centrifugal and axial flow fans.

Centrifugal Fans

Centrifugal fans produce pressure by changing the magnitude and direction of the velocity of the inlet air. Different types of centrifugal fans are distinguished by their impeller (blade) shape and other features. Operating characteristics of several types of centrifugal fans are presented in Table 4.3.11. The fan types are described below in order of increasing efficiency.

Radial — Because the fan wheel is rugged and simple, radial fans are usually used in industrial systems where the air stream may be contaminated or may be designed to carry solid materials in the air stream.

Forward curved — Forward curved fans generally produce less pressure than backward inclined fans, but are relatively easy to fabricate. They also tend to operate at a lower speed and produce less noise than other centrifugal fan types. Because of these characteristics, forward curved fans are used for low-pressure HVAC applications such as residential systems or packaged air handling systems.

Airfoil — Backward curved fans are of higher efficiency, can produce high static pressures and are *power limiting*. With power-limiting fans the power draw of the fan will climb to a maximum at a maximum efficiency (and particular flow rate) and then decrease as the flow rate increases. Radial and forward curved fans do not exhibit a power-limiting characteristic and, therefore, may overload motors.

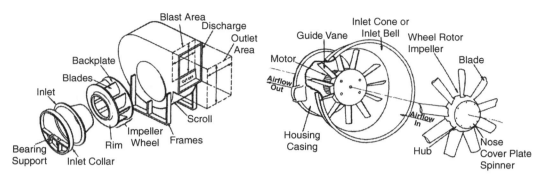

FIGURE 4.3.10 Centrifugal and axial fan components.

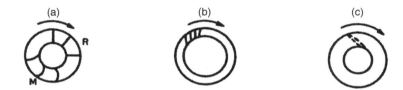

FIGURE 4.3.11 Types of centrifugal fans include: (a) radial, (b) forward curved, and (c) airfoil.

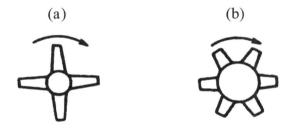

FIGURE 4.3.12 Types of axial fans include: (a) propeller and (b) tubeaxial.

Backward inclined — Backward inclined fans are similar to airfoil fans in both configuration and performance. The primary difference is that unlike airfoil fan blades, backward-inclined blades are simply flat rather than foils. Consequently, their efficiency is slightly lower. Airfoil and backward curved centrifugal fans are most commonly used in commercial HVAC systems due to their efficiency and pressure head capabilities.

Axial Flow Fans

Axial flow fans impart energy to the air by giving it a swirling motion. The different types of axial fans are distinguished by their blade shape, blade pitch, and hub diameter to blade tip diameter ratio (hub ratio). Two types of axial flow fans are presented in Figure 4.3.12. The types of axial flow fans in order of increasing efficiency are described below.

Propeller — Propeller fans generate low pressure differentials and have low efficiency and low cost. They are most often installed unducted for low pressure, high volume applications such as air circulation, spot cooling, or transfer of air from one space to another with very little need for pressure differential. Propeller fans are characterized by an impeller with single thickness blades attached to a small diameter hub/shaft and a simple circular ring housing. Propeller fans essentially have a zero hub ratio (hub diameter/fan diameter), having enough hub only to satisfy the mechanical requirements to drive the fan.

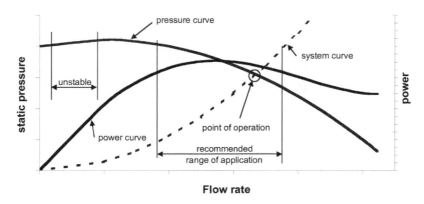

FIGURE 4.3.13   Typical fan performance curve.

Tubeaxial — Tubeaxial fans generally have four to eight airfoil or single thickness blades attached to a larger diameter hub mounted in a cylindrical housing. Tubeaxial fans generally have a hub ratio of 0.25 or greater. Tubeaxial fans are of higher efficiency than propeller fans and are able to generate more pressure.

Vaneaxial — In order to improve the flow vector components and fan efficiency, straightening vanes can be added to the fan housing. Vaneaxial fans are simply tubeaxial units equipped with guide vanes. Vaneaxial fans may be used in general HVAC applications for low, medium, and high pressure systems. They are more compact than centrifugal fans and are advantageous when space restrictions exist.

Fan Arrangement and Classification

Fan specifications include fan class (1, 2, or 3) which reflects the sturdiness of the fan — the higher the class, the higher the fan structural strength, allowing it to operate at higher impeller speeds and in more adverse conditions. Fans are also specified by their arrangement, which indicates the motor location, air discharge orientation, and drive train type (direct drive or pulley drive). Centrifugal fans can be specified as either single width, single inlet (SWSI) or double width, double inlet (DWDI). These designations indicate the width of the impeller wheel and whether the air may enter the fan on only one side or from two sides.

Fan Performance Curves

A typical fan performance curve is shown in Figure 4.3.13. Fan curves are a basic tool used in the design and analysis of performance. As can be seen in the figure, fan curves are presented with air flow rate (CFM) along the horizontal axis and differential pressure (psi or feet of head) along the vertical axis. The relationship between pressure and flow for a particular fan, operating at a particular speed, is presented as a single line. For the particular speed and fan, a given flow and pressure differential operating point will lie on the performance (pressure curve) line. The power curve, also shown, presents the fan power required at a particular flow rate and pressure. Variation of air density or fan speed will change the fan curves in a way that is predictable by fan laws discussed later in this section.

For a particular air distribution system, a *system curve* can be overlaid on a fan curve to predict and visualize how a fan will interact with the system. A system curve is generated by knowing the pressure drop that will occur in a system at a particular flow rate and invoking the second fan law (described below) to generate points at other flow rates. Note in Figure 4.3.13 that two different ranges are shown, the recommended application range and the unstable range. In selecting a fan, it is important for the fan to operate in the recommended application range to ensure good efficiency. If the fan/system combination results in operation in the unstable range, there is a high risk that the flow will exhibit undesirable pulsations.

In lieu of a fan curve, manufacturers very often present their fan performance data in a table. A fan performance table will provide information similar to that found in a fan curve.

Fan Power

A fan rotor adds energy to the air stream in the form of static energy (pressure), kinetic energy (velocity), and heat. The quantity and form (kinetic, static, or heat) of energy added to the air stream is a function of power input, fan efficiency, motor efficiency, and whether the fan motor is situated in the air stream or out of the air stream.

A useful formulation for the calculation of power required to develop a particular flow rate at a particular differential pressure is as follows:

$$Power = C \cdot Flow \cdot \Delta P / (\eta_{fan} \cdot \eta_{motor})$$

where

C = system constant
ΔP = air pressure rise
η = efficiency

Fan efficiency is the ratio of the useful work added to the flowing air to the shaft work input; motor efficiency is the ratio of the shaft work to the motor electric power input.

Fan Laws

Fan laws relate performance variables for fans of similar type and geometry. Standardized fan performance curves or tables are generated for a particular fan operating at a particular speed at standard atmospheric conditions. The performance of the fan can be predicted at other speeds, flow rates, pressures, and air densities using fan laws. There are three fundamental fan laws that can be used to predict changes in these variables.

Fan Law Number 1: *In a fixed system, fan flow rate is proportional to fan speed.*

$$CFM_{new} = CFM_{old} \cdot \frac{RPM_{new}}{RPM_{old}}$$

A fan acts as a constant volume device — the faster the blades operate the higher the volumetric flow rate will be. Since a fan is a constant volume scoop, an important axiom to the first fan law is that changes in density have *no effect* on the volumetric flow rate generated by a fan at a particular speed. However, the mass flow rate increases with density.

Fan Law Number 2: *In a fixed system, the static pressure rise across the fan varies as the square of the fan speed.*

$$SP_{new} = SP_{old} \cdot \left[\frac{RPM_{new}}{RPM_{old}}\right]^2$$

Since more air mass creates a higher pressure differential across a fan, an important axiom to the second fan law is that the pressure change across a fan is directly proportional to the density of the air. This relationship can be expressed as

$$SP_{new} = SP_{old} \cdot \frac{\rho_{new}}{\rho_{old}}$$

Fan Law Number 3: *In a fixed system, the power of a fan varies as the cube of the fan speed (RPM).*

$$HP_{new} = HP_{old} \cdot \left[\frac{RPM_{new}}{RPM_{old}}\right]^3$$

As in the second fan law, an axiom to the third fan law is that fan power varies directly with air density:

$$HP_{new} = HP_{old} \cdot \frac{\rho_{new}}{\rho_{old}}$$

The fan laws are expressions of the fact that the fan curves of similar fans are homologous. This implies that at the same point of rating, geometrically similar fans have the same efficiency.

The most common use of the fan laws is to determine the performance of a specific fan at altitudes and/or air temperatures other than those at which a particular fan is rated. As altitude or temperature increases, the density of air decreases and the pressure differential and work required will reduce proportionally per fan laws 2 and 3.

System Curve

For a fixed HVAC air distribution system, a given air flow rate requires a specific total pressure in the system. The distribution system components — including dampers, filters, coils, ductwork, and diffusers — represent a resistance that must be overcome. In HVAC applications, the relationship between pressure and flow rate for a system of fixed resistance follows the second fan law.

Using the second fan law, the system curve can be determined from one known operating point (usually design) for a given system. Once determined, the relationship described by the second fan law can be used to calculate other system operating points and plot the system curve. By superimposing the system curve on the fan curve, the operating point of a given fan in a given system can be determined (see Figure 4.3.13). Since fan performance data are generally published for a specific fan speed, the fan laws can be used to generate additional fan curves. The intersection between the system curve and the fan curve identifies the flow and pressure differential at which the fan will operate for a particular system.

System Effect

In the duct design section of this chapter, methods for calculating pressure drop in distribution system equipment are discussed. However, even when these methods are followed, the installed fan performance measured by field tests may differ from the designer's calculated performance. This not uncommon occurrence results because the published fan performance data are based on standardized tests conducted in laboratories using specific fan entry and exit configurations. The fan tests are conducted under ideal conditions where the flow into and out of the fan has no air swirling, and the air stream has a uniform velocity. In actual systems, the inlet or outlet conditions of the fan may be less than ideal in that there may be air swirl or significant velocity gradients at the fan inlet or outlet. These velocity gradients or swirling actions are what cause the system effect to occur. Since testing and system configurations differ, *system effect* factors must be taken into account to properly understand and predict the fan-system operation.

Figure 4.3.14 shows the impact of the system's effect on performance. The system effect causes the system flow coefficient to be higher and the system curve to be steeper than expected. If the system effect is not taken into account during system design, the installed flow rate will be less than the design flow rate. To increase flow, the fan speed and system pressure must increase, which increases power requirements and operating costs as well.

The system effect factor cannot be measured in the field but can be predicted and accounted for using methods outlined in Chapter 32 of the 1997 *ASHRAE Fundamentals Handbook* (ASHRAE 1997) and in AMCA Publication 201 (AMCA 1990). The data account for differences in velocity profiles between fans as tested and fans as installed. The method consists of identifying a flow configuration that most closely

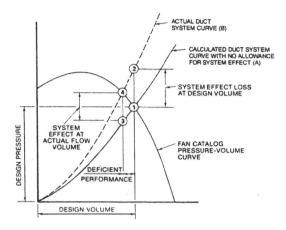

FIGURE 4.3.14 System effect (courtesy of SMACNA, 1990, *HVAC Systems Duct Design*).

matches the particular configuration being considered. Once the flow configuration is identified, the proper system effect curve can be selected and applied to estimate the effect as a pressure loss. An otherwise accurate design can be significantly in error if system effects are not considered.

Fan Flow Modulation

In variable air volume distribution systems, the volume of air delivered by the fan varies as building loads vary. To modulate the flow, the characteristics of either the fan or the system must change. The system performance changes by changing the system resistance, by use of dampers, for example. Fan characteristics may be changed by one of several methods, including a variable-speed drive motor, fan discharge dampers, and fan inlet vane dampers.

To change system resistance, branch duct dampers must accomplish large pressure drops since fan speed remains constant. As the damper resistance increases, the system constant decreases, and the system curve shifts to the left. To achieve low flow rates, extreme pressures may be exerted over branch dampers. This results in unstable flow and noise generation. Thus, most distribution system flow control occurs at the fan.

For fan control with a discharge damper, a damper absorbs excess pressure near the fan while a constant static pressure is maintained in the duct. As the damper closes, the fan curve shifts to the left. It is common to represent fan performance with capacity control dampers as a family of curves, each passing through the origin.

Fan control with inlet-vane dampers has characteristics similar to control with discharge dampers. For this configuration though, the capacity dampers are an integral part of the fan equipment. A properly designed unit can accomplish capacity control with a lower power requirement than discharge dampers.

Changing the fan performance through changing fan speed is the most energy efficient method for accomplishing capacity control because, as illustrated by the fan laws, fan power requirements drop with the cube of the fan flow rate. However, variable speed drive controls are more complex and expensive.

Fan Selection

After the air distribution system has been defined and the system performance curve evaluated, the fan can be selected to meet the system requirements. Fan selection involves choosing the size, type, class, arrangement, and capacity control to accomplish the job most economically.

The most efficient operating area for a fan is usually clearly presented in graphic and tabular presentations of manufacturer data. For variable air volume systems, it is important to know the frequency at which the fan will be operating at different part loads to select the most efficient fan. In general, the less expensive the fan capacity control, the more expensive the operating costs.

4.3.4 Ducts

Ducts are conduits used to carry air from air handling units to or from conditioned or otherwise ventilated spaces. Supply, return, or exhaust air ducts are sized to deliver or remove the amount of air required to meet zone loads or ventilation requirements at the design conditions. Duct sizing and application is influenced by space availability, desired location of room diffusers or returns, and operating versus capital cost tradeoffs. Proper designs must consider allowable noise levels, fire code requirements, duct leakage, operation and maintenance accessibility requirements, and heat conduction rates to or from the duct.

Construction and Codes

Ducts are often constructed of galvanized steel, but other materials such as black carbon steel, stainless steel, aluminum, copper, fiberglass reinforced plastic, and concrete are used in some applications, depending on factors such as corrosion, durability, purpose, price, and pressure requirements. Application considerations prescribe the best material choice. Different duct materials have different properties, such as roughness (which affects air pressure drop), ease of modification, weight, cost, thermal expansion, rigidity, porosity, strength, weldability, and corrosion resistance.

The minimum requirement for duct material strength and thickness is dictated by code and affected by the system air pressure and duct air velocity. Duct friction is a function of surface roughness and velocity. In general, ducts are classified into two velocity regimes — low velocity (below 2500 ft/min) and high velocity (up to 4500 ft/min). Recommended duct friction rates differ between the two regimes. In general, low air velocity applications include constant volume systems and duct sections near spaces where high noise levels are unacceptable. High air velocity systems are used primarily in the main trunks of VAV or industrial systems to reduce duct capital costs and space requirements. Ducts can be either internally or externally insulated to reduce the transfer of both heat and noise.

Industrial Ventilation: A Manual of Recommended Practice (ACGIH 1998) provides an excellent resource outlining the application of ventilation systems in industrial settings. Two publications, *HVAC Systems Duct Design* (SMACNA 1990) and *HVAC Duct Construction Standards* (SMACNA 1995), provide detailed design information on duct construction, installation methods, design, and application for all common systems in use today.

Theory of Air Flow in Ducts

Fluid flow is measured indirectly by a pressure differential measurement. Pressure is lost and flow is hindered by friction arising from the interaction of the fluid with the conduit. This interaction occurs because of both static losses and dynamic losses. Static pressure loss results from the friction of the air on the wall of the duct work. Dynamic losses occur under turbulent flow conditions whenever there is a sudden change in direction or magnitude of the velocity of the air flow.

Bernoulli's equation, which is a specialized form of the first law energy balance, can be applied for the analysis of air flow in ducts. Written in terms of pressure, the equation can be expressed in the following form:

$$SP_1 + VP_1 + HP_1 + \Delta P_{fan} = SP_2 + VP_2 + HP_2 + PL_2$$

States (1) and (2) are points in the ducted air stream, as shown in Figure 4.3.15, where

SP = static pressure — pressure normal to flow

VP = velocity pressure — pressure arising from flow velocity that equals $\rho V^2/2g_c$ where g_c is (32.2 lb_m ft/s^2)/lb$_f$

HP = potential pressure — pressure arising from columns of air occurring because of a height change (normally a small effect)

ΔP = total pressure increase across fan (if present)

PL = pressure loss due to friction

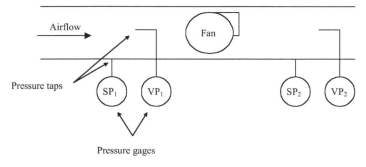

FIGURE 4.3.15 Ducted air flow.

Summing the pressure losses (*PL*) occurring across each duct section and system component (such as dampers, fitting, filters, and coils) determines the pressure loss that must be overcome by the fan. Knowing the pressure loss in the various branches in a complex ducting system allows for the prediction of the fraction of supply air that will follow each branch. Proper application of Bernoulli's equation and friction loss equations allows the designer to size the ducting such that each section will have the proper pressure drop at the design flow rate, resulting in the desired branch flow.

Although duct pressure drop is much lower than component pressure drop, its evaluation is important to minimize the use of dampers to pressure balance the system. The total pressure decrease across a section of ductwork is due to friction and dynamic losses. The theory underlying the frictional and dynamic duct losses is outlined below.

Friction Losses

For fluid flow in conduits, the friction pressure drop (PL or ΔP) can be determined from the Darcy equation

$$\Delta P_f = f\left(\frac{L}{D_h}\right)(VP)$$

where

ΔP_f = pressure loss arising from friction
f = friction factor, a function of Reynolds number and duct internal roughness
L = duct length over which the friction loss occurs
D_h = hydralic cross-sectional diameter
VP = velocity pressure

For circular ducts, the hydraulic diameter is the same as the duct diameter. For other ducts, the hydraulic diameter is equal to

$$D_h = \frac{4A}{P}$$

where

A = the cross-sectional area
P = the perimeter of the ducting

The value of the friction factor is dependent on the flow regime characterized by the Reynolds number. The dimensionless Reynolds number represents the ratio of fluid inertial to viscous forces, specifically

$$\text{Re} = \frac{D_h V}{\nu}$$

where

ν = the kinematic viscosity of the fluid.

V = the fluid velocity

For laminar flow (Re < 2300), the friction factor is only a function of the Reynolds number. For turbulent flow, the friction factor is a function of the Reynolds number and duct surface roughness. The Moody friction factor chart presents the friction factor graphically for both laminar and turbulent flow.

Because use of the Darcy equation can be cumbersome, specialized charts have been developed for determining air pressure loss in galvanized ducts based on the Moody diagram. Figure 4.3.16 presents duct friction loss per unit of duct length as a function of duct volumetric flow rate.

The shaded chart area delimits recommended duct sizes for low and high velocity applications. Duct sizes are specified for circular ducts. Rectangular ducts having equivalent hydraulic characteristics may be substituted for the circular size. Rectangular duct equivalents for circular duct sizes are published in *ASHRAE* (2001) and *SMACNA* (1990). To evaluate friction loss for ducts constructed of materials other than galvanized steel, roughness factors available from SMACNA (1990) can be applied to the pressure losses from Figure 4.3.16.

By invoking the second fan law, the pressure drop for various flow rates can be determined for a fixed duct system by the following relationship:

$$\frac{\Delta P_{new}}{\Delta P_{old}} = \frac{V_{new}^2}{V_{old}^2} = \frac{CFM_{new}^2}{CFM_{old}^2}$$

where

CFM = the air volumetric flow rate

V = air velocity

ΔP = the differential pressure measurement in the duct section of interest.

This relationship is the basis for the system curve equation commonly used in HVAC design.

Dynamic Losses

Dynamic losses occur in fittings when there is a change in flow direction or velocity. Dynamic losses occur across duct fittings, such as elbows, tees, entries, exits, transitions, and junctions. The pressure loss associated with each type of fitting is proportional to the fluid velocity pressure. This relationship can be expressed as

$$\Delta P_d = C \cdot DP$$

where

C = the local loss or dynamic loss coefficient

DP = dynamic pressure ($\rho V^2/2g_c$)

Coefficient values for different fittings have been determined by laboratory testing and are published in several HVAC reference manuals, such as *SMACNA* (1990) and *ASHRAE* (2001). For converging or diverging flow junctions, two loss coefficients are reported — one for the main branch pressure loss and

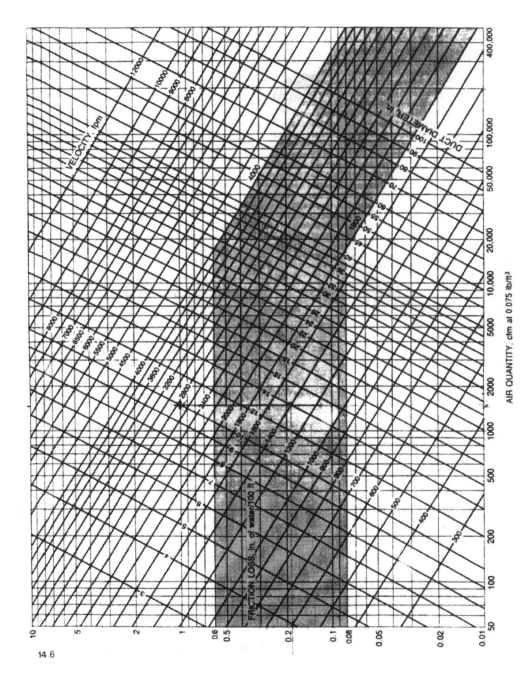

FIGURE 4.3.16 Duct friction loss chart (courtesy of SMACNA, 1990, *HVAC Systems Duct Design*).

14.6

the other for the branch duct pressure loss. Since there are many fitting configurations for which there is not an exact match in the literature, the designer needs to use judgement in the estimation of a *C* value for unique or unpublished configurations.

Duct System Components and Considerations

General construction details representing acceptable practice are presented in *HVAC Duct Construction Standards* (SMACNA 1995). While a detailed discussion of construction standards and methods is beyond the scope of this chapter, some of the more important ducting elements are introduced below.

Duct Joints, Seams, Sealing, Hanging, and Reinforcement

Normally, installers are given latitude in the selection of duct seams, hangers, and reinforcements, but the construction must meet applicable codes. Particular construction details needed to meet requirements for static pressure, sealing, materials, duct support, and other provisions are outlined in the *HVAC Duct Construction Standards* (SMACNA 1995).

Turning Vanes

A construction feature that reduces pressure loss in system components and fittings is turning vanes. Used in system components such as elbows and tees, turning vanes reduce the dynamic friction loss by directing the air flow parallel to the duct walls.

Fire Dampers

Fire dampers are required where ducts penetrate a firewall. They must be accessible for checking and resetting, so a duct door is normally installed in the duct adjacent to the fire damper. Fire dampers have a fusible link that melts and separates at a particular temperature, causing the spring loaded damper to slam shut. Similar to fire dampers are smoke and fire dampers that normally include a fusible link, but they may also be actuated by a control system to control smoke contamination in buildings.

Balance Dampers

An important activity that occurs during the construction process is the testing and balancing (TAB) of air systems. TAB professionals make field adjustments to air systems to ensure that proper flow is delivered to or from each zone or space, and that the overall flow of the systems is as specified.

In designing and installing ducting systems, it is important to make adequate provisions for TAB. One of the most important components that must be included to allow TAB is the balance damper. Balance dampers can be of either single or multiple blade construction but generally have a locking mechanism so that the person performing TAB on the system can set and lock the balance dampers in the proper position. Further information on TAB of air systems is available through the American Air Balance Council (AABC) and the National Environmental Balance Bureau (NEBB).

Duct Air Leakage

Duct leakage can significantly increase CFM air flow requirements and energy loss, decreasing overall system efficiency and effectiveness. Duct leakage impacts system performance by increasing fan power requirements, cooling coil loads, and reheat coil loads. The level of impact depends on the leakage rate, fan type, fan control, and system type. Typical leakage levels have been measured by the Florida Solar Energy Center (FSEC) for residential and small commercial buildings. For residential duct systems, leaks are 10–20% of fan flow on each side of the fan. For light commercial buildings, field studies suggest that duct leakage is actually higher than that found in residences — the average leakage in the supply ducts being over 20% of fan flow. Lawrence Berkeley National Laboratory (LBNL) has performed limited measurements of duct leakage in large commercial buildings. The performance data are inconclusive, but it is clear that some fraction of these buildings have significant (i.e., >10%) duct leakage. Even more unclear for this class of buildings is the impact of duct leakage on energy use. Some analyses of this question have been published; however, because the interactions between different factors are so complex, consensus on the appropriate analysis procedure has not been reached. Nevertheless, a conservative evaluation of the cost effectiveness of duct sealing in large commercial buildings warrants sealing ducts in most locations analyzed.

The functional form usually used to describe the relationship between the pressure in a duct and the flow through the leaks in that duct is as follows:

$$CFM_{leaks} = C_{leak} \Delta P^n_{\text{Duct to Space}}$$

where the pressure differential is the difference between the pressure in the ducts and the pressure of the space surrounding the ducts. When testing ducts for leakage, a known measured pressure differential is applied, and the flow required to maintain that pressure differential is determined by using a calibrated fan. By using several data points for Q and ΔP, one can solve for C_{leak} and n. For leaks that look like orifices (e.g., holes), n is 0.5, whereas, for leaks with some length (lap joints between duct sections) n is approximately 0.6 to 0.65.

Duct Conduction and Insulation

Duct heat gains and losses can be significant, particularly for single-story buildings with ducts located outside the conditioned building space. Insulating ducts reduces unwanted heat transfer, condensation on cold surfaces, and noise. Duct heat gains and losses must be determined to size distribution system fans and cooling coils. Methods for estimating duct heat transfer based on entering air temperature, exiting air temperature, and surrounding air temperature are outlined in Chapter 2.1 of this handbook.

The duct heat transfer coefficient is predominantly a function of duct construction, duct insulation level, and duct air flow rate. For example, experimental measurements of duct heat transfer coefficients for a 10-in square metal duct with 2-in $3/4$ lb/ft^3 faced fibrous glass range from 0.134 to 0.148 Btu/(hr ft^2 F) for duct velocities ranging from 780 to 3060 feet per minute. For a 10 in round duct with the same insulation and velocity range, the duct heat transfer coefficients range from 0.157 to 0.163 Btu/(hr ft^2 F) (Lauvray, 1978).

Duct System Design

The primary concern of the designer is that the desired amount of air be delivered to the zones. Other important considerations that must be accounted for in the duct system design include

- Available space for installation and access
- Meeting noise criteria
- Air leakage rates to or from the ducting
- Heat gain or loss
- Testing and balancing
- Smoke and fire control
- First costs versus operating costs

Duct Design Methods

Duct design methods are used to determine the size of main and branch duct sections. Several methods have been developed for achieving acceptable designs that balance capital and operating costs. While in many systems, the pressure drop of the duct is only a small fraction of the total system pressure drop, it is important to accurately estimate duct losses so that flow will be properly balanced among zones and rooms served by an AHU.

There are several duct design methods, including (1) equal friction, (2) static regain, (3) velocity reduction, and (4) the T-method. The two most common of the four, equal friction and static regain, are discussed below in detail. In general, the approach of each of the four methods is as follows.

In the *equal friction* method, the goal is to maintain the same pressure gradient throughout the system.

In the *static regain* method, the duct sections are sized so that the friction losses are offset by converting available velocity pressure to static pressure at junctions.

In the *velocity reduction method*, specific duct velocities (which dictate duct size) are specified and designed for various parts of the system.

The *T-method* is an optimization procedure for evaluating the economic trade-off between duct size (capital cost) and fan energy (operating cost).

No matter the methodology used for duct sizing, it is best to create a design that delivers the proper amount of air to each space with little need for throttling through the use of balance dampers. If a balance damper must be significantly closed during TAB, it is often an indication that the ducting has not been designed or constructed carefully.

Equal Friction

The most common manual method for duct design is the equal friction method, in which all duct sections are sized to maintain an equal pressure gradient in the system (thus the term *equal friction*). The method is normally applied to low pressure/low velocity systems. The gradient commonly chosen is 0.1 in water gauge of pressure drop per 100 ft duct length. With this method, branch ducts of different lengths will not be balanced. Achieving a balanced system to ensure appropriate zone flow rates relies on the proper adjustment of the balance dampers in the zone boxes.

Static Regain

The static regain method is the most commonly used computerized method for duct sizing. When precisely applied, the static regain method results in a system that is self-balancing. The method is based on the requirement that the static pressure losses occurring in a duct section are regained by a decrease in velocity pressure in the following section. Velocity pressure decreases through a decrease in velocity. A decrease in velocity occurs through an increase in duct size. This design approach results in each duct section having nearly the same static pressure, thereby creating a system that is self balancing.

Design Software

While the equal friction method can be completed by hand calculations, the static regain method requires a computerized analysis. Both methods are available in computerized form from major HVAC equipment manufacturers. Although the static regain method is complicated and requires several calculation iterations, in computerized form it is as straightforward as the equal friction method.

4.3.5 Terminal Units

Terminal units (also known as terminal "boxes") receive conditioned air from a central air handling unit and vary the volume and/or temperature of the air delivered to the conditioned space to maintain the zone setpoint. A single terminal box serves one thermal zone. Each box may distribute air to several zone supply diffusers. Air distribution systems can be distinguished by their terminal type, and the basic control of a central air handling system is dependent on the terminal type associated with the unit. Terminal unit types are described below.

Constant Volume Reheat

Constant volume reheat terminal units are simply reheat coils located in the supply ducts near the zones served. A constant volume of supply air from the central air handling unit flows through the reheat coil and into the zone. If the supply air is too cold to maintain the zone setpoint temperature, the reheat coil, which usually uses electricity or hot water as its heat source, will modulate to reheat the supply air, thereby maintaining the zone setpoint. Because constant volume reheat systems often operate in a mode where the air is cooled by a vapor compression cycle, only to be reheated at the terminal units, they tend to use a large amount of energy when compared to other system types.

Variable Volume without Reheat

Variable volume (VAV) terminal units without reheat are common. The terminals consist of a modulating damper that changes the quantity of air into the space in response to a zone thermostat to maintain the zone

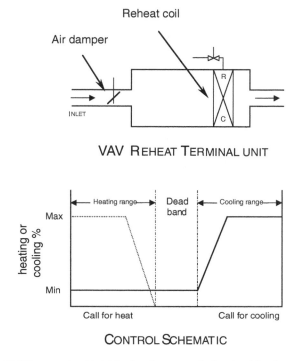

FIGURE 4.3.17   Variable air volume terminal unit with reheat.

setpoint temperature. Often, variable volume terminal units will have a minimum damper position or volume to assure that adequate air ventilation and circulation occurs in the space. In such units, overcooling can occur since no provision is made for the tempering of the supply air when zone cooling loads are small.

Variable Volume Reheat

Variable volume (VAV) terminal units with reheat have an air damper that modulates the quantity of supply air from the central air handler in response to the zone thermostat. Additionally, a reheat coil is provided to temper the supply air when there is no zone cooling load, or if heat is required in the space. Sometimes, the variable volume box and the reheat coil are separated in a zone. In this case, the system is composed of standard variable volume terminal units (without reheat) and zone baseboard heating coils. The coils add heat to the room during low cooling loads or when the zone requires heating.

Figure 4.3.17 shows a schematic drawing of a VAV terminal unit with a reheat coil. In most cases, the units are controlled as shown on the right side of the figure. Specifically, when there is a call for heat, as shown on the x-axis, the VAV damper will be closed to its minimum position, and the reheat coil will be in full heating mode. As the call for heat decreases, the controller will begin to reduce the output to the heating coil until the heating coil is fully off, while the primary air damper remains at its minimum position. On a call for cooling, the heating coil will remain off, and the primary air damper will modulate open until the box goes to maximum cooling.

Dual Duct Constant Volume

Dual duct constant volume terminal units are served with warm and cool primary air from the central air handling unit (Figure 4.3.18). The terminal unit blends these two air streams to produce the necessary supply air temperature to maintain zone setpoint. The terminal unit contains a damper, or set of dampers, that varies the percentage of cool and warm air, but not total volume, supplied to the zone to maintain the zone setpoint temperature. Like their constant volume reheat terminal unit counterparts, these units can use a great deal of energy because the supply air streams are heated in the hot deck and cooled in the cold deck, only to be blended to some intermediate temperature to satisfy the zone loads.

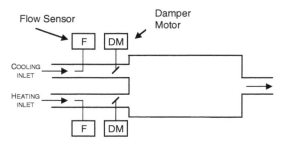

VAV Dual Duct Terminal Unit

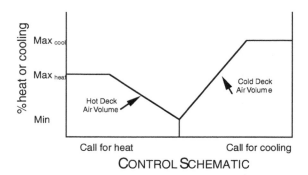

Control Schematic

FIGURE 4.3.18   Dual duct terminal unit.

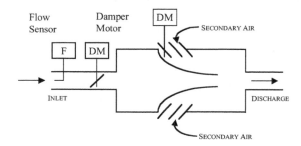

FIGURE 4.3.19   Induction terminal unit.

Dual Duct Variable Volume

Like the dual duct constant volume units, dual duct variable volume terminals are served by both warm and cool air. Unlike the constant volume units, the variable volume units vary both the percentage and the total volume of air into the space to maintain the zone temperature setpoint.

Other Terminal Unit Variations

Two primary terminal units types, single duct and dual duct, are outlined above. Variations and enhancements on these terminal units are described below.

1. Induction terminal units — A VAV induction system has a terminal unit that blends primary air from the central unit with recirculated (secondary) return air from the zone. The higher volume of blended air is then introduced to the space via the supply diffusers. The increased air flow is advantageous to increase the mixing of air in the space.

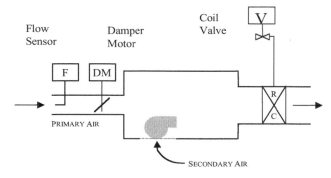

FIGURE 4.3.20 Parallel fan-powered terminal unit.

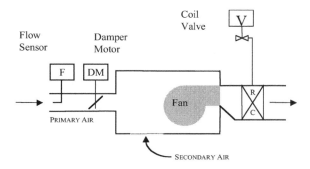

FIGURE 4.3.21 Series fan-powered terminal unit.

Induction boxes, in general, require a higher supply duct static pressure than other terminal unit types in order to induce the secondary air flow and operate properly. For this reason, operating costs using induction VAV boxes are usually higher. Figure 4.3.19 illustrates an induction terminal unit.

2. Fan-powered terminal units — Secondary air may also be introduced to the terminal boxes with a *fan-powered induction* terminal unit. In these units, the fan draws air from the plenum or zone. The induction fan may be installed in a parallel or series arrangement (Figures 4.3.20 and 4.3.21) with the primary air flow. In series, it operates continuously and provides a constant air flow to the zone. In parallel, it operates intermittently to induce the secondary air flow as needed to meet heating demand.

Because fan-powered terminal units have a local fan, the pressure requirements are lower than other terminal types, which can reduce the load on the main air handling unit fan. However, since each of the terminal unit fans require energy, the overall energy use of a system with fan-powered terminals is often greater than systems that have terminals not fan-powered.

Induction units and fan-powered boxes are often used in VAV systems since they allow higher zone air flow rates with a negligible increase in coil loads. A higher flow rate into the zone ensures higher air velocities and greater occupant comfort.

3. Bypass terminal units — Smaller central air handling units (generally below 20 tons of cooling capacity) used in VAV applications usually do not include provisions for the modulation of the primary air volume. In these situations, a bypass terminal unit (Figure 4.3.22) is sometimes used. Bypass units receive a constant volume of air from the central air handling unit but supply a variable air volume to the zone as needed to maintain setpoint. The remainder of the primary air is diverted, via a modulating damper, to a return air plenum or duct.

These units can be more efficient than constant volume units since the need to reheat this air is avoided. Because the supply air flow is constant, this is not a true VAV system; hence supply air temperature reset is

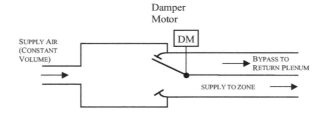

FIGURE 4.3.22   Bypass terminal unit.

FIGURE 4.3.23   A square ceiling-mounted diffuser, a register, and a grille (courtesy of Carnes Company, Inc.).

often used to prevent a high percentage of the supply air from bypassing the space and reducing air circulation and occupant comfort.

4.3.6   Diffusers

This section describes the equipment and design fundamentals for supply-air outlets and return-air and exhaust-air inlets of air distribution systems. Just before entering or exiting a conditioned building space, conditioned air passes through a *diffuser*, *register*, or *grille*. These devices are not interchangeable since each varies in the way it transitions air to or from the conditioned space. Diffusers are designed to entrain room air with high velocity supply air to induce room air circulation. Registers have slotted or perforated openings and are equipped with a damper to provide direction or volume control of supply air. Grilles are registers without dampers and do not provide any mixing or flow control. Generally, supply air is pushed through diffusers and return air is pulled through grilles. A typical example of each type of device is presented in Figure 4.3.23.

Supply Air Diffusers

Diffusers are installed at supply-air outlets and come in a variety of configurations, including square, circular, and slot. Generally, square and circular ceiling- mounted diffusers consist of a series of concentric rings or louvers. These diffusers discharge air radially in all directions. A slot diffuser is an elongated outlet with an aspect ratio of 25:1 or greater and a maximum height of approximately 3 in (SMACNA, ASHRAE 1988). Figure 4.3.24 shows a slot diffuser in a ceiling mount.

Diffusers are important for maintaining a safe and comfortable indoor environment. They promote proper mixing of the supply air with the room air. Generally, conditioned air is supplied to the outlet at velocities higher and temperatures lower than those acceptable in the occupied zone. Supply-air outlet diffusers slow and temper the supply air by entraining room air into the primary air stream. The *entrainment* of secondary air results in a surface or *Coanda effect*, illustrated in Figure 4.3.25.

The surface effect is caused by the primary air stream moving adjacent to a ceiling or wall, creating a low pressure area adjacent to the surface, and causing the air stream to flow parallel to the surface throughout the length of throw. This effect inhibits the horizontal drop of the cold, primary air stream. Diffusers which result in larger areas of surface spread tend to have a larger surface effect. The surface effect permits the temperature differentials between the primary and secondary air to be large while still maintaining occupant comfort.

FIGURE 4.3.24 Ceiling-mounted slot diffuser (courtesy of Carnes Company, Inc.).

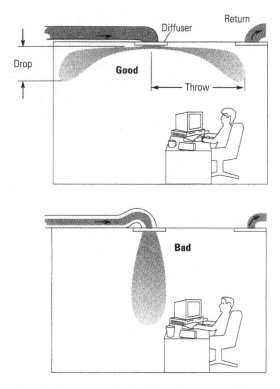

FIGURE 4.3.25 Coanda effect of supply air diffuser (courtesy of E-source, Boulder, CO).

Outlet Fixture Design Procedure

The general procedure for selecting and locating supply-air outlet devices can be summarized as follows.

1. Calculate zone supply air flow rate from design load calculations.
2. Select type and quantity of room outlets by evaluating outlet air flow rate, outlet flow pattern, and building structural characteristics.
3. Locate outlets to provide uniform room temperature through use of a uniform distribution pattern with modifications due to high gain or loss parts of a zone, e.g., windows.
4. Select proper outlet size from manufacturer's literature, based on outlet air flow, discharge velocity and throw, distribution pattern, pressure loss, and sound level requirements.

TABLE 4.3.3   Sidewall, Sill Throw, and Total Pressure for a Slot Diffuser

CFM/FT	Neck Size	2.0 in	2.5 in	3.0 in	4.0 in	5.0 in	6.0 in	8.0 in	10.0 in	12.0 in
100	Total Pressure (iwg)	.19	.081	.063	.032	.020	.013			
	Sidewall Throw (ft.)	20–27	18–26	17–24	15–23	13–22	11–19			
	Sill Throw (ft.)	18–22	16–20	15–18	12–15	9–13	7–10			
200	Total Pressure (iwg)			.25	.13	.080	.052	.026	.017	.012
	Sidewall Throw (ft.)			30–40	27–38	25–36	23–33	20–29	16–25	14–21
	Sill Throw (ft.)			25–32	23–28	20–24	17–22	15–19	13–17	11–14

TABLE 4.3.4   Characteristic Room Length Definitions

Diffuser Type	Characteristic Length (L)
High sidewall grille	Distance to wall perpendicular to jet
Circular ceiling diffuser	Distance to closest wall or intersecting jet
Ceiling slot diffuser	
Perforated, louvered ceiling diffuser	
Sill grille	Length of room in direction of jet flow
Light troffer diffuser	Distance to midplane between outlets plus distance from ceiling to occupied zone

Proper selection and spacing of supply-air outlet devices will reduce the occurrence of excess fluctuation of conditions and maintain comfort within the room. The physiological effect of temperature, humidity, and air motion on the human body is measured in terms of the *effective draft temperature*. The percent of locations in a room where the effective draft temperature is within the comfort range is defined as the air diffusion performance index (ADPI). A higher ADPI means more desirable conditions are achieved. To simplify diffuser specification and completing step 4 above, guidelines have been developed for selecting supply-air outlets based on maximizing ADPI in the building zone.

An understanding of several terms is required to apply the guidelines. The diffuser *throw* is the maximum distance from the outlet device to a point in the air stream where the velocity in the stream cross-section equals a specified terminal velocity. *Sidewall throw* is the horizontal distance from the diffuser. *Sill throw* is the horizontal plus vertical distance from the diffuser mounted in the sill or floor. The *drop* is the vertical distance from the diffuser.

For most devices the *terminal velocity*, V_t, for which the throw is defined is specified as 50 fpm. For ceiling slot-diffusers the value has been set at 100 fpm. The throw distance for a particular terminal velocity is denoted as T_v. The subscript refers to the terminal velocity for which it is defined, such as T_{50}.

The throw of the device is impacted by the flow rate through the device and its neck area. The pressure drop of the device is also impacted by these two parameters. Table 4.3.3 presents sample catalog data for a ceiling-mounted slot diffuser similar to the one depicted in Figure 4.3.24. The neck dimensions are listed as diameter inches. Flow rate is presented in terms of active diffuser length. The total pressure is the static and velocity pressure drop through the device. Sidewall and sill throw minimum and maximum values are based on T_{100} and T_{50}, respectively.

In selecting and spacing diffusers, the throw is compared to the *characteristic room length*, L, the distance from the outlet device in the principal horizontal direction of the air flow to the nearest boundary wall or intersecting air jet. Definitions of characteristic room length for several diffuser types are listed in Table 4.3.4.

Guidelines

Table 4.3.5 recommendations for T_v/L values for room loads ranging from 20–80 Btu/h ft^2 for several types of supply-air outlets. Recommended values are given for meeting a maximum ADPI. Also, a range of T_v/L values is listed that meet a minimum ADPI. To use the selection guidelines, the room dimensions, load, and air volume requirements must be known. With this information available, proceed as follows.

TABLE 4.3.5 ADPI Selection Guide

Terminal Device	Room Load, Btu/(h · ft²)	T_{50}/L for Max. ADPI	Maximum ADPI	For ADPI Greater Than	Range of $T_{0.25}/L$
High	80	1.8	68	—	—
sidewall	60	1.8	72	70	1.5–2.2
grilles	40	1.6	78	70	1.2–2.3
	20	1.5	85	80	1.0–1.9
Circular	80	0.8	76	70	0.7–1.3
ceiling	60	0.8	83	80	0.7–1.2
diffusers	40	0.8	88	80	0.5–1.5
	20	0.8	93	90	0.7–1.3
Sill grille	80	1.7	61	60	1.5–1.7
straight	60	1.7	72	70	1.4–1.7
vanes	40	1.3	86	80	1.2–1.8
	20	0.9	95	90	0.8–1.3
Sill grille	80	0.7	94	90	0.8–1.5
spread	60	0.7	94	80	0.6–1.7
vanes	40	0.7	94	—	—
	20	0.7	94	—	—
Ceiling	80	0.3*	85	80	0.3–0.7
slot diffusers	60	0.3*	88	80	0.3–0.8
(for T_{100}/L)	40	0.3*	91	80	0.3–1.1
	20	0.3*	92	80	0.3–1.5
Light	60	2.5	86	80	<3.8
troffer	40	1.0	92	90	<3.0
diffusers	20	1.0	95	90	<4.5
Perforated and	11–51	2.0	96	90	1.4–2.7
louvered ceiling				80	1.0–3.4
diffusers					

Source: ASHRAE (2001).

* The column value is actual T_{100}/L.

1. Make a preliminary selection of outlet type, number, and location.
2. Determine room characteristic length, L.
3. Select the recommended T_v/L ratio from the table.
4. Calculate the throw distance, T_v.
5. Select the appropriate supply outlet size from manufacturer's literature.
6. Verify that selection meets other design criteria for noise and pressure drop.

Return and Exhaust Air Grilles

Grilles are located at return-air or exhaust-air inlets and consist of a framed set of vertical or horizontal vanes. The vanes control the air flow in the vertical or horizontal plane and may be fixed or adjustable. They return air to the central system or exhaust air to the outside. They ensure proper, unrestricted air flow in the space and maintenance of building pressure. Return-air inlets may be connected to a duct or to another space (often the plenum above the conditioned space). Exhaust-air inlets remove air directly from the building and are always ducted.

In general, the same type of equipment used for outlets can be used for inlets. However flow outlets require accessory devices to deflect, equalize, and turn the air stream from the duct approach to produce a uniform air flow into the outlet device. For the return air, these accessories are not required, but return-air branch ducts should be equipped with volume dampers to balance the air flow. In some applications, exhaust intakes are designed and positioned to remove contaminants or heat directly from the source. Some examples of such applications include laboratory fume hoods and kitchen exhaust canopies.

The major concern with return and exhaust inlet devices is that there be a sufficient number of inlets to maintain inlet velocities within the recommended range. In general, return- and exhaust-air diffusers do not affect the air patterns in the space. However, do not mount return-air inlets close to supply-air outlets or flow short-circuiting will occur.

The location of the return and exhaust inlets can help increase HVAC efficiency. For HVAC systems predominately operating in the cooling mode, improved performance is achieved when heat is removed at the source rather than having it distributed through the space. Due to the nature of some loads, such as solar, it may be difficult to remove them at the source. For lighting loads, mounting return-air inlets near ceiling-mounted fixtures keeps the heat from dispersing into the space.

Design Procedure

The general procedure for selecting and locating inlet devices is as follows.

1. Calculate room return- and exhaust-air flow rates from design load calculations.
2. Select type and quantity of room inlets by evaluating inlet air flow rate, inlet velocity, pressure loss, and sound level.
3. Locate inlets to enhance room air circulation and removal of undesirable loads and contaminants.
4. Select proper inlet size from manufacturer's literature based on inlet airflow, inlet velocity, pressure drop, and sound level.

4.3.7 Air Handling System Control

Air handling systems are sized to meet the design peak capacity. Since the design capacity is needed during only a very small percentage of the year, it is necessary to reduce system capacity to meet the particular setpoints when operating at conditions less than full system capacity. Control systems as applied to air handling systems function to reduce system capacity and meet off-peak loads.

Typically, mechanical designers provide a sequence of operations in the design documents that outline how the system is supposed to function. The sequence of operation is used as a basis for the selection and application of control systems to air handling systems. Once the systems are installed and commissioned, the duty of system operation and maintenance falls upon the building operator.

This section lists the types of control systems normally found in air handling units and discusses air handler sequences of operation and the control philosophies normally used in controlling commonly used air system components and arrangements.

Control System Types

There are three basic types of control systems:

- *Pneumatic control* — systems that use air pressure for sensing system states, such as temperature, and for controlling and actuating devices such as valves and dampers.
- *Electric* — systems that use electric volts or current for system sensing and actuation.
- *Digital/electronic* — systems that use volts or current for system sensing and actuation and digital systems for system programming, sensing, and actuation.

Regardless of the specific type of control system used for a particular system, the system is designed to meet the requirements of the sequence of operations.

Control of the Major Components of Air Handling Systems

Following is an outline of the philosophies used in control of air handling system components.

The fans in CAV systems may operate *intermittently* or *continuously*. With intermittent fan operation or cycling, the system fans run only as needed to meet zone loads. Intermittent operation is commonly found in residential buildings and small commercial buildings. In continuous fan operation, air is

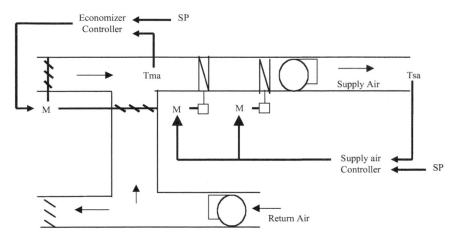

FIGURE 4.3.26 AHU control schematic.

continuously provided to the zones during occupied hours. This type of operation improves indoor air quality since outdoor ventilation air is constantly supplied. Continuous fan operation is ubiquitous in commercial buildings and may be mandatory by code to meet ventilation and other requirements.

Air handling systems used for comfort control are often shut off during unoccupied periods to conserve energy and to reduce system wear due to long run hours. Start/stop control of such systems can be accomplished through the use of time clocks. Clocks can be of the 24-hour, 7-day type, or, in the case of digital or other electronic control systems, be programmed to start or stop a system at any time during an entire year.

When a system is commanded into stop mode, it is always important to protect against potential system damage, such as the freezing of coils. To avoid such situations, outside air dampers will normally be commanded to their fully closed positions, and other steps will be taken to guard against freezing, such as the opening of hot water coil valves when sensed temperatures in critical areas are too close to freezing.

Control of Ventilation and Economizers

Airside economizers, so called to differentiate them from other types of economizers such as waterside economizers, normally consist of a set of dampers that work together to blend return and outside air to maintain a mixed or supply air, or space setpoint. Figure 4.3.26 shows a typical airside economizer control setup.

In this system the controller provides an output signal to modulate the return and outside air dampers to maintain the mixed air temperature at setpoint. Often, an economizer will be designed so that it maintains a supply-air setpoint. Use of the mixed-air control point, however, can be advantageous since it can reduce system hunting and stability problems due to the mass of the heating and cooling coils.

The use of temperature as the controlled variable results in a *dry-bulb economizer*. Another commonly used economizer scheme is the *enthalpy economizer* which is used to maximize the amount of free cooling by using the air stream (return or outside air) that has the lowest enthalpy. Enthalpy economizers can be difficult to control and maintain since they rely on accurate determination of air humidity, a notoriously problematic state to measure.

Ventilation Control

The control of ventilation rates is important for maintaining acceptable indoor air quality while maintaining efficient operation. The maintenance of the minimum ventilation rate is often accomplished through the use of a ventilation damper that is fully open when the system is on, and is fully closed when the system is off. Such a minimum outside air damper is particularly effective in a CAV system, where the system can be balanced during system startup to ensure that a specified quantity of outside air is always introduced into the supply air stream.

In VAV systems, where the flow of supply air is not constant, maintenance of a specified quantity of outside air is more difficult. Several methods to maintain constant ventilation rates while the supply air rate modulates have been used, with varying degrees of success. Some of the methods include

- Supply air/return air fan tracking, where the return-air fan is controlled to move slightly less air than the supply fan, thereby, at least in theory, ensuring a certain quantity of ventilation air.
- Use of mixed air, return air, and outside air temperature sensors to calculate the fraction of outside air entering the supply fan based on first law mixing equations.
- Use of 100% outside air fans (injection fans) to positively deliver a constant quantity of ventilation air into the mixing box. This method works well at the cost of added capital equipment expenditures.
- Use of differential pressure sensors, situated across a fixed orifice. In this configuration, a negative pressure with respect to the outside is maintained in the mixing box, and the outside air damper is modulated to maintain a constant pressure across the fixed orifice.
- Use of CO_2 monitors in the conditioned space or return air ducting to maintain the CO_2 concentrations at acceptable levels.

Control of Air Handling Unit Fans

In CAV systems, fans operate at a constant speed to provide a constant volume of air to and from the conditioned space. In this case, the supply and return fans are balanced during system commissioning to maintain the constant flow rates. The only control, then, of the fans is start/stop control.

In VAV systems, the supply and usually return fans are modulated using variable inlet vanes or variable speed drives to vary the air flow to the required rate. In nearly all cases, the supply fan is modulated to maintain a duct static pressure setpoint at some location in the supply air ducting. As the associated VAV terminal boxes modulate open to provide more cooling to the spaces, the pressure in the supply air ducting will decrease. In response to the decreased duct pressure, the control system will increase the air flow into the ducting to maintain the pressure setpoint. Conversely, as the associated VAV boxes close to reduce cooling to the spaces, the duct pressure will tend to increase, causing the control system to reduce the air flow rate to maintain the static pressure setpoint.

Control of Main Air Handling Unit Heating and Cooling Coils

The coils in the air handling units are modulated to maintain temperature and, in some cases, humidity setpoints. For air handlers that serve multiple spaces, the most common approach is to modulate the coil capacity to maintain a supply-air temperature setpoint. The controller modulates a control valve in the case of steam, hot water, and chilled water cooling coils. The controller modulates or stages the source in systems with vapor-compression air conditioning, a furnace, or electric resistance heating coils. Figure 4.3.26 shows a control schematic for the maintenance of a supply-air setpoint in a central air handling unit with hot water heating coils and chilled water cooling coils. As shown in the figure, the controller receives a signal of the supply air temperature, compares it to the setpoint value, and sends the appropriate signal to adjust the coil valve position.

Control of Terminal Units

VAV terminal units modulate the supply air volume, the supply air temperature, or both, delivered to the conditioned space to maintain the space setpoint temperature. Two basic types of local control loops are commonly used on VAV terminal units. One type results in *pressure dependent* operation, and the other results in *pressure independent* operation.

Terminal units where the space thermostat controls the primary damper directly, as shown in Figure 4.3.27, are pressure dependent. In the figure, T_z is the actual zone temperature, *setpoint* is the desired zone temperature, C is the local controller, and DM is the air damper motor. In this arrangement, the controller output directly operates the air damper actuator. Pressure dependent boxes are so called because the flow rate of the primary air through the terminal depends on the static air pressure at the inlet of the unit and the primary air damper position.

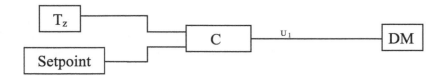

FIGURE 4.3.27   Pressure dependent terminal box local control system arrangement.

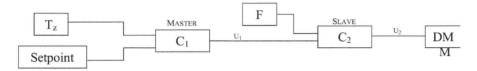

FIGURE 4.3.28   Pressure independent terminal unit control system arrangement.

Pressure dependent VAV terminal units are usually less expensive than pressure independent terminal units because they have simpler control systems and monitoring equipment.

To achieve pressure independence, the control system is arranged in a master/slave configuration, as shown in Figure 4.3.28, where:

T_z is the zone temperature
Setpoint is the desired zone temperature
C_1 is the master controller that resets the flow rate setpoint
F is the measured air flow of the primary air stream
C_2 is the slave controller modulating to maintain the setpoint flow rate from C_1
DM is the primary air damper position

The master-slave control system arrangement results in pressure independence because the flow rate into the zone is directly controlled with the slave controller, so it is not a function of the pressure at the inlet of the terminal unit. Pressure independent terminal units are more stable under varying pressure conditions but are more expensive to purchase and maintain because of the need for a flow measurement device (F) and a second controller.

Sequences of Operation for Air Handing Systems

Using the fundamental control schemes outlined above, a sequence of control operations that governs the control of the entire air handling system can be developed. The sequence of operation describes the method of control, control points, control circuits, control sequence, control features, and operation. Specific procedures for the following operations are generally included:

- Start-up
- Fan-speed control
- Supply- and mixed-air temperature control
- Room-air temperature control
- Equipment safety interlocks

The start-up describes the manual and automatic methods to start the system fans and the sequence of control contacts used to power other system components. Fan speed is actively controlled in VAV systems. Thus, the sequence of operation for VAV systems includes a section on fan speed control. Equipment safety interlocks include safety switches, limit switches, and smoke detection. A sequence of operation for a number of different HVAC systems is presented in Appendix G of *Air-Conditioning Systems Design Manual* (ASHRAE 1993).

4.3.8 Secondary Air System Design

The previous sections of this chapter outline the various components found in typical air systems. Selection of each component and integrating it into a working system that satisfies all design requirements and parameters is the designer's task. Following is a step-by-step process that can be used in system design.

1. Define system requirements and design conditions.

For general HVAC systems, the requirements include the following items:

- Inside air temperature and humidity based on comfort requirements as outlined in ASHRAE Standard 55 — *Thermal Environmental Conditions for Human Occupancy* (ASHRAE 1992a).
- Design values for outside air temperature and humidity.
- Design sensible and latent loads for each conditioned zone as determined from the building load analysis.
- Ventilation requirements as outlined in ASHRAE Standard 62.1 – *Ventilation for Acceptable Indoor Air Quality* (ASHRAE 1989). Ventilation requirements are usually driven by ventilation standards for the maintenance of indoor air quality but may be based on pressurization requirements, exhaust makeup air requirements, or other needs.
- Noise criteria.
- Space and system configurations requirements.

2. Select a supply air temperature and calculate the zone design flow rates.

A value of 55°F is typical (lower temperatures usually result in more efficient system operation and lower energy costs).

3. Select and arrange diffusers and return air grilles in each space.

Select number, spacing, and layout to provide adequate flow, ventilation, and air movement. In VAV systems, be sure to consider both design conditions and low flow rates when selecting the diffusers.

4. Select terminal units.

If designing a system that serves multiple zones, terminal units should be selected so they supply the design amount of air at or near their full open positions. Each unit should allow for adequate turn-down for off-peak conditions. To avoid circulation and ventilation problems, avoid sizing VAV terminal units to supply less than 0.6 CFM/ft^2 in the minimum damper position.

5. Layout and size duct work.

Start duct layout from the terminal units to the diffusers, then from the air handling unit(s) to the terminal units. Be sure to consider duct air leakage, heat loss, and noise in sizing the ducts. Provide adequate provisions for test and balance. Place fire and smoke dampers where appropriate.

6. Size the main air handling unit and components.

- Select heating coils, cooling coils, humidification devices. The coils must be adequate to offset the load arising from the space, the outside ventilation air, and the gains or losses in the supply and return air paths.
- Size mixing box and dampers. Include return-air dampers, mixed-air dampers, outside-air dampers. Dampers must be sized to provide good control of the air streams without causing too much pressure drop.
- Select and arrange filtration systems for the application.
- Select the fans to offset the pressure losses in each of the system components.

Because of duct air leakage, the amount of supply air leaving the air handler will be less than the amount of air introduced to the zones. This should be accounted for in sizing the air handling unit components or measures should be taken to seal all duct circuits.

7. Provide for mounting and maintenance.
Be mindful of weight and clearances. Be sure to provide access for component replacement, maintenance, and cleaning.

8. Prepare a sequence of operations.

9. Generate plans and specifications.

4.3.9   Air System Commissioning and Operation

After an air secondary system is designed and installed, it is particularly important that the system be commissioned to ensure its effective operation. Commissioning involves identifying building system equipment, control, and operational problems and fixing them so that the building performs according to the design intent. To ensure efficient system operation and long life, it is also important that the system operators and service technicians be trained to have a strong fundamental understanding of the system they are charged with operating.

Commissioning is discussed in Chapter 7.1.

Definition of Terms

Airside economizer: An air system control option that maximizes the use of outdoor air for cooling. The economizer consists of dampers, temperature and humidity sensors, actuators, and controls.

Coanda effect: The diversion of an air stream from its normal flow path due to its attachment to an adjacent surface (such as a ceiling or wall). The effect results from a low pressure region between the fluid and the surface. In supply air streams, the effect prevents cold air from dropping in a narrow column into the space.

Constant air volume system: CAV systems supply a constant flow rate of air when the fans are on. Single duct or dual duct and single zone or multiple zone systems may be constant volume systems. Single duct CAV systems commonly include terminal reheat.

Diffuser: An air distribution system outlet comprised of deflecting members designed to discharge air in various directions and promote mixing of primary air with secondary air.

Direct expansion (DX) coils: The evaporator coil in a refrigeration compression cycle. The refrigerant, the primary working fluid, absorbs heat from the secondary fluid and changes phase from liquid to gas.

Drop: The vertical distance between the supply air outlet and the lower edge of a horizontally projected air stream at the end of its throw.

Dual duct system: One type of a category of systems that supplies heating and cooling in separate ducts to multiple zones. In this type, the warm and cool air streams are mixed close to the zone served as compared to multizone systems that mix the two air streams close to the central supply fan.

Effectiveness: A measure of heat exchanger efficiency that equals the ratio of the actual amount of heat transferred to the maximum heat transfer possible between the fluid streams. The theoretical maximum is dependent on the fluids' entering state (i.e., temperature) and heat capacity.

Entrainment: The capture of surrounding air (secondary air) by the supply air stream (primary air) as it is discharged from a zone diffuser.

Evaporative cooling: The adiabatic exchange of heat between an air stream and a wetted surface or water spray. Sensitive cooling of the air stream occurs as it becomes saturated and approaches the wet-bulb temperature. It is an effective cooling method for dry climates.

Grille: A louvered or perforated device for air passage which can be located on the ceiling, wall, or floor.

Multizone system: One type of a category of systems that supplies heating and cooling in separate ducts to multiple zones. In this type, the warm and cool air streams are mixed close to the central supply fan as compared to dual duct systems that mix the two air streams close to the zone served. This system type is poorly named since many systems serve multizones although they are not necessarily *multizone* systems.

Pressure dependent terminal boxes: A type of VAV terminal box where the space thermostat controls the box damper position directly. For a given damper position, the actual flow rate through the box is system pressure dependent.

Pressure independent terminal boxes: A type of VAV terminal box where the space thermostat and a supply air flow sensor control the box damper position. The flow sensor compensates for changes in system pressure making the box pressure independent.

Register: A grille equipped with a movable damper to control the direction of flow and/or volume of flow.

Sensitive heat ratio (SHR): The ratio of sensible cooling to total cooling in an air cooler. The SHR establishes a line of constant slope on a psychrometric chart for a graphical solution to moist air cooling analysis.

System effect: An increase in system pressure drop resulting from the ducting configuration entering and exiting the air-handling unit fan. Abrupt duct transitions can result in unintentional air twirling that negatively impacts fan performance.

System curve: A graphic presentation showing the relationship between air flow rate and pressure drop for a particular system.

Throw: Upon leaving a supply air outlet, the distance the maximum velocity air stream travels before being reduced to a specified terminal velocity, usually defined as 50 or 100 feet per minute.

Terminal velocity: The maximum velocity of an air stream leaving a supply air outlet at the end of its throw.

Variable air volume systems: VAV systems modulate the flow of air supplied to the zones. Single duct or multiple duct multiple zone systems may be VAV systems.

Zone: A building thermal space that has comfort conditions controlled by a single thermostat.

References

ACGIH. 1998. *Industrial Ventilation: A Manual of Recommended Practice*. American Council of Governmental Industrial Hygienists, Cincinnati, OH.

AMCA. 1990. *Publication 201 Fans and Systems*. Air Movement and Control Association International, Arlington Heights, IL.

ASHRAE. 2001. *ASHRAE Fundamentals Handbook*. American Society of Heating, Ventilation, and Air-Conditioning Engineers, Atlanta, GA.

ASHRAE. 1996. *1996 ASHRAE Systems and Equipment Handbook*. American Society of Heating, Ventilation, and Air-Conditioning Engineers, Atlanta, GA.

ASHRAE. 1992a. *Thermal Environmental Conditions for Human Occupancy*. ANSI/ASHRAE 55-1992. American Society of Heating, Ventilation, and Air-Conditioning Engineers, Atlanta, GA.

ASHRAE. 1992b. *Gravimetric and Dust-Spot Procedures for Testing Air-Cleaning Devices Used in General Ventilation for Removing Particulate Matter*. ANSI/ASHRAE 52.1-1992. American Society of Heating, Ventilation, and Air-Conditioning Engineers, Atlanta, GA.

ASHRAE. 1989. *Ventilation for Acceptable Indoor Air Quality*. ANSI/ASHRAE 62-1989. American Society of Heating, Ventilation, and Air-Conditioning Engineers, Atlanta, GA.

Lauray, T. L. 1978. Experimental heat Transmission Coefficients for Operating Air Duct Systems. *ASHRAE Journal*, pp. 68–73, June.

SMACNA. 1995. *HVAC Duct Construction Standards*. Sheet Metal and Air Conditioning Contractors' National Association, Chantilly, VA.

SMACNA. 1990. *HVAC Systems Duct Design*. Sheet Metal and Air Conditioning Contractors' National Association, Chantilly, VA.

Additional Information

Information regarding the design requirements of HVAC systems can be found in the following resources. For comfort, refer to ASHRAE Standard 55 — *Thermal Environmental Conditions for Human Occupancy* (ASHRAE 1992a). For design values of outdoor design conditions for cities in the U.S. and worldwide, refer to values published in *ASHRAE Fundamentals*, Chapter 24 (ASHRAE 1997). Outdoor air ventilation requirements are outlined in ASHRAE Standard 62.1 — *Ventilation for Acceptable Indoor Air Quality* (ASHRAE 1989). For a sample sequence of control operations written for several different types of HVAC systems, refer to Appendix G of *Air-Conditioning Systems Design Manual* (ASHRAE 1993).

Industrial Ventilation: A Manual of Recommended Practice (ACGIH 1998) provides an excellent resource outlining the application of ventilation systems in industrial settings. *HVAC Systems Duct Design* (SMACNA 1990) and *HVAC Duct Construction Standards* (SMACNA 1995) provide detailed design information on duct construction, installation methods, design, and application for all common systems in use today.

4.4   Electrical Systems

Moncef Krarti

This chapter outlines some general concepts related to electric systems for HVAC applications in buildings. First, a review of basic characteristics of an electric system operating under alternating current is provided. Then, electric equipment commonly used in HVAC systems is described. Finally, design procedures for electrical distribution systems specific to motors are illustrated step-by-step with specific examples. Throughout this chapter, several measures of improving the energy efficiency of electrical systems are provided. Moreover, simplified calculation methods are presented to evaluate of the cost effectiveness of the proposed energy efficiency measures.

In most buildings and industrial facilities, electric systems consume a significant part of the total energy use. Table 4.4.1 compares electricity consumption in three sectors (residential, commercial, and industrial) for both the U.S. and France, which is representative of most western European countries. It is clear that in the U.S. electric energy is used more significantly in commercial and residential buildings than in the industrial facilities where fossil fuels (such as coal, oil, and natural gas) are typically used.

For residential buildings, lighting and heating, ventilating, and air conditioning (HVAC) each account for approximately 20% of total U.S. electricity use. Refrigerators represent another important energy end-use in the residential sector with about 16% of electricity. For the commercial sector as a whole, lighting accounts for over 40% while HVAC accounts for only 11% of the total electricity use. However, for commercial buildings with space conditioning, HVAC is one of the major electricity end-uses and can be more energy intensive than lighting. Moreover, computers and other office equipment (such as printers, copiers, and facsimile machines) are becoming an important electric energy end-use in office buildings.

To ensure that all electric equipment operate safely, it is important to design a reliable distribution system. In the U.S., the National Electric Code (NEC) provides specific requirements for a safe design of electrical installations. For buildings, a typical electrical installation includes the following equipment:

- A unit substation with a transformer to step down the voltage
- A set of lighting panelboards and motor control centers that house circuit breakers, fuses, disconnect switches, and overload loads
- A set of wiring distribution systems including feeders and branch circuits consisting of electrical conductors and conduits

To properly design an electric installation, it is important to first estimate the load associated with all the electric utilization equipment, including lighting fixtures, appliances, and motors.

TABLE 4.4.1   Percentage Share of Electricity in Total
Energy Use in Three Sectors for U.S.[a] and France[b]

Sector	U.S.	France
Residential buildings	61%	52%
Commercial buildings	52%	68%
Industrial facilities	12%	52%

[a] *Source:* Office of Technology Assessment (1995).
[b] *Source:* Electricité de France (1997).

4.4.1   Review of Basics

Alternating Current Systems

For a linear electrical system subject to an alternating current (AC), the time variation of the voltage and current can be represented as a sine function:

$$v(t) = V_m \cos \omega t \tag{4.4.1}$$

$$i(t) = I_m \cos(\omega t - \phi) \tag{4.4.2}$$

where

V_m and I_m are the maximum instantaneous values of voltage and current, respectively. These maximum values are related to the effective or root mean square (rms) values as follow:

$$V_m = \sqrt{2} * V_{rms} = 1.41 * V_{rms}$$

$$I_m = \sqrt{2} * I_{rms} = 1.41 * I_{rms}$$

In the U.S., the values of V_{rms} are typically 120 V for residential buildings or plug-load in the commercial buildings, 277 V for lighting systems in commercial buildings, and 480 V for motor loads in commercial and industrial buildings. Higher voltages can be used for certain industrial applications. ω is the angular frequency of the alternating current and is related to the frequency f as follows:

$$\omega = 2\pi f$$

In the U.S., the frequency f is 60 Hz, that is 60 pulsations in one second. In other countries, the frequency of the alternating current is $f = 50$ Hz.

ϕ is the phase lag between the current and the voltage. In this case, the electrical system is a resistance (such as incandescent lamp), the phase lag is zero, and the current is on phase with the voltage. If the electrical system consists of a capacitance load (such as a capacitor or a synchronous motor), the phase lag is negative and the current is in advance relative to the voltage. Finally, when the electrical system is dominated by an inductive load (such as a fluorescent fixture or an induction motor), the phase lag is positive and the current lags the voltage.

Figure 4.4.1 illustrates the time variation of the voltage for a typical electric system. The concept of root mean square (also called effective value) for the voltage, V_{rms}, is also indicated in Figure 4.4.1. It should be noted that the cycle for the voltage waveform repeats itself every 1/60 s (since the frequency is 60 Hz).

The instantaneous power, $p(t)$, consumed by the electrical system operated on one-phase AC power supply can be calculated using Ohm's law:

$$p(t) = v(t) \cdot i(t) = V_m I_m \cos \omega t \cdot \cos(\omega t - \phi) \tag{4.4.3}$$

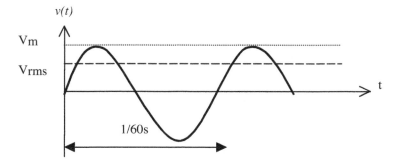

FIGURE 4.4.1 Illustration of the voltage waveform and the concept of V_{rms}.

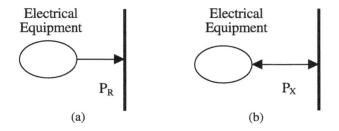

FIGURE 4.4.2 Illustration of the direction of electricity flow for (a) real power, and (b) reactive power.

The above equation can be rearranged using some basic trigonometry and the definition of the rms values for voltage and current:

$$p(t) = V_{rms} \cdot I_{rms}(\cos\phi \cdot (1 + \cos2\omega t) + \sin\phi \cdot \sin2\omega t) \qquad (4.4.4)$$

Two types of power can be introduced as a function of the phase lag angle ϕ: the real power P_R and the reactive power P_X as defined below:

$$P_R = V_{rms} \cdot I_{rms}\cos\phi \qquad (4.4.5)$$

$$P_X = V_{rms} \cdot I_{rms}\sin\phi \qquad (4.4.6)$$

Note that both types of power are constant and are not a function of time. To help understand the meaning of each power, it is useful to note that the average of the instantaneous power consumed by the electrical system over one period is equal to P_R:

$$\bar{P} = \frac{1}{T}\int_0^T p(t)dt = P_R \qquad (4.4.7)$$

Therefore, P_R is the actual power consumed by the electrical system over its operation period (which consists typically of a large number of periods T). P_R is typically called real power and is measured in kW. Meanwhile, P_X is the power required to produce a magnetic field to operate the electrical system (such as induction motors) and is stored and then released; this power is typically called reactive power and is measured in kVAR. A schematic is provided in Figure 4.4.2 to help illustrate the meaning of each type of power.

While the user of the electrical system actually consumes only the real power, the utility or the electricity provider has to make available to the user both the real and reactive power. The algebraic sum of P_R and

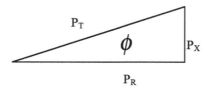

FIGURE 4.4.3   Power triangle for an electrical system.

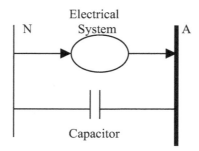

FIGURE 4.4.4   The addition of a capacitor can improve the power factor of an electrical system.

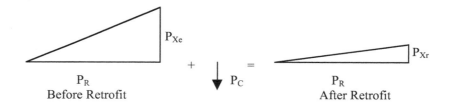

FIGURE 4.4.5   Effect of adding capacitors on the power triangle of the electrical system.

P_X constitutes the total power, P_T. Therefore, the utility has to know, in addition to the real power needed by the customer, the magnitude of the reactive power, and thus the total power.

As mentioned earlier, for a resistive electrical system, the phase lag is zero, and, thus, the reactive power is also zero (see Equation 4.4.6). Unfortunately, for commercial buildings and industrial facilities, the electrical systems are not often resistive and the reactive power can be significant. In fact, the higher the phase lag angle ϕ, the more important the reactive power P_X. To illustrate the importance of the reactive power relative to the real power P_R and the total power P_T consumed by the electrical system, a power triangle is used to represent the power flow, as shown in Figure 4.4.3.

Power Factor Improvement

As mentioned in the previous section, the reactive power must be supplied by the utility even though it is not actually registered by the power meter (as real power used). The magnitude of this reactive power increases as the power factor decreases. To account for the loss of energy due to the reactive power, most utilities have established rate structures that penalize any user who has a low power factor. Therefore, a significant savings in utility costs can be achieved by improving the power factor. As illustrated in Figure 4.4.4, this power factor improvement can be obtained by adding a set of capacitors to the entire electrical system. The size of these capacitors, P_C, is typically measured in kVAR (the same unit as the reactive power) and can be determined as indicated in Figure 4.4.5 using the power triangle analysis:

$$P_C = P_{Xe} - P_{Xr} = P_R \cdot (\tan\phi_e - \tan\phi_r) \tag{4.4.8}$$

where,

P_{Xe} and P_{Xr} are reactive power, respectively, before retrofit (existing conditions) and after retrofit (retrofitted conditions).

P_C is the reactive power of the capacitor to be added.

ϕ_e and ϕ_r are phase lag angle, respectively, before retrofit (existing conditions) and after retrofit (retrofitted conditions).

Using the values pf_e of power factor before and after Pf_r the retrofit, the size of the capacitors can be determined:

$$P_C = P_R \cdot [\tan(\cos^{-1} pf_e) - \tan(\cos^{-1} pf_r)] \qquad (4.4.9)$$

The calculations of the cost savings due to power factor improvement depend on the utility rate structure. In most rate structures, one of three options, summarized below, is used to assess the penalty for low power factor. Basic calculation procedures are typically needed to estimate the annual cost savings in the utility bills:

- *Modified billing demand:* In this case, the demand charges are increased in proportion to a fraction by which the power factor is less than a threshold value. The size for the capacitors should be selected so the system power factor reaches at least the defined threshold value.
- *Reactive power charges:* In this rate, charges for reactive power demand are included as part of the utility bills. In this option, the size of the capacitors should ideally be determined to eliminate this reactive power (i.e., so that the power factor is unity).
- *Total power charges:* This rate is similar to the rate described above but the charges are set for the building/facility total power. Again, the capacitors should be sized so the power factor is equal to unity.

The calculations of the cost savings due to power factor improvement are illustrated in Example 4.4.1.

Example 4.4.1

Problem: Consider a building with a total real power demand of 500 KW with a power factor of $pf_e = 0.70$. Determine the required size of a set of capacitors to be installed in parallel with the building service entrance so that the power factor becomes at least $pf_r = 0.90$.

Solution

The size in kVAR of the capacitor is determined using Equation 4.4.9:

$$P_C = 500 \cdot [\tan(\cos^{-1} 0.70) - \tan(\cos^{-1} 0.90) = 268 kVAR$$

Thus a capacitor rated at 275 kVAR can be selected to ensure a power factor of the building electrical system of 0.90.

4.4.2 Electrical Motors

In the U.S., there were 125 million operating motors in the range of 1 to 120 hp in 1991. These motors consumed approximately 55% of the electricity generated in the U.S. (Andreas, 1992). In large industrial facilities, motors can account for as much as 90% of the total electrical energy use. In commercial buildings, motors can account for more than 50% of the building electrical load.

Motors convert electrical energy to mechanical energy and are typically used to drive machines. The driven machines serve several purposes in the building, including moving air (supply and exhaust fans), moving liquids (pumps), moving objects or people (conveyors, elevators), compressing gases (air compressors, refrigerators), and producing materials (production equipment). To select the type of motor to be used for a particular application, several factors have to be considered, including

- The form of electrical energy that can be delivered to the motor: direct current (DC) or alternating current (AC), single or three phase.
- The requirements of the driven machine, such as motor speed and load cycles.
- The environment in which the motor is to operate: normal (where a motor with an open-type ventilated enclosure can be used), hostile (where a totally enclosed motor must be used to prevent outdoor air from infiltrating the motor), or hazardous (where a motor with an explosion-proof enclosure must be used to prevent fires and explosions).

The basic operation and the general characteristics of AC motors are discussed in the following sections. In addition, simple measures are described to improve the energy efficiency of existing motors.

Overview of Electrical Motors

There are basically two types of electric motors used in buildings and industrial facilities: (1) induction motors and (2) synchronous motors. Induction motors are the more common type, accounting for about 90% of the existing motor horsepower. Both types use a motionless stator and a spinning rotor to convert electrical energy into mechanical power. The operation of both types of motor is relatively simple and is briefly described below.

Alternating current is applied to the stator, which produces a rotating magnetic field in the stator. A magnetic field is also created in the rotor. This magnetic field causes the rotor to spin in trying to align with the rotating stator magnetic field. The rotation of the magnetic field of the stator has an angular speed that is a function of both the number of poles, N_P, and the frequency, f, of the AC current, as expressed in Equation 4.4.10:

$$\omega_{mag} = \frac{4\pi \cdot f}{N_P} \tag{4.4.10}$$

The above expression is especially useful when we discuss the use of variable frequency drives for motors with variable loads.

One main difference between the two motor types (synchronous and induction) is how the rotor field is produced. In an induction motor, the rotating stator magnetic field induces a current, thus a magnetic field, in the rotor windings which are typically of the squirrel-cage type. Because its magnetic field is induced, the rotor cannot rotate with the same speed as the stator field (if the rotor spins with the same speed as the stator magnetic field, no current can be induced in the rotor since the stator magnetic field remains unchanged relative to the rotor). The difference between the rotor speed and the stator magnetic field rotation is called the *slip factor*.

In a synchronous motor, the rotor field is produced by application of direct current through the rotor windings. Therefore, the rotor spins at the same speed as the rotating magnetic field of the stator, and, thus, the rotor and the stator magnetic field are synchronous in their speed.

Because of their construction characteristics, the induction motor is basically an inductive load and thus has a lagging power factor, while the synchronous motor can be set so it has a leading power factor (i.e., acts like a capacitor). Therefore, it is important to remember that a synchronous motor can be installed both to provide mechanical power and to improve the power factor of a set of induction motors. This option may be more cost-effective than just adding a bank of capacitors.

Three parameters are important to characterize an electric motor during full-load operation:

- The mechanical power output of the motor, P_M. This power can be expressed in kW or horsepower (hp). The mechanical power is generally the most important parameter in selecting a motor.
- The conversion efficiency of the motor, η_M. This efficiency expresses the mechanical power as a fraction of the real electric power consumed by the motor. Due to various losses (such as friction,

FIGURE 4.4.6 Definition of the efficiency of a motor.

core losses due to the alternating of the magnetic field, and resistive losses through the windings), the motor efficiency is always less than 100%. Typical motor efficiencies range from 75 to 95% depending on the size of the motor.

- The power factor of the motor, pf_M. As indicated earlier in this chapter, the power factor allows the estimation of the reactive power needed by the motor.

Using the schematic diagram of Figure 4.4.6, the real power used by the motor can be calculated as follows:

$$P_R = \frac{1}{\eta_M} \cdot P_M \tag{4.4.11}$$

Therefore, the total power and the reactive power needed to operate the motor are, respectively,

$$P_T = \frac{P_R}{pf_M} = \frac{1}{pf_M \cdot \eta_M} \cdot P_M \tag{4.4.12}$$

$$P_X = P_R \tan \phi = \frac{1}{\eta_M} \cdot P_M \cdot \tan(\cos^{-1} pf_M) \tag{4.4.13}$$

Energy Efficient Motors

General Description

Based on their efficiency, motors can be classified into two categories: (1) standard efficiency motors, and (2) high or premium efficiency (i.e., energy efficient) motors. The energy efficient motors are 2 to 10 percentage points more efficient than standard efficiency motors, depending on the size. Table 4.4.2 summarizes the average efficiencies for both standard and energy efficient motors that are currently available commercially. The improved efficiency for the high or premium motors are mainly due to better design with use of better materials to reduce losses. However, this efficiency improvement comes with a higher price of about 10 to 30% more than standard efficiency motors. These higher prices are partially the reason that only one fifth of the motors sold in the U.S. are energy efficient.

However, the installation of premium efficiency motors is becoming a common method of improving the overall energy efficiency of buildings. The potential for energy savings from premium efficiency motor retrofits is significant. In the U.S. alone, it was estimated that replacing the 125 million operating motors (in the range of 1 to 120 hp) with premium efficiency models would save approximately 60 Thw of energy per year (Nadel et al., 1991).

To determine the cost effectiveness of motor retrofits, there are several tools available, including the MotorMaster developed by the Washington State Energy Office (WSEO, 1992). These tools have the advantage of providing large databases for cost and performance information for various motor types and sizes.

Adjustable Speed Drives (ASDs)

With more emphasis on energy efficiency, an increasing number of designers and engineers are recommending the use of variable speed motors for various HVAC systems. Indeed, the use of adjustable speed

TABLE 4.4.2 Typical Motor Efficiencies

Motor mechanical power output kW (hp)	Average nominal efficiency for standard efficiency motor	Average nominal efficiency for premium efficiency motor
0.75 (1.0)	0.730	0.830
1.12 (1.5)	0.750	0.830
1.50 (2.0)	0.770	0.830
2.25 (3.0)	0.800	0.865
3.73 (5.0)	0.820	0.876
5.60 (7.5)	0.840	0.885
7.46 (10)	0.850	0.896
11.20 (15)	0.860	0.910
14.92 (20)	0.875	0.916
18.65 (25)	0.880	0.926
22.38 (30)	0.885	0.928
29.84 (40)	0.895	0.930
37.30 (50)	0.900	0.932
44.76 (60)	0.905	0.933
55.95 (75)	0.910	0.935
74.60 (100)	0.915	0.940
93.25 (125)	0.920	0.942
111.9 (150)	0.925	0.946
149.2 (200)	0.930	0.953

Source: Adapted from Hoshide, 1994.

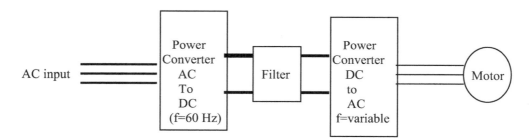

FIGURE 4.4.7 Basic concept of ASD power inverter.

drives (ASDs) is now becoming common, especially for supply and return fans in variable air volume (VAV) systems and for hot and chilled water pumps in central heating and cooling plants.

Electronic ASDs convert the fixed-frequency AC power supply (50 or 60 Hz) first to a DC supply and then to a variable frequency AC power supply, as illustrated in Figure 4.4.7. Therefore, the ASDs can change the speed of AC motors with no moving parts presenting high reliability and low maintenance requirements.

In order to achieve the energy savings potential for any HVAC application, the engineer needs to know the actual efficiency of the motors. For ASD applications, it is important to separate the losses between the drive and the motor to achieve an optimized HVAC system with the lowest operating cost. Specifically, the engineer would need to know the loss distribution, including the iron losses, copper losses, friction and windage losses, and the distribution of losses between stator and rotor. To determine these losses and thus the motor efficiency, accurate measurements are needed. Unfortunately, existent power measurement instruments are more suitable for sinusoidal rather than distorted waveforms (which are typical for ASD applications).

The use of low-cost solid state power devices and integrated circuits to control speed have made most of the commercially available ASDs draw power with extremely high harmonic content. Some investigators (Filipski and Arseneau 1990; Domijan et al. 1995; Czarkowski and Domijan, 1997) have studied the behavior of commercially available power measurement instruments subjected to voltage and current waveforms typical of ASD motor connections. In particular, three drive technologies used in HVAC industry were investigated by Czarkowski and Domijan (1997) as part of ASHRAE project 770-RP (Domijan et al., 1996), namely PWM induction, switched reluctance, and brushless DC drives. The main finding of their investigation is that existing power instruments fail to accurately measure power losses (and thus motor efficiency) due to the high harmonic content in the voltage spectra, especially for brushless DC and switched reluctance motors which represent a substantial portion of the HVAC market.

Typically, the motor losses were measured indirectly by monitoring the power input and the power output then by taking the difference. This traditional approach requires an extremely high accuracy for the power measurements to achieve "reasonable" estimation of motor losses, especially for premium efficiency motors (with low losses). However, a new approach to measure the ASD motor losses *directly* has been proposed (Fuchs and Fei, 1996).

Energy Savings Calculations

There are three methods to calculate the energy savings due to energy efficient motor replacement. These three methods are outlined below.

Method 1: *Simplified Method*

This method has been and is still used by most energy engineers to determine the energy and cost savings incurred by motor replacement. Inherent to this method, two assumptions are made: (1) the motor is fully loaded, and (2) the change in motor speed is neglected.

The electric power savings due to the motor replacement is computed as follows:

$$\Delta P_R = P_M \cdot \left(\frac{1}{\eta_e} - \frac{1}{\eta_r} \right) \tag{4.4.14}$$

Where,

P_M is the mechanical power output of the motor.

η_e is the design (i.e., full-load) efficiency of the existing motor (e.g., before retrofit).

η_r is the design (i.e., full-load) efficiency of the energy-efficient motor (e.g., after retrofit).

The electric energy savings incurred from the motor replacement is thus

$$\Delta kwh = \Delta P_R \cdot N_h \cdot LF_M \tag{4.4.15}$$

where,

N_h is the number of hours per year during which the motor is operating.

LF_M is the load factor of the motor's operation during one year.

Method 2: *Mechanical Power Rating Method*

In this method, the electrical peak demand of the existing motor is assumed to be proportional to its average mechanical power output:

$$P_{R,e} = \frac{P_M}{\eta_{op,e}} \cdot LF_{M,e} \cdot PDF_{M,e} \tag{4.4.16}$$

where,

$\eta_{op,e}$ is the motor efficiency at the operating average part-load conditions. To obtain this value, the efficiency curve for the motor can be used. If the efficiency curve for the specific existing motor is not available, a generic curve can be used.

$LF_{M,e}$ is the load factor of the existing motor and is the ratio between the average operating load of the existing motor and its rated mechanical power. In most applications, the motor is oversized and operates at less than its capacity.

$PDF_{M,e}$ is the peak demand factor and represents the fraction of typical motor load that occurs at the time of the building peak demand. In most applications, $PDF_{M,e}$ can be assumed to be unity since the motors often contribute to the total peak demand of the building.

Since the mechanical load does not change after installing an energy efficient motor, it is possible to consider a smaller motor with a capacity $P_{M,r}$ if the existing motor is oversized with a rating of $P_{M,e}$. In this case, the smaller energy efficient motor can operate at a higher load factor than the existing motor. The new load factor, LF_r, of the energy efficient motor can be calculated as follows:

$$LF_r = LF_e \cdot \frac{P_{M,r}}{P_{M,e}} \qquad (4.4.17)$$

Moreover, the energy efficient motors often operate at a higher speed than the standard motors they replace since they have lower internal losses. This higher speed actually has a negative impact since it reduces the effective efficiency of the energy efficient motor by a factor called the slip penalty. The slip penalty factor, $SLIP_P$, is defined as shown in Equation 4.4.18:

$$SLIP_P = \left(\frac{\omega_{M,r}}{\omega_{M,e}}\right)^3 \qquad (4.4.18)$$

where,

$\omega_{M,e}$ is the rotation speed of the existing motor
$\omega_{M,r}$ is the rotation speed of the energy-efficient motor

Using an equation similar to Equation 4.4.16, the peak electrical demand for the retrofitted motor (e.g., energy efficient motor) can be determined:

$$P_{R,r} = \frac{P_{M,r}}{\eta_{op,r}} \cdot LF_{M,r} \cdot PDF_{M,r} \cdot SLIP_P \qquad (4.4.19)$$

The electrical power savings due to the motor replacement can thus be estimated:

$$\Delta P_R = P_{R,e} - P_{R,r} \qquad (4.4.20)$$

The electric energy savings can be therefore calculated using Equation 4.4.15.

Method 3: *Field Measurement Method*

In this method, the motor electrical power demand is measured directly on-site. Typically, current, I_M, voltage, V_M, and power factor, pf_M, readings are recorded for the existing motor to be retrofitted. For three-phase motors (which are common in industrial facilities and in most HVAC systems for commercial buildings), the electrical power used by the existing motor can be either directly measured or calculated from current, voltage, and power factor readings as follows:

$$P_{R,E} = \sqrt{3} \cdot V_M \cdot I_M \cdot pf_M \qquad (4.4.21)$$

The load factor of the existing motor can be estimated by taking the ratio of the measured current over the nameplate full-load current, I_{FL}, as expressed in Equation 4.4.22:

$$LF_{M,E} = \frac{I_M}{I_{FL}} \qquad (4.4.22)$$

TABLE 4.4.3 Level of Participation in Lighting Conservation Programs by U.S. Commercial Buildings

Lighting Retrofit	Percent Participation in Number of Buildings	Percent Participation in Floor Area of Spaces
Energy efficient lamps and ballasts	31	49
Specular reflectors	18	32
Time clock	10	23
Manual dimmer switches	10	23
Natural lighting control sensors	7	13
Occupancy sensors	5	11

Source: From EIA, 1995

A study by Biesemeyer and Jowett (1996) has indicated that Equation 4.4.20 more accurately estimates the motor load ratio than an approach based on the ratio of the motor speeds (i.e., measured speed over nominally rated speed) used by BPA (1990) and Lobodovsky (1994). It should be noted that Equation 4.4.20 is recommended for load ratios that are above 50% since, for these load ratios, a typical motor draws electrical current proportional to the imposed load.

The methodology for the calculation of the electrical power and energy savings is the same as described for the Mechanical Power Rating Method using Equations 4.4.17 through 4.4.20.

4.4.3   Lighting Systems

Lighting accounts for a significant portion of the energy use in commercial buildings. For instance, in office buildings, 30 to 50% of electricity consumption is used to provide lighting. In addition, heat generated by lighting contributes to the thermal load to be removed by the cooling equipment. Typically, energy retrofits of lighting equipment are very cost effective with payback periods of less than two years in most applications. In the U.S., lighting system conservation features were the most often installed measures to reduce energy costs in commercial buildings, as shown in Table 4.4.3. The data for Table 4.4.3 are based on the results of a survey (EIA, 1995) to determine the level of participation of commercial buildings in a variety of specific types of conservation programs and energy technologies.

To better understand the retrofit measures that need to be considered in order to improve the energy efficiency of lighting systems, a simple estimation of the total electrical energy use due to lighting can be provided by Equation 4.4.23:

$$Kwh_{Lit} = \sum_{j=1}^{J} N_{Lum,j} \cdot WR_{Lum,j} \cdot N_{h,j} \tag{4.4.23}$$

where,

$N_{Lum,j}$ is the number of lighting luminaires of type j in the building to be retrofitted. Recall that a luminaire consists of the complete set of a ballast, electric wiring, housing, and lamps.
$WR_{Lum,j}$ is the wattage rating for each luminaire of type j. In this rating, the energy use due to both the lamps and ballast should be accounted for.
$N_{h,j}$ is the number of hours per year when the luminaires of type j are operating.
J is the number of luminaire types in the building.

It is clear from Equation 4.4.23 that there are three options to reduce the energy use due to lighting, as briefly discussed below:

(a) Reduce the wattage rating for the luminaires including both the lighting sources (e.g., lamps) and the power transforming devices (e.g., ballasts), which would therefore decrease the term $WR_{Lum,j}$ in Equation 4.4.23. In the last decade, technological advances, such as compact fluorescent lamps and electronic ballasts, have increased the energy efficiency of lighting systems.

(b) Reduce the time of use of the lighting systems through lighting controls, which would therefore decrease the term $N_{h,j}$ in Equation 4.4.23. Automatic controls have been developed to decrease the use of a lighting system, so illumination is provided only during times when it is actually needed. Among energy efficient lighting controls are occupancy sensing systems and light dimming controls through the use of daylighting.

(c) Reduce the number of luminaires, which would therefore decrease the term $N_{Lum,j}$ in Equation 4.4.23. This goal can be achieved only in cases where delamping is possible due to over-illumination.

In this section, only measures related to the general actions described in items (a) and (b) are discussed. To estimate the energy savings due to any retrofit measure for the lighting system, Equation 4.4.23 can be used. The energy use due to lighting has to be calculated before and after the retrofit, and the difference between the two estimated energy uses represents the energy savings. Throughout the section, examples of lighting retrofit are presented.

Energy Efficient Lighting Systems

Improvements in the energy efficiency of lighting systems have provided several opportunities to reduce electrical energy use in buildings. This section discusses the energy savings calculations for the following technologies:

- High efficiency fluorescent lamps
- Compact fluorescent lamps
- Compact halogen lamps
- Electronic ballasts

First a brief description is provided for the factors that an auditor should consider to achieve and maintain an acceptable quality and level of comfort for the lighting system. Second, the design and the operation concepts are summarized for each available lighting technology. Then, the energy savings that can be expected from retrofitting existing lighting systems using any of the new technologies are estimated and discussed.

Typically, three factors determine the proper level of light for a particular space. These factors include age of the occupants, speed and accuracy requirements, and background contrast (depending on the task being performed). It is a common misconception to consider that overlighting a space provides higher visual quality. Indeed, it has been shown that overlighting can actually reduce the illuminance quality and the visual comfort level within a space, in addition to wasting energy. Therefore, it is important, when upgrading a lighting system, to determine and maintain the adequate illuminance level as recommended by the appropriate authorities. Table 4.4.3 summarizes the lighting levels recommended for various activities and applications in selected countries, including the U.S., based on the most recent illuminance standards.

High Efficiency Fluorescent Lamps

Fluorescent lamps are the most commonly used lighting systems in commercial buildings. In the U.S., fluorescent lamps illuminate 71% of the commercial space. Their relatively high efficacy, diffuse light distribution, and long operating life are the main reasons for their popularity.

A fluorescent lamp consists generally of a glass tube with a pair of electrodes at each end. The tube is filled at very low pressure with a mixture of inert gases (primarily argon) and with liquid mercury. When the lamp is turned on, an electric arc is established between the electrodes. The mercury vaporizes and radiates in the ultraviolet spectrum. This ultraviolet radiation excites a phosphorous coating on the inner surface of the tube which emits visible light. High efficiency fluorescent lamps use a krypton-argon mixture which increases the efficacy output by 10 to 20% from a typical efficacy of 70 lumens/watt to about 80 lumens/watt. Improvements in phosphorous coating can further increase the efficacy to 100 lumens/watt.

TABLE 4.4.4 Recommended Lighting Levels for Various Applications in Selected Countries (in Lux Maintained on Horizontal Surfaces)

Application	France AEF (92&93)	Germany DIN5035 (90)	Japan JIS (89)	U.S./Canada IESNA (93)
Offices				
General	425	500	300–750	200–500
Reading Tasks	425	500	300–750	200–500
Drafting (detailed)	850	750	750–1500	1000–2000
Classrooms				
General	325	300–500	200–750	200–500
Chalkboards	425	300–500	300–1500	500–1000
Retail Stores				
General	100–1000	300	150–750	200–500
Tasks/Till Areas	425	500	750–1000	200–500
Hospitals				
Common Areas	100	100–300	—	—
Patient Rooms	50–100	1000	150–300	100–200
Manufacturing				
Fine Knitting	850	750	750–1500	1000–2000
Electronics	625–1750	100–1500	1500–300	1000–2000

The handling and the disposal of fluorescent lamps is highly controversial due to the fact that mercury inside the lamps can be toxic and hazardous to the environment. A new technology is being tested to replace the mercury with sulphur to generate the radiation that excites the phosphorous coating of the fluorescent lamps. The sulphur lamps are not hazardous and would present an environmental advantage to the mercury-containing fluorescent lamps.

The fluorescent lamps come in various shapes, diameters, lengths, and ratings. A common labeling method used for fluorescent lamps is

$$F.S.W.C - T.D.$$

where

> F stands for fluorescent lamp.
> S refers to the style of the lamp. If the glass tube is circular, then the letter C is used. If the tube is straight, no letter is provided.
> W is the nominal wattage rating of the lamp (such as 4, 5, 8, 12, 15, 30, 32, 34, 40, etc.)
> C indicates the color of the light emitted by the lamp: W for white, CW for cool white, BL for black light
> T refers to tubular bulb.
> D indicates the diameter of the tube in eighths of one inch (1/8 in = 3.15 mm) and is, for instance 12 (D = 1.5 in = 38 mm) for the older and less energy efficient lamps and 8 (D = 1.0 in = 31.5 mm) for more recent and energy efficient lamps.

Thus, F40CW-12 designates a fluorescent lamp that has a straight tube, uses 40W electric power, has a cool white color, and is tubular with 38 mm (1.5 in) diameter.

Among the most common retrofit in lighting systems is the upgrade of the conventional 40W T12 fluorescent lamps to more energy efficient lamps such 32W T8 lamps. For a lighting retrofit, it is recommended that a series of tests be conducted to determine the characteristics of the existing lighting system. For instance, it is important to determine the illuminance level at various locations within the space especially in working areas such as on benches and/or desks.

Compact Fluorescent Lamps

These lamps are miniaturized fluorescent lamps with small diameters and shorter lengths. The compact lamps are less efficient than full-sized fluorescent lamps with only 35 to 55 lumens/Watt. However, they are more energy efficient and have longer lives than incandescent lamps. Currently, compact fluorescent lamps are being heavily promoted as an energy saving alternative to incandescent lamps, even though they may have some drawbacks. In addition to their high cost, compact fluorescent lamps are cooler and thus provide less pleasing contrast than incandescent lamps.

Compact Halogen Lamps

Compact halogen lamps are adapted for use as direct replacements for standard incandescent lamps. Halogen lamps are more energy efficient, produce whiter light, and last longer than incandescent lamps. Indeed, incandescent lamps convert typically only 15% of their electrical energy input into visible light — 75% is emitted as infrared radiation, and 10% is used by the filament as it burns. In halogen lamps, the filament is encased inside a quartz tube which is contained in a glass bulb. A selective coating on the exterior surface of the quartz tube allows visible radiation to pass through but reflects the infrared radiation back to the filament. This recycled infrared radiation permits the filament to maintain its operating temperatures with 30% less electrical power input.

Halogen lamps can be dimmed and present no power quality or compatibility concerns as can be the case for the compact fluorescent lamps.

Electronic Ballasts

Ballasts are integral parts of fluorescent luminaires since they provide the voltage level required to start the electric arc and regulate the intensity of the arc. Before the development of electronic ballasts in the early 1980s, only magnetic or "core and coil" ballasts were used to operate fluorescent lamps. While the frequency of the electrical current is kept at 60 Hz (in countries other than the U.S., the frequency is set at 50 Hz) by the magnetic ballasts, electronic ballasts use solid state technology to produce high frequency (20–60 MHz) current. The use of high frequency current increases the energy efficiency of the fluorescent luminaires since the light is cycling more quickly and appears brighter. When used with high efficiency lamps (T8 for instance), electronic ballasts can achieve 95 lumens/Watts as opposed to 70 lumens/Watts for conventional magnetic ballasts. It should be mentioned however that efficient magnetic ballasts can achieve the same lumen/Watt ratios as electronic ballasts.

Other advantages that electronic ballasts have relative to their magnetic counterparts include

- *Higher power factor.* The power factor of electronic ballasts is typically in the 0.90 to 0.98 range. Meanwhile, conventional magnetic ballasts have a low power factor (less than 0.80) unless a capacitor is added, as discussed in Section 4.4.2.
- *Less flicker.* Since magnetic ballasts operate at 60 Hz current, they cycle the electric arc about 120 times per second. As a result, flicker may be perceptible, especially if the lamp is old, during normal operation or when the lamp is dimmed to less than 50% capacity. However, electronic ballasts cycle the electric arc several thousand times per second and flicker problems are avoided, even when the lamps are dimmed to as low as 5% of capacity.
- *Less noise.* Magnetic ballasts use electric coils and generate an audible hum which can increase with age. Such noise is eliminated by the solid state components of the electronic ballasts.

Lighting Controls

As illustrated by Equation 4.4.23, energy savings can be achieved by not operating the lighting system at full capacity when illumination becomes unnecessary. The control of the lighting system can be achieved by several means, including manual on/off and dimming switches, occupancy sensing systems, and automatic dimming systems using daylighting controls.

While energy savings can be achieved by manual switching and manual dimming, the results are typically unpredictable since they depend on occupant behavior. Scheduled lighting controls provide a more efficient approach to energy savings but can also be affected by the frequent adjustments from occupants. Only automatic light switching and dimming systems can respond in real-time to changes in occupancy and climatic changes. Some of the automatic controls available for lighting systems are briefly discussed next.

TABLE 4.4.5   Energy Savings Potential with Occupancy Sensor Retrofits

Space Application	Range of Energy Savings
Offices (Private)	25–50%
Offices (Open Space)	20–25%
Rest Rooms	30–75%
Conference Rooms	45–65%
Corridors	30–40%
Storage Areas	45–65%
Warehouses	50–75%

Occupancy Sensors

Occupancy sensors save energy by automatically turning off the lights in spaces that are not occupied. Generally, occupancy sensors are suitable for most lighting control applications and should be considered for lighting retrofits. It is important to properly specify and install the occupancy sensors to provide reliable lighting during periods of occupancy. Indeed, most failed occupancy sensor installations result from inadequate product selection and improper placement. In particular, the auditor should select the proper motion sensing technology used in occupancy sensors. Two types of motion sensing technologies are currently available in the market:

Infrared sensors — register the infrared radiation emitted by various surfaces in the space, including the human body. When the controller connected to the infrared sensors receives a sustained change in the thermal signature of the environment (as is the case when an occupant moves), it turns the lights on. The lights are kept on until the recorded changes in temperature are not significant. The infrared sensors operate adequately only if they are in direct line-of-sight with the occupants and thus must be used in smaller enclosed spaces with regular shapes and without partitions.

Ultrasound sensors — operate on a sonar principle like submarine and airport radars. A device emits a high frequency sound (25–40 KHz) beyond the hearing range of humans. This sound is reflected by the surfaces inside a space (including furniture and occupants) and is sensed by a receiver. When people move inside the space, the pattern of sound waves changes. The lights remain on until no movement is detected for a preset period of time (e.g., 5 minutes). Unlike infrared radiation, sound waves are not easily blocked by obstacles such as wall partitions. However, the ultrasound sensors may not operate properly in large spaces which tend to produce weak echoes.

Based on a study by EPRI, Table 4.4.5 shows typical energy savings expected from occupancy sensor retrofits. Significant energy savings can be achieved in spaces where occupancy is intermittent, such as conference rooms, rest rooms, storage areas, and warehouses.

Light Dimming Systems

Dimming controls allow the variation of the intensity of lighting system output based on natural light level, manual adjustments, and occupancy. A smooth and uninterrupted decrease in the light output is defined as a continuous dimming as opposed to stepped dimming in which the lamp output is decreased in stages by preset amounts.

Computer software, such as RADIANCE (LBL, 1991), can accurately estimate the energy savings from dimming systems that use natural light controls (e.g., daylighting). With such computer tools, an engineer can predict the percentage of time when natural light is sufficient to meet all lighting needs.

Example 4.4.3 provides a simple calculation procedure to estimate the energy savings from a lighting retrofit project.

Example 4.4.3

Problem: Consider a building with a total of 500 luminaires of four 40W lamps/luminaire. Determine the energy saving after replacing those with two 32W high efficacy lamps/luminaire. This building is operated 8 hours/day, 5 days/week, 50 weeks/year.

Solution

The energy saving in kWh is

$$\Delta kWh = 500 \cdot (4*40 - 4*32) \cdot 8.5.50 \cdot \frac{1}{1000} = 32,000 \ kWh/yr$$

Thus, the energy saving is 116,800 kWh/year.

4.4.4 Electrical Distribution Systems

All electrical systems have to be designed in order to provide electrical energy as safely and reliably as economically possible. Figure 4.4.8 shows a typical one-line diagram of an electrical system for a building. The main distribution panel includes the switchgear-breakers, to distribute the electric power, and the unit substation, to step down the voltage. The unit substation consists typically of a high voltage disconnect switch, a transformer, and a set of low voltage breakers. The circuit breakers for lighting and plug-connected loads are housed in lighting panelboards while the protective devices for motors are assembled typically into motor control centers (MCCs). Specifically, an MCC consists of: overload relays, to prevent the current from the motor from exceeding dangerous levels, and fuse disconnect switches or breakers to protect the motor from short circuit currents.

An important part of any electrical system is the electrical wiring that connects all the system components. Three types of connecting wires can be identified:

- Service entrance conductors are those electrical wires that deliver electricity from the supply system to the facility. For large facilities, electricity is typically supplied by an electric utility at a relatively high voltage (13.8 kV) requiring a transformer (part of a unit substation) to step down the voltage to the utilization level.

- Feeders are the conductors that deliver electricity from the service entrance equipment location to the branch-circuits. Two types of feeders are generally distinguished: the main feeders that originate at the service entrance (or main distribution panel) and the subfeeders that originate at distribution centers (lighting panelboards or motor control centers).

- Branch circuits are the conductors that deliver electricity to the utilization equipment from the point of the final over-current device.

Transformers

The transformer is the device that allows change to the voltage level of an alternating current. In particular, it is common to use transformers at generating stations to increase the transmission voltages to high levels (13,800 volt) and near or inside buildings to reduce the distribution voltages to low levels for utilization (480 or 208 volt).

A typical transformer consists of two windings: primary and secondary. The primary winding is connected to the power source, while the secondary winding is connected to the load. Between the primary and the secondary windings, there is no electrical connection. Instead, the electric energy is transferred through inductance within the core, which is generally made of laminated steel. Therefore, transformers operate only on alternating current.

There are basically two types of transformers: liquid-filled and dry-type. In liquid-filled transformers, the liquid acts as a coolant and as insulation dielectric. Dry-type transformers are constructed so that the core and coils are open to allow for cooling by the free movement of air. In some cases, fans may be installed to increase the cooling effect. The dry-type transformers are widely used because of their lighter weight and simpler installation, compared to liquid-filled transformers.

A schematic diagram for a single-phase transformer is provided in Figure 4.4.9. A three-phase transformer can be constructed from a set of three single-phase transformers electrically connected so that

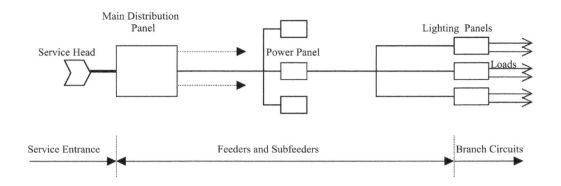

FIGURE 4.4.8 A schematic one-line diagram for a basic electrical distribution system within a building.

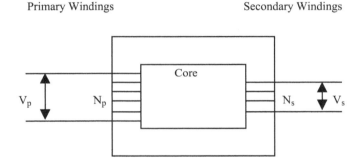

FIGURE 4.4.9 Simplified model for a single-phase transformer.

the primary and the secondary windings can be either wye or delta configurations. For buildings, delta-connected primary and wye-connected secondary is the most common arrangement for transformers. It can be shown that the primary and the secondary voltages, V_p and V_s, are directly proportional to the respective number of turns, N_p and N_s, in the windings:

$$\frac{V_p}{V_s} = \frac{N_p}{N_s} = a \qquad (4.4.24)$$

where a is the turns ratio of the transformer. As indicated by Equation 4.4.24, the turns ratio can be determined directly from the voltages without knowing the actual number of turns on the transformer windings.

Transformers are rated by their volt-ampere capacity of the secondary windings. For large transformers, the power output in kilo volt-ampere, or kVA, rating is generally used as expressed in Equation 4.4.25:

$$kVA = \frac{\sqrt{3} \cdot V_s \cdot I_s}{1000} \qquad (4.4.25)$$

where V_s and I_s are respectively the rated line-to-line voltage and the rated line current of the secondary.

Transformers are typically very efficient with energy losses (in the core and windings) representing only 1–2% of the transformer capacity. It may be cost effective to invest in more efficient transformers, especially if they are used continuously at their rated capacity, as illustrated in Example 4.4.4.

Example 4.4.4

Problem

Determine the cost effectiveness of selecting a unit with an efficiency of 99.95% rather than 99.90% for a 500-kVA rated transformer. Assume the following:

- The cost of electricity is $0.10/kWh.
- The installed costs of 99.0% and 99.5% efficient transformers are respectively $7000 and $9500.
- The average power factor of the load is 0.90.
- The no-load losses are the same for both transformers.

For the analysis, consider two cases for the length of time during which the transformer is used at its rated capacity:

(a) 10 hours/day and 250 days/year
(b) 16 hours/day and 300 days/year

Solution

To determine the cost effectiveness of installing an energy efficient transformer, a simplified economic analysis based on the payback period is used. The savings in energy losses in kWh for the high efficiency transformer can be calculated as follows:

$$kWh_{saved} = N_h \cdot kVA \cdot pf \cdot \left(\frac{1}{\eta_{std}} - \frac{1}{\eta_{eff}} \right)$$

where

N_h is the total number of hours (per year) during which the transformer is operating at full-load [for case (a) N_h = 10*250 = 2500 hrs/yr; for case (b) N_s = 16*300 = 4800 hrs/yr].
kVA is the rated transformer power output [500 kVA].
pf is the annual average power factor of the load [pf = 0.90].
η_{std} and η_{eff} are the efficiency of the standard transformer and the efficient transformer [0.990 and 0.995], respectively.

The energy savings and the payback period for each case are presented below.

For case (a) N_h = 2500 hrs/yr

The energy savings in kWh is calculated as follows:

$$kWh_{saved} = 2500 * 500 * 0.90 * \left(\frac{1}{0.990} - \frac{1}{0.995} \right) = 5710 \ kWh/yr$$

Therefore, the payback period, *PB*, for investing on the efficient transformer is

$$PB = \frac{\$9000 - \$7000}{5710 kWh * \$0.07/kWh} = 5.0 \ years$$

For case (b) N_h = 4800 hrs/yr

The energy savings in kWh is calculated as follows:

$$kWh_{saved} = 4800 * 500 * 0.90 * \left(\frac{1}{0.990} - \frac{1}{0.995} \right) = 10,964 \ kWh/yr$$

Thus, the payback period, *PB*, for investing on the efficient transformer is

$$PB = \frac{\$9000 - \$7000}{10,964(10964)kWh * \$0.07/kWh} = 2.6 \; years$$

It is clear that it can be cost-effective to consider investing in a more energy efficient transformer, especially when the load is supplied during longer periods of time. It should be noted that additional energy savings can be expected during no-load conditions for the energy efficient transformer.

Electrical Wires

The term *electrical wire* is generic and refers typically to both a conductor and a cable. The conductors are copper or aluminum wires that actually carry electrical current. Cable refers generally to the complete wire assembly, including the conductor, insulation, and any shielding and/or protective covering. A cable can have more than one conductor, each with it own insulation.

The size of an electrical conductor represents its cross-sectional area. In the U.S., two methods are used to indicate the size of a conductor: the American wire gauge (AWG) for small sizes and thousand of circular mils (MCM). For the AWG method, the available sizes are from number 18 to number 4/0 — the higher the number the smaller the conductor size. For buildings, the smallest size of copper conductor that can be used is number 14 which is rated for a maximum loading of 15 amperes. AWG size designation became inadequate soon after its implementation in the early 1900s due to the ever-increasing electrical load in buildings. For larger conductors, the cross-sectional area is measured in circular mils. A circular mil corresponds to the area of a circle that has a diameter of 1 mil or 1/1000th of an inch. For instance, a conductor with a diameter of .5 in (500 mils) has a circular mil area of 250,000 which is designated by 250 MCM.

To determine the correct size of conductors to be used for feeders and branch circuits in buildings, three criteria generally need to be considered:

- *The rating of the continuous current under normal operating conditions.* The National Electric Code (NEC, 1996) refers to the continuous current rating as the ampacity of the conductor. The main parameters that affect the ampacity of a conductor include the physical characteristics of the wire, such as its cross-sectional area (or size) and its material, and the conditions under which the wire operates, such as the ambient temperature and the number of conductors installed in the same cable. Table 4.4.5 indicates the ampacity rating of copper and aluminum conductors with various sizes. Various derating correction and factors may need to be applied to the ampacity of the conductor to select its size.

- *The rating of short circuit current under fault conditions.* Indeed, high short circuit currents can impose significant thermal or magnetic stresses not only on the conductor but also on all the components of the electrical system. The conductor has to withstand the relatively high short circuit current since the protective device requires some finite time before detecting and inter-rupting the fault current.

- *The maximum allowable voltage drop across the length of the conductor.* Most electrical utilization equipment is sensitive to the voltage applied to it. It is therefore important to reduce the voltage drop that occurs across the feeders and the branch circuits. The NEC recommends a maximum voltage drop of 3% for any one feeder or branch circuit, with a maximum voltage drop from the service entrance to the utilization outlet of 5%.

For more details, the reader is referred to section 220 of the NEC that covers the design calculations of both feeders and branch circuits.

Two conductor materials are commonly used for building electrical systems: copper and aluminum. Because of its highly desirable electrical and mechanical properties, copper is the preferred material used for conductors of insulated cables. Aluminum has some undesirable properties and its use is restricted. Indeed, an oxide film, which is not a good conductor, can develop on the surface of aluminum and can

TABLE 4.4.5 Ampacity of Selected Insulated Conductors Used in Buildings

Conductor Size (AWG or MCM)	THW (Copper)	THHN (Copper)	THW (Aluminum)	THHN (Aluminum)
18	—	14	—	—
16	—	18	—	—
14	20	25	—	—
12	25	30	20	25
10	35	40	30	35
8	50	55	40	45
6	65	75	50	60
4	85	95	65	75
3	100	110	75	85
2	115	130	90	100
1	130	150	100	115
1/0	150	170	120	135
2/0	175	195	135	150
3/0	200	225	155	175
4/0	230	260	180	205
250	255	290	205	230
300	285	320	230	255
350	310	350	250	280
400	335	380	270	305
500	380	430	310	350
600	420	475	340	385
700	460	520	375	420
750	475	535	385	435
800	490	555	395	450
900	520	585	425	480
1000	545	615	445	500
1250	590	665	485	545
1500	625	705	520	585
1750	650	735	545	615
2000	665	750	560	630

Source: Adapted from NEC Table 310-16.

cause poor electrical contact, especially at the wire connections. It should be noted that aluminum can be considered in cases when cost and weight are important criteria for the selection of conductors. However, it is highly recommended, even in these cases, to use copper conductors for the connections and the equipment terminals to eliminate poor electrical contact.

To protect the conductor, several types of insulation materials are used. The cable (which is the assembly that inludes the conductor, insulation, and any other covering) is identified by letter designations depending on the type of insulation material and the conditions of use. In buildings, the following letter designations are used.

- *For the insulation material type:* A (asbestos), MI (mineral insulation), R (rubber), SA (silicone asbestos), T (thermoplastic), V (varnished cambric), and X (cross-linked synthetic polymer).
- *For the conditions of use:* H (heat up to 75°C), HH (heat up to 90°C), UF (suitable for underground), W (moisture resistant).

Thus, the letter designation THW refers to a cable that has a thermoplastic insulation rated for maximum operating temperature of 75°C and suitable for use in dry as well as wet locations.

Moreover, some types of electrical cables have outer coverings that provide mechanical/corrosion protection, such as lead sheath (L), nylon jacket (N), armored cable (AC), metal-clad cable (MC), and nonmetallic sheath cable (NM).

For a full description of all types of insulated conductors, their letter designations, and their uses, the reader is referred to the NEC, article 310 and table 310-13.

In general, electrical cables are housed inside conduits for additional protection and safety. The types of conduits commonly used in buildings are listed below:

- Rigid metal conduit (RMC) can be of either steel or aluminum and has the thickest wall of all types of conduits. Rigid metal conduit is used in hazardous locations such as areas of high exposure to chemicals.
- Intermediate metal conduits (IMC) has a thinner wall than the rigid metal conduit but can be used in the same applications.
- Electrical metallic tubing (EMT) is a metal conduit but with a very thin wall. The NEC restricts the use of EMT to locations where it is not subjected to severe physical damage during installation or after installation.
- Electrical nonmetallic conduit (ENC) is made of nonmetallic material such as fiber or rigid PVC (polyvinyl chloride). Generally, rigid nonmetallic conduit cannot be used where subject to physical damage.
- Electrical nonmetallic tubing (ENT) is a pliable corrugated conduit that can be bent by hand. Electrical nonmetallic tubing can be concealed within walls, floors, and ceilings.
- Flexible conduit can be readily flexed and thus is not affected by vibration. Therefore, a common application of the flexible conduit is for the final connection to motors or recessed lighting fixtures.

It should be noted that the number of electrical conductors that can be installed in any one conduit is restricted to avoid any damage of cables (especially when the cables are pulled through the conduit). The NEC restricts the percentage fill to 40% for three or more conductors. The percentage fill is defined as the fraction of the total cross-sectional area of the conductors — including the insulation — over the cross-sectional area of the inside of the conduit.

When selecting the size of the conductor, the operating costs and not only the initial costs should be considered. As illustrated in Example 4.4.5, the cost of energy encourages the installation of larger conductors than are required by the NEC, especially for the smaller sized conductors (i.e., numbers 14, 12, 10, and 8). Unfortunately, most designers do not consider the operating costs in their design for several reasons, including the uncertainties in electricity prices.

Example 4.4.5

Problem

Determine if it is worthwhile to install number 10 (AWG) copper conductor instead of number 12 (AWG) on a 400 ft branch circuit that feeds a load of 16 amperes. Assume that

- The load is used 10 hours/day and 250 days/year.
- The cost of electricity is $0.10/kWh.
- The installed costs of No. 12 and No. 10 conductors are, respectively, $60.00 and $90.00 per 1000 ft cable.

Solution

In addition to the electric energy used to meet the load, which is independent of the conductor size, there is an energy loss in the form of heat generated by the flow of current, I, through the resistance of the conductor, R. The heat loss in Watts can be calculated as follows:

$$Watts = R \cdot I^2$$

Using the information by the NEC (Table 8), the resistance of both conductors No. 12 and 10 can be determined to be, respectively, 0.193 ohm and 0.121 ohm per 100 ft. Thus, the heat loss for the 400-ft branch circuit if No. 12 conductor is used can be estimated as follows:

$$Watts_{12} = 0.193 * 400/100 \cdot (16)^2 = 197.6 \ W$$

Similarly, the heat loss for the 400-ft branch circuit when No. 10 conductor is used is found to be

$$Watts_{10} = 0.121 * 400/100 \cdot (16)^2 = 123.9\ W$$

The annual cost of copper losses for both cases can be easily calculated:

$$Cost_{12} = 197.6\ W * 250\ days/yr * 10\ hrs/Day * 1\ kW/1000\ W * \$0.10/kWh = \$49.4/yr$$

$$Cost_{10} = 123.9\ W * 250\ days/yr * 10\ hrs/Day * 1\ kW/1000\ W * \$0.10/kWh = \$31.0/yr$$

Therefore, if No. 10 is used instead of No. 12, the payback period, *PB*, for the higher initial cost for the branch circuit conductor is

$$PB = \frac{(\$90/1000\ ft - \$60/1000\ ft) * 400\ ft}{(\$49.4 - \$31.0)} = 0.68\ yr = 8\ months$$

The savings in energy consumption through the use of larger conductors can thus be cost-effective. Moreover, it should be noted that the larger size conductors reduce the voltage drop across the branch circuit which permits the connected electrical utilization equipment to operate more efficiently. However, the applicable code has to be carefully consulted to determine if a larger size conduit is required when larger size conductors are used.

Branch Circuits

In general, branch circuits originate at the panelboards and/or motor control centers to serve lighting fixtures, general use receptacles, specific purpose equipment, and motors. In commercial buildings, branch circuits for lighting and receptacles are common, and their design requirements are detailed in Articles 210 and 220 of the NEC. The design requirements for motor branch circuits are generally more involved and are considered in article 430 of the NEC.

Branch Circuits for Lighting

Branch circuits for fluorescent lighting and for smaller wattage medium-based incandescent lamps (up to 300 watts) are restricted to 15 or 20 amperes. Fixed lighting units with heavy-duty lampholders can be connected to circuits rated up to 50 amperes when installed in other-than dwelling units. In general, the lighting used to illuminate areas such as offices and schools is considered to be a continuous load. Since, the NEC restricts the maximum loading on a circuit supplying a continuous load to 80% of the circuit rating, the maximum loading on a 20-ampere lighting circuit is 16 amperes.

Branch Circuits for Receptacles

The NEC defines a receptacle as a contact device installed at the outlet for the connection of a single attachment plug. The minimum load for an outlet is set by the ampere rating of the appliance served by the outlet. However, the majority of receptacles are installed for general purpose use. Therefore, the exact loads of receptacles are generally unknown. To compute the loads on a receptacle branch circuit, a minimum loading of 180 volt-amperes should be allowed for each general use receptacle outlet, regardless of whether a single, duplex, or triplex receptacle is installed. Thus, the maximum number of general use receptacles allowed on a 15- and 20-ampere branch circuit is 10 and 13, respectively (assuming a power supply source of 120 volt).

Branch Circuit for Motors

Electric motors have unique starting and running characteristics. Therefore, the branch circuits for motors have to be designed with special considerations. In particular, the starting currents of motors can be as high as six times that of their rated full-load running currents. To avoid the motor from shutting down, protective devices have to be properly designed to account for the transient starting current that can last up to 15 seconds. Moreover, the protective devices have to be able to react accurately to any

TABLE 4.4.6 Maximum Size of Overcurrent Protection Device as Required by the NEC

Wire size	Copper wire	Aluminum or copper-clad aluminum wire
No. 14 AWG	15 amperes	No. 14 AWG aluminum is not permitted
No. 12 AWG	20 amperes	15 amperes
No. 10 AWG	30 amperes	25 amperes

overloads and to protect the motor from being damaged. Thus, the branch circuit of motors includes the following components:

- Protection device for short circuit and ground fault protection
- Conductors to supply electric power to the motor
- Motor controller for overload protection
- Disconnection means to safely isolate the motor from the power source supply

Protective Devices

One of the main requirements in the design of electrical systems is to minimize power outages and damage in cases of fault conditions. Protective devices provide the means to isolate the faulted segment of the electrical system as quickly and safely as possible. Specifically, a protective device has two major functions: the detection of the fault condition and disconnection of the faulted section from the remainder of the electrical system. Some protective systems combine both functions, such as fuses, while other types separate the two actions, such as high voltage circuit breakers. Article 240 of the NEC covers the over-current protective devices.

Abnormal or fault conditions can occur on an electrical system for several reasons, including the following:

- Overloads occur when electrical equipment draws excessive current demands. Fault currents are considered overload currents when they are up to 600% of the full-load capacity of the electrical system.
- Short circuits result in considerably large flows of current in excess of 600% of the full-load current rating. Typically, short circuits are due to electrical failures, such as breakdown in the conductor insulation (arcing fault) or an accidental connection of two phases (bolted fault).
- Single phasing on three-phase systems such as motors.
- Over-voltages and transient surges that occur when the electrical system is subject to lightning.

Protective devices are characterized and rated using the following parameters:

- Maximum continuous voltage that can be applied to the electrical system without causing the conductor insulation to fail.
- Maximum continuous current that can flow in the electrical system without resulting in overheating
- Interrupting current defined as the maximum current up to which the protective device can safely operate to disconnect the electrical system.
- Short-time ratings, including the momentary current (the maximum current that the protective device can withstand without failure) and the specified time current (the current that the protective device can withstand for a specified time — typically 0.5 seconds — without failure).

In general, the size of the protective device should be less than the ampacity of the conductor being protected. Table 4.4.6 provides the maximum size of the overcurrent protective device required by the NEC depending on the conductor size of the branch circuit or feeder.

Two types of devices are commonly used to protect electrical systems in buildings: fuses and circuit breakers.

Fuses

The NEC defines the fuse as an over-current protective device with a circuit opening and fusible part that is heated and severed by the passage of current through it. Currently, there are several types of fuses suitable for various applications. The basic construction of all fuses has remained essentially the same over the years. However, for current-limiting fuses, the fusible element is made of silver, packed in a quartz filler, and hermetically sealed inside a ceramic case. For motor applications with high starting current, dual-element time-delay fuses are used to prevent the protective device from tripping each time the motor is operated.

The main advantages of fuses compared to other types of protective devices are

- Low initial cost
- Little maintenance since fuses are simple to construct
- Generally compact and require little space to be installed
- High current interrupting capabilities
- Inherently fail-safe devices since when fuses fail, they automatically open the circuit

However, fuses also present several disadvantages, including

- Can cause single phasing in three phase systems
- Are not flexible since the time response of the fuses are fixed and not adjustable
- Must be replaced after each operation

Circuit Breakers

The NEC defines the circuit breaker as a device designed to open and close a circuit by non automatic means and to open the circuit automatically on a predetermined matter without injury to itself when properly applied within its rating. Circuit breakers are available with various voltage and continuous current ratings, as well as interrupting current rating, response characteristics, and methods of operation. For instance, molded-case breakers are compact and relatively inexpensive, but they have generally low interrupting ratings and thus cannot be applied to large systems. In addition, electronic solid state trip units are currently commonly used, especially for power circuit breakers. Indeed, solid state trip units provide more flexibility and accuracy than the mechanical dual-magnetic types.

It should be noted that circuit breakers can be installed either in single pole or multi-pole. Multi-pole breakers are generally gang-operated so that all the poles are closed and opened simultaneously by one common operating mechanism (such as a handle). Therefore, circuit breakers cannot cause single phasing in three-phase systems, as is the case with fuses.

Compared to fuses, the circuit breakers provide the following advantages:

- Can serve as means of both protecting and switching an electrical circuit
- Do not cause single phasing
- Can be remotely operated
- Can easily incorporate ground-fault protection

However, breakers have some disadvantages compared to fuses. In particular, circuit breakers

- Have higher initial cost
- Require more space since they are larger
- Require more maintenance because of their complexity in construction and operation
- Do not limit fault current and thus the electrical system is subject to higher thermal and magnetic stresses under fault conditions
- Are not a fail-safe device since the trip mechanism can be damaged and the breaker can be left in a closed position

Design Requirements for HVAC Systems

The NEC includes a number of articles specific to electrically driven heating, air conditioning, and refrigeration equipment. Some of the design requirements based on the NEC that apply to the HVAC systems are summarized below.

Article 424 applies for fixed electric space heating equipment. In particular, the following design requirements should be considered for electric heating equipment:

- The branch circuit conductor and the protective device should not be smaller than 125% of the total load (i.e., heater and motor).
- A disconnect means is required to disconnect the controller and the heater.
- The disconnect must be located within sight of the heater but may not be readily accessible.

Article 422 applies to any air conditioning and refrigerating equipment that does not have a hermetic refrigerant motor compressor, such as fan-coil units and evaporator coils. In fact, the scope of article 422 includes appliances that are fastened in place, permanently connected, or cord-and-plug-connected in any occupancy. Some of the design requirements of article 422 are listed below:

- The conductor ampacity to any individual appliance shall not be less than required by the appliance marking or instruction.
- The overcurrent protection for appliances must not exceed the protective device rating marked on the appliance.
- Circuit breakers or switches can serve as the disconnect means for permanently connected equipment rated over 300 VA.

Article 440 is specific to electrically driven air conditioning and refrigeration equipment that has hermetic refrigerant motor compressors. In particular, the NEC states that:

- The branch circuit conductors to a single motor compressor must have an ampacity not less than 125% of the motor compressor current. For several motors, conductors must have an ampacity of not less than 125% of the highest rated motor compressor current of the group, plus the sum of the other motor compressor currents of the group.
- The protection device rating must not exceed the manufacturer's values marked on the equipment. For instance, if the nameplate specifies "HACR Circuit Breaker," then the equipment must be protected by a circuit breaker that is rated for heating, air conditioning, or refrigeration equipment.
- The disconnecting means should be readily accessible and within sight of the equipment. Only room air conditioners, household refrigerators and freezers, drinking water coolers, and beverage dispensers are permitted to use the attachment plug and receptacle as the disconnecting means.

For motors, starters need to be specified in addition to the conductors, protective devices, and disconnecting means. A starter is a controller whose primary function is to start and stop the operation of the motor either manually or automatically. However, starters can have additional features, such as overload, under-voltage, single-phasing protection, reversal of direction of rotation, and reduced voltage starting. In addition to the manual starters that can be used only to start and stop small motors, magnetic starters are widely used because of their proven reliability. Some of the common types and features of magnetic starters are summarized below:

- Under-voltage protection to prevent motors from restarting whenever the power is restored
- Combination starters to provide disconnecting means and overcurrent protection to the electrical system
- Reversing starters to reverse the direction of the motor rotation. Reversing starters can also be used to plug a motor to a rapid stop.

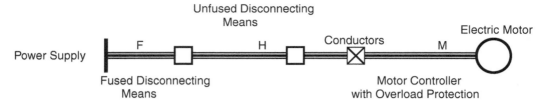

FIGURE 4.4.10 Typical branch circuit components for a single motor.

- Full-voltage starters to be used when the motor starting current does not cause serious disturbances on the electrical system
- Reduced-voltage starters to be used whenever full-voltage starting can cause unacceptable disturbances on the electrical system. Different mechanisms are used to reduce the starting voltage, including autotransformer arrangements and wye-delta connections.

Typically, motor control centers (MCCs) are used to house the motor starters in addition to protection devices, disconnecting means, and overload relays. MCCs are available in modular plug-in units allowing flexibility in arrangement and ease of maintenance.

In the U.S., the design of branch circuits, feeders, protective devices, and motor control centers for motors is based on NEC requirements. The following sections provide the design steps, through a series of examples, to illustrate the procedure to select the adequate conductor size, fuse rating, and MCC layout.

Branch Circuit for one Motor

To design the branch circuit for a single motor, four essential components need to be sized: the conductor, the protective device with the disconnecting means, the overload protection, and the unfused disconnect means, as shown in Figure 4.4.10. The following steps are suggested to size all the components of motor branch circuits. To better illustrate the design procedure, we consider the specific case of a branch circuit that supplies electrical power to a 40 hp, 460 V, three-phase squirrel-cage induction motor, with nameplate full-load current of 50 A. The motor is protected with a non-time-delay fuse and is supplied with THW conductor. Both the motor and its controller are out of sight of the branch circuit source of supply.

Step 1: Motor load

The rated current for a motor is determined from the NEC (1996) using Table 430-18 (for single-phase motors) or Table 430-150 (for three-phase motors). For a 40 hp, three-phase motor, the rated current is 52 A.

Step 2: Conductor size

Using the NEC requirement for motor branch circuits, the minimum ampacity of the conductor is 125% of 52 A or 65 A. Using Table 4.4.5 (Table NEC 310-16), the size of a THW conductor is No. 6 AWG (rated at 65 A). Since the motor required 3 conductors (for a three-phase motor with a delta connection), the conduit size is 1 in (Table 3A of chapter 9 from the NEC).

Step 3: Protective device with a disconnecting means

Since the protection device is a non-time delay fuse, a factor of 300% should be used to determine the rating of the fuse — that is 300% of 52 A or 156 A. It should be noted that NEC Table 430-152 should be used to obtain the rating factors. A standard size of fuse should be selected (the largest standard size after 156 A) or 175 A. The disconnecting size is based on the horsepower and the type of the protective device for the motor. Table 4.4.7 presents the standard rating of switch

TABLE 4.4.7 Standard Rating of Switch (in Amperes) for Three-Phase Motors (Rated at 480 Volt)

Horsepower Rating Range	Fused Switch With Non-time Delay Fuses	Fused Switch With Time Delay Fuses	Unfused Switch
Below 7.5 hp (5.6 kW)	30	30	30
7.5–15 hp (5.6–11.2 kW)	60	30	30
15–20 hp (11.2–14.9 kW)	100	60	60
20–25 hp (14.9–18.7 kW)	100	60	60
25–30 hp (18.7–22.4 kW)	200	60	60
30–50 hp (22.4–37.3 kW)	200	100	100
50–60 hp (37.3–44.8 kW)	400	100	100
60–100 hp (44.8–74.6 kW)	400	200	100
Above 100 hp (74.6 kW)	400	400	200

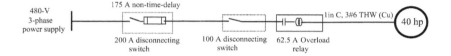

FIGURE 4.4.11 The branch circuit specifications for the 40 hp motor.

for 480-volt three-phase motors. For a 40 hp motor protected with a non-time delay fuse, the switch size is 200 A.

Step 4: Overload protection

The rating of the overload protection is 125% of 50 A (nameplate full-load current) or 62.5 A.

Step 5: Unprotected disconnecting means

The rating of unfused switch depends on the size of the motor. For a 40 hp motor, a 100 A rated unfused switch is needed (see Table 4.4.7).

A one-line diagram for the branch circuit is shown in Figure 4.4.11.

Feeder for Several Motors

In general, two components need to be sized to design a motor feeder: the conductor and the protective device with the disconnecting means. Typically, overload protection is not provided to the feeder since individual motors are cleared by their own overload relays in case any excessive overloads occur. The design procedure for motor feeders is illustrated below using a feeder that supplies electric power to the following three-phase, 460-V motors: one 40 hp, two 30 hp, three 20 hp, and six 10 hp. All the motors have full voltage nonreversing starters (FVNR) except the 40 hp motor with full voltage reversing (FVR) starter. The feeder and all the branch circuits are protected with non-time delay fuses and are supplied with THW conductors.

Step 1: Motor load

The rated load current for all three-phase motors is determined from the NEC Table 430-150. The rated current loads for 40 hp, 30 hp, 20 hp, and 10 hp 460-V motors are 52 A, 40 A, 27 A, and 14 A, respectively.

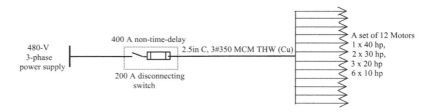

FIGURE 4.4.12   The feeder specifications to supply several motors.

Step 2: Conductor size

Using the NEC requirement for motor branch circuit, the minimum ampacity of the conductor is 125% of 52 A plus the sum of 40 A × 2, 27 A × 3, and 14 A × 6. Thus the ampacity of the conductor is 297 A. Therefore, the size of THW feeder conductors is No. 350 MCM (rated at 310 A) based on Table 4.4.5 or Table NEC 310-16. Since the motor required 3 conductors (for a three-phase motor with a delta connection), the conduit size is 2.5 in (Table 3A of chapter 9 from the NEC).

Step 3: Protective device with a disconnecting means

Since the protection device is a non-time delay fuse, a factor of 300% should be used for the largest motor. According to the NEC (1996), the fuse should be set at a rating no greater than the value calculated by taking the largest rating of the largest motor current multiplied by the appropriate factor (here, 300%) plus the sum of the full-load currents of all the remaining motors — that is, the nearest standard fuse size (300% × 52 A + 40 A × 2 + 27 A × 3 + 14 A × 6 = 401 A) or 400 A. From Table 4.4.7, the switch size is 200 A.

A one-line diagram for the feeder is shown in Figure 4.4.12.

Motor Control Centers

Motor control centers (MCCs) are recommended when several motors need to be controlled from one location. The centralized location of the control units can offer several benefits, including convenience of operation and ease of maintenance. For commercial buildings, MCCs are typically located in the mechanical rooms that house the fans, pumps, and HVAC equipment.

The overall size of MCCs depends on several factors including the number of motors, the type of the motor starters, and the rating of the protective devices. In the U.S., MCCs are typically arranged in modules of 20 in × 20 in × 90 in. The procedure for determining the space requirements of MCCs is illustrated below, using a motor control center that feeds the motors defined in the previous section.

Step 1: Space requirements for motor starters

Each module (20 in × 20 in × 90 in) of a motor control center can hold a variable number of drawout control units. Each drawout control unit houses the starter, the protection device (circuit breaker or fused switch), and a disconnecting means for one motor. The drawout control units have standard incremental dimensions. Commonly, the smallest unit has a dimension of 12 in (30 cm) with an increment of 6 in (15 cm) for larger units. Other increments such as 6.5 or 7 in (16.5 or 17.7 cm) are also available. Typically, the increments are referred to as space factors. Table 4.4.8 indicates the number of space factors required for motor control units protected by either circuit breakers or fusible switches for full voltage nonreversing (FVNR) and full voltage reversing (FVR) controllers for 480-volt three-phase motors.

Table 4.4.9 provides the space factor requirements for various drawout control units used in the MCC that serve the motors considered in the previous section.

TABLE 4.4.8 Typical Space Factors Required for MCCs (Based on a Space Factor = 6 in = 15 cm)

Motor Horsepower Range	Circuit Breaker (FVNR)	Fusible Switch (FVNR)	Circuit Breaker (FVR)	Fusible Switch
Below 10 hp (7.5 kW)	2	2	3	3
10–25 hp (7.5–19 kW)	2	2	3	3
25–50 hp (19–37 kW)	4	4	4	5
50–100 hp (37–75 kW)	4	6	4	6
100–200 hp (75–149 kW)	6	7	10	12
200–400 hp (149–298 kW)	12	12	12	12

Note: Manufacturers' specifications should be consulted.

TABLE 4.4.9 Number of Space Factors Required by the Motor Control Units

Motor hp	Type of Starter	Space Factor per One Motor	Number of Motors	Number of Space Factors
40	FVR	5	1	5
30	FVNR	4	2	8
20	FVNR	2	3	6
10	FVNR	2	6	12

Therefore, the total number of space factors required for all the motor control units is 31. In addition, one space factor is typically allocated to the incoming feeder cables. Thus, 32 space factors are needed for the MCC that serves 40 hp, 2 × 30 hp, 3 × 20 hp, and 6 × 10 hp motors. Since each module can hold 12 space factors, the number of modules needed for the MCC is 32/12 or 3 modules with 4 spare space factors.

Step 2: Layout of the MCC

Figure 4.4.13 shows a possible layout for the MCC, including the position of the drawout control units for all the motors.

The MCC can be free standing or mounted against a wall. To allow for easy access to the drawout control units, sufficient working space in the front of the MCC should be available. For an MCC rated at 480 volt, the NEC requires a minimum clearance of 3.5 ft (1.07 m) from the front face of the MCC to the nearest grounded surface such as a wall.

4.4.5 Power Quality

Under ideal operation conditions, the electrical current and voltage vary as a sine function of time. However, utility generator or distribution system problems such as voltage drops, spikes, or transients can cause fluctuations in the electricity, which can reduce the life of electrical equipment including motors and lighting systems. Moreover, an increasing number of electrical devices operating on the system can cause distortion of the sine waveform of the current and/or voltage. This distortion leads to poor power quality which can waste energy and harm both electrical distribution and devices operating on the systems.

Total Harmonic Distortion

The power quality can be defined as the extent to which an electrical system distorts the voltage or current sine waveform. The voltage and current for an electrical system with ideal power quality vary as a simple sine function of time, often referred to as the fundamental harmonic, and are expressed by Equations 4.4.1 and 4.4.2, respectively. When the power is distorted, due for instance to electronic ballasts (which change the frequency of the electricity supplied to the lighting systems), several harmonics need to be

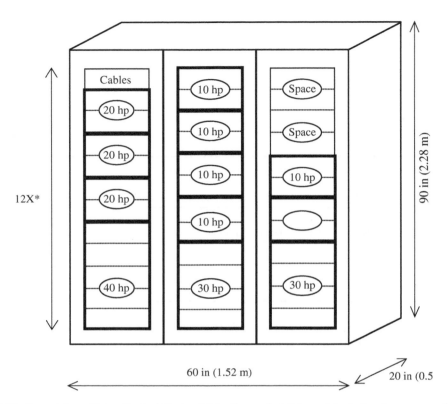

FIGURE 4.4.13 Layout for a Motor Control Center. *Note: the symbol X is used to designate one space factor [that is 6 in (15 cm)].

used in addition to the fundamental harmonic to represent the voltage or current time-variation as shown in Equations 4.4.26 and 4.4.27:

$$v(t) = \sum_{k=1}^{N_V} V_k \cos(k\omega - \theta_k) \tag{4.4.26}$$

$$i(t) = \sum_{k=1}^{N_I} I_k \cos(k\omega - \phi_k) \tag{4.4.27}$$

Highly distorted waveforms contain numerous harmonics. While the even harmonics (i.e., second, fourth, etc.) tend to cancel each other's effects, the odd harmonics (i.e., third, fifth, etc.) have peaks that coincide and significantly increase the distortion effects. To quantify the level of distortion for both voltage and current, a dimensionless number, referred to as the total harmonic distortion (THD), is determined through a Fourier series analysis of the voltage and current waveforms. The THD for voltage and current are respectively defined as follows:

$$THD_V = \sqrt{\frac{\displaystyle\sum_{k=2}^{N_V} V_k^2}{V_1^2}} \tag{4.4.28}$$

TABLE 4.4.10 Typical Power Quality Characteristics (Power Factor and Current THD) for Selected Electrical Loads

Electrical Load	Real Power Used (W)	Power Factor	Current THD (%)
Incandescent Lighting Systems			
100 W incandescent lamp	101	1.0	1
Compact Fluorescent Lighting Systems			
13 W lamp with magnetic ballast	16	0.54	13
13 W lamp with electronic ballast	13	0.50	153
Full-Sized Fluorescent Lighting Systems			
(2 lamps per ballast)	87	0.98	17
T12 40 W lamp with magnetic ballast	72	0.99	5
T12 40 W lamp with electronic ballast	93	0.98	22
T10 40 W lamp with magnetic ballast	75	0.99	5
T10 40 W lamp with electronic ballast	63	0.98	6
T8 32 W lamp with electronic ballast			
High Intensity Discharge Lighting Systems			
400 W high-pressure sodium lamp	425	0.99	14
with magnetic ballast	450	0.94	19
400 W metal halide lamp			
with magnetic ballast			
Office Equipment			
Desktop computer without monitor	33	0.56	139
Color monitor for desktop computer	49	0.56	138
Laser printer (in standby mode)	29	0.40	224
Laser printer (printing)	799	0.98	15
External fax/modem	5	0.73	47

Source: Adapted from NLPIP (1995).

$$THD_I = \sqrt{\frac{\sum_{k=2}^{N_I} I_k^2}{I_1^2}} \qquad (4.4.29)$$

Table 4.4.10 provides current THD for selected but specific lighting and office equipment loads (NLPIP, 1995). Generally, it is found that devices with high current THD contribute to voltage THD in proportion to their share of the total building electrical load. Therefore, the engineer should consider the higher-wattage devices before the lower devices to reduce the voltage THD for the entire building or facility. Example 4.4.6 shows a simple calculation procedure that can be followed to assess the impact of an electrical device on the current THD. Thus, the engineer can determine which devices to correct first to improve the power quality of the overall electric system. Typically, harmonic filters are added to electrical devices to reduce the current THD values.

Example 4.4.6

Problem

Assess the impact on the current THD of a building of two devices: the 13 W compact fluorescent lamp (CFL) with an electronic ballast and the laser printer while printing. Use the data from Table 4.4.13.

Solution

Both devices have an rms voltage of 120 V (i.e., $V_{rms} = 120$ V); their rms current can be determined using the real power used and the the power factor given in Table 4.4.10 and Equation 4.4.30 (see Equation 4.4.5):

$$I_{rms} = \frac{P_R}{V_{rms} \cdot pf} \tag{4.4.30}$$

The above equation gives an rms current of 0.22 A for the CFL and 6.79 A for the printer. These values are the rms of each device's fundamental current waveform and can be used in the THD equation, Equation 4.4.27, to estimate the total harmonic current of each device:

$$I_{tot} = I_{rms} \cdot THD_I \tag{4.4.31}$$

The resultant values of 0.33 A for the CFL and 1.02 A for the printer show that although the printer has relatively low current THD (15%), the actual distortion current produced by the printer is more than three times that of the CFL because the printer uses more power.

IEEE (1992) recommends a maximum allowable voltage THD of 5% at the building service entrance (i.e., point where the utility distribution system is connected to the building electrical system). Based on a study by Verderber et al. (1993), the voltage THD reaches the 5% limit when about 50% of the building electrical load has a current THD of 55% or when 25% of the building electrical load has a current THD of 115%.

It should be noted that when the electrical device has a power factor of unity (i.e., $pf = 1$), there is little or no current THD (i.e., $THD_I = 0\%$) since the device has only a resistive load and effectively converts input current and voltage into useful electric power. As shown in Table 4.4.10, the power factor and the current THD are interrelated, and both define the characteristics of power quality. In particular, Table 4.4.10 indicates that lighting systems with electronic ballast have typically high power factor and low current THD. This good power quality is achieved using capacitors to reduce the phase lag between the current and voltage (thus improving the power factor as discussed in Section 4.4.2) and filters to reduce harmonics (and therefore increase the current THD value).

The possible problems that have been reported due to poor power quality include:

- Overload of neutral conductors in three-phase with four wires. In a system with no THD, the neutral wire carries no current if the system is well balanced. However, when the current THD becomes significant, the currents due to the odd harmonics do not cancel each other but rather add up on the neutral wire which can overheat and be a fire hazard.
- Reduction in the life of transformers and capacitors. This effect is mostly caused by distortion in voltage.
- Interference with communication systems. Electrical devices that operate with high frequencies, such as electronic ballasts (that operate at frequencies ranging from 20 to 40 kHz), can interfere and disturb the normal operation of communication systems such as radios, phones, and energy management systems (EMS).

4.4.6 Summary

This chapter provides an overview of the basic characteristics of electrical systems in HVAC applications for buildings. In particular, the operation principles of motors are emphasized. Throughout the chapter, several measures are described to improve the energy performance of existing or new electrical installations. Moreover, illustrative examples are presented to evaluate the cost effectiveness of selected energy efficiency measures. For instance, it is shown that the use of larger conductors for branch circuits can be justified based on the reduction of energy losses and thus operating costs. Moreover, the chapter provides suggestions to improve the power quality, increase the power factor, and reduce lighting energy use in buildings. These suggestions are presented to illustrate the wide range of issues that an engineer should address when designing, analyzing, or retrofitting electrical systems for buildings.

References

Andreas, J., 1992, *Energy-Efficient Motors: Selection and Application*, Marcel Dekker, New York, NY.

Biesemeyer, W.D. and Jowett, J., 1996, Facts and fiction of HVAC motor measuring fo energy savings, Proceedings of the ACEEE 1996 Summer Study on Energy Efficiency in Buildings, ACEEE, Washington, D.C.

BPA, 1990, *High Efficiency Motor Selection Handbook*, Bonneville Power Administration, Portland, OR.

Czarkowski, D. and Domijan, A., 1997, Performance of Electric Power Meters and Analyzers in Adjustable Speed Drive Applications, *ASHRAE Transactions*, Vol. 103, Part 1.

Domijan, A., Embriz-Sander, E., Gilani, A.J., Lamer, G., Stiles, C., and Williams, C.W., 1995, Watthour Meter Accuracy under Controlled Unbalanced Harmonic Voltage and Current Conditions, *IEEE Transactions Power Delivery*, Winter Meeting of IEEE Power Engineering Society.

Domijan, A., Czarkowski, D., abu-Aisheh, A., and Embriz-Sander, E., 1996, Measurements of Electrical Power Inputs to Variable Speed Motors and Their Solid State Power Converters: Phase Ii, *ASHRAE Final Report* RP-770.

Domijan, A., Abu-Aisheh, A., and Czrakowski, D., 1997, Efficiency and Separation of Losses of an Induction Motor and Its Adjustable Speed Drive at Different Loading/Speed Combinations, *ASHRAE Transactions*, Vol. 103, Part 1.

Filipski, P.S. and Arseneau, R., 1990, Behavior of wattmeters and watthour meters under distorted waveform conditions, IEEE tutorial course, 13–22.

Fuchs, E.F. and Fei, R., 1996, A new computer-aided method for the efficiency measurement of low-loss transformers and inductors under non-sinusoidal operation, *IEEE Transactions on Power Delivery*, 11(1), 292–304.

IEEE, 1992, Guide for Harmonic Control and Reactive Compensation of Static Power Converters, *IEEE*, 519–1992. Piscataway, NJ: Institute of Electrical and Electronics Engineers.

Lobodovsky, K.K., 1994, Motor efficiency management, *Energy Engineering*, 91(2), 32–43.

Nadel, S., Shepard, M., Greenberg, S., Katz, G., and de Almeida, A., 1991, *Energy-Efficient Motor Systems: A Handbook of Technology, Program, and Policy Opportunities*, American Coucil for Energy-Efficient Economy, Washington D.C.

NEC, 1996, *National Electrical Code*, published by the National Fire Protection Association, Quincy, MA.

NLPIP, 1995, Power Quality, Lighting Answers, Newsletter of the National Lighting Product Information Program, 2(2):5.

Verderber, R.R., Morse, O.C., and Alling, W.R., 1993, Harmonics From Compact Fluorescent Lamps, *IEEE Transactions on Industry Applications*, 29(3):670–674.

5

Controls

Peter Armstrong
Pacific Northwest National Laboratory

Michael R. Brambley
Pacific Northwest National Laboratory

Peter S. Curtiss
Kreider & Associates

Srinivas Katipamula
Pacific Northwest National Laboratory

5.1 Control Fundamentals

Peter S. Curtiss

A building energy system is the combination of the HVAC plant, heating and cooling distribution paths, process loop controllers, and building energy management systems (EMSs). All of these must act in concert to produce a comfortable and healthy environment for the people who work in these buildings. The physical HVAC system must achieve the goals listed here:

- The building must maintain an internal temperature within a range acceptable by the occupants.
- Fresh air must be drawn into the building and distributed efficiently.
- Conditioned air, water, and gas must travel throughout the building to locations where they are needed.
- Internally generated air- and water-borne pollutants must be flushed from the building.
- Temperatures, pressures, flows, light levels, and energy use must all be properly controlled and any adverse interactions accounted for.

To accomplish this, a control system should be used to provide adequate feedback of the many complex processes in the building and to satisfy a number of objectives.

The main objective of the control system is *process stabilization*. HVAC processes can generally be divided into three categories: *self-stabilizing, moderate self-stabilizing*, and *unstable*. An example of the former is a well-designed coal furnace in which the amount of heat given off is the same as that produced by the burning of the coal. A moderately self-stabilizing process is usually stable, but small perturbations can lead to unstable response. A cooling coil at design conditions can be operated with a simple on-off control on the valve, but, once the load decreases, the room temperatures downstream of the coil will not be controlled adequately. An electric boiler is a good example of an unstable process, as it needs to have controls in place to prevent the boiler from reaching dangerous pressures. Unstable processes usually require some way of measuring the process, some way of assigning a setpoint to the process, and some method of controlling the process.

Another objective is the *suppression of disturbances*. Buildings are inherently dynamic, subject to constant occupant- and weather-driven load disturbances. Feedback, feedforward, and anticipating controls are used to maintain desired setpoints during periods of external disturbances.

The control system should ideally also perform some degree of *process optimization* to minimize system energy use and operating cost. Process optimization and energy savings in HVAC systems are achieved in a number of ways. The easiest is through scheduling, that is, to simply turn off devices such as pumps and fans when not in use. Another method is supervisory control, where the setpoints of the various processes are modified depending on the current load conditions.

Finally, the control system should provide a high degree of *labor saving and personal safety* features. Manually changing setpoints or adjusting valves can be boring jobs and are tasks much better suited to a continually attendant control system. Also, an automatic control system will not "lose interest" or not notice when a process is approaching dangerous levels.

Overview of Control Systems

As in any science, the field of HVAC controls has its own jargon and definition of events. Before discussing control systems in any detail, it is helpful to know the nomenclature. Formal definitions of some of the control components follow.

Process: a coil, damper, fan, or other piece of equipment that produces a motion, temperature change, pressure, etc. as a function of the actuator position and external disturbances. The output of the process is called the process value. If a positive action in the actuator causes an increase in the process value, then the process is called ***direct-acting***. A heating coil is direct-acting. If positive action in the actuator decreases the process value, for example a cooling coil, it is called ***reverse-acting***.

Sensor: a device that produces some kind of signal indicative of the process value. Sensors use pneumatic, fluidic, or electric impulses to transmit information.

Setpoint: the desired value for a process output. The difference between the setpoint and the process value is called the ***process error***.

Controller: sends signals to an actuator to effect changes in a process. The controller compares the setpoint and the process value to determine the process error. It then uses this error to adjust the output and bring the process back to the setpoint. The controller *gain* dictates the amount that the controller adjusts its output for a given error.

Actuator: a pneumatic, fluidic, or electric device that moves a damper or a valve, activates a relay, or performs any physical action that will control a process.

External disturbance: any driving force that is unmeasured or unaccounted for by the controller.

Open-loop system: one in which there is no feedback. A whole-house attic fan is an example. It will continue to run even though the house may have already cooled. Also, timed on/off devices are open loops.

Feedback (closed-loop) system: contains a process, a sensor, and a controller. Figure 5.1.1 shows some of the components and terms used when discussing feedback loop systems. In this diagram, a setpoint is compared with the measured process value. The difference between the two values is the error. The controller uses the error to generate an output signal that is sent to an actuator. The actuator, in turn, translates the control signal into a physical change of the process. A sensor measures the change of the process, and the cycle begins anew.

These feedback loop components are illustrated in the reservoir level control example shown in Figure 5.1.2. Here a float sensor adjusts the flow of water out of the tank via an armature that acts like the controller and actuator.

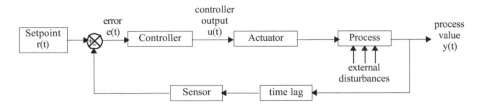

FIGURE 5.1.1 Typical components of a feedback loop.

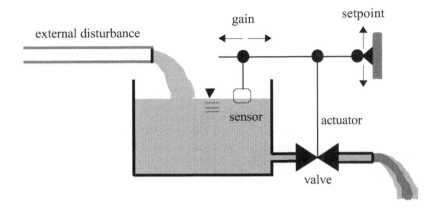

FIGURE 5.1.2 Control system example showing components.

Sensor and Process Characteristics

Sensors and processes have various characteristics that play a role in the ability of a feedback loop to maintain stable control of a system.

The *time constant* of a sensor or process is a quantity that describes the dynamic response of the device or system. Usually the time constant is related to the mass of the object. For example, the physical mass of a heating coil must first heat up before it can heat a stream of air passing through it. Likewise, a temperature sensor recording this change will probably have a protective sheaf around it that must first heat up before the sensor registers a change of temperature. The time constant of a coil can be several minutes, while for sensors it is typically tens of seconds.

The *dead time* or *lag time* of a process is the time between the change of a process and the time this change arrives at the sensor. The delay time is not related to the time constant of the sensor, although the effects of the two are similar. Large dead times must be properly treated by the control system to prevent unstable control.

Hysteresis is the nonlinear response of an actuator that results in different valve or damper positions depending on whether the control signal is increasing or decreasing. That is, a control signal of 50% may result in a valve position of 45% when the control signal is increasing from zero, but a valve position of 55% when the control signal is decreasing from 100%.

The *dead band* of a process is a range of the process value in which no control action is taken. A dead band is usually used in a 2-position control to prevent "chattering" or in split-range systems to prevent sequential control loops from fighting each other.

The *control point* is the actual, measured value of a process (i.e., the setpoint + steady state offset + compensation).

The *stability* of a feedback loop is an indication of how well the process is controlled or, alternatively, how controllable the process is. Stability is determined by any number of criteria, including overshoot, settling time, correction of deviations due to external disturbances, etc.

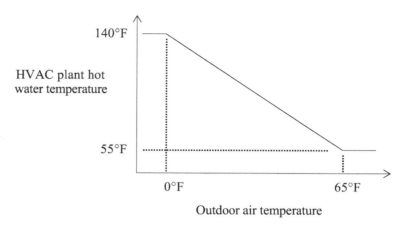

FIGURE 5.1.3   Outdoor air reset control.

Other Types of Control Loops

If the cause of a process disturbance is known, a control loop can be designed to counteract disturbances before their effects are seen at the end process. *Compensation control* (or *reset control*) is when the control point of a process is shifted upward or downward depending on the input from a second sensor. This is a subset of *feedforward control*, an example of which is outdoor-air reset control, as shown in Figure 5.1.3. In this example, as the ambient air temperature increases, the temperature of the hot water is reset upward in anticipation of a greater heating load.

Another type of feedforward control uses a model of the process to predict what the process value will be at some point in the future, based upon the current and past conditions. The controller then specifies a control action to be taken in the present that will reduce the future process error. This is called *predictive control*. Note that predictive controllers are different than *adaptive controllers*. Adaptive controllers are essentially feedback loop processes in which the gains are modified dynamically to adapt to the current process conditions.

Control Signals

There are several methods for passing control signals from one location to another. *Electric control* uses low voltages (typically 24 VAC) or line voltages (110 VAC) to measure values and effect changes in controlled variables. *Electronic control* is associated with the use of solid state, electronic components used for measurement and amplification of measured signals and the generation of proportional control signals. *Pneumatic controls* use compressed air as the medium for measuring and controlling processes. *Fluidic controls* are similar to pneumatic controls except that hydraulic fluid is used instead of compressed air.

5.1.1   Sensors

Closed-loop control of building systems is possible only if the control system is able to accurately measure the process. This section discusses some of the different types of sensors and sensing mechanisms that can be found in a building. As with the control loops, there are terms which must be defined prior to any discussion of the sensors.

> **Actual value:** the true or actual value of a process. This value can never be known with absolute certainty since it must be determined by measurements that will always incorporate some error.
> **Measured value** (or **process value**): the estimate of the actual value. The measurement error is the difference between the actual and measured values.
> **Uncertainty:** the possible value of the error. The uncertainty range is the probable range of the errors (for example, the process value = actual value ± uncertainty).

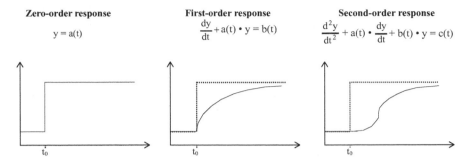

FIGURE 5.1.4 Different types of sensor response.

Process range: a process will vary between some lower and upper bounds relating to the dynamics of the physical process. This is called the process range.

Accuracy of a sensor: the expected uncertainty of measurement, usually specified by the manufacturer of the instrument. This can be given in engineering units (e.g., ± 1°F) or as a function of the range (e.g., ± 5% full scale).

Repeatability: the ability of a sensor to measure the same value during successive measurements. The systematic error is a constant error due to an inherent problem in the process or in the sensor. Sometimes a process or instrument will behave differently when the process value increases as opposed to when the process value is decreasing. This is called the hysteresis of the process.

Sensitivity: the ratio of the change in the sensor output corresponding to a unit change in the measured variable.

Calibration: the relationship between the sensor output and the corresponding engineering units is called the calibration of the sensor.

Resolution (of a sensor): the smallest readable change of the value of the measured variable.

Sensor Response

The sensor may not respond immediately if it has much mass or a small surface area in contact with the measured process. The sensor response also depends on the type of process measured. Figure 5.1.4 shows some of the different types of sensor response that may be experienced. A zero-order response is typical of sensors such as voltage and current transducers where there is an immediate response to a change in the measured process, while higher order responses are usually seen in the measurements of other processes. The first-order response is quite common and is quantified by the *time constant* of the sensor.

Sensor Time Constant — The time constant of a sensor is usually found from experimentation, although if the physical characteristics of the sensor are well known then it can be calculated. For example, consider a temperature sensor with the following known properties: mass m, specific heat c_p, surface area A, and surface conductance coefficient h (in units of Btu/hr·ft²·°F or W/cm²·°C). The overall heat capacitance of this sensor can be found from

$$C_{sensor} = m \cdot c_p$$

The heat capacity has units of energy per temperature change. Energy (in the form of heat) can pass into this sensor at a rate of

$$UA_{sensor} = A \cdot h$$

with units of power per temperature difference.

Suppose this sensor is allowed to reach a steady state temperature and is then placed into a large container of water at temperature T_{water}. The amount of energy transferred between the water and the sensor in a given time is

$$Q_{water \rightarrow sensor} = UA_{sensor} \cdot (T_{water} - T_{sensor}) \cdot \Delta t$$

The sign convention is that energy *into* the sensor is positive. The change of energy stored in the sensor over a given interval is

$$Q_{storage} = C_{sensor} \cdot [T_{sensor}(t) - T_{sensor}(t - \Delta t)] = C_{sensor} \cdot \Delta T_{sensor}$$

The energy balance is $Q_{storage} = Q_{water \rightarrow sensor}$ or

$$C_{sensor} \cdot \Delta T_{sensor} - UA_{sensor} \cdot (T_{water} - T_{sesnsor}) \cdot \Delta t = 0$$

For an infinitesimally small time interval, this can be written as

$$C_{sensor} \cdot \frac{dT_{sensor}}{dt} + UA_{sensor} \cdot (T_{sensor} - T_{water}) = 0$$

The *time constant* can now be defined. It is given as $\tau = C_{sensor}/UA_{sensor}$ in units of seconds. The energy balance is then

$$\frac{dT_{sensor}}{dt} + \frac{1}{\tau} T_{sensor} = \frac{1}{\tau} T_{water}$$

The general solution of this first-order, nonhomogeneous equation yields

$$T_{sensor}(t) = T_{water} + [T_{sensor}(0) - T_{water}]e^{-t/\tau}$$

Note that when $t = \tau$, the sensor will be e^{-1}, or 37%, of the total temperature change away from its final value.

Electronic Temperature Sensors

Almost all electronic temperature sensors use thermocouples, thermistors, or RTDs. The output of these sensors is often amplified or otherwise modified to provide a more meaningful signal to the control system.

Thermocouple — When any two dissimilar metals are in contact, a current is generated that corresponds to the temperature of the junction. This is the principle behind thermocouples. Advantages of thermocouples include

- Self-powered — no excitation voltage is necessary
- Simple — no electronics other than the constituent metals
- Rugged — difficult to break or damage
- Inexpensive — typically a few dollars per point
- Wide variety and temperature ranges — see thermocouple types in Table 5.1.1

Some of the disadvantages of thermocouples are

- Non-linear — polynomial conversion equations are required for full temperature range
- Low voltage — amplification of the signal may be necessary
- Reference required — independent measurement of the voltmeter temperature may be necessary
- Least sensitive — accuracy may be plus or minus several degrees

TABLE 5.1.1   Thermocouple Types and Ranges

Type	Metals +	Metals −	Average Seebeck Coefficient μV/°F	Average Seebeck Coefficient ref. °F	Std. Error °F	Range °F
B	94% Pt/6% Rh	70% Pt/30% Rh	3.3	1112	7.9–15.5	32 to 3200
E	90% Ni/10% Cr	Constantan	32.5	32	3.1–7.9	−450 to 1800
J	Iron	Constantan	27.9	32	2.0–5.2	−350 to 1400
K	90% Ni/10% Cr	Ni	21.9	32	2.0–5.2	−450 to 2400
R	87% Pt/10% Rh	Pt	6.4	1112	2.5–6.8	−60 to 3100
S	90% Pt/10% Rh	Pt	5.7	1112	2.5–6.8	−60 to 3100
T	Cu	Constantan	21.1	32	1.4–5.2	−450 to 800

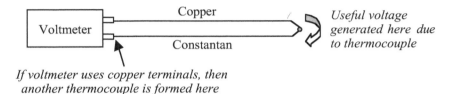

FIGURE 5.1.5   Measuring thermocouple with voltmeter.

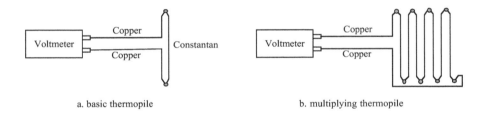

a. basic thermopile   b. multiplying thermopile

FIGURE 5.1.6   Examples of thermopiles.

Care must be exercised when using thermocouples. A standard copper-constantan thermocouple will produce a voltage proportional to the temperature at the junction of the two metals. However, it is not possible to measure this voltage directly because the connection to a voltmeter will also result in a thermocouple, as shown in Figure 5.1.5. However, if the temperature of the voltmeter terminal is known, then the necessary correction factor can be easily calculated and applied to the resulting signal.

If one wishes to measure the temperature *difference* between two points, then any even number of thermocouples can be arranged to produce a *thermopile* that can be measured with a standard voltmeter without a reference temperature. Figure 5.1.6a shows how a basic thermopile works. The two junctions produce opposing voltages; the temperature difference is proportional to the voltage difference. Figure 5.1.6b shows a thermopile that could be used for averaging purposes or to amplify the signal of a point measurement.

Resistance Temperature Device (RTD) — The resistance of most metals changes as the temperature of the metal changes. This principle is used in resistance temperature devices. Typically a thin wafer of platinum is laser-etched to provide a known resistance at a reference temperature. These types of sensors are quite stable and accurate and provide a more linear response than thermocouples. Disadvantages include

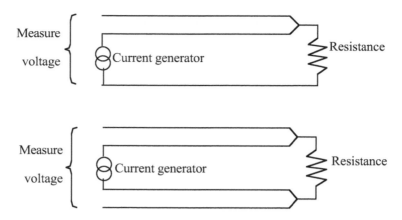

FIGURE 5.1.7   Three- and four-wire resistance measurements.

- Cost — RTDs are typically $50–100 per point
- Current source required — an excitation voltage is required to measure the resistance
- Small resistance change — total range of temperature may be represented by 30 to 40 Ω resistance range
- Low absolute resistance — reference resistance typically 100 Ω
- Self-heating — excitation voltage can cause sensor to warm and provide false readings

The temperature of an RTD is calculated as

$$°C = (\Omega_{RTD} - \Omega_{ref})/(100 \cdot \alpha)$$

where α is the average slope of the Ω_{RTD} versus temperature line between 0 and 100°C. Typical values are $\alpha = 0.00385$ or $\alpha = 0.00392$ for a Ω_{ref} of 100 Ω at 0°C.

RTD sensors have nominal resistance values of 100 to 120 ohms under normal conditions. Standard 18-gauge wire has a resistance of about 0.67 ohms per 100 feet. It is therefore necessary to use 3- or 4-wire resistance measurements to avoid introducing too much error into the resulting signal. In such a measurement, a power source is used to send a small current (typically 1 to 10 mA) through the resistor, and the resulting voltage is measured (Figure 5.1.7). If the current is accurately controlled, then the resistance can be found through Ohm's law.

Thermistor — Certain semiconductors or metal oxides can be packaged into a small probe in which the electrical resistance through the probe varies inversely with the temperature. This is the basic principle behind thermistors. The sensors often have high reference resistances (in the kΩ range) and very fast response times. Disadvantages include a strong nonlinearity, limited operational temperature range, fragility of the sensor, and the same current requirements and self-heating problems found in RTDs. Thermistor temperatures are calculated from

$$1/T = A + B \cdot \ln(\Omega_{thermistor}) + C \cdot [\ln(\Omega_{thermistor})]^3$$

where T is the temperature in Kelvin, $\Omega_{thermistor}$ is the resistance of the thermistor, and A, B, and C are curve fitting constants. Typical resistance is 5000 ohms at 25°C.

Note that with both RTDs and thermistors, it is important not to allow current to flow through the sensor continuously, as this will lead to heating of the resistor and erroneous readings. This self heating is a function of both the current and the sensor resistance where the power converted to heat is given as i^2R. The self-heating effect will be attenuated by the mass of the sensor and any factors in the local

FIGURE 5.1.8   Schematic of a wood-based humidity sensor.

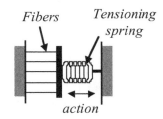

FIGURE 5.1.9   Schematic of a fiber-based humidity sensor.

FIGURE 5.1.10   Schematic of a thin-film humidity sensor.

environment that may transport heat away from the sensor, but usually it is best to provide power to the sensor only when a reading is taken.

Humidity Measurements

To determine all moist air properties, one needs only to know the two of the following: dry-bulb temperature, wet-bulb temperature, relative humidity, and humidity ratio. Some traditional ways of measuring moist air properties are listed below.

Two pieces of different types of wood can be used to determine the humidity ratio (Figure 5.1.8). A given humidity ratio will cause both pieces to absorb water, but in different amounts. The sensor will bend accordingly and generate a physical control action that can be transmitted to a controller.

In a similar fashion, fibers will contract or expand depending on the local humidity ratio and the corresponding absorption of water by the fibers (Figure 5.1.9). Both natural and synthetic fibers are available for this use. The tension of the fibers can be measured by sensors on the fiber supports.

Thin-film capacitors or resistors can be used to determine the relative humidity (Figure 5.1.10). These devices consist of a thin wafer or piece of foil that changes electrical properties as the relative humidity changes.

Chilled mirror systems use an electronically cooled reflective surface to determine the dew-point of an airstream (Figure 5.1.11). The mirror is cooled until it is no longer a specular surface (that is, until moisture in the air begins to condense on the mirror surface).

Perhaps the most standard method for measuring humid air properties is the use of a sling psychrometer (Figure 5.1.12). This device is simply two thermometers on a single base with a moistened absorbent material around the bulb of one of the thermometers. Air is forced across the absorbent material and, through evaporation, is forced to bring it to the wet-bulb temperature.

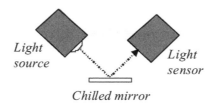

FIGURE 5.1.11   Schematic of a chilled mirror dew-point sensor.

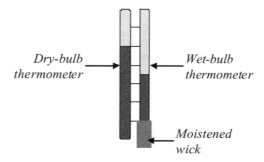

FIGURE 5.1.12   Schematic of a sling psychrometer.

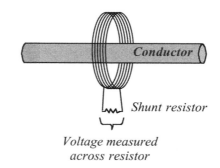

FIGURE 5.1.13   Schematic of a current transducer.

Current Measurements

Electrical current is usually measured with current transducers (sometimes called "current donuts" — Figure 5.1.13). This kind of transducer is simply a long, continuous winding of wire that uses the induced magnetic field from current flow in a power line to generate a proportional measurement signal.

The current transducers can be solid core or split core. Solid core transducers are single coils of wire that must be installed on a conductor before the conductor is connected to the load. Split core transducers can be opened on one side to allow installation around a wire (see Figure 5.1.14) and are used for short-term monitoring or in situations where the existing conductor cannot be broken or disconnected.

Pressure Measurements

Pressure will often be cited in either *gauge* (sometimes *gage*) or *absolute* pressure. Gauge pressure refers to the pressure above the ambient atmospheric pressure, while absolute pressure uses a complete vacuum as the zero reference.

A *bulb and capillary* arrangement is often used to transmit temperature signals (Figure 5.1.15). The bulb is filled with a refrigerant or other material that changes pressure as a function of the temperature.

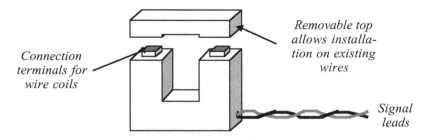

FIGURE 5.1.14 Schematic of a split core current transducer.

FIGURE 5.1.15 Schematic of a bulb and capillary sensor.

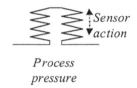

FIGURE 5.1.16 Schematic of a bellows pressure sensor.

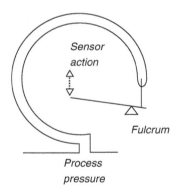

FIGURE 5.1.17 Schematic of a bourdon tube.

This kind of sensor is useful for taking average readings across a wide surface area and also for freeze protection.

A *bellows* sensor uses a flexible coupling to amplify changes in pressure and translate the pressure change into a physical motion (Figure 5.1.16). The opposite end of the bellows is attached to some kind of armature that will perform an action depending on the displacement of the bellows.

Bourdon tubes (Figure 5.1.17) are flattened pieces of pipe, capped at one end, which flex slightly when a pressure is applied to the open end. This motion is then translated into an actuator motion or dial adjustment through an armature connected to the pipe. Bourdon tubes are used in many dials and gauges.

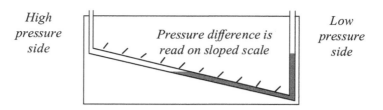

High pressure side   *Pressure difference is read on sloped scale*   *Low pressure side*

FIGURE 5.1.18   Schematic of a liquid manometer.

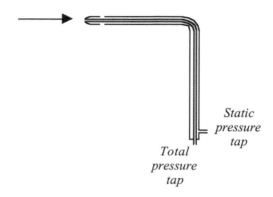

Static pressure tap

Total pressure tap

FIGURE 5.1.19   Schematic of a pitot tube.

A *manometer* (Figure 5.1.18) is often used to measure the difference between dynamic and static pressures. This is an inclined tube containing a fluid of a known and consistent specific gravity. The pressure measurements can be made by attaching a differential pressure probe to the ends of the tube.

Electronic sensors use a couple of different methods for determining the pressure applied to the sensor. Some incorporate a grid of thin wires with very specific electrical resistances. The grid is attached to a membrane that can develop a concavity under pressure, thus flexing the grid. As the grid is flexed, the resistance across the wires changes. This resistance can be measured by passing a small, constant current through the mesh and measuring the resulting voltage. *Piezoelectric* sensors rely on the physical properties exhibited by specific materials, specifically, crystals that produce a small current under pressure. This current can be measured and used as the basis for determining the pressure on the crystal.

Air Flow Measurements

Pitot tubes are most often used for hand measurements of the air flow rate in a duct. These probes consist of concentric tubes, one that is open to the oncoming air flow and the other with openings perpendicular to the air flow (Figure 5.1.19). The former measures the total pressure and the latter measure the static pressure. The difference between the two is the dynamic pressure from which the air speed can be determined from

$$v = \sqrt{\frac{2 \cdot P_d}{\rho} g_c}$$

where P_d is the dynamic pressure, ρ is the density of the air, and g_c is the acceleration of gravity (32.17 lbm·ft/lbf·s²). Since there is no air flow through the tube, the length of the tube does not matter, and long pitot tubes can be used to take measurements deep inside very large ductwork.

The profile of the air velocity through a duct is usually not uniform across the face of the duct. For this reason, it is preferable to take multiple readings of the air flow in a duct and average the result. To

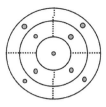

Equal area regions for
rectangular (left) and
round ducts (right)

FIGURE 5.1.20 Examples of equal area duct measurements.

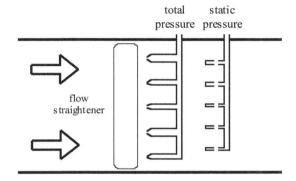

FIGURE 5.1.21 Schematic of an air flow station.

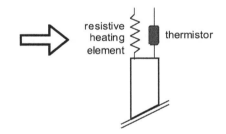

FIGURE 5.1.22 Schematic of a hot wire anemometer.

ensure that a true average is taken, the readings should be taken at equal area sections of a duct, as shown in Figure 5.1.20.

A *pitot-tube traverse* is often used to measure the air velocity because it is easy to navigate a pitot tube across the face of a duct.

Air flow stations are used to measure the average total and static pressures in a duct (Figure 5.1.21). In essence, they act like permanently installed pitot traverse measurements. The measurements are taken with many modified pitot tubes that span the cross section of the duct. Since the pressure will be adversely affected by any turbulence in the duct, flow straighteners are usually included just upstream of the pitot arrays. In addition, it is important to situate an air flow station in a long, straight region of duct, several duct diameters downstream of any kind of obstruction or elbow in the duct.

Hot wire anemometers (Figure 5.1.22) rely on heating an airstream to determine the air flow rate. The current passing through a resistive heating element is varied in order to maintain a constant temperature (around 200°F) at a downstream thermistor. Since the response of the thermistor will also depend on the air temperature, it is necessary to measure this value as well. A simpler device measures the heating element current while the resistance is held constant by a feedback controller using a wheatstone bridge.

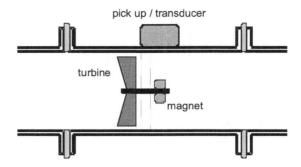

FIGURE 5.1.23   Schematic of a turbine flow meter.

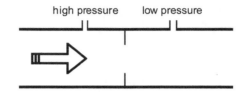

FIGURE 5.1.24   Schematic of an orifice plate meter.

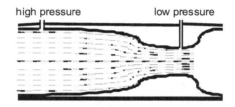

FIGURE 5.1.25   Schematic of a venturi meter.

Fluid/Gas Flow Measurements

A *turbine meter* (Figure 5.1.23) is a volumetric flow meter that consists of a propeller with a magnet attached to the shaft. The propeller makes a certain number of turns per volume of fluid that passes through the blades. A magnetic pick-up counts the number of times the propeller turns over a given time interval. The turns-to-volume ratio is called the k-factor of the turbine meter. The k-factor should remain constant but can tend to drop at low flow values due to friction in the turbine bearings.

Both *orifice plates* (Figure 5.1.24) and *venturi meters* (Figure 5.1.25) use constrictions in a fluid stream to induce a pressure drop. The pressure drop can then be correlated to a flow rate given the fluid's density and kinematic viscosity. The main difference between the two types of flow meters is that venturi meters attempt to preserve laminar flow while orifice plates usually produce turbulence. Orifice plates, therefore, create higher total system pressure drops than venturi meters, but they are also much less expensive.

Radiation Sensors

Typically, solar radiation is measured with a *pyranometer*. One type of pyranometer uses alternating black and white fields to measure radiation (see Figure 5.1.26). Sunlight striking the sensor causes the black surfaces to become warmer than the adjacent white surfaces; the temperature difference is then measured using a thermopile. The sensor is calibrated according to the correlation between the temperature difference and the intensity of the sunlight.

Other types of pyranometers use a *photovoltaic* chip (or *photodiode*) to produce (or allow) a current corresponding to the ambient radiation signal (Figure 5.1.27).

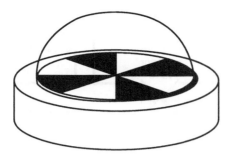

FIGURE 5.1.26   Temperature-based pyranometer.

FIGURE 5.1.27   Illustration of a photocell-based radiation sensor.

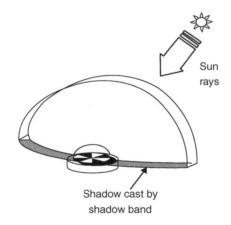

FIGURE 5.1.28   Pyranometer with shadow band.

A *pyrheliometer* is a pyranometer that tracks the sun and is shaded so that it views only a very small angle of the sky corresponding to the angular extent of the sun. These devices are used to measure only the beam radiation coming directly from the solar disk.

A *shadow band pyranometer* uses a suspended band that blocks out direct beam radiation (Figure 5.1.28). The reading from a shadow band pyranometer is usually used in conjunction with a standard pyranometer to determine both the beam and diffuse radiation.

A *multi-pyranometer array* uses several fixed pyranometers to measure the solar radiation on several surfaces simultaneously. A knowledge of solar geometry is then used to determine the individual radiation components.

5.1.2   Actuators

Electric and Electronic Actuators

Electronic actuators use a series of motors and reduction gears to move valves and dampers. They will accept control signals up to 20 VDC or 20 mA and translate a signal into an actuator position. Because

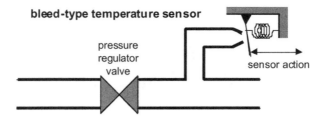

FIGURE 5.1.29   Schematic of a bleed type pneumatic sensor.

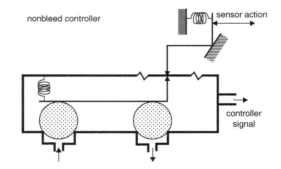

FIGURE 5.1.30   Schematic of a non-bleed controller.

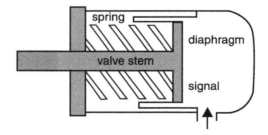

FIGURE 5.1.31   Schematic of a pneumatic valve actuator.

the motors are often geared down significantly to achieve the desired torque at the valve or damper shaft, the travel time of electronic actuators can be tens of seconds or even minutes.

Pneumatic Actuators

Pneumatic sensors and controllers are an important part of HVAC systems, particularly in older buildings. Pneumatic actuators use compressed air to generate force on diaphragms located within the actuators. With a relatively small surface area and modest air pressures, it is possible to generate sufficient forces to move valves, dampers, etc. Pneumatic actuators are generally much faster acting than electronic actuators since the full force is applied as soon as the signal is received by the actuator. It is, however, difficult to implement complex control algorithms using pneumatic components. Pneumatic actuators are also often more bulky than electronic actuators.

Pneumatic controllers can usually be divided into two different types: *bleed* (Figure 5.1.29) and *non-bleed* (Figure 5.1.30). Typical air pressures used in such systems vary from 3 to 20 psi. The signals from the controllers can be used to create a mechanical action. The force required by that action determines both the working pressure used and the size of the diaphragm used in the actuator.

Pneumatic actuators (Figure 5.1.31) take advantage of the energy inherent in the compressed air signal. The actuator body is sized so that the required torque or force will be achieved.

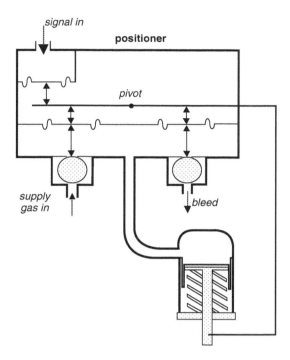

FIGURE 5.1.32 Schematic of a pneumatic positioner.

Of course, most actuators, valves, and dampers are subject to friction and sticking. This can be annoying if you are trying to obtain exact control of a process. To overcome this problem, many valves use a *positioner* to convert the pneumatic signal from the controller into the proper pressure sent to the actuator. Figure 5.1.32 shows the relation between an actuator and the positioner mechanics.

5.1.3   Steady State and Dynamic Processes

A process is basically a group of mechanical equipment in which something is put in that is changed or transformed somehow to produce an output. We are familiar with the action-reaction of processes from our every-day life experience, but to quantify these responses we need to put them in more mathematically rigorous terms. Many processes in a building will be at steady state, while others may be in a more-or-less constant state of change. Steady state processes are addressed first.

Steady State Operation

The true response of even the simplest function is actually quite complex. It is very difficult to identify and quantify every single input due to the stochastic nature of life. However, practically any process can be approximated by an equation which takes into account the known input variables and produces a reasonable likeness to the actual process output.

It is convenient to use differential equations to describe the behavior of processes. For this reason, the "complexity" of the function is denoted by the number of terms in the corresponding differential equation (i.e., the *order* or *degree* of the differential equation). In steady state analysis, we usually consider a step change in the control signal and observe the ensuing response. The following descriptions assume a step input to the function, as illustrated in Figure 5.1.33.

A step change such as this is highly unlikely in the field of HVAC controls and can be applied only to a digital event, such as a power supply being switched on or a relay being energized. Zero-order functions (also mostly theoretical) have a one-to-one correspondence to the input,

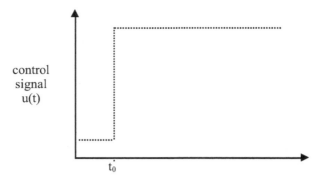

FIGURE 5.1.33   Step change function used in the explanations.

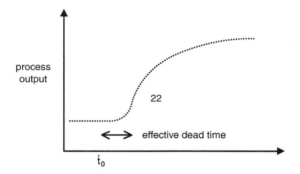

FIGURE 5.1.34   Response of second-order process to step change showing dead time.

$$y(t) = a_0 \cdot u(t)$$

First-order functions produce a curved line in the output with a step-change as input,

$$\frac{dy(t)}{dt} + a_1 \cdot y(t) = b_1 \cdot u(t)$$

and higher order functions produce lines with multiple curves.

The function that relates the process value to the controller input is called the *transfer function* of the process. The time between the application of the step change t_0 and the time at which the complete change in the process value has been achieved is called the *transfer period*. If there is a sufficient distance between the process output and the sensor, then you can observe the dead time during which the process is seemingly not affected by the control signal (see Figure 5.1.34).

The *process gain* (or *static gain* or *steady state gain*) is the ratio of the percentage change of the process to the percentage change of the control signal for a given response. The gain can be positive (as in a heating coil) or negative (as in a cooling coil).

To summarize, the transfer function of a process has several components: dead time, transfer function of the physical system, transfer period, and process gain.

Dynamic Response

In practice, very few processes are controlled in a steady-state fashion, i.e., by a series of step changes. Usually, the control signal is constantly modulating much like the way you constantly make small changes to the steering wheel of your car when driving down the highway. We now consider dynamic process changes by returning to buckets filled with water. Figure 5.1.35 shows two reservoirs of water connected by a thin tube. When water is added to the reservoir on the left, the water level in the reservoir on the

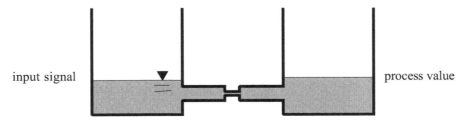

FIGURE 5.1.35 Example used in dynamic response explanation.

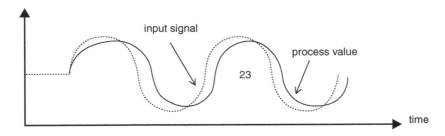

FIGURE 5.1.36 Dynamic response of water bucket example.

right will begin to rise, albeit slowly as the water must first pass through the thin constriction. In this sense, the level of water in the reservoir on the left can be considered the control signal that effects a process, i.e., the level of water on the right.

It is obvious that a step change in the control signal will bring about a first-order response of the process value. Suppose, however, that a periodic signal is applied to the level of the bucket on the left, that is, the water level increases and decreases at a constant time interval. If the frequency of the signal is small enough, we see a response in the level in the bucket on the right which varies as a function of this driving force, but with a delay and a small decrease in the amplitude, as shown in Figure 5.1.36.

Here the *dynamic process gain* is less than one even though the static process gain is 1. There is no dead time in this process; as soon as we begin to increase the control signal, the process value will also begin to increase. The dynamic process gain, therefore, can be defined similarly to that of the static gain — it is the ratio of the amplitude of the two signals, comparable to the normalized ranges used in the static gain definition.

The dynamic gain, as its name suggests, is truly dynamic. It changes not only according to the transfer function but also to the frequency of the control signal. As the frequency increases, the output lags even farther behind the input and the gain continues to decrease. At one point, the frequency may be exactly right to cancel any past effects of the input signal (i.e., the phase shift is 180°) and the dynamic gain approaches zero. If the frequency rises any higher, the process output may be decreasing as the control signal is increasing (this can easily be the case with a cooling or heating coil due to the mass effects) and the dynamic gain is negative.

5.1.4 Feedback Loops

In most houses, the furnace comes on based upon a relay activated in the thermostat. When the temperature falls below some user-set value, the furnace fires up and delivers heat to the space. This situation is an example of a feedback loop. Now, suppose the system designed knew the heat loss in the house as a function of the outdoor air temperature and the amount of heat that could be delivered by the furnace. In this case, the design may incorporate only the outside air temperature into the staging of the furnace. Such a situation is an example of feedforward control. This example illustrates the main differences between feedback and feedforward loops:

- Feedback loops use a direct measurement of the process under control.
- Feedforward (compensation) control measures the external disturbances and uses these values to control the process in anticipation of the process value.

The Control Loop

Recall that a control loop must have at least the following:

- An actuator that affects the process
- The process being controlled
- A sensor to measure the process value
- A controller that calculates the error and sends a signal to the actuator
- A setpoint input

The objective of the control loop is to maintain the process at the setpoint when

- The setpoint is changed
- The load on the process is changed
- The transfer function of the process is changed (e.g., clogged filters, fouled heat exchanged, degradation of equipment, and changes in the external disturbances)

In practice, most controllers look at three components of the error: the actual value of the error, the running sum of the error, and the change of error over time. Each of these components is multiplied by some gain, the products are summed, and the result is sent to the actuator.

Single feedback loops consist of one each of the elements of a control loop. This type of control can be applied to relatively simple control algorithms where the biggest disturbance is usually a load change, such as pressure and flow controls.

If a process consists of several subprocesses, each with a relatively different transfer function, it is often useful to use *cascaded control loops*. For example, consider a manufacturing line in which 100% outside air is used but which must also have very strict temperature control of the room air temperature. The room temperature is controlled by changing the position of a valve on a coil at the main air handling unit which feeds the zone. Typically, the time constant of the coil will be much smaller than the time constant of the room. A single feedback loop would probably result in poor control since there is so much dead time involved with both processes. The solution is to use two controllers: the first (the *master*) compares the room temperature to the thermostat setting and sends a signal to the second (the *slave*), which uses that signal as its own setpoint for controlling the coil valve. The slave controller measures the output of the coil, not the temperature of the room. The controller gain on the master can be set lower than that of the slave to prevent excessive cycling.

Sometimes control action is needed at more than one point in a process. The best example of this is an air-handling unit that contains both heating and cooling coils in order to maintain a fixed outlet air temperature no matter what the season is. Typically, a *sequential* (or *split range*) system in an air handling unit will have three temperature ranges of operation — the first for heating mode, the last for cooling mode, and a middle dead-band region where neither the cooling nor heating coils are operating. Most sequential loops are simply two different control loops acting off the same sensor. The term *sequential* refers to the fact that in most of these systems the components are in series in the air or water stream.

As noted earlier, feedforward loops (Figure 5.1.37) can be used when the effects of an external disturbance on a system are known. It can be useful to combine feedback and feedforward controllers in systems where the proper method for obtaining good control is, in part, to attack the problem at its root. An example of this is outside air temperature reset control used to modify supply air temperatures. The control loop contains both a discharge air temperature sensor (the *primary* sensor) and an outdoor air temperature sensor (the *compensation* sensor). The designer should have some idea about the influence of the outside temperature on the heating load and can then assign an *authority* to the effect of the

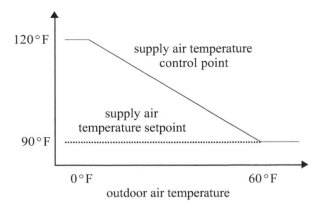

FIGURE 5.1.37 Example of feedforward control.

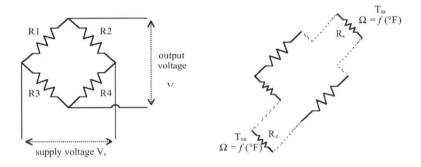

FIGURE 5.1.38 Bridge circuit used to bias the control point in feedforward loop example. Left shows standard bridge circuit; right shows location of outdoor and supply air temperature sensors.

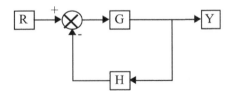

FIGURE 5.1.39 Block diagram of feedback loop.

outside air temperature on the controller setpoint. As the outdoor temperature increases, the control point decreases and vice versa.

The *authority* of each input can be defined using a bridge circuit (Figure 5.1.38).

If all resistances are equal, then $V_o = 0$. If R3 is the compensation resistor, R_c, plus some fixed resistors that bring the total up to approximately the discharge air sensor R2, and R1 and R4 are more or less the same as R2, then the authority of the compensation sensor is $R_c \div R2$.

Mathematical Representation of a Feedback Loop

For the purpose of analysis it is convenient to define a feedback loop mathematically. A general feedback loop is shown in Figure 5.1.39. The controller, actuator, and process have all been combined into the *forward transfer function* (or *open-loop transfer function*) **G** and the sensor and dead time have all been combined into the *feedback path* transfer function **H**. The *closed-loop transfer function* is defined as

$$\frac{Y}{R} = \frac{G}{1 + G \cdot H}$$

The right side of this equation is usually a fraction expressed in polynomials, in which case the roots of the numerator are called the "zeros" of the transfer function, and the roots of the denominator are called the "poles."

The denominator of the closed-loop transfer function, $1 + G \cdot H$, is called the *characteristic function*. Setting the characteristic function equal to zero gives the *characteristic equation*

$$1 + G \cdot H = 0$$

which can be used for determining the stability of the overall transfer function.

Sometimes it is necessary to express transfer functions in either the frequency or discrete time domains. Laplace and z-transforms are used, respectively, to do so.

Laplace Transforms — The transfer function of a closed-loop process is often written using Laplace transforms, a mapping of a continuous time function to the frequency domain and defined as

$$F(s) = \int_0^\infty f(t)e^{-st}dt$$

This formulation can greatly simplify problems involving ordinary differential equations that describe the behavior of systems. A transformed differential equation becomes a purely algebraic term that can be easily manipulated. The result needs to then be inverted back to the continuous time domain. The inverse Laplace transform is given by

$$f(t) = \frac{1}{2\pi j}\int_{\sigma - j\infty}^{\sigma + j\infty} F(s)e^{st}ds$$

where s is a real constant integer greater than the real part of any singularity of $F(s)$.

If $F(s)$ is the Laplace transform of $f(t)$, then the Laplace transform of the nth derivative of $f(t)$ is

$$L\left\{\frac{d^n f}{dt^n}\right\} = s^n F(s) - s^{n-1}f(0^+) - s^{n-2}f(0^+) - \ldots - f^{n-1}f(0^+)$$

where $f(0^+)$ is the initial value of $f(t)$ evaluated as $t \to 0$ from the positive region.

The Laplace transform of a time function $f(t)$ delayed in time by T equals the Laplace transform of $f(t)$ multiplied by e^{-sT}. This is applicable any time there is dead time between the process and the sensor. When investigating stability criteria, however, there are times when we wish to preserve the polynomial expression of the Laplace transform. For this reason, the time lag is often given by the Padé approximation:

$$e^{-sT} \approx \frac{2 - sT}{2 + sT}$$

If the Laplace transform of $f(t)$ is $F(s)$ and if there exists a limit of $sF(s)$ as s goes to infinity, then the initial value of the time function is

$$\lim_{t \to 0} f(t) = \lim_{s \to \infty} sF(s)$$

Likewise, if $sF(s)$ is analytic on the imaginary axis and in the right half-plane, then the final value of the time function is

$$\lim_{t \to \infty} f(t) = \lim_{s \to 0} sF(s)$$

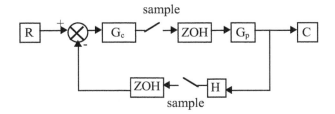

FIGURE 5.1.40 Feedback loop block diagram for discrete time controllers.

Example of Laplace Transform — Laplace transforms enable the calculations of complex differential equations to be reduced to algebraic manipulation. For example, consider a differential equation used to describe a process in the time domain:

$$\frac{d^n y}{dt^n} + a_1 \frac{d^{n-1} y}{dt^{n-1}} + a_2 \frac{d^{n-2} y}{dt^{n-2}} + \ldots + a_{n-1} \frac{dy}{dt} + a_n y = b_0 \frac{d^m u}{dt^m} + b_1 \frac{d^{m-1} u}{dt^{m-1}} + \ldots + b_{m-1} \frac{du}{dt} + b_m u$$

To express this relationship in the frequency domain, it is convenient to have a reference value as a constant offset and to use only the dynamic portion of the driving signal. This allows all but the first term of the Laplace transform of the derivative to be set to zero. The Laplace transform of the previous equation is then

$$s^n Y(s) + A_1 s^{n-1} Y(s) + \ldots + A_{n-1} s Y(s) + A_n Y(s) = B_0 s^m U(s) + B_1 s^{n-1} U(s) + \ldots + B_{n-1} s U(s) + B_n U(s)$$

which can be rewritten as

$$Y(s)\cdot(s^n + A_1 s^{n-1} + \ldots + A_{n-1} s + A_n) = U(s)\cdot(B_0 s^m + B_1 s^{n-1} + \ldots + B_{n-1} s + B_n)$$

and the transfer function is found from

$$\frac{Y(s)}{U(s)} = \frac{s^m + B_1 s^{m-1} + \ldots + B_{m-1} s + A_m}{s^n + A_1 s^{n-1} + \ldots + A_{n-1} s + A_n}$$

Z-Transforms — A significant number of HVAC control applications are accomplished by computers and digital control systems. In such systems, the sampling is not continuous as required for something like a Laplace transform. The control loop schematic looks similar to Figure 5.1.40.

It would be prohibitively expensive to include a voltmeter or ohmmeter on each loop, so the controller employs what is called a *zero-order hold*. This basically means that the value read by the controller is "latched" until the next value is read. This discrete view of the world precludes the use of Laplace transforms for analyses, and it is therefore necessary to find some other means of simplifying the simulation of processes and controllers. The z-transform is used for this purpose.

Recall that the Laplace transform is given as

$$L\{f(t)\} = \int_0^\infty f(t)e^{-st}dt$$

Now suppose a process is sampled at a discrete, constant interval T. The index k will be used to describe the number of the interval, that is,

at time $t = 0, k = 0$,
at time $t = T, k = 1$,
at time $t = 2T, k = 2$,
at time $t = 3T, k = 3$,

and so forth. The equivalent Laplace transform of a process that is sampled at a constant interval T can be represented as

$$L\{f^*(t)\} = \sum_{k=0}^{\infty} f(kT)e^{-s \cdot kT}$$

By substituting the *backward-shift operator* z for e^{Ts} we get the definition of the z-transform:

$$Z\{f(t)\} = \sum_{k=0}^{\infty} f(kT)z^{-k}$$

Example of Using the Discrete Time Domain — The conversion from the continuous time domain to the discrete time domain is best seen through example. Consider a process described by

$$y(k) = a_1 y(k-1) + a_2 y(k-2) + a_3 y(k-3) + \ldots$$

$$+ b_1 u(k-1) + b_2 u(k-2) + b_3 u(k-3) + \ldots$$

The equivalent z-transform is given by

$$y(1 - a_1 z^{-1} - a_2 z^{-2} - a_3 z^{-3} + \ldots) = u(b_1 z^{-1} - b_2 z^{-2} - b_3 z^{-3} + \ldots)$$

and the transfer function can now be found as

$$\frac{y}{u} = \frac{b_1 z^{-1} - b_2 z^{-2} - b_3 z^{-3} + \ldots}{1 - a_1 z^{-1} - a_2 z^{-2} - a_3 z^{-3} + \ldots}$$

Table 5.1.2 lists some of the more common transforms used in the analysis of building systems. More extensive tables can be found in most mathematics and numerical analysis reference books.

5.1.5   Controllers

Controllers are like processes in that they have gains and transfer functions. Generally, there is no dead time in a controller, or it is so small as to be negligible. Recall that the process static gain can be viewed as the total change in the process value due to a 100% change in the controller output. A proportional controller acts like a multiplier between an *error signal* and this process gain. Under stable conditions, therefore, there must be some kind of error to yield any controller output. This is called the steady state or static offset.

Ideally, a controller gain is chosen that compensates for the dynamic gain of the process under normal operating conditions. The total loop dynamic gain can be considered as the product of the process, feedback, and controller gains. If the total dynamic loop gain is one, the process will oscillate continuously at the natural frequency of the loop with no change in amplitude of the process value. If the loop gain is greater than one, the amplitude will increase with each cycle until the limits of the controller or process are reached, or something breaks. If the dynamic loop gain is less than one, the process will eventually settle into stable control.

The *controller bias* is a constant offset applied to the controller output. It is the output of the controller if the error is zero, $u = K \cdot e + M$, where M is the bias. This is useful for processes that become nonlinear at the extremes.

PID Control

Most control systems in HVAC processes use proportional-integral-derivative (PID) control where the controller output is given by

Control output = $K_p \cdot$ [error + $K_i \cdot$ integral of error + $K_d \cdot$ derivative of error]

where K_p, K_i, and K_d are the controller gains.

TABLE 5.1.2   List of Some S- and Z-Transforms

Continuous Time Domain	Frequency Domain	Discrete Time Domain
1 $t = 0$ 0 $t \neq 0$	n/a	1
1 $t = k$ 0 $t \neq k$	n/a	z^{-k}
1	$\dfrac{1}{s}$	$\dfrac{z}{z-1}$
t	$\dfrac{1}{s^2}$	$\dfrac{Tz}{(z-1)^2}$
e^{-at}	$\dfrac{1}{(s+a)^2}$	$\dfrac{z}{z-e^{-aT}}$
$t\,e^{-at}$	$\dfrac{1}{s+a}$	$\dfrac{Tze^{-aT}}{(z-e^{-aT})^2}$
$1 - e^{-at}$	$\dfrac{a}{s(s+a)}$	$\dfrac{z(1-e^{-aT})}{(z-1)(z-e^{-aT})}$
$e^{-at} - e^{-bt}$	$\dfrac{b-a}{(s+a)(s+b)}$	$\dfrac{z(e^{-aT}-e^{-bT})}{(z-e^{-aT})(z-e^{-bT})}$

Proportional control action is the amount that the controller output changes for a given error. The proportional term, $K_p \cdot e$, has the greatest effect when the process value is far from the desired setpoint. Very large values of K_p will tend to force the system into oscillatory response. The proportional gain effect of the controller goes to zero as the process approaches setpoint. Purely proportional control should therefore be used only when (1) the time constant of the process is small so a large controller gain can be used, (2) the process load changes are relatively small so that the offset is limited, and (3) the offset is within an acceptable range.

Integral action is the rate at which the controller output changes for a given error sum. The integral term K_i is the reciprocal of the reset time, T_r, of the system. Integral control is used to cancel any steady state offsets that would occur using purely proportional control. This is sometimes called *reset* control.

Derivative action is the amount that the controller output changes for a given rate of change of the error. Derivative control is typically used in cases where there is a large time lag between the controlled device and the sensor used for the feedback. This term has the overall effect of preventing the actuator signal from going too far in one direction or another and can be used to limit excessive overshoot.

PID Controller in the Time Domain — The PID controller can be represented in a variety of ways. In the time domain, the controller output is given by

$$u(t) = K_p\left[e(t) + K_i \int_0^t e(t)\,dt + K_d \frac{de(t)}{dt} \right]$$

PID Controller in the Frequency Domain — It is relatively straightforward to derive the Laplace transform of the time domain PID equation. Recall that the Laplace transforms for the integral and derivative of a function are

$$L\left[\int_0^t f(u)du\right] = \frac{F(s)}{s} \quad \text{and} \quad L\left[\frac{df(t)}{dt}\right] = sF(s) - f(0)$$

The output of the PID controller can therefore be expressed as

$$U(s) = K_p\left\{E(s) + K_i\frac{E(s)}{s} + K_d[sE(s) - e(t_0)]\right\}$$

Note that we are using upper case letters for s-domain functions. If we assume that the error at time $t = t_0$ is zero (before any perturbations), this equation is equivalent to

$$U(s) = E(s)\left[K_p + \frac{K_pK_i}{s} + K_pK_ds\right]$$

and the transfer function of the controller is therefore

$$\frac{U(s)}{E(s)} = \left[K_p + \frac{K_pK_i}{s} + K_pK_ds\right]$$

PID Controller in the Z-Domain — If the data is measured discretely at time intervals Δt, the PID controller can be represented by

$$u(k) = K_p\left[e(k) + K_i\Delta t\sum_{i=0}^k e(i) + K_d\frac{e(k) - e(k-1)}{\Delta t}\right]$$

The change of the output from one time step to the next is given by $u(k) - u(k-1)$, so the PID *difference equation* looks like this:

$$u(k) - u(k-1) = K_p\left[\left(1 + \frac{K_d}{\Delta t}\right)e(k) + \left(K_i\Delta t - 1 - 2\frac{K_d}{\Delta t}\right)e(k-1) + \left(\frac{K_d}{\Delta t}\right)e(k-2)\right]$$

which simplifies as

$$u(k) - u(k-1) = q_0e(k) + q_1e(k-1) + q_2e(k-2)$$

where

$$q_0 = K_p\left(1 + \frac{K_d}{\Delta t}\right); \quad q_1 = K_p\left(K_i\Delta t - 1 - 2\frac{K_d}{\Delta t}\right); \quad q_2 = K_p\left(\frac{K_d}{\Delta t}\right)$$

The difference equation can then be written as

$$u(1 - z^{-1}) = e(q_0 + q_1z^{-1} + q_2z^{-2})$$

and the z-domain transfer function of the PID controller is therefore

$$\frac{u(z)}{e(z)} = \frac{q_0 + q_1z^{-1} + q_2z^{-2}}{1 - z^{-1}} = \frac{q_0z^2 + q_1z + q_2}{z^2 - z}$$

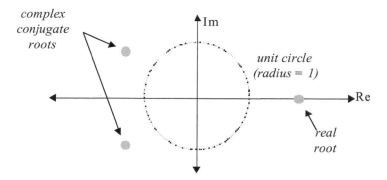

FIGURE 5.1.41 Plotting the roots of the characteristic equation.

5.1.6 Stability of Feedback Loops

Obviously, in HVAC feedback control one wishes to reduce the error. Control systems engineers can use different cost functions in the analysis of a given controller depending on the criteria for the controlled process. Some of these cost functions (or *performance indices*) are listed below:

ISE	integral of the square of the error	$\int e^2$
ITSE	integral of the time and the square of the error	$\int te^2$
ISTAE	integral of the square of the time and the absolute error	$\int t^2\lvert e\rvert$
ISTSE	integral of the square of the time and the square of the error	$\int t^2 e^2$

Stability in a feedback loop means that the feedback loop will tend to converge on a value as opposed to exhibiting steady state oscillations or divergence. Recall that the closed loop transfer function is given by

$$\frac{Y}{R} = \frac{G}{1 + GH}$$

and that the denominator, $1 + GH$, is called the characteristic equation. Typically, this equation will be a polynomial in s or z depending on the method of analysis of the feedback loop. Two necessary conditions for stability are that all powers of s or z must be present in the characteristic equation from zero to the highest order, and all coefficients in the characteristic equation must have the same sign.

The roots of the characteristic equation play an important role in determining the stability of a process. These roots can be real and/or imaginary and can be plotted as shown in Figure 5.1.41. In the s-domain, if all the roots are in the left-hand plane (i.e., to the left of the imaginary axis), then the feedback loop is guaranteed to be asymptotically stable and will converge. If one or more roots are in the right-hand plane, then the process is unstable. If one or more roots lie on the imaginary axis and none are in the right-hand plane, then the process is considered to be marginally stable. In the z-domain, if all the roots lie within the unit circle about the origin, then the feedback loop is asymptotically stable and will converge. If one or more roots lie outside the unit circle, then the process is unstable. If one or more roots lie on the unit circle and none are outside the unit circle, then the process is marginally stable.

Routh-Hurwitz Criteria — The Routh-Hurwitz method is a tabular manipulation of the characteristic equation. If the characteristic equation is given by

$$a_0 s^n + a_1 s^{n-1} + \ldots + a_{n-1} s + a_n = 0$$

then the Routh-Hurwitz method constructs a table as follows:

$$s^n \quad a_0 \ a_2 \ a_4 \ \ldots$$
$$s^{n-1} \quad a_1 \ a_3 \ a_5 \ \ldots$$
$$s^{n-2} \quad X_1 \ X_2 \ X_3 \ \ldots$$
$$s^{n-3} \quad Y_1 \ Y_2 \ Y_3 \ \ldots$$
$$\vdots \quad \vdots \ \vdots \ \vdots \ \vdots$$

where

$$X_1 = \frac{a_1 a_2 - a_0 a_3}{a_1}; \qquad X_2 = \frac{a_1 a_4 - a_0 a_5}{a_1}; \qquad X_3 = \frac{a_1 a_6 - a_0 a_7}{a_1} \ldots$$

$$Y_1 = \frac{X_1 a_3 - a_1 X_2}{X_1}; \qquad Y_2 = \frac{X_1 a_5 - a_1 X_3}{X_1} \ldots$$

and so forth. Now consider the first column of coefficients. The number of roots in the right-hand plane is equal to the number of sign changes in the first column. In other words, if all the elements in the first column have the same sign, then there are no roots in the right-hand plane and the process is stable. A few considerations about the Routh-Hurwitz method:

- If the first element of any row is zero but the remaining elements are not, then use some small value ε and interpret the final column results as $\varepsilon \to 0$.
- If one of the rows before the final row is entirely zeros, then (1) there is at least one pair of real roots of equal magnitude but opposite signs, or (2) there is at least one pair of imaginary roots that lie on the imaginary axis, or (3) there are complex roots symmetric about the origin.

Tuning Feedback Loops — One method of obtaining the desired critically damped response of HVAC processes is to determine the closed-loop transfer function in the form

$$\frac{Y}{R} = \frac{(s + A_1)(s + A_2)\ldots(s + A_m)}{(s + B_1)(s + B_2)\ldots(s + B_n)}$$

The coefficients A and B depend on both the process characteristics and the controller gains. The objective of pole-zero cancellation is to find values for the controller gains that will set some numerator coefficients equal to those in the denominator, effectively canceling terms. As can be imagined, however, this can be a very difficult process, particularly when working with complex roots of the equations. This method is typically used only with very simple system models.

Often it is more convenient to test a feedback loop *in situ*. The *reaction curve technique* has been developed which allows one to field tune PID constants using open-loop response to a step change in the controller output. Consider the process response shown in Figure 5.1.42 where Δc is the change of process output, Δu is the change of controller, L is the time between change and intersection, and T is the time between lower intersection and upper intersection. We can define the following variables:

$$A = \Delta u \div \Delta c, \qquad B = T \div L, \qquad \text{and} \qquad R = L \div T$$

which can be used with the values in Table 5.1.3 to estimate "decent" control constants.

The *ultimate frequency* test involves increasing the proportional gain of a process until it begins steady state oscillations. K_p^* is defined as the proportional gain that results in steady oscillations of the controlled system, and T^* is the period of the oscillations. The desired controller gains are given in Table 5.1.4.

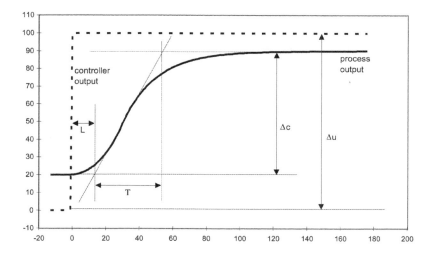

FIGURE 5.1.42 Response of process to step change in controller output.

TABLE 5.1.3 Recommended Control Constants from Reaction Curve Tests

Controller Components	Zeigler-Nichols			Cohen and Coontz		
	K_p	$\frac{K_p}{K_i}$	$\frac{K_d}{K_p}$	K_p	$\frac{K_p}{K_i}$	$\frac{K_d}{K_p}$
P	AB	—	—	$AB\left(1 + \dfrac{R}{3}\right)$	—	—
P + I	0.9AB	3.3L	—	$AB\left(1.1 + \dfrac{R}{12}\right)$	$L\dfrac{30 + 3R}{9 + 20R}$	—
P + D	—	—	—	$AB\left(1.25 + \dfrac{R}{6}\right)$	—	$L\dfrac{6 - 2R}{22 + 3R}$
P + I + D	1.2AB	2L	0.5L	$AB\left(1.33 + \dfrac{R}{4}\right)$	$L\dfrac{32 + 6R}{13 + 8R}$	$L\dfrac{4}{11 + 2R}$

TABLE 5.1.4 Recommended Control Constants from Ultimate Frequency Tests

Controller Components	K_p	$\frac{K_p}{K_i}$	$\frac{K_d}{K_p}$
P	$0.5\ K_p^*$	—	—
P + I	$0.45\ K_p^*$	$0.8T^*$	—
P + I + D	$0.6\ K_p^*$	$0.5T^*$	$0.125T^*$

5.1.7 Control Diagrams

Line diagrams are pictorial representations of components of a control loop. These schematics are useful for presenting information about subsystems or simple control diagrams. Figure 5.1.43 shows an example control line diagram for a motor starter. When starting a large motor, it is desirable to minimize the initial current draw. This circuit in the diagram is a *wye-delta* starter, common in many large motors. At start-up, the relays 1S and 1M1 are energized and the motor windings are energized in a wye configuration. This allows the motor to start with low voltage and current draw. After an acceleration period, relay 1A

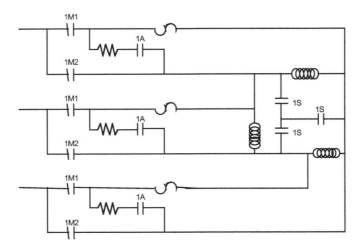

FIGURE 5.1.43   Control line diagram for a motor starter.

is energized allowing current to be shunted across the resistors, then 1S is de-energized, 1M2 is energized, and the motor is running in a delta configuration.

Ladder diagrams are used frequently to describe the logic flow of complex control systems. Figure 5.1.44 shows a ladder diagram for the start-up and safety features of a chiller and condenser combination. Each of the rungs of the ladder is labeled to indicate the function of that rung.

5.1.8   Control of Air Distribution Systems

In larger buildings, air handlers are used to push air through ductwork until it arrives at zones that need to be conditioned. The zones use their own equipment to provide localized heating or cooling to meet the thermostat setpoint. To ensure that there is sufficient air pressure at the zones to provide adequate control and air circulation, the pressure in the supply air ducts is usually controlled to a fixed setpoint. The supply air duct static pressure is typically measured at a point about 75% of the total duct length downstream from the fan, although some researchers have indicated that this is not the ideal location (Figure 5.1.45). The pressure in the supply duct will be controlled to between 3 and 20 in of equivalent water pressure depending on the building size. In optimal control, the duct pressure setpoint will be dynamically adjusted to ensure that at least one terminal box supply air damper is fully open. That is, the duct pressure will be set to the minimum value allowed by the zone or zones with the maximum supply air demand.

Supply fans usually control to maintain duct static pressure with a high-limit cutout safety. Return fans are controlled as either speed tracking (open loop), direct building static (reset to zero when no OA), air flow tracking (requires air flow sensor), or return duct static pressure. Relief fans are used to control direct building static pressure or to track the amount of outside air brought into the building.

Most buildings are pressurized, which helps control infiltration and dirt intake. The pressure rise across the fan is controlled using inlet vanes or outlet dampers. Certain zones are depressurized, such as laboratories and zones under *smoke evacuation mode*. With these strategies, the exhaust fan tracks above the supply fan to guarantee a net loss of conditioned air from a zone.

There are several different ways to vary the pressure rise across a fan. *Outlet dampers* restrict the air leaving the fan, while *inlet vanes* restrict the air entering the fan and can also "pre-swirl" the air to improve fan efficiency. *Variable frequency drives* (VFDs or sometimes *variable speed drives*, *VSDs*) change the actual speed of the motor by decomposing the standard 50 or 60 Hz signal and rebuilding it at a desired frequency through a process called *pulse width modulation*. Since the motor is not operating at the design frequency,

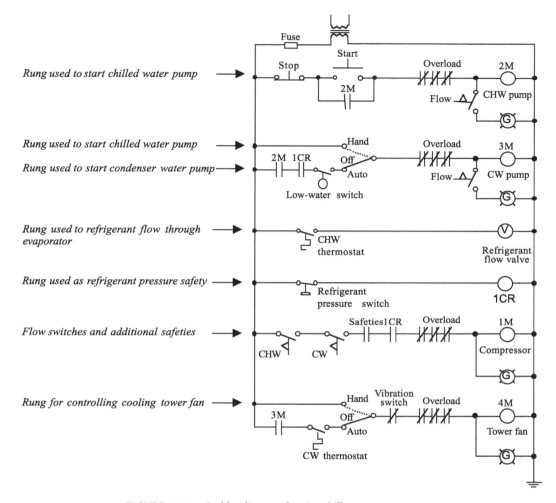

FIGURE 5.1.44 Ladder diagram showing chiller start-up sequence.

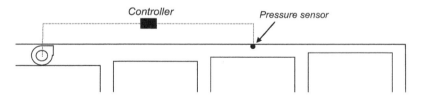

FIGURE 5.1.45 Supply air duct showing location of static pressure sensor.

however, there may be certain frequencies that cause undue vibration or harmonics in the fan. Most VFDs have *critical speed step-over* circuits that are adjustable ranges that can be locked out to avoid any resonant speeds. Two step-over ranges of individually adjustable width will be sufficient for nearly all applications. Figure 5.1.46 shows the location of each of these methods of pressure modulation.

Figure 5.1.47 shows the principle behind the operation of a variable frequency drive. The standard line frequency (left) is decomposed and then used to generate a series of impulses that "blend together" to create a new periodic wave at a different frequency. Figure 5.1.47 shows how 60 Hz power is converted to a 40 Hz signal.

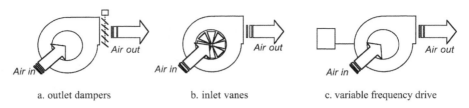

a. outlet dampers   b. inlet vanes   c. variable frequency drive

FIGURE 5.1.46   Different methods of fan pressure modulation.

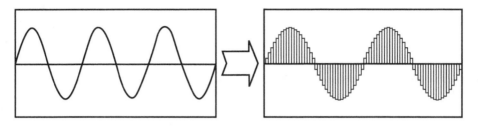

FIGURE 5.1.47   Variable frequency modulation.

Terminal Box Control

The terminal box provides the endpoint control of the zone temperature. It is connected to the local thermostat or central control system and provides the fine adjustment of the local temperature (see Figure 5.1.48). Terminal boxes can have many different configurations, including induced air return from the plenum, reheat coils (usually only in perimeter zones), and fans to pull air in from the plenum. In the case of fan-powered mixing boxes with reheat, the control of the boxes is similar to that shown in Figure 5.1.49. The damper closes as the room cools until it reaches a minimum value allowed for indoor air quality. The fan and heaters come on in succession to provide recirculation and heating of room air, respectively.

Air Temperature Control

The supply air temperature in a single-deck system is usually maintained at about 50 to 60°F. One controller can be used to modulate both the CHW and HW valves on the air handler. Valves on the chilled and hot water coils are modulated to maintain the air setpoint temperature (see Figure 5.1.50). If the outdoor air is below 32°C, then a preheat coil might be activated to prevent water in the cooling or heating coils from freezing.

If the outdoor air temperature is close to the desired supply air temperature, then the exhaust and outdoor air dampers are fully opened and the return air damper is completely closed, as shown in Figure 5.1.51. Under these conditions, 100% outside air enters the building and the chiller can be shut off because there is no need for cooling the air. This is called *economizer mode*. Strictly speaking, the economizer mode should be used whenever the enthalpy difference between the outdoor air and the supply air is less than the enthalpy difference between the return air and the supply air. However, it is difficult to measure enthalpy, so usually only the air temperatures are used.

A similar damper configuration is used at night to cool a building mass in preparation for the next day's cooling load. This is called *night purging*. The idea with night purging is to precool the building so the peak cooling loads the next day are reduced. However, in some very humid locations it is claimed that the introduction of humid air into the building leads to the absorption of water by the interior building materials, office paper, etc. and that this can actually lead to an increase in the latent load during the next day.

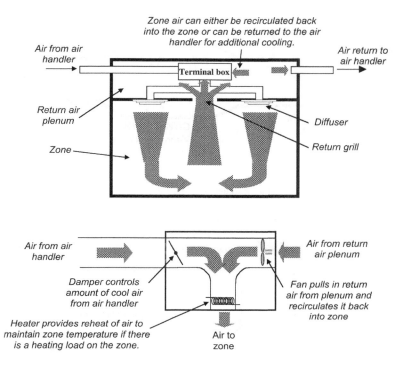

FIGURE 5.1.48 Components of terminal boxes.

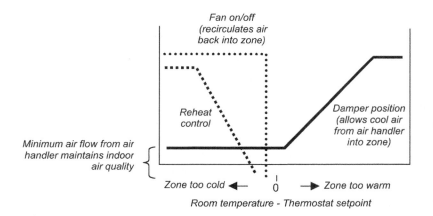

FIGURE 5.1.49 Typical control of fan-powered terminal box.

5.1.9 Control of Water Distribution Systems

Water distribution systems are the combination of pumps, pipes, and other apparatus that move hot and chilled water throughout a building. The control schemes implemented in water distribution systems must maintain controllable pressure across control valves, maintain required flow through a heating or cooling source, maintain desired water temperature to terminal units, maintain minimum flow through pumps, properly stage multipump systems, and prevent cavitation in pumps.

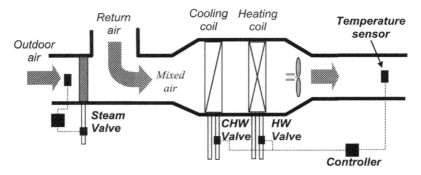

FIGURE 5.1.50 Typical air handler showing preheat, cooling, and reheat coils.

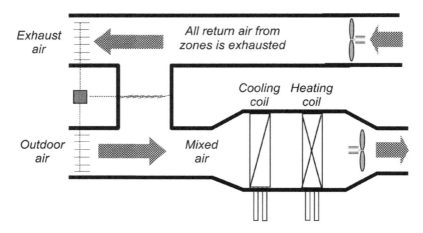

FIGURE 5.1.51 Air handler showing dampers in economizer mode.

Water flows in HVAC systems are controlled through valves. Valves have many different arrangements, depending on the needs and function of the system. Three typical valve types are shown in Figure 5.1.52.

Linear valves are good for steam flow because the steam is at a constant temperature, and the latent heat of condensation is uniform with the change of pressure. Equal percentage valves are good for cooling and heating coils because the combination of the typical coil with an equal percentage valve provides effectively linear control of the coil heating or cooling. When multiple valves are installed in a system, some will be responsible for causing greater pressure drops in the system than others. The *authority* of a valve is the ratio of the valve pressure drop to the total system pressure drop.

While, ideally, a valve will exhibit linear behavior even at very low flows, it is more likely that there is a significant step change in the flow when the valve plug first separates from the seal. The actual minimum flow when a valve is cracked open is about 3 to 5% of maximum possible flow. This value is known as the *turndown ratio*. A typical commercial valve might have a turndown ratio of 5% (20:1), while industrial process control valves may have a turndown ratio of 50:1 to 100:1. Such fine precision, however, is generally not needed in HVAC applications.

The actual flow rate through a valve is determined by the *flow coefficient*, C_v, which is the number of gallons per minute of 60°F water that will flow through the valve with a pressure drop of 1 psi when the valve is fully open. The flow under any other condition can be found from GPM = $C_v \cdot \Delta P^{0.5}$ where ΔP is the pressure drop across the valve.

Often a valve's purpose is not only to shut off the flow but to redirect flow from one pipe to another. This is accomplished through *three-way mixing and diverting valves*. The two types of valves are constructed differently as shown in Figure 5.1.53, so they should not be used interchangeably. Mixing and

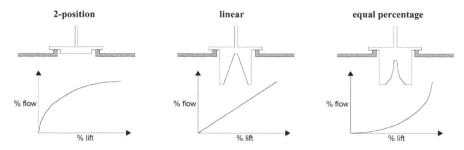

FIGURE 5.1.52   Valve configurations and corresponding flow versus lift charts.

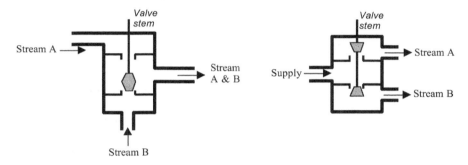

FIGURE 5.1.53   Schematics of three-way mixing (left) and diverting (right) valves.

diverting valves are particularly useful on coils because it allows the flow to remain relatively constant through a particular water loop branch over the operating range of the coil. This helps maintain the water distribution system pressure at a fixed value. A *balancing valve* is used to make the pressure drop through the bypass loop similar to the pressure drop through the coil.

Water Temperature and Pressure Control

Most water distribution systems use a primary/secondary loop configuration as shown in Figure 5.1.54. In such a system, the primary loop is often used to obtain gross control of the water temperature, and the valve to the secondary loop is used to maintain more precise temperature control of the distributed water.

The pressure across the supply and return lines of a water distribution system must be maintained above a minimum threshold in order for the control valves in that system to work properly. Consider the extreme: if there were no pressure drop in the system (i.e., no water flow), the valves would do absolutely nothing. The problem is compounded if three-way valves are not used on the system and the pressure control must be accomplished through other means. One way to aid the maintenance of a minimum pressure drop across the system is to use reverse-return plumbing. Compared with a direct return system, the reverse-return provides for a roughly equal pressure drop across all loads. In large systems, however, the extra capital cost of the piping may need to be considered. Figure 5.1.55 illustrates direct and reverse return piping.

With a fixed-speed pump and widely varying loads with two-way valves, a pressure drop across the water distribution supply and return can be maintained using a bypass valve controlled by a differential pressure sensor (Figure 5.1.56). Care must be taken when choosing the gains for the control loop, however, since water is not compressible and small changes in the valve position can lead to large changes in the system pressure drop.

In a variable-flow system (Figure 5.1.57), the pump speed is varied to maintain a fixed pressure drop across the supply and return branches. As with the bypass loop, however, care must be taken not to set the control loop gains too high because excessive system pressure oscillations can occur rapidly.

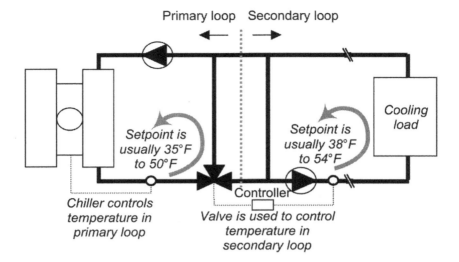

FIGURE 5.1.54   Schematic of chilled water primary and secondary loops.

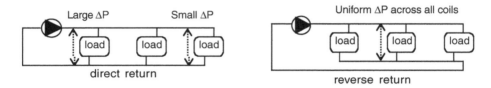

FIGURE 5.1.55   Direct and reverse return piping schematics.

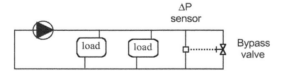

FIGURE 5.1.56   Flow bypass sensor position.

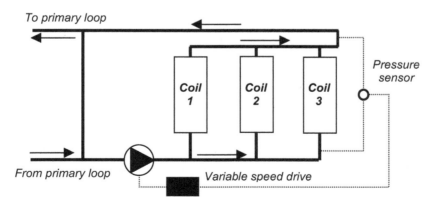

FIGURE 5.1.57   Variable flow water distribution system showing location of a pressure sensor.

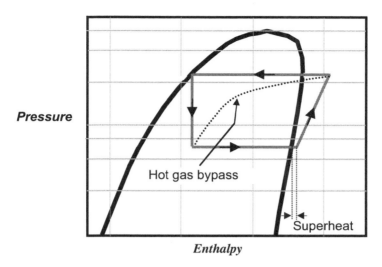

FIGURE 5.1.58   Pressure-enthalpy diagram showing superheat and hot gas bypass.

5.1.10   Control of Chillers

Chillers are used to cool water or air by means of a vapor compression cycle. Methods for modulating the capacity of a chiller depend on the type of compressor used in the cycle.

A *centrifugal* compressor acts like a water pump. Refrigerant enters the compressor and is subject to the centrifugal forces of an impeller, increasing both the velocity and the pressure of the refrigerant. Centrifugal compressors can also use inlet vanes to preswirl the refrigerant or to impede the refrigerant flow. The chiller controller operates a pneumatic or electric actuator to reposition the inlet vanes as a function of the chiller water temperature. If speed control is available, the controller sequences the motor or rotor speed with the position of the inlet vanes.

A *reciprocating* compressor behaves like a car engine in reverse. Refrigerant enters the cylinders and is compressed by pistons that are driven from a central shaft. Valves at the top of the cylinders allow the refrigerant to enter or leave the cylinders at the proper stages of the cycle. To reduce the capacity of the chiller, some of the valves are forced open so that no compression takes place. These types of compressors, therefore, have very distinct operating capacities (for example, 0, 25%, 50%, 75%, and 100% of rated peak capacity)

A *screw-type* compressor forces the refrigerant into a series of interlocking screws. The farther along the meshing screws the refrigerant travels, the greater the compression. Modulation of capacity occurs with a pneumatic or electric actuator used to position a sliding bypass valve. The valve allows refrigerant to leave the mesh at various points along the screws. As a result, these kinds of compressors have good modulating characteristics.

With all types of compressors, it is desirable to have only gaseous refrigerant enter the compressor. Liquid refrigerant in the compressor cannot be compressed and can cause pitting of the compressor surfaces. To avoid this *slugging* of the compressor, chillers usually operate with some amount of *superheat* where the evaporator outlet temperature is slightly outside the saturation curve. A typical value for superheat is about 5 to 10°F, meaning that the refrigerant is 5 to 10°F above the saturation temperature at the compressor inlet pressure. The thermal expansion valve on a chiller is often controlled to maintain the superheat setpoint.

Hot gas bypass is sometimes used to modulate the capacity of reciprocating chillers. A solenoid valve is used to let hot refrigerant flow directly from the compressor outlet back into the evaporator inlet. Figure 5.1.58 shows both the superheat and hot gas bypass processes on a pressure-enthalpy diagram.

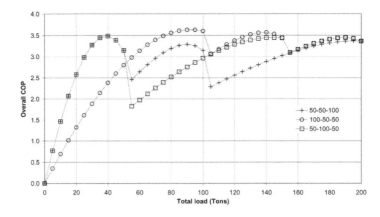

FIGURE 5.1.59   Total cooling plant COP as a function of cooling load and different chiller staging orders.

Most chillers are operated on a primary/secondary distribution loop system which guarantees flow through the evaporator barrel and insures against freezing. There are a number of other safety controls on a chiller designed to protect it from freezing or high refrigerant pressures. Some of these safeties are listed here. Most if not all of these safeties will cause the chiller to shut off and will not reset automatically; human intervention is usually required.

- High condenser pressure
- Low refrigerant pressure or temperature
- High motor temperature
- Motor current overload
- Low oil sump temperature
- High oil sump temperature
- Evaporator water flow interlock
- Condenser water flow interlock

Large buildings may have several chillers of different sizes that are brought on-line according to the load on the cooling plant. Since the part-load performance of a chiller may not be very good, it is often desirable to operate the chillers at 80 to 100% of their rated capacities to ensure that the kW/ton ratio is sufficiently small. A number of issues are associated with multiple chiller operation, however, particularly if the chillers are of different capacities. If the morning load is small but the afternoon load is expected to be large, it may be advantageous to bring the large chiller on-line in the morning even at a low part-load ratio. This procedure will prevent the large chiller from doing a "warm-start" in the afternoon, when the start-up transient power consumption can be high enough to reset the demand load for the building.

If the building has more than two chillers, it may be necessary to include some intelligence in the start-up algorithm that can identify the proper combination of chillers to obtain the highest overall plant COP. Figure 5.1.59 shows the effects of different starting dispatch orders for a building with two 50-ton chillers and one 100-ton chiller. Clearly, some combinations are better than others, particularly at low loads.

The CHW outlet setpoint should be the same on each chiller in a multiple chiller plant. The condenser and evaporator flows through each chiller should be in proportion to the relative capacity of each chiller.

Some chillers have built-in control circuitry that limits the current draw. This is useful if one is trying to manage the utility demand charges. Under current limiting control, the chiller will unload capacity rather than trying to meet the setpoint if the current limit is reached.

Some variable speed drives also have circuitry that automatically controls the maximum output current of the drive. This is necessary to protect the current carrying components. Typically, a drive's rating is

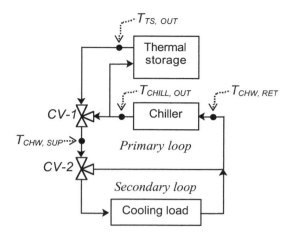

FIGURE 5.1.60 Schematic of thermal storage flows.

at 100% current. Constant torque drives typically have a maximum current limit of 150% and variable torque drives have a maximum current limit of 100 to 115%.

Thermal Energy Storage

It is becoming more common to use thermal energy storage (TES) to shed daytime peaking loads to the evening hours when overall electrical demand is low and the cost of electricity is usually cheaper. The "coolth" is stored either by creating ice or by cooling a large volume of water. A simplified schematic of ice storage is shown in Figure 5.1.60. At night, the three-way mixing valve CV-2 is closed to force all the water from the chiller back into the primary loop. If ice is used as the thermal storage medium, then the chiller setpoint $T_{CHW, SUP}$ is set somewhere around 25 to 30°F. The water is used to cool the thermal storage tank. During the daytime the CV-2 will modulate to maintain a desired setpoint temperature in the secondary loop. To retrieve the stored energy, one of the following methods is usually used.

The simplest method for thermal storage control is the *chiller-priority* strategy. The chiller is sized for a capacity less than the building peak load. During the evenings, the chiller is used to make chilled water or ice. During the day the chiller operates at full capacity and the difference between the chiller output and the building load is made up with the cold storage. This technique maintains a high chiller efficiency and provides for a known demand curve in case demand limiting is implemented. The main disadvantage of this method is that the chiller still operates during the daytime hours and incurs both consumption and demand charges.

Storage-priority control attempts to use as much of the stored energy as possible and to minimize the chiller operation during the hours when electricity is most expensive. The clear advantages of this strategy are that the chiller can quite possibly be turned off during these on-peak hours and the cost savings are significant. The main disadvantage is that a control scheme is required that can predict the next-day building loads and generate enough ice to handle that load. If the storage is depleted during the day, then the chiller may need to be placed on-line during periods of high energy cost.

Constant proportional control of thermal storage refers to a technique in which the ratio of the cooling provided by the chiller and the storage is a constant. In other words, the proportion of cooling provided by either device remains constant throughout the day regardless of the load. This combines the simplicity of chiller-priority control with the demand limiting potential of storage-priority control. When using constant proportion control, the outlet temperatures of the chiller and storage tank remain constant. The chiller outlet temperature setpoint is given as

$$T_{CHILL, OUT} = T_{CHW, RET} - f \cdot (T_{CHW, RET} - T_{SUP})$$

where f is the fraction of cooling that is to be handled by the chiller.

Cooling Tower Control

Cooling towers are typically controlled to maintain the condenser return water temperature, i.e., the temperature of the water that goes back into the chiller condenser. This is done by either staging multiple cooling tower fans or by varying the fan speeds with variable frequency drives. Since cooling towers take advantage of evaporative cooling, *the setpoint of the cooling tower can be no lower than the wet-bulb temperature of the ambient air.* Thus, ambient temperature data is needed both in the design and control of cooling towers. Also, note that cooling towers tend to create their own microclimates, so any measurement of the ambient conditions must be made in the vicinity of the cooling tower. The *approach temperature* of a cooling tower is the difference between the design outlet temperature and the design wet-bulb temperature. As the approach temperature decreases, the available cooling also decreases. If the approach temperature is below about 5°F, the cooling capacity of the tower will be effectively negligible.

If the cooling tower has multiple cells (i.e., fan and draft columns) and the fans are controlled using variable frequency drives, then the fans in all cells should be operated at the same speed. This will provide the correct amount of cooling at the minimum fan energy cost. If the cooling tower has multiple cells and the fans are controlled with multiple speed fans, then the lowest speed fans should be put on-line first as additional capacity is needed.

Tower ("Free") Cooling — In tower cooling mode, the chiller is turned off and the water bypasses the chiller altogether. If the building cooling load is small and the outside conditions are within the right range, it is possible to achieve "free" cooling by sending the cooling tower water directly to the load (Figure 5.1.61).

Heat Recovery

Some chiller models allow for the recovery of condenser heat for relatively low-temperature hot water applications (Figure 5.1.62). Heat from the condenser can be used to provide some or all of the heating energy for hot water distribution loops in the range of 100 to 130°F. Temperatures above this are not possible since the condenser would then be too hot for efficient chiller operation. The temperature controls for the heat recovery loop are usually installed by the manufacturer. The condenser heat recovery can be placed before or after the conventional heating source. Note that it is always necessary to have conventional refrigerant cooling equipment in case the chiller is operating when the demand for hot water is low.

5.1.11 Control of Boilers and Steam Systems

Boilers are used to generate steam for use in larger buildings. Steam is a convenient "prime mover" since it does not require an additional pumping power to circulate. The boiler produces steam, and the steam expands into the distribution system. Of course, there will be some condensation within the distribution system as some of the steam cools. This condensate is removed using steam traps that feed to a condensate receiver tank. A float switch in this tank closes when the tank is full and activates a pump that returns the condensate back to the boiler. Figure 5.1.63 illustrates a steam distribution system and condensate return.

Boilers use electricity, natural gas, or oil to heat water and produce steam. *Hydronic* boilers do not produce steam but rather pressurized hot water. Most of the control circuitry for boilers is installed by the manufacturer, although often the user will have the ability to change the outlet pressure setpoint. Low pressure steam systems operate in the 10 to 20 psig range.

As with all large HVAC equipment, boilers have safety controls that will shut off the boiler (or prevent ignition) if a condition exists that is hazardous to the equipment or the facilities personnel. Some boiler combustion safeties include shutting off the gas flow if the flame is unintentionally extinguished, purging the combustion chamber of any unburnt gases before ignition, purging the combustion chamber of any unburnt gases after flame shut-off, and ensuring flame integrity upon startup before increasing gas flow to maximum. Of course, there will always be a water flow interlock to prevent boiler operation when there is no water flow. In addition, many large boilers (greater than 1 million Btu/hour) monitor flue gas conditions to estimate combustion efficiency. Ideally, CO_2 concentrations in the flue gas will be

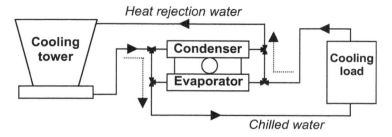

FIGURE 5.1.61 Chilled water plumbing needed for "free" cooling.

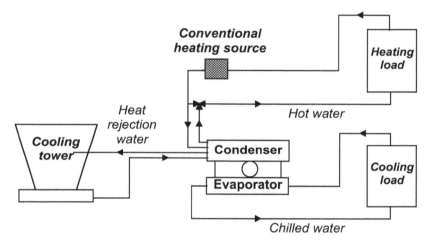

FIGURE 5.1.62 Piping schematic for chiller with heat recovery.

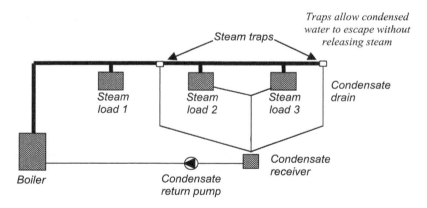

FIGURE 5.1.63 Schematic of steam distribution system and condensate return.

10 to 12 percent, while O_2 concentrations should be around 3 to 5 percent. Lower concentrations of O_2 are impractical and unsafe, while higher concentrations imply that too much air is entering the combustion chamber and must be warmed.

Boilers are rated in terms of their efficiency or in terms of the annual fuel utilization efficiency (AFUE). The efficiency is a simple, steady state ratio of input energy to output energy while the AFUE efficiency takes cycling into account. The problem with cycling a boiler is that the pressure vessel must first reach boiler temperatures before steam is produced. During cycling (see Figure 5.1.64), a considerable amount of energy is lost during the transient conditions.

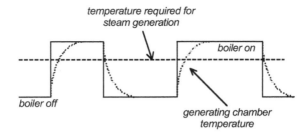

FIGURE 5.1.64   Boiler pressure chamber temperature versus time during boiler cycling.

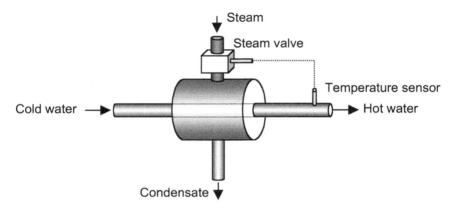

FIGURE 5.1.65   Schematic of a steam/hot water converter showing steam valve.

Steam/Hot Water Converter

A converter is like a heating coil, except that it is much smaller and heats water, not air. The water and steam never mix; they just exchange energy. Converters are used for generating hot water at remote points in a building. They are used instead of multiple water heaters. The steam flow is controlled to maintain the hot water temperature setpoint. Figure 5.1.65 shows a schematic of a steam/hot water converter and its steam valve.

Radiant Heating

Floor panels are often controlled by outside air temperature. Maximum allowable temperatures are about 85°F. Wall and ceiling panels have less mass and can operate at temperatures up to 100°F and 120°F, respectively. Walls and ceilings can also be used for radiant cooling, but care must be taken to avoid condensation. Note that radiant heating temperatures are good matches to solar collector temperatures. Temperature sensors for radiant heating should be located away from the panels so that the control maintains the proper air temperature and is not biased by direct heating from the panels.

Building Warm-up/Cool-down

Many buildings are allowed to "float" at night, which means that the HVAC equipment is turned off and the building temperature can drift up or down depending on the difference between the inside and outside temperature. Of course, the building must be at a comfortable temperature when the occupants arrive, so it is necessary to start the HVAC system beforehand. The key is to have an understanding of the building time constant so that the morning warm-up (or pull-down, if in cooling mode) can start so the building reaches the desired setpoint without wasting energy. Figure 5.1.66 shows the effect of different start times on the building temperature. During warm-up mode, any exhaust fans and relief fans are turned off, the building pressure control and the air flow tracking differential are reset to zero (if used), and the thermostat control is overridden or reset.

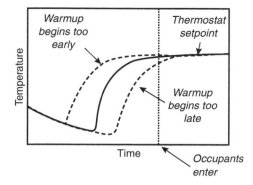

FIGURE 5.1.66 Effect of different HVAC warm-up start times on building temperature.

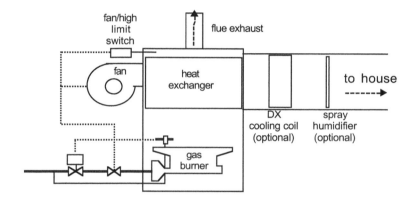

FIGURE 5.1.67 Schematic of residential furnace showing location of components and sensors.

Control of Furnaces

Furnaces are gas- or oil-fired devices (most often residential) that combine combustion elements with a fan to circulate warm air throughout a house. The flame safeguard controls on most residential furnaces are thermocouples protected by a metal sheath. Other methods of flame detection include bimetal sensors (slow response) and ultraviolet flame sensing (fast and reliable but expensive). Figure 5.1.67 shows a residential furnace and its components.

5.1.12 Supervisory Control

Supervisory controllers are used to govern the operation of the entire HVAC plant and/or building climate control. This type of controller attempts to minimize a cost function associated with the operation of the building under the current conditions. The controller will vary setpoints, perform load shedding, switch cooling modes from mechanical to storage, etc. A basic diagram of a supervisory controller is given in Figure 5.1.68.

The principle behind the plant-wide optimization is to have a supervisory controller that can be used to predict the behavior of the plant over a wide range of operating conditions. This is accomplished by any number of modeling techniques from simple regression to neural networks to sophisticated building simulations. Once the model has been developed and calibrated, it can be used to examine a number of "what-if" scenarios to determine the economically optimum operating conditions. The two examples in Figure 5.1.69 show the results of a building model that has been used to look at the effects of (a) chilled water temperature control, and (b) chilled water and supply air temperature control on the hourly cost of operating the HVAC plant.

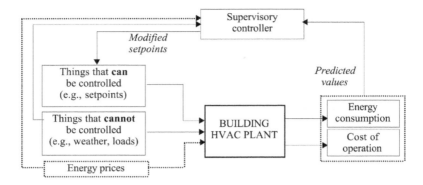

FIGURE 5.1.68   Supervisory controller information paths.

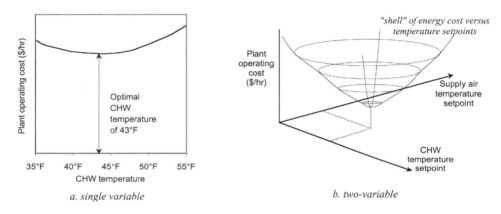

a. single variable   *b. two-variable*

FIGURE 5.1.69   Supervisory optimization decision charts.

5.1.13   Advanced System Design Topics — Neural Networks for Commercial Building Controls

Neural networks (NNs), offer considerable opportunity to improve control achievable in standard PID systems. This section provides a short introduction to this novel approach to control which has been used in a number of commercial products. For the basis of NNs, see Section 6.2.3.

A proof of concept experiment in which NNs were used for both local and global control of a commercial building HVAC system was conducted, in the JCEM laboratory at the University of Colorado, in which full-scale testing of multizone HVAC systems can be done repeatably. Data collected in the laboratory were used to train NNs for both the components and the full systems involved (Curtiss, 1993a, 1993b). Any neural network-based controller will be useful only if it can perform better than a conventional PID controller. Figures 5.1.70 and 5.1.71 show typical results for the PID and NN control of a heating coil. The difficulty that the PID controller experienced is due to the highly nonlinear nature of the heating coil. A PID controller tuned at one level of load is unable to control acceptably at another whereas the NN controller does not have this difficulty. With the NN controller we see excellent control — minimal overshoot and quick response to the setpoint changes.

In an affiliated study, Curtiss et al. (1993b) showed that NNs offered a method for global control of HVAC systems as well. The goal of such controls could be to reduce energy consumption as much as possible while meeting comfort conditions as a constraint. Energy savings of over 15% were achieved by the NN method vs. standard PID control.

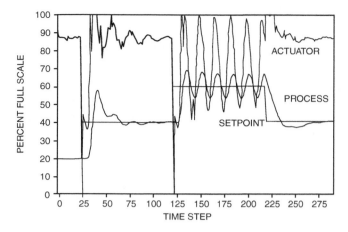

FIGURE 5.1.70   PID controller response to step changes in coil load. Proportional gain of 2.0 (from Curtiss et al., 1993a.)

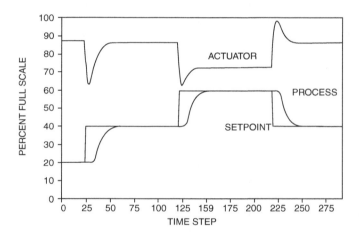

FIGURE 5.1.71   NN controller with learning rate of 1.0 and window of 15 time steps (from Curtiss et al., 1993a.)

References

ASHRAE, 1989. *Handbook of Fundamentals*. American Society of Heating Refrigeration and Air Conditioning Engineers, 1791 Tullie Circle NE, Atlanta, GA.

Curtiss, P.S., Kreider, J.F., and Brandemuehl, M.J., 1993a. Adaptive Control of HVAC Processes Using Predictive Neural Networks. *ASHRAE Transactions,* Vol. 99, Pt. 1.

Curtiss, P.S., Brandemuehl, M.J., and Kreider, J.F., 1993b. Energy Management in Central HVAC Plants using Neural Networks. *ASHRAE Transactions,* Vol. 100, Pt. 1.

Haines, R.W., 1987. *Control Systems for Heating, Ventilating, and Air Conditioning*, 4th ed., Van Nostrand Reinhold, New York, NY.

Kreider, J.F., Rabl, A., and Curtiss, P.S., 2001. *Heating and Cooling of Buildings*, 2nd ed., McGraw-Hill, New York, NY.

McClelland, J.L. and Rumelhart, D.E., 1988. *Exploration in Parallel Distributed Processing*, MIT Press, Cambridge, MA.

Shinners, S., 1972. *Modern Control System Theory and Application*, 2nd ed., Addison-Wesley Publishing Company, Reading, MA.

5.2   Intelligent Buildings

Michael R. Brambley, Peter Armstrong, Michael Kintner-Meyer, Robert G. Pratt, and Srinivas Katipamula

The topic of "intelligent buildings" (IBs) emerged in the early 1980s. Since then, the term has been used to represent a variety of related yet differing topics, each with a slightly different focus and purpose. Wiring and networking infrastructure companies emphasize the cabling requirements for communication in intelligent buildings and the need to accommodate future needs for higher speed broadband. Lucent (Lucent 2000) for example, defines an IB as "… one with a completely integrated wiring architecture. A single cabling system that handles all information traffic — voice, data, video, even the big building management systems."

Developers focusing on advanced technologies for use as parts of buildings, furnishings, or structural members emphasize intelligent building technology, such as smart windows, active walls, ergonomic furniture, or furnishings with built-in computing technology or network/communication connections (see, for example, Ambient 2000).

Other organizations define intelligent buildings in the context of the overall performance of the building or how successfully the occupants of the building conduct business. The European Intelligent Buildings Group (EIBG 2000) uses the following definition: "An intelligent building incorporates the best available concepts, materials, systems, and technologies, integrating these to achieve a building that meets or exceeds the performance expectations of the building's stakeholders. These stakeholders include the building's owners, managers, and users, as well as the local and global community."

Dexter (1996) provided an excellent discussion of heating, ventilation, and air conditioning systems for intelligent building systems.

Finally, the U.S. Department of Energy (2000) defines intelligent building systems as follows:

Intelligent building systems (IBSs) will use data from design, together with sensed data, to automatically configure controls, commission (i.e., start up and check out) and operate buildings. Control systems will use advanced, robust techniques and will be based on smaller, cheaper, and more abundant sensors than today. Intelligent devices will use this wealth of data to ensure optimal building performance continuously by controlling building systems and continuously recommissioning them using automated tools that detect and diagnose performance anomalies and degradation. Intelligent building systems will provide much more than today's rudimentary control. They will optimize operation across building systems, inform and implement energy purchasing, guide maintenance activities, and report building performance, while ensuring that occupant needs for comfort, health, and safety are met at lowest possible cost.

Although these definitions differ, they share a common goal of providing better services to building occupants to make them more productive, more comfortable, healthier, safer, more secure, and better able to conduct business using modern technology to provide better building services.

Webster's Dictionary (1999) defines *intelligent* as "having intelligence" and *intelligence* as "1.a. The capacity to acquire and apply knowledge. b. The faculty of thought and reason." By these definitions, few, if any, building technologies are intelligent. Yet, the infrastructure in which intelligence could be deployed has grown rapidly: high speed communication by audio, video, and other data, high speed processing, soon to exceed 1 GHz in home computers, and enormous data storage capabilities at lower cost than ever before. Application of building automation technology has become commonplace, yet building automation systems do not satisfy these definitions of "intelligent."

For purposes of this chapter, we adopt a slight variation of the definitions provided by Dexter and the USDOE. We define intelligent building technologies (IBT) as technologies incorporated into the building structure, external shell, or building systems that contribute to progress in moving buildings and building systems toward an ultimate vision of buildings that possess intelligence and display at least some characteristics that satisfy Webster's definition of intelligent. We also adapt the definition provided by the USDOE as the ultimate vision of intelligent buildings — buildings that set up and operate themselves. Intelligent buildings will provide

- High-quality, personalized, and localized environmental control, including all aspects of the environment — heat, cooling, light, ventilation, and acoustics.
- Highly reliable, cost effective building services and environmental control — levels of reliability that far exceed today's at minimal costs, as automated learning systems that adapt to changing conditions are introduced into buildings.
- High speed multimode communication for voice, data, graphics, audio, and video — these technologies will provide services for tenants but will also provide the foundation on which intelligent building technology will be built.
- Flexible, reconfigurable workspaces and services — the increasing pace and changing styles of work in today's and tomorrow's rapidly growing economy require building spaces that can accommodate change. Flexible work spaces and the services provided to them (telephony, data, video, electricity, comfort conditioning, ventilation, and lighting) that can be reconfigured overnight or during lunch will make the space more adaptable to the needs of business.
- Efficient, robust, cost effective building operation — better control, provided by intelligent technology, will optimize building operation with respect to cost, while meeting occupants needs.
- Sustainable building practices — intelligent buildings will provide solutions for making our building stock higher performing and more sustainable.

Despite the advancement of technology, particularly in information technology, over the last 20 years, buildings and the technologies in them still do not display intelligent characteristics. While the concept of an intelligent building (IB) has been evolving for almost 20 years, relatively few buildings completed by the year 2000 would be considered intelligent. IB-enabling building automation (BA) technologies, however, have developed at an unprecedented rate in these same two decades. This seeming irony is easily explained by industry observers: the building industry is very conservative and is slow to adopt any new technology, let alone one that is rapidly changing. There is, nevertheless, little doubt that building automation technologies will be widely adopted — it is only a question of when. The fact that we can, at this time, paint a rather clear picture of "what" and "how" may be taken as evidence that "when" is not too far off.

This section, therefore, differs from others in this handbook because it summarizes the best estimate of IB technologies expected to be seen before long. It does not represent the certainty of historical experience and scientific inevitability embodied in other chapters that treat more mature technologies. However, because wide adoption appears to be imminent, we have decided to discuss the advanced concepts involved in IB technology.

The remainder of this section provides a discussion of drivers for the continued emergence of IBT, more detailed descriptions of technologies that the authors believe will be important to realizing the vision of intelligent buildings, and suggestions for how building designers, owners, and operators can prepare for intelligence in their buildings. Our focus is on building environmental systems that ensure comfortable and healthy indoor environments.

5.2.1 Why Intelligent Buildings Are Needed — Demand and Benefits

A survey of office tenants by the Building Owners and Managers Association International in 1999 revealed building features that tenants consider important (BOMA 1999). The most highly desired services and amenities were

- Comfortable temperatures
- Good indoor air quality
- High-quality building maintenance

- Responsive building management
- Building management's ability to meet tenant needs
- Effective noise control

BOMA also asked which of a set of thirteen intelligent building features were considered most important. The results were

- HVAC systems that provide comfortable temperatures
- Ability for tenants to control comfort
- Built-in wiring for the Internet
- High-speed LAN/WAN connectivity
- Fiber optics capability
- Conduits for power, data, and voice cabling
- Controlled access security system with monitoring
- Redundant power source

A number of political, economic, sociologic, and technologic factors contribute to the evolving demand for higher quality, more sophisticated, more reliable building services. These drivers include

- *Building owners searching for ways to cut costs of ownership and increase value.* Buildings owners are concerned with retaining tenants, and employers are concerned with retaining employees, especially in a tight labor market. Providing quality office space is more important than ever before for retaining tenants and employees. Tenants are demanding premium services, like high-speed Internet access, that were not required even five years ago. Owners are seeking ways to reduce costs, often looking to operations and maintenance (O&M) for cost reductions. This often means decreases in O&M staffing, even when tenants are demanding better service. Intelligent building technologies should provide solutions to this problem by providing tools that help operation and maintenance staff target their efforts more effectively. For example, if a sensor has failed and is causing poor control of an air-handler, an automated diagnostician could identify that a specific sensor has failed, provide estimates of the cost and comfort impacts of this problem, and direct staff to replace that specific sensor. In some cases, like poor control, an intelligent system could correct itself, learning from a record of complaints as well as past performance resulting in decreased costs while increasing performance. In addition to these evident costs, the risk of litigation related to indoor environmental conditions has increased in significance over the last decade. Indoor air quality and its effects on health have been the cause of notable legal action. Intelligent monitoring and control systems will help alleviate problems in the future.
- *The information technology revolution spurred by growth of the Internet.* Owners are struggling to meet tenant demands for access to rapidly changing information technology (IT) services. Keeping up with changes in information technology and providing for upgrades as technology evolves is increasingly important for retaining tenants. From the perspective of providing enhanced building environmental control, advances in IT (e.g., wireless, broadband, distributed processes, data mining, rapid widespread information sharing, etc.) provide new technologies by which high quality, flexible environmental control could be ensured. Information technology provides the new foundation on which intelligent building technologies can be built.
- In recent years, a global trend in *utility deregulation* has occurred. Deregulation will open new opportunities to building owners by providing increased customer choice in energy supply. New full service energy and O&M performance contracts are emerging in the energy services sector, and new, nontraditional services, such as air quality, comfortable temperatures, and continuous building commissioning, are likely to provide a new way of valuing energy and the services it provides.
- A shift is also occurring from central generating plants to *distributed power generation*. The economies of scale that existed for decades are giving way to economies of mass production. Smaller,

modular power generation systems, tailored, on-site, power solutions, and "green" power are emerging alternatives for building owners. These technologies may provide the answer to tenant desires for reliable, backed-up power sources, while also potentially leading to lower energy costs.

- In recent years, *carbon management* has emerged as a global scientific and political issue. After decades of investigation and discussion among the scientific community, the potential dangers of global climate change resulting from anthropogenic carbon emissions have become recognized as an important global issue. Sustainable energy policy is viewed by many as the solution to this problem. Currently, international political pressure is driving the U.S. towards policy that will address this issue. Premium tenants increasingly want "sustainable" or "green" buildings. Intelligent building technologies are likely to provide a major contribution toward using energy more efficiently in buildings and controlling the contribution of the buildings sector to atmospheric carbon concentrations. Government is likely to be a catalyst in this area through research, incentives, taxes, or potentially even regulation.

- In the U.S. and many of the industrialized countries of the world, *changing demographics* will influence the needs and desires of building occupants. This is particularly true for the U.S. where the "baby boomers" are now reaching middle age and will become seniors in the next decade or two. Meeting the needs of an aging population, such as increased demand for health care, changing needs with respect to housing, and changing work arrangements, could drive the need for changes to the building stock and create a role for intelligent buildings to help meet these needs.

Other factors make the time opportune for the development of intelligent building technologies. Building automation systems are likely to provide the mechanism for delivery of intelligent building technology for indoor environmental control. They currently provide a network of control and monitoring equipment that coordinates (or could coordinate) the functions of HVAC, security, fire protection, and other building services, while facilitating and automating the building management. These capabilities are clearly desirable. Increasing interest in IBs is driven by falling costs and growing capabilities of the underlying information technology. Let's consider some of these costs and functions.

Hardware costs. Moore's Law (Intel 2000) rings true for most of the technologies upon which building automation relies. Modems, bit drivers, hardware protocol converters (e.g., RS232 to RS485), communication protocol converters (bridges, routers, gateways), microprocessors, memory, computers-on-a-chip, analog components, PLAs, mixed ASICs, etc. have faithfully doubled in power and speed every 18 months for about three decades. Costs have generally followed a downward trend all the while. Lower cost and higher speed, capacity, and computing power mean that increased functionality is ever more cost effective.

Network media are well established. TCP/IP, CEBus™, BACnet™, Lon Talk™, Profibus™, and Fieldbus™ protocol standards are used at the supervisory level, field panel level, and in some cases at the device level in building automation systems. HART, one-wire, and similar multidrop standards are used for simple devices. All of these standards have achieved widespread acceptance in terms of their hardware/electrical requirements.

Communications standards have passed the critical threshold of multivendor acceptance. Third-party providers of control components and systems compatible with existing proprietary standards (e.g., Delta for Honeywell) have been very successful for a couple of decades. BACnet and Lon Works now provide open standards which show that the trend towards multi-vendor interoperability continues.

Enterprise management has shifted from cautious acceptance to embracing network information sharing technologies. The networks that tie a BAS with the rest of the enterprise and the intelligent software applications that manage the BAS are the keys for the next generation of distributed facilities management systems (Bayne 1999). Controls manufacturers, engineers, and researchers are developing software solutions that take advantage of integrated networks to provide easy access to operating and control data (Olken et al., 1996; O'Neill 1998; Brambley et al., 1998; Katipamula et al., 1999; Chassin 1999). Use of state-of-the-art controls that facilitate distributed processing, coupled with gateways that provide interfaces between the control networks and the data networks (Internet and intranet, respectively), will provide better monitoring and control of building systems and enable management of distributed facilities from either a central or remote location.

Operational efficiency is being pushed ever harder as skilled labor costs rise and the competition to provide services increases. This provides an opportunity to bring benefits by embedding knowledge and the ability to reason in building automation systems.

Improved operation is also a strong driver. Conditions that lead to complaints should be avoided, and response, when complaints are voiced, should be fast and their resolution appropriate and permanent. Occupants expect good lighting, thermal comfort, and a clean and adequate supply of fresh and recirculated air that is free of odors as well as contaminants. Automated systems can respond quicker than a human operator or engineer manually identifying problems and compensating for them. Meeting occupant's needs in a timely, consistent manner may require intelligent systems.

Maintenance managers are beginning to expect BASs to include tracking and scheduling features. These capabilities have been a part of BASs for years, but they have been difficult or inconvenient to use, and the knowledge required to convert raw data to information has been largely absent. These problems are likely to be addressed by the next generation of BASs that should include the capability to interpret data into useful information. Automatic fault detection is also becoming an expected, if limited, BAS feature; extension to automated diagnosis promises substantial labor savings for operation and maintenance.

Efficient infrastructure is the corollary to operational efficiency. Owners prefer to invest in building systems that are efficient in the first place and adaptable to changing building use. Standards, such as BACnet, emerged from building owners' dissatisfaction for years with being "hostage" to the vendor of their BAS. Standards are a step toward remedying this problem. True "plug and play" is the ultimate Intelligent Building solution to this problem, where system components can be removed, new ones installed and automatically set up and operated.

5.2.2   Intelligent Building Technologies

Intelligent building HVAC technologies will include advanced control systems, automated diagnostics, and flexible systems that can be easily reconfigured to adapt to reconfiguration of building spaces. Adaptive control systems will require plentiful, accurate, reliable, long-lived sensors. Today, sensors in buildings are generally low quality and unreliable. Sensor drifting, complete failure, and even improper placement are commonplace in buildings. Intelligent building systems will require better, more abundant sensors. But cost is often cited as a major impediment to adding more sensors to a new design or retrofit. This will require sensor costs to decrease, so the intelligence of control systems can correspondingly increase. Intelligence without information (data) is of little use — sensor technology will be a key to realizing IBT.

Intelligent environmental control systems will provide localized, personalized comfort control. Initially this might be done by providing work stations or seating stations in conference rooms with individualized controls for ventilation and air temperature. In the longer term, this will involve the system automatically recognizing that a specific individual is sitting at a specific location and then providing the conditions necessary to keep that person comfortable and productive.

Advanced control systems will provide intelligent power management. Energy will not be used when not needed. Environmental control systems will automatically respond to occupancy levels — not just whether a space is occupied or not, but how many occupants are present in each space and what activities are taking place (e.g., exercising compared to sitting at a desk working at a computer).

To provide these capabilities, advanced control systems will need levels of intelligence not present in today's systems. They will need to recognize changes in conditions both inside the building and outside (e.g., outdoor air temperature, humidity levels, solar insulation, wind conditions) and adapt accordingly to minimize the cost of operation while meeting the requirements of the occupants. This will require capabilities that are adaptive, predictive, and learning.

Automated diagnostics will also play an important role in intelligent building systems. These systems can "observe" the operation of environmental control systems and identify when problems in operation occur and then identify the causes of a problem. This capability is much more than a simple alarm. It requires embedded knowledge of the building systems, both control systems and physical components, how they should operate, and how they behave when degraded or failing. Automated diagnostics will lead to better maintenance that helps continuously ensure proper indoor conditions and minimal costs to meet those conditions. Automated diagnostics will also play an important role in automating the commissioning process. Coupled with "plug and play," automated diagnostics will enable automatic set up and operation of many building systems.

Flexible building systems will be important to buildings of the future. They will include rapidly reconfigurable spaces, reconfigurable services for those spaces, robust, self-balancing HVAC systems, reconfigurable lights, electric, and communication services, and "plug and play" components that can be inserted and rapidly placed into service. Part of providing flexible spaces may involve changes in the types of systems used to provide comfort. An example is the potential use of modular, miniature heat pumps in place of large, centralized HVAC systems. New technologies, like this, will most likely find a role in intelligent buildings of the future.

This introduction has provided an overview of some generic types of intelligent building technologies. The following sections discuss specific intelligent building technologies in more detail.

Smart Windows

The window has a truly profound effect on many aspects of a building's existence. Operable windows admit air, light, and sound. Windows contribute a large, if not dominant share of the cooling load and, depending on climate and other aspects of building design, can also represent a large share of the heating load.

Careful, large sample studies have demonstrated very convincingly, if not conclusively, that daylighting increases occupant performance and satisfaction. Retail sales were found to be up 30–50% per unit floor area at the 99.9% confidence level in a recent analysis of sales in 108 almost identical stores (Heschong 1999a). Elementary school students' performance in fully daylit classrooms was found to be 7–15% better than in windowless classrooms in a recent analysis of 21,000 students in three geographically and climatologically disperse school districts (Heschong 1999b). Operable windows were also shown to have a positive effect on students' standardized test scores. Another "finding suggests that *control* of light and/or diffusion of direct sunlight are important features to include in a classroom skylight system" (Heschong 1999b).

Poorly designed fenestration can cause many problems. Even well-designed window systems usually involve tradeoffs. However, proper control of window light transmittance and other window aspects — *smart windows* — can eliminate most of the design compromises while retaining all of the potential benefits. Note that fixed shading and diffusing elements are often called "solar control elements"; here we refer only to active elements as controls.

Daylight control mechanisms include chromogenic glazings, manual and automatic window shades, moveable insulation, and moveable diffusing screens. Thermochromic and photochromic glazings provide a simple form of control based on just one variable, that is, of preventing excessive daylight under direct or bright sky, thus preventing excessive heat gain. Electrochromic windows with electrically controllable transmission characteristics enable greater control and can become part of an integrated building temperature, lighting, and energy control strategy.

Much of the benefit of daylighting is attributed to the subtle visual stimulation of light levels, patterns of light and shade, and spectral distribution that change throughout the day. A similar effect may explain occupants' preference for operable windows. Mechanical delivery of fresh air provides the basic requirement for reasonably clean air with sufficient oxygen and no objectionable odors. The stimulation of naturally occurring changes in air movement, however, is missing with mechanical ventilation.

SWITCHABLE WINDOW TECHNOLOGIES

For this discussion, we refer to a switchable technology as one that uses materials applied to (coated on) glass and has the ability to change tinting level (transmission) or opacity when subjected to an outside influence, specifically heat, light, or electricity. We then split this broader category into two groups: electrochromic (EC) and non-electrochromic (non-EC). The non-EC variety includes liquid crystal and suspended particle technologies. These technologies physically operate quite differently from EC, and in some cases their application is also different. For instance, "polymer dispersed liquid crystal" has been used in windows where privacy is desired because it turns opaque in its un-electrified state. Otherwise, in its electrified state PDLC is clear.

Electrochromic technologies are part of the larger chromogenic family which includes photo-chromic and thermochromic materials. These are materials that change their tinting level or opacity when exposed to light (*photo*chromic), heat (*thermo*chromic), and electricity (*electro*chromic). Within the EC category, there are several types. The primary three are

- Inorganic thin film
- Organic polymer
- Organic solution phase

While these three are all considered electrochromic, the materials and processes that comprise and form the EC systems as well as the resulting performance, appearance, durability, and application vary greatly. For example, with organic polymer electrochromics, the EC films are applied to the inside surfaces of facing panes of glass, and the panes are then adhered together sandwich-style using a polymer electrolyte material between the films. The polymer material must perform not only as one of the layers critical to a functioning electrochromic system, i.e., the electrolyte, but it also must hold the two pieces of glass together.

On the other hand, inorganic ceramic thin-film EC is made of ceramic materials that are sputter-coated onto one pane of glass (similar to how low-E glass is formed) and fired at high temperatures. The heating "bakes" the thin films onto the glass something like the way glazing is fired onto pottery. The pane can then be fabricated into industry-standard dual-pane windows, or, if desired, into a conventional laminated glass structure for situations where extra strength and safety are required.

Source: From SAGE Electrochromics Inc., 2000. With permission.

Operable windows also present the designer with potential problems. Filtration and intelligent controls that ensure energy efficient operation (especially when the occupant goes home without shutting windows) are essential elements of the smart windows concept.

In summary, design with smart windows provides occupants stimulation and personal control of light and ventilation while also providing the potential for energy efficiency. Proper integration with all other building systems and their controls ensures realization of that energy potential and long-term, reliable, and economic building operation.

Plug and Play Control Concepts

Plug and play functionality is broadly forecast to be a key feature of building controls in the future. The term *plug and play* is adopted from personal computer operating systems that detect newly installed hardware, establish communications with it, determine its type and function, and select and install the drivers it requires from a library of possibilities. The analog for building HVAC systems and controls is a system by which the hardware is installed and networked, the hardware announces its presence and preferred operating conditions over the network, and the control system automatically develops the algorithms and control code needed to operate the systems. Ultimately, the only inputs required from the designer might be the set of desired operating modes, or overrides to default decisions such as utilization of night/weekend setbacks, selection of the basis for controlling the economizer (temperature or enthalpy), utilization of a supply temperature reset schedule, etc.

Achieving this vision of plug and play controls will bring a number of benefits to building owners, contractors, and operators. By automatically configuring the equipment, controllers, sensors, and actuators, inappropriate or incomplete control strategies can be virtually eliminated. Proactive, continuous commissioning procedures can be automatically generated that calibrate sensors against one another (or manual readings where necessary), test all operating modes, check for proper equipment installation (backwards flows, etc.), ensure proper actuation of controlled devices, and detect unacceptable valve and damper leakage. Cross-wiring of sensors and actuators can be discovered and corrected automatically by simply remapping them. These and related capabilities will

- Reduce the manual labor in setting up control systems and crafting control algorithms
- Ensure compatibility of control strategies with equipment characteristics
- Utilize the best appropriate control strategies
- Provide a degree of standardization for control strategies and algorithms that assists with their maintenance
- Reduce callbacks by detecting errors at the time of installation
- Generally result in a higher quality product

Intelligent buildings with plug and play controls will exhibit lower startup costs, fewer problems, increased comfort, lower O&M costs, greater adaptability to changing needs, and increased energy efficiency.

This vision of plug and play controls will not be achieved overnight. At its most primitive level, initial plug and play capability might be achieved by quasi-manual methods. For example, equipment manufacturers would specify equipment model numbers, characteristics, embedded sensors, optimal and limiting operating conditions, and preferred and alternative operating strategies on a Web site. Similarly, controls manufacturers would specify controller model numbers and characteristics and maintain a library of typical control strategies and algorithms on their Web site. The installed equipment and controllers would have bar codes, identifying the make and model, that are scanned by the installer and mapped to an electronic blueprint of the HVAC system schematic. This would link the equipment and associated controllers and their characteristics with the HVAC system topology. The controls designer would specify operating strategies and modes in electronic form also mapped to the equipment to be controlled via the HVAC blueprints. The plug and play system would then retrieve all relevant information and algorithms from the Web sites, automatically assemble the control software (specifying algorithms and setpoints), and generate automated and manual test procedures for commissioning and on-going diagnostics. The system also could retrieve and generate all relevant documentation for the control system and HVAC equipment that would be needed for on-going maintenance.

At the next level of plug and play functionality, the equipment and controllers might be networked via an intranet so that they automatically announce their presence and even might contain their own Web sites with all the relevant information. They also could utilize some form of location system so that their physical location could be mapped to the HVAC system layout, and hence the system topology automatically determined, eliminating most of the manual labor for creating this mapping.

Ultimately, plug and play controls might truly mimic computer system functionality in that the topology of the equipment, the controllers, and their sensors would be created and tested automatically through a mechanism of automatically exploring what is connected to what. This would provide added diagnostic capability (such as detecting that "the pump is installed backwards") and would eliminate the need for a direct link between the system physical layout and the system schematic.

Wireless Controls

As the tremendous growth[*] over the last several years in the wireless telephony area has affected almost everyone, there is very little reason to believe that this trend will not change the traditional HVAC controls industry. Major inroads of wireless technologies have been observed in the automatic meter reading (AMR) industry where increasing competition in retail electricity and gas markets has spurred the adoption of low cost radio frequency (RF) applications for reading electrical and gas meters. Wireless sensor technology has recently appeared on the HVAC controls market. The first application was a wireless sensor for a variable air volume (VAV) distribution system commercially available in 1997. At the AHR Exhibition held in February 2000, another controls vendor showcased a wireless temperature sensor for VAV and other controls applications. Both technologies are intended to communicate sensor data relatively short distances within a building. They were not intended for use as wireless communication applications across office buildings or other concrete and steel structures.

Recently, a new school of thought for wireless controls for buildings proposed to use wireless communications exclusively for entire buildings or sections of buildings. This is a departure from the previously adopted notion that wireless communications bridges only short distances to a control device. Novel radio frequency techniques would be used to achieve communications of approximately 1500 ft in typical commercial buildings while still complying with Federal Communications Commission (FCC) regulations[**] that limit RF signal power output of license-free RF operations. While this technology may provide a viable and cost effective option to wired control networks, the earliest adoption of wireless controls in building automation are expected to be the low cost RF technology designed to communicate short distances within a building's structure.

Wireless control technologies are becoming increasingly more attractive on the basis of cost savings compared to a wired control system. Other benefits include the following.

Extensibility of Control System — Wireless sensors and control devices can be readily added as retrofits to accommodate the changing needs of the occupants. As interior office space layout undergoes frequent changes to respond to organizational changes of the tenant, the controls technology can simply move with the walls. For instance, wall thermostats might be attached to a wall using Velcro. This thermostat could then easily be re-attached to a new wall of a reconfigured space. Likewise, the lighting control could be reconfigured easily as small office spaces are consolidated into larger conference rooms, or larger rooms are subdivided into small individual offices.

Enabling Advanced Diagnostics and Controls — Wireless controls technology is also an enabling technology that provides new economically viable controls enhancements or retrofit opportunities in those cases where using conventional wired technology would not be economically justifiable. Low-cost wireless sensors could significantly improve the information on thermal and environmental conditions in buildings, which would enable advanced diagnostics procedures and optimal control techniques to be developed and deployed. For instance, if several IAQ sensors could be inexpensively deployed to take real-time measurements, root causes of indoor air problems could be more easily detected.

First Cost Savings Opportunities — It is generally the communication and integration of many sensors into the existing control architecture that is cost-prohibitive when done using existing wiring practice. In particular, the labor component of the installation of additional wires is expensive. According to RS Means Mechanical Cost Data, communication cables for HVAC control applications installed in cable conduits are approximately $3.50 per linear foot of wiring, with the cost for ANSI/TIA/EIA category 5

---

[*]28 million subscribers to wireless telephony services in June of 1995, 61 million in mid 1998, *Progress Report: Growth and Competition in U.S. Telecommunications 1993–1998*, Council of Economic Advisers, February 1999. National Telecommunications and Information Administration, U.S. Department of Commerce.
[**]Federal Communications Commission Regulation Part 15. Radio Frequency Devices. Federal Communications Commission, Washington, DC.

cables for telecommunication at $0.15/ft and approximately $0.07/ft for #18 AWG cable used for thermostat connections (RS Means 1999).

Wireless controls in commercial buildings contribute to cost savings opportunities in the following three ways:

- *Cost for the physical wires and wiring conduits.* This is by far the least significant cost component. Communication cables for horizontal copper wiring in plenums are $0.15–0.20/linear ft or less. Noncommunication wiring for devices such as thermostats is approximately half the cost of communication wiring.
- *Cost of accessibility to wiring conduits.* This cost component is site specific. Issues may arise, for example in older buildings with asbestos ceilings, which may require specialists to be at the site to supervise or perform some of the work. Furthermore, there could be a high premium for inconveniencing tenants or customers in office buildings where the wiring work can be done only during the regular work schedule.
- *The cost of labor.* Accessibility to the wiring conduits or places in which wires can be laid determines the labor hours for wiring and, thus, the labor cost. Complicated wiring with short runs is generally more time intensive than long, straight runs. Depending on the location, labor cost, and whether or not union labor is employed, the labor for a linear foot may vary significantly. Means suggests $3/ft for labor as an average cost estimate.

Special Considerations with Wireless Controls

Wireless controls have been deployed successfully for several years. Most applications represented building-to-building communications or combined telemetry and control tasks with outdoor transceivers. The environment in which the wireless communication was performed was generally not subject to changes. Furthermore, outdoor environments generally cause less interference with wireless communications.

Early attempts to deploy wireless controls in commercial building environments with concrete structures yielded mixed results. Although there are no comprehensive studies published on the engineering issues for successful deployment of RF controls in office environments, there is anecdotal evidence of interference, signal attenuation, and battery lifetime problems. Interference problems have been addressed with frequency hopping and spread-spectrum techniques. Attenuation problems may still exist and require some engineering tasks for properly locating the RF transceivers. Depending on the signal strength, frequency, and surrounding environments, repeater stations have been deployed to robustly communicate the control signals to the intended receiver.

Full advantage of wireless controls can be achieved if the device is battery powered or, in other words, does not require electrical wiring for the power supply. A lifetime of 5 years or more is generally required for building controls applications to be acceptable. Data rate requirements and the RF signal power output determine the energy consumption of the control device. The power requirements for the device processing circuitry are generally significantly less than the RF communication energy consumption. Therefore, sophisticated power management techniques have been deployed to conserve electric energy whenever possible. The RF transceivers can be switched into a "sleep" mode and can "wake up" at a set time to communicate updated information, or they can be device-initiated when, for instance, the system properties above a threshold value are sensed. Therefore, applications such as remote thermostats or temperature sensors are now early wireless applications. The power requirements for closed loop controls with update rates of a second or less are relatively power intensive and place a real challenge to meet the acceptable lifetime requirement.

Automated Diagnostics in Intelligent Buildings

Commissioning is traditionally viewed as the process of manually testing an HVAC system to ensure that it performs as designed and as expected, in terms of function, capacity, efficiency, and ability to maintain thermal comfort. Diagnosis is commonly viewed as a passive process of observing system performance after installation and identifying problems. Although increasing numbers of buildings are being commissioned today, overall the number of commissions remains small. The number of buildings that receive either post-construction diagnostic or re-commissioning services is miniscule. As a result, when combined with the complexity of modern HVAC systems and a lack of routine maintenance, operational problems in commercial buildings are rampant.

Widespread adoption of automated diagnostic and commissioning procedures will help drive the cost of these services down and increase availability of the expertise in software packages. This is the subject of Chapters 7.1 and 7.2 of this handbook.

Automated diagnostic and commissioning procedures are directly linked to plug and play controls' functionality, with mutual and synergistic benefits. The automated procedures benefit from plug and play because the available sensors to support diagnostics are known; the control actions that should result under any conditions can be checked against the intended control strategy. This vastly reduces or eliminates the need to manually obtain and enter this information when setting up diagnostic or commissioning tools. Further, special operating modes, sometimes called proactive diagnostics, can be used to create operating conditions under which sensors can be checked against one another for consistency or to highlight correct or incorrect performance. These operating modes can be both designed and promulgated through plug and play functions, reducing cost. Where automated procedures must be supplemented by manual measurements and observations, the plug and play control system in intelligent buildings could generate the procedures and forms automatically, as well as set the system to the appropriate operating conditions on command.

Automated diagnostic procedures also bring added value to plug and play controls' functionality. The plug and play controls could automatically adjust or compensate for some problems, for example by changing a setpoint, substituting one sensor for another that is out of service, or adjusting PID loop parameters to enhance stability. These actions might be temporary until a repair is made or permanent if the adjustment is in response to changed building needs. Further, intelligent buildings could maintain records of design intent, control strategies, maintenance actions, and equipment specifications that will greatly simplify and reduce the cost of correcting or repairing problems found by automated diagnostic procedures.

Integrating Alternate Generation Technologies: Fuel Cells, Microturbines, and Solar Power

On-site generation technologies provide redundancy and power quality in almost all installations. In some installations, energy cost is an additional driver, and in others the "green" statement itself motivates the building owner. Deriving the full potential benefits of on-site power generation usually requires a level of integration and control coordination not found in conventional buildings. Important differences among the available solar, fuel-cell, and microturbine technologies should be considered during initial planning of on-site generation. These technologies for distributed generation are described in Chapter 3.1.

Automated Real-Time Energy Purchasing Capabilities

The Electric Power Research Institute (EPRI), in its *Electricity Technology Roadmap*, describes a future technology scenario, in which smart homes of the not-too-distant future will be equipped with Internet devices and controllers that could automatically search the Internet for access to the lowest cost power or seek power with other valued attributes, such as "green" power or high reliability power for critical applications (EPRI, 1998). While the scenario for the residential customer may appear to be far in the future, the earliest adopters will most likely be the commercial and industrial customers in deregulated energy markets. To date, the regulatory framework for the dynamic procurement of electric power already exists. California, for instance, offers nonresidential end-users the option to bid their load directly into the California Power Exchange for the purchase of wholesale electric power at hourly spot-market prices (demand bidding) [PBEP, 1997; PSCP, 1997]. Realizing this future scenario would require the integration of the facility management system with the bidding communication technology. With an integration of facility management, company accounting, and the bidding system, bid scenarios could be generated automatically based on load forecasts, the flexibility to manage electric power demands, and the assessment of the economic value of consuming electric energy at each hour of the day.

The growth of the Internet has drastically increased the interest in and relevance of electronic commerce. There are already electronic auction servers operating on the Internet, such as APX (Automated Power Exchange) providing opportunities to buy and sell power. Web-based intelligent agents are emerging to facilitate the searching task for an optimal power portfolio that meets the customer's cost and risk requirements. These agents work autonomously, scanning the Web for power offerings that meet a set of requirements — cost, date of delivery, and other characteristics that are important to the customer (EPRI, 1999; Reticular, 1999). It remains uncertain whether or not residential customers will ultimately use this technology and engage in power e-commerce over the Internet or whether this commerce will remain in the domain of energy service companies and large corporate energy managers.

Optimized Dynamic Building Systems

Optimal use of part-load efficiency, building thermal capacitance, and off-peak utility rates requires use of a number of simple concepts that collectively represent a very challenging problem. The key to this challenge is the high level coordinated control of various plant and distribution elements based on robust on-line, building-specific (i.e., adaptive) models of building thermal response, internal gain profiles, and equipment performance. A corollary requirement is on-line fault detection and diagnosis (FDD), described in Chapter 7.2.

Programmed (fixed) schedules for HVAC start/stop are no longer acceptable. To optimize performance, start-stop must function at zone level, must learn occupancy schedules at the zone level, and must use thermal set-up and set-back response models that adapt to changing building characteristics, including rezoning, and changing schedules of the occupants in any given zone.

In a well-designed building that has very efficient lighting and office equipment, life-cycle cost-optimized insulation, solar gain/shading/reflecting elements, and personal control of local temperature and humidity, the daily cooling loads may be quite modest. The possibility of reducing these loads further, or at least reducing the cost of meeting them, by nocturnal precooling of available building mass has been studied by a number of researchers.

Precooling can be achieved at night by operating a chiller near its most efficient part-load capacity, or by operating air handlers in economizer mode. Occupied zones are generally cooled at least to the lowest comfortable temperature during the night. Best use of thermal mass is obtained by cooling below the comfort band and then allowing zones to recover to just the minimum acceptable temperature by the time occupants arrive. Zones are maintained at the minimum temperature until the marginal cost of power crosses some threshold at which time cooling is modulated to stop or reduce the rate of rise at the building electric meter. This threshold must be carefully and dynamically selected (changing even up to the instant that it is crossed) to maximize savings at the meter without allowing zone conditions to rise above the comfort band. Modulation of the plant during the building demand-limiting period is also critical to success of the precooling strategy. Precooling also may involve significant latent load by absorption/desorption of water from the building fabric and contents.

Achieving the full potential of such dynamic control requires short-term (12–24 hour) forecasts of weather and occupant loads, a realistic and seasonally adaptive model of the building's thermal response to indoor and outdoor conditions, and real-time communication of utility rates.

Control of ice or chilled water storage is conceptually much simpler than control involving the thermal mass of the building fabric, but the need to justify the higher cost of discrete storage, and the desire to save (or at least minimize any increase) in plant energy use while pursuing the "easy" demand-related savings can also be quite challenging. Real-time pricing (RTP) adds yet another layer of complexity (Henze, et al., 1997).

5.2.3   How to Prepare for IB Technologies

This chapter has explained how improved control and monitoring can reduce labor costs, improve occupant satisfaction and productivity, and save energy. Future buildings will take maximum advantage of IB enabling technologies by using fundamentally new design criteria and processes. Existing buildings, however, can also benefit.

Upgrading Communication and Information Infrastructure

Two important elements of IBs can be retrofit incrementally to existing buildings: networks and sensors.

Communications Infrastructure — Typical office buildings have multiple tenants who maintain their own internal networks. The building owner, however, provides — and frequently must upgrade — high speed Internet access. The communications infrastructure needed for IBs is, ideally, designed and installed as part of the tenant-driven Internet access upgrades. The design should ensure that the appropriate network topology is installed and that sufficient data bandwidth is extended to key locations, including mechanical and electrical rooms and other equipment locations (e.g., rooftop, elevator, penthouse). Collection nodes serving HVAC and lighting distribution must also be considered, however. Multidrop lines will eventually extend to every room, and the drop density over open-plan office areas will at least equal that extended to small individual offices. In individual offices a single local controller will provide, minimally, the interface to an occupant control device, lighting sensors and actuators, HVAC sensors (IAQ and MRT as well as air temperature), and an HVAC terminal box.

Information infrastructure — Plans for upgrades should be reviewed to ensure adequate provisions for sensors whenever system upgrades of any type — communications, HVAC, lighting, tenant finish — are made. The decision to install a local controller should be based on cost, features, and technically useful life. Local controllers must be fully compatible with standard communications protocols, such as LonTalk, BACnet, or TCP/IP, and should also be compatible with application level software standards (XML, Java, OLE). Compatibility means that all features should be available and fully functional without requiring use of the manufacturer's software.

Upgrading HVAC Infrastructure

Current models of most air handlers and packaged equipment include microprocessor-based controls as a standard or optional feature. Most controller original equipment manufacturers (OEMS) offer a line of generic controllers that are configured to each equipment manufacturers specifications by firmware. The trend is towards control boards that (1) conform to BACnet, LonTalk, TCP/IP, or one of the half dozen widely accepted industrial control or building automation communications standards, and (2) allow firmware upgrades via the communications port. It is desirable to select equipment that uses controller hardware/firmware/communications protocols so that migration toward full integration of building controls can proceed smoothly and economically. It is also a good idea to consider the sensor suite included in each type of package and the provision for adding additional sensors and monitoring equipment in the future. The number and type of expansion inputs provided by the controller are factors that must be considered in selecting competing equipment. The ability of the controller microprocessor and associated architecture to support future FDD algorithms appropriate to the equipment type is also important but much more difficult to assess.

References

Brambley, M.R. et al., 1998. Automated diagnostics for outdoor air, ventilation, and economizers, *ASHRAE Journal*, vol. 40, no. 10, 49–55.

Braun, J.E., 1992, A comparison of chiller-priority, storage-priority and optimal control of an ice-storage system, *ASHRAE Trans.* vol. 98, part 1, AN-92-8-1, 893–902.

Building Owners and Managers Association International (BOMA), 1999, *What Tenants Want: 1999 BOMA/ULI Office Tenant Survey Report*, BOMA, Washington, DC.

Chassin, D.P., Computer software architecture to support automated building diagnostics, *Proceedings of the 1999 CIB W78 Workshop on Information Technologies in Construction*, Vancouver, Canada, vol.4, May 31–June 3, 1999.

Dexter, A.L., 1996, Intelligent buildings: fact or fiction?, *Int. J. HVAC&R Res.*, vol. 2, no. 2.

Electric Power, Revision 1, August 1999, Reticular Systems, Inc., San Diego.

EPRI, 1998, The electricity technology roadmap. *EPRI J.* November/December 1998, 25–31.

EPRI, 1999, Prototype intelligent software agents for trading electricity: competitive/cooperative power scheduling in an electronic marketplace, *EPRI Report* TR-113366.

European Intelligent Buildings Group (EIBG), June, 2000, The intelligent building, available at URL: http://www.eibg.org./.

GMD. June, 2000, Ambiente workspaces of the future, GMD — German National Research Center for Information Technology, Darmstadt, Germany, available at URL: http://www.darmstadt.gmd.de/ambiente/.

Henze, G.P., Dodier, R.H., and Krarti, M. 1997, Development of a predictive optimal controller for thermal energy storage systems, *Int. J. HVAC&R Res.*, ASHRAE, vol. 3, no. 3, 233–264.

Heschong, L., 1999, *Daylighting in Schools: An Investigation into the Relationship Between Daylighting and Human Performance*, Heschong Mahone Group for PG&E/CBEE Third Party Program, Fair Oaks, CA.

Heschong, L., 1999, *Skylighting and Retail Sales: An Investigation into the Relationship Between Daylighting and Human Performance*, Heschong Mahone Group for PG&E/CBEE Third Party Program, Fair Oaks, CA.

Intel Corporation, 2000, Intel's processor hall of fame, what is Moore's law. Available at URL: http://www.intel.com/intel/museum/25anniv/hof/moore.htm.

Katipamula, S. et al., Facilities management in the 21st century, *HPAC Heating/Piping/Air Conditioning Engineering*, July 1999, pp. 51–57.

Keeney, K.R., and Braun, J.E., 1997, Application of building precooling to reduce peak cooling requirements, *ASHRAE Trans.*, vol. 103, part 1 (PH-97-4-1), 463–469.

Keeney, K., and Braun, J.E., 1996. A simplified method for determining optimal cooling control strategies for thermal storage in building mass, *Int. J. HVAC&R Res.*, ASHRAE, vol. 2, no. 1, 59–78.

Lucent Technologies, 2000, SYSTIMAX intelligent buildings, Available at URL: http://www.lucent.com/systimax/buildings.html.

Morris, F.B., Braun, J.E., and Treado, S.J., 1994, Experimental and simulated performance of optimal control of building thermal performance, *ASHRAE Trans.*, vol.100, part 1, paper no. 3776, 402–414.

Olken, F. et al., 1996. Remote building monitoring and control, *Proceedings of the ACEEE 1996 Summer Study on Energy Efficiency in Buildings*, American Council for an Energy-Efficient Economy, Washington, D.C.

O'Neil, P., 1998. Opening up the possibilities, *Engineered Systems*, vol 15, no. 6.

PBEP, 1997, Power exchange bidding and bid evaluation protocol (PBEP), PX Tariff, Californian Power Exchange, 1997.

PSCP, 1997, Power exchange scheduling and control protocol (PSCP), PX Tariff, Californian Power Exchange, 1997.

Reticular, 1999, Using intelligent agents to implement an electronic auction for buying and selling electric power, Revision 1, August 1999, Reticular Systems, Inc., San Diego, CA.

Rossi, T.M., and Braun, J.E., 1996, Minimizing operating costs of vapor compression equipment with optimal service scheduling, *Int. J. HVAC&R Res.*, ASHRAE, vol. 2, no. 1, 3–26.

RS Means, 1999, *Mechanical Cost Data 1999*. 15 Mechanical, 157 — Air-Conditioning and Ventilation, 420–5000, RS Means Company, Inc., Kingston, MA.

SAGE Electrochromics, Inc. 2000. Sage Glass® Explained. Available at URL: http://sage.ec.com/pages/.sgexplained.html.

Seem, J.E., and Braun, J.E., 1992, The impact of personal environmental control on building energy use, *ASHRAE Trans.*, vol. 98, part 1, AN-92-8-2, 903-909.

U.S. Dept. of Energy (DOE) 2000. Recommended Future Directions for R&D in Intelligent Buildings Systems. Available at URL: http://www.eren.doe.gov/buildings/systemsfuture.html.

Webster's II New College Dictionary, Houghton Mifflin Company, Boston, 1999, p. 576.

6

HVAC Design Calculations

Peter S. Curtiss
Kreider & Associates, LLC

Jeffrey S. Haberl
Texas A&M University

Joe Huang
Lawrence Berkeley Laboratory

David Jump
Lawrence Berkeley Laboratory

Jan F. Kreider
Kreider & Associates, LLC

Ari Rabl
École des Mines de Paris and University of Colorado

T. Agami Reddy
Drexel University

Max Sherman
Lawrence Berkeley Laboratory

6.1 Energy Calculations — Building Loads

Ari Rabl and Peter S. Curtiss

Heating and cooling loads are the thermal energy that must be supplied to or removed from the interior of a building in order to maintain the desired comfort conditions. That is the demand side of the building, addressed in this chapter and the next. Once the loads have been established, one can proceed to the supply side and determine the performance of the required heating and cooling equipment, as discussed in Chapters 4.1–4.4.

Of primary concern to the designer are the maximum or peak loads because they determine the capacity of the equipment. They correspond to the extremes of hot and cold weather, called design conditions. But while in the past it was common practice to limit oneself to the consideration of peak loads, examination of annual performance has now become part of the designer's job. The oil crises sharpened our awareness of energy, and the computer revolution has given us the tools to optimize the design of the building and to compute the cost of energy. In this chapter we address the calculation of peak loads. Methods for the determination of annual energy requirements are presented in Chapter 6.2.

A load calculation consists of a careful accounting of all the thermal energy terms in a building. While the basic principle is simple, a serious complication can arise from storage of heat in the mass of the building: In practice, this is very important for peak cooling loads, even in lightweight buildings typical in the U.S. For peak heating loads, the heat capacity can be neglected unless one insists on setting the thermostat back even during the coldest periods. For annual energy consumption, the effect of heat capacity depends on the control of the thermostat: it is negligible if the indoor temperature is constant but can be quite significant with thermostat set back or up.

This chapter begins with models for air exchange in Section 6.1.1. Section 6.1.2 discusses the design conditions, heat loss coefficient, and thermal balance of a building. Section 6.1.3 examines the limitations of a steady state analysis. The need for zoning, i.e., the separate treatment of different parts of a building where the loads are too dissimilar to be lumped together, is discussed in 6.1.4, and a steady-state method for peak heating loads is presented in Section 6.1.5. Peak cooling loads are calculated in Section 6.1.6 by using a modified steady state method, CLF-CLTD. To provide an algorithm for dynamic load calculations, the transfer function method is presented in Section 6.1.7; using this method has become relatively simple, thanks to computers with spreadsheets.

The calculation of loads presented here does not take into account the losses in the distribution system. These losses can be quite significant, especially in the case of uninsulated ducts, and they should be taken into account in the analysis of the HVAC system. Distribution systems are the province of Chapter 4.3.*

6.1.1   Air Exchange

Fresh air in buildings is essential for comfort and health, and the energy for conditioning this air is an important term. Not enough air, and one risks sick-building syndrome; too much air, and one wastes energy. The supply of fresh air, or air exchange, is stated as the flow rate $\dot{V}$ of the outdoor air that crosses the building boundary and needs to be conditioned [ft³/min (m³/s or L/s)]. Often it is convenient to divide it by the building volume, as $\dot{V}/V$, expressing it in units of air changes per hour. Even though it is customary to state the air flow as the volumetric rate, the mass flow $\dot{m} = \rho\dot{V}$ would be more relevant for most applications in buildings. The relation between mass flow and volume flow depends on the density ρ, which varies quite significantly with temperature and pressure.

To estimate the air exchange rate, the designer has two sources of information: data from similar buildings and models. The underlying phenomena are complicated, and a simple comparison with other buildings may not be reliable. The modeling approach can be far more precise, but may require a fair amount of effort.

It is helpful to distinguish two mechanisms that contribute to the total air exchange:

- *Infiltration* — uncontrolled airflow through all the little cracks and openings in a real building
- *Ventilation* — natural ventilation through open windows or doors and mechanical ventilation by fans

Data for Air Exchange

Air exchange rates can be measured directly by means of a tracer gas. Sulfur hexafluoride (SF_6) has been a favorite because it is inert and harmless, and it can be detected at concentrations above 1 part per billion (ppb). The equipment is relatively expensive but allows determination at hourly or even shorter intervals (Sherman et al., 1980; Grimsrud et al., 1980). More recently a low-cost alternative has been developed that uses passive perfluorocarbon sources and passive samplers (Dietz et al., 1985). Each source can cover up to several hundred cubic meters of building volume, at a cost of about $50, but the method

---

* For future reference, we note that the loads of each zone, as calculated in this chapter, include the contribution of outdoor air change. However, for the analysis of air-based central distribution systems, it is convenient to exclude the contribution of ventilation air from the zone loads and to count it instead as load at the air handler. Keeping separate the load due to ventilation air is straightforward because this load is instantaneous.

yields only averages over sampling periods of several weeks (in fact, it averages the inverse of the air exchange rate). Carbon dioxide is interesting as a tracer gas because it is produced by the occupants and can be used for monitoring indoor air quality; as a measure of air exchange, it is uncertain to the extent that the number of occupants and their metabolism are not known.

An entirely different method of determining the airtightness of a building is pressurization with a blower door (a special instrumented fan that is mounted in the frame of a door for the duration of the test). To obtain accurate data, one needs fairly high pressures, around 0.2 to 0.3 in WG (50 to 75 Pa), which are higher than natural conditions in most buildings. The extrapolation to lower values requires assumptions about the exponent in the flow-pressure relation, and it is not without problems (see Chapter 23 of ASHRAE, 1989a). In buildings with mechanical ventilation, one could bypass the need for a blower door by using the ventilation system itself, if the pressure is sufficiently high.

In the past, not much attention was paid to airtight construction, and older buildings tend to have rather high infiltration rates, in the range of 1 to 2 air changes per hour. With current conventional construction in the U.S., one finds lower values, around 0.3 to 0.7. These values are seasonal averages; instantaneous values vary with wind and indoor-outdoor temperature difference. When infiltration is insufficient to guarantee adequate indoor air quality, forced ventilation becomes necessary. The required air exchange rate depends, of course, on the density of occupants. In residential buildings, the density is relatively low, and with conventional U.S. construction, infiltration is likely to be sufficient. But it is certainly possible to make buildings much tighter than 0.3 air changes per hour of infiltration. In fact, Swedish houses are standardly built to such high standards that uncontrolled infiltration rates are around 0.1 air changes per hour; mechanical ventilation supplies just the right amount of outdoor air, and an air-to-air heat exchanger minimizes the energy consumption. In the U.S. for buildings with forced ventilation, ASHRAE ventilation Standard 62-99 applies. Good sites for mechanical exhaust are kitchen and bathrooms, to remove indoor air pollution and excessive humidity.

Uncontrolled air exchange is highly dependent on wind and on temperatures. Even with closed windows, it can vary by a factor of 2 or more, being lower in summer than in winter. The variability of air exchange is indicated schematically in Figure 6.1.1 as the relative frequency of occurrence for three types of house: a leaky house and a moderately tight house, both with natural infiltration, and a very tight house with mechanical ventilation. The last guarantees adequate supply of fresh air at all times, without the energy waste of conditioning unnecessary air. With open windows, the air exchange rate is difficult to predict accurately. It varies with the wind, and it is highly dependent on the aerodynamics of the building and its surroundings. The designer needs data on ventilation rates with open windows to assess comfort conditions in buildings with operable windows during the transition season between heating and cooling. Figure 6.1.2 presents data for natural ventilation in a two-story house. Depending on which windows are open and where the ventilation is measured, the air change rate varies between 1 and 20 per hour.

Models for Air Leakage

For a more precise model of air exchange, let us recall from Chapter 2.1 that the flow through an opening is proportional to the area and to some power of the pressure difference:

$$\dot{V} = Ac\Delta p^n \tag{6.1.1}$$

where A = area of opening, ft² (m²)

$\Delta p = p_o - p_i$ = pressure difference between outside and inside, inWG (Pa)

c = flow coefficient, ft/(min · inWG^n) [m/(s · Pa^n)]

n = exponent, between 0.4 and 1.0 and usually around 0.65 for buildings

In general, different openings may have different coefficients and exponents. This equation is an approximation, valid only for a certain range of pressures and flows; different n and c may have to be used for other ranges. There is another problem in applying this equation to buildings: the width of an

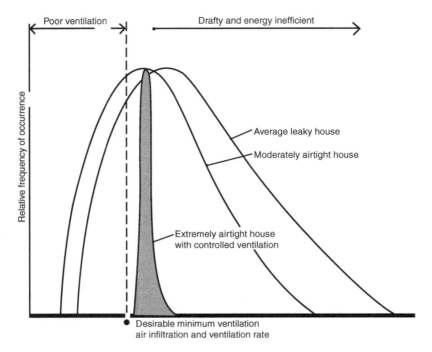

FIGURE 6.1.1 Variability of air exchange for three types of houses, plotted as relative frequency of occurrence versus air change rate (from Nisson and Duff, 1985).

FIGURE 6.1.2 Measured ventilation rates, as a function of wind speed, in a two-story house with windows open on lower floor (from Achard and Gicquel, 1986). The curves are labeled according to the location of open windows and measuring point as follows:

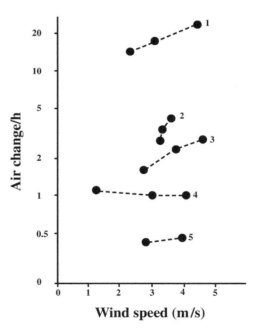

Open Windows	Measuring Point
1. All upper floor	Upper floor
2. Upper floor windward	Upper floor
3. All upper floor	Lower floor
4. Upper floor leeward	Upper floor
5. None	Whole house

opening can change with pressure. Blower door tests have shown that the apparent leakage area can be significantly higher for overpressurization than for underpressurization; external pressure tends to compress the cracks of a building (Lydberg and Honarbakhsh, 1989). In buildings without mechanical ventilation, the pressure differences under natural conditions are positive over part of the building and negative over the rest; here it is appropriate to average the blower door data over positive and negative

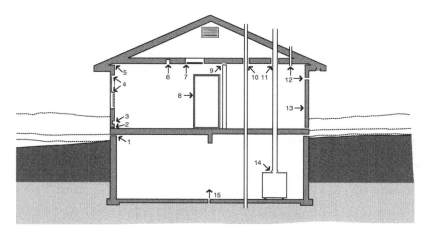

FIGURE 6.1.3 Typical air leakage sites in a house. (1) Joints between joists and foundation; (2) joints between sill and floor; (3) electrical boxes; (4) joints at windows; (5) joints between wall and ceiling; (6) ceiling light fixtures; (7) joints at attic hatch; (8) cracks at doors; (9) joints at interior partitions; (10) plumbing-stack penetration of ceiling; (11) chimney penetration of ceiling; (12) bathroom and kitchen ventilation fans; (13) air/vapor barrier tears; (14) chimney draft air leaks; (15) floor drain (air enters through drain tile) (from Nisson and Duff, 1985).

pressure differences. In buildings where over- or underpressure is maintained, the leakage area data should correspond to those conditions.

The total flow is obtained by summing over all openings k as

$$\dot{V} = \sum_k A_k c_k \Delta p_k^n \qquad \text{(include only terms with } \Delta p_k > 0) \qquad (6.1.2)$$

where A_k = leakage area
        c_k = flow coefficient
        n_k = exponent
        $\Delta p_k = p_o - p_i$ = local pressure difference

If one sums naively over all openings of a building, the result (averaged over momentary fluctuations) is zero because the quantity of air in a building does not change. The flow into the building must equal the flow out; the former corresponds to positive terms in the sum, the latter to negative terms. What interests us here is the energy needed for conditioning the air that flows into the building. Therefore, the sum includes only the terms with $p_o > p_i$.

As indicated by the subscript k, all the terms can vary from one point to another in the building. Therefore, a fairly detailed calculation may be required. Typical air leakage sites in houses are shown in Figure 6.1.3, and data for leakage areas can be found in Table 6.1.1 for a wide variety of building components.

Actually, in many applications, one need not worry about this. For relatively small buildings without mechanical ventilation, the LBL model can be used; this model, discussed shortly, bypasses Equation 6.1.2 by correlating the flow directly with wind speed, temperature difference, and total leakage area. In many (if not most) buildings with mechanical ventilation, one maintains a significant pressure difference between the interior and exterior[*]. If this pressure difference is larger than the pressures induced by wind and temperature, the latter can be neglected and all terms in Equation 6.1.2 have the same sign. Before proceeding to these applications, we have to discuss the origin of the pressure differences.

---

[*] Overpressure in the building allows better control and comfort. Underpressure can be maintained with smaller ducts and lower cost, but at the risk of condensation, freezing, and possibly draft.

TABLE 6.1.1   Data for Effective Leakage Areas of Building Components at 0.016 in WG (4 Pa), cm^2

Component	Best Estimate	Maximum	Minimum
Sill foundation-Wall			
Caulked, in^2/ft of perimeter	0.04	0.06	0.02
Not caulked, in^2/ft of perimeter	0.19	0.19	0.05
Joints between ceiling and walls			
Joints, in^2/ft of wall			
(only if not taped or plastered and no vapor barrier)	0.07	0.12	0.02
Windows			
Casement			
Weather-stripped, in^2/ft^2 of window	0.011	0.017	0.006
Not weather-stripped, in^2/ft^2 of window	0.023	0.034	0.011
Awning			
Weather-stripped, in^2/ft^2 of window	0.011	0.017	0.006
Not weather-stripped, in^2/ft^2 of window	0.023	0.034	0.011
Single-hung			
Weather-stripped, in^2/ft^2 of window	0.032	0.042	0.026
Not weather-stripped, in^2/ft^2 of window	0.063	0.083	0.052
Double-hung			
Weather-stripped, in^2/ft^2 of window	0.043	0.063	0.023
Not weather-stripped, in^2/ft^2 of window	0.086	0.126	0.046
Single-slider			
Weather-stripped, in^2/ft^2 of window	0.026	0.039	0.013
Not weather-stripped, in^2/ft^2 of window	0.052	0.077	0.026
Double-slider			
Weather-stripped, in^2/ft^2 of window	0.037	0.054	0.02
Not weather-stripped, in^2/ft^2 of window	0.074	0.110	0.04
Doors			
Single door			
Weather-stripped, in^2/ft^2 of door	0.114	0.215	0.043
Not weather-stripped, in^2/ft^2 of door	0.157	0.243	0.086
Double door			
Weather-stripped, in^2/ft^2 of door	0.114	0.215	0.043
Not weather-stripped, in^2/ft^2 of door	0.16	0.32	0.1
Access to attic or crawl space			
Weather-stripped, in^2 per access	2.8	2.8	1.2
Not weather-stripped, in^2 per access	4.6	4.6	1.6
Wall-Window frame			
Wood frame wall			
Caulked, in^2/ft^2 of window	0.004	0.007	0.004
No caulking, in^2/ft^2 of window	0.024	0.038	0.022
Masonry wall			
Caulked, in^2/ft^2 of window	0.019	0.03	0.016
No caulking, in^2/ft^2 of window	0.093	0.15	0.082
Wall-Door frame			
Wood wall			
Caulked, in^2/ft^2 of door	0.004	0.004	0.001
No caulking, in^2/ft^2 of door	0.024	0.024	0.009
Masonry wall			
Caulked, in^2/ft^2 of door	0.0143	0.0143	0.004
No caulking, in^2/ft^2 of door	0.072	0.072	0.024
Domestic hot water systems			
Gas water heater (only if in conditioned space), in^2	3.1	3.9	2.325
Electric outlets and light fixtures			
Electric outlets and switches			
Gasketed, in^2 per outlet and switch	0	0	0
Not gasketed, in^2 per outlet and switch	0.076	0.16	0
Recessed light fixtures, in^2 per fixture	1.6	3.1	1.6
Pipe and duct penetrations through envelope			

TABLE 6.1.1 (continued)   Data for Effective Leakage Areas of Building Components at 0.016 in WG (4 Pa), cm²

Component	Best Estimate	Maximum	Minimum
Pipes			
Caulked or sealed, in² per pipe	0.155	0.31	0
Not caulked or sealed, in² per pipe	9.3	1.55	0.31
Ducts			
Sealed or with continuous vapor barrier, in² per duct	0.25	0.25	0
Unsealed and without vapor barrier, in² per duct	3.7	3.7	2.2
Fireplace			
Without insert			
Damper closed, in² per fireplace	10.7	13	8.4
Damper open, in² per fireplace	54	59	50
With insert			
Damper closed, in² per fireplace	5.6	7.1	4.03
Damper open or absent, in² per fireplace	10	14	6.2
Exhaust fans			
Kitchen fan			
Damper closed, in² per fan	0.775	1.1	0.47
Damper open, in² per fan	6	6.5	5.6
Bathroom fan			
Damper closed, in² per fan	1.7	1.9	1.6
Damper open, in² per fan	3.1	3.4	2.8
Dryer vent			
Damper closed, in² per vent	0.47	0.9	0
Heating ducts and furnace-Forced-air systems			
Ductwork (only if in unconditioned space)			
Joints taped or caulked, in² per house	11	11	5
Joints not taped or caulked, in² per house	22	22	11
Furnace (only if in conditioned space)			
Sealed combustion furnace, in² per furnace	0	0	0
Retention head burner furnace, in² per furnace	5	6.2	3.1
Retention head plus stack damper, in² per furnace	3.7	4.6	2.8
Furnace with stack damper, in² per furnace	4.6	6.2	3.1
Air conditioner			
Wall or window unit, in² per unit	3.7	5.6	0

Note: For conversion to SI units: 1 in² = 6.45 cm², 1 ft² = 0.0929 m², and 1 in²/ft² = 69 cm²/m².
Source: From ASHRAE, 1989a.

Pressure Terms

The pressure difference $\Delta p = p_o - p_i$ is the sum of three terms:

$$\Delta p = \Delta p_{wind} + \Delta p_{stack} + \Delta p_{vent} \tag{6.1.3}$$

the first due to wind, the second due to the stack effect (like the flow induced in a heated smokestack), and the third due to forced ventilation, if any. We take the pressure differences to be positive when they cause air to flow toward the interior. The flow depends only on the total Δp, not on the individual terms. The relative contribution of the wind, stack, and ventilation terms varies across the envelope, and because of the nonlinearity, one cannot calculate separate airflows for each of these effects and add them at the end.

The wind pressure is given by Bernoulli's equation (in SI units)

$$p_{wind} = \frac{\rho}{2}(v^2 - v_f^2) \tag{6.1.4SI}$$

where v = wind speed (undisturbed by building), m/s
v_f = final speed of air at building boundary

ρ = air density, kg/m³

In USCS units, we have

$$p_{wind} = \frac{\rho}{2g_c}(v^2 - v_f^2) \qquad lb_f/ft^2 \qquad (6.1.4US)$$

with g_c = 32.17 $(lb_m \cdot ft)/(lb_f \cdot s^2)$, the wind speed being in feet per second, and the air density in pound-mass per cubic foot.

Under standard conditions of 14.7 psi (101.3 kPa) and 68°F (20°C), the density is

$$\rho = 0.075 \ lb_m/ft^3 \qquad (\rho = 1.20 \ kg/m^3)$$

but one should note that the density of outdoor air can deviate more than 20% above (winter at sea level) or below (summer in the mountains) these values. In USCS units, the ratio of

$$\frac{\rho}{g_c} = 0.00964 \ inWG/(mi/h)^2$$

under standard conditions if pressure is in inches water gauge and wind speed in miles per hour.

The wind speed is strongly modified by terrain and obstacles, being significantly higher far above the ground (see ASHRAE, 2001). Since the final speed v_f is awkward to determine, a convenient shortcut is to use Equation 6.1.4 with $v_f = 0$, multiplying it instead by a pressure coefficient C_p:

$$p_{wind} = C_p \frac{\rho}{2} v^2 \qquad (6.1.5)$$

The quantity $p_{wind}/C_p = (\rho/2) v^2$ is plotted versus wind speed v in Figure 6.1.4a. Numerical values for C_p can be gleaned from Figure 6.1.4b, where this coefficient is plotted as a function of the angle between the wind and the surface normal. Typical values are in the range from approximately -0.6 to 0.6, depending on the direction of the wind.

Actually we are interested in the pressure difference between the interior and exterior of a building. If the interior of an entire floor offers no significant flow resistance, one can find the indoor pressure due to wind by averaging the flow coefficient over all orientations of the surrounding wall. Since that average is approximately -0.2, the local pressure difference $p_o - p_i$ at a point of the wall is, in that case,

$$\Delta p_{wind} = \Delta C_p \frac{\rho}{2} v^2 \qquad (6.1.6)$$

with $\Delta C_p = C_p - (-0.2)$ being the difference between the local pressure coefficient and the average.

The *stack effect* is the result of density differences between air inside and outside the building. In winter, the air inside the building is warmer, hence less dense than the air outside. Therefore, the indoor pressure difference (bottom versus top) is less than the outdoor pressure difference between the same heights. Consequently, there is an indoor-outdoor pressure difference. It varies linearly with height, and the level of neutral pressure is at the midheight of the building, as suggested by Figure 6.1.5, if the leaks are uniformly distributed. During the cooling season when indoor air is colder than the outside, the effect is reversed.

The pressure difference is given by

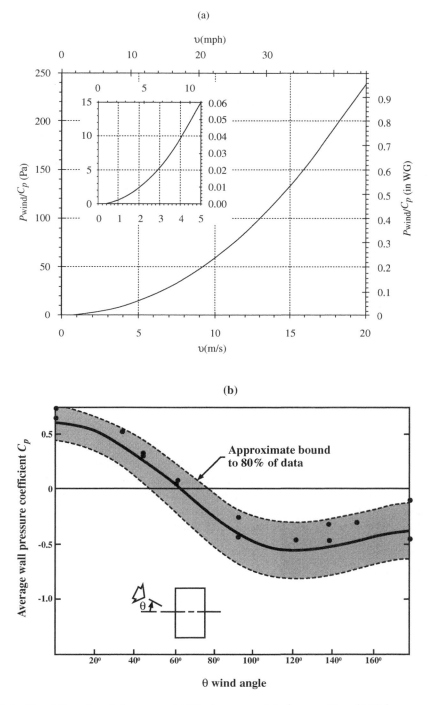

FIGURE 6.1.4 Correlations for wind pressure. (a) Wind pressure plotted as $p_{wind}/C_p = (\rho/2)v^2$ versus wind speed v. (b) Typical values of pressure coefficient C_p of Equation 6.1.5 for a rectangular building as a function of wind direction (from ASHRAE, 1989a; dots indicate the values from Figure 14.6 of that reference).

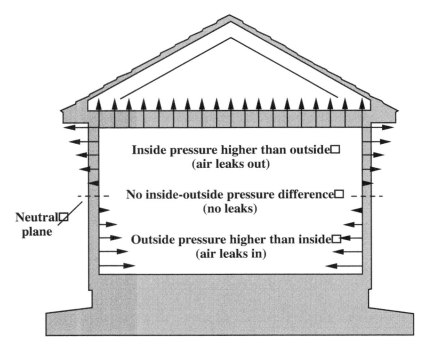

Inside pressure higher than outside☐
(air leaks out)

No inside-outside pressure difference☐
(no leaks)

Neutral☑
plane

Outside pressure higher than inside☐
(air leaks in)

FIGURE 6.1.5   Air leakage due to stack effect during heating season (from Nisson and Vutt, 1985).

$$\Delta p_{stack} = -C_d \rho_i g \Delta h \frac{T_i - T_o}{T_o} \tag{6.1.7SI}$$

$$\Delta p_{stack} = -C_d \frac{\rho_i g}{g_c} \Delta h \frac{T_i - T_o}{T_o} \tag{6.1.7US}$$

where ρ_i = density of air in building = 0.075 lb_m/ft^3 (= 1.20 kg/m^3)
  Δh = vertical distance from neutral pressure level, up being positive, ft (m)
  g = 32.17 ft/s^2 (9.80 m/s^2) = acceleration due to gravity [g_c = 32.17 ($lb_m \cdot ft$)/($lb_f \cdot s^2$)]
  T_i and T_o = indoor and outdoor absolute temperatures, °R (K)
  C_d = draft coefficient, a dimensionless number to account for the resistance to air flow
       between floors

The draft coefficient ranges from about 0.65 for typical modern office buildings to 1.0 if there is no resistance at all. Equation 6.1.7 is plotted in Figure 6.1.6 as a function of $\Delta T = T_i - T_o$ and Δh, assuming air at 75°F (24°C). Since the relation is linear in Δh, this figure can be read outside the range shown by simply changing the scales of the axes. For a brief summary, one can say that the stack pressure amounts to

$$\frac{\Delta p_{stack}}{C_d \Delta h \Delta T} = 0.04 \quad Pa/(m \cdot K) \tag{6.1.8SI}$$

$$\frac{\Delta p_{stack}}{C_d \Delta h \Delta T} = 0.00014 \quad lb_f/(ft^2 \cdot ft \cdot °R) \tag{6.1.8US}$$

The stack effect tends to be relatively small in low-rise buildings, up to about five floors, but in high-rise buildings it can dominate and should be given close attention.

Finally, in buildings with *mechanical ventilation*, there is the pressure difference Δp_{vent} if the intake and exhaust flows are not equal. The resulting pressure difference depends on the design and operation of

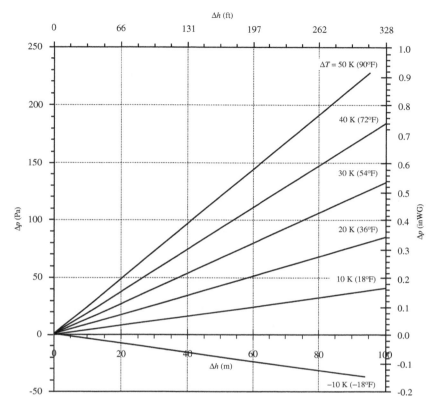

FIGURE 6.1.6 Pressure difference due to stack effect.

the ventilation system and on the tightness of the building. In addition, there is some coupling to the wind and stack terms. Thus, the determination of Δp_{vent} may be somewhat difficult. However, as we now show, the situation is simple when Δp_{vent} is larger in magnitude than the wind and stack terms. This is an important case since many designers aim for slight overpressurization of commercial buildings by making the outdoor air intake larger than the exhaust flow.

Consider how the pressures and flows are related in a fairly tight building where mechanical ventilation maintains overpressure or underpressure Δp_{vent} relative to the outside; for simplicity, we assume it uniform in the entire building. The law of conservation of mass implies that the net air flow provided by the ventilation system equals the net leakage $\dot{V}$ across the envelope, as calculated, according to Equation 6.1.2, by summing over all leakage sites k of the building envelope:

$$\dot{V} = \sum A_k c_k \Delta p_k^{n_k} \qquad \text{with} \qquad \Delta p_k = \Delta p_{wind,k} + \Delta p_{stack,k} + \Delta p_{vent} \qquad (6.1.9)$$

For simplicity, let us assume at this point one single value n for the exponent. Then the pressure term on the right-hand side of Equation 6.1.9 can be rewritten in the form

$$\Delta p_k^n = \Delta p_{vent}^n (1 + x_k)^n \qquad \text{with} \qquad x_k = \frac{\Delta p_{wind,k} + \Delta p_{stack,k}}{\Delta p_{vent}} \qquad (6.1.10)$$

TABLE 6.1.2   Stack Coefficient a_s

	Number of Stories		
	One	Two	Three
Stack coefficient a_s, $(\text{ft}^3/\text{min})^2/(\text{in}^4 \cdot {}^\circ\text{F})$	0.0156	0.0313	0.0471
Stack coefficient a_s, $(\text{L/s})^2/(\text{cm}^4 \cdot \text{K})$	0.000145	0.000290	0.000435

Source: From ASHRAE, 1989a.

As long as $|x_k| < 1$, the binomial expansion can be used, with the result[*]

$$\dot{V}_{vent} = \Delta p_{vent}^n \sum A_k c_k \left[1 + n x_k + \frac{n(n-1)}{2} x_k^2 + \dots \right]$$

The quantity x_k is positive in some parts of the building, negative in others. In fact, if the distribution of cracks is approximately symmetric (top-bottom and windward-leeward), then for each term with positive x_k there will also be one with approximately the same coefficient but negative x_k. Thus the linear terms in the expansion tend to cancel. The higher-order terms are small, beginning with x_k^2 which is multiplied by $n(n-1)/2$, a factor that is always less than 0.125 in absolute value since $0 < n < 1$. Thus the contributions of the x_k-dependent terms are much smaller than that of the leading term. Therefore, *if a building is pressurized to Δp_{vent} by mechanical ventilation and if wind and stack pressures are smaller than Δp_{vent} then it is indeed a fair approximation to neglect them altogether and write*

$$\dot{V}_{vent} \approx \Delta p^n \sum A_k c_k \qquad \text{with } \Delta p = \Delta p_{vent} \qquad (6.1.11)$$

the sum covering the entire envelope of the building. Had we allowed for different exponents n_k in Equation 6.1.9, Δp with its exponent would remain inside the sum, but the conclusion about the negligibility of stack and wind terms continues to hold.

LBL Model for Air Leakage

To apply Equation 6.1.2, one needs data for leakage areas and flow coefficients of all the components of a building. Much research has been done to obtain such data, both for components and for complete buildings, e.g., by pressurizing a building with a blower door. Data were presented in Table 6.1.1.

The total leakage area is obtained by adding all the leakage areas of the components as illustrated in the following example. Once the total leakage area has been found, either by such a calculation or by a pressurization test, the airflow $\dot{V}$ can be estimated by the following model, developed at Lawrence Berkeley Laboratory (LBL) as reported by ASHRAE (2001):

$$\dot{V} = A_{leak} \sqrt{a_s \Delta T + a_w v^2} \quad \text{L/s} \qquad (6.1.12)$$

where A_{leak} = total effective leakage area of building, cm^2
    a_s = stack coefficient of Table 6.1.2, $(\text{L/s})^2/(\text{cm}^4 \cdot \text{K})$
    ΔT = $T_i - T_o$, K
    a_w = wind coefficient of Table 6.1.3, $(\text{L/s})^2/(\text{cm}^4 \cdot (\text{m/s})^2]$
    v = average wind speed, m/s

This model is applicable to single-zone buildings *without* mechanical ventilation.

---

[*] The ith power of x_k in this series is multiplied by the binomial coefficient

$$\binom{n}{i} = \frac{n!}{i!(n-i)!}$$

TABLE 6.1.3 Wind Coefficient a_w

Shielding class	Description	Wind Coefficient a_w, $(L/s)^2/[cm^4 \cdot (m/s)^2]$			Wind Coefficient a_w, $(ft^3/min)^2/[in^4 \cdot (mi/h)^2]$		
		Number of Stories			Number of Stories		
		One	Two	Three	One	Two	Three
1	No obstructions or local shielding	0.000319	0.000420	0.000494	0.0119	0.0157	0.0184
2	Light local shielding; few obstructions, a few trees or small shed	0.000246	0.000325	0.000382	0.0092	0.0121	0.0143
3	Moderate local shielding; some obstructions within two house heights, thick hedge, solid fence, or one neighboring house	0.000174	0.000231	0.000271	0.0065	0.0086	0.0101
4	Heavy shielding; obstructions around most of perimeter, buildings or trees within 10 m in most directions; typical suburban shielding	0.000104	0.000137	0.000161	0.0039	0.0051	0.0060
5	Very heavy shielding; large obstructions; typical downtown shielding	0.000032	0.000042	0.000049	0.0012	0.0016	0.0018

Source: From ASHRAE, 1989a.

Further Correlations for Building Components

Most commercial buildings have features that are not included in the model of the previous section:

- Mechanical ventilation
- Revolving or swinging doors
- Curtain walls (i.e., the non load-bearing wall construction commonly employed in commercial buildings)

These features can be analyzed by using the method presented in the *Cooling and Heating Load Calculation Manual* (1979, 1992), published by ASHRAE. Here we give a brief summary. The air flows are determined from the following correlations, where Δp is the total pressure difference at each point of the building, calculated as described above.

For residential-type doors and windows, the flow per length l_p of perimeter is given by an equation of the form

$$\frac{\dot{V}}{l_p} = k(\Delta p)^n \qquad \dot{V}, \text{ ft}^3/\text{min}; \; l_p, \text{ ft}; \; \Delta p, \text{ in WG} \qquad (6.1.13\text{US})$$

Numerical values can be found in Figure 6.1.7 for windows and residential-type doors, and in Figure 6.1.8 for swinging doors when they are closed, as functions of the pressure difference Δp, for several values of k corresponding to different types of construction. The exponent n is 0.5 for residential-type doors and windows, and 0.65 for swinging doors. The larger values for n and for k for the latter account for the larger cracks of swinging doors. Since the equations are dimensional, all quantities must be used with the specified units. We have added dual scales to the graphs, so Figures 6.1.7–6.1.10 can be read directly in both systems of units.

Obviously the air flow increases markedly when doors are opened. Figure 6.1.9 permits an estimate of the air flow through swinging doors, both single-bank and vestibule-type, as a function of traffic rate. The coefficient C $[(ft^3/min)/(in\ WG)^{-0.5}]$ for Figure 6.1.9a is found from Figure 6.1.9b for the number of people passing the door per hour. The equation for the flow in part a is

$$\dot{V} = C(\Delta p)^{0.5} \qquad \dot{V}, \text{ ft}^3/\text{min}; \; \Delta p, \text{ in WG} \qquad (6.1.14\text{US})$$

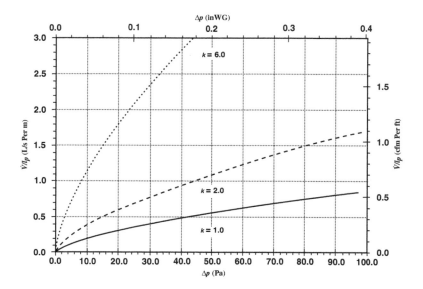

FIGURE 6.1.7   Window and residential-type door air infiltration $\dot{V}$ per perimeter length l_p. The curves correspond to Equation 6.1.13, with n = 0.65 and coefficient k [(ft³/min)/(inWG)$^{0.65}$] according to construction type, as shown in the following table:

Coefficient	Windows		Doors (Residential Type)
	Wood Double-Hung (locked)	Other Types	
k = 1.0 "tight"	Weather-stripped, small gap width 0.4 mm (1/64 in)	Weather-stripped: wood casement and awning windows, metal casement windows	Very small perimeter gap and perfect-fit weather-stripping — often characteristic of new doors
k = 2.0 "average"	Non weather-stripped, small gap width 0.4 mm (1/64 in); or weather-stripped, large gap width 2.4 mm (3/32 in)	All types of sliding windows, weather-stripped (if gap width is 0.4 mm, this could be "tight"; or non weather-stripped metal casement windows (if gap width is 2.4 mm, this could be "loose")	Small perimeter gap with stop trim, good fit around door, and weather-stripping
k = 6.0 "loose"	Non weather-stripped, large gap width 2.4 mm (3/32 in)	Non weather-stripped vertical and horizontal sliding windows	Large perimeter gap with poor-fitting stop trim and weather-stripping, or small perimeter gap without weather-stripping

Analogous information on flow per unit area of curtain wall can be determined by the equation

$$\frac{\dot{V}}{A} = K(\Delta p)^{0.65} \qquad \dot{V}, \text{ft}^3/\text{min}; A, \text{ft}^2; \Delta p, \text{in WG} \qquad (6.1.15\text{US})$$

It is presented in Fig. 6.1.10 for three construction types, corresponding to the indicated values of the coefficient K (ft³/min)/(ft² · inWG$^{0.65}$).

6.1.2   Principles of Load Calculations

Design Conditions

Loads depend on the indoor conditions that one wants to maintain and on the weather, the latter of which is not known in advance. If the HVAC equipment is to guarantee comfort at all times, it must be designed for peak conditions. What are the extremes? For most buildings it would not be practical to aim for total protection by choosing the most extreme weather on record and adding a safety margin.

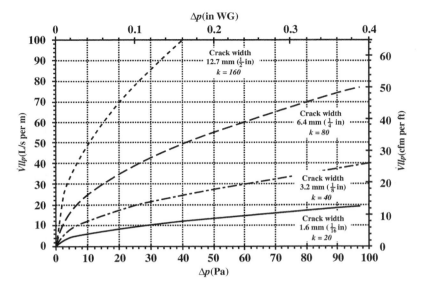

FIGURE 6.1.8 Infiltration through closed swinging door cracks, $\dot{V}$ per perimeter length l_p. The curves correspond to Equation 6.1.13 with n = 0.5 and k [(ft³/min)/(inWG)^0.5].

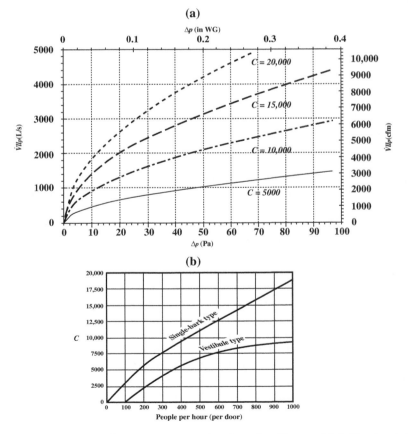

FIGURE 6.1.9 Infiltration due to door openings as a function of traffic rate. (a) Infiltration (with n = 0.5). (b) Coefficient C [(ft³/min)/(inWG)^0.5].

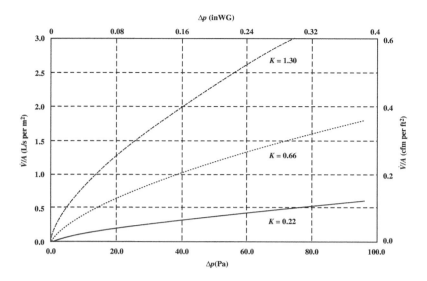

FIGURE 6.1.10   Infiltration per area of curtain wall for one room or one floor. The curves correspond to Equation 6.1.15, with coefficient k $[(ft^3/min)/(ft^2 \cdot inWG)^{0.65}]$ according to construction type.

Coefficient	Construction
K = 0.22 "tight"	Close supervision of workmanship; joints are redone when they appear inadequate
K = 0.66 "average"	Conventional
K = 1.30 "loose"	Poor quality control, or older building where joints have not been redone

Such oversizing of the HVAC equipment would be excessive, not just in first cost but also in operating cost; most of the time, the equipment would run with poor part-load efficiency. Therefore compromise is called for, reducing the cost of the HVAC equipment significantly while accepting the risk of slight discomfort under rare extremes of weather. The greater the extreme, the rarer the occurrence.

To help with the choice of design conditions, ASHRAE has published weather statistics corresponding to several levels of probability. They are the conditions that are exceeded at the site in question during a specified percentage of time of an average season. For warm conditions, the *ASHRAE Handbook of Fundamentals* lists design conditions for the 0.4%, 1.0%, and 2% levels. For cooling, these percentage probabilities refer to 12 months (8760 hours). To see what these statistics imply, consider Washington, D.C., where the 1% level for the dry-bulb temperature is 92°F (33°C). Here the temperature is above 92°F during 0.010 × 8760 h = 87.6 h of an average summer. Since the hottest hours are concentrated during afternoons rather than spread over the entire day, the corresponding number of days can be considerably higher than 1% of the 122 days of the summer season (June – September).

The cold weather design conditions are listed as 99.0% and 99.6% conditions because that is the percentage of the typical year when the temperature is *above* these levels.

A certain amount of judgment is needed in the choice of design conditions. For ordinary buildings, it is customary to base the design on the level of 1.0% in summer and 99.0% in winter. For critical applications such as hospitals or sensitive industrial processes, or for lightweight buildings, one may prefer the more stringent level of 0.4% in summer (99.6% in winter). Thermal inertia can help in reducing the risk of discomfort; it delays and attenuates the peak loads, as will be explained in the following sections. Therefore one may move to a less stringent level for a given application if the building is very massive.*

---

* As a guide for the assessment of the relation between persistence of cold weather and thermal inertia, we note that according to studies at several stations, temperatures below the design conditions can persist for up to a week (ASHRAE, 2001).

For peak heating loads, one need not bother with solar radiation because the extremes occur during winter nights. For cooling loads, solar radiation is crucial, but its peak values are essentially a function of latitude alone. For opaque surfaces, the effect of solar radiation is treated by means of the sol-air temperature, for glazing, by means of the solar heat gain factor. Design values of the solar heat gain factor for a set of surface orientations and latitudes can be found in (ASHRAE 2001).

As for humidity and latent loads, the ASHRAE tables include design wet-bulb temperatures, also at the 0.4%, 1.0%, and 2% levels along with the coincident dry-bulb temperature. Alternatively, the Tables also show the mean coincident wet-bulb temperatures, defined as the average wet-bulb temperature at the corresponding dry-bulb values (also at the 0.4%, 1.0%, and 2% levels). For winter no wet-bulb temperature data are given. Usually this poses no serious problem because latent loads during the heating season are zero if one does not humidify. If one does humidify, uncertainties in the value of the outdoor humidity have little effect on the latent load because the absolute humidity of outdoor air in winter is very low.

Wind speed is another weather-dependent variable that has a bearing on loads. Traditionally the ASHRAE (2001) value

$$v_{win} = 15 \text{ mi/h (6.7 m/s)} \tag{6.1.16a}$$

has been recommended for heating loads, if there is nothing to imply extreme conditions (such as an exposed hilltop location). For cooling loads, a value half as large is recommended

$$V_{sum} = 7.5 \text{ mi/h (3.4 m/s)} \tag{6.1.16b}$$

because wind tends to be less strong in summer than in winter. Of particular interest is the surface heat transfer coefficient (radiation plus convection) h_o for which ASHRAE (2001) recommends the design values.

$$h_{o,win} = 6.0 \text{ Btu/(h} \cdot \text{ft}^2 \cdot \text{°F)} \quad [34.0 \text{ W/(m}^2 \cdot \text{K)}] \tag{6.1.17a}$$

$$h_{o,sum} = 4.0 \text{ Btu/(h} \cdot \text{ft}^2 \cdot \text{°F)} \quad [22.7 \text{ W/(m}^2 \cdot \text{K)}] \tag{6.1.17b}$$

The better a building is insulated and tightened, the less its heat transmission coefficient depends on wind. With current practice for new construction in the U.S., typical wind speed variations may change the heat transmission coefficient by about 10% relative to the value at design conditions. Temperature and humidity for normal indoor activities should be within the comfort region delineated in Chapter 2.2. The comfort chart indicates higher indoor temperatures in summer than in winter because of the difference in clothing.

Building Heat Transmission Coefficient

One of the most important terms in the heat balance of a building is the heat flow across the envelope. As discussed in Chapter 2.1, heat flow can be assumed to be linear in the temperature difference when the range of temperatures is sufficiently small; this is usually a good approximation for heat flow across the envelope. Thus one can calculate the heat flow through each component of the building envelope as the product of its area A, its conductance U, and the difference $T_i - T_o$ between the interior and outdoor temperatures. The calculation of U (or its inverse, the R_{th} value) is described in Chapter 2.1. Here we combine the results for the components to obtain the total heat flow.

The total conductive heat flow from interior to exterior is

$$\dot{Q}_{cond} = \sum_k U_k A_k (T_i - T_o), \tag{6.1.18}$$

with the sum running over all parts of the envelope that have a different composition. It is convenient to define a total *conductive heat transmission coefficient* K_{cond}, or *UA* value, as

$$K_{cond} = \sum_k U_k A_k \qquad (6.1.19)$$

so that the conductive heat flow for the typical case of a single interior temperature T_i can be written as

$$\dot{Q}_{cond} = K_{cond}(T_i - T_o) \qquad (6.1.20)$$

In most buildings, the envelope consists of a large number of different parts; the greater the desired accuracy, the greater the amount of detail to be taken into account.

As a simplification, one can consider a few major groups and use effective values for each. The three main groups are glazing, opaque walls, and roof. The reason for distinguishing the wall and the roof lies in the thickness of the insulation; roofs tend to be better insulated because it is easier and less costly to add extra insulation there than in the walls. With these three groups one can write

$$K_{cond} = U_{glass} \cdot A_{glass} + U_{wall} \cdot A_{wall} + U_{roof} \cdot A_{roof} \qquad (6.1.21)$$

if one takes for each the appropriate effective value. For instance, the value for glazing must be the average over glass and framing.

In the energy balance of a building, there is one other term that is proportional to $T_i - T_o$. It is the flow of sensible heat [in watts (Btu/h)] due to air exchange:

$$\dot{Q}_{air} = \rho c_p \dot{V}(T_i - T_o) \qquad (6.1.22)$$

where ρ = density of air
        c_p = specific heat of air
        $\dot{V}$ = air exchange rate ft³/h (m³/s)

At standard conditions, 14.7 psia (101.3 kPa) and 68°F (20°C), the factor ρc_p has the value

$$\rho c_p = 0.018 \text{ Btu/(ft}^3 \cdot °\text{F)} \qquad [1.2 \text{ kJ/(m}^3 \cdot \text{K})] \qquad (6.1.23)$$

In USCS units, if $\dot{V}$ is in cubic feet per minute, it must be converted to ft³/h by multiplying by 60 (ft³/h)/(ft³/min). It is convenient to combine the terms proportional to $T_i - T_o$ by defining the total heat transmission coefficient K_{tot} of the building as the sum of conductive and air change terms:

$$K_{tot} = K_{cond} + \rho c_p \dot{V} \qquad (6.1.24)$$

A more refined calculation would take surface heat transfer coefficients into account, as well as details of the construction. In practice, such details can take up most of the effort.

Heat Gains

Heat gains affect both heating and cooling loads. In addition to familiar solar gains, there are heat gains from occupants, lights, and equipment such as appliances, motors, computers, and copiers. Power densities for lights in office buildings are around 20–30 W/m². For lights and for resistive heaters, the nominal power rating (i.e., the rating on the label) is usually close to the power drawn in actual use. But for office equipment, that would be quite misleading; the actual power has been measured to be much lower, often by a factor of 2 to 4 (Norford et al., 1989). Some typical values are indicated in Table 6.1.4. In recent years, the computer revolution has brought a rapid increase in electronic office equipment, and the impact on loads has become quite important, comparable to lighting. The energy consumption for office equipment is uncertain — will the occupants turn off the computers between uses or keep them running nights and weekends?

TABLE 6.1.4 Typical Heat Gain Rates for Several Kinds of Equipment

Equipment	Heat Gain Btu/h	Heat Gain W	Comments
Television set	170–340	50–100	
Refrigerator	340–680	100–200	Recent models more efficient
Personal computer (desktop)	170–680	50–200	Almost independent of use while turned on
Impact printer	34–100	10–30 standby	Increases about twofold during printing
Laser printer	510	150 standby	Increases about twofold during printing
Copier	500–1000	150–300 standby	Increases about twofold during printing

Note: Measured values are often less than half of the nameplate rating.
Source: Based on ASHRAE, 1989a, Norford et al., 1989, and updates.

TABLE 6.1.5 Nominal Heat Gain Values from Occupants

Activity	Total Btu/h	Total W	Sensible Btu/h	Sensible W	Latent Btu/h	Latent W
Seated at rest	340	100	240	70	100	30
Seated, light office work	410	120	255	75	150	45
Standing or walking slowly	495	145	255	75	240	70
Light physical work	850	250	310	90	545	160
Heavy physical work	1600	470	630	185	970	285

Source: Based on ASHRAE, 1989a.

For special equipment such as laboratories or kitchens, it is advisable to estimate the heat gains by taking a close look at the inventory of the equipment to be installed, paying attention to the possibility that much of the heat may be drawn directly to the outside by exhaust fans.

Heat gain from occupants depends on the level of physical activity. Nominal values are listed in Table 6.1.5. It is instructive to reflect on the origin of this heat gain. The total heat gain must be close to the caloric food intake, since most of the energy is dissipated from the body as heat. An average of 100 W corresponds to

$$100 \text{ W} = 0.1 \text{ kJ/s} \times \left(\frac{L \, kcal}{4.186 \, kJ} \right) \times (24 \times 3600 \text{ s/day}) = 2064 \text{ kcal/day} \tag{6.1.25}$$

indeed a reasonable value compared to the typical food intake (note that the dieticians' calorie is really a kilocalorie). The latent heat gain must be equal to the heat of vaporization of the water that is exhaled or transpired. Dividing 30 W by the heat of vaporization of water, we find a water quantity of 30 W/(2450 kJ/kg) = 12.2 × 10⁻⁶ kg/s, or about 1.1 kg/24 h. That also appears quite reasonable.

The latent heat gain due to the air exchange is

$$\dot{Q}_{air,lat} = \dot{V}\rho h_{fg} \left(W_o - W_i \right) \tag{6.1.26}$$

where $\dot{V}$ = volumetric air exchange rate, ft³/min (m³/s or L/s)
ρ = density, lb$_m$/ft³ (kg/m³)
ρh_{fg} = 4840 Btu/(h · ft³/min) [3010 W/(L/s)] at standard conditions
W_i, W_o = humidity ratios of indoor and outdoor air

Heat Balance

Loads are the heat that must be supplied or removed by the HVAC equipment to maintain a space at the desired conditions. The calculations are like accounting. One considers all the heat that is generated in the space or that flows across the envelope; the total energy, including the thermal energy stored in

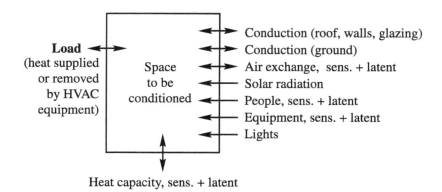

FIGURE 6.1.11   The terms in a load calculation.

the space, must be conserved according to the first law of thermodynamics. The principal terms are indicated in Figure 6.1.11. Outdoor air, occupants, and possibly certain kinds of equipment contribute both sensible and latent heat terms.

Load calculations are straightforward in the static limit, i.e., if all input is constant. As discussed in the following section, that is usually an acceptable approximation for the calculation of peak heating loads. But for cooling loads, dynamic effects (i.e., heat storage) must be taken into account because some of the heat gains are absorbed by the mass of the building and do not contribute to the loads until several hours later. Dynamic effects are also important whenever the indoor temperature is allowed to float.

Sometimes it is appropriate to distinguish several aspects of the load. If the indoor temperature is not constant, the instantaneous load of the space may differ from the rate at which heat is being supplied or removed by the HVAC equipment. The load for the heating or cooling plant is different from the space load if there are significant losses from the distribution system, or if part of the air is exhausted to the outside rather than being returned to the heating or cooling coil.

It is convenient to classify the terms of the static energy balance according to the following groups. The sensible energy terms are

1. Conduction through the building envelope other than ground,

$$\dot{Q}_{cond} = K_{cond}\left(T_i - T_o\right) \tag{6.1.27}$$

2. Conduction through the floor, $\dot{Q}_{floor}$
3. Heat due to air exchange (infiltration and/or ventilation), at rate $\dot{V}$,

$$\dot{Q}_{air} = \dot{V}\rho c_p\left(T_i - T_o\right) \tag{6.1.28}$$

4. Heat gains from solar radiation, lights, equipment (appliances, computers, fans, etc.), and occupants,

$$\dot{Q}_{gain} = \dot{Q}_{sol} + \dot{Q}_{lit} + \dot{Q}_{equ} + \dot{Q}_{occ} \tag{6.1.29}$$

Combining the heat loss terms and subtracting the heat gains, one obtains the total *sensible load*

$$\dot{Q} = \dot{Q}_{cond} + \dot{Q}_{air} + \dot{Q}_{floor} - \dot{Q}_{gain} + \dot{Q}_{stor} \tag{6.1.30}$$

where we have added a term $\dot{Q}_{stor}$ on the right to account for storage of heat in the heat capacity of the building (the terms *thermal mass* and *thermal inertia* are also used to designate this effect). A dynamic analysis includes this term; a static analysis neglects it.

We have kept $\dot{Q}_{floor}$ as a separate item because it should not be taken proportional to $T_i - T_o$ except in cases such as a crawl space or uninsulated slab on grade, where the floor is in fairly direct contact with outside air. More typical is conduction through massive soil (see Chuangchid and Krarti, 2000; Kreider, Rabl, and Curtiss, 2001) are appropriate. In traditional construction, the floor term has usually been small, and often it has been neglected altogether. But in superinsulated buildings it can be relatively important.

Using the total heat transmission coefficient K_{tot},

$$K_{tot} = K_{cond} + \dot{V}\rho c_p \tag{6.1.31}$$

one can write the sensible load in the form

$$\dot{Q} = K_{tot}\left(T_i - T_o\right) + \dot{Q}_{floor} - \dot{Q}_{gain} \pm \dot{Q}_{stor} \tag{6.1.32}$$

For signs, we take the convention that $\dot{Q}$ is positive when there is a heating load and negative when there is a cooling load. Sometimes, however, we prefer a positive sign for cooling loads. In that case, we will add subscripts *c* and *h* with the understanding that

$$\dot{Q}_c = -\dot{Q} \quad \text{and} \quad \dot{Q}_h = \dot{Q} \tag{6.1.33}$$

The latent heat gains are mainly due to air exchange, equipment (such as in the kitchen and bathroom), and occupants. Their sum is

$$\dot{Q}_{lat} = \dot{Q}_{lat,air} + \dot{Q}_{lat,occ} + \dot{Q}_{lat,equ} \tag{6.1.34}$$

The total load is the sum of the sensible and the latent loads.

During the heating season, the latent gain from air exchange is usually negative (with the signs of Equation 6.1.26) because the outdoor air is relatively dry. A negative Q_{lat} implies that the total heating load is greater than the sensible heating load alone — but this is relevant only if there is humidification to maintain the specified humidity ratio W_i. For buildings without humidification, one has no control over W_i, and there is not much point in calculating the latent contribution to the heating load at a fictitious value of W_i.

6.1.3 Storage Effects and Limits of Static Analysis

The storage term $\dot{Q}_{stor}$ in Equation 6.1.32 is the rate of heat flow into or out of the mass of the building, including its furnishings and even the air itself. The details of the heat transfer depend on the nature of the building, and they can be quite complex.

One of the difficulties can be illustrated by considering an extreme example: a building that contains in its interior a large block of solid concrete several meters thick. The conductivity of concrete is relatively low, and diurnal temperature variations do not penetrate deeply into the block; only in the outer layer, to a depth of roughly 0.20 m, are they appreciable. Thus the bulk of the block does not contribute any storage effects on a diurnal time scale. The static heat capacity, defined as the product of the mass and the specific heat, would overestimate the storage potential because it does not take into account the temperature distribution under varying conditions.

To deal with this effect, some people (e.g., Sonderegger, 1978) have used the concept of *effective heat capacity* C_{eff}, defined as the periodic heat flow into and out of a body divided by the temperature swing at the surface. It depends on the rate of heat transfer and on the frequency. The effective heat capacity is smaller than the static heat capacity, approaching it in the limit of infinite conductivity or infinitely long charging and discharging periods. As a rule of thumb, for diurnal temperature variations, the effective heat capacity of walls, floors, and ceilings is roughly 40 to 80% of the static heat capacity, assuming typical construction of buildings in the U.S. (wood, plaster, or concrete, 3–10 cm thick). For items, such as furniture, that are thin relative to the depth of temperature variations, the effective heat capacity approaches the full static value.

Further complications arise from the fact that the temperatures of different parts of the building are almost never perfectly uniform. For instance, sunlight entering a building is absorbed by the floor, walls, and furniture and raises their temperature. The air itself does not absorb any appreciable solar radiation and is warmed only indirectly. Thus the absorbed radiation can cause heat to flow from the building mass to the air, even if the air is maintained thermostatically at uniform and constant temperature.

Heat capacity tends to be more important for cooling than for heating loads, for a number of reasons. Summer heat flows are more peaked than those in winter. Peak heating loads correspond to times without sun and the diurnal variation of $T_i - T_o$ is small compared to its maximum in most climates. By contrast, for peak cooling loads, the diurnal variation of $T_i - T_o$ is comparable to its maximum, and solar gains are crucial. Also, in climates with cold winters, heating loads are larger than cooling loads, and the storage terms, for typical temperature excursions, are relatively less important in winter than in summer.

Consequently, the traditional steady-state calculation of peak heating loads was well justified for buildings with a constant thermostat setpoint. However, thermostat setback can have a sizable impact on peak heating loads because setback recovery occurs during the early morning hours on top of the peak heat loss; section 6.1.5 discusses this point in more detail. Peak cooling loads, by contrast, are usually not affected by thermostat setup because recovery is not coincident with the peak gains.

Storage effects for latent loads are difficult to analyze (see e.g., Fairey and Kerestecioglu, 1985; and Kerestecioglu and Gu, 1990), and most of the current computer programs for building simulation, such as DOE2.1 and BLAST (see Chapter 6.2), do not account for moisture exchange with the building mass. In practice, this neglect of moisture storage is usually not a serious problem. Precise humidity control is not very important in most buildings. Where it is important, e.g., in hospitals, temperature and humidity are maintained constant around the clock. When the air is at constant conditions, the moisture in the materials does not change much and storage effects can be neglected; but in buildings with intermittent operation, these effects can be large, as shown by Fairey and Kerestecioglu (1985) and by Wong and Wang (1990).

From this discussion emerge the following recommendations for the importance of dynamic effects:

- They can significantly reduce the peak cooling loads, with or without thermostat setup.
- They can be neglected for peak heating loads, except if thermostat setback recovery is to be applied even during the coldest periods of the year.
- For the calculation of annual consumption, they can have an appreciable effect if the indoor temperature is not kept constant.

Storage of *latent* heat is neglected for most applications.

Thus a simple static analysis is sufficient for some of the problems the designer is faced with, but not for the peak cooling load. To preserve much of the simplicity of the static approach in a method for peak cooling loads, ASHRAE has developed the CLF-CLTD method which modifies the terms of a static calculation to account for thermal inertia. This method is presented in section 6.1.6; it can be used for standard construction if the thermostat setpoint is constant.

6.1.4   Zones

So far we have considered the interior as a single zone at uniform temperature — a fair approximation for simple houses, for certain buildings without windows (such as warehouses), or for buildings that are dominated by ventilation. But in large or complex buildings, one usually has to calculate the loads separately for a number of different zones. There may be several reasons. An obvious case is a building where different rooms are maintained at different temperatures, e.g., a house with an attached sunspace.

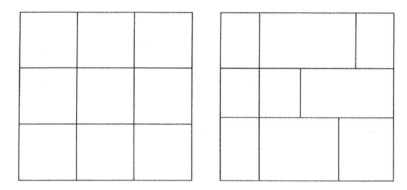

FIGURE 6.1.12 Variation of schematic floor plan to show that surface area of interior walls is independent of arrangement.

Here the heat balance equation is written for each zone, in the form of Equation 6.1.32 but with an additional term

$$\dot{Q}_{j-k} = U_{j-k} A_{j-k} \left(T_j - T_k \right) \tag{6.1.35}$$

for the heat flow between zones j and k.

However, even when the entire building is kept at the same temperature, multizone analysis becomes necessary if the spatial distribution of heat gains is too nonuniform. Consider, for example, a building with large windows on the north and south sides, during a sunny winter day when the gains just balance the total heat loss. Then neither heating nor cooling would be required, according to a one-zone analysis. But how can the heat from the south get to the north?

Heat flow is the product of the heat transfer coefficient and the temperature difference, as in Equation 6.1.35. Temperature differences between occupied zones are small, usually not more than a few Kelvins; otherwise there would be complaints about comfort. The heat transfer coefficients between zones are often not sufficiently large for effective redistribution of heat, especially if there are walls or partitions.

The basic criterion for zoning is the ability to control the comfort conditions; the control is limited by the number of zones one is willing to consider. To guarantee comfort, the HVAC plant and distribution system must be designed with sufficient capacity to meet the load of each zone. In choosing the zones for a multizone analysis, the designer should try to match the distribution of heat gains and losses. A common and important division is between interior and perimeter zones, because the interior is not exposed to the changing environment. Different facades of the perimeter should be considered separately for cooling load calculations, as suggested in Figure 6.1.12. Corner rooms should be assigned to the facade with which they have the most in common; usually this will be the facade where a comer room has the largest windows. Corner rooms are often the critical rooms in a zone, requiring more heating or cooling (per unit floor area) than single-facade rooms of the same zone while being most distant from AHUs.

Actually there are different levels to a zoning analysis, corresponding to different levels of the HVAC system. In an air system, there are major zones corresponding to each air handler. Within each air handler zone, the air ducts, air outlets, and heating or cooling coils must have sufficient capacity and sufficient controllability to satisfy the loads of each sub zone; the design flow rates for each room are scaled according to the design loads of the room. For best comfort (and if cost were no constraint), each zone should have its own air handler and each room its own thermostat. There is a tradeoff between equipment cost and achievable comfort, and the best choice depends on the circumstances. If temperature control is critical, one installs separate air handlers for each of the five zones in Figure 6.1.12 and separate thermostats for each room. To save equipment cost, one often assigns several zones to one air handler and

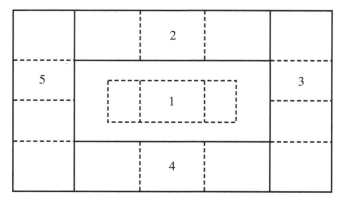

FIGURE 6.1.13   Example of recommended zoning. Thick lines represent zones, labeled 1 through 5. Dashed lines represent subzones.

several rooms to one thermostat, but the more divergent the loads, the more problematic the control. For the building of Figure 6.1.12, a single air handler and five thermostats may be adequate if the distribution of heat gains is fairly uniform and if the envelope is well insulated with good control of solar gains.

Another example is a house whose air distribution system has a single fan (typical of all but the largest houses). Even though there is only one major zone, the detailed design of the distribution system demands some attention to subzones. Within each room, the peak heating capacity should match the peak heat loss. Also, it is advisable to place heat sources close to points with large heat loss, i.e., under windows (unless they are highly insulating).

The choice of zones is not always clear-cut, and the design process may be iterative. Depending on the distribution of gains and losses, one may want to assign several rooms to a zone, one room to a zone, or even several zones to a room (if it is very large). With finer zonal detail one improves the control of comfort, but at the price of greater calculational effort and higher HVAC system cost. In an open office space, there is no obvious boundary between interior and perimeter; here a good rule is to make the perimeter zone as deep as the penetration depth of direct solar radiation, typically a few meters. Spaces connected by open doors, e.g., offices and adjacent hallways, can sometimes be treated as a single zone. Separate zones are advisable for rooms with large computers or energy-intensive equipment. In multistory buildings, one may want to treat the top floor apart from the rest.

The calculation of peak heating loads and capacities can often be done without considering separate perimeter zones, because peak heating loads occur when there is no sun; with uniform internal gains, the corresponding thermal balance is uniform around the perimeter. But while the calculation can be carried out for a single zone, the operation requires multiple zones: The heating system must allow separate control of different facades to compensate for the variability of solar gains during the day. For cooling loads, a multizone analysis is essential, because the peak loads occur when the sun is shining.

As discussed in the previous section, peak cooling loads require a dynamic analysis whereas peak heating loads can be estimated quite well by static models (at least in the absence of thermostat setback). Compared to heating loads, the calculation of cooling loads of large buildings is thus doubly complicated: It requires multiple zones and dynamic analysis if one wants reasonable accuracy.

A related issue is the coincidence between peak loads of different zones. To determine the capacity of the central plant, one needs to know the peak load of the totality of zones served by the plant. This is usually less than the simple sum of the individual peak loads because of noncoincidence. The term diversity is used to designate the ratio of the actual system peak to the sum of the individual peak loads. In practice, one often finds a diversity around 0.6 to 0.8 for large buildings or groups of buildings (e.g., university campuses); for better estimates at the design stage, computer simulations are recommended (see Chapter 6.2).

6.1.5 Heating Loads

Since the coldest weather may occur during periods without solar radiation, it is advisable not to rely on the benefit of solar heat gains when calculating peak heating loads (unless the building contains long-term storage). If the indoor temperature T_i is constant, a static analysis is sufficient and the calculation of the peak heating load $\dot{Q}_{h,max}$ is very simple. Find the design heat loss coefficient K_{tot}, multiply by the design temperature difference $T_i - T_o$, and subtract the internal heat gains on which one can count during the coldest weather

$$\dot{Q}_{h.max} = K_{tot}\left(T_i - T_o\right) - \dot{Q}_{gain} \tag{6.1.36}$$

to find the design heat load. However, it is also necessary to warm a space that has had night setback. In a given situation, the required extra capacity, called the *pickup load*, depends on the amount of setback $T_i - T_o$, the acceptable recovery time, and building construction. For reasonable accuracy, a dynamic analysis is recommended. Optimizing the capacity of the heating system involves a tradeoff between energy savings and capacity savings, with due attention to part load efficiency. As a general rule for residences, ASHRAE (1989a) recommends oversizing by about 40% for a night setback of 10°F (5.6 K), to be increased to 60% oversizing if there is additional setback during the day. In any case, some flexibility can be provided by adapting the operation of the building. If the heating capacity turns out insufficient, one can reduce the depth and duration of the setback during the very coldest periods.

In commercial buildings with mechanical ventilation, the demand for extra capacity during setback recovery is reduced if the outdoor air intake is closed during unoccupied periods. In winter that should always be done for energy conservation (unless air quality problems demand high air exchange at night).

6.1.6 CLTD/CLF Method For Cooling Loads[*]

Because of thermal inertia, it is advisable to distinguish several heat flow rates. The heat gain is the rate at which heat is transferred to or generated in a space. The cooling load is the rate at which the cooling equipment would have to remove thermal energy from the air in the space in order to maintain constant temperature and humidity. Finally, the heat extraction rate is the rate at which the cooling equipment actually does remove thermal energy from the space.[**]

Conductive heat gains and radiative heat gains do not enter the indoor air directly; rather they pass through the mass of the building, increasing its temperature relative to the air. Only gradually are they transferred to the air. Thus their contribution to the cooling load is delayed, and there is a difference between heat gain and cooling load. Averaged over time, these rates are, of course, equal, by virtue of the first law of thermodynamics.

The heat extraction rate is equal to the cooling load only if the temperature of the indoor air is constant (as assumed in this section). Otherwise, the heat flow to and from the building mass causes the heat extraction rate to differ from the cooling load.

ASHRAE, which sets standard load calculation procedures, is in a transition period regarding its load estimation as this book goes to press. Therefore, what follows is the long-standing CLTD/CLF method used for at least two decades by HVAC engineers. The final section of this chapter summarizes the most recent developments in load calculation procedures even though not all are finalized.

To account for transient effects without having to resort to a full-fledged dynamic analysis, a special shorthand method has been developed that uses the cooling load temperature difference (CLTD) and

---

[*] An updated version of the CLTD/CLF method, the CLTD/SCL/CLF method, is described in Section 6.1.8.

[**] In Chapter 4.3 on HVAC systems, we encountered yet another rate, the coil load; it is the rate at which the cooling coil removes heat from the air, and it can be different from the heat extraction rate due to losses in the distribution system.

cooling load factor (CLF). To explain the principles, note that the cooling load due to conduction across an envelope element of area A and conductance U would be simply

$$\dot{Q}_{cond} = UA(T_o - T_i) \tag{6.1.37}$$

under static conditions, i.e., if indoor temperature T_i and outdoor temperature T_o were both constant. When the temperatures vary, this is no longer the case because of thermal inertia. But if the temperatures follow a periodic pattern, day after day, $\dot{Q}_{c,cond}$ will also follow a periodic pattern. Once $\dot{Q}_{c,cond}$ has been calculated, one can define a CLTD as the temperature difference that gives the same cooling load when multiplied by UA. If such temperature differences are tabulated for typical construction and typical temperature patterns, they can be looked up for quick determination of the load. Thus the conductive cooling load is

$$\dot{Q}_{cond} = UA \cdot CLTD_t \tag{6.1.38}$$

where the subscript t indicates the hour t of the day.

Likewise, if there is a constant radiative heat gain in a zone, the corresponding cooling load is simply equal to that heat gain. If the heat gain follows a periodic pattern, the cooling load also follows a periodic pattern. The cooling load factor (CLF) is defined such that it yields the cooling load at hour t when multiplied by the daily maximum $\dot{Q}_{max}$ of the heat gain:

$$\dot{Q}_{c,rad,t} = \dot{Q}_{max} \cdot CLF_t \tag{6.1.39}$$

The CLFs account for the fact that radiative gains (solar, lights, etc.) are first absorbed by the mass of the building, becoming a cooling load only as they are being transferred to the air. Only convective gains can be counted as cooling load without delay. Some heat gains, e.g., from occupants, are partly convective and partly radiative; the corresponding CLFs take care of that.

The CLTDs and CLFs of ASHRAE have been calculated by means of the transfer functions discussed in the next section. To keep the bulk of numerical data within reasonable limits, only a limited set of standard construction types and operating conditions has been considered. Some correction factors are provided to extend the applicability, however, without escaping the constraint that the indoor temperature T_i be constant.

If one has to do a CLTD/CLF calculation by hand, it is advisable to use a worksheet such as the one reproduced in Figure 6.1.14 to make sure that nothing is overlooked[*]. The calculation needs to be done for the hour when the peak occurs. That hour can be guessed if a single load dominates because in that case it is the hour with the largest value of CLTD or CLF. If several loads with noncoincident peaks are of comparable importance, the hour of the combined peak may not be entirely obvious, and the calculation may have to be repeated several times. In most buildings, peak cooling loads occur in the afternoon or early evening. Figures 6.1.15 to 6.1.18 give an indication when the components of the cooling load are likely to reach their peak.

The steps of the calculation are summarized in the worksheet of Figure 6.1.14. We now proceed to discuss these steps, illustrating them by filling out the worksheet for a zone of an office building. The procedure has to be carried out for each zone of the building.

For *walls* and *roofs*, the conductive cooling load at time t is calculated by inserting the appropriate CLTD into Equation 6.1.38. We have plotted CLTD versus time for three roof types in Figure 6.1.15. The heavier the construction, the smaller the amplitude and the later the peak. Figure 6.1.16 shows analogous results for sunlit walls having the four cardinal orientations.

---

[*] The tables of CLTD and CLF values are too voluminous to include in this book. They are included in the "HCB Software" available from Kreider & Associates (jfk@well.com) in electronic form or from ASHRAE (1989a) in tabular form.

Job ID	Date			Initials	
Site	Latitude			Longitude	
Design conditions	Indoor temp.		Rel. humid.	Outdoor temp.	Rel. humid.
Room	Identification			Dimensions	

Latent loads					Instantaneous
	$\dot{V}$	W_o	W_i	$\Delta W = W_o - W_i$	$\dot{Q}_{lat} = p \times h_{fg} \times \dot{V} \times \Delta W$
Air exchange					
	N = number			$\dot{Q}_{lat/unit}$	$\dot{Q}_{lat} = N \times \dot{Q}_{lat/unit}$
Appliances					
People					
TOTAL LATENT					

Sensible loads					hour t	hour t
Component and orientation	Construction type		U	A	$CLTD_t$	$\dot{Q}_t = U \times A \times CLTD_t$
Walls						
Roof						
Glazing conduction						
Glazing solar		A	SC	$SHGF_{max}$	CLF_t	$\dot{Q}_t = A \times SC \times SHGF_{max} \times CLF_t$
Air exchange		V	$\dot{V}$	T_i	T_o	$\dot{Q} = p \times c_p \times \dot{V} \times (T_o - T_i)$ (instantaneous)
Internal partitions			U	A	ΔT across partition	$\dot{Q} = U \times A \times \Delta T$ (instantaneous)
Ceiling						
Floor						
Sides						
Ducts						
Internal gains		num-ber	gain /unit	$\dot{Q}$	CLF_t	$\dot{Q}_t = \dot{Q} \times CLF_t$
Appliances						
Fans						
Lights						
Motors						
People						
TOTAL SENSIBLE						

FIGURE 6.1.14 Worksheet for CLTD/CLF method for a specific zone. At sea level ρc_p = 1.08 Btu/(h · °F)]/(ft³/min) [1.2 (W/K)/(L/s)] and ρh_{fg} = 4840 (Btu/h)/(ft³/min) [3010 W/(L/s)].

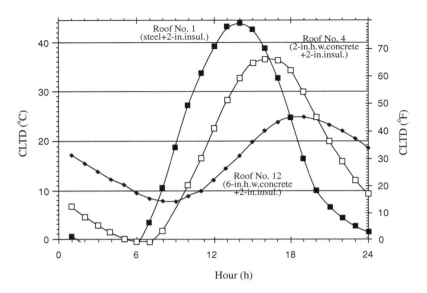

FIGURE 6.1.15   CLTDs for three roof types.

For these CLTDs, the following conditions have been assumed:

- High absorptivity for solar radiation ("dark").
- Solar radiation for 40°N on July 21.
- $T_i = 25.5°C$ (78°F).
- T_o has a mean of $T_{o,av} = (T_{o,max} + T_{o,min})/2 = 29.4°C$ (85.0°F) and a daily range = $T_{o,max} - T_{o,min} = 11.7°C$ (21.0°F), with $T_{o,max} = 35.0°C$ (95.0°F) being the design temperature.
- Outdoor convective heat transfer coefficient ho = 17 W/(m² · K) [3.0 Btu/(h · ft² · °F)].
- Indoor convective heat transfer coefficient hi = 8.3 W/(m² · K) [1.46 Btu/(h · ft² · °F)].
- No forced ventilation or air ducts in the ceiling space.

When conditions are different, one should correct the CLTDs according to the formula

$$\text{CLTD}_{cor} = (\text{CLTD} + \text{LM})K + (25.5°C - T_i) + (T_{o,av} - 29.4°C) \qquad (6.1.40\text{SI})$$

$$\text{CLTD}_{cor} = (\text{CLTD} + \text{LM})K + (78°C - T_i) + (T_{o,av} - 85°C) \qquad (6.1.40\text{US})$$

where LM   = correction factor for latitude and month
       K = color adjustment factor
       $T_i, T_{o,av}$ = actual values for application

And $T_{o,av}$ is obtained by subtracting 0.50 × daily range from $T_{o,max}$, the design temperature of the site. The color correction K is 1.0 for dark and 0.5 for light surfaces; values less than 1.0 should be used only when one is confident that the surface will permanently maintain low absorptivity.

How about other construction types? Two factors are affected: the U value and the CLTD. One should always use the correct U value for the actual construction in Equation 6.1.38. As for the CLTD, one should select the construction type that is closest in terms of mass and heat capacity.

For *windows*, one treats conductive and solar heat gains separately, according to the decomposition:

Heat gain through glass = conduction due to $T_i - T_o$

$$+ \text{ heat gain due to solar radiation transmitted through or absorbed by glass} \qquad (6.1.41)$$

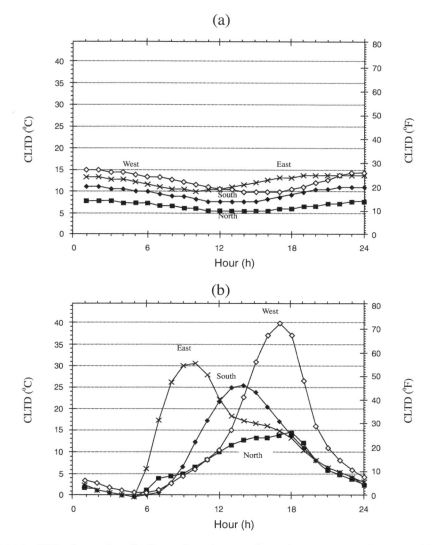

FIGURE 6.1.16 CLTDs for sunlit walls. Four orientations are shown for two construction types: (a) Group A walls [heavy, for example, 8-in (200 mm) concrete with insulation]; (b) Group G walls (light, for example, frame or curtain wall).

The conductive part is calculated as in Equation 6.1.38:

$$\dot{Q}_{c,cond,glaz,t} = UA \cdot CLTD_{glaz,t} \qquad (6.1.42)$$

Solar gains through windows are treated by means of the solar heat gain factor SHGF. It is defined as the instantaneous heat gain [Btu/(h · ft²)(W/m²)] due to solar radiation through reference glazing. There are two components in this solar gain: the radiation absorbed in the glass and the radiation transmitted through the glass. The latter is assumed to be totally absorbed in the interior of the building, a reasonable assumption in view of the cavity effect. The radiation absorbed in the glass raises its temperature, thereby changing the conductive heat flow. The SHGF combines this latter contribution with the radiation transmitted to the interior. For glazing types other than the reference glazing, one multiplies the SHGF by the shading coefficient SC.

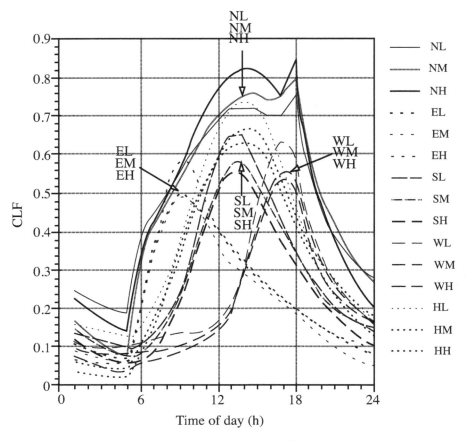

FIGURE 6.1.17 Cooling load factors for glass without interior shading for 5 orientations (E = east, S = south, W = west, N = north, H = horizontal) and 3 construction types (L = light, M = medium, H = heavy).

To calculate the contribution to the cooling load, the daily maximum of the solar heat gain is multiplied by the cooling load factor. Thus the actual cooling load at time t due to solar radiation is given by the formula

$$\dot{Q}_{c,sol,t} = A \cdot SC \cdot SHGF_{max} \cdot CLF_t \tag{6.1.43}$$

where A = area, ft^2 (m^2)
        SC = shading coefficient
        $SHGF_{max}$ = maximum solar heat gain factor, Btu/(h $\cdot$ ft^2) (W/m^2)
        CLF_t = cooling load factor for time t

SHGFmax is the value of SHGF at the hour when the radiation attains its maximum for a particular month, orientation, and latitude. The CLF takes into account the variation of the solar radiation during the day, as well as the dynamics of its absorption in the mass of the building and the gradual release of this heat. A separate set of CLFs is available (ASHRAE, 1989a) for each orientation and for each of three construction types, characterized in terms of the mass of building material per floor area: light = 30 lb/ft^2 (146 kg/m^2), medium = 70 lb/ft^2 (341 kg/m^2), and heavy = 130 lb/ft^2 (635 kg/m^2). Each set comprises all hours from 1 to 24. A subset is plotted in Figure 6.1.17.

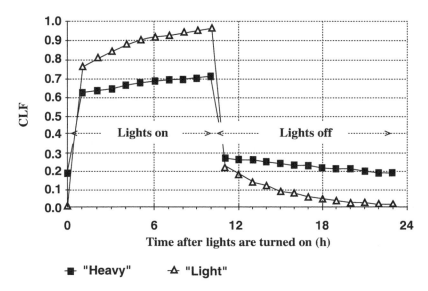

FIGURE 6.1.18 Cooling load factors for lights that are on 10 h/day, for two extreme construction types: light and heavy (a coefficient and b classification, as per Table 6.1.14A, are 0.45 D for heavy and 0.75 A for light).

In analogous fashion, CLFs have been computed for heat gains from *internal heat sources*. There are different factors for each of the three major categories:

Occupants

$$\dot{Q}_{occ,t} = \dot{Q}_{occ}\text{CLF}_{occ,t} \qquad (6.1.44)$$

Lights

$$\dot{Q}_{lit,t} = \dot{Q}_{lit}\text{CLF}_{lit,t} \qquad (6.1.45)$$

Equipment such as appliances

$$\dot{Q}_{app,t} = \dot{Q}_{app}\text{CLF}_{app,t} \qquad (6.1.46)$$

In these equations, the $\dot{Q}$ Btu/h (W) on the right side is the rate of heat production, assumed constant for a certain number of hours and zero the rest of the time. The $\dot{Q}_t$ on the left side is the resulting cooling load at hour t, for t from 1 to 24. For each load profile there is a different set of CLFs.

Figure 6.1.18 plots a set of CLFs for lights. The lights are on for 10 h/day, and the time axis shows the number of hours after the lights have been turned on. Two construction types are shown, representing the highest and the lowest thermal inertia in the tables. Once again, the CLF curves show how the actual loads are attenuated and delayed by the thermal inertia.

Equations 6.1.38 to 6.1.46, together with the corresponding tables, as summarized in Figure 6.1.13, are what is known as the *ASHRAE CLTD/CLF method for cooling load calculations*. Note the assumptions that have been made:

- *The indoor temperature T_i is assumed to be constant*
- *Periodic conditions*, corresponding to a series of identical design days

Such limitations are the price of simplicity. If one wants to analyze features like variable occupancy (weekday or weekend) or thermostat setup, one must resort to a dynamic analysis.

Since thermostat setup and reduced weekend heat gains are frequently encountered in commercial buildings, one may wonder about the applicability of the CLF-CLTD method for such cases. If the building is kept at constant T_i, the method can indeed be used (for identical weather there is a relatively small increase in cooling load from Monday to Friday, and, since it is preceded by four days with identical conditions, the prediction of the peak is reliable). Thermostat setup, on the other hand, necessitates a dynamic analysis, as presented in the next section.

This traditional CLTD/CLF method has been changed by ASHRAE researchers recently to include a larger selection of wall types and other features. This new method is described in Section 6.1.8 along with tables of values needed for its use. We have not included tables for the traditional CLTD/CLF method in this handbook in the interest of space. The coefficients are available for this original method as indicated in an earlier footnote.

6.1.7 Transfer Functions for Dynamic Load Calculations

Basis of the Method

The instantaneous load $\dot{Q}$ can be considered the response of the building or room to the driving terms $\{T_i, T_o, \dot{Q}_{sol}$, etc.) that act on it. The transfer function method calculates the response of a system by making the following basic assumptions:

- Discrete time steps (all functions of time are represented as series of values at regular time steps, hourly in the present case)
- Linearity (the response of a system is a linear function of the driving terms and of the state of the system)
- Causality (the response at time t can depend only on the past, not on the future)

As an example, suppose there is a single driving term $u(t)$ and the response is $y(t)$. To make the expressions more readable, let us indicate the time dependence as a subscript, in the form $y(t) = y_t$, $u(t) = u_t$, and so on. Then according to the transfer function model, the relation between the response and the driving term is of the form

$$y_t = -(a_1 y_{t-1\Delta t} + a_2 y_{t-2\Delta t} + \cdots + a_n y_{t-n\Delta t}) + (b_0 u_t + b_1 u_{t-1\Delta t} + b_2 u_{t-2\Delta t} + \cdots + b_m u_{t-m\Delta t}) \quad (6.1.47)$$

with time step

$$\Delta t = 1\mathrm{h} \quad (6.1.48)$$

where a_1 to a_n and b_0 to b_m are coefficients that characterize the system; they are independent of the driving term or response. Equation 6.1.47 is obviously linear. It satisfies causality because y_t depends only on the past values of the response ($y_{t-1\Delta t}$ to $y_{t-n\Delta t}$) and on present and past values of the driving terms[*] (u_t to $u_{t-m\Delta t}$).

The past state of the system enters because of the coefficients a_1 to a_n and b_1 to b_m; this is how thermal inertia is taken into account. The response is instantaneous only if these coefficients are zero. The greater their number and magnitude, the greater the weight of the past.

The accuracy of the model increases as the number of coefficients is enlarged and as the time step is reduced. For load calculations, hourly time resolution and a handful of coefficients per driving term will suffice. The coefficients are called *transfer function coefficients.*

Incidentally, the relation between u and y could be written in symmetric form

$$a_0 y_t + a_1 y_{t-1\Delta t} + a_2 y_{t-2\Delta t} + \cdots + a_n y_{t-n\Delta t} = b_0 u_t + b_1 u_{t-1\Delta t} + b_2 u_{t-2\Delta t} + \cdots + b_m u_{t-m\Delta t} \quad (6.1.49)$$

---

[*] A series such as Equation 6.1.47 is also known as a time series.

which is equivalent because one can divide both sides of the equation by a_0. Since the roles of u and y are symmetric, one can use the same model to find, e.g., the load (i.e., the heat $\dot{Q}$ to be supplied or removed) as a function of T_i, or T_i as a function of $\dot{Q}$.

Equation 6.1.49 can be readily generalized to the case where there are several driving terms. For instance, if the response T_i is determined by two driving terms, heat input $\dot{Q}$ and outdoor temperature T_o, then one can write the transfer function model in the form

$$a_{i,0}T_{i,t} + a_{i,1}T_{i,t-1\Delta t} + \ldots + a_{i,n}T_{i,t-n\Delta t} = a_{o,0}T_{o,t} + a_{o,1}T_{o,t-1\Delta t} + \ldots + a_{o,m}T_{o,t-m\Delta t}$$

$$+ a_{Q,0}\dot{Q}_t + a_{Q,1}\dot{Q}_{t-1\Delta t} + a_{Q,2}\dot{Q}_{t-2\Delta t} + \ldots + a_{Q,r}\dot{Q}_{t-r\Delta t} \qquad (6.1.50)$$

with three sets of transfer function coefficients: $a_{i,0}$ to $a_{i,n}$, $a_{o,0}$ to $a_{o,m}$, and $a_{Q,0}$ to $a_{Q,r}$. This equation can be considered an algorithm for calculating $T_{i,t}$, hour by hour, given the previous values of T_i and the driving terms T_o and $\dot{Q}$. Likewise, if T_i and T_o were given as driving terms, one could calculate $\dot{Q}$ as response.

Any set of response and driving terms can be handled in this manner. Thus loads can be calculated hour by hour, for any driving terms (meteorological data, building occupancy, heat gain schedules, etc.), and it is, in fact, the method used by the computer simulation program DOE2.1 (Birdsall et al., 1990; see also Chapter 6.2 of this handbook).

Once the necessary numerical values of the transfer function coefficients have been obtained, the calculation of peak loads is simple enough for a spreadsheet. One specifies the driving terms for the peak day and iterates an equation like Equation 6.1.50 until the result converges to a steady daily pattern. Transfer function coefficients have been calculated and listed for a wide variety of standard construction types (ASHRAE, 1989a), and some excerpts will be presented here. PREP (1990) can be used to calculate transfer function coefficients for walls and roofs not in the standard ASHRAE database.

The remainder of this section discusses the transfer function method in detail; it is also included in the software. The method involves three steps:

1. Calculation of the conductive heat gain (or loss) for each distinct component of the envelope, by Equation 6.1.51.
2. Calculation of the load of the room at constant temperature, based on this conductive heat gain (or loss) as well as any other heat source in the room, by Equation 6.1.56.
3. Calculation of the heat extraction (or addition) rate for the cooling (or heating) device and thermostat setpoints of the room, by Equation 6.1.61.

Conductive Heat Gain

The conductive heat gain (or loss) $\dot{Q}_{cond,t}$ at time t through the roof and walls is calculated according to the formula

$$\dot{Q}_{cond,t} = -\sum_{n \geq t} d_n \dot{Q}_{cond,t-n\Delta t} + A\left(\sum_{n \geq 0} b_n T_{os,t-n\Delta t} - T_i \sum_{n \geq 0} c_n\right) \qquad (6.1.51)$$

where A = area of roof or wall, m^2 (ft^2)

Δt = time step = 1 h

$T_{o,st}$ = sol-air temperature of outside surface at time $_t$

b_n, c_n, d_n = coefficients of conduction transfer function

The indoor temperature T_i is multiplied by the sum of the c_n values, so the individual c_n coefficients are not needed (this is because T_i is assumed constant at this point; the extension to arbitrary T_i comes shortly). In general, the initial value $\dot{Q}_{cond,t}$ = O is not known; its value does not matter if the calculation is repeated over a sufficient number of time steps until the resulting pattern becomes periodic within the desired accuracy. Usually 4–7 days' worth will be sufficient.

Numerical values of the coefficients of the conduction transfer function are listed in Table 6.1.6: roofs in Table 6.1.6a and walls in Table 6.1.6b. If the room in question is adjacent to rooms at a different temperature, the heat gain across the partitions is also calculated according to Equation 6.1.51.

It is instructive to establish the connection of the transfer function coefficients with the U value. In the steady-state limit, i.e., when $\dot{Q}_{cond}$, T_{os}, and T_i are all constant, Equation 6.1.51 becomes

$$\dot{Q}_{cond} \sum_{n \geq 1} d_n = A \left(T_{os} \sum_{n \geq 0} b_n - T_i \sum_{n \geq 0} c_n \right) \qquad \text{where } d_0 = 1 \qquad (6.1.52)$$

Since in that limit we also have

$$\dot{Q}_{cond} = AU\left(T_{os} - T_i\right) \qquad (6.1.53)$$

the coefficients of T_{os} and T_i must be equal,

$$\sum_{n \geq 0} b_n = \sum_{n \geq 0} c_n \qquad (6.1.54)$$

and the U value is given by

$$U = \frac{\displaystyle\sum_{n \geq 0} c_n}{\displaystyle\sum_{n \geq 0} d_n} \qquad (6.1.55)$$

The Load at Constant Temperature

The above calculation of the conductive heat gain (or loss) is to be repeated for each portion of the room envelope that has a distinct composition. The relation between these conductive gains and the total load depends on the construction of the entire room. For example, a concrete floor can store a significant fraction of the heat radiated by lights or by a warm ceiling, thus postponing its contribution to the cooling load of the room.

For each heat gain component $\dot{Q}_{gain}$, the corresponding cooling load $\dot{Q}_c$ (or reduction of the heating load) at constant T_i is calculated by using another set of coefficients, the coefficients v_n and w_n, of the *room transfer function*,

$$\dot{Q}_{c,t} = v_0 \dot{Q}_{gain,t} + v_1 \dot{Q}_{gain,t-1\Delta t} + v_2 \dot{Q}_{gain,t-2\Delta t} + \cdots$$

$$- w_1 \dot{Q}_{c,t-1\Delta t} - w_2 \dot{Q}_{c,t-2\Delta t} - \cdots \qquad (6.1.56)$$

with the subscript t indicating time, as before. The coefficient w_0 of $\dot{Q}_{c,t}$ is not shown because it is set equal to unity. The coefficients for a variety of room construction types are listed in Tables 6.1.7 and 6.1.8. In these tables, all coefficients with index 2 or higher are zero. Since w_0 is unity, Table 6.1.7 shows only a single coefficient w_1. Again, it is instructive to take the steady-state limit and check the consistency with the first law of thermodynamics. It requires that the sum of the v_n values equal the sum of the w_n values:

$$\sum_{n \geq 0} v_n = \sum_{n \geq 0} w_n \qquad (6.1.57)$$

The entries of Tables 6.1.7 and 6.1.8 do indeed satisfy this condition.

TABLE 6.1.6   Coefficients of Conduction Transfer Function[a]

(a) Roofs

(Layer sequence left to right = inside to outside)		$n=0$	$n=1$	$n=2$	$n=3$	$n=4$	$n=5$	$n=6$	Σc_n	U	δ	λ
Layers E0 A3 B25 E3 E2 A0	b_n	0.00487	0.03474	0.01365	0.00036	0.00000	0.00000	0.00000	0.05362	0.080	1.63	0.97
Steel deck with 3.33-in insulation	d_n	1.00000	−0.35451	0.02267	−0.00005	0.00000	0.00000	0.00000				
Layers E0 A3 B14 E3 E2 A0	b_n	0.00056	0.01202	0.01282	0.00143	0.00001	0.00000	0.00000	0.02684	0.055	2.43	0.94
Steel deck with 5-in insulation	d_n	1.00000	−0.60064	0.08602	−0.00135	0.00000	0.00000	0.00000				
Layers E0 E1 B15 E4 B7 A0	b_n	0.00000	0.00065	0.00339	0.00240	0.00029	0.00000	0.00000	0.00673	0.043	4.85	0.82
Attic roof with 6-in insulation	d_n	1.00000	−1.34658	0.59384	−0.09295	0.00296	−0.00001	0.00000				
Layers E0 B22 C12 E3 E2 C12 A0	b_n	0.00059	0.00867	0.00688	0.00037	0.00000	0.00000	0.00000	0.01652	0.138	5.00	0.56
1.67-in insulation with 2-in h.w. concrete RTS	d_n	1.00000	−1.11766	0.23731	−0.00008	0.00000	0.00000	0.00000				
Layers E0 E5 E4 B12 C14 E3 E2 A0	b_n	0.00000	0.00024	0.00217	0.00251	0.00055	0.00002	0.00000	0.00550	0.057	6.32	0.60
3-in insulation w/4-in l.w. conc. deck and susp. clg.	d_n	1.00000	−1.40605	0.58814	−0.09034	0.00444	−0.00006	0.00000				
Layers E0 E5 E4 C5 B6 E3 E2 A0	b_n	0.00001	0.00066	0.00163	0.00049	0.00002	0.00000	0.00000	0.01477	0.090	7.16	0.16
1-in insul. w/4-in h.w. conc. deck and susp. clg.	d_n	1.00000	−1.24348	0.28742	−0.01274	0.00009	0.00000	0.00000				
Layers E0 E5 E4 C13 B20 E3 E2 A0	b_n	0.00001	0.00060	0.00197	0.00086	0.00005	0.00000	0.00000	0.00349	0.140	7.54	0.15
6-in h.w. deck w/0.76-in insul. and susp. clg.	d_n	1.00000	−1.39181	0.46337	−0.04714	0.00058	0.00000	0.00000				
Layers E0 E5 E4 B15 C15 E3 E2 A0	b_n	0.00000	0.00000	0.00002	0.00014	0.00024	0.00011	0.00002	0.00053	0.034	10.44	0.30
6-in insul. w/6-in l.w. conc. deck and susp. clg.	d_n	1.00000	−2.29459	1.93694	−0.75741	0.14252	−0.01251	0.00046				
Layers E0 C13 B15 E3 E2 C12 A0	b_n	0.00000	0.00000	0.00007	0.00024	0.00016	0.00003	0.00000	0.00050	0.045	10.48	0.24
6-in h.w. deck w/6-in ins. and 2-in h.w. RTS	d_n	1.00000	−2.27813	1.82162	−0.60696	0.07696	−0.00246	0.00001				

TABLE 6.1.6 (continued)   Coefficients of Conduction Transfer Function[a]

(b) Walls

(Layer sequence left to right = inside to outside)		$n = 0$	$n = 1$	$n = 2$	$n = 3$	$n = 4$	$n = 5$	$n = 6$	Σc_n	U	δ	λ
Layers E0 A3 B1 B13 A3 A0 Steel siding with 4-in insulation	b_n	0.00768	0.03498	0.00719	0.00006	0.00000	0.00000	0.00000	0.04990	0.066	1.30	0.98
	d_n	1.00000	-0.24072	0.00168	0.00000	0.00000	0.00000	0.00000				
Layers E0 E1 B14 A1 A0 A0 Frame wall with 5-in insulation	b_n	0.00016	0.00545	0.00961	0.00215	0.00005	0.00000	0.00000	0.01743	0.055	3.21	0.91
	d_n	1.00000	-0.93389	0.27396	-0.02561	0.00014	0.00000	0.00000				
Layers E0 C3 B5 A6 A0 A0 4-in h.w. concrete block with 1-in insulation	b_n	0.00411	0.03230	0.01474	0.00047	0.00000	0.00000	0.00000	0.05162	0.191	3.33	0.78
	d_n	1.00000	-0.76963	0.04014	-0.00042	0.00000	0.00000	0.00000				
Layers E0 A6 C5 B3 A3 A0 4-in h.w. concrete with 2-in insulation	b_n	0.00099	0.00836	0.00361	0.00007	0.00000	0.00000	0.00000	0.01303	0.122	5.14	0.41
	d_n	1.00000	-0.93970	0.04664	0.00000	0.00000	0.00000	0.00000				
Layers E0 E1 C8 B6 A1 A0 8-in h.w. concrete block with 2-in insulation	b_n	0.00000	0.00061	0.00289	0.00183	0.00018	0.00000	0.00000	0.00552	0.109	7.11	0.37
	d_n	1.00000	-1.52480	0.67146	-0.09844	0.00239	0.00000	0.00000				
Layers E0 A2 C2 B15 A0 A0 Face brick and 4-in l.w. conc. block with 6-in insulation	b_n	0.00000	0.00000	0.00013	0.00044	0.00030	0.00005	0.00000	0.00093	0.043	9.36	0.30
	d_n	1.00000	-2.00875	1.37120	-0.37897	0.03962	-0.00165	0.00002				
Layers E0 C9 B6 A6 A0 A0 8-in common brick with 2-in insulation	b_n	0.00000	0.00005	0.00064	0.00099	0.00030	0.00002	0.00000	0.00200	0.106	8.97	0.20
	d_n	1.00000	-1.78165	0.96017	-0.16904	0.00958	-0.00016	0.00000				
Layers E0 C11 B6 A1 A0 A0 12-in h.w. concrete with 2-in insulation	b_n	0.00000	0.00001	0.00019	0.00045	0.00022	0.00002	0.00000	0.00089	0.112	10.20	0.13
	d_n	1.00000	-2.12812	1.53974	-0.45512	0.05298	-0.00158	0.00002				

[a] U, b_n, and, c_n are in Btu/(h · ft² · °F); d_n and A are dimensionless; and δ is in hours [1 Btu/(h · ft² · °F) = 5.678 W/(m² · K)]. For definition of layer codes and thermal properties, see Table 6.1.1A.

Source: From ASHRAE, 1989a, with permission.

TABLE 6.1.7 The w_1 Coefficient of the Room Transfer Function ($w_0 = 1.0$ and higher terms are zero)

	Room Envelope Construction[b]				
	2-in (51-mm) Wood floor	3-in (76-mn) Concrete floor	6-in (152-mm) Concrete floor	8-in (203-mm) Concrete floor	12-in (305-mm) Concrete floor
Room air[a] circulation and S/R type	Specific mass per unit floor area, lb/ft²				
	10	40	75	120	160
Low	−0.88	−0.92	−0.95	−0.97	−0.98
Medium	−0.84	−0.90	−0.94	−0.96	−0.97
High	−0.81	−0.88	−0.93	−0.95	−0.97
Very high	−0.77	−0.85	−0.92	−0.95	−0.97
	−0.73	−0.83	−0.91	−0.94	−0.96

[a] Circulation rate:

Low: Minimum required to cope with cooling load from lights and occupants in interior zone. Supply through floor, wall, or ceiling diffuser. Ceiling space not used for return air, and $h = 0.4$ Btu/(h · ft² · °F) [2.27 W/(m² · K)], where h = inside surface convection coefficient used in calculation of w_1 value.

Medium: Supplied through floor, wall, or ceiling diffuser. Ceiling space not used for return air, and $h = 0.6$ Btu/(h · ft² · °F) [3.41 W/(m² · K)].

High: Room air circulation induced by primary air of induction unit, or by room fan and coil unit. Ceiling space used for return air, and $h = 0.8$ Btu/(h · ft² · °F) [4.54 W/(m² · K)].

Very high: Used to minimize temperature gradients in a room. Ceiling space used for return air, and $h = 1.2$ Btu/(h · ft² · °F) [6.81 W/(m² · K)].

[b] Floor covered with carpet and rubber pad; for a bare floor or if covered with floor tile, take next w_1 value down the column.

Source: From ASHRAE, 1989a, with permission.

Therefore, Equation 6.1.56 has to be applied separately to each of the heat gain types in Table 6.1.8, and the resulting cooling load components $\dot{Q}_{c,t}$ are added to obtain the total cooling load of the room at time t. The heat gain types are as follows:

- Solar gain (through glass without interior shade) and the radiative component of heat from occupants and equipment
- Conduction through envelope and solar radiation absorbed by interior shade
- Lights
- Convective gains (from air exchange, occupants, equipment)

For lights, the coefficients depend on the arrangement of the lighting fixture and the ventilation system. While specific numbers vary a great deal with the circumstances, the general pattern is common to all peak cooling loads: *thermal inertia attenuates and delays the peak contributions of individual load components.* The total peak is usually less than the result of a steady state calculation, although it could be more if the time delays act in the sense of making the loads coincide. Daily average loads, in contrast to peak loads, can be determined by a static calculation, if the average indoor temperature is known; that follows from the first law of thermodynamics. But if the thermostat allows floating temperatures, the indoor temperature is, in general, not known without a dynamic analysis. With the transfer functions described so far, one can calculate peak loads when the indoor temperature T_i is constant. That is how the CLFs and CLTDs of Section 6.1.6 were determined. We now address the generalization to variable T_i.

Variable Indoor Temperature and Heat Extraction Rate

The indoor temperature T_i may vary, not only because of variable thermostat setpoints but also because of limitations of the HVAC equipment (capacity, throttling range, imperfect control). The extension to variable T_i requires one additional transfer function. Recall that the behavior of a room can be described

TABLE 6.1.8   The v Coefficients of the Room Transfer Function[a] (Only v_0 and v_1 Are Nonzero)

Heat Gain Component	Room Envelope Construction[b]	v_0	v_1
		\ Dimensionless	
Solar heat gain through glass[c] with no	Light	0.224	$1 + w_1 - v_0$
interior shade; radiant heat from	Medium	0.197	$1 + w_1 - v_0$
equipment and people	Heavy	0.187	$1 + w_1 - v_0$
Conduction heat gain through exterior	Light	0.703	$1 + w_1 - v_0$
walls, roofs, partitions, doors, windows	Medium	0.681	$1 + w_1 - v_0$
with blinds, or drapes	Heavy	0.676	$1 + w_1 - v_0$
Convective heat generated by equipment	Light	1.000	0.0
and people, and from ventilation and	Medium	1.000	0.0
infiltration air	Heavy	1.000	0.0

Heat Gain from Lights[d]

Furnishings	Air Supply and Return	Type of Light Fixture	v_0	v_1
Heavyweight simple furnishings, no carpet	Low rate; supply and return below ceiling (V ≤ 0.5)[e]	Recessed, not vented	0.450	$1 + w_1 - v_0$
Ordinary furnishings, no carpet	Medium to high rate, supply and return below or ceiling (V ≥ 0.5)	Recessed, not vented	0.550	$1 + w_1 - v_0$
Ordinary furnishings, with or without carpet on floor	Medium to high rate, or induction unit or fan and coil, supply and return below, or through ceiling, return air plenum (V ≥ 0.5)	Vented	0.650	$1 + w_1 - v_0$
Any type furniture, with or without carpet	Ducted returns through light fixtures	Vented or freehanging in air stream with ducted returns	0.750	$1 + w_1 - v_0$

[a] The transfer functions in this table were calculated by procedures outlined in Mitalas and Stephenson (1967) and are acceptable for cases where all heat gain energy eventually appears as cooling load. The computer program used was developed at the National Research Council of Canada, Division of Building Research.

[b] The construction designations denote the following:

> *Light construction:* such as frame exterior wall, 2-in (51-mm) concrete floor slab, approximately 30 lb of material/ft^2 (146 kg/m^2) of floor area.
> *Medium construction:* such as 4-in (102-mm) concrete exterior wall, 4-in (102-mm) concrete floor slab, approximately 70 lb building material/ft^2 (341 kg/m^2) of floor area.
> *Heavy construction*: such as 6-in (152-mm) concrete exterior wall, 6-in (152-mm) concrete floor slab, approximately 130 1b of building material/ft^2 (635 kg/m^2) of floor area.

[c] The coefficients of the transfer function that relate room cooling load to solar heat gain through glass depend on where the solar energy is absorbed. If the window is shaded by an inside blind or curtain, most of the solar energy is absorbed by the shade and is transferred to the room by convection and long-wave radiation in about the same proportion as the heat gain through walls and roofs; thus the same transfer coefficients apply.

[d] If room supply air is exhausted through the space above the ceiling, and lights are recessed, such air removes some heat from the lights that would otherwise have entered the room. This removed light heat is still a load on the cooling plant if the air is recirculated, even though it is not a part of the room heat gain as such. The percent of heat gain appearing in the room depends on the type of lighting fixture, its mounting, and the exhaust airflow.

[e] V is room air supply rate in (ft^3/min)/ft^2 of floor area.

Source: From ASHRAE, 1989a, with permission.

by a relation like Equation 6.1.50 which links the output (room temperature T_i) to all the relevant input variables (outdoor temperature T_o, heat input or extraction by the HVAC system $\dot{Q}$, solar heat gains, etc.)

$$a_{i,0}T_{i,k} + a_{i,1}T_{i,k-1} + \ldots + a_{i,l}T_{i,k-l} = a_{o,0}T_{o,k} + a_{o,1}T_{o,k-1} + \ldots + a_{o,m}T_{o,k-m}$$

$$+ a_{Q,0}\dot{Q}_k + a_{Q,1}\dot{Q}_{k-1} + \ldots + a_{Q,n}\dot{Q}_{k-n} \qquad (6.1.58)$$

TABLE 6.1.9 Normalized Coefficients of Space Air Transfer Function

Room envelope construction	P_0	P_1	g_0^*	g_1^*	g_2^*	g_0^*	g_1^*	g_2^*
	Dimensionless		Btu/(h · ft² · °F)			W/(m² · K)		
Light	1.00	−0.82	1.68	−1.73	0.05	9.54	−9.82	0.28
Medium	1.00	−0.87	1.81	−1.89	0.08	10.28	−10.73	0.45
Heavy	1.00	−0.93	1.85	−1.95	0.10	10.50	−11.07	0.57

Source: From ASHRAE, 1989a, with permission.

A separate set of transfer function coefficients is needed for each input variable with different time-delay characteristics; here we have indicated only T_o and $\dot{Q}$ explicitly. Now consider two different control modes, mode 1 with the constant value $T_{i,ref}$ assumed and mode 2 with arbitrary T_i, all input being the same except for $\dot{Q}$. Let

$$\delta T_i = T_{i,ref} - T_i \qquad (6.1.59)$$

and

$$\delta \dot{Q} = \dot{Q}_{ref} - \dot{Q} \qquad (6.1.60)$$

Designate the differences in T_i and $\dot{Q}$ between these two control modes. Taking the difference between Equation 6.1.50 for mode 1 and for mode 2, we see that all variables other than δT_i and $\delta \dot{Q}$ drop out. The transfer function between δT_i and $\delta \dot{Q}$ is called *space air transfer function*, and, following ASHRAE practice, its coefficients are designated by $p_n \, (= a_{Q,n})$ and $g_n \, (= a_{i,n})$

$$\sum_{n \geq 0} p_n \delta \dot{Q}_{t-n\Delta t} = \sum_{n \geq 0} g_{n,t} \delta T_{i,t-n\Delta t} \qquad (6.1.61)$$

A subscript t has been added to g_n to allow the transfer function to vary with time if the air exchange rate varies. Numerical values can be obtained from Table 6.1.9. While p_n is listed directly, g_n is given in terms of g_n^* from which g_n is calculated according to

$$g_{0,t} = g_0^* A + P_0 K_{tot,t}$$

$$g_{1,t} = g_1^* A + P_1 K_{tot,t-\Delta t}$$

$$g_{2,t} = g_2^* A \qquad (6.1.62)$$

where A = floor area and $K_{tot,t}$ W/K [Btu/(h · °F)] is the total heat transmission coefficient of the room. The latter is the sum of conductive and air change terms according to Equation 6.1.24:

$$K_{tot,t} = K_{cond} + \rho c_p \dot{V} \qquad (6.1.24)$$

and a subscript t for time dependence has been added to allow for the possibility of variable air change. Of course, K_{cond} is the sum of the conductance-area products for the envelope of the room.

To verify the consistency of these coefficients with the first law of thermodynamics, let us take the steady state limit where $\delta \dot{Q}$, δT_i, and $K_{tot,t}$ are constant and can be pulled outside the sum. Replacing the g_n by Equation 6.1.62, we find

$$\delta \dot{Q} \, \Sigma p_n = \delta T_i \, [A \Sigma g_n^* + K_{tot} \, \Sigma p_n] = \text{steady state limit} \qquad (6.1.63)$$

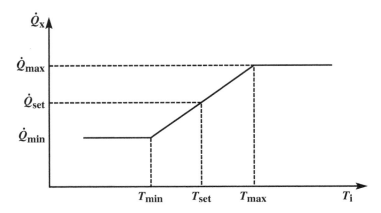

FIGURE 6.1.19 Control law of Equation 6.1.66 for heat extraction rate $\dot{Q}_x$ (solid line) as function of room temperature T_i.

A look at the numerical values of g_n^* in Table 6.1.9 shows that their sum vanishes. Thus the equation reduces to

$$\delta\dot{Q} = \delta T_i K_{tot} \tag{6.1.64}$$

as it should. Since K_{tot} is positive, $\delta\dot{Q}$ and δT_i have the same sign; this means that $\delta\dot{Q}$ is positive for positive heat input to the room. If we want to state cooling loads $\dot{Q}_c$ as positive quantities, we should therefore take $\dot{Q}_c = -\dot{Q}$ and $\delta\dot{Q}_c = -\delta\dot{Q}$. In particular, if we call the cooling load at temperature T_i the heat extraction rate $\dot{Q}_x$ (because that is the rate at which the HVAC equipment must extract heat to obtain the temperature T_i), we can write Equation 6.1.61 in the form

$$\sum_{n\geq0} p_n \left(\dot{Q}_{x,t-n\Delta t} - \dot{Q}_{c,ref}\right) = \sum_{n\geq0} g_{n,t}\left(T_{i,ref} - T_{i,t-n\Delta t}\right) \tag{6.1.65}$$

where $\dot{Q}_{c,ref}$ is the cooling load at the constant temperature $T_{i,ref}$.

Using Equation 6.1.61, one can calculate $\delta\dot{Q}$ for any δT_i, or δT_i for any $\delta\dot{Q}$. It can also be used for a mixed regime where δT_i is specified for certain hours and $\delta\dot{Q}$ for others. The calculation proceeds from one hour to the next, solving for δT_i or $\delta\dot{Q}$ as appropriate. The daily cycle is iterated until the result converges to a stable pattern.

Capacity limitations of the HVAC equipment can also be included. One checks, at each hour with thermostatic control, whether the actual load (at T_i = thermostat setpoint) exceeds the capacity. If it does, one solves for T_i instead, setting $\dot{Q}_x$ equal to the capacity for this hour.

Likewise, one can account for the throttling range of a control system which modulates the heat extraction rate according to the control law shown in Figure 6.1.19. Stated as an equation, this means that the heat extraction rate $\dot{Q}_x$ is determined by the room temperature T_i according to

$$\dot{Q}_{x,t} = \begin{cases} \dot{Q}_{max} & \text{for } T_{i,t} > T_{max} \\ \dot{Q}_{set} + \dot{Q}'\left(T_{i,t} - T_{set}\right) & \text{for } T_{min} < T_{i,t} < T_{max} \\ \dot{Q}_{min} & \text{for } T_{i,t} < T_{min} \end{cases} \tag{6.1.66}$$

where $\dot{Q}' = (\dot{Q}_{max} - \dot{Q}_{min})/(T_{max} - T_{min})$ and T_{set} is the thermostat setpoint; we have added the subscript t to indicate that this equation applies instantaneously at each hour t. At each new hour t, Equations 6.1.65 and 6.1.66 can be considered as a system of two equations for two unknowns: $\dot{Q}_{x,t}$ and $T_{i,t}$. After finding the solution, one repeats the process for the next hour.

TABLE 6.1.10   Summary of Load Calculations

Item	Method and Comments
Zones	Define zones Zone = part of building that can be assumed to have uniform loads. For each zone carry out the steps below
Design conditions	Determine appropriate values of temperatures (T_i and T_o) and humidities (W_i and W_o) for peak conditions at site in question
Conduction	$K_{cond} = \sum_k U_k A_k$
Air change	$\dot{V}\, \rho c_p$ For relatively simple buildings, use LBL model, Equation 6.1.12; otherwise use correlations
Heat gains	Solar, lights, equipment, occupants
Heating load, sensible	$\dot{Q}_{h,max} = K_{tot}(T_i - T_o) - \dot{Q}_{gain}$, with $K_{tot} = K_{cond} + \dot{V}\,\rho c_p$
Cooling load, "static," sensible	$\dot{Q}_{c,cond,t} = UA\ CLTD_t$ $\dot{Q}_{c,rad,t} = \dot{Q}_{max} CLF_t$
Cooling load, dynamic, sensible	Transfer function method Equations 6.1.51 to 6.1.66
Latent loads	Latent gain from air change $\dot{Q}_{air,lat} = \dot{V}\, h_{fg}(W_o - W_i)$, also latent gains from occupants and from equipment

A general load calculation procedure is summarized in Table 6.1.10 to summarize the chapter to this point.

6.1.8   New Methods for Load Calculations

In the ever-present quest for more accurate and versatile load prediction methods, new procedures are being developed all the time. In the past few years, two new calculation procedures have been developed that are relatively similar to the existing methods but implement enough differences to warrant mention. The CLTD/CLF method has been updated to include a new term, the solar cooling load (SCL), and is now termed the CLTD/SCL/CLF method. The radiant transfer series (RTS) method is a noniterative modification of the transfer function method. An overview of both of these new methods is presented in this section.

CLTD/SCL/CLF Method for Calculating Cooling Loads

The CLTD/CLF method presented in section 6.1.6 has been modified in a number of important ways. The selection of roof and wall CLTD values requires a number of look-up tables but allows the use of essentially any arbitrary wall construction. In addition, the solar heat gain factors have been replaced with a term called the solar cooling load. While the overall methodology of the CLTD/CLF method has been preserved, a full description of the necessary computations of the new method is presented here to avoid confusion with the original method in Section 6.1.6.

Roof CLTD Value Selection

The CLTD/SCL/CLF method uses 10 types of roofs. The roof types are numbered 1, 2, 3, 4, 5, 8, 9, 10, 13, and 14. The roof type chosen depends on the principal roof material, the location of the mass in the roof, the overall R-value of the roof, and whether or not there is a suspended ceiling. Table 6.1.11 below shows the cross reference chart used to select a roof type.

The tables of new roof CLTD values are calculated based on an indoor temperature of 78°F, maximum and mean outdoor temperatures of 95°F and 85°F, respectively, and a daily range of 21°F. Once the 24 CLTD values are selected, they are each adjusted by

$$Corrected\ CLTD = CLTD + (78 - T_i) + (T_{om} - 85) \tag{6.1.67}$$

Where T_i is the actual inside design dry-bulb temperature and T_{om} is the mean outside design dry-bulb temperature,

TABLE 6.1.11 Cross Reference Table Used to Determine Roof Type

Mass Location	Principal Roof Material Description	ASHRAE code	Susp. Ceiling	R Value (ft²·h·°F/Btu) 0–5	5–10	10–15	15–20	20–25	25–30
Inside insulation	2 in HW Concrete	C12	No	2	2	4	4	5	—
			Yes	5	8	13	13	14	—
Evenly spaced	1 in Wood	B7	No	1	2	2	4	4	—
			Yes	—	4	5	9	10	10
	2 in HW Concrete	C12	No	2	—	—	—	—	—
			Yes	3	—	—	—	—	—
	Steel deck	A3	No	1	1	1	2	2	—
			Yes	1	1	2	2	4	—
	Attic-Ceiling comb.	n/a	No	1	2	2	2	4	—
Outside insulation	2 in HW Concrete	C12	No	2	3	4	5	5	—
			Yes	3	3	4	5	—	—

Source: Adapted from McQuiston, F. and Spitler, J., (1992).

$$T_{om} = \text{Outside design dry-bulb temperature} - \frac{\text{Daily range}}{2} \qquad (6.1.68)$$

No adjustments to the CLTD are recommended for color or ventilation. The CLTD charts are usually published for several different latitudes; interpolation between the latitudes for an exact site is acceptable.

Wall CLTD Value Selection
The CLTD/SCL/CLF uses 15 wall types numbered sequentially 1 through 16 with no wall 8. The wall type is chosen based on the principal wall material, the secondary wall material, the location of the mass in the wall, and the overall wall R-value. Table 6.1.12 below shows an example cross-reference chart used to select a wall type. The tables of wall CLTD values are divided by latitude. The wall CLTDs were calculated using the same conditions as the roof CLTD values and may require adjustments based on the actual inside and ambient conditions. Interpolation between the tables may be necessary to obtain the correct values for a given site.

Once the roof and wall CLTD values have been selected and adjusted as necessary, the conductive heat flow through the roof and walls is calculated for each hour as in the original CLTD/CLF method,

$$q(hr) = U \cdot A \cdot CLTD(hr) \qquad (6.1.69)$$

where

U = overall heat transfer coefficient for the surface (Btu/hr·ft²·°F)
A = area of the surface, and
$CLTD$ = the cooling load temperature difference.

Glass CLTD Value Selection
The glass CLTD values remain the same as they were in the original method. As with the roof and wall CLTDs, the fenestration CLTD values may need to be corrected based on Equations 6.1.67 and 6.1.68. The conductive load calculation from the glass uses the same method as for the roof and walls. The CLTD values for the glass are given in Table 6.1.13.

Solar Cooling Load
The new method replaces the maximum solar heat gain factor with the solar cooling load (SCL). This new value is used to calculate the radiative (solar) heat gain through any glass surface in the building. The radiative solar gains are then given by

$$q(hr) = A \cdot SC \cdot SCL \qquad (6.1.70)$$

where A is the area of the glass surface, SC is the shading coefficient, and SCL is the solar cooling load factor. The shading coefficient is the ratio of the actual solar heat gain to that from the reference window used to calculate the SCL.

TABLE 6.1.12 Example Wall Type Selection Table. Values Are Shown for Mass Located Inside Insulation.

R Value ft2·h·°F/Btu	Principal Wall Material (ASHRAE Material Code)										
	A2	C1	C2	C3	C4	C5	C6	C7	C8	C17	C18
Stucco and/or plaster											
2.0 to 2.5	5	—	—	—	—	5	—	—	—	—	—
2.5 to 3.0	5	3	—	2	5	6	—	—	5	—	—
3.0 to 3.5	5	4	2	2	5	6	—	—	6	—	—
3.5 to 4.0	5	4	2	3	6	6	10	4	6	—	5
4.0 to 4.75	6	5	2	4	6	6	11	5	10	—	10
4.75 to 5.5	6	5	2	4	6	6	11	5	10	—	10
5.5 to 6.5	6	5	2	5	10	7	12	5	11	—	10
6.5 to 7.75	6	5	4	5	11	7	16	10	11	—	11
7.75 to 9.0	6	5	4	5	11	7	—	10	11	—	11
9.0 to 10.75	6	5	4	5	11	7	—	10	11	4	11
10.75 to 12.75	6	5	4	5	11	11	—	10	11	4	11
12.75 to 15.0	10	10	4	5	11	11	—	10	11	9	12
15.0 to 17.5	10	10	5	5	11	11	—	11	12	10	16
17.5 to 20.0	11	10	5	9	11	11	—	15	16	10	16
20.0 to 23.0	11	10	9	9	16	11	—	15	16	10	16
23.0 to 27.0	—	—	—	—	—	—	—	16	—	15	—
Steel or other light-weight siding											
2.0 to 2.5	3	—	—	2	3	5	—	—	—	—	—
2.5 to 3.0	5	2	—	2	5	3	—	—	5	—	—
3.0 to 3.5	5	3	1	2	5	5	—	—	5	—	—
3.5 to 4.0	5	3	2	2	5	5	6	3	5	—	5
4.0 to 4.75	6	4	2	2	5	5	10	4	6	—	5
4.75 to 5.5	6	5	2	2	6	6	11	5	6	—	6
5.5 to 6.5	6	5	2	3	6	6	11	5	6	—	6
6.5 to 7.75	6	5	2	3	6	6	11	5	6	—	10
7.75 to 9.0	6	5	2	3	6	6	12	5	6	—	11
9.0 to 10.75	6	5	2	3	6	6	12	5	6	4	11
10.75 to 12.75	6	5	2	3	6	7	12	6	11	4	11
12.75 to 15.0	6	5	2	4	6	7	12	10	11	5	11
15.0 to 17.5	10	6	4	4	10	7	—	10	11	9	11
17.5 to 20.0	10	10	4	4	10	11	—	10	11	10	11
20.0 to 23.0	11	10	4	5	11	11	—	10	11	10	16
23.0 to 27.0	—	—	—	—	—	—	—	10	—	11	16

Source: Adapted from McQuiston, F. and Spitler, J., (1992).

TABLE 6.1.13 CLTD Values for Fenestration

Hour	CLTD	Hour	CLTD	Hour	CLTD	Hour	CLTD
1	1	7	−2	13	12	19	10
2	0	8	0	14	13	20	8
3	−1	9	2	15	14	21	6
4	−2	10	4	16	14	22	4
5	−2	11	7	17	13	23	3
6	−2	12	9	18	12	24	2

Source: Adapted from McQuiston, F. and Spitler, J., (1992).

Using the SCL value tables requires that you know the number of walls, floor covering, inside shading, and a number of other variables for the zone. The tables are also broken down by building type, with different tables for zones in

- Single story buildings
- Top floor of multistory buildings
- Middle floors of multistory buildings
- First floor of multistory buildings

Table 6.1.14 gives the zone types for the SCL for the first story of multistory buildings. The zone type listed here is for the SCL *and is not necessarily the same zone type used for the CLF Tables*. Once the zone type has been determined, the SCL can be found from tables such as shown in Table 6.1.15.

Accounting for Adjacent Zones

The CLTD/SCL/CLF method treats the conductive heating load from any adjacent spaces through internal partitions, ceilings, and floors as a simple steady state energy flow

$$q(hr) = U \cdot A \cdot (T_a - T_r) \tag{6.1.71}$$

where T_a is the temperature in the adjacent space and T_r is the temperature of the room in question.

Occupant Loads

People within a space add both sensible and latent loads to the space. The heating load at any given hour due to the occupants is given as

$$q(hr) = N \cdot F_d \cdot [q_s \cdot CLF(hr) + q_l] \tag{6.1.72}$$

where N is the number of people in the space and F_d is the diversity factor. As implied by the preceding equation, the latent load is assumed to immediately translate into a cooling load on the system, while the sensible load is subject to some time delay as dictated by the mass of the room, i.e., its capability to absorb heat and release it at a later time. The diversity factor, F_d, takes into account the variability of the actual number of occupants in the space and has typical values as given in Table 6.1.16.

The CLF values are read from tables. To find the *CLF* it is first necessary to determine the zone type. This is done in a similar fashion as for the solar cooling loads. That is, the building type, room location, and floor coverings must be known before the zone type can be found. Table 6.1.17 gives the zone types for people, equipment, and lights for interior (nonperimeter) zones. Note that the zone type for occupants and equipment is not the same as for the lighting. The same holds true for the solar cooling load: the zone types for occupants is not the same as the zone type for the SCL.

Once the zone type has been determined, the occupant CLF is found from the lookup tables such as shown in Table 6.1.18. This table shows values for Type A zones only; the zones get progressively more massive for types B, C, and D. Figure 6.1.20 shows the cooling load factors for type A and D zones that are occupied for twelve hours.

Note that the occupant CLF will be 1.0 for all hours in building with high occupant density (greater than 1 person per 10 ft^2), such as auditoriums and theaters. The CLF will also be 1.0 in buildings where there is 24 hour per day occupancy.

Lighting Loads

At any given hour the load due to the lighting is approximated as

$$q(hr) = Watts \cdot F_d \cdot F_{sa} \cdot CLF(hr) \tag{6.1.73}$$

where *Watts* is the total lamp wattage in the space, F_d is the diversity factor, and F_{sa} is a ballast special allowance factor. The diversity factor *i* takes into account the variability of the actual wattage of lights on at any given time and has typical values as given in Table 6.1.19.

The lighting CLF values come from tables and are found in a fashion similar to that for the occupants. It should be remembered that the zone types for lighting are not necessarily the same zone types for the solar cooling load or the occupants. Note that the lighting *CLF* will be 1.0 for buildings in which the lights are on 24 hours per day or where the cooling system is shut off at night or on the weekends.

If the calculations are done in IP units, then the result from Equation 6.1.73 is multiplied by 3.41 to convert watts to Btu/hr.

TABLE 6.1.14 Zone Type for Solar Cooling Load for First Story of Multistory Buildings

Mid-Floor Type	Ceiling Type	Floor Covering	Partition Type	Inside Shade	Zone Type
		1 or 2 Walls			
2.5 in Concrete	With	Carpet	Gypsum	Full	A
				Half to None	B
			Concrete block	Full	B
				Half to None	C
		Vinyl	Gypsum	Full	C
				Half to None	C
			Concrete block	Full	D
				Half to None	D
	Without	Carpet	Gypsum	—	B
			Concrete block	Full	C
				Half to None	C
		Vinyl	Gypsum	Full	C
				Half to None	D
			Concrete block	Full	C
				Half to None	D
1 in Wood	—	Carpet	Gypsum	Full	A
				Half to None	B
			Concrete block	Full	B
				Half to None	C
		Vinyl	Gypsum	Full	B
				Half to None	C
			Concrete block	Full	C
				Half to None	D
		3 Walls			
2.5 in Concrete	With	Carpet	Gypsum	Full	A
		Carpet	Gypsum	Half to None	B
		Carpet	Concrete block	—	B
		Vinyl	Gypsum	Full	C
		Vinyl	—	Half to None	C
		Vinyl	Concrete block	Full	C
	Without	Carpet	Gypsum	—	B
		Carpet	Concrete block	Full	B
		Carpet	Concrete block	Half to None	C
		Vinyl	Gypsum	Full	C
		Vinyl		Half to None	C
		Vinyl	Concrete block	Full	C
1 in Wood	—	Carpet	Gypsum	Full	A
				Half to None	B
			Concrete block	—	B
		Vinyl	Gypsum	Full	B
				Half to None	C
			Concrete block	Full	C
				Half to None	C
		4 Walls			
2.5 in Concrete	With	Carpet	Gypsum	Full	A
				Half to None	B
		Vinyl	Gypsum	—	C
	Without	Carpet	Gypsum		B
		Vinyl	Gypsum		B
1 in Wood	—	Carpet	Gypsum	Full	A
				Half to None	A
		Vinyl	Gypsum	Full	B
				Half to None	C

Source: Adapted from McQuiston, F. and Spitler, J., (1992).

TABLE 6.1.15 Solar Cooling Load (SCL) for Sunlit Glass at 36° North Latitude for July

Zone Type	Orientation	1	2	3	4	5	6	7	8	9	10	11	12	13	14	15	16	17	18	19	20	21	22	23	24
A	N	0	0	0	0	0	25	29	28	32	36	39	40	41	39	36	32	33	36	12	6	3	1	1	0
	NE	0	0	0	0	0	79	129	139	120	84	58	50	45	41	37	32	26	17	7	3	2	1	0	0
	E	0	0	0	0	0	86	153	184	182	155	107	67	54	45	39	33	26	17	7	3	2	1	0	0
	SE	0	0	0	0	0	42	90	125	142	140	119	86	58	48	40	34	27	17	7	3	2	1	0	0
	S	0	0	0	0	0	8	17	24	36	53	70	80	79	68	52	38	29	18	7	3	2	1	0	0
	SW	0	0	0	0	0	8	17	24	30	35	38	57	90	122	141	144	127	85	32	15	8	4	2	1
	W	1	0	0	0	0	8	17	24	30	35	38	40	66	115	159	188	191	149	53	25	12	6	3	2
	NW	1	0	0	0	0	8	17	24	30	35	38	40	40	56	93	129	148	127	43	21	10	5	2	1
	HOR	0	0	0	0	0	20	66	120	171	215	246	263	265	251	221	178	124	66	28	13	7	3	2	1
B	N	2	2	2	1	1	21	25	29	32	33	36	38	38	38	35	32	33	35	15	10	7	5	4	3
	NE	2	1	1	1	1	68	109	120	108	81	61	54	50	46	42	37	30	22	12	9	6	5	3	3
	E	2	2	2	1	1	73	130	158	161	143	106	75	63	55	48	41	34	25	14	10	7	5	4	3
	SE	2	2	2	1	1	36	77	107	124	125	111	85	64	55	48	41	33	24	14	10	7	5	4	3
	S	2	2	2	1	1	7	14	21	31	47	61	71	72	65	52	41	33	24	13	9	7	5	5	3
	SW	6	4	3	3	2	8	15	21	27	31	35	51	80	108	126	131	119	86	43	29	20	14	11	8
	W	8	6	5	4	3	9	16	22	27	32	35	37	60	101	140	166	172	141	63	42	29	20	15	11
	NW	6	5	4	3	2	8	15	21	27	31	35	37	38	52	84	115	132	117	49	32	22	16	11	8
	HOR	8	6	5	4	3	19	57	103	148	188	218	237	244	237	215	182	137	88	53	37	26	19	14	11
C	N	5	5	4	4	3	24	25	24	27	31	33	35	35	34	32	29	31	34	16	14	10	8	6	6
	NE	7	6	6	5	5	71	106	111	95	68	51	48	46	44	41	37	32	24	16	13	11	10	9	8
	E	9	8	8	7	6	77	128	148	145	124	89	62	56	52	47	43	37	29	20	17	15	13	12	11
	SE	8	8	7	6	5	40	77	102	114	112	97	73	55	49	45	40	35	27	18	15	13	12	11	9
	S	6	6	5	4	4	10	17	22	31	45	58	65	65	57	45	35	30	22	14	11	10	9	8	7
	SW	13	12	10	9	8	14	20	25	29	32	35	50	77	102	116	118	105	74	34	26	21	18	16	14
	W	16	15	13	12	11	16	22	27	31	34	36	37	59	98	132	154	155	122	48	34	28	24	21	18
	NW	12	11	10	9	8	14	20	25	29	32	35	36	38	50	84	108	122	104	37	26	21	18	15	14
	HOR	24	22	19	17	16	31	66	107	145	178	203	217	220	212	192	161	122	81	53	44	38	34	30	27
D	N	8	7	6	6	5	21	22	21	25	27	30	31	32	32	31	29	30	32	17	14	12	11	10	9
	NE	11	10	9	8	7	59	87	93	82	63	51	49	47	46	43	40	35	29	28	19	17	15	14	12
	E	15	13	12	11	10	65	105	123	124	110	84	65	60	57	53	48	43	36	28	25	22	20	18	16
	SE	13	12	11	10	9	36	65	85	96	97	87	70	56	52	49	45	40	33	25	22	20	18	16	15
	S	9	9	8	7	6	11	16	20	27	39	49	56	57	52	43	36	31	26	19	16	15	13	12	11
	SW	20	18	16	14	13	17	21	25	28	31	33	45	67	87	100	103	95	72	41	35	30	27	24	22
	W	25	22	20	18	16	20	24	27	30	33	34	36	53	84	112	131	134	111	55	45	39	34	31	28
	NW	18	17	15	14	12	16	20	24	27	30	32	34	34	45	69	92	104	92	42	34	29	26	23	21
	HOR	37	33	30	27	24	35	62	94	125	153	174	189	195	191	179	157	128	95	72	63	56	51	46	41

Solar Time (Hour of Day)

Source: Adapted from McQuiston, F. and Spitler, J., (1992).

TABLE 6.1.16 Typical Diversity Factors for Occupants in Large Buildings

Building Type	F_d
Apartment	0.40 to 0.60
Industrial	0.85 to 0.95
Hotel	0.40 to 0.60
Office	0.75 to 0.90
Retail	0.80 to 0.90

Source: Adapted from McQuiston, F. and Spitler, J., (1992).

TABLE 6.1.17 Zone Types for Use in Determining the CLF — Interior (i.e., Nonperimeter) Zones Only

Zone Parameters			Zone Type	
Middle Floor	Ceiling Type	Floor Covering	People and Equipment	Lights
Single Story				
N/A	N/A	Carpet	C	B
N/A	N/A	Vinyl	D	C
Top Floor				
2.5 in Concrete	With	Carpet	D	C
2.5 in Concrete	With	Vinyl	D	D
2.5 in Concrete	Without	*	D	B
1 in Wood	*	*	D	B
Bottom Floor				
2.5 in Concrete	With	Carpet	D	C
2.5 in Concrete	*	Vinyl	D	D
2.5 in Concrete	Without	Carpet	D	D
1 in Wood	*	Carpet	D	C
1 in Wood	*	Vinyl	D	D
Mid-Floor				
2.5 in Concrete	N/A	Carpet	D	C
2.5 in Concrete	N/A	Vinyl	D	D
1 in Wood	N/A	*	C	B

* The effect of this parameter is negligible in this case.
Source: Adapted from McQuiston, F. and Spitler, J., (1992).

Appliance and Equipment Loads

Equipment can add heat either through resistive heating or from electrical motors operating in the equipment. The CLTD/SCL/CLF method accounts for both types of equipment heat separately. In addition, the equipment loads are further broken down into sensible or latent components. The latent components are assumed to become immediate loads on the cooling system. The latent loads are found in tables devoted to hospital equipment, restaurant equipment, and office equipment; latent loads are cited only for the hospital and restaurant equipment. An example of these kinds of loads is given in Table 6.1.20.

The sensible component of the loads is adjusted by

$$q(hr) = q_{sa} \cdot CLF(hr) \tag{6.1.74}$$

where q_{sa} is the sensible heat gain per appliance as found from the tables. The cooling load factor is found by first determining the zone type and then by looking up the CLF in a table appropriate for that zone type as was done for the occupants and lighting. While the zone type is similar for occupants and equipment, it may not be the same as for lighting.

TABLE 6.1.18 Occupant Cooling Load Factors for Type A Zones

		\multicolumn{9}{c}{Total Hours That Space Is Occupied}

Hour After Occupants Enter Space		2	4	6	8	10	12	14	16	18
	1	0.75	0.75	0.75	0.75	0.75	0.75	0.76	0.76	0.77
	2	0.88	0.88	0.88	0.88	0.88	0.88	0.88	0.89	0.89
	3	0.18	0.93	0.93	0.93	0.93	0.93	0.93	0.94	0.94
	4	0.08	0.95	0.95	0.95	0.95	0.96	0.96	0.96	0.96
	5	0.04	0.22	0.97	0.97	0.97	0.97	0.97	0.97	0.97
	6	0.02	0.10	0.97	0.97	0.97	0.98	0.98	0.98	0.98
	7	0.01	0.05	0.33	0.98	0.98	0.98	0.98	0.98	0.98
	8	0.01	0.03	0.11	0.98	0.98	0.98	0.99	0.99	0.99
	9	0.01	0.02	0.06	0.24	0.99	0.99	0.99	0.99	0.99
	10	0.01	0.02	0.04	0.11	0.99	0.99	0.99	0.99	0.99
	11	0.00	0.01	0.03	0.06	0.24	0.99	0.99	0.99	0.99
	12	0.00	0.01	0.02	0.04	0.12	0.99	0.99	0.99	1.00
	13	0.00	0.01	0.02	0.03	0.07	0.25	1.00	1.00	1.00
	14	0.00	0.01	0.01	0.02	0.04	0.12	1.00	1.00	1.00
	15	0.00	0.00	0.01	0.02	0.03	0.07	0.25	1.00	1.00
	16	0.00	0.00	0.01	0.01	0.02	0.04	0.12	1.00	1.00
	17	0.00	0.00	0.01	0.01	0.02	0.03	0.07	0.25	1.00
	18	0.00	0.00	0.00	0.01	0.01	0.02	0.05	0.12	1.00
	19	0.00	0.00	0.00	0.01	0.01	0.02	0.03	0.07	0.25
	20	0.00	0.00	0.00	0.01	0.01	0.02	0.03	0.05	0.12
	21	0.00	0.00	0.00	0.00	0.01	0.01	0.02	0.03	0.07
	22	0.00	0.00	0.00	0.00	0.01	0.01	0.02	0.03	0.05
	23	0.00	0.00	0.00	0.00	0.00	0.01	0.01	0.02	0.03
	24	0.00	0.00	0.00	0.00	0.00	0.01	0.01	0.02	0.03

Source: Adapted from McQuiston, F. and Spitler, J., (1992).

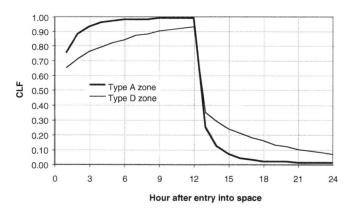

FIGURE 6.1.20 Occupant cooling load factors for Type A and Type D zones for a space that is occupied for 12 hours.

TABLE 6.1.19 Typical Diversity Factors for Lighting in Large Buildings

Building Type	F_d
Apartment	0.30 to 0.50
Industrial	0.80 to 0.90
Hotel	0.30 to 0.50
Office	0.70 to 0.85
Retail	0.90 to 1.00

Source: Adapted from McQuiston, F. and Spitler, J., (1992).

TABLE 6.1.20   Partial List of Recommended Heat Gain for Restaurant Equipment

| Appliance Type | Size | Maximum Input Rating | | Heat Gain Rate (Btu/hr) | | |
| | | | | Unhooded | | Hooded |
		Watts	Btu/hr	Sensible	Latent	Sensible
Barbeque (pit), per 5 lbs of food capacity	80–300 lbs	200	680	440	240	210
Barbeque (pressurized), per 5 lbs of food capacity	45 lbs	470	1600	550	270	260
Blender, per gal of capacity	0.25–1.0 gals	1800	6140	4060	2080	1980
Braising pan, per gal of capacity	27–35 gals	400	1360	720	380	510
Cabinet (large hot holding)	16.3–17.3 ft³	2080	7100	610	340	290
Cabinet (large hot serving)	37.6–40.5 ft³	2000	6820	610	310	280
Cabinet (large proofing)	16.0–17.0 ft³	2030	6930	610	310	280
Cabinet (small hot holding)	3.3–6.5 ft³	900	3070	270	140	130
Cabinet (very hot holding)	17.3 ft³	6150	20,980	1880	960	850
Can opener		170	580	580		0
Coffee brewer	12 cups/2 burners	1660	5660	3750	1910	1810
Coffee heater, per boiling burner	1–2 burners	670	2290	1500	790	720
Coffee heater, per warming burner	1–2 burners	100	340	230	110	110
Coffee/hot water holding urn, per gal of capacity	3.0 gal	460	1570	580	200	260
Coffee urn (large), per gal of capacity	6.0–10.0 gal	2500	8530	2830	1430	1.36
Coffee urn (small), per gal of capacity	3.0 gal	1580	5390	1770	920	850
Cutter (large)	18 in bowl	750	2560	2560		820
Cutter (small)	14 in bowl	370	1260	1260		410
Cutter and mixer (large)	7.5–11.3 gal	3730	12,730	12,730		4060

Source: Adapted from McQuiston, F. and Spitler, J., (1992).

The total cooling load in the space is then found from the sum of the sensible and latent loads. If there is a cooling load due to equipment with electrical motors that run equipment in the space then the space cooling load is incremented by

$$q(hr) = 2545 \cdot \frac{HP}{\eta} \cdot F_l \cdot F_u \cdot CLF(hr) \tag{6.1.75a}$$

where HP is the rated horsepower of the motor, η is the efficiency, F_l is the load factor (power used divided by rated horsepower, typically around 12), and F_u is the motor use factor (accounting for intermittent use). The term 2545 converts from HP to Btu per hour. Equation 6.1.75a assumes that both the equipment and the motor are located within the space. If the equipment is in the space but the motor is located outside the space, then this equation is de-rated by the motor efficiency:

$$q(hr) = 2545 \cdot HP \cdot F_l \cdot F_u \cdot CLF(hr) \tag{6.1.75b}$$

Conversely, if the motor is inside the space but it acts on equipment outside the space, the cooling load is incremented by

$$q(hr) = 2545 \cdot HP \cdot \frac{1-\eta}{\eta} \cdot F_l \cdot F_u \cdot CLF(hr) \tag{6.1.75c}$$

As with the lighting, the CLF is always 1.0 when the cooling system does not operate 24 hours per day.

Air Infiltration
The sensible and latent cooling loads introduced by infiltration are treated the same way in the CLTD/SCL/CLF method as they were in the original CLTD/CLF method. Specifically, the infiltrating air is assumed to immediately become a load on the cooling system.

TABLE 6.1.21   Period Response Factors for Representative Roof Types 1 Through 8, Btu/h·ft²·°F

	Roof Type							
	1	2	3	4	5	6	7	8
Y_{P0}	0.004870	0.000556	0.006192	0.000004	0.000105	0.003675	0.001003	0.003468
Y_{P1}	0.036463	0.012356	0.044510	0.000658	0.002655	0.034908	0.009678	0.022622
Y_{P2}	0.026468	0.020191	0.047321	0.004270	0.007678	0.054823	0.017455	0.045052
Y_{P3}	0.008915	0.012498	0.035390	0.007757	0.008783	0.050193	0.017588	0.047168
Y_{P4}	0.002562	0.005800	0.026082	0.008259	0.007720	0.041867	0.015516	0.042727
Y_{P5}	0.000708	0.002436	0.019215	0.006915	0.006261	0.034391	0.013169	0.037442
Y_{P6}	0.000193	0.000981	0.014156	0.005116	0.004933	0.028178	0.011038	0.032544
Y_{P7}	0.000053	0.000388	0.010429	0.003527	0.003844	0.023078	0.009213	0.028228
Y_{P8}	0.000014	0.000152	0.007684	0.002330	0.002982	0.018900	0.007678	0.024472
Y_{P9}	0.000004	0.000059	0.005661	0.001498	0.002309	0.015478	0.006397	0.021212
Y_{P10}	0.000001	0.000023	0.004170	0.000946	0.001787	0.012675	0.005328	0.018386
Y_{P11}	0.000000	0.000009	0.003072	0.000591	0.001383	0.010380	0.004437	0.015937
Y_{P12}	0.000000	0.000003	0.002264	0.000366	0.001070	0.008501	0.003696	0.013814
Y_{P13}	0.000000	0.000001	0.001668	0.000225	0.000827	0.006962	0.003078	0.011973
Y_{P14}	0.000000	0.000001	0.001229	0.000138	0.000640	0.005701	0.002563	0.010378
Y_{P15}	0.000000	0.000000	0.000905	0.000085	0.000495	0.004669	0.002135	0.008995
Y_{P16}	0.000000	0.000000	0.000667	0.000052	0.000383	0.003824	0.001778	0.007797
Y_{P17}	0.000000	0.000000	0.000491	0.000032	0.000296	0.003131	0.001481	0.006758
Y_{P18}	0.000000	0.000000	0.000362	0.000019	0.000229	0.002564	0.001233	0.005858
Y_{P19}	0.000000	0.000000	0.000267	0.000012	0.000177	0.002100	0.001027	0.005077
Y_{P20}	0.000000	0.000000	0.000196	0.000007	0.000137	0.001720	0.000855	0.004401
Y_{P21}	0.000000	0.000000	0.000145	0.000004	0.000106	0.001408	0.000712	0.003815
Y_{P22}	0.000000	0.000000	0.000107	0.000003	0.000082	0.001153	0.000593	0.003306
Y_{P23}	0.000000	0.000000	0.000079	0.000002	0.000063	0.000945	0.000494	0.002866

Source: Adapted from Spitler, J.D. and Fisher, D.E., (1999).

The Radiant Time Series Method for Hourly Cooling Load Calculations

The radiant time series method is a new method currently under development by ASHRAE. This method is similar to the transfer function method except that the several-day-long iterative computations for the conductive heat flows and room radiative transfer functions have been replaced by a set of 24 response factors that are used directly for the calculations at each hour. The principal differences between the transfer function method and the RTS method are outlined in this section.

Surface Conduction Heat Transfer

The transfer function method uses an iterative process to calculate the conductive heat flow across the roof and wall surfaces of a building. Depending on the driving forces and the wall material, this may require several repetitions of each day's values before the iteration converges. The RTS method replaces this iteration with a simple summation,

$$q_{cond}(hr) = A \cdot \sum_{j=0}^{23} Y_{Pj} \cdot \left(T_{e,\theta-j\delta} - T_r \right) \tag{6.1.76}$$

where A is the surface area, Y_{Pj} is the jth response factor, $T_{e,\theta-j\delta}$ is the sol-air temperature from j hours ago, and T_r is the space temperature, assumed to be constant. The response factors for the walls and roof can be found from lookup tables (such as Table 6.1.21) similar to those created for the transfer function coefficients.

Once the conductive loads have been calculated, the transmitted solar heat and window conductive heat gains through each window, the absorbed solar gain, and the internal gains are calculated the same as with the transfer function method.

TABLE 6.1.22   Typical Radiative Fraction of Building Heat Gains

Heat Gain Type	Typical Radiative Fraction
Occupants	0.7
Suspended fluorescent lighting, unvented	0.67
Recessed fluorescent lighting, vented to return air	0.59
Recessed fluorescent lighting, vented to supply & return air	0.19
Incandescent lighting	0.71
Equipment	0.2–0.8
Conductive heat gain through walls	0.63
Conductive heat gain through roofs	0.84
Transmitted solar radiation	1.0
Solar radiation absorbed by window glass	0.63

Source: Adapted from McQuiston, F. and Spitler, J., (1992).

Radiative and Convective Fractions

When all the heat flows into the building have been calculated, the loads must be further broken down into the radiative and convective components. Table 6.1.22 shows recommended values for the radiative fraction; the convective fraction is simple

$$Load\ convective\ fraction = 1 - Load\ radiative\ fraction \qquad (6.1.77)$$

The convective fraction immediately becomes a cooling load on the building HVAC system. The radiative portion is absorbed by the building materials, furniture, etc. and is convected into the space as a time-lagged and attenuated cooling load as described in the next section.

Conversion of Radiant Loads

The radiant loads are converted to hourly cooling loads through the use of radiant time factors. Similar to the response factors, the time factors estimate the cooling load based on past and present heat gains.

$$q_{cool}(hr) = \sum_{j=0}^{23} r_j \cdot q_{\theta-j\delta} \qquad (6.1.78)$$

where r_0 is the fraction of the load convected to the space at the current time, r_1 is the fraction at the previous hour, and so forth. This step replaces the zone transfer function of the transfer function method.

Two sets of radiative time factors must be determined for each zone: one for the transmitted solar heat gain and one for all other types of heat gain. The difference between the two is that the former is assumed to be absorbed by the floor only while the latter is assumed to be evenly distributed throughout the space. The radiant time factors are determined through a zone heat balance model as described by Spitler et al (1997).

6.1.9   Summary

We have described the tools for calculating heating and cooling loads. The focus has been on peak loads (annual loads and energy consumption are addressed in Chapter 6.2). The procedure begins with the definition of the zones and the choice of the design conditions, followed by a careful accounting of all thermal energy terms, including conduction, air change, and heat gains. The formulas for the load calculation depend on whether the thermostat setpoint is constant or variable. The first case is much simpler, allowing a static calculation (for heating loads) or a quasistatic calculation (for cooling loads). Correct analysis of loads for variable setpoints requires a dynamic method; in such a case, the transfer function method can be used both for heating and for cooling. For latent loads, a static calculation is usually considered sufficient. Further detail on load calculations is contained in ASHRAE (1998).

References

Achard, P. and R. Gicquel (1986). *European Passive Solar Handbook.* Commission of the European Communities, Directorate General XII for Science, Research and Development, Brussels, Belgium.

ASHRAE (1979). *Cooling and Heating Load Calculation Manual.* GRP 158. American Society of Heating, Refrigerating and Air-Conditioning Engineers, Atlanta, GA.

ASHRAE (1999). *Standard 62-99. Ventilation for Acceptable Indoor Air Quality.* American Society of Heating, Refrigerating and Air-Conditioning Engineers, Atlanta, GA.

ASHRAE (1989a). *Handbook of Fundamentals.* American Society of Heating, Refrigerating and Air-Conditioning Engineers, Atlanta, GA.

ASHRAE (1989b). *Standard 90.1-1989: Energy Efficient Design of New Buildings, Except Low-Rise Residential Buildings.* American Society of Heating, Refrigerating and Air-Conditioning Engineers, Atlanta, GA.

ASHRAE (1998). *Cooling and Heating Load Calculation Principles*, American Society of Heating, Refrigerating and Air-Conditioning Engineers, Atlanta, GA.

ASHRAE (2001). *Handbook of Fundamentals.* American Society of Heating, Refrigerating and Air-Conditioning Engineers, Atlanta, GA.

Birdsall, B., W. F. Buhl, K. L. Ellingtop, A. E. Erdem, and F. C. Winkelmann (1990). *Overview of the DOE2.1 Building Energy Analysis Program* Report LBL-19735, rev. 1., Lawrence Berkeley Laboratory, Berkeley, CA, 94720.

Chuangchild, P. and Krarti, M. (2000). Parametric Analysis and Development of a Design Tool for Foundation Heat Gain for Refrigerated Warehouses, *ASHRAE Trans.*, vol. 106, pt. 2.

Dietz, R. N., T. W. Ottavio, and C. C. Cappiello (1985). Multizone Infiltration Measurements in Homes and Buildings Using Passive Perfluorocarbon Tracer Method. *ASHRAE Trans.*, vol. 91, pt. 2.

Fairey, P. W. and A. A. Kerestecioglu (1985). Dynamic Modeling of Combined Thermal and Moisture Transport in Buildings: Effects on Cooling Loads and Space Conditions. *ASHRAE Trans.*, vol. 91, pt. 2A, p. 461.

Grimsrud, D. T., M. H. Sherman, I. E. Ianssen, A. N. Pearman, and D. T. Hanje (1980). An Intercomparison of Tracer Gases Used for Air Infiltration Measurements. *ASHRAE Trans.*, vol. 86, pt. 1.

Kerestecioglu, A. A. and L. Gu (1990). Theoretical and Computational Investigation of Simultaneous Heat and Moisture Transfer in Buildings: "Evaporation and Condensation" Theory. *ASHRAE Trans.*, vol. 96, pt. I.

Kreider, J.F., Rabl, A., and Curtiss, P. (2001). *Heating and Cooling of Buildings: Design for Efficiency, 2nd edition*, McGraw-Hill, New York, NY.

Lydberg, M. and A. Honarbakhsh (1989). *Determination of Air Leakiness of Building Envelopes Using Pressurization at Low Pressures.* Swedish Council for Building Research. Document DI9:1989, Giivle, Sweden.

McQuiston, F. and Spitler, J. (1992). *Cooling and Heating Load Calculation Manual.* ASHRAE.

Mitalas, G.P. and Stephenson, D.G. (1967). Room Thermal Response Factors, *ASHRAE Trans.*, vol. 73, pt 2.

Nisson, I. D. N. and G. Duff (1985). *The Superinsulated Home Book.* John Wiley & Sons, New York, NY.

Norford, L. K., A. Rabl, I. P. Harris, and I. Roturier (1989). Electronic Office Equipment: The Impact of Market Trends and Technology on End Use Demand. In T. B. Iohansson et al., Eds. *Electricity: Efficient End Use and New Generation Technologies, and Their Planning Implications.* Lund University Press, Lund, Sweden, 1989, pp. 427–460.

PREP (1990). Included in *TRNSYS-A Transient System Simulation Program.* Solar Energy Laboratory, Engineering Experiment Station Report 38-12, University of Wisconsin, Madison.

Sherman, M. H., D. T. Grimsrud, P. E. Condon, and B. V. Smith (1980). Air Infiltration Measurement Techniques. Proc. 1st lEA Symp. Air Infiltration Centre, London, and Lawrence Berkeley Laboratory Report LBL 10705, Berkeley, CA.

Sonderegger, R. C. (1978). Diagnostic Tests Determining the Thermal Response of a House. *ASHRAE Trans.*, vol. 84, pt. 1, p. 691.

Spitler, J.D. and Fisher, D.E. (1999). On the relationship between the radiant time series and transfer function methods for design cooling load calulations, *International Journal of HVAC&R Research*, vol. 5, no. 2, pp. 125–138.

Wong, S. P. W. and S. K. Wang (1990). Fundamentals of Simultaneous Heat and Moisture Transfer between the Building Envelope and the Conditioned Space Air. *ASHRAE Trans.*, vol. 96, pt. 2.

6.2 Simulation and Modeling — Building Energy Consumption

Joe Huang, Jeffrey S. Haberl, and Jan F. Kreider

The advent of the oil embargo in 1973 and the subsequent oil shocks led to an increased awareness of the cost of energy consumed to heat and cool buildings. Whereas previous advances in building heating and cooling had been focused mostly on increased comfort and convenience, attention also went to the amount of energy wasted and the need to develop ways to make energy use more efficient.

The first step to understanding building energy utilization is to know how much energy is being used in the building and how that is divided among space conditioning and other uses. Unfortunately, this is not as easy to do as it sounds. Even if that information were readily available, such as how much natural gas and electricity are consumed by a boiler and air conditioner, it would still be difficult to evaluate in terms of energy efficiency, which would require accounting for differences in building size, usage patterns, internal conditions, and climate. As interest has grown in improving building energy efficiency, engineers, energy experts, and analysts have gone more and more to computer-based methods that simulate in detail the energy flows in a building over long extended periods, typically of a year or more.

6.2.1 Steady State Energy Calculation Methods

Degree-Day Method

The earliest method used to estimate the heating energy consumption of a building is the degree-day method first developed in the 1930s by the gas utilities to predict gas consumption. A degree-day is the sum of the number of degrees that the average daily temperature (technically the average of the daily maximum and minimum) is above (for cooling) or below (for heating) a base temperature times the duration in days. Thus, a day where the average daily temperature is 12 degrees lower than the base temperature would accumulate 12 degree-days, as would three days, each of which was 4 degrees below the base temperature. Summed over an entire year, the number of heating or cooling degree-days remains a convenient single number for indicating climate severity.

Historically, gas companies found 65°F to be the most appropriate base temperature for estimating fuel deliveries in the 1930s. The concept behind the degree-day method is that the base temperature represents the "balance point" of a building at which the building's internal heat gains are just sufficient to counterbalance the heat losses to the outside, so that the building requires neither heating nor cooling. Below that balance temperature the building requires heating, while above that temperature the building requires cooling, in proportion to the difference from the base temperature.

Since its invention in the 1930s, base 65°F heating degree-days, and to less of an extent, base 65°F cooling degree-days, have become widely accepted as the most convenient, simple indicators of climate severity. In the U.S., heating degree-days vary from less than 500 in Miami, 1000–3000 in the south, 3000–7000 in the north, to extremes of over 8000 in Bismarck, ND and 10,000 in Anchorage, AK. Correspondingly, cooling degree-days vary from 0 in Anchorage, less than 100 in Seattle, 500–1200 in the north, 1200–3000 in the south, and over 3000 in Phoenix and Miami. Figures 6.2.1 and 6.2.2 show the heating and cooling degree-days for the U.S. averaged over 30 years from 1950 to 1980.

In the degree-day method, the building heat load, i.e., the amount of heating energy input or cooling energy extraction, is estimated as the number of degree-days times 24 (to convert to degree hours), times the overall building heat loss coefficient (Btu/hr °F). The overall heat loss coefficient is the sum of the (U-value x area) for all external surfaces, such as walls, windows, doors, and roof, plus the heat losses or gains due to infiltration. The heating load equation is

$$HL = HDD_{65} * 24 \left[\sum U_i A_i + (\text{Infiltration Air Changes per Hour} \times \text{Volume}) \times 0.018^* \right]$$

---

[*] The constant 0.018 is the product of density and specific heat for air at sea level. For other altitudes this constant must be adjusted by the density ratios. For example, at the altitude of Denver the density is 0.06 lb/ft³ and the constant has a value of 0.0144.

FIGURE 6.2.1   U.S. heating degree-days base 65°F (1950–1980, 30-year averages).

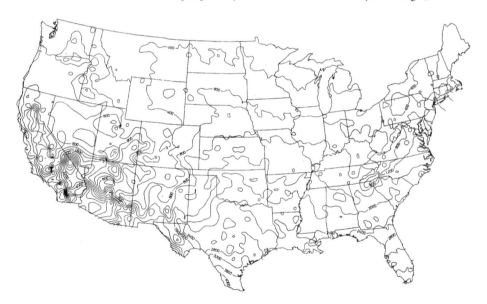

FIGURE 6.2.2   U.S. cooling degree-days base 65°F (1950–1980, 30-year averages).

The cooling load equation is similar except that cooling degree-days are used in place of heating degree-days.

To derive the heating or cooling energy consumption, the heating or cooling loads need to be multiplied by the efficiency of the HVAC system. For example, for a 1800 ft² house in Denver (HDD=5940) with an overall heat loss coefficient of 400 Btu/hr-°F, 0.7 air change rate per hour, 14,400 ft³ volume, heated by a furnace with an AFUE of 0.78, and having an average duct loss factor of 0.10, the total energy use would be

$$HE = [5940 * 24 * (400 + 0.7 * 14000 * 0.0144)]/(0.78 * 0.90 * 1000000) = 110 \; MMBtu$$

Similarly, for the same house in Denver (CDD = 630) cooled by an air conditioner with a SEER of 9.0, the cooling energy use in kWh would be

$$CE = [630 * 24 * (400 + 0.7 * 14000 * 0.0144)]/(9.0 * 0.90 * 1000) = 1010 \text{ kWh}$$

The above example shows the degree-day method in its simplest form. Its limitations are easy to list. The method does not consider the effects of solar heat gain or building thermal mass, nor can it account for variations in infiltration rates, thermostat settings (such as night setback), or occupant actions such as window venting on cool summer nights or during the spring and fall seasons.

When building energy calculations started to attract more attention in the late 1970s and early 1980s, efforts were made to improve the degree-day method by using variable base temperatures, calculating degree-hours instead of days, etc. These are described in the following section. Although these modified degree-day methods have gained in accuracy, this has come at the expense of computational ease, and in most cases calculations can be done conveniently only using a computer program. As the capability of personal computers has grown, degree-day methods have fallen increasingly out of favor because they remain fundamentally steady state calculations that are unable to capture fully the transient heat flows that dominate building thermal processes.

Despite its limitations, there are situations where the simple degree-day method can still be of use. By virtue of its simplicity, it provides a quick answer that can be used as a starting point or check for more detailed calculations. For estimating heating energy use in light construction residential buildings with low solar gain in cold climates, the simple degree-day method may be adequate when gathering the additional data on climate and building conditions may be unwarranted or impractical. The simple and clear-cut formulation of the degree-day method is also valuable as a way to distill the results from more detailed calculations that are often hard to interpret. For example, in the correlation methods described later, detailed hourly simulation results can be presented as the imputed heating degree-days that a building "sees" when the variations in building operations and conditions are considered. This not only helps in visualizing the detailed results, but permits interpolation using the degree-day method for small changes in the building shell.

Variable Base Degree-Day Method

The variable base degree-day method was developed to account for the fact that balance point temperatures vary between buildings and even within a building depending on the time of day. The original 65°F base temperature developed in the 1930s implied that if a building were maintained at 70°F, the heat gains from the sun and internal processes would contribute on average 5°F of "free heat," so that heating was required only when the outside temperature dropped below 65°F. As buildings are now better insulated and more air-tight, their balance point temperatures should be lower. For residential buildings, several studies have found that the average balance point temperature is now 55–57°F, instead of 65°F. For commercial buildings, the combination of low surface-to-volume ratios, high window-to-wall ratios, and high internal gains have caused their balance point temperatures to drop even further to 50°F or lower.

The balance point temperature for a building differs markedly between daytime and nighttime. During the day, the building receives heat gain from the sun as well as from human activity, including equipment and lights. At night, there is, of course, no solar heat gain and human activity is reduced. In a typical residential building, the balance point temperature depression may be around 15°F, while at night it is only 3°F. In a commercial building, the difference can be even greater because of higher internal heat gains during the day and very low heat gains at night.

The variable base degree-day method attempts to account for these different building conditions by calculating first the balance point temperature of a building and then the heating and cooling degree hours at that base temperature. Since the method subdivides the day into daytime and nighttime periods, degree-hours have to be used in place of degree-days. Whereas degree-days are calculated from the average between the daily maximum and minimum temperatures, degree-hours are calculated from the temperature for each hour. In general, the number of (degree-hours)/24 can be from slightly to significantly larger than the number of degree-days at the same base temperature. The differences are particularly significant for cooling. This can be explained by considering what happens on a spring day when the average daily temperature may be below the base temperature, but several afternoon hours are above it. Such a day would have no cooling degree-days but a number of cooling degree-hours. In Washington, the (heating degree-hours)/24 and (cooling degree-hours)/24 are 4% and 5% greater than their respective degree-day values. However, in Sacramento, the differences are 24% and 26%, respectively.

In this method, the degree-day calculations are repeated for daytime and nighttime conditions for each month of the year. The balance point temperature depression (BPD) for any building is calculated as the total heat gains divided by the overall building conductance, i.e.,

$$\text{BPD} = \frac{(\text{solar heat gain} + \text{internal heat gain})_{\text{average per hour}}}{\text{overall building conductance}}$$

For example, a building in Denver with 50ft^2 of double-pane windows (solar heat gain coefficient of 0.79) in each orientation receives average total daily solar heat gains of 9219 Btu on the northside, 24,526 Btu on the eastside, 60,388 Btu on the southside, and 24,846 Btu on the westside. The average hourly solar heat gain (assuming 12 hours as daytime hours) is 9915 Btu/hour. If the building has other internal heat gains of 3200 Btu/hour and an overall building conductance of 576.4 Btu/hr-°F, its BPD would be 13,115/576.4 = 23°F. During the nighttime hours, the building has internal gains of only 1400 Btu/hour, and BPD of 1400/576.4 = 2°F. If the thermostat were maintained at 70°F during the day and 65°F at night, base 47°F (70 − 23°F) heating degree-hours should be used for the daytime hours, while base 63°F (65 − 2°F) heating degree-hours should be used for the nighttime hours.

On the cooling side of the equation, the variable base degree day (VBDD) method follows a similar logic to identify the number of cooling degree-hours that a building actually "sees." The cooling balance point changes dramatically depending on whether the building is being vented. If the windows are open, the heat gains are flushed out of the building and have no effect on its cooling load. However, if the windows are closed, then the solar and internal gains create a balance temperature significantly lower than the thermostat setpoint for cooling. The way these two conditions are handled in a VBDD method is that the cooling degree-hours are calculated using the balance temperature with the windows closed, but they are not accumulated for the hours when the temperatures are below the thermostat setpoint, when the windows are assumed to be open. This is illustrated in Figure 6.2.3, which shows the one-to-one relationship between cooling degree-hours and the temperature difference between the balance and outdoor air temperature. However, those degree-hours occurring in the shaded triangle when the outdoor temperatures are below the thermostat setpoint are considered "vented" and not added to the running total of cooling degree-hours.

Although the simple example shown earlier for the variable degree-day method may not seem very onerous, one must keep in mind that the balance point calculations have to be repeated for each month, and that it requires calculating the solar heat gain through windows (and skylights), which in turns requires calculating sun angles and the distribution of solar gain by orientation. After the balance point temperatures have been derived, the procedure then requires the calculation of heating and cooling degree-hours at different base temperatures.

The net result of this extra detail is to make the method too tedious for either hand calculations or implementation on a spreadsheet program. In the early 1980s, several PC programs were written using the variable base degree-day method. Researchers at LBL developed the CIRA (Computerized Instrumented Residential Audit) program, that was licensed to Burt Hill Kosan Rittelman (www.burthill.com) in 1984, who marketed it under the name EEDO (Energy Efficient Design Options). The CIRA/EEDO program is a DOS program written in BASIC for quick analysis of retrofit potentials and options in residential buildings. After the inspector or analyst has entered basic information about the location, building geometry and thermal conditions, and already installed retrofit measures, the program uses the variable base degree-day method to calculate the base case energy use of the house and the energy savings for 100 or more potential energy retrofits. The program then compares these energy savings to the costs

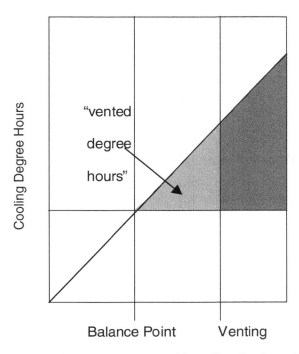

FIGURE 6.2.3 Schematic representation of "vented" cooling degree-hours.

of the measures, and produces a report listing the recommended retrofit measures in order of their cost effectiveness.

The CIRA/EEDO program contains weather data for 200 U.S. locations, including monthly solar heat gain by orientation, and a regression-based technique to interpolate monthly heating and cooling degree-hours at any base temperature. In calculating the effects of night setback on heating loads, the program uses correlations to hourly simulations done with the BLAST program (see section on Simulations) to account for the thermal lag due to the building thermal mass.

Variable base degree-day programs like CIRA represent the best that can be expected of this steady state method. The CIRA/EEDO program was very appropriate at the time it was developed given the limited capabilities of personal computers at the time, and its intended use for analysis of retrofit measures for residential buildings. Comparisons of CIRA results to those using the DOE-2 hourly simulation program done by one author (Huang) in 1984 showed good agreement for heating energy use, but some discrepancies in cooling energy use. The comparisons are particularly divergent in places such as Miami, since CIRA does not calculate latent cooling loads or the energy used for dehumidification. Further refinement of the degree-day method does not seem warranted because (1) it remains a steady state method that does not recognize a building's thermal history, and (2) additional refinement of degree-day terminology would require detailed processing of hourly weather data at a complexity approaching their use in hourly simulations.

Bin and Modified Bin Methods

The next evolution in building energy calculations from the various degree-day methods are the bin methods, which have also developed into several variations. The underlying assumption of the bin method is that for a given temperature at the same general time of day (morning, afternoon, evening, etc.), the heating and cooling loads of a building should be roughly the same. Therefore, one can derive a building's annual heating and cooling loads by calculating its loads for a set of "snapshots" defined by temperature "bins," multiplying the calculated loads by the number of hours represented by each bin, and then totaling the sums to derive the building's annual heating and cooling loads. In the original formulation of the bin method, there was no accounting of the effects of solar gain or wind on the calculated loads, which were done simply as the difference between the outdoor and indoor balance temperatures times the building conductance, divided by the efficiency of the heating or cooling system.

Later versions of the bin method accounted for these effects by using more detailed binned data that gave the average wind speeds and solar gains by month, and the number of hours within bins separated by month as well as time of day. When doing the calculation, the solar and wind effects are taken into account for each month and time of day period. The bin method is described in Kreider, Rabl, and Curtiss (1994).

6.2.2 Dynamic Hourly Simulation Methods

The previous methods described are all steady state calculations, the differences being the number of snapshots used to characterize the energy use of the building over the entire year. Even the most complex of these methods still misses the dynamic response of the building to changes in the weather or the building controls. As public interest in building energy use increased in the late 1970s, a number of general purpose computer programs have been developed to simulate the energy flows of a building, including its system and plant, on an hourly or even subhourly basis for an entire year. These efforts have been largely funded by branches of the federal government, notably the Department of Energy and Department of Defense, and some state government offices, such as the California Energy Commission.

The simplest simulation programs use networks that are the thermal equivalents to electrical RC circuits. Temperatures are represented by voltages, heat flows by currents, and thermal masses by capacitances. Network programs are generally limited to smaller, one-zone buildings such as residences or small office buildings that have shell-dominant loads and simple heating and cooling systems. Network programs that are still widely used today include the *CalRes* program, mandated by the California Energy Commission as the official program for showing compliance to California's Title-24 Building Energy Standard in residential buildings, *SeriRes* and *ENERGY-10*, both developed by the National Renewable Energy Laboratory (NREL, formerly known as the Solar Energy Research Institute, or SERI).

Energy-10

ENERGY-10 is a recent software product completed in 1996 through a partnership of the Passive Solar Industries Council (e-mail address: psicdc@aol.com), NREL, LBNL, and the Berkeley Solar Group with funding from DOE. The aim of the program is to provide a user-friendly simulation tool for the design of passive solar strategies in small and medium-sized buildings under 10,000 ft^2. The CNE simulation engine of *ENERGY-10* is a two-zone network model that runs on an hourly time-step. Substantial effort was made to add capabilities to CNE to model passive solar and energy-efficient strategies as daylighting, solar orientation, thermal mass, ventilation, and ground-coupled cooling.

Because the objective of *ENERGY-10* is to encourage architects and engineers to incorporate passive solar design strategies in the early design phase of a project, the user interface requires a minimum number of inputs and has an Auto-Build feature that automatically generates two building files at once — one for the proposed design and the other for a generic reference design of the same size and usage pattern. The Auto-Build feature assists users in quickly evaluating the merits of a proposed design or design strategies. Figure 6.2.4 is a sample input screen showing the Windows-based input procedure of *ENERGY-10* which, at a minimum, requires only five inputs — location, building use category, size, HVAC, and utility rates — to make an initial simulation. Figure 6.2.5 shows a second level, more detailed input screen, once the user has more specific data on the proposed building. In keeping with the philosophy for a quick and simple-to-use design tool, *ENERGY-10* also presents the program output in a highly graphical manner. Figures 6.2.6 and 6.2.7 show sample output for the total heating and cooling energy costs and heat flows for a proposed design compared to the reference case.

DOE-2 and BLAST

For larger and more complex buildings, the two most widely used public-domain whole-building simulation programs are DOE-2 and BLAST. In contrast to *ENERGY-10*, these two are much more like standard engineering programs rather than design tools; the primary work has gone into algorithm

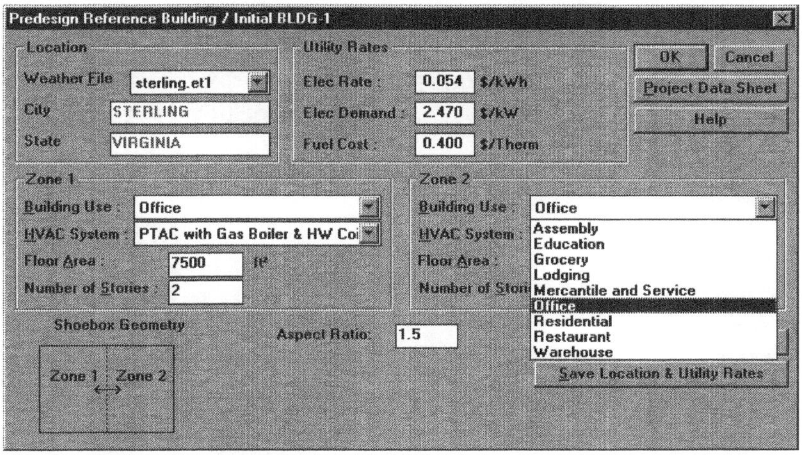

FIGURE 6.2.4 *ENERGY-10* input screen (1).

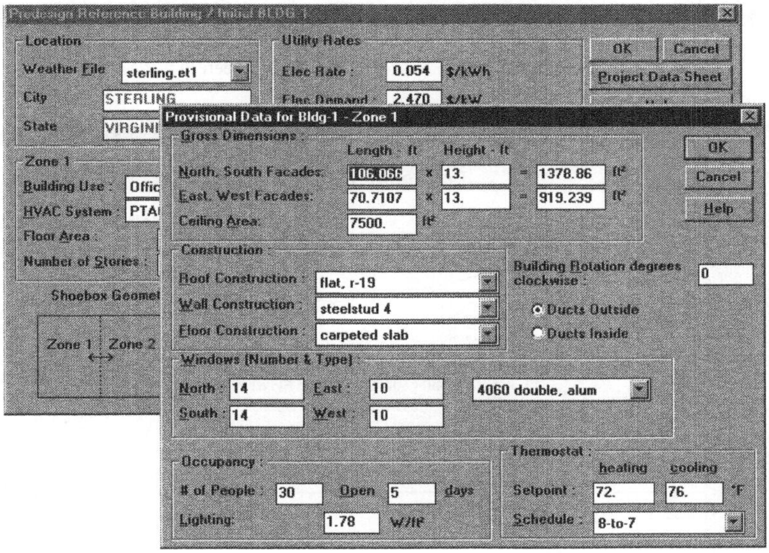

FIGURE 6.2.5 *ENERGY-10* input screen (2).

development and numerical analysis, rather than the user interface (although private vendors have since developed several graphical interfaces). In addition to DOE-2 and BLAST, there are also a number of private sector programs, developed primarily for practicing engineers to design HVAC systems, that can also do annual simulations, including TRACE (developed by the Trane Company) and HAP (developed by Carrier Corporation). This review will not cover these proprietary programs since their calculational routines are not publicly available. The general sense of these proprietary programs is that they are less detailed than the two primary public-domain programs in their loads calculation but are comparable in their system simulations.

DOE-2 is a public domain program originally started by the Lawrence Berkeley National Laboratory (LBNL) in 1979 in collaboration with Los Alamos Scientific Laboratory and Argonne National Laboratory, with support from the U.S. Department of Energy (DOE). For the past 20 years, LBNL has continued to develop and maintain the program, the current (and probably last) public version of the program

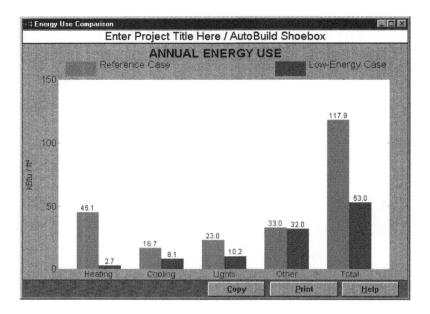

FIGURE 6.2.6 *ENERGY-10* output screen (1).

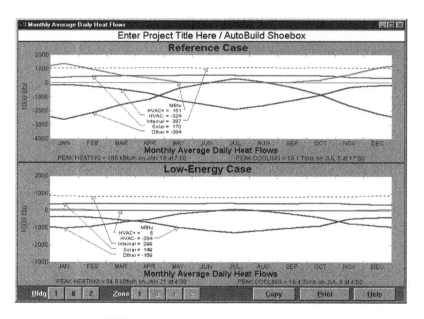

FIGURE 6.2.7 *ENERGY-10* output screen (2).

being DOE-2.1E, released in 1993. The basic DOE-2 program uses a text-based input and output procedure that is quite powerful, but unfortunately difficult and time-consuming to learn.

Mainframe and PC versions of DOE-2, as well as the FORTRAN source code, are available from the Energy Science and Technology Software Center (ESTSC) in Oak Ridge, TN (e-mail address: estsc@adonis.osti.gov). There are also more than a dozen PC versions or derivative software packages based on DOE-2 being sold by private vendors, some of whom have added graphical user interfaces to make the program easier to use. However, these interfaces generally come at the cost of some loss of modeling capability and a fundamental understanding of how the program works. They may be very helpful in introducing DOE-2 at a basic level, but may at some point hinder experienced users from tapping into more advanced features such as user-defined functions, macro expressions, or purposely tweaking the model for specific applications. A good source of information on DOE-2 (as well as BLAST) software resources, consultants, and Web sites is the *Building Energy Simulation User News*, a newsletter published quarterly by the Simulation Research Group of LBNL (e-mail address: KLEllington@lbl.gov).

The BLAST program was developed by the U.S. Army Construction Engineering Research Laboratory (CERL) with funding from various U.S. Department of Defense agencies, and first released in 1977. Since 1983, BLAST has been maintained and supported by the Building Systems Laboratory at the University of Illinois at Urbana-Champaign. The current version is BLAST 3.0, which was completed in 1980. The BLAST program contains three main subprograms — *Space Load Prediction* which calculates the building's space conditioning loads, *Air Distribution System Simulation* which models the performance and control of the air handling system to meet the previously calculated loads, and *Central Plant Simulation* which models the performance and energy usage of the boilers, chillers, and other equipment that supply the heating and cooling needed by the air handling system. In recent years, the Building System Laboratory has also developed a Windows-based graphical interface, the *Heat Balance Loads Calculator* (HBLC), which allows users to visualize the building model as it is being developed. PC versions of the BLAST program and HBLC are both available from the BSL at the University of Illinois (e-mail address: www.bso.uiuc.edu). As with DOE-2, new developments and current information about BLAST appear in the *Building Energy Simulation User News* newsletter published quarterly at LBNL.

DOE-2 and BLAST have both been maintained for close to 20 years, and can be regarded as mature software products if one accepts their 1970s software architecture and modeling techniques.

Starting in 1996, DOE began developing a new building simulation program, EnergyPlus, with a core development team consisting of staff from LBNL, the University of Illinois BSL, and CERL, thus combining the experience and expertise of the original DOE-2 and BLAST development teams. In addition to improving the simulation techniques and capabilities, other goals and characteristics of the EnergyPlus program are

- To create a modular software platform to facilitate future enhancements by other researchers and thus eliminate the bottleneck in software development found with DOE-2 and BLAST due to their arcane structure and coding
- To maintain EnergyPlus as a calculational engine and leave the development of user-interfaces to third-party private ventures. As of 2001, two beta versions of EnergyPlus were to be released to interested reviewers and a number of licensing agreements signed with potential interface developers.

The following discussion of building energy simulation programs focuses on their underlying calculation techniques, input data requirements, and output variables, using DOE-2 as an example. When talking about DOE-2, we refer to the original program with its text-based batch input procedure. This is not meant to downplay the benefits of commercial graphical interfaces, but more a reflection of the author's own experience; it also reflects more directly the operations of the basic program itself. There is no getting around the fact that making a credible computer simulation of a large building requires a sizeable amount of information and knowledge about building physics and operations, as well as equipment performance and controls, that cannot be avoided. Each portion of this input information may seem reasonable to the professionals in that discipline — the building geometry and construction materials to the architects, the HVAC equipment characteristics and operations to the mechanical engineer, the lighting schedules and operating hours to the building managers, the weather data to the meteorologist, etc. — but *in toto* may be difficult when requested of a single person attempting to put together a DOE-2 computer input file. A graphical user interface may simplify this task by using ready-made input templates, but users should at least be aware of the range of input assumptions that are required.

General Modeling Technique and Capabilities of the DOE-2 Program

Despite many similarities, there are some significant differences between DOE-2 and BLAST. Both programs simulate building energy use in sequential fashion, modeling first the building's heating and cooling loads, then the actions of the HVAC system to meet those loads, and finally the actions of the central plant to provide the energy required by the system. DOE-2 adds a further economics module that calculates the cost of the consumed energy, which can be difficult to do in places that have complicated utility tariffs. This sequential modeling approach results in a weak coupling between the three modeling steps and necessitates some simplifications. For example, because the actual zone temperature is calculated only in the systems subprogram, the loads calculations are done using a constant zone temperature and then adjusted using a steady state approximation during the systems simulation.

Whereas in the simpler degree-day or bin methods, the calculations are done for an average or a number of average conditions and then aggregated to yearly totals, in the hourly simulation method the heat flows, building conditions, system operations, etc. are tracked hour by hour through the entire year. This includes the hourly variations in heat conduction through the walls, roof, windows, doors, and floor, solar heat gain on the walls and roof and through the windows, convection due to air infiltration through cracks and leaks in the building envelope, and internal heat gain from people, lights, and equipment.

To calculate heat conduction through building surfaces, both DOE-2 and BLAST use the transfer function or response factor technique, but in somewhat different formulations. This analytical technique was developed in the 1980s to model dynamic heat flow by characterizing it as a time series of thermal responses at different faces of a building surface to a unit excitation at either the inside or outside face, either a heat flux pulse, as in DOE-2, or a temperature pulse, as in BLAST. Once the transfer functions or response factors for a building surface have been calculated, they are then used to calculate the dynamic heat flows hour by hour based on the varying excitations for each hour. Transfer functions or response factors capture the effects of thermal mass in dampening heat flows through building structures. The impact of wind on heat conduction is taken into account by varying the outside air surface coefficient depending on the wind speed shown in the weather file, with adjustments for neighborhood wind shielding effects.

To calculate solar heat gain, hourly simulation programs such as DOE-2 or BLAST need to track the position of the sun, the amount of direct and diffuse solar radiation, the orientation of the building surface, the relative position and size of shading surfaces such as overhangs, fins, neighboring buildings, as well as self-shading from other parts of the building, and in the case of windows, the transmission characteristics of the glazing depending on the sun angle. To calculate convective heat flows due to infiltration, various models are used to estimate the amount of air leakage based on temperature differences, wind speeds, and local shielding factors. To calculate internal heat gain, DOE-2 or BLAST relies on user input schedules, e.g., the number of people in a building depending on time of day and the day type, energy intensities, e.g., the amount of sensible and latent heat gain per person, and, for certain types of equipment such as stoves or boilers, the fraction of heat gain that actually remains in the space.

Perhaps the most substantial difference between DOE-2 and BLAST is in how the two programs calculate the zone heating or cooling loads. After the heat flows to a zone have been determined, DOE-2 uses the weighting factor method to compute the zone's cooling load. Weighting factors are similar to response factors but relate the thermal response of an entire space rather than that of an individual building surface. BLAST, on the other hand, uses a heat balance method that models the energy exchange between all the surfaces making up the zone. The weighting factor method is quicker but has the disadvantage that the zone properties must be constant throughout a simulation. For this reason, DOE-2 has difficulty in modeling strategies that affect the zone properties, such as movable night insulation or increased thermal coupling to the air with ventilative cooling.

The DOE-2 program was designed as a whole-building energy analysis program for large commercial buildings but has been used to model anything from single-zone residential houses to large skyscrapers with up to 128 thermal zones, hundreds of surfaces, and dozens of schedules for occupancy, equipment use, and equipment controls.

The DOE-2 *Systems* subprogram has a library of 26 system types, each with an assumed configuration and default characteristics. These are listed in Table 6.2.1. Different pieces of HVAC equipment are modeled using two performance curves, one giving its full-load performance (efficiency or coefficient of performance) as a function of outdoor air conditions, and the other its part-load performance as a

TABLE 6.2.1   HVAC Systems Types Modeled in DOE-2

Code-Word	Description of System
	Single Supply Duct Types
SZRH	Single-Zone Fan with Optional Sub-Zone Reheat
PSZ	Packaged Single-Zone
SZCI	Single-Zone Ceiling Induction
RHFS	Constructed Done, Reheat Fan System
VAVS	Variable Volume Fan
PIU	Power Induction Unit
PVAVS	Packaged Variable-Air Volume
PVVT	Packaged Variable-Volume, Variable-Temperature
PTGSD	Packaged Total Gas Solid Desiccant
CBVAV	Ceiling Bypass
EVAP-COOL	Evaporative Cooling
	Air Mixing Types
MZS	Multi-Zone Fan
PMZS	Packaged Multi-Zone Fan
DDS	Dual Duct Fan
	Terminal Units
TPFC	Two-Pipe Fan Coil
FPFC	Four-Pipe Fan Coil
TPIU	Two-Pipe Induction Unit
FPIU	Four-Pipe Induction Unit
PTAC	Packaged Terminal Air-Conditioner
HP	Heat Pump
	Residential
RESYS	Residential
RESVVT	Residential Variable-Volume, Variable-Temp
	Heating Zone
FPH	Floor Panel Heating
HVSYS	Heating and Ventilating
UHT	Unit Heater
UVT	Unit Ventilator
	Diagnostics
SUM	Sum Zone Loads

function of the part-load ratio (fraction of full-load operation during the hour). The performance curves, as well as the control and operation of the configured system, can be modified by the user, but the modeling of innovative systems or nonstandard configurations would require changes to the original source code.

The DOE-2 *Plant* subprogram models the performance and operation of large plant heating and cooling equipment, including boilers; electric, gas-fired, or engine-driven chillers; cooling towers; thermal storage systems; electric generators; and the parasitic energy use of pumps and fans. Like in the *Systems* subprogram, the full- and part-load performance of the plant equipment are modeled using various curves.

The DOE-2 *Economics* subprogram allow users to input utility rate structures, first costs, and maintenance and overhaul costs in order to compute the operational costs, energy savings, investment statistics, and overall life-cycle costs.

TABLE 6.2.2 Comparison of General Features and Capabilities of DOE-2, BLAST, IBLAST, and EnergyPlus

	DOE-2	BLAST	IBLAST	EnergyPlus
Integrated Simultaneous Solution	No	No	Yes	Yes
Integrated loads/system/plant				
Iterative solution				
Tight coupling				
Multiple Time Step Approach	No	No	Yes	Yes
User-defined time step for interaction between zones and environment (15-minute default)				
Variable time step for interactions between zone air mass and HVAC system (>1 minute)				
Input Functions	Yes	No	No	Yes
User can modify code with reprogramming				
New Reporting Mechanism	No	No	No	Yes
Standard reports				
User-definable report with graphics				

Source: From Crawley, D.B. et al. (2000). With permission.

In terms of ability to model specific heat flows, equipment types, or control strategies, in some cases DOE-2 may be more accurate, while in other cases BLAST might be more accurate. Tables 6.2.2, 6.2.3, and 6.2.4 compare a number of salient features and modeling capabilities of DOE-2, two versions of BLAST, and the new EnergyPlus program.

Overall Structure of the DOE-2 Program

DOE-2 consists of three separate programs:

doebdl — an input processor that reads the input file, checks for syntax, logic, and data completeness, supplies defaults when no input values are given, computes response factors and weighting factors, and produces an output ASCII file for debugging; if no errors are found, *doebdl* produces binary files used as input by *doesim*.

doesim — the main simulation program that models the energy use of a building for specified run-periods of up to a year; *doesim* consists of four subprograms that are run sequentially: LOADS, SYSTEMS, PLANT, and ECONOMICS.

doewth — a stand-alone weather processing program to convert raw weather data into DOE-2's required binary format.

A schematic of the DOE-2 program is shown in Figure 6.2.8. The left half of the figure shows the input processing in *doebdl*, while the right half shows the simulation steps in *doesim* starting at the top. For each subprogram, i.e., LOADS, SYSTEMS, PLANT, and ECONOMICS, there is a parallel section in the *doebdl* processor. DOE-2 can be stopped after any of the subprograms. Conversely, it can skip directly to a later subprogram if the output binary files from the previous subprogram have been saved.

Weather Data for Hourly Simulations

Hourly simulation programs require detailed hourly weather data. Required are 8760 hourly observations, at the minimum: dry-bulb temperature, wind speed, and direct and diffuse solar radiation. DOE-2 and BLAST also require some moisture measure, i.e., wet-bulb or dewpoint temperature, absolute or relative humidity, along with atmospheric pressure. Useful also are wind direction and sky cover. With one notable exception, all these data are reported at major airport weather stations. Measured solar radiation, however, is available only from very few research sites, which, moreover, tend not to be major urban centers. For most sites, the only alternative is to estimate the amount of solar radiation based on the reported cloud cover and sky conditions. Estimated solar radiation on an annual basis compares well, but hourly values can be off substantially.

Weather data used in hourly simulations can be categorized as either typical or actual year. Typical year data are likely to be available in the format needed by the individual simulation programs, either from software venders or institutions maintaining the programs, while actual year data exist only in raw

TABLE 6.2.3   Comparison of Loads Features and Capabilities of DOE-2, BLAST, IBLAST, and EnergyPlus

	DOE-2	BLAST	IBLAST	EnergyPlus
Heat Balance Calculation	No	Yes	Yes	Yes
Simultaneous calculation of radiation and convection processes each time step				
Interior Surface Convection				
Dependent on temperature and air flow	No	Yes	Yes	Yes
Internal thermal mass	Yes	Yes	Yes	Yes
Moisture Absorption/Desorption	No	No	Yes	Yes
Combined heat and mass transfer in building envelopes				
Thermal Comfort	No	Yes	Yes	Yes
Human comfort model based on activity, inside dry-bulb, humidity, and radiation				
Anisotropic Sky Model	Yes	No	No	Yes
Sky radiance depends on sun position for better calculation of diffuse solar on tilted surfaces				
Advanced Fenestration Calculations	Yes	No	No	Yes
Controllable window blinds				
Electrochromic glazing				
WINDOW 4 Library	Yes	Yes	Yes	Yes
More than 200 window types — conventional, reflective, low-E, gas-filled, electrochromic				
User defined using WINDOW 4				
Daylighting Illumination and Controls	Yes	No	No	Yes
Interior illuminance from windows and skylights				
Step, dimming, on/off luminaire controls				
Glare simulation and control				
Effects of dimming on heating and cooling				

Source: From Crawley, D.B. et al. (2000). With permission.

TABLE 6.2.4   Comparison of HVAC Features and Capabilities of DOE-2, BLAST, IBLAST, and EnergyPlus

	DOE-2	BLAST	IBLAST	EnergyPlus
Fluid Loops	No	Yes	Yes	Yes
Connect primary equipment and coils				
Hot water, loops, chilled water and condenser loops, refrigerant loops				
Air Loops				
Connects fans, coils, mixing boxes, zones	No	No	No	Yes
User-Configurable HVAC Systems	No	No	No	Yes
Hardwired Template HVAC Systems	Yes	Yes	Yes	No
High-Temperature Radiant Heating	No	Yes	No	Yes
Gas/electric heaters, wall radiators				
Low-Temperature Radiant Heating	No	No	Yes	Yes
Heated floor/ceiling				
Cooled ceiling				
Atmospheric Pollution Calculation	Yes	Yes	No	Yes
CO_2, SO_x, NO_x, CO, particulate matter and hydrocarbon production				
On-site and at power station				
Calculate reductions in greenhouse gases				
SPARK Connection	No	No	No	Yes
TRNSYS Connection	No	No	No	Yes

Source: From Crawley, D.B. et al. (2000). With permission.

form from archival sources and must be processed into the formats needed by the simulation program. Typical year weather data are useful for evaluating expected building energy performance or complying with building energy standards, but actual year data must be used for reconciling actual energy consumption records.

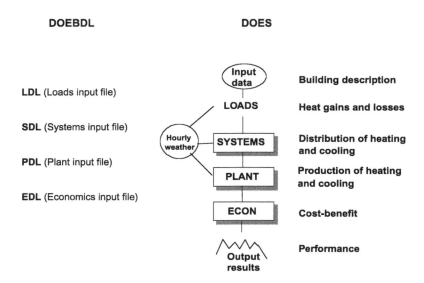

FIGURE 6.2.8   Data flow in DOE-2. (From Winkelmann, F. *The User's News*. With permission.)

Typical year weather files attempt to represent average weather conditions for a location over many years. These are often a synthetic year made up of 12 actual but typical months, as in the TMY2 (Typical Meteorological Year, 2nd version) files produced by the National Renewable Energy Laboratory (NREL, e-mail address: www.nrel.gov) for 239 U.S. sites, or the WYEC2 (Weather Year for Energy Calculations, 2nd version) files produced by the American Society of Heating, Refrigerating, and Air-Conditioning Engineers (ASHRAE, e-mail address: www.ashrae.org) for 55 U.S. and Canadian locations. Both the TMY2 and WYEC2 weather files have hourly solar radiation data from different cloud and sky models. An older, now less commonly used, form of typical year data are the Test Reference Year (TRY) files, available from the National Climatic Data Center (NCDC, e-mail address: www.ncdc.noaa.gov) which are actual years chosen as the most representative for nearly 60 U.S. locations. TRY weather files have no solar radiation data, only cloud cover information. TRY data are widely used in Europe.

Actual year weather data are routinely available for several hundred major airports from the NCDC going back many decades. These weather files also have no solar radiation, which must be estimated from the cloud and sky cover information either by hand or, more typically, by the weather processing utility programs that accompany a simulation program.

Starting in 1997, a new problem has arisen with U.S. and Canadian weather data due to the decision by the meteorological authorities, such as the National Weather Service and the Federal Aviation Authority in the U.S., to replace manual observations with Automated Surface Observing Stations (ASOS). The new ASOS weather stations promise more consistent and reliable weather recordings, but these data lack even the cloud and sky condition records needed for estimating solar radiation. Consequently, until new procedures or additional instrumentation are added that record solar radiation, the new ASOS weather data will not be usable in hourly building energy simulations. The installation of ASOS in major U.S. airport locations was planned to be completed by the end of the year 2000.

DOE-2 Building Input File

The primary interaction between the user and DOE-2 is through a building input file written in DOE-2's Building Description Language, or BDL. BDL is a pseudo-English text-based input format consisting of a DOE-2 keyword followed by the user input, which could be either numeric, such as *HEIGHT*=10 or *HEIGHT* 10 (the = sign is optional), or a user-defined character name followed by a DOE-2 keyword, such as WEST_WALL= *EXTERIOR-WALL*. All DOE-2 input commands are terminated by a double period (..). For example, the inputs for an example building wall in DOE-2 BDL is shown next (in English units), with the DOE-2 keywords indicated in italics:

STUCCO *MATERIAL*
   THICKNESS 0.073 *CONDUCTIVITY* 0.4167 *DENSITY* 166 *SPECIFIC-HEAT* 0.20 ..
CONCRETE *MATERIAL*
   THICKNESS 0.633 *CONDUCTIVITY* 0.80 *DENSITY* 144 *SPECIFIC-HEAT* 0.14 ..
EXT-WALL-LAY *LAYER*
   MATERIALS (STUCCO, CONCRETE) ..
EXT-WALL-CON *CONSTRUCTION*
   LAYER = EXT-WALL-LAY ..
WEST_WALL *EXTERIOR WALL*
   HEIGHT 10 *WIDTH* 35
X 10 Y 120 Z 0
   CONSTRUCTION EXT-WALL-LAY ..

The first two inputs describe the thermal properties of stucco and concrete. The third composes these two materials into a layer, which in turn is referenced as a construction. The remaining inputs describe the dimensions of the wall, locates it in the building's coordinate system, and assigns it to the wall construction just mentioned.

Figure 6.2.9 shows a sample complete DOE-2 input file for a simple box-shaped house. For any DOE-2 simulation, the input file has to define not only the thermal characteristics and geometrical layout of the physical building, but also the hour-by-hour variations in the internal conditions and the operational characteristics and control of the HVAC system and plant. Key input items include the internal loads or heat gains produced by occupants, lights, and equipment, the thermostat settings and schedules for the HVAC system, and its full-load and part-load performance.

It is not the intention of this discussion to explain the DOE-2 input file in detail, but to give a general sense of what inputs are needed to do a DOE-2 simulation of a building. The "$" indicates comments which are concluded with another "$."

DOE-2 Outputs

A DOE-2 simulation can produce more than 20 verification and nearly 50 different output files. The verification reports summarize the input building parameters, such as the number, orientation, area, and U-values of walls and windows in the building, and useful primarily for checking that the input data have been correctly entered.

The output files in *Loads* give the peak and monthly heating and cooling loads of the building, and their breakdown by building component or heat flow path. Figure 6.2.10 shows a sample *Loads* output report giving the peak heating and cooling loads of a building, while Figure 6.2.11 shows the monthly breakdown of heating and cooling loads by building component. These loads are approximate, as they are calculated at an assumed constant zone temperature, and categorized as heating whenever there is a net heat loss from the building, and as cooling whenever there is a net heat gain to the building. These loads bear only approximate similarity to the actual heating and cooling loads computed in the *Systems* subprogram, which takes into account actual thermostat settings and deadband and free cooling through economizers or natural ventilation. These *Loads* reports, however, are still useful in showing the relative magnitude of heat flows through different parts of a building and identifying possible areas of concern.

The output files in *Systems* give, among others, the design specifications for the space conditioning system, the actual peak and seasonal heating and cooling loads imposed on the system, and the energy used to meet these loads. If the building is modeled with a central plant, most of the *System* loads will be passed to the plant, and only energy used for zone-level heating and cooling will appear in the *System* reports. Since the *System* simulation considers only the zone- or building-level loads, the load breakdowns by building component shown in the *Loads* report do not appear in the *Systems* reports.

Figure 6.2.12 shows a sample *System* report listing the design parameters of the space conditioning system. The capacities of the heating and cooling equipment and the fan system are determined in one of three ways — (1) specified by the user, (2) sized by DOE-2 based on design-day conditions specified by the user, or (3) sized by DOE-2 based on the peak loads from the *Loads* simulation. Because DOE-2 is a

```
INPUT LOADS ..
TITLE      LINE-1 *SINGLE FAMILY RESIDENCE*
           LINE-2 *SAMPLE DOE-2.1E INPUT FILE *  ..

ABORT                    ERRORS ..
           DIAGNOSTIC         WARNINGS  ..
           RUN-PERIOD         JAN 1 2000 THRU DEC 31 2000 ..
           LOADS-REPORT       SUMMARY=(LS-C,LS-D)  ..

           BUILDING-LOCATION
                      SHIELDING-COEF=.19          $Site parameters for$
                      TERRAIN-PAR1=.85            $building obstructions$
                      TERRAIN-PAR2=.20            $and ground roughness$
                      WS-TERRAIN-PAR1=.85         $used in Sherman-Grimsrud$
                      WS-TERRAIN-PAR2=.20 ..      $infiltration method$

$ ------Schedule----- $
SCH-1      =DAY-SCHEDULE       (1) (.024)  (2) (.022)  (3,5) (.021)
                               (6) (.026)  (7) (.038)  (8) (.059)
                               (9) (.056) (10) (.060) (11) (.059)
                              (12) (.046) (13) (.045) (14) (.030)
                              (15) (.028) (16) (.031) (17) (.057)
                           (18,19) (.064) (20) (.052) (21) (.050)
                              (22) (.055) (23) (.044) (24) (.022)  ..
INT-LDS-1 =SCHEDULE          THRU DEC 31 (ALL) SCH-1 ..

$ -----Materials----- $
STUD-2    =MAT     TH=.4583 COND=.0667 DENS=32  S-H=.33 $2X6 Stud$ ..
R19WALL   =MAT     TH=.4583 COND=.0337 DENS=7.6 S-H=.23 $R-19 w/0.20 Stud$ ..
INSUL-1   =MAT     TH=.2917 COND=.0265 DENS=1.5 S-H=.2  $R-11 Insulation$ ..
INSUL-3   =MAT     TH=.7917 COND=.0264 DENS=1.5 S-H=.2  $R-30 Insulation$ ..
SHEATH-1  =MAT     TH=.0417 COND=.0382 DENS=22  S-H=.31 $1/2-in Sheathing$ ..
DRYWALL-1 =MAT     TH=.0417 COND=.0926 DENS=50  S-H=.26 $1/2-in Drywall$ ..
AL-SIDE-1 =MAT     TH=.0104 COND=.0171 DENS=170 S-H=.29 $Alum Siding$ ..
AS-SHG-1  =MAT     TH=.0208 COND=.0473 DENS=70  S-H=.30 $Asphalt Shingle$ ..
PLYW-1    =MAT     TH=.0417 COND=.0667 DENS=34  S-H=.29 $1/2-in Plywood$ ..
AT-AIR-1  =MAT     RES=1.3                               $Attic Air Space$ ..
EXP-POLY-1=MAT     TH=.0833 COND=.0177 DENS=2.2 S-H=.29 $1-in Polystyrene$ ..
EXP-POLY-2=MAT     TH=.1667 COND=.0177 DENS=2.2 S-H=.29 $2-in Polystyrene$ ..
CONCRETE-1=MAT     TH=.3333 COND=.7576 DENS=140 S-H=.2  $4-in concrete$ ..
CONCRETE-2=MAT     TH=.6667 COND=.7576 DENS=140 S-H=.2  $8-in Concrete$ ..
CARP/PAD-1=MAT     RES=2.08                              $Carpet and pad$ ..
DRYSOIL   =MAT     TH=.5    COND=.5    DENS=125 S-H=.2  $Dry Soil$ ..

$ ------Glazing----- $

WINDOW-1  =GLASS-TYPE       GLASS-TYPE-CODE=2002
                           FRAME-ABS=0.9
                           SPACER-TYPE-CODE=0 ..

$ ------Constructions----- $

IWLAY-1   =LAYERS       MAT=(AL-SIDE-1,SHEATH-1,R19WALL,DRYWALL-1) ..
INS-WL-1  =CONS         LAYERS=IWLAY-1   ROUGHNESS=3 ..
IRLAY-1   =LAYERS       MAT=(AS-SHG-1,PLYW-1,AT-AIR-1,INSUL-3,
                            DRYWALL-1) I-F-R=.61 ..
INS-RF-1  =CONS         LAYERS=IRLAY-1   ABS=.86 ..
SRLAY-1   =LAYERS       MAT=(AS-SHG-1,PLYW-1,AT-AIR-1,INSUL-1,STUD-2,
                            DRYWALL-1) I-F-R=.61 ..
STUD-RF-1 =CONS         LAYERS=SRLAY-1   ABS=.86 ..
CRLAY-1   =LAYERS       MAT=(PLYW-1,PLYW-1,CARP/PAD-1)
                            I-F-R=0.92 ..
```

FIGURE 6.2.9 Sample DOE-2 input file for single-family house (part 1).

```
CRL-1    =CONS              LAYERS=CRLAY-1 ..
BWLAY-1      =LAYERS        MAT=(DRYSOIL,EXP-POLY-2,CONCRETE-2)
                           I-F-R=0.68 ..
BASE-WL-1 =CONS            LAYERS=BWLAY-1 ..
DIRLAY-1 =LAYERS           MAT=(DRYSOIL) ..
DIRT-1   =CONS             LAYERS=DIRLAY-1 ..
DR-1     =CONS             U=.7181  ABS=.78  ROUGHNESS=4 ..


$ ------Space Description----- $
COND-1    =SPACE-CONDITIONS   SOURCE-SCHEDULE=INT-LDS-1
                             SOURCE-TYPE=PROCESS
                             SOURCE-BTU/HR=56000
                             SOURCE-SENSIBLE=1
                             SOURCE-LATENT=.225
                             INF-METHOD=S-G
                             HOR-LEAK-FRAC=.4
                             FRAC-LEAK-AREA=.0005
                             FLOOR-WEIGHT = 0
                             FURNITURE-TYPE=LIGHT
                             FURN-FRACTION=.29
                             FURN-WEIGHT=8
                             EQUIP-SCHEDULE=INT-LDS-1
                             EQUIPMENT-W/SQFT=2.51
                             EQUIP-SENSIBLE=0.0
                             EQUIP-LATENT=0.0 ..


$ -----Zone-1 - Crawl Space (unvented)---- $
CRAWL-1      =SPACE           A=1540  V=3080
                             INF-METHOD=S-G
                             FRAC-LEAK-AREA=.0005
                             FLOOR-WEIGHT=0
                             Z-TYPE=UNCONDITIONED  T=(65) ..
GROUND-1     =U-F            CONS=DIRT-1  A=1540  U-EFF=.196 ..


$ -----Zone-2 - House (conditioned)---- $
HOUSE-1      =SPACE           A=1540  V=12320  S-C=COND-1 ..
SOUTH-WL-1 =E-W             H=8  W=42  AZ=180  CONS=INS-WL-1 ..
SOUTH-WIN-1 =WI             W=1.7292  H=3.6667  G-T=WINDOW-1
                           FRAME-WIDTH=0.1667  X=12  Y=2.875  M=7 ..
S-DOOR-1   =DOOR           H=7  W=1.4285  CONS=DR-1  X=30.5 ..
EAST-WL-1    =E-W           H=8  W=36.667  AZ=90  X=42  Y=0
                           CONS=INS-WL-1 ..
EAST-WIN-1  =WI            LIKE SOUTH-WIN-1  ..
E-DOOR-1    =DOOR          LIKE S-DOOR-1 ..
NORTH-WL-1 =E-W            LIKE SOUTH-WL-1  X=42  Y=36.667  AZ=0 ..
NORTH-WIN-1 =WI           LIKE SOUTH-WIN-1  ..
N-DOOR-1    =DOOR          LIKE S-DOOR-1 ..
WEST-WL-1    =E-W           LIKE EAST-WL-1  X=0  Y=36.667  AZ=270 ..
WEST-WIN-1  =WI            LIKE SOUTH-WIN-1  ..
W-DOOR-1    =DOOR          LIKE S-DOOR-1 ..
SOUTH-RF-1  =ROOF         H=18.333  W=42  Z=8  AZ=180  TILT=22.62
                           M=.9  CONS=INS-RF-1 ..
SOUTH-RF-2  =ROOF         H=18.333  W=42  Z=8  AZ=180  TILT=22.62
                           M=.1  CONS=STUD-RF-1 ..
NORTH-RF-1  =ROOF         LIKE SOUTH-RF-1  X=42  Y=36.667  AZ=0 ..
NORTH-RF-2  =ROOF         LIKE SOUTH-RF-2  X=42  Y=36.667  AZ=0 ..
FLOOR-1      =I-W           CONS=CRL-1  H=36.667  W=42  Z=0  TILT=180
                           N-T=CRAWL-1  ..
END ..
COMPUTE LOADS ..
INPUT SYSTEMS ..
           SYSTEMS-REPORT    VERIFICATION=(SV-A)
                             SUMMARY=(SS-A,SS-H) ..
```

FIGURE 6.2.9 (continued)   Sample DOE-2 input file for single-family house (part 2).

```
                       $ ------Schedules----- $
HEAT-1      =SCHEDULE       THRU DEC 31 (ALL) (1,6) (70) (7,24) (70) ..

COOL-1      =SCHEDULE       THRU DEC 31 (ALL) (1,24) (78) ..
VENT-1      =SCHEDULE       THRU DEC 31 (ALL) (1,24) (-1) ..
VTEMP-1     =SCHEDULE       THRU MAY 14 (ALL) (1,24) (-4)
                           THRU SEP 30 (ALL) (1,24) (-4)
                           THRU DEC 31 (ALL) (1,24) (-4) ..
                             $Natural ventilation type - enthalpic$
                             $temp. based on previous 4 day's loads$
VOPEN-1     =SCHEDULE       THRU DEC 31 (ALL) (1,6)   (0)
                                             (7,23)  (1)
                                             (24)    (0) ..


HOUSE-1     =ZONE              DESIGN-HEAT-T=70
                              DESIGN-COOL-T=78
                              ZONE-TYPE=CONDITIONED
                              THERMOSTAT-TYPE=TWO-POSITION
                              HEAT-TEMP-SCH=HEAT-1
                              COOL-TEMP-SCH=COOL-1 ..


CRAWL-1     =ZONE              ZONE-TYPE=UNCONDITIONED ..

            $ ------Air Conditioner and Furnace parameters----- $
AIR-1       =SYSTEM-AIR        VENT-METHOD=S-G
                              FRAC-VENT-AREA=.011
                              MAX-VENT-RATE=20
                              NATURAL-VENT-SCH=VENT-1
                              VENT-TEMP-SCH=VTEMP-1
                              OPEN-VENT-SCH=VOPEN-1 ..
SYS-1       =SYSTEM            SYSTEM-TYPE=RESYS
                              ZONE-NAMES=(HOUSE-1,CRAWL-1)
                              SYSTEM-AIR=AIR-1
                              FURNACE-HIR=1.42      $Eff.=0.78 +10% duct losses
                              COOLING-EIR=.370      $SEER=10, COP=3.0 +10% duct$
                              MAX-SUPPLY-T=110
                              MIN-SUPPLY-T=55 ..
END ..
COMPUTE SYSTEMS ..
STOP ..
```

FIGURE 6.2.9 (continued)   Sample DOE-2 input file for single-family house (part 3).

dynamic simulation that takes into account the building's thermal inertia and the noncoincidence of peak load components, the design parameters from either the second or third procedures tend to be small compared to those derived from standard engineering sizing calculations. Although "right sizing" can lead to higher operational efficiencies and capital cost savings, the designer should evaluate whether the building and weather files contain the appropriate design conditions. There is an optional input in the *System* subprogram for a SIZING-RATIO if a safety factor is desired.

Figure 6.2.13 shows a sample *System* report of the heating, cooling, and electrical loads by month, as well as the monthly peak loads and the coincident outdoor air conditions. In contrast to the approximate loads shown in the *Loads* report, these are the true heating and cooling loads being met by the system. Since the example shown is for a residential house, there is no central plant and the heating gas and cooling electricity energy uses appear on a *System* report shown in Figure 6.2.14. For large buildings with central plants, this *System* report would show only the zone-level energy use, with additional *Plant* output reports showing the monthly and peak energy uses of plant equipment such as boilers, chillers, cooling towers, and pumps.

Figure 6.2.15 shows the summary BEPS (Building Energy Performance Summary) report that gives the annual energy used by the building broken down by major end uses. In addition to these summary reports, DOE-2 also allows users to select from several hundred hourly variables at the global and subprogram levels and print out hourly reports for a selected time period in either ASCII or binary

```
SINGLE FAMILY RESIDENCE    SAMPLE DOE-2.1E INPUT FILE    DOE-2.1E-097 Tue May  2 04:18:59
REPORT- LS-C BUILDING PEAK LOAD COMPONENTS                          WEATHER FILE- STERLING, VA
WYEC2
------------------------------------------------------------------------------------------------
-


*** BUILDING ***


                    FLOOR  AREA     1540  SQFT       143  M2
                    VOLUME         12320  CUFT       349  M3

                           COOLING  LOAD                      HEATING  LOAD
                    ============================          ========================

      TIME                       JUL 24  5PM                    JAN  5  5AM

      DRY-BULB TEMP            93 F          34 C           18 F         -8 C
      WET-BULB TEMP           82 F          28 C           14 F        -10 C
      TOT HORIZONTAL SOLAR RAD  199 BTU/H.SQFT  627 W/M2      0 BTU/H.SQFT    0
W/M2
      WINDSPEED AT SPACE      6.3 KTS       3.2 M/S        13.1 KTS      6.8 M/S
      CLOUD AMOUNT 0(CLEAR)-10   5                             6


                      SENSIBLE        LATENT                    SENSIBLE
                   (KBTU/H)  ( KW ) (KBTU/H) ( KW )          (KBTU/H)  ( KW )
                   --------  ------ -------- ------          --------  --------
WALL CONDUCTION      2.039   0.597  0.000   0.000           -2.722    -0.797
ROOF CONDUCTION      3.391   0.993  0.000   0.000           -2.226    -0.652
WINDOW GLASS+FRM COND 4.352  1.275  0.000   0.000           -6.086    -1.783
WINDOW GLASS SOLAR   8.232   2.412  0.000   0.000            0.125     0.037
DOOR CONDUCTION      0.985   0.288  0.000   0.000           -1.193    -0.349
INTERNAL SURFACE COND -1.419 -0.416 0.000   0.000           -1.419    -0.416
UNDERGROUND SURF COND 0.000  0.000  0.000   0.000            0.000     0.000
OCCUPANTS TO SPACE   0.000   0.000  0.000   0.000            0.000     0.000
LIGHT     TO SPACE   0.000   0.000  0.000   0.000            0.000     0.000
EQUIPMENT TO SPACE   0.000   0.000  0.000   0.000            0.000     0.000
PROCESS   TO SPACE   2.602   0.762  0.718   0.210            1.197     0.351
INFILTRATION         2.907   0.852  6.741   1.975          -12.978    -3.803
                   --------  ------ -------- ------          --------  --------
TOTAL               23.088   6.765  7.459   2.186          -25.301    -7.413
TOTAL / AREA         0.015   0.047  0.005   0.015           -0.016    -0.052

TOTAL LOAD          30.547 KBTU/H   8.950  KW         -25.301 KBTU/H   -7.413  KW
TOTAL LOAD / AREA   19.84 BTU/H.SQFT 62.559 W/M2       16.430 BTU/H.SQFT 51.816
W/M2


         ********************************************************************
         *                                                                  *
         *   NOTE  1)THE ABOVE LOADS EXCLUDE OUTSIDE VENTILATION AIR         *
         *   ----    LOADS                                                   *
         *         2)TIMES GIVEN IN STANDARD TIME FOR THE LOCATION           *
         *           IN CONSIDERATION                                        *
         *         3)THE ABOVE LOADS ARE CALCULATED ASSUMING A               *
         *           CONSTANT INDOOR SPACE TEMPERATURE                       *
         *                                                                  *
         ********************************************************************
```

FIGURE 6.2.10   Loads output report on peak load components.

format. These are useful for providing a detailed look at the energy performance of the building, space-conditioning system, or control strategy on an hourly level. For example, Figure 6.2.16 shows a plot of hourly DOE-2 results for the conduction heat gains through the wall and window and the solar heat gain through the window of a typical office room over several days.

Accuracy of the DOE-2 Program

Because of the amount of and flexibility in the input data needed to do an hourly simulation with programs such as DOE-2, it is difficult to distinguish the accuracy of the program algorithms from the accuracy of the input data. In typical engineering applications, as distinguished from a research project, experienced users are able to achieve accuracies of 10–12% in monthly peak demand, 8–10% in monthly energy use, 10–15% in annual peak demand, and 3–5% in annual energy use for large commercial

```
SINGLE FAMILY RESIDENCE          SAMPLE DOE-2.1E INPUT FILE          DOE-2.1E-097   Tue May  2 04:18:59 2000LDL RUN  1
REPORT- LS-D BUILDING MONTHLY LOADS SUMMARY                                        WEATHER FILE- STERLING, VA  WYEC2
```

MONTH	COOLING ENERGY (MBTU)	TIME OF MAX DY	TIME OF MAX HR	DRY-BULB TEMP	WET-BULB TEMP	MAXIMUM COOLING LOAD (KBTU/HR)	HEATING ENERGY (MBTU)	TIME OF MAX DY	TIME OF MAX HR	DRY-BULB TEMP	WET-BULB TEMP	MAXIMUM HEATING LOAD (KBTU/HR)	ELECTRICAL ENERGY (KWH)	MAXIMUM ELEC LOAD (KW)
JAN	0.20217	25	15	63.F	48.F	9.638	-7.324	5	5	18.F	14.F	-25.301	119.	0.247
FEB	0.22431	5	15	63.F	44.F	9.304	-5.675	11	6	26.F	25.F	-19.151	108.	0.247
MAR	0.93053	20	16	71.F	48.F	15.963	-4.504	22	6	28.F	24.F	-18.527	119.	0.247
APR	2.42814	27	17	86.F	64.F	21.600	-2.081	16	3	36.F	31.F	-13.724	115.	0.247
MAY	4.01552	1	14	94.F	71.F	20.610	-0.874	7	4	46.F	45.F	-9.702	119.	0.247
JUN	5.74028	27	17	85.F	64.F	20.791	-0.093	10	5	57.F	52.F	-3.581	115.	0.247
JUL	6.50512	24	16	93.F	82.F	23.088	-0.041	10	5	58.F	55.F	-3.077	119.	0.247
AUG	5.93819	16	16	93.F	73.F	23.080	-0.043	14	6	58.F	51.F	-2.902	119.	0.247
SEP	4.18373	1	16	91.F	72.F	22.172	-0.362	30	5	49.F	46.F	-6.790	115.	0.247
OCT	2.34292	10	14	81.F	62.F	17.284	-1.650	28	5	37.F	35.F	-10.594	119.	0.247
NOV	0.71645	21	13	63.F	46.F	10.815	-3.427	30	2	37.F	30.F	-14.891	115.	0.247
DEC	0.15903	3	15	57.F	46.F	8.763	-6.270	24	24	31.F	26.F	-17.740	119.	0.247
TOTAL	33.386						-32.344						1404.	
MAX						23.088						-25.301		0.247

FIGURE 6.2.11 Loads output report on monthly load components.

REPORT- SV-A SYSTEM DESIGN PARAMETERS   SYS-1   WEATHER FILE- STERLING, VA WYEC2

SYSTEM NAME	SYSTEM TYPE	ALTITUDE MULTIPLIER	FLOOR AREA (SQFT)	MAX PEOPLE
SYS-1	RESYS	1.000	3080.0	0.

SUPPLY FAN (CFM)	ELEC (KW)	DELTA-T (F)	RETURN FAN (CFM)	ELEC (KW)	DELTA-T (F)	OUTSIDE AIR RATIO	COOLING CAPACITY (KBTU/HR)	SENSIBLE (SHR)	HEATING CAPACITY (KBTU/HR)	COOLING EIR (BTU/BTU)	HEATING EIR (BTU/BTU)
929.	0.119	0.4	0.	0.000	0.0	0.000	32.018	0.609	-39.709	0.37	0.37

ZONE NAME	SUPPLY FLOW (CFM)	EXHAUST FLOW (CFM)	FAN (KW)	MINIMUM FLOW RATIO	OUTSIDE AIR FLOW (CFM)	COOLING CAPACITY (KBTU/HR)	SENSIBLE (SHR)	EXTRACTION RATE (KBTU/HR)	HEATING CAPACITY (KBTU/HR)	ADDITION RATE (KBTU/HR)	MULTIPLIER
HOUSE-1	929.	0.	0.000	1.000	0.	0.00	0.00	23.09	0.00	-40.15	1.0
CRAWL-1	0.	0.	0.000	0.000	0.	0.00	0.00	0.00	0.00	0.00	1.0

FIGURE 6.2.12 System output report on system design parameters.

SINGLE FAMILY RESIDENCE SAMPLE DOE-2.1E INPUT FILE DOE-2.1E-097 Tue May 2 04:18:59 2000SDL RUN 1
REPORT- SS-A SYSTEM MONTHLY LOADS SUMMARY FOR SYS-1 WEATHER FILE- STERLING, VA WYEC2

	C O O L I N G						H E A T I N G						E L E C	
MONTH	COOLING ENERGY (MBTU)	TIME OF MAX DY	TIME OF MAX HR	DRY-BULB TEMP	WET-BULB TEMP	MAXIMUM COOLING LOAD (KBTU/HR)	HEATING ENERGY (MBTU)	TIME OF MAX DY	TIME OF MAX HR	DRY-BULB TEMP	WET-BULB TEMP	MAXIMUM HEATING LOAD (KBTU/HR)	ELEC-TRICAL ENERGY (KWH)	MAXIMUM ELEC LOAD (KW)
JAN	0.00000					0.000	-9.098	5	5	18.F	14.F	-28.900	180.	0.368
FEB	0.00000					0.000	-7.404	12	2	24.F	20.F	-22.876	162.	0.352
MAR	0.01458	25	16	76.F	53.F	6.497	-5.896	22	6	28.F	24.F	-21.582	167.	1.044
APR	0.54407	28	18	84.F	67.F	17.291	-2.619	16	4	36.F	31.F	-16.186	210.	2.224
MAY	1.00029	31	17	86.F	68.F	16.767	-0.877	7	5	46.F	45.F	-11.256	258.	2.296
JUN	3.44213	24	17	90.F	76.F	20.775	-0.027	10	6	57.F	52.F	-2.876	554.	2.628
JUL	5.51768	24	17	93.F	82.F	26.914	-0.008	9	6	59.F	56.F	-1.826	820.	3.307
AUG	4.99961	29	17	91.F	76.F	24.810	-0.002	14	7	58.F	51.F	-1.036	751.	3.069
SEP	2.12490	9	17	88.F	74.F	23.646	-0.159	30	6	49.F	46.F	-5.932	384.	2.868
OCT	0.50496	10	15	81.F	62.F	12.686	-1.080	28	6	37.F	35.F	-9.968	196.	1.630
NOV	0.00000					0.000	-3.381	30	2	37.F	30.F	-16.186	147.	0.335
DEC	0.00000					0.000	-7.341	24	24	31.F	26.F	-19.953	175.	0.341
TOTAL	18.148						-37.892						4004.	
MAX						26.914						-28.900		3.307

FIGURE 6.2.13 System output report on system monthly and peak loads.

SINGLE FAMILY RESIDENCE   NFRC DOE-2.1E INPUT FILE   DOE-2.1E-091 Wed May 3 02:49:02 2000SDL RUN 1
REPORT- SS-H SYSTEM MONTHLY LOADS SUMMARY FOR   SYS-1   WEATHER FILE- STERLING, VA WYEC2

| | --FAN ELEC-- | | --FUEL HEAT-- | | --FUEL COOL-- | | --ELEC HEAT-- | | --ELEC COOL-- | |
MONTH	FAN ENERGY (KWH)	MAXIMUM FAN LOAD (KW)	GAS OIL ENERGY (MBTU)	MAXIMUM GAS OIL LOAD (KBTU/HR)	GAS OIL ENERGY (MBTU)	MAXIMUM GAS OIL LOAD (KBTU/HR)	ELECTRIC ENERGY (KWH)	MAXIMUM ELECTRIC LOAD (KW)	ELECTRIC ENERGY (KWH)	MAXIMUM ELECTRIC LOAD (KW)
JAN	27.	0.087	14.704	42.802	0.000	0.000	0.	0.000	33.	0.050
FEB	22.	0.069	12.141	34.823	0.000	0.000	0.	0.000	32.	0.050
MAR	18.	0.065	9.956	33.064	0.000	0.000	0.	0.000	30.	0.891
APR	10.	0.081	4.900	25.616	0.000	0.000	0.	0.000	84.	1.900
MAY	7.	0.076	2.119	18.602	0.000	0.000	0.	0.000	132.	2.000
JUN	16.	0.090	0.635	6.235	0.000	0.000	0.	0.000	423.	2.317
JUL	25.	0.115	0.615	4.639	0.000	0.000	0.	0.000	676.	2.972
AUG	22.	0.110	0.600	3.427	0.000	0.000	0.	0.000	610.	2.738
SEP	10.	0.104	0.894	10.801	0.000	0.000	0.	0.000	258.	2.543
OCT	6.	0.063	2.493	17.031	0.000	0.000	0.	0.000	71.	1.459
NOV	10.	0.049	6.139	25.611	0.000	0.000	0.	0.000	21.	0.050
DEC	22.	0.060	12.229	30.850	0.000	0.000	0.	0.000	34.	0.050
TOTAL	196.		67.425		0.000		0.		2405.	
MAX		0.115		42.802		0.000		0.000		2.972

FIGURE 6.2.14 System output report on system monthly and peak energy use.

```
SINGLE FAMILY RESIDENCE          NFRC DOE-2.1E INPUT FILE          DOE-2.1E-091  Wed May  3 02:49:02 2000PDL RUN  1
REPORT- BEPS  BUILDING ENERGY PERFORMANCE SUMMARY                                WEATHER FILE- STERLING, VA  WYEC2
-----------------------------------------------------------------------------------------------------------------

                         ENERGY TYPE:   ELECTRICITY   NATURAL-GAS
                         UNITS: MBTU


                         CATEGORY OF USE
                         ---------------
                           MISC EQUIPMT        4.8           0.0
                           SPACE HEAT          0.0          67.4
                           SPACE COOL          7.6           0.0
                           PUMPS & MISC        0.6           0.0
                           VENT FANS           0.7           0.0
                                           --------      --------
                               TOTAL         13.7          67.4


TOTAL SITE ENERGY        81.09 MBTU     52.7 KBTU/SQFT-YR GROSS-AREA      52.7 KBTU/SQFT-YR NET-AREA
TOTAL SOURCE ENERGY     108.43 MBTU     70.4 KBTU/SQFT-YR GROSS-AREA      70.4 KBTU/SQFT-YR NET-AREA

PERCENT OF HOURS ANY SYSTEM ZONE OUTSIDE OF THROTTLING RANGE =   0.0
PERCENT OF HOURS ANY PLANT LOAD NOT SATISFIED               =   0.0
NOTE: ENERGY IS APPORTIONED HOURLY TO ALL END-USE CATEGORIES.
```

FIGURE 6.2.15 Plant output report on building energy performance summary.

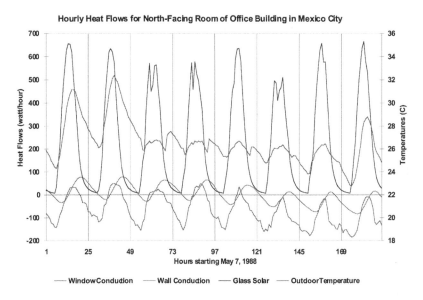

FIGURE 6.2.16 Hourly plot of DOE-2 calculated conductive heat flows and solar heat gains in a typical north-facing office module in Mexico City.

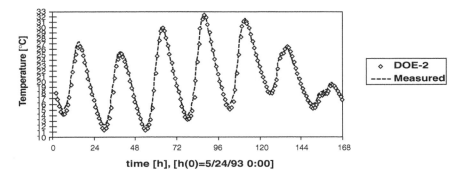

FIGURE 6.2.17 Comparison of DOE-2 simulated and measured zone temperatures in an experimental test chamber in the U.K. (From Winkelmann, F. *The User's News*. With permission.)

buildings with sufficient information on building conditions and utility bills. In an International Energy Agency project that compared the results from various computer simulation programs to measured indoor temperatures for a well-monitored test cell in the U.K., the simulated indoor temperatures from DOE-2 were virtually identical to the measured results (see Figure 6.2.17).

In another project done at LBNL, DOE-2 simulations were done for several small unoccupied test houses in the San Diego area that were monitored in detail over several summers. The discrepancy in indoor temperatures was less than 0.3°C on average, and around 1°C maximum. In general, it seems that given sufficient time and effort to gather input parameters and calibrate (more truthfully, adjust) the building model, quite close agreements to measured data can be achieved. However, the same cannot be said for cases when there is no measured data for comparison. In such instances, there can be substantial discrepancies due to input errors, or user bias.

Future Developments in Building Energy Calculation Methods
As mentioned previously, since 1996 the U.S. Department of Energy has been supporting the development of a new building energy simulation program, EnergyPlus, that aims to be an analysis platform for the next

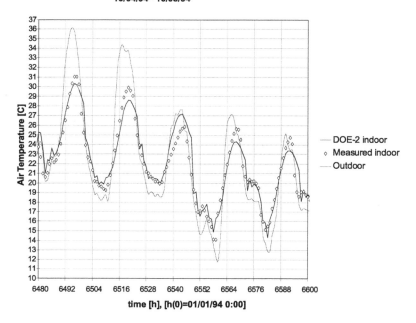

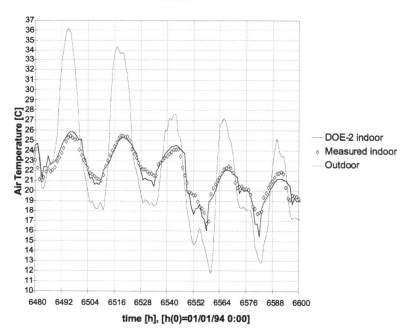

FIGURE 6.2.18 Comparison of DOE-2 simulated and measured zone temperatures in unoccupied test houses in San Diego. (From Winkelmann, F. and Meldem, R. (1995). With permission.)

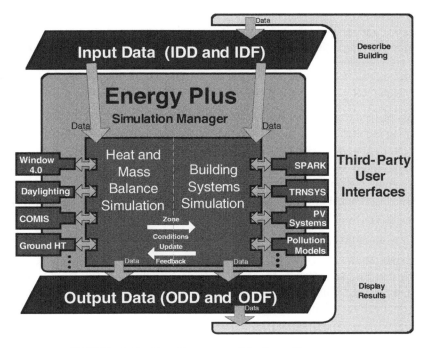

FIGURE 6.2.19 Overall structure of the EnergyPlus program.

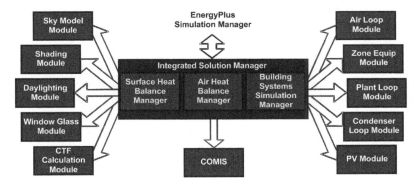

FIGURE 6.2.20 EnergyPlus integrated simulation manager. (From Crawley, D.B. et al. (2000). With permission.)

decade or more. Figures 6.2.19 and 6.2.20 illustrate the structure and calculation sequence of this new simulation tool. A general description of this new program is given in an article by Crawley et al. (1999).

6.2.3 Inverse Modeling

At the present time there are many different methods for analyzing energy use in buildings. In general, the applications of these methods are driven by the motivation behind the investigation of a building's energy use, including retrofit energy savings analysis, diagnosing equipment malfunctions, energy auditing indices, component efficiency testing, and demand side management (DSM) evaluation. These methods can be divided into two basic approaches — forward modeling and inverse modeling — with a third approach that represents methods that contain aspects of both definitions — forward plus inverse (Rabl, 1988; Rabl and Riahle, 1992). Forward modeling was the subject of Sections 6.2.1 and 6.2.2.

In the inverse approach, the analysis is conducted on the empirical behavior of the building as it relates to one or more driving forces or parameters. This approach is referred to as a system identification, parameter identification, or *inverse modeling*. In the inverse modeling approach, one assumes a structure or physical configuration of the building or system being studied and then attempts to identify the important parameters through the use of a statistical analysis (Rabl and Rialhe, 1992). In general, there are two basic types of inverse models: steady state inverse models and dynamic inverse models. A third category, hybrid models, includes models that have characteristics of both forward and inverse models.

Inverse modeling techniques have been successful in the following cases:

- Identifying the energy savings from building retrofits
- Estimating the performance of an existing building under future weather and occupancy conditions
- Predicting hourly (or subhourly) loads and energy use levels for optimal operation of HVAC systems under demand or real time pricing utility rates
- Constructing a model of HVAC subsystems for the optimal, adaptive control of that subsystem
- Faulting diagnosis of HVAC systems

Steady State Inverse Models

The simplest form of an inverse model is a *steady state inverse model* of a building's energy use. The simplest steady state inverse model can be calculated by statistically regressing monthly utility consumption data against average billing period temperatures. Although simple in concept, the most accurate methods use sophisticated change-point regression procedures that simultaneously solve for several parameters including a weather independent base-level parameter, one or more weather dependent parameters, and the point or points at which the model switches from weather dependent to weather independent dependent behavior. In its simplest form, the 65°F (18.3°C) degree-day model is a change-point model that has a *fixed* change point at 65°F. Other examples include the three- and five-parameter Princeton Scorekeeping Methods* (PRISM) (Fels 1986) and a four-parameter model** (4P) developed by Ruch and Claridge (1991). An inverse bin method has also been proposed to handle more than four changepoints (Thamiseran and Haberl, 1995).

Models Using One Independent Variable

Figure 6.2.21 shows several types of steady state, single variable inverse models. Figure 6.2.21a shows a simple one-parameter, or "average," model whose equation is given in the first line of Table 6.2.5. Figure 6.2.21b shows a steady state two-parameter (2P) model where β_0 is the y-axis intercept, and β_1 is the slope of the regression line for positive values of x, where x represents the ambient air temperature. The 2P model represents cases when either heating or cooling is always required.

Figure 6.2.21c shows a three-parameter, change point model. This is typical of natural gas energy use in a single family residence that utilizes gas for space heating and domestic water heating. In the equation in Table 6.2.5, β_0 represents the baseline energy use, and β_1 is the slope of the regression line for values of ambient temperature less than the change point β_2. In this type of notation, the exponent (+) indicates that only positive values of the parenthetical expression are considered. Figure 6.2.21d shows a three-parameter model for cooling energy.

Figures 6.2.21e and 6.2.21f illustrate four-parameter models for heating and cooling, respectively. In a four-parameter model, β_0 represents the baseline energy exactly at the change point β_3. β_1 and β_2 are the lower and upper region regression slopes for ambient air temperature below and above the change point β_3. Figure 6.2.21g illustrates a five-parameter model (Fels, 1986). Such a model is useful for modeling buildings that are electrically heated and cooled. The five-parameter model has two change points and a base level consumption value as shown in the final equation in Table 6.2.5.

---

  * The three parameters include a weather independent base-level use, a change-point, and a temperature dependent parameter or slope of a line that is determined by regression.

  ** The four parameters include a change point, a slope above the change point, a slope below the change point, and the energy use associated with the change point.

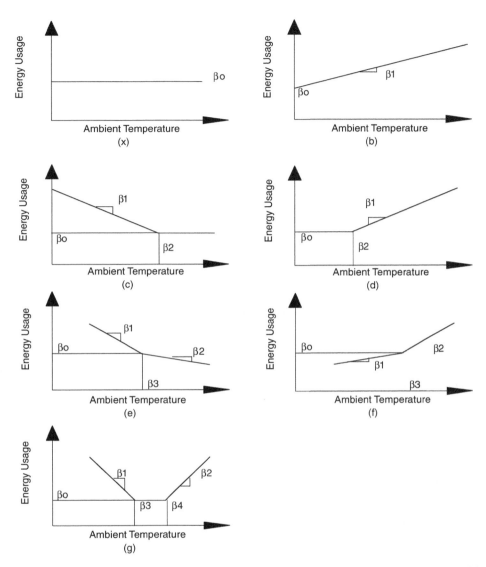

FIGURE 6.2.21 Steady state, single variable models appropriate for commercial building energy use: (a) one-parameter model, (b) two-parameter model shown for cooling energy use, (c) three-parameter heating energy use model, (d) three-parameter cooling energy use model, (e) four-parameter heating energy use model, (f) four-parameter cooling energy use model, and (g) five-parameter heating and cooling model with separate change points.

The advantage of these simple steady state inverse models is that their use can be easily automated and applied to a large numbers of buildings where monthly utility billing data and average daily temperatures for the billing period are available. Steady state inverse models can also be applied to daily data, which allows one to compensate for differences in weekday and weekend use (Claridge et al., 1992).

The disadvantages of the steady state inverse models include

- An insensitivity to dynamic effects (i.e., thermal mass)
- Insensitivity to variables other than temperature (for example humidity and solar gain)
- Inappropriateness for certain building types, such as buildings that have strong on/off schedule dependent loads, or buildings that display multiple change points. In such cases, alternative models must be used.

TABLE 6.2.5   Change Point (CP) Model Equations

$$E_{period} = \beta_0$$
(average model)

$$E_{period} = \beta_0 + \beta_1 T$$
(single CP model)

$$E_{period} = \beta_0 + \beta_1(\beta_2 - T)^+$$
(double CP model — heating)

$$E_{period} = \beta_0 + \beta_1(T - \beta_2)^+$$
(double CP model — cooling)

$$E_{period} = \beta_0 + \beta_1(\beta_3 - T)^+ - \beta_2(T - \beta_3)^+$$
(triple CP model — heating)

$$E_{period} = \beta_0 - \beta_1(\beta_3 - T)^+ + \beta_2(T - \beta_3)^+$$
(triple CP model — cooling

$$E_{period} = \beta_0 - \beta_1(\beta_3 - T)^+ + \beta_2(T - \beta_4)^+$$
(five-parameter model — cooling)

Steady State Inverse Models Using More Than One Independent Variable

Multiple regression techniques allow the analyst to investigate the influence of more than one independent variable (such as outdoor air temperature and humidity, solar radiation, and indicators of scheduling) on a response variable (such as building energy use). The form of the general linear regression model is

$$Y = \beta_0 + \beta_1 X_1 + \beta_2 X_2 + \dots + \beta_P X_P + \varepsilon$$

where Y is the response variable; $X_1, X_2, \dots X_P$ are the independent variables; $\beta_0, \beta_1, \beta_2, \dots, \beta_P$ are the p regression parameters; and ε is the error term. When $p=2$, the response surface is a plane. When $p > 2$, the response surface is called a hyperplane.

Interactions between independent variables can be considered by using the product of two independent variables. Curvature in the response surface can be introduced through the use of independent polynomial variables. The equation below demonstrates a model with two independent variables, each in quadratic form, with an interaction term:

$$Y = \beta_0 + \beta_1 X_1 + \beta_2 X_1^2 + \beta_3 X_2 + \beta_4 X_2^2 + \beta_5 X_1 X_2 + \varepsilon$$

The choice of the model should be guided by the analyst's understanding of the physical system and its expected response. Several standard statistical tests exist for evaluating the goodness-of-fit of the model and the degree of influence that each of the independent variables exerts on the response variable (Draper and Smith, 1981; Neter et al., 1989).

When modeling building energy use data, the independent variables are often linearly correlated with each other (as in the case of outdoor air temperature, humidity, and solar radiation). When multicolinearity exists, the regression coefficients may not indicate the relative importance of the independent variables. In addition, the uncertainties of the estimates of the regression parameters (reported as the standard error of each parameter estimate) may be so large that the model's usefulness for prediction purposes is comprised.

Dynamic Inverse Models — Neural Networks

More advanced forms of inverse models include *dynamic inverse models*. Examples of dynamic inverse models include equivalent thermal network analysis (Sonderegger, 1977;), ARMA models (Subbarao, 1990; Reddy, 1989), Fourier series models (Dhar et al., 1995), and artificial neural networks (Kreider

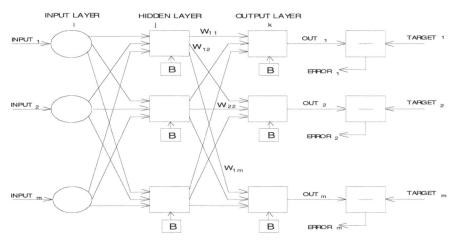

FIGURE 6.2.22   Schematic diagram of a neural network showing input layer, hidden layers, and output along with target training values. Hidden and output layers consist of connected neurons; the input layer does not contain neurons.

and Wang, 1991; Kreider and Haberl, 1994). Neural networks seem to hold considerable promise among those listed and are described in detail below.

The use of the term *dynamic* refers to the fact that these models are capable of capturing dynamic effects, such as mass dynamics, which traditionally have required the solution of a set of differential equations. These models are better suited for handling intercorrelated forcing functions or independent parameters. The advantages of dynamic inverse models include the ability to model complex systems which are dependent on more than one independent parameter. The disadvantages of dynamic inverse models include their increasing complexity and the need for more detailed measurements to "tune" the model. Unlike steady state inverse models, dynamic inverse models usually require a high degree of user interaction and knowledge of the building or system being modeled.

Neural Network Construction

An artificial neural network is a massively parallel, dynamic system of interconnected, interacting parts based loosely on some aspects of the brain. Neural networks are considered to be intuitive because they learn by example rather than by following programmed rules. The ability to "learn" is one of the key aspects of neural networks. A neural network consists of several layers of neurons that are connected to each other. A *connection* is a unique information transport link from one sending neuron to one receiving neuron. The structure of part of an NN is schematically shown in Figure 6.2.22. Any number of input, output, and hidden layer (only one hidden layer is shown) neurons can be used. One of the challenges of this technology is to construct a net with sufficient complexity to learn accurately without imposing a burden of excessive computational time.

The neuron is the fundamental building block of a neural network. A set of inputs is applied to each. Each element of the input set is multiplied by a weight, indicated by the W in Figure 6.2.22, and the products are summed at the neuron. The symbol for the summation of weighted inputs is termed *INPUT* and must be calculated for each neuron in the network. In equation form, this process for one neuron is

$$INPUT = \sum_i O_i W_i + B$$

where

> O_i are inputs to a neuron, i.e., outputs of the previous layer
> W are weights, and
> B are the biases.

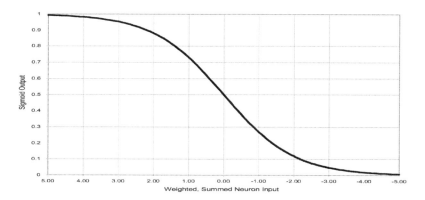

FIGURE 6.2.23 Sigmoid function used to process the weighted sum of network inputs.

After *INPUT* is calculated, an activation function *F* is applied to modify it, thereby producing the neuron's output as described shortly.

Artificial networks have been trained by a wide variety of methods (McClelland and Rumelhart, 1988; Wasserman, 1989). Back-propagation is one systematic method for training multilayer neural networks. The weights of a net are initiated with small random numbers. The objective of training the network is to adjust the weights iteratively so that application of a set of inputs produces the desired set of outputs matching a training data set. Usually a network is trained with a data set that consists of many input-output pairs; these data are called a training set. Training the net using back-propagation requires the following steps:

1. Select a training pair from the training set and apply the input vector to the network input layer.
2. Calculate the output of the network, OUT_i.
3. Calculate the error $ERROR_i$ between the network output and the desired output (the target vector from the training pair).
4. Adjust the weights of the network in a way that minimizes the error.
5. Repeat steps 1 through 4 for each vector in the training set until the error for the entire set is lower than the user specified, preset training tolerance.

Steps 1 and 2 are the *forward pass*. The following expression describes the calculation process in which an activation function *F* is applied to the weighted sum of inputs *INPUT*:

$$OUT = F(INPUT) = F\left(\sum O_i W_i + B\right)$$

where *F* is the activation function, and *B* is the bias of each neuron.

A common activation function is the sigmoid function

$$F(INPUT) = \frac{1}{(1 + e^{-INPUT})}$$

A sigmoid function is shown in Figure 6.2.23. It has a value of 0.0 when *INPUT* is a large negative number and a value of 1.0 for large and positive *INPUT*, making a smooth transition between these limiting values. The bias *B* is the activation threshold for each neuron. The bias avoids the tendency of a sigmoid function to get "stuck" in the saturated, limiting value area.

Steps 3 and 4 above comprise the *reverse pass* in which the delta rule (McClelland and Rumelhart, 1988) is used as follows. For each neuron in the output layer, the previous weight *W(n)* is adjusted to a new value *W(n+1)* to reduce the error by the following rule:

$$W(n + 1) = W(n) + (n\delta)OUT$$

where $W(n)$ is the previous value of a weight
  $W(n+1)$ is the weight after adjusting
  η is the training rate coefficient
  δ is calculated from

$$\delta = \left(\frac{\partial INPUT}{\partial OUT}\right)(TARGET - OUT) = OUT(1 - OUT)(TARGET - OUT)$$

in which *TARGET* (see Figure 6.2.22) is the training set target value. This method of correcting weights bases the magnitude of the correction on the error itself.

Of course, hidden layers have no target vector; therefore, back-propagation trains these layers by propagating the output error back through the network layer by layer, adjusting weights at each layer. The delta rule adjustment is calculated from

$$\delta_j = OUT(1 - OUT)\sum \delta_{j+1}W_{j+1}$$

This overall method of adjusting weights belongs to the general class of steepest descent algorithms. The weights and bias after training contain meaningful system information; before training the initial, random biases and random weights have no physical meaning.

Neural Networks Applied to Buildings — Some Examples
This section describes a number of applications of NNs to residential and commercial building systems. Two energy prediction cases are presented and one commercial building control demonstration is included. Anstett and Kreider (1993), Kreider and Wang (1991), and Wang and Kreider (1992) report additional case studies.

Commercial Buildings — There are a number of reasons that an NN prediction of building energy use has been found to be useful in commercial buildings; among them are

- Prediction of what a properly operating building should be doing compared to actual operation — if there is a difference, it can be used in an expert system to produce early diagnoses of building operation problems.
- Prediction of what a building, prior to an energy retrofit, would have consumed under present conditions — when compared to the measured consumption of the retrofitted building, the difference represents a good estimate of the energy savings due to the retrofit. This represents one of the few ways that actual energy savings can be determined after the pre-retrofit building configuration has ceased to exist.

Figure 6.2.24 shows results typical of several hundred networks constructed on a number of commercial buildings. This building is an academic engineering center located in central Texas. The cooling load is created by solar gains, internal ("free") gains, outdoor air sensible heat, and outdoor air humidity loads. The NN is used to predict the pre-retrofit energy consumption for comparison with measured consumption of the retrofitted building. Six months of pre-retrofit data were available with which to train a network.

The known building consumption data are shown in the figure by the solid line while the NN predictions are shown by the dashes. The figure shows that NNs trained for one period (here, September 1989) can predict energy consumption well into the future (January 1990).

The network used for this prediction has two hidden layers. The input layer contains eight neurons that receive eight different types of input data as listed below. The output layer consists of one neuron that gives the output datum (chilled water consumption). Each training fact (i.e., training data set),

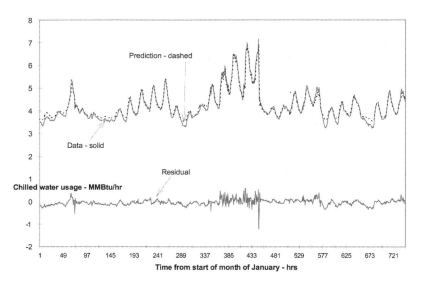

FIGURE 6.2.24 NN prediction of whole building, hourly chilled water consumption, for example, commercial building.

therefore, contains eight input data (independent variables) and one pattern datum (dependent variable). The eight hourly input data used in each hour's data vector were selected on physical bases (Kreider, Rabl and Curtiss, 2001) as follows:

- Hour number (0–2300)
- Ambient dry-bulb temperature
- Horizontal insolation
- Humidity ratio
- Wind speed
- Weekday/weekend binary flag (0,1)
- Past hour's chilled water consumption
- Second past hour's chilled water consumption

These easily measured independent variables were able to predict the chilled water use to an RMS error of less than 4% (JCEM, 1992a, 1992b, 1992c).

The choice of an optimal network's configuration for a given problem remains an art. The number of hidden neurons and layers must be sufficient to meet the requirement of the given application. However, if too many neurons and layers are employed, the network becomes "brittle" and tends to memorize data rather than learning, that is, finding the underlying patterns within the data. Further, choosing an excessively large number of hidden layers significantly increases the required training time for certain learning algorithms.

The error and training time are two important training criteria. One measure of the accuracy of a network prediction is the root mean square error defined as

$$RMSE = \left\{ \frac{\sum (OUT - TARGET)}{N} \right\}^{0.5}$$

where the network output *OUT* and target *TARGET* are both normalized before training to the closed interval [0-1]. The RMS error is dimensionless.

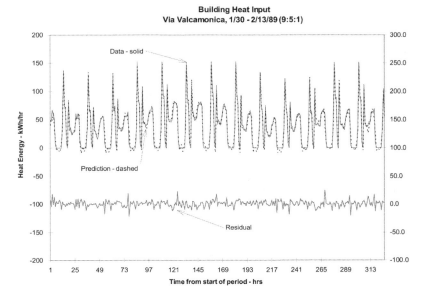

FIGURE 6.2.25   NN prediction of residential heating energy for the first six weeks of 1989.

Residential Buildings — Neural networks also have been applied to residential buildings. Figure 6.2.25 shows a typical prediction for a building in Varese, Italy. Hourly data collected by the Joint Research Center of the European Community included outdoor temperature and insolation, indoor temperature and heating energy consumption. The weather data and time information (hour of day, day of year) were used to make the predictions shown in the figure. It is seen that the diurnal patterns of heating energy usage are predicted well by the NN model.

Application of Inverse Models

The principal applications of inverse models appear to be in the following areas:

- Evaluation of energy conservation programs
- Prescreening indices for energy auditing
- Building energy management
- Optimal control
- *In situ* characterization of HVAC

In each of these applications both steady state and dynamic inverse models have been applied. In general, steady state inverse models are used with monthly and daily data containing one or more independent variables. Dynamic inverse models are usually used with hourly or sub-hourly data in cases where the thermal mass of a building is significant enough to delay the heat gains or losses.

Evaluation of Energy Conservation Programs

Aside from simply regressing energy use against temperature (e.g., often a two-parameter model with a slope and y-axis intercept), other widely-used steady state inverse methods for the evaluation of energy conservation retrofits include three-, four- and five-parameter change points models previously described (Fels, 1986; Ruch and Claridge, 1993). Such models have been shown to be useful for statistically determining average weather dependent and weather independent energy use for buildings. Three-parameter change point models can yield baseline energy use, the temperature at which weather dependent energy use begins to increase energy use above the baseline (i.e., the change point), and the linear slope of the temperature dependency above (cooling model) or below (heating model) the temperature change point.

The existence of a change point in heating or cooling data that is plotted against ambient temperature can be physically justified since most HVAC systems use a thermostat that turns systems on or off above or below a setpoint temperature. Change point regressions work best with heating data from buildings with systems that have few or no part-load nonlinearities (i.e., systems that become less efficient as they begin to cycle on-off with part-loads). In general change point regressions for cooling loads exhibit less of a good fit because of changes in outdoor humidity which influence latent coil loads. Other factors that decrease the goodness of fit of change point models include solar effects, thermal lags, and on-off HVAC schedules. In buildings with continuous, year-round cooling or heating, four-parameter models exhibit a better statistical fit over three-parameter models (i.e., grocery stores and office buildings with high internal loads). However, results of every modeling effort should be inspected for reasonableness (i.e., make sure that the regression is not falsely indicating an unreasonable relationship).

One of the main advantages of using a steady state inverse model to evaluate the effectiveness of energy conservation retrofits lies in its ability to factor out year-to-year weather variations. This can be accomplished by using a Normalized Annual Consumption or NAC (Fels, 1986). Basically, once the regression parameters have been calculated for both pre-retrofit and post-retrofit periods, the annual energy conservation savings can be calculated by comparing the difference one obtains by multiplying the pre-retrofit and post-retrofit parameters by the weather conditions for the average year. Typically, ten to twenty years of average daily weather data from a nearby National Weather Service site are used to calculate 365 days of average weather conditions which are then used to calculate the average pre-retrofit and post retrofit conditions.

Energy Management

Steady state and dynamic inverse models can be used by energy management and control systems to predict energy use (Kreider and Haberl, 1994). Hourly or daily comparisons of measured energy use against predicted energy use can be used to determine if systems are being left on unnecessarily or are in need of maintenance. Combinations of predicted energy use and a knowledge-based system have been shown to be capable of indicating above normal energy use and diagnosing the possible cause of the malfunction if sufficient historical information about malfunction signatures has been previously gathered (Haberl and Claridge, 1987). Hourly systems that utilize artificial neural networks have also been constructed (Kreider and Wang, 1991).

6.2.4　Hybrid Modeling

Forward plus inverse models or hybrid models encompass everything that does not neatly fit into the exact definition of forward or inverse models. For example, when a traditional fixed-schematic simulation program such as DOE-2 or BLAST (or even a component based model) is used to simulate the energy use of an existing building, then one has a *forward* analysis method that is being used in an *inverse* application, i.e., the forward simulation model is being calibrated or fit to the actual energy consumption data from a building in much the same way that one fits a linear regression of energy use to temperature. Such an application is a *hybrid model*.

Although at first this might appear to be a simple process, there are several practical difficulties in achieving a "calibrated simulation," including the measurement and adaptation of weather data for use by the simulation programs (i.e., converting global horizontal solar into beam and diffuse solar radiation), the choice of methods used to calibrate the model, and the choice of methods used to measure the required input parameters for the simulation (i.e., the weight of the building, infiltration coefficients, and shading coefficients). In the scientific sense, truly "calibrated" models have been achieved only in very few applications since they require a very large number of input parameters, a high degree of expertise, and enormous amounts of computing time, patience, and financial resources — much more than most practical applications would allow. However, examples exist in the literature of different methods employed to calibrate simulation models, including Bronson et al. (1992), Haberl et al. (1995), Kaplan et al. (1990), Corson (1992), Bou Saada and Haberl (1995a, 1995b), and Hsieh (1988).

TABLE 6.2.6 Classification of Methods for the Thermal Analysis of Buildings

Method	Forward	Inverse	Hybrid	Comments:
Steady State Methods				
Simple linear regression		X		One dependent parameter, one independent parameter. May have slope and y-intercept.
Multiple linear regression		X	X	One dependent parameter, multiple independent parameters.
Modified degree-day method	X			Based on fixed reference temperature of 65°F.
Variable base degree-day method	X			Variable reference temperatures.
ASHRAE bin method and inverse bin method	X	X	X	Hours in temperature bin times load for that bin.
Change point models: 3-parameter (PRISM CO, HO), 4-parameter, 5-parameter (PRISM HC).		X	X	Uses daily or monthly utility billing data and average period temperatures.
ASHRAE TC 4.7 modified bin method	X		X	Modified bin method with cooling load factors.
Dynamic Methods				
Thermal network (Sonderegger, 1977)	X	X	X	Uses equivalent thermal parameters (inverse mode).
Response factors (Stephenson and Mitalas, 1967)	X			Tabulated or as used in simulation programs.
Fourier Analysis (Shurcliff, 1984; Dhar, 1995)	X	X	X	Frequency domain analysis convertible to time domain.
ARMA Model (Subbarao, 1986)		X		Autoregressive Moving Average model.
ARMA Model (Reddy, 1989)		X		Multiple-input autoregressive moving average model.
BEVA, PSTAR (Subbarao, 1986)	X	X	X	Combination of ARMA and Fourier series, includes loads in time domain.
Modal analysis (Bacot et al., 1984)	X	X	X	Bldg. described by diagonalized differential equation using nodes.
Differential equation (Rabl, 1988)		X		Analytical linear differential equation.
Computer simulation (DOE-2, BLAST)	X		X	Hourly simulation programs with system models.
Computer emulation (HVACSIM+, TRNSYS)	X		X	Sub-hourly simulation programs.
Artificial neural networks (Kreider and Wang, 1991; Kreider, 1992; Kreider and Haberl, 1994)		X	X	Connectionist models.

6.2.5 Classification of Methods

In Table 6.2.6 different methods of analyzing building energy use are classified using an expanded version of Rabl's definitions (Rabl, 1988). Simple linear regression and multiple linear regression are the most widely used forms of inverse analysis. In the proper application, multiple linear regression must adequately address intercorrelations among the independent parameters as discussed above.

6.2.6 How to Select an Approach

Table 6.2.7 presents a decision diagram for selecting an inverse model where usage of the model (diagnostics — D, energy savings calculations — ES, design — DE, and control — C), degree of difficulty in understanding and applying the model, time scale for the data used by the model (hourly — H, daily — D, monthly — M, and subhourly — S), calculation time, and input variables used by the models (temperature — T, humidity — H, solar — S, wind — W, time — t, thermal mass — tm) are the criteria used to determine the choice of a particular model.

TABLE 6.2.7　Decision Diagram for Selection of Inverse Models

Method	Usage[a]	Difficulty	Time[b] Scale	Calc. Time	Variables[c]	Accuracy
Simple linear regression	ES	Simple	D,M	Very Fast	T	Low
Multiple linear regression	D,ES	Moderate	D,M	Fast	T,H,S,W,t	Medium
ASHRAE bin method and inverse bin method	ES	Moderate	H	Fast	T	Medium
Change point models.	D,ES	Moderate	H,D,M	Fast	T	Medium
ASHRAE TC 4.7 modified bin method	ES,DE	Moderate	H	Medium	T,S,tm	Medium
Thermal network	D,ES,C	Complex	S,H	Fast	T,S,tm	High
Fourier Series Analysis	D,ES,C	Complex	S,H	Medium	T,H,S,W,t,tm	High
ARMA Model	D,ES,C	Complex	S,H	Medium	T,H,S,W,t,tm	High
Modal analysis	D,ES,C	Complex	S,H	Medium	T,H,S,W,t,tm	High
Differential equation	D,ES,C	Very Complex	S,H	Fast	T,H,S,W,t,tm	High
Computer Simulation (Component-based)	D,ES,C, DE	Very Complex	S,H	Slow	T,H,S,W,t,tm	Medium
Computer simulation (Fixed schematic)	D,ES,DE	Very Complex	H	Slow	T,H,S,W,t,tm	Medium
Computer emulation	D,C	Very Complex	S,H	Very Slow	T,H,S,W,t,tm	High
Artificial Neural Networks	D,ES,C	Complex	S,H	Fast	T,H,S,W,t,tm	High

[a] Usage shown includes diagnostics (D), energy savings calculations (ES), design (DE), and control (C).
[b] Time scales shown are hourly (H), daily (D), monthly (M), and subhourly (S).
[c] Variables include temperature (T), humidity (H), solar (S), wind (W), time (t), thermal mass (tm).

References

Anstett, M. and J.F. Kreider, (1993). Application of Artificial Neural Networks to Commercial Building Energy Use Prediction, *ASHRAE Trans.*, 99, (Part. 1), 505–517.

Bacot, P., A. Neveu, and J. Sicard, (1984). Analyse Modale Des Phenomènes Thermiques en Regime Variable Dans le Batiment, *Revue Generale de Thermique*, No. 267, 189.

Bou Saada, T. and J. Haberl, (1995a). A Weather-Daytyping Procedure for Disaggregating Hourly End-Use Loads in an Electrically Heated and Cooled Building from Whole-Building Hourly Data, *Proc. 30th IECEC*, 349–356.

Bou Saada, T., and J. Haberl, (1995b). An Improved Procedure for Developing Calibrated Hourly Simulation Models, *Proc. Int. Building Performance Simulation Assoc.*

Bronson, D., S. Hinchey, J. Haberl, and D. O'Neal, (1992). A Procedure for Calibrating the DOE-2 Simulation Program to Non-Weather Dependent Loads, 1992 *ASHRAE Trans.*, 98 (Part 1), 636–652.

Clarke, J.A., (1985). *Energy Simulation in Building Design*, Adam Hilger Ltd., Boston, MA.

Claridge, D.E., M. Krarti, and M. Bida, (1987). A Validation Study of Variable-Base Degree-Day Cooling Calculations, *ASHRAE Trans.*, 93(2), 90–104.

Claridge, D.E., J.S. Haberl, R. Sparks, R. Lopez, and K. Kissock, (1992). Monitored Commercial Building Energy Data: Reporting the Results. *ASHRAE Trans.*, 98 (Part 1), 636–652.

Clark, D.R., (1985). *HVACSIM+ Building Systems and Equipment Simulation Program: Reference Manual.* NBSIR 84-2996, U. S. Department of Commerce, Washington, D.C.

Cole, R.J., (1976). The Longwave Radiation Incident Upon the External Surface of Buildings. *The Building Services Engineer*, 44, 195–206.

Cooper, K.W. and D.R. Tree, (1973). A Re-evaluation of the Average Convection Coefficient for Flow Past a Wall. *ASHRAE Trans.*, 79, 48–51.

Corson, G.C. (1992). Input-Output Sensitivity of Building Energy Simulations, *ASHRAE Trans.*, 98 (Part 1), 618.

Crawley, D.B. et al., (1999). EnergyPlus: a New Generation Building Energy Simulation Program, in *Proceedings of the Renewable and Advanced Energy Systems for the 21st Century Conference*, April 11–15, 1999, Lahaina, Maui, Hawaii.

Crawley, D.B. et al., (2000). EnergyPlus: Energy simulation program, *ASHRAE Journal*, April.

Draper, N. and H. Smith, (1981). *Applied Regression Analysis*, 2nd edition, John Wiley & Sons, New York, NY.

Davies, M.G., (1988). Design Models to Handle Radiative and Convective Exchange in a Room. *ASHRAE Trans.*, 94 (Part. 2), 173–195.

Erbs, D. G., S. A. Klein, and W. A. Beckman, (1983). Estimation of Degree-Days and Ambient Temperature Bin Data from Monthly-Average Temperatures. *ASHRAE J.* 25(6), 60.

Fels, M., Ed. (1986). Measuring Energy Savings: The Scorekeeping Approach, *Energy and Buildings*, vol. 9.

Fels, M. and M. Goldberg, (1986). Refraction of PRISM Results in Components of Saved Energy, *Energy and Buildings*, 9:169.

Haberl, J.S. and P. Komor, (1990b). Improving Commercial Building Energy Audits: How Daily and Hourly Data Can Help, *ASHRAE J.*, 32 (9), 26–36.

Haberl, J.S., D. Bronson, D. O'Neal, (1993). An Evaluation of the Impact of Using Measured Weather Data Versus TMY Weather Data in a DOE-2 Simulation of an Existing Building in Central Texas. *ASHRAE Trans.*

Haberl, J.S. and D.E. Claridge, (1987). An Expert System for Building Energy Consumption Analysis: Prototype Results, *ASHRAE Trans.*, 93 (Part. 1), 979–998.

Huang, Y.J. et al., (1984), Home Energy Rating Systems: Sample Approval Methodology for Two Tools, LBL-18669, Lawrence Berkely Laboratory, Berkeley, CA.

Joint Center for Energy Management, (1992). *Interim Report: Artificial Neural Networks Applied to Loan-STAR Data.* Report No. TR/92/10, June.

Joint Center for Energy Management, (1992). *Second Interim Report: Artificial Neural Networks Applied to LoanSTAR Data.* Report No. TR/92/11, June.

Joint Center for Energy Management, (1992). *Final Report: Artificial Neural Networks Applied to Loan-STAR Data.* Report No. TR/92/15, September.

Kaplan, M., J. McFerran, J. Jansen, and R. Pratt, (1990). Reconciliation of a DOE2.1c Model with Monitored End-Use Data From a Small Office Building, *ASHRAE Trans.*, 96, (Part 1), 981.

Katipamula, S. and D.E. Claridge, (1993). Use of Simplified Models to Measure Retrofit Energy Savings, *ASME J. Solar Energy Engineering*, 115, 77–84.

Kreider, J.F. and X.A. Wang, (1992). Improved Artificial Neural Networks for Commercial Building Energy Use Prediction, *Solar Engineering*, 92, 361–366, ASME, New York.

Kreider, J.F. and J.S. Haberl, (1994). Predicting Hourly Building Energy Usage: The Great Predictor Shootout — Overview and Discussion of Results, *ASHRAE Trans.*

Kreider, J.F., A. Rabl, and P.S. Curtiss, (1994). *Heating and Cooling of Buildings: Design for Efficiency*, McGraw-Hill, New York, 890.

Kreider, J.F. and X.A. Wang, (1991). Artificial Neural Networks Demonstration for Automated Generation of Energy Use Predictors for Commercial Buildings. *ASHRAE Trans.*, 97, (Part 1).

Liu, M., and D. Claridge, (1995). Application of Calibrated HVAC System Models to Identify Component Malfunctions and To Optimize the Operation and Control Strategies, *Proc. of the 1995 ASME Solar Engineering Conf.*, 1, 209–218.

MacDonald, J.M. and D.M. Wasserman, (1989). *Investigation of Metered Data Analysis Methods for Commercial and Related Buildings*, ORNL Rept., ORNL/CON-279, May.

McClelland, J. L. and D. E. Rumelhart, (1988). *Exploration in Parallel Distributed Processing*. MIT Press, Cambridge, MA.

Neter, J., W. Wasserman, and M. Kutner, (1989). *Applied Linear Regression Models*, 2nd edition, Richard C. Irwin, Inc., Homewood, IL.

Rabl, A. (1988). Parameter Estimation in Buildings: Methods for Dynamic Analysis of Measured Energy Use, *J. Solar Energy Engineering*, 110, 52–66.

Rabl, A., A. Riahle, (1992). Energy Signature Model for Commercial Buildings: Test With Measured Data and Interpretation, *Energy and Buildings*, 19, 143–154.

Reddy, T., (1989). Application of Dynamic Building Inverse Models to Three Occupied Residences Monitored Non-Intrusively, *Proc. Thermal Performance of Exterior Envelopes of Buildings IV*, ASHRAE/DOE/BTECC/CIBSE.

Reddy, T. and D. Claridge, (1994). Using Synthetic Data to Evaluate Multiple Regression and Principle Component Analyses for Statistical Modeling of Daily Building Energy Consumption, *Energy and Buildings*. 24, 35–44.

Ruch, D., L. Chen, J. Haberl, and D. Claridge, (1993). A Change-Point Principal Component Analysis (CP/PCA) Method for Predicting Energy Usage in Commercial Buildings: The PCA Model, *J. Solar Energy Engineering*, 115, No. 2, May.

Ruch, D., and D. Claridge, (1991). A Four Parameter Change-Point model for Predicting Energy Consumption in Commercial Buildings, *Proc. ASME-JSES-JSME International Solar Energy Conf.*, 433–440.

Sonderegger, R.C., (1977). Dynamic Models of House Heating Based on Equivalent Thermal parameters, Ph.D. Thesis, Center for Energy and Environmental Studies Report No. 57, Princeton University, Princeton, NJ.

Thamilseran, S and J. Haberl, (1995). A Bin Method for Calculating Energy Conservation Retrofit Savings in Commercial Buildings, *Proc. 1995 ASME/JSME/JSES Intl. Solar Energy Conf.*, 111–124.

U.S. Air Force, (1978). *Engineering Weather Data, AF Manual* AFM 88-29, U.S. Government Printing Office, Washington, D.C.

U.S. Army, (1979). *BLAST, The Building Loads Analysis and System Thermodynamics Program — Users Manual*. U.S. Army Construction Engineering Research Laboratory Report E-153.

U.S. Department of Energy, (1981). *DOE Reference Manual Version 2.1A*. Los Alamos Scientific Laboratory Report LA-7689-M, Version 2.1A. Lawrence Berkeley Laboratory, Report LBL -8706 Rev. 2.

Wang, X.A. and J.F. Kreider, (1992). Improved Artificial Neural Networks for Commercial Building Energy Use Prediction, *Solar Engineering*.

Wasserman, P.D., (1989). *Neural Computing, Theory and Practice*. Van Nostrand Reinhold, New York, NY.

Winkelmann, F. and Meldem, R., Comparison of DOE-2 with measurements in the Pala Test Houses, *LBNL Report 37979*, July 1995.

Yuill, G.K., (1990). *An Annotated Guide to Models and Algorithms for Energy Calculations Relating to HVAC Equipment*. American Society of Heating, Refrigerating, and Air-Conditioning Engineers, Inc., Atlanta, GA.

6.3   Energy Conservation in Buildings

Max Sherman and David Jump

6.3.1   The Indoor Environment

The building sector is an important part of the energy picture. In the United States about 40% of all U.S. energy expenses are attributable to buildings. While buildings may consume about one-third of the fuel resources in the country, they consume over 65% of the electricity.

Residential buildings accounted for 17 quads of energy at a cost of 104 billion dollars, while commercial buildings (offices, stores, schools, and hospitals) accounted for about 13 quads and $68 of energy consumption in 1989. A breakdown by sector reveals where most of the energy is used in buildings and provides background for improving building energy efficiency. The OTA's estimate suggests that by 2010, more that 10 quads of energy could be saved with improved energy efficiency.

The energy consumed in residential and commercial building provides many services, including weather protection, thermal comfort, communications, facilities for daily living, esthetics, a healthy work environment, and so on. Since in a modern society, people spend the vast majority of their time inside buildings, the quality of the indoor (or built) environment is important to their comfort, and good thermal performance of buildings is important for energy efficiency as well as productivity of workers in commercial buildings.

This chapter first discusses issues that influence the quality of the indoor environment in terms of the occupants' comfort, health, and productivity. Most of the facets of providing an acceptable indoor environment involve energy-intensive services. Factors influencing space conditioning energy needs, such as the building envelope's thermal properties and ventilation requirements, are then discussed. Next, the means of providing space conditioning to the building interior is examined. Thermal distribution systems are discussed in this section; the actual heating and cooling equipment is covered in Chapters 4.1–4.3. Where applicable, opportunities for energy conservation through improved efficiency are demonstrated. Finally, some building simulation tools are reviewed to provide the reader with the means necessary to complete the complicated analysis of building energy consumption.

Thermal Comfort

The most fundamental building service is to protect the occupants of the building from the outdoor environment. The building structure keeps the wind and rain out, but energy is required to provide an acceptable thermal environment for the occupants. The amount of energy required depends in part on the optimum comfort conditions required by the occupants activity levels and clothing (Figure 6.3.1). Thermal comfort is discussed in Chapter 2.2.

Lighting and Visual Comfort

After the HVAC system, lighting accounts for the largest energy consumption of the building services. Building occupants require sufficient light levels to go about their normal activities. Just as the building envelope protects the occupants from weather, it also reduces the amount of usable sunlight. The desire to continue activities at any time or place necessitates the use of electric lighting.

In the perimeter of buildings and throughout small buildings natural lighting or daylighting can provide a high-quality visual environment. Care must be taken in the design phase to control the admission of daylight as it varies over the day and throughout seasons. This is done primarily with the use of exterior overhangs, fins, awnings, blinds, and interior shades, drapes, or blinds. In the core areas of buildings, electric lighting must be continuously provided to meet the occupants' lighting requirements.

Achieving visual comfort requires more than providing average light levels. Glare from high-intensity sources, poor color rendition, or flickering can all cause discomfort or reduce visual performance.

Indoor Air Quality

Good indoor air quality may be defined as air that is free of pollutants that cause irritation, discomfort or ill health to occupants, or premature degradation of the building materials, paintings, and furnishings or equipment. Thermal conditions and relative humidity also impact the perception of air quality in addition to their effects on thermal comfort. Focus on indoor air quality issues increased as reduced-ventilation energy-saving strategies, and consequently increased pollution levels, were introduced. A poor indoor environment can manifest itself as a sick building in which some occupants experience mild illness symptoms during periods of occupancy. More serious pollutant problems may result in long-term or permanent ill-health effects.

An almost limitless number of pollutants may be present in a space, of which many are at immeasurably low concentrations and have largely unknown toxicological effects. Sources of indoor air pollutants in the home and in offices and their typical concentrations are given in Table 6.3.1. The task of identifying and assessing the risk of individual pollutants has become a major research activity in the past 20 years. Some pollutants can be tolerated at low concentrations, while irritation and odor often provide an early warning of deteriorating conditions. Health-related air quality standards are typically based on risk assessment and are specified in terms of a maximum permitted exposure, which is determined by exposure time and pollutant concentration. Higher concentrations of pollutants are normally permitted for shorter term exposures.

Air quality needs for comfort are highly subjective and dependent on circumstances. Some occupations allow higher exposures than would be allowed for the home or office. Health-related air quality standards are normally set at minimum safety requirements and may not necessarily provide for adequate comfort or energy efficiency at work or in the home.

Pollution-free environments are a practical impossibility. Optimum indoor air quality relies on an integrated approach to managing exposures by the removal and control of pollutants and ventilating the occupied space. It is often useful to differentiate between unavoidable pollutants (such as human bioef-

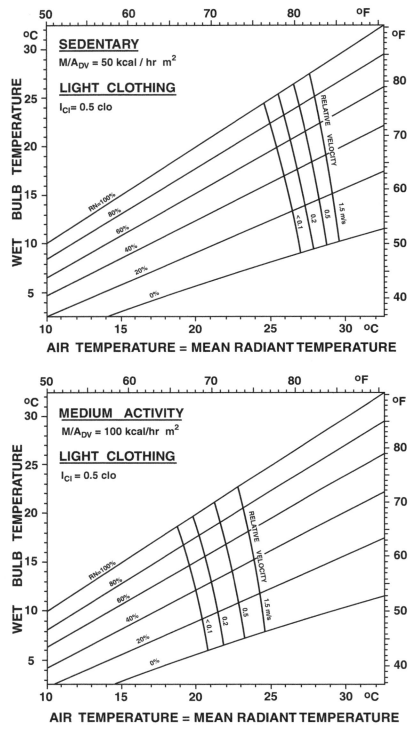

FIGURE 6.3.1 Comfort lines for persons with light and medium clothing at two different activity levels. (From Fanger, P.O., *Thermal Comfort Analysis and Applications in Environmental Engineering,* McGraw-Hill, New York, 1970.)

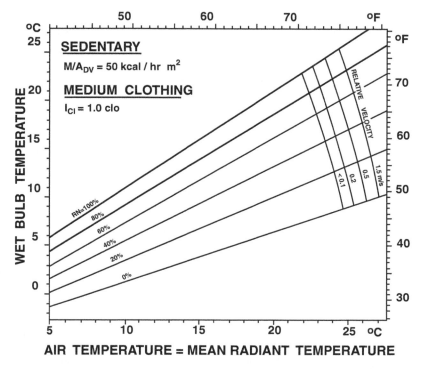

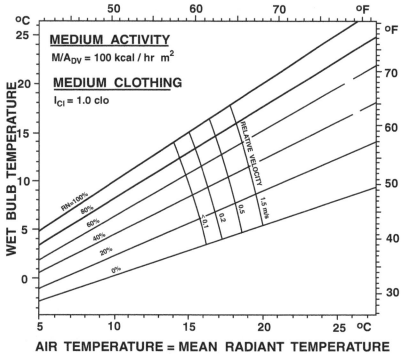

FIGURE 6.3.1 (*continued*)

TABLE 6.3.1 Principal Indoor Pollutants, Sources, and Typical Concentrations

Pollutant	Source	Concentrations
Respirable particles	Tobacco smoke, unvented kerosene heaters, wood and coal stoves, fireplaces, outside air, occupant activities, attached facilities	>500 μg/m³ bars, meetings, waiting rooms with smoking 100–500 μg/m³ smoking sections of planes 10 to 100 μg/m³ homes 1,000 μg/m³ burning food or fireplaces
NO, NO₂	Gas ranges and pilot lights, unvented kerosene and gas space heaters, some floor heaters, outside air	25 to 75 ppb homes with gas stoves 100 to 500 ppb peak values for kitchens with gas stoves or kerosene gas heaters
CO	Gas ranges, pilot lights, unvented kerosene and gas space heaters, tobacco smoke, back drafting water heater, furnace, or wood stove, attached garages, street level intake vents, gasoline engines	>50 ppm when oven used for heating >50 ppm attached garages, air intakes 2 to 15 ppm cooking with gas stove
CO₂	People, unvented kerosene and gas space heaters, tobacco smoke, outside air	320 to 400 ppm outdoor air 2,000 to 5,000 ppm crowded indoor environment, inadequate ventilation
Infectious, allergenic, irritating biological materials	Dust mites and cockroaches, animal dander, bacteria, fungi, viruses, pollens	>1,000 cfu/m³ homes with mold problems, offices with water damage 500 ± 200 cfu/m³ homes and offices without obvious problems
Formaldehyde	Urea Formaldehyde Foam Insulation (UFFI), glues, fiberboard, pressed board, plywood, particle board, carpet backing fabrics	0.1 to 0.8 ppm homes with UFFI 0.5 ppm average in mobile homes
Radon and radon daughters	Ground beneath a home, domestic water, some utility natural gas	1.5 pCi/l estimated average in homes >6 pCi/l in 3 to 5% homes
Volatile organic compounds: benzene, styrene, tetrachloroethylene, dichlorobenzene, methylene chloride, chloroform	Outgassing from water, plasticizers, solvents, paints, cleaning compounds, mothballs, resins, glues, gasoline, oils, combustion, art materials, photocopiers, personal car products	Typical concentrations of selected compounds: benzene — 15 μg/m3; 1,1,1 trichloroethylene — 20 μg/m3; chloroform — 2 μg/m3; tetrachloroethylene — 5 μg/m3; styrene — 2 μg/m3; m, p-dichlorobenzene — 4 μg/m3; m,p-xylene — 15 μg/m3
Semivolatile organics: chlorinated hydrocarbons, DDT heptachlor, chlordane, polycyclic compounds	Pesticides, transformer fluids, germicides, combustion of wood, tobacco, kerosene and charcoal, wood preservatives, fungicides, herbicides, insecticides	limited data
Asbestos	Insulation on building structural components, asbestos plaster around pipes and furnaces	>1,000 ng/m³ when friable asbestos, otherwise no systematic measurements

Note: A cfu is a colony forming unit.
Source: From Samet, J.M. et al. (1988).

fluents) over which little source control is possible, and avoidable pollutants (such as emissions of volatile organic compounds) for which control is possible. Whole-building ventilation usually provides an effective measure to deal with the unavoidable emissions, but source control is the preferred and sometimes only practical method to address avoidable pollutant sources. Examples of source control are given in Table 6.3.3.

TABLE 6.3.2 Methods of Controlling Sources of Indoor Pollution

Use of building materials, furnishings, and consumer products with low emissions rates
Physical removal of emitted pollutant
Isolating, encapsulating, or controlling emission sources
Local venting of pollutants at the point of emission (e.g., range hood, substructure radon control system)

Source: From Nero, A.V. Jr., (1992). With permission.

Productivity

The indoor environment contains and is affected by a variety of other issues that have indirect effects on thermal conservation in buildings. One of these that is of primary importance is worker productivity. Productivity is the workers' efficiency in performing their duties and responsibilities, which ultimately result in the economic well-being of the organization. In commercial buildings productivity has traditionally been viewed as the monetary return on employee compensation. Efforts to increase worker productivity have evolved from improving job satisfaction by various means (Stokes, 1978) and improving worker incentives (Lawler and Porter, 1967), to focusing on factors that negatively influence productivity, such as poor indoor environments.

Poor indoor environments can be generally described in three categories: inadequate thermal comfort, unhealthy environments, and poor lighting. Manifestations of poor productivity can be characterized by worker illness, absenteeism, distractions to concentration, and drowsiness or lethargy at work as well as by defects and mistakes in manufacturing and routine office work, and so forth. Primarily because of inadequate productivity measures, direct relationships between productivity and environmental factors are difficult to quantify (Daisey, 1989).

Examination of the cost of improving energy efficiency in buildings reveals that while significant energy cost savings are being achieved through retrofits, the relative savings may be dwarfed by savings due to increased worker productivity. Romm and Browning (1994) presented data based upon a national survey of office building stock in the United States showing that while energy costs are roughly $1.8/ft² yr, the office workers' salaries amount to approximately $130/ft² yr. As the authors state, "a 1% gain in productivity is equivalent to the entire annual energy cost." The point, often overlooked when considering an energy efficiency measure's cost-effectiveness, is that increased worker productivity can dramatically reduce the payback time of the retrofit.

Envelope Thermal Properties

Some of the most important properties of building materials are their strength, weight, durability, and cost. In terms of energy conservation, their most important properties are their ability to absorb and transmit heat. The materials' thermal properties govern the rate of heat transfer between the inside and outside of the building, the amount of heat that can be stored in the material, and the amount of heat that is absorbed into the surface by heat conduction and radiation. The rate of heat transfer through the building materials in turn determines the magnitude of heat losses and gains in the building. This information is important in order to determine the proper and most efficient design of space heating equipment required to maintain the desired indoor environmental conditions.

Heat loss and gain through the building envelope is a complex process involving four main mechanisms: heat conductance through solid and porous parts of the building envelope; heat convection from air to walls, ceilings, floors, and exteriors; solar radiation absorbed on exterior surfaces and transmitted through windows; and heat transport through ventilation or infiltration of air. This section discusses the thermal properties associated with the first three mechanisms.

In order to take full advantage of different materials' thermal properties for energy conservation purposes, it is necessary first to determine the nature of building loads in each building sector. Residential, lightly loaded small commercial buildings and warehouses typically have low internal loads (e.g., heat from appliances, office equipment, lights, people, etc.), high infiltration loads, and high envelope transmission loads. The heat losses in these buildings are roughly proportional to the indoor-outdoor temperature difference. Depending on orientation and shading, solar heat gains can also be large. In large commercial, industrial, and institutional buildings, envelope transmission loads are relatively lower than in houses and affect only the peripheral zones, not the building core. In these buildings, internal loads are dominant. Chapters 6.1 and 6.2 discuss additional details of thermal loads.

Above-Grade Opaque Surfaces

Steady-state heat transfer through the walls, floors and ceilings of a building depends on the indoor-outdoor temperature difference and the heat transmittance through each envelope component. Nothing can be done about the weather, and indoor conditions are constrained by occupant thermal comfort, but the conductance of the building envelope can be advantageously manipulated. Equation 6.3.1 shows the calculation required to determine envelope transmission heat losses, where the summation is taken over each component of the building envelope that separates the interior from the exterior.

$$Q_{tr} = \sum_i (UA)_i \Delta T \qquad (6.3.1)$$

where

ΔT is the indoor-outdoor air temperature difference, in K (°F)
A is the component's surface area, in m^2 (ft^2)
U is the thermal transmittance of the component, in W/m^{2}K (Btu/hr ft^2°F)

The inverse of the transmittance U, is the component's resistance to heat flow, $R = 1/U$. Thermal resistance is analogous to electrical resistance when the heat transfer is one-dimensional, which is often the case in buildings. Thus, a very good approximate method of determining wall resistance to heat flow, for example, is to use electric circuit analogs. This is particularly useful when analyzing composite walls, ceilings, or floors made up of supporting framework, insulation, interior wallboard, exterior facing and so on, where the total resistance to heat flow can be determined from the individual resistances of each component. As an example, consider the composite wall of Figure 6.3.2(a), which represents the structure of a wall in a house, with insulation between wood studs on 46 cm (18 inches) centers. The interior wallboard is gypsum and the exterior facing is Douglas fir. The wall is represented by the electrical circuit shown in Figure 6.3.2(b).

The total thermal resistance of the wall is given by

$$R_{tot} = R_1 + T_2 + (R_5 R_6)/(R_5 + R_6) + R_3 + R_4$$

where $\quad R_1 = 1/h_i, \ R_2 = a/k_1, \ R_3 = c/k_2, \ R_4 = 1/h_o, \ R_5 = b/(0.11 \cdot k_3), \ R_6 = 1/(0.89 \cdot k_4)$

R_1 and R_4 are convective resistances, while all others are conductive resistances. R_5 and R_6 have been corrected for the fractional amount of area perpendicular to heat flow that they occupy. To show the effect of insulation, the calculation of the total wall resistance will be done first by assuming the insulation space is occupied by air. The calculation will be repeated with fiberglass installed in the insulation space.

Using thermal conductivity data from Table 6.3.3, and assuming typical values for the convection coefficients as shown:

$$h_i = 7.5 \ \text{W/m}^2\text{K}, \ h_o = 15 \ \text{W/m}^2\text{K},$$
$$k_1 = 0.48 \ \text{W/m K}, \ a = 0.0127 \ \text{m}$$
$$k_2 = 0.11 \ \text{W/m K}, \ c = 0.0254 \ \text{m}$$
$$k_3 = 0.17 \ \text{W/m K}, \ b = 0.1016 \ \text{m}$$

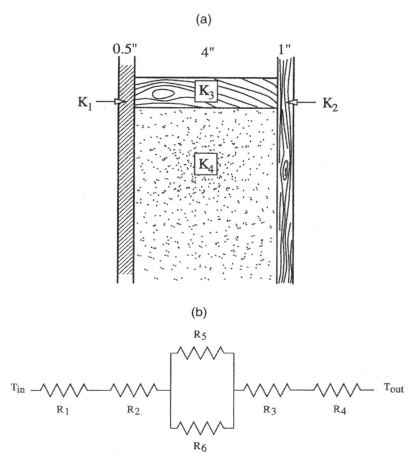

FIGURE 6.3.2 (a) Schematic representation of a composite wall section. (b) Equivalent thermal resistance for wall section of (a).

for the airspace, $R_6 = 0.44$ Km²/W. Using these values, the total thermal resistance is

$$R_{tot} = 0.01 + 0.06 + (5.43 \cdot 0.44)/(5.43 + 0.44) + 0.23 + 0.07 = 0.78 \text{ Km}^2/\text{W} \ (4.43°\text{F ft}^2 \text{ hr}/\text{Btu})$$

If the air space is filled with cellulose insulation, $k_4 = 0.043$ W/m K, $R_4 = 0.1016/(.89*0.042) = 2.72$ K m²/W, and

$$R_{tot} = 0.01 + 0.06 + (5.43 \cdot 2.72)/(5.43 + 2.72) + 0.23 + 0.07 = 2.18 \text{ Km}^2/\text{W} \ (12.38°\text{F ft}^2 \text{ hr}/\text{Btu})$$

which is more than double the thermal resistance of the wall with air between the studs.

The wood frame in the previous example acts as a thermal bridge between the wallboard and external facing. To further reduce heat losses a 2.54 cm (1 inch) layer of open-cell rigid foam can be added between the wood frame and exterior facing. This would result in a new total resistance of

$$R_{tot} = 2.18 + 0.0254/0.033 = 2.95 \text{ Km}^2/\text{W} \ (16.75°\text{F ft}^2 \text{ hr}/\text{Btu})$$

TABLE 6.3.3 Thermal Properties of Some Common Building Materials

Material	Density kg/m³ (lb/ft³)	Conductivity W/m K (Btu/hr ft°F)	Specific Heat J/kg K (Btu/lb°F)	Emissivity Ratio
	Wallboard			
Douglas fir plywood	140 (8.7)	0.11 (0.06)	2,720 (0.65)	
Gypsum board	1,440 (90)	0.48 (0.27)	840 (0.20)	
Particle board	800 (50)	0.14 (0.08)	1,300 (0.31)	
	Masonry			
Red brick	1,200 (75)	0.47 (0.27)	900 (0.21)	0.93
White brick	2,000 (125)	1.10 (0.64)	900 (0.21)	
Concrete	2,400 (150)	2.10 (121)	1,050 (0.25)	
Hardwoods			1,630 (0.39)	
Oak	704 (44)	0.17 (0.10)		0.09 (planed)
Birch	704 (44)	0.17 (0.10)		
Maple	671 (42)	0.16 (0.09)		
Ash	642 (40)	0.15 (0.09)		
Softwoods			1,630 (0.39)	
Douglas fir	559 (35)	0.14 (0.08)		
Redwood	420 (26)	0.11 (0.06)		
Southern pine	614 (38)	0.15 (0.09)		
Cedar	375 (23)	0.11 (0.06)		
Steel (mild)	7,830 (489)	45.3 (26.1)	500 (0.12)	0.12 (cleaned)
	Aluminum			
Alloy 1100	2,740 (171)	221 (127.7)	896 (0.21)	0.09 (commercial sheet)
Bronze	8,280 (517)	100 (57.8)	400 (0.10)	
Rigid Foam Insulation	32.0 (2.0)	0.033 (0.02)		
Glass (soda-lime)	2,470 (154)	1.0 (0.58)	750 (0.18)	0.94 (smooth)

Sources: ASHRAE Handbook of Fundamentals, 2001; Holman, J.P., 1976.

which is a 35% improvement. It is common practice in cold climates and in superinsulated houses to use an external insulation layer. However adding more insulation to the house exterior is not always beneficial. Adding insulation decreases envelope heat transmission losses; however, the percentage reduction diminishes quickly with increased insulation. After a certain point, the energy cost savings do not justify the cost of the added insulation. This point represents the economic optimum insulation thickness, and is determined by minimizing the life cycle and installation costs. This optimum varies by climate zone.

The application of insulation in internally loaded buildings must be analyzed on a case-by-case basis. For example, in a commercial building in a cool climate with outdoor temperatures below room temperatures more than half the time, adding insulation to the envelope would unnecessarily increase cooling costs. The optimum insulation thickness depends on the amount of internal and solar gains, hours of use, and so on. These factors vary between zones of the building.

The overall conductance of the building can be found by analyzing each component separately, then summing over all components. The process is cumbersome because the bookkeeping of all conductivities, thicknesses, resistances, and so on must be accurate. However, calculated values have been shown to agree well with measured data if the quality of installation is high.

Heat transmission through the building envelope is one of the major loss mechanisms in residences. Increasing the thermal resistance of the envelope by adding insulation reduces space heat loss on a long-term basis. This increases the building's thermal efficiency and also improves the occupants' thermal comfort by providing a more constant indoor temperature. Insulating walls and ceilings also keeps the inside surface of the exterior wall above the dewpoint temperature, thereby preventing condensation.

There are many types of insulation materials. Table 6.3.4 lists some of the various types of insulation, the various forms available, and the approximate thermal resistance per unit thickness.

TABLE 6.3.4 Available Building Envelope Insulation

Insulation Type	Blanket	Batt	Loose Fill	Rigid Panels	Formed-in-Place	R-Value/thickness m² K/W/m (°F ft² hr/Btu/in.)
Cellulose	√	√	√			25.7 (3.7)
Rock wool and fiberglass	√	√	(pellets)	(semirigid board)	22.9 (3.3)	
Perlite	√					18.7 (2.7)
Polystyrene				√		24.3 (3.5)
Urea formaldehyde, urethane					√	31.2, 36.7 (4.5, 5.3)

TABLE 6.3.5 Appropriate Application of Insulation Forms

Batts	Between joists, on unfinished attic floors or basement/crawl space ceilings
Blankets	Same as batts, but with longer continuous coverage
Loose fill	Poured in unfinished attic floor, useful around obstructions and hard-to-reach corners
Rigid boards	Interior or exterior basement walls
Foam	Between framing studs in wall, virtually anywhere in building

Sources: From Jones, P. (1979).

While each type of insulation shows a high R-value, there are some disadvantages to some insulation types that must be mentioned. Both cellulose and perlite will pack down and lose their insulation value when they get wet. Rock wool and fiberglass irritate the skin. Polystyrene is moisture resistant, but combustible. Urea formaldehyde and urethane give off noxious gases during fire, even though they're fire resistant. They require specialized equipment to inject the foam into wall cavities, which makes them the most expensive type. If not installed correctly, they'll leave a lingering odor.

Reflective-type insulations are not mentioned in Table 6.3.4. These insulations generally have smooth and shiny surfaces and are installed with this surface facing the source of heat. Reflective insulations are generally installed over exposed studs in walls, attics, and floors, often enclosing an air space underneath. The thermal resistance of reflective insulations so installed depends on the orientation of the insulation and direction of heat flow. Some blanket and batt insulations have reflective backing, which also serves as a vapor barrier.

Currently the highest thermal resistance per unit thickness insulation is the blown polymer foam types as shown in Table 6.3.4. These foam insulations contain chlorofluorocarbons (CFCs), which have been identified with the depletion of the earth's ozone layer. Use of CFCs was discontinued in 1996 in the United States. The performance of blown foam insulations with CFC substitutes is expected to decrease by up to 25%.

Insulation technology is still evolving. A new type of insulation showing great promise is gas-filled panels (GFPs). These insulations are made up of a thin walled baffle structure with low-emittance coatings. The minimal solid construction prevents heat conduction; the baffle structure and coating reduce convection and radiation heat transfer. High-performance GFPs use low-conductivity gases such as argon and krypton in place of air in the baffle spaces. Originally designed for refrigerators and freezers, GFPs can be applied as a building insulation. The R-value per inch for argon filled units, R7.5/in, is twice as high as fiberglass insulation, and more expensive krypton filled units have achieved R12/in. Projected costs in 1992 for 3-inch R22 GFPs is $0.60/ft². GFPs are in development (Griffith and Arashteh, 1992).

The various forms of insulation are convenient to use in different parts of the building. Table 6.3.5 shows where each form is typically installed.

Figure 6.3.3 shows a map of the United States roughly divided into climate zones. The recommended R-value has of residential ceiling, wall and floor insulation corresponding to the zones on the map are given in Table 6.3.6.

The thermal mass of a building material is the product of the material's specific heat and density. Judicious use of a building's thermal mass can also be used to increase its energy efficiency. For example, in winter, a building's thermal mass can store heat from solar radiation during the day and be made to

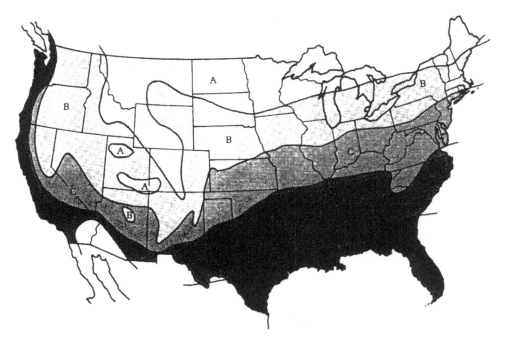

FIGURE 6.3.3   U.S. zones based on insulation needs. (From Jones, P., *How to Cut Heating and Cooling Costs*, Butterick Publishing, New York, 1979.)

TABLE 6.3.6   Recommended R-Value for U.S. Zones

Zone	Ceiling	Wall	Floor
A	R-38	R-19	R-22
B	R-33	R-19	R-22
C	R-30	R-19	R-19
D	R-26	R-19	R-13
E	R-26	R-13	R-11

Source: Jones, P., *How to Cut Heating and Cooling Costs*, Butterick Publishing, New York, 1979.

release its heat to the interior at night, thereby reducing the need for space heating. These strategies are reviewed in the chapter on passive solar energy. Peak power shifting can also be accomplished with the use of a building's thermal mass by cooling the mass overnight and circulating indoor air over it to provide cooling during the day. Such strategies make use of lower utility rates during the night. Examples of materials with a high thermal mass are concrete, masonry, and cinder block. Gypsum board also serves as thermal mass in houses, though to a lesser degree. To use a building's thermal mass effectively, it should have a large exposed surface area to the interior.

Foundations and Basements

Foundations and basements of buildings raise other issues besides those of the rest of the building envelope. Unlike the ambient air, the ground temperature does not undergo large daily temperature swings. Problems with foundations and basements arise primarily in winter. For houses with uninsulated concrete slab foundations, the floor of the house can become very cold and water vapor diffusing through carpets can condense on the slab surface. This causes damage to the floors and carpets. The same problems can occur on uninsulated basement floors and walls. Also, pipes can freeze in uninsulated basements and the space is generally unusable. Moisture diffusion through concrete and cinder block and exposed soil in basements can be a problem. These problems can be mitigated by the proper placement of insulation and vapor barriers.

Insulating foundations and basements also has energy benefits. In a handbook on foundations and basements from Oak Ridge National Laboratory (ORNL, 1988), it was shown that both interior and exterior insulated masonry basement walls saved approximately the same amount of energy, while basement walls insulated only from the ground level to the frost line saved less energy. In regard to moisture problems, a vapor barrier between the soil and foundation can be used to keep condensing water vapors outside the building shell.

Albedo and Shading

In summer, solar radiation striking roofs and walls of houses and other buildings can be absorbed and transmitted to the building interior. For houses, attic temperatures can soar as a result of sunshine being absorbed on the roof and transmitted to the attic as heat. This increases the heat transmittance through the ceiling. One strategy to reduce ceiling heat gains is to increase attic insulation levels appropriately. In warmer climates such as the sun belt of the United States, attic insulation levels are being increased to the same levels as those found in the coldest climates of Table 6.3.6. This strategy can be expensive. Another strategy for preventing solar heat gains from penetrating the building envelope is to reflect the incoming solar radiation from building surfaces or by shading the surfaces from the sun with deciduous trees. These strategies are much more cost-effective.

Reflecting solar gains from building surfaces is a method with some precedence in history. In ancient Greece, whitewashed building walls reflected solar radiation to keep the interiors cool. This practice is being revisited in modern times. Recent studies have shown that increasing the albedo of the building roofs and walls can reduce air conditioner energy consumption. It was shown that a 22% reduction in the air conditioning bill was achieved by increasing the albedo from 0.2 to 0.6 of a residential roof in Sacramento (Akbari, et al. 1992).

Albedo is a measure of a surface's reflectance. Generally the higher the surface's reflectance, the lower its emissivity, but this does not always hold for paints, which can reflect visible light but absorb infrared radiation. The scale for both reflectivity and emissivity is from 0 to 1. Many paint manufacturers have begun listing the reflective properties on their products. Surface roughness and color also have an impact: rougher, darker surfaces generally have a lower albedo. Costs for retrofitting buildings with higher-albedo paints are minimal if the paint is used at the time of regular building maintenance.

Planting deciduous trees (i.e., broad-leaved trees that lose their leaves in the fall) near a house can provide the needed shading of buildings in summer, while allowing needed solar gains into the building in winter. There are many beneficial side effects as well: trees cool their surroundings by evapotranspiration, absorbing heat from the ambient air and evaporating up to 100 gallons of water per day, trees can filter air pollution, provide a windbreak for houses, reduce street noise, prevent soil erosion and provide habitat for wildlife. A study of the cooling effect of trees has shown that annual cooling costs of air conditioning were reduced by 10% by the addition of shade trees (Akbari, et al. 1994).

Air conditioning energy and cost savings can be further increased by combining the benefits of the use of shade trees, shrubs, and other greenery with raised albedo building surfaces and other surfaces in the urban environment as well. Figure 6.3.4 shows the typical albedo of various surfaces found in communities. When combining the shading properties of trees with high albedo surfaces, researchers estimate cooling costs to be reduced by 15 to 35% for the whole community. An excellent information source on the cooling effect of trees and high-albedo surfaces is listed in the bibliography (USEPA, 1992).

Windows

Windows, or building fenestration systems, influence building energy use in four ways. Heat is transmitted through windows in the same manner as through walls. Windows absorb and transmit direct and scattered solar radiation into the building, where it is absorbed on surfaces and convected to the inside air. Air leakage around windows increases infiltration loads. Windows also let in visible light, reducing the need for electric lighting.

Heat transmittance through building window systems is generally larger than through the opaque part of the building shell. A typical single pane of window glass has an insulating value of R-1. This value is low in comparison to typical R-15 wall thermal resistances. Adding a second or third pane of glass doubles

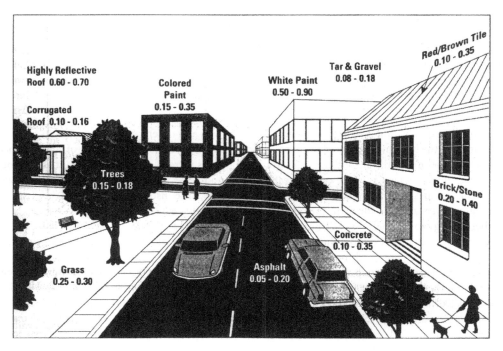

FIGURE 6.3.4 Surface albedo values in a typical urban environment. (From the United States Environmental Protection Agency, "Coding Our Communities: A Guidebook on Tree Planting and Light-Colored Surfaces," USEPA Report No. 22p-2001, Office of Policy Analysis, Climate Change Division, Washington, D.C., January 1992.)

or triples the R-value respectively. Alternatively, filling the gap space between the window panes with argon gas and adding a transparent low emissivity coating can increase the resistance to R4. A triple layer window with two low-emissivity coatings and argon gas fill can achieve R8. Naturally, windows become more expensive with each step taken to increase the R-value.

The overall heat conductance of windows is made up of three parts: through the center of the glass, through the edge of the glass and through the window frame, as in Eq. (6.3.2)

$$U_0 = \left(U_{cg} A_{cg} + U_{eg} A_{eg} + U_f A_f \right) \big/ A_{pf} \tag{6.3.2}$$

where U is the heat conductance, A is the area of the surface, cg refers to the center of the glass, eg refers to the edge of the glass (defined by the area $2\frac{1}{2}$ inches from the frame), f refers to the frame and A_{pf} refers to the opening area of the wall. This methodology of determining the window U value is preferred because of the different combinations of window and frame technologies possible in fenestration systems.

The National Fenestration Rating Council (NFRC, 1995) publishes a directory listing for over 20,000 commercially available windows.

Windows also allow radiative heat transmission through the glass. The amount of solar radiation passing through the glass during the day depends on the window's optical properties. Generally, clear glass is not spectrally selective and allows most incident solar radiation to pass through, with a small amount of absorptance (depending on thickness) and with about 8% reflected from each layer of glass. However, most architectural-quality glass is opaque to long wave radiation from surfaces at temperatures below 120°C. This tends to produce the greenhouse effect, in which radiation passing through the glass is absorbed on interior room surfaces and then re-radiated with long wave radiation to other interior surfaces, which warm the room air by convection. This re-radiated energy is unable to exit through the glass.

Figure 6.3.5 shows the relative intensity of solar radiation as a function of wavelength. Also shown in the figure is the region the human eye perceives as visible light. To make optimum use of daylighting,

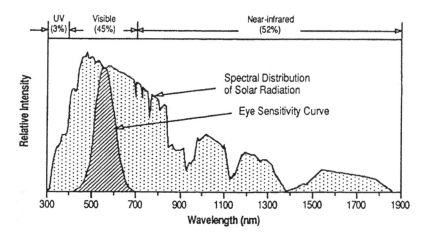

FIGURE 6.3.5 Solar spectrum. (From Davids, B.J., "Taking the Heat out of Sunlight — New Advances in Glazing Technology for Commercial Buildings," Proc. ACEEE Summer Study on Energy Efficiency in Buildings, August 26–September 1, Asilomar Conference Center, Pacific Grove, CA, 1990.)

an ideal spectrally selective coating on the surface of a window would block out all solar radiation with wavelengths below about 450 nm and above about 690 nm. Such a coating would pass virtually all visible light through the window while blocking out approximately 50% of all solar radiation. Multilayer metal and dielectric coatings on glass approach this behavior, allowing more than 80% of visible light through and reflecting most of the rest of the sun's spectrum. Some types of green, blue-green, and blue absorbing glasses also provide spectral selectivity, although they are generally not as effective as reflective coatings.

Some of the nonreflected radiation is absorbed in the glass; the rest is transmitted through it. Part of the absorbed energy is conducted to the inside surface. Thus, the solar heat gain from a window is the sum of the transmitted radiation and the fraction of absorbed radiation that is conducted to the inside surface:

$$q_i = E_t\left(\tau_s + N_i\alpha_s\right) = E_t \times \text{SHGC} \tag{6.3.3}$$

where E_t is the total solar radiation incident on the window, τ_s is the glass transmittance, N_i is the fraction conducted inward and α_s is the glass absorbance. The fraction of the total radiation that finds its way to the interior is called the Solar Heat Gain Coefficient (SHGC). The SHGC and the window's overall U-value should be used in any description of a window's energy performance.

The total rate of heat transfer through a window is the sum of the normal heat transmittance due to the indoor-outdoor temperature difference, the solar gain, and an infiltration term as shown in Eq. (6.3.4)

$$q_{tot_i} = \text{SHGC}_i E_t + U_0 \Delta T + K Q_{\text{inf}} \Delta T \tag{6.3.4}$$

Cracks around window frames or loose sliding window sashes are sites for air infiltration in buildings. This topic is covered in the next section. In building design or retrofit, appropriate window selection depends on the climate and nature of the building. For heating-dominated buildings, it is desirable to have a high transmittance over the entire solar spectrum, but a high reflectivity (low emissivity) over the long wavelength infrared portion of the spectrum. This arrangement allows most solar radiation into the building, but traps low-temperature radiant heat inside the building. Many commercially available windows with low-emittance (low-e) coatings approach this performance. The opposite performance is desirable for cooling dominated buildings. The ideal in this case is to have a high reflectance (low emittance) over all solar wavelengths outside the visible portion of the spectrum. Depending on the spectral selectivity of the window, this would eliminate from 50 to 70% of the solar radiation from entering the building, without loss in light transmission. Windows approximating this ideal behavior are available. Many absorbing glass and reflective coatings also block some portion of the visible spectrum which may be desirable to control glare from direct sun. These glazings may also change the color rendering properties of the transmitted light. This must also be considered in the selection of window systems.

Windows provide daylight, which aids visual comfort and reduces the need for artificial lighting, thus saving energy. An accurate evaluation of daylighting is beyond the purposes of this chapter. As a rule of thumb, properly designed windows can provide daylight adequate for typical indoor tasks for a depth $2^{1}/_{2}$ times the height of the window, based on normal sill height. In typical office buildings, if the effective aperture (visible transmittance of glass times fractional area of wall that is glazed) of a building is in the range of 0.2 to 0.35, daylighting can provide approximately 50% of annual electric lighting needs in perimeter zones adjacent to windows. In a skylighted building, an effective aperture of 0.04 can provide 50 to 70% of lighting needs. For more on the subject, the reader is referred to "Recommended Practices of Daylighting" (IES, 1979).

Infiltration and Ventilation

Ventilation is the building service most associated with controlling the indoor air quality to provide a healthy and comfortable indoor environment. In large buildings ventilation is normally supplied through mechanical systems, but in smaller buildings such as single-family homes it is principally supplied by leakage through the building envelope (i.e., *infiltration*). Most U.S. buildings with mechanical ventilation systems also use the system for thermal energy distribution. We restrict our discussion here to ventilation of *outdoor* air.

Ventilation is the process by which clean air is provided to a space. It is needed to meet the metabolic requirements of occupants and to dilute and remove pollutants emitted within a space. Usually ventilation air must be conditioned by heating or cooling it to maintain thermal comfort and then it becomes an energy liability. Ventilation energy requirements can exceed 50% of the space conditioning load; thus excessive or uncontrolled ventilation can be a major contributor to energy costs and global pollution. Thus, in terms of cost, energy, and pollution, efficient ventilation is vital, but inadequate ventilation can cause comfort or health problems for the occupants.

Mechanically Dominated Ventilation

Most medium and large buildings are ventilated by mechanical systems designed to bring in outside air, filter it, supply it to the occupants, and then exhaust an approximately equal amount of stale air.

Ideally these systems should be based on criteria that can be established at the design stage. After the systems are designed and installed, attempts to mitigate problems may lead to considerable expense and energy waste, and may not be entirely successful. The key factors that must be included in the design of ventilation systems are given in Table 6.3.7, along with suggested sources for more information.

These factors differ for various building types and occupancy patterns. For example, in office buildings pollutants tend to come from occupancy, office equipment, and automobile fumes. Occupant pollutants typically include metabolic carbon dioxide emission, odors, and sometimes smoking. When occupants are the prime source, carbon dioxide acts as a surrogate and can be used to cost-effectively modulate the ventilation, forming what is known as a *demand controlled ventilation* system.

TABLE 6.3.7   Design Considerations for Mechanical Ventilation

Code requirement, regulations or standards	ASHRAE Standard 55
Ventilation strategy and system sizing	ASHRAE Standard 62-1989
Climate and weather variations	ASHRAE *Handbook of Fundamentals*, (2001)
Air distribution, diffuser location, and local ventilation	ASHRAE *Handbook of Fundamentals*, (2001) ASHRAE *Handbook of Systems and Equipment*, (2000)
Location of outdoor air inlets and outlets	ASHRAE *Handbook of Fundamentals*, (2001)
Ease of operations and maintenance	Equipment Manufacturer, ASHRAE *Handbook of Applications*, (1999)
Impact of system on occupants (e.g., acoustics and vibration)	ASHRAE *Handbook of Fundamentals*, (2001)

Schools are dominated by high occupant loads, transient occupancy, and high levels of metabolic activity. Design ventilation in hospitals must aim at providing fresh air to patient areas, combined with clean room design for operating theaters. Ventilation in industrial buildings poses may special problems, which frequently have to be assessed on an individual basis. Contaminant sources are varied, but often well-defined, and limiting values are often determined by occupational standards. Poorly designed, operated or maintained ventilation systems, rather than the ventilation rate itself, may cause sick building syndrome (SBS). The causes of SBS were summarized earlier.

Infiltration

Infiltration is the process of air flowing in (or out) of leaks in the building envelope, thereby providing ventilation in an uncontrolled manner. All buildings are subject to infiltration, but it is more important in smaller buildings. In larger buildings there is less surface area to leak for a given amount of building volume, so the same leakage matters less. More important, the pressures in larger buildings are usually dominated by the mechanical ventilation system and the leaks in the building envelope have only a secondary impact on the ventilation rate. However, infiltration in larger buildings may affect thermal comfort, control, and system balance.

In low-rise residential buildings (most typically, single-family houses) infiltration is the dominant force. In these buildings mechanical systems contribute little (intentionally) to the ventilation rate.

Infiltration is made up of two parts: weather-induced pressures and envelope leakage. Since little of practical import can be done about the weather, it is the envelope leakage, or *air tightness*, that is the variable factor in understanding infiltration. Virtually all knowledge about the air tightness of buildings comes through making *fan pressurization* measurements, done most typically with a *blower door*.

Blower door is the popular name for a device that is capable of pressurizing or depressurizing a building and measuring the resultant air flow and pressure. The name comes from the common utilization of the technology, where there is a fan (i.e., blower) mounted in a door; the generic term is "fan pressurization." Blower door technology was first used around 1977 to test the tightness of building envelopes (Blomsterberg, 1977), but the diagnostic potential of the technology soon became apparent. Blower doors helped uncover hidden bypasses that accounted for a much greater percentage of building leakage than did the presumed culprits of window, door, and electrical outlet leakage. The use of blower doors as part of retrofitting and weatherization became known as *House Doctoring*. This led to the creation of instrumented audits and computerized optimizations (Blasnik and Fitzgerald, 1992). A brief description of a typical blower door test follows in the measurements section.

While it is understood that blower doors can be used to measure air tightness, the use of blower door data alone cannot be used to estimate real-time air flows under natural conditions or to estimate the behavior of complex ventilation systems. However, a rule of thumb relating blower door data to seasonal air change data exists (Sherman, 1987): namely, the seasonal air exchange can be estimated from the flow required to pressurize the building to 50 Pascal divided by 20. Ventilation and infiltration air flows are generally measured with a tracer gas, as described in the measurements section.

A more accurate description of infiltration rates can be found by separating the leakage characteristics of the building from the driving forces, which are wind- and temperature-induced pressures on the building shell. Modeling the leakage data from blower door tests as orifice flow, Sherman and Grimsrud (1980) developed the LBL infiltration model, Eq. (6.3.5):

$$Q = L\left(f_s \Delta T + f_w V_w^2\right)^{0.5} \tag{6.3.5}$$

Here Q is the volumetric air flow rate, L is the effective leakage area of the house at 4 Pa, ΔT is the indoor-outdoor temperature difference, V_w is the time-averaged wind speed, and f_s and f_w are the stack and wind coefficients, respectively, as determined from a fit of the data. The model was validated by Sherman and Modera (1984) and incorporated into the *ASHRAE Handbook of Fundamentals* (2001). Much of the subsequent work on quantifying infiltration is based on that model, including ASHRAE Standard 119 (1988) and ASHRAE Standard 136 (1993).

Blower doors are used to find and fix the leaks. A common method of locating leak sites is to hold a smoke source near the leak and to watch where the smoke exits the house. Depending on the leak site, different methods are used to stop the leakage.

Often, the values generated by blower door measurements are used to estimate infiltration for both indoor air quality and energy consumption analyses. These estimates in turn are used for comparison to standards or to provide program or policy decisions. Each specific purpose has a different set of associated blower door issues.

Compliance with standards, for example, requires that the measurement protocols be clear and easily reproducible, even if this reduces accuracy. Public policy analyses are more concerned with getting accurate aggregate answers than reproducible individual results. Measurements that might result in costly actions are usually analyzed conservatively, but "conservatively" for IAQ is diametrically opposed to "conservatively" for energy conservation.

Because infiltration depends on the weather, buildings that have much of it can have quite variable ventilation rates. Determining when there is insufficient infiltration to provide adequate indoor air quality or energy-wasteful excess infiltration is not a simple matter. The trade-off in determining optimal levels depends on various economic and climactic factors.

Individual variations notwithstanding, Sherman and Matson (1993) have shown that the stock of housing in the United States is significantly overventilated from infiltration and that there are 2 exajoules (1.9 quads) of potential annual savings that could be captured. Much of this savings could be captured by simple tightening, but a significant portion requires installation of a ventilation system or strategy to assure adequate ventilation levels.

Natural Ventilation

Natural ventilation is a strategy suitable for use in mild climates or during mild parts of the year. As commonly interpreted, natural ventilation is the use of operable parts of the building envelope (e.g., windows, etc.) to allow natural airflow at the discretion of the occupants.

Natural ventilation shares many of the same properties as infiltration: it depends on weather for driving forces; it is a function of the leakage area of the buildings; and so on. The distinguishing feature of natural ventilation, however, is that it is under the control of the building occupants.

From the point of view of the HVAC designer, natural ventilation is quite bothersome, because a conservative ventilation designer cannot count on it, but one must consider its potential effects on the building load. From the perspective of the occupants, however, natural ventilation gives them more control of their environment and usually makes it more acceptable. Studies have shown that those in naturally ventilated buildings tend to suffer less sick building syndrome and less respiratory disease (e.g., colds) than buildings that are fully mechanically ventilated.

The designs of new commercial buildings have been curtailing the availability of natural ventilation as an option by removing operable windows. Natural ventilation still dominates in the residential sector.

Ventilative Cooling

In dwellings, natural ventilation serves more than just a means to provide clean air; it serves to cool the building and its occupants and reduce the requirement for mechanical cooling. Fans are used to assist ventilative cooling by increasing the air change rates. These *whole house fans* are of much larger capacity than is needed for ventilation. A standard such fan may provide as much as 10 air changes per hour, compared with the ventilation requirement of no more than half an air change per hour.

Ventilative cooling removes internally generated heat as long as the outdoor temperature is less than the indoor temperature. When thermally massive elements are included in the structure, night ventilation can be used to store "coolth" and reduce the cooling requirements the following day.

The air motion caused by ventilative cooling provides additional cooling to the occupants of the building by removing more heat and lowering the apparent temperatures. Increased air motion raises the upper limit of acceptable air temperature from a thermal comfort perspective and therefore also reduces cooling demand.

In commercial buildings ventilative cooling is accomplished principally by the use of an *economizer*. Economizers are nothing more than dampers that allow outdoor air to be used instead of recycled (i.e., return) air in the building's thermal distribution system. This is usually done when there is a cooling load and the outdoor air is cooler than the indoor air.

A simple economizer works in lieu of the cooling system of the building. Because of the large internal loads of some commercial buildings it can be necessary to supply some mechanical cooling and also use outdoor air. Such a system is called an *integrated economizer* because it can do both. Not all systems are capable of running in this way. In buildings using cooling towers, *waterside economizers* can effectively use evaporative cooling to substitute for chiller operation.

Heat Recovery

Other than periods when ventilative cooling is useful, considerable energy is lost from a building through the departing air stream. When air change is dominated by infiltration, little can be done to recapture this energy, but if the departing air stream is centrally collected, a variety of methods for recovering or recycling the waste heat become possible (AIVC TN 45, 1994).

Ventilation heat recovery is the process by which sensible and latent heat is recovered from the air stream. Methods have been developed for air-to-air systems, in which the energy in the departing air stream is transferred directly to the entering air stream. Heat recovery can also be accomplished with heat-pump systems, in which heat from the exhaust air is pumped to another system such as the domestic hot water.

While the heat recovery process can be efficient at collecting the heat, benefits must always be weighed against the energy needed to drive the recovery, the capital costs of the equipment, and the maintenance. Induced losses such as infiltration or duct leakage must be understood. Without careful design and construction of the building envelope the system total performance can be considerably impaired and in some cases could increase the total energy costs.

The efficiency of all systems can be defined in terms of the proportion of outgoing ventilation energy that is recovered. Quoted efficiencies can be quite high (e.g., 65–75%), and the attractiveness therefore quite strong. For various reasons, however, field studies do not always come up to expectations. Basically, if poor attention is given to planning and installation, then the level of heat recovery can be quite disappointing.

Measurement Methods

A brief description of a blower door test follows. The reader is referred to ASTM Standard E779 for a complete description. Shown in Figure 6.3.6 is a sketch of a fan mounted in a doorway of a single-family house. A means of measuring the pressure difference between the house and outside is provided in this case by a digital manometer. Other pressure measuring instruments are acceptable as long as they are accurate over the measured pressure range: 0 to 100 Pascal. Volumetric air flow through the fan must also be determined. The air flow through most blower door fans is calibrated against fan speed in revolutions per minute or the pressure drop across the fan (range of approximately 0 to 500 Pa). The latter method is shown in Figure 6.3.6. The test is performed after ensuring that all windows and fireplace dampers in the house are closed. Some protocols also require exhaust vents in the kitchen and bathrooms to be sealed. The general procedure is to depressurize the house in steps of about 12 Pascal to about 50 Pascal, recording the house pressure and fan air flow at each step. The air flow direction through the fan is then reversed and the procedure is repeated for house pressurization measurements.

The air flow is plotted against the house pressure and a power law relationship of the type $Q = c(\Delta P)^n$ is fitted to the data (c and n are determined from the chosen curve fitting procedure). Using this relationship and modeling the house leakage as orifice flow, the effective leakage area of the house is determined by Eq. (6.3.6)

$$\mathrm{ELA} = CQ_r \left(\rho / (2\Delta P) \right)^{0.5} \tag{6.3.6}$$

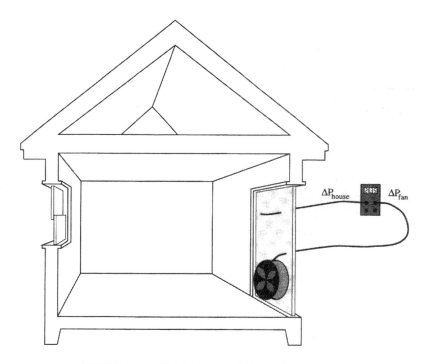

FIGURE 6.3.6   Sketch of a typical blower door test setup.

where ELA is the effective leakage area at the reference pressure ΔP_r, usually 4 Pa, Q_r is the reference pressure air flow, ρ is the air density, and C is a conversion constant. At 4 Pascal, the ELA is a good estimate of the equivalent area of holes in the envelope that provide the same leakage. Note that the ELA at 4 Pa is determined by extrapolation of measured data.

Other indications of a house's air tightness are often used. Two of these are CFM50 and ACH50. CFM50 is the flow in cubic feet per minute at a house pressure of 50 Pa. ACH 50 is the air change rate at 50 Pa, which is obtained by dividing CFM50 by the house volume in cubic feet. Unlike ELA measurements, which require an entire range of air flow and house pressure data, both CFM50 and ACH50 can be determined from one measurement. This simplifies testing. On the other hand, normalizing ELAs with house floor area enables comparisons between houses—to determine if one house is unusually leaky, for example. These air tightness indications are used for comparisons to standards, to provide background for program or policy decisions, or to estimate the energy load caused by the infiltration.

Tracer gas techniques have become widely used to measure the ventilation rates in buildings. The basic principle involved is that of conservation of mass (of tracer gas) as expressed in the continuity equation. By monitoring the injection and concentration of the tracer, one can infer the exchange of air. Although there is only one continuity equation, there are many different experimental injection strategies and analytical approaches. These different techniques may result in different estimates of infiltration due to uncertainties and biases of the procedures.

An ideal tracer gas is one which is inert, safe, mixes well, does not adhere to surfaces and its concentration in air is easily measurable. A mass balance of tracer gas within the building, assuming the outdoor concentration is zero, takes the form of Eq. (6.3.7):

$$\frac{dc(t)}{dt} = F(t) - Q(t)c(t) \qquad (6.3.7)$$

where

> V is the volume of the interior space
> $c(t)$ is the tracer gas concentration at time t
> $dc(t)/dt$ is the time rate of change of tracer concentration
> $F(t)$ is the tracer injection rate at time t
> $Q(t)$ is the airflow rate of the building at time t

The development of Eq. (6.3.7) made use of a number of assumptions: the airflow out of the building accounts for the removal of tracer gas, not chemical reactions or absorption onto surfaces, and the tracer gas concentration is uniform throughout the interior space.

There are a number of methods to use in determining infiltration and ventilation rates with a tracer gas. The choice of method in a given situation will depend on the practical details of the experiment as well as the reason for measuring the air change in the first place. A standardization method (ASTM 1990) is the decay method, which requires the least time and usually the least preparation.

In the decay method, a small amount of tracer gas is mixed with the interior air. The injection is then stopped and the concentration of tracer gas is monitored. The air change rate is then determined from the solution of Eq. (6.3.7):

$$c(t) = c_0 e^{-\text{ACH}_t} \qquad (6.3.8)$$

where c_0 is the concentration at time $= 0$ and ACH $= Q/V$ is the air change rate.

The advantages of the decay method are numerous: the injection rate F need not be measured, the concentration of tracer gas can be measured on site or collected in sample containers and analyzed elsewhere, and the test can be performed with a minimum of equipment and time. Disadvantages include errors introduced by the nonuniform mixing of tracer gas with air, large uncertainties in the air change rate unless the precision of the measuring equipment is high, and biased estimates of the average air change rate.

Other experimental techniques, such as the constant concentration and constant injection techniques are used. The constant concentration technique can be both accurate and precise, but it requires the most equipment as well as sophisticated control systems and real-time data acquisition. The constant injecting technique (without charge-up) can be considered a somewhat simpler version of the constant concentration technique, in that no active control of the injection rate is needed.

As more detailed information is required for both energy and indoor air quality purposes, researchers are turning to complex, multizone tracer strategies. Both single-gas and multigas techniques are being utilized, but only multigas techniques are capable of uniquely determining the entire matrix of air flows.

Tracer gases are used for a wide range of diagnostic techniques including leak detection and atmospheric tracing. In cases (e.g., mechanically ventilated rooms) in which ventilation rates are known, tracer gases can be used to measure the ventilation efficiency within the zone. Age of air concepts are often used to describe the spatial variation of ventilation. Sandberg (1983) summarizes the definitions and some of the tracer techniques for determining the efficiency (e.g., by seeding inlet streams or monitoring exhaust streams). Further discussion of intrazonal air flows and ventilation efficiency is beyond the scope of this chapter.

6.3.2 Review of Thermal Distribution Systems

Thermal distribution systems are the ductwork, piping or other means used to transport heat or cooling effect from the heating or cooling equipment (furnaces, boilers, compressors, etc.) to the building space where it is needed. This section focuses on the distribution system connected to the heating/cooling equipment, rather than on the equipment itself. For a review of air distribution equipment, refer to

Chapter 4.3. Energy efficiency research in buildings has been primarily focused on the building shell, lighting, appliances, or the space-conditioning equipment. Although the need for improved efficiency in thermal distribution systems has been often cited (Modera, 1989, Cummings and Tooley, 1989), this need has received more attention only in recent years.

Thermal distribution systems are primarily characterized by two transport mediums, air and water. Andrews and Modera (1991) classify the type of distribution system used in residential and small commercial buildings and estimate the potential energy savings possible. In residential applications they found that 85% of the primary energy used for space conditioning was in forced-air systems, with the remainder in hydronic systems. In small commercial buildings the authors reported that 69% of all small commercial buildings in 1986 were heated or cooled with forced air systems, and that the fraction was continually increasing. The focus of this section will be on forced air systems, because this type of system is the largest fraction in buildings and because they have the most potential for efficiency improvements.

There are three primary modes of energy loss associated with air distribution. One is direct air leakage from the ducts to unconditioned spaces or to outside via holes or cracks in the ductwork. This mechanism is mainly a function of the quality and durability of the duct installation. Another is heat conduction through the duct walls resulting from inadequate insulation. The third mode is increased infiltration resulting from increased pressures across building shell leakage sites, pressures which are generated by the operating forced air system. Infiltration is also increased when the system is not operating because the leakage sites in the duct system add to the building's overall leakage area.

The magnitude of the energy loss depends on many factors: level and location of duct leakage sites, insulation level, location of ducts, space conditioning system sizing, and climate region are a few examples. Typically, the inefficiencies of ducts are at their worst at the time of day when they are needed most. This is due to the extreme temperature differences between the ducts and their surroundings, which increase the conduction losses through the duct walls and worsen return-side leakage losses. This is true in both residential and commercial buildings where the ducts can be located in attics or on rooftops, for example. Because the demand for space conditioning is highest during these peak hours, inefficient ducts exacerbate the problem electric utilities have in meeting the power demand.

Measurement Methods

Measurement techniques available to use in the analysis of distribution efficiencies and to characterize the existing stock of forced air systems are outlined here. Duct leakage area (DLA) is an important parameter used to characterize both direct duct leakage losses and additional envelope leakage area. The most common measurement methods are listed in Table 6.3.8.

The blower door subtraction method utilizes two blower door tests, a normal blower door test as described in a previous section, and a second test with the duct system sealed from the house by taping over the duct registers. The DLA is then determined from the difference between the ELAs measured in each test. This method yields duct leakage to outside values only, as the total leakage area of ducts may also include leakage sites to inside the conditioned space. The accuracy of this method is low because the DLA is determined by the subtraction of two large numbers.

In the blower door plus flow hood measurement, the method is similar to a normal blower door test except that a flow capture hood is placed over one unsealed register, usually a return register, and the airflow into the ducts is measured during the test. Simultaneous measurements of house and duct pressure are made, and with the measured flow through the flow capture hood an orifice model is used to determine the DLA. This method has been shown to be more accurate than the subtraction technique.

TABLE 6.3.8 Duct Leakage Area Measurement Methods

No.	Primary Equipment	Result	Accuracy	Reference
1	Blower door	DLA (to outside)	low	(Proctor et al., 1993)
2	Blower door + flow hood	DLA (to outside)	medium	(Proctor et al., 1993)
3	Duct pressurizing fan	DLA (total)	high	(Energy Conservatory, 1995)
4	Duct fan + blower door	DLA (to outside)	high	(Energy Conservatory, 1995)

The third measurement of DLA is the most versatile and is gaining acceptance as the preferred method in field measurements: the fan pressurization technique. The procedure is similar to the determination of the house ELA. In this method, all registers are taped except one, where a fan is connected. The fan has been previously calibrated for volumetric flow rate. With this technique, airflow and duct pressurization measurements are made and the total duct leakage of the supply side, return side, or entire system is determined. Measurements of duct leakage to the outside are made with the help of a blower door. Besides being the most accurate of the three methods, this method has advantages when sealing ducts in retrofit applications.

Actual duct air leakage rates during system operation can be measured by individually measuring the fan air flow rate and air flow through the supply and return registers. The volumetric leakage rate of supply air is then the difference between the fan air flow rate and the sum of the air flow out the supply registers. The method is similar on the return side. Fan air flow rates can be measured with a tracer gas technique (ASTM Standard E741). Air flow rates out of registers can be measured with a flow capture hood or a modified version of a flow capture hood as done in June and Modera (1994).

An indication of duct conduction losses can be obtained by measuring temperatures in the supply and return plenums and in each supply register. The percentage of energy lost due to duct conduction is the temperature difference between the supply plenum and register temperature divided by the difference between the supply and return plenum temperatures. Measurement of the distribution system's impact on the building infiltration rate can be measured by the tracer gas or blower door techniques outlined previously.

Alternatively, a duct system's efficiency can be determined by an electric coheating method. This method compares the energy used by the heating equipment (furnace, heat-pump, etc.) and duct system to maintain a set indoor temperature with the energy consumed by electric heaters placed inside the conditioned space to maintain the same indoor temperature. This method requires extensive data acquisition and measurement equipment, but has revealed much insight on the factors that influence a duct system's energy impacts. An example of this method is found in Palmiter and Francisco (1994).

As research into the energy efficiency of air distribution systems has only recently begun, development of standard methods of testing to determine distribution system efficiencies is not yet complete. ASHRAE is currently sponsoring the development of such tests. There are three test pathways currently in development. The first is the design pathway, and it will rely on computer modeling of the building, equipment, and duct system. The other two pathways involve field measurements. The research pathway is intended to obtain the most complete description of the *in situ* performance of a duct system. The diagnostic pathway is intended to rely on field measurements which may be obtained quickly and at minimal cost. Each field measurement pathway will contain descriptions of the recommended procedures to analyze an air distribution system's performance. First versions of the ASHRAE test methods were available in 1997.

Residential Ducts

Forced air heating and cooling systems are used in approximately 50% of existing single family housing in the United States (Andrews and Modera, 1991). In the Northeast and Midwest, 52% of all homes have forced air systems, with 44% of those homes having ducts in unconditioned spaces such as attics or crawl spaces, and 50% having them in partly conditioned spaces, such as basements. In the southern and western United States, 46% of all homes have forced air systems, with 82% of their ducts located in unconditioned spaces.

Field work on existing housing has revealed the potential for efficiency improvements. Modera et al. (1992) measured DLAs (at 4 Pa) of 0.90 cm^2/m^2 (normalized for house floor area) in 19 California houses built before 1980, and 0.92 cm^2/m^2 in post-1980 construction (12 houses). Jump and Modera (1994) measured 1.57 cm^2/m^2 leakage area (at 25 Pa) in 13 houses in Sacramento, California. In terms of conduction losses, Modera et al. reported that 23% of the energy delivered to the air at the coil was lost due to conduction before it arrived at the registers. In the Sacramento houses, Jump and Modera reported a 13% loss. In terms of distribution system induced air infiltration, Gammage (1986) reported an average increase of 80% in the infiltration rate of 31 Tennessee houses when the air handler fan was operating. In five Florida houses, Cummings and Tooley (1989) determined that the infiltration rate tripled when the air handler fan was on and internal doors in the house were open, and that rate further tripled when internal doors were closed. The former effect was attributed to duct system leakage, while the latter effect was attributed to pressure imbalances in the house due to inadequate return-air pathways.

Palmiter and Francisco (1994) measured the system and delivery efficiencies in 24 all-electric homes in the Pacific Northwest. They found that delivery efficiencies averaged 56% for 22 of those homes, which had ducts in unconditioned spaces. The corresponding system efficiency was 71%. In the 2 homes with 50% or more of the ducts in conditioned spaces, they found that delivery efficiencies were 67%, with system efficiencies of 98% on average.

The majority of new single-family housing construction is housing with solid concrete foundations. This practice usually results in duct placement in the attic. This is the most unfavorable local for the distribution system because of the extreme temperature differences that exist between ducts and attics, particularly in the summer. In designing homes for energy efficiency, care must be taken to either seal and insulate the ducts well, or locate them in a less harsh environment, preferably inside the conditioned space. Efficiency-conscious home builders use techniques such as fan pressurization to verify minimum leakage levels in the duct system. Stum (1993) has other advice for duct installation in new construction.

Efficiency improvements in existing housing is mainly accomplished by duct sealing and insulation retrofits. Monitoring of duct system retrofits has shown a reduction of space conditioning energy use of up to 20% for the houses studied (Jump and Modera, 1994; Palmiter and Francisco, 1994). The average cost per house in the Jump and Modera study was $600. This cost is approximately $1\frac{1}{2}$ as much as adding R19 attic insulation.

In new construction, the savings potential is much larger. Actual energy savings are much larger if care is taken at the time of installation, ensuring airtight duct systems and adequate insulation levels. Other efficiency improvements which can be incorporated into new construction are the inclusion of zoned air distribution systems and installing the ducts inside the conditioned space. In the latter case, energy losses from the duct system will not be lost to outside the home.

Commercial Buildings

Commercial building air distribution systems have received far less attention as compared to residential systems. Small commercial buildings have received some attention because of their similarities with residential systems. Andrews and Modera (1991) determined that 69% of small commercial buildings used forced air systems for heating and cooling.

Large commercial building air distribution systems are characterized by extensive networks of ducts, mixing boxes, dampers, in-line fans, and controls. They operate at higher pressures and serve many different building zones and often the building's ventilation system is included. Unlike residential systems, the duct location in large commercial buildings is not typically in a severe environment. Two sources of energy losses in these systems come from warming the conditioned air with the in-line fans and leaky air dampers. Efficiency problems with large commercial duct systems has not received as much attention as residential and small commercial systems.

Advanced Systems

A new application of an existing technology is being developed for energy-efficient cooling, thermal comfort, and high indoor air quality in commercial buildings (Feustel, 1993). These systems provide cooling effect through radiative exchange between humans and water-cooled ceilings or ceiling panels. There are several advantages to radiant hydronic systems. The first is that water is a far better thermal medium than air. Water systems in general do not leak, and when they do, the problem is quickly noted and repaired. Second, the preferred thermal comfort arrangement of a cool head and warm feet is maintained in these systems. Overhead plenum space need no longer be provided. High indoor air quality is maintained by the continual supply of fresh air, eliminating the need to mix fresh and recirculated air as is the case in all air systems. Such systems are more common in Europe, but small studies have shown their high application potential in the United States. (Feustel, 1993; Feustel and Stetiu, 1995).

6.3.3 Tools

A variety of tools and reference materials are available to help estimate the issues discussed in this chapter.

Whole-Building Simulation Tools

Chapters 6.1 and 6.2 describe building thermal analysis tools in detail. Computer-based building energy simulation tools allow architects, engineers, and researchers much-needed flexibility in analyzing a buildings energy performance while preserving accuracy. These tools serve a variety of functions, from relatively uncomplicated algorithms that are used to predict a building's peak loads and system requirements to comprehensive room-by-room analysis packages that yield information about the load impacts of specific building components, the performance of heating or cooling equipment, and life cycle costs.

Both BLAST and DOE-2, as well as many other energy simulation tools are now being used as the energy analysis engines behind user-friendly computer interfaces. This is an attempt to make these valuable tools more accepted among architects and design engineers, as well as researchers. A more recent innovation involves linking these engines with CAD software.

Air Flow and Air Quality Simulation Tools

Air flow models are used to simulate the rates of incoming and outgoing air flows for a building with known leakage under given weather and shielding conditions. Air flow models can be divided into two main categories: single-zone models and multizone models. Single-zone models assume that the structure can be described by a single, well-mixed zone. The major application for this model type is the single-story, single-family house with no internal partitions. A large number of buildings, however, have structures that would characterize them more accurately as multizone structures. Therefore, more detailed models have been developed, which also take internal partitions into account.

Multizone airflow network models deal with the complexity of flows in a building by recognizing the effects of internal flow restrictions. They require extensive information about flow characteristics and pressure distribution.

As for their single-zone counterparts, these models are based on the mass-balance equation. Unlike the single-zone approach, where there is only one internal pressure to be determined, the multizone model must determine one pressure for each of the zones. This adds considerably to the complexity of the numerical solving algorithm, but by the same token, the multizone approach offers great potential in analyzing infiltration and ventilation airflow distribution.

The advantage of multizone models, besides being able to simulate air flows for larger buildings, is their ability to calculate mass flow interactions between the different zones. Understanding the air mass flow in buildings is important for several reasons:

- Exchange of outside air with inside air necessary for building ventilation
- Energy consumed to heat or cool the incoming air to inside comfort temperature
- Air needed for combustion
- Airborne particles and germs transported by air flow in buildings
- Smoke distribution in case of fire

Literature reviews undertaken in 1985 and 1992 (Fuestel and Kendon, 1985; Fuestel and Dieris, 1992) showed a large number of different multizone airflow models. One of the first models found was already published in 1970 (Jackman, 1970). Newer airflow models (e.g., CONTAM 93 [Walton, 1993] and COMIS [Feustel and Raynor-Hoosen, 1990]) include pollutant transport models, are more user-friendly and run on personal computers. Furthermore, faster and more robust solvers guarantee shorter calculation times and allow integration of all kinds of flow resistance (e.g., large vertical openings with two-way flows) besides the basic crack flow resistance. The limits of zones and flow paths per zone a model can handle depends on the computer storage.

One of the most bothersome exercises for the modeler is to provide the characteristic parameters for the flow resistance and the outside pressure field for different wind directions. In COMIS, the wind pressure distribution for rectangular shaped buildings can be calculated using an algorithm derived from a parametric study based on wind tunnel results (Feustel and Raynor-Hoosen, 1990).

Thermal Distribution Simulation Tools

Air Distribution Systems

Air distribution system leakage, conduction losses, and the associated impact on whole-building air infiltration has received little attention in building energy simulation tools. The widely known building simulation program DOE-2 uses a simple efficiency multiplier of approximately 0.9 (Lawrence Berkeley Laboratory, 1984) to compensate for duct energy losses. Other simulation tools such as the Thermal Analysis Research Program (TARP) (Walton, 1983) and TRNSYS (University of Wisconsin, 1978) ignored the space conditioning equipment and associated distribution system when calculating building thermal loads.

In more recent times, researchers have begun to consider air distribution systems and their impact on building thermal loads in simulations. An ASHRAE special projects committee (Jakob et al., 1986) showed a reduction of up to 40% in the overall system efficiency in the heating mode. Parker et al. (1993) used a simulation tool, FSEC 2.1 (Kerestecioglu and Gu, 1990), coupled with a detailed duct model to predict duct system impacts on building loads and associated energy consumption. Details considered in the duct model were duct leakage flows based upon duct leakage areas and operating pressures, infiltration impacts across the return ducts and building envelope and heat storage and heat transfer losses across the duct walls. They found that the impacts of duct leakage were of the largest magnitude and that electrical demand during summer peak hours were significantly increased.

Modera and Jansky (1992) developed a simulation tool to analyze air distribution system energy impacts in residences. The tool is based upon the DOE-2 thermal simulation code (Birdsall et al. 1990), the COMIS airflow network code (Feustel and Raynor-Hoosen, 1990), and a duct performance model developed specifically for the simulation tool. The duct performance model calculates the combined impacts of duct leakage and conduction on duct performance and also acts as the interface between COMIS and DOE-2. One of the major findings of their study was the identification of a thermalsiphon loop with a heat exchange rate of more than four times larger than that due to system-off duct leakage. Modera and Treidler (1995) improved the simulation tool in order to look at the thermal siphon effect in more detail and improve the modeling of the duct thermal mass and its effects on duct losses and model duct impacts on multispeed space conditioning equipment. They estimated thermal siphon loads to be between 5 and 16% of the heating load, and that duct thermal mass effects decrease the energy delivery efficiency of the duct system by 1 to 6%. Most significantly, multispeed air conditioners were shown to be more sensitive to duct efficiency than single-speed equipment, because their efficiency decreases with increasing building load. Subsequent field measurements have shown the simulation tool to be accurate.

Hydronic Systems

Energy savings and peak load impacts of radiant hydronic systems have not been studied as systematically. Stetiu and Feustel (1995), developed a simulation tool to perform sensitivity analyses of nonresidential radiantly cooled buildings. The model is based on a methodology for describing and solving the dynamic, nonlinear equations that correspond to complex physical systems as found in buildings. Accurate simulation of the dynamic performance of hydronic radiant cooling systems is described. The model calculates loads, heat extraction rates, room air temperature and room surface temperature distributions, and can be used to evaluate issues such as thermal comfort, controls, system sizing, system configuration, and dynamic response. The authors present favorable comparisons with available field data.

Window Thermal Analysis and Daylighting/Fenestration Tools

Fenestration software programs are used to determine the windows thermal performance and daylighting capabilities. The software can be either stand alone or used as a front end of a whole building energy simulation program. To facilitate their use in the building design stage, window analysis software often has CAD-compatible inputs for geometric details or graphics displays. Software developers are continually

TABLE 6.3.9 Sample of Fenestration Software

Software	Analysis	CAD	Building Simulation	Platform	Developer
ADELINE	thermal/ lighting	yes	yes	DOS	Lawrence Berkeley National Laboratory, International Energy Agency
AGI	lighting	yes	no	DOS	Lighting Analysis Corporation
CALA	lighting	yes	no	DOS	Holophane Corporation
Building Design Advisor	thermal/ lighting	yes	yes	WINDOWS	Lawrence Berkeley National Laboratory
Lumen-Micro	lighting	yes	no	DOS	Lighting Technologies Inc.
LUXICON	lighting	yes	no	WINDOWS	Cooper Lighting
Radiance	lighting	yes	no	UNIX	Lawrence Berkeley National Laboratory
RESFEN	thermal	no	yes	DOS	Lawrence Berkeley National Laboratory
SUPERLITE	lighting	no	no	DOS	Lawrence Berkeley National Laboratory, International Energy Agency
WINDOW 4.1	thermal	no	no	DOS	Lawrence Berkeley National Laboratory

improving the user interface to make the programs more accessible to building designers, who may not have time or design fees to use less friendly tools.

The thermal performance of a window is characterized by its overall *U*-value, solar heat gain coefficient, and shading coefficient. These factors depend on the number of panes, gas-filled spaces, low-emittance coatings, frame material and construction, and so on. Window thermal performance software does the required bookkeeping and calculations to determine the parameters important to a window's impact on the thermal loads in a building. Such programs can be used to design and develop new window products and compare the performance characteristics of different types of windows.

The daylighting capabilities of windows are important to consider in the design phase. Some fenestration programs are designed to demonstrate a window's ability to illuminate a space. Some of these programs can calculate the interreflection of light from surfaces in the space and present the results in high-resolution photorealistic graphical displays using a variety of ray-tracing and radiosity techniques. This offers a significant advantage over traditional software that provides simple numerical results or isolux contours. The results tell the designer where the room is under- or overilluminated and where glare problems may exist. These programs can also be integrated into whole building energy simulation programs and thus present a complete picture of the impacts of the windows on the building's energy efficiency and comfort. Table 6.3.9 presents a sample of available lighting or fenestration software. A more complete review of the lighting or daylighting design software can be found in (IESNA, 1995).

Conclusion

This chapter presented the issues that determine the quality of the indoor environment and the energy issues that affect them. Many facets of building properties and energy services that affect comfort levels of occupants, their health, and their productivity were reviewed. These included conductive, convective, and radiant heat transmission through the building envelope, ventilation and infiltration, and thermal distribution. It was demonstrated how improvements in the building envelope or thermal distribution system can provide the same services, but much more efficiently. Technical advances in construction materials, insulation, windows, and paint provide the means to control building loads or use them to advantage. This reduces the requirements of heating or cooling in the building, thus reducing energy consumption, operating costs, and peak power demand. Whole building simulation programs allow the building designer or retrofitter to evaluate an energy service or shell technology's impact on a building's energy efficiency. These tools give architects the information and means necessary to evaluate a proposed building's thermal and visual comfort, heating and cooling equipment, energy budgets, design cost-effectiveness, and so on. The benefit to society is reduced pressure on limited natural resources, independence from foreign fuel supplies, less demand for new power plants, and reduced air pollution and groundwater contamination.

References

Akbari, H., S. Bretz, H. Taha, D. Kurn, and J. Hanford, "Peak Power and Cooling Energy Savings of High-Albedo Roofs," Submitted to *Energy and Bulidings*. Excerpts from Lawrence Berkely National Laboratory Report LBL-34411, Berkeley, CA, 1992.

Akbari, H., S. Bretz, H. Taha, D. Kurn, and J. Hanford, "Peak Power and Cooling Energy Savings of Shade Trees," Submitted to *Energy and Bulidings*. Excerpts from Lawrence Berkely National Laboratory Report LBL-34411, Berkeley, CA, 1994.

AIVC, "A Review of Building Airtightness and Ventilation Standards," TN 30, Air Infiltration and Ventilation Centre, UK, 1990.

AIVC, "Air-to-Air Heat Recovery in Ventilation," TN 45, Air Infiltration and Ventilation Centre, UK, 1990.

ASHRAE Standard 119, "Air Leakage Performance for Detached Single-Family Residential Buildings," American Society of Heating, Refrigerating and Air conditioning Engineers, 1988.

ASHRAE Standard 62, "Air Leakage Performance for Detached Single-Family Residential Buildings," American Society of Heating, Refrigerating and Air conditioning Engineers, 1989.

ASHRAE Handbooks, American Society of Heating, Refrigerating and Air conditioning Engineers, 2001.

ASHRAE Standard 55, "Thermal Environmental Conditions for Human Occupancy," American Society of Heating, Refrigerating and Air conditioning Engineers, 1992.

ASHRAE Standard 119, "Air Leakage Performance for Detached Single-Family Residential Buildings," American Society of Heating, Refrigerating and Air conditioning Engineers, 1988.

ASHRAE Standard 136, "A Method of Determining Air Change Rates in Detached Dwellings," American Society of Heating, Refrigerating and Air conditioning Engineers, 1993.

ASTM STP 1067, "Air Change Rate and Airtightness in Buildings, American Society of Testing and Materials," M.H. Sherman, Ed., 1990.

ASTM Standard E741-83, "Standard Test Method for Determining Air Leakage Rate by Tracer Dilution," *ASTM Book of Standards*, American Society of Testing and Materials, Vol. 04.07, 1994.

ASTM Standard E779-87, "Test Method for Determining Air Leakage by Fan Pressurization," *ASTM Book of Standards*, American Society of Testing and Materials, Vol. 04.07, 1991.

ASTM Standard E1186-87, "Practices for Air Leakage Site Detection in Building Envelopes," *ASTM Book of Standards*, American Society of Testing and Materials, Vol. 04.07, 1991.

Andrews, J.W. and M.P. Modera, "Energy Savings Potential for Advanced Thermal Distribution Technology in Residential and Small Commercial Buildings," Prepared for the Building Equipment Division, Office of Building Technologies, U.S. Dept. of Energy, Lawrence Berkeley Laboratory Report, LBL-31042, 1991.

Andrews, J.W. and M.P. Modera, "Thermal Distribution in Small Buildings: A Review and Analysis of Recent Literature," Brookhaven National Laboratory Report, BNL-52349, September 1992.

Birdsall, B., W.F. Buhl, K.L. Ellington, A.E. Erdem, and F.C. Winkelmann, "Overview of the DOE-2 Building Energy Analysis Program, Version 2.1D," Lawrence Berkeley Laboratory Report LBL-19735 Rev.1, Lawrence Berkeley Laboratory, Berkeley, CA, July 1992.

Blasnik, M. and J. Fitzgerald, "In Search of the Missing Leak," *Home Energy*, Vol. 9, No. 6, November/December 1992.

Blomsterberg, A., "Air Leakage in Dwellings," Dept. Bldg. Constr. Report No. 15, Swedish Royal Institute of Technology, 1977.

Cummings, J.R. and J.J. Tooley Jr., "Infiltration and Pressure Differences Induced by Forced Air Systems in Florida Residences," *ASHRAE Trans.*, Vol. 95, Pt. 2, 1989.

Daisey, J.M., "Buildings of the 21st Century: A Perspective on health and Comfort and Work Productivity," presented at the International Energy Agency's Workshop on Buildings of the 21st Century: "Developing Innovative Research Agendas," Gersau, Switzerland, May 16–18, 1979.

Davids, B.J., "Taking the Heat Out of Sunlight—New Advances in Glazing Technology for Commercial Buildings," Proc. ACEEE Summer Study on Energy Efficiency in Buildings, August 26–September 1, Asilomar Conference Center, Pacific Grove, CA, 1990.

Energy Conservatory, "Minneapolis Duct Blaster Operation Manual," The Energy Conservatory, 5158 Bloomington Ave, S, Minneapolis, MN 55417. 1995.

Fuestel, H.E., "Hydronic Radiant Cooling—Preliminary Performance Assessment," Lawrence Berkeley Laboratory Report, LBL-33194, 1993.

Fuestel, H.E. and C. Stetiu, "Hydronic Radiant Cooling—Preliminary Assessment," *Energy and Buildings,* Vol. 8, 1985.

Feustel, H.E. and V.M. Kendon, "Infiltration Models for Multicellular Structures—A Literature Review," *Energy and Bulidings,* Vol. 8, 1985.

Feustel, H.E. and A. Raynor-Hoosen, "Fundamentals of the Multizone Airflow Model COMIS," Technical Note 29, Air Infiltration and Ventilation Centre, Warwick, UK, 1990, also Lawrence Berkeley National Laboratory Report LBNL-28560, 1990.

Feustel, H.E. and J. Dieris, "A Survey of Airflow Models for Multizone Structures," *Energy and Buildings,* Vol. 18, 1992.

Gammage, R.B., A.R. Hawthorne, and D.A. White, "Parameters Affecting Air Infiltration and Air Tightness in Thirty-One East Tennessee Homes," In *Measured Air Leakage of Buildings, ASTM STP 904,* H.R. Trechsel and P.L. Lagus, Eds., pp. 61–69, American Society for Testing and Materials, Philadelphia, 1986.

Griffith, B. and D. Arasteh, "Gas Filled Panels: A Thermally Improved Building Insulation," Proc. ASHRAE/DOE/BTECC Conference: Thermal Performance of the Exterior Envelopes of Buildings V, Clearwater Beach, FL, December 1992.

Holman, J.P., *Heat Transfer,* 4th Edition, McGraw-Hill, New York, 1976.

IES, "Recommended Practices of Daylighting," IES RP-5, Illuminating Engineering Society of North America, 120 Wall St., 17th Floor, New York, NY 10005, 1979.

IESNA, "1995 Software Summary, Lighting Design and Applications,"

Vol. 25, No. 7, Illuminating Engineering Society of North America, 120 Wall St., 17th Floor, New York, NY 10005, July 1995.

Jackman P.J. "A Study of Natural Ventilation of Tall Office Buildings," *J. Inst. Heat. Vent. Eng.,* 38, 1970.

Jakob, F.E., D.W. Locklin, P.E. Fisher, L.J. Flanigan, and R.A. Cudnick, "SP43 Evaluation of Systems Options for Residential Forced Air Heating," *ASHRAE Transactions,* Vol. 92, Pt. 2, Atlanta, GA, 1986.

Jones, P., *How to Cut Heating and Cooling Costs,* Butterick Publishing, New York, 1979.

Jump, D.A. and M.P. Modera, "Energy Impacts of Attic Duct Retrofits in Sacramento Houses," Proc. ACEEE 1994 Summer Study, Pacific Grove, CA, August 28–September 3, 1994, American Council for an Energy Efficient Economy, 1001 Connecticut Av., NW, Suite 801, Washington, DC 20036, 1994.

Kerestecioglu, A. and L. Gu, "Theoretical and Computational Investigation of Heat and Moisture Transfer in Buildings: Evaporation and Condensation Theory," *ASHRAE Trans.,* Vol. 96, Pt. 1, 1990.

Lawler, E.E. III and L.W. Porter, "The Effect of Performance on Job Satisfaction," *Industrial Relations,* Vol. 7, pp. 20–28, 1967.

Lawrence Berkeley Laboratory, "DOE 2.1 Supplement, Version 2.1C," Building Energy Simulation Group, Lawrence Berkeley Laboratory, Berkeley, CA, 1984.

Modera, M.P., "Residential Duct System Leakage: Magnitude, Impacts, and Potential for Reduction," *ASHRAE Trans.,* Vol. 95, Pt. 2, 1989, also LBL-26575.

Modera, M.P., D.J. Dickerhoff, R.E. Jansky, and B.V. Smith, "Improving the Energy Efficiency of Residential Air Distribution Systems in California—Final Report: Phase 1," Lawrence Berkeley Laboratory Report, LBL-30886, 1991.

Modera, M.P. and E.B. Treidler, "Improved Modeling of HVAC System/Envelope Interactions in Residential Buildings," Proc. 1995 ASME International Solar Energy Conference (March 19–24), 1995.

Modera, M.P. and R. Jansky, "Residential Air-Distribution Systems: Interactions with the Building Envelope," Lawrence Berkeley Laboratory Report LBL-31311, UC-350, Lawrence Berkeley Laboratory, Berkeley, CA, July 1992.

Modera, M.P., J.C. Andrews, and E. Kweller, "A Comprehensive Yardstick for Residential Thermal Distribution Efficiency," Proc. ACEEE 1992 Summer Study, Pacific Grove, CA, August 30–September 5, 1992, American Council for an Energy Efficient Economy, 1001 Connecticut Ave., NW, Suite 801, Washington, DC 20036, 1992.

Modera, M.P. and D.A. Jump, "Field Measurement of the Interactions Between Heat Pumps and Duct Systems in Residential Buildings," Proc. 1995 ASME International Solar Energy Conference (March 19–24), 1995; also LBL-36047.

National Fenestration Rating Council (NFRC), "Certified Products Directory," 4th Edition, Silver Springs, MD, January 1995.

Nero, A.V. Jr., "Personal Methods of Controlling Exposure to Indoor Air Pollution," *Principles and Practice of Environmental Medicine,* Plenum Medical Book Company, New York, 1992.

Oak Ridge National Laboratory (ORNL), "Building Foundation Design Handbook," ORNL/SUB/86-72143/1, Oak Ridge National Laboratory, Oak Ridge, TN, 37831, May 1988.

Palmiter, L.E. and P.W. Francisco, "Measured Efficiency of Forced-Air Distribution Systems in 24 Homes," Proc. ACEEE 1994 Summer Study, Pacific Grove, CA, August 28–September 3, 1994, American Council for an Energy Efficient Economy, 1001 Connecticut Ave. NW, Suite 801, Washington, DC 20036, 1992.

Parker, D., P. Fairey, and L. Gu, "Simulation of the Effects of Duct Leakage and Heat Transfer on Residential Space-Cooling Energy Use," *Energy and Buildings,* Vol. 20, No. 2, 1993.

Proctor, J., M. Blasnik, B. Davis, T. Downey, M.P. Modera, G. Nelson, and J.J. Tooley, Jr., "Leak Detectors: Experts Explain the Techniques," *Home Energy,* Vol. 10, No. 5, pp. 26–31, September/October 1993.

Romm, J.J. and W.D. Browning, "Greening the Building and the Bottom Line: Increasing Productivity Through Energy Efficient Design," Proc. ACEEE 1994 Summer Study on Energy Efficiency in Buildings, Panel 9, Demonstrations and Retrofits, Pacific Grove, CA, September 1994.

Samet, J.M., M.C. Marbury, and J.D. Spengler, "Health Effects and Sources of Indoor Air Pollution. Part II," *Am. Rev. Respir. Dis.,* Vol. 137, pp. 221–242, 1988.

Sandberg, M. and M. Sjoberg, "The Use of Moments for Assessing Air Quality in Ventilated Rooms," *Building & Environment,* Vol. 18, p. 181, 1983.

Sherman, M.H., "Estimation of Infiltration from Leakage and Climate Indicators," *Energy and Buildings,* Vol. 10, No. 1, p. 81, 1987.

Sherman, M.H., D.T. Grimsrud, "The Measurement of Infiltration using Fan Pressurization and Weather Data," Proc. First International Air Infiltration Centre Conference, London, England. Lawrence Berkeley Laboratory Report, LBL-10852, October 1980.

Sherman, M.H., N.E. Matson, "Ventilation-Energy Liabilities in U.S. Dwellings," Proc. 14th AIVC Conference, pp. 23–41, 1993, LBL Report No. LBL-33890, 1994.

Sherman, M.H., M.P. Modera, "Infiltration Using the LBL Infiltration Model," Special Technical Publication No. 904, "Measured Air Leakage Performance of Buildings," pp. 325–347. ASTM, Philadelphia, PA, 1984.

Stetiu, C. and H.E. Feustel, "Development of a Model to Simulate the Performance of Hydronic Radiant Cooling Ceilings," presented at the ASHRAE Summer Meeting, San Diego, CA, June 1995.

Stokes, B., "Worker Participation—Productivity and the Quality of Work Life," Worldwatch Paper 25, Worldwatch Institute, December 1978.

Stum, K., "Guidelines for Designing and Installing Tight Duct Systems," *Home Energy,* Vol. 10, No. 5, pp. 55–59, September/October 1993.

Treidler, E.B. and M.P. Modera, "Peak Demand Impacts of Residential Air-Conditioning Conservation Measures," Proc. ACEEE 1994 Summer Study, Pacific Grove, CA, August 28–September 3, 1994, American Council for an Energy Efficient Economy, 1001 Connecticut Ave. NW, Suite 801, Washington, DC20036, 1992.

United States Congress, Office of Technology Assessment, "Building Energy Efficiency," OTA-E-518, Washington, D.C.: U.S. Government Printing Office, May, 1992.

United States Environmental Protection Agency, "Cooling Our Communities: A Guidebook on Tree Planting and Light-Colored Surfaces," USEPA Report No. 22P-2001, Office of Policy Analysis, Climate Change Division, Washington, D.C., January 1992.

University of Wisconsin, "TRNSYS, A Transient Simulation Program," University of Wisconsin Experiment Station, Report 38-9, Madison WI, 1978.

Walton, G.N., "Thermal Analysis Research Program," National Bureau of Standards, NSBIR 83-2655, Washington, D.C., 1983.

Walton, G. "CONTAM 93, User Manual," NISTIR 5385, National Institute of Standards and Technology, 1993, Chapter 53. "Thermal Insulation and Airtightness Building Regulations," Royal Ministry of Local Government and Labor, Norway, 27 May, 1987.

6.4   Solar Energy System Analysis and Design

T. Agami Reddy

Successful solar system design is an iterative process involving consideration of many technical, practical, reliability, cost, code, and environmental considerations (Mueller Associates, 1985). The success of a project involves identification of and intelligent selection among trade-offs, for which a proper understanding of goals, objectives, and constraints is essential. Given the limited experience available in the solar field, it is advisable to keep solar systems as simple as possible and not be lured by the promise of higher efficiency offered by more complex systems. Because of the location-specific variability of the solar resource, solar systems offer certain design complexities and concerns not encountered in traditional energy systems.

The objective of this chapter is to provide energy professionals with a fundamental working knowledge of the scientific and engineering principles of solar collectors and solar systems relevant to both the feasibility study and schematic design of a solar project. Conventional equipment such as heat exchangers, pumps, and piping layout is but briefly described. Because of space limitations, certain equations/correlations had to be omitted, and proper justice could not be given to several concepts and design approaches. Effort has been made to provide the reader with pertinent references to textbooks, manuals, and research papers.

A detailed design of solar systems requires in-depth knowledge and experience in (1) the use of specially developed computer programs for detailed simulation of solar system performance, (2) designing conventional equipment, controls, and hydronic systems, (3) practical aspects of equipment installation, and (4) economic analysis. These aspects are not addressed here, given the limited scope of this chapter. Readers interested in acquiring such details can consult manuals such as SERI (1989) or Mueller Associates (1985).

It is obvious that the rather lengthy process just outlined pertains to large solar installations. The process is much less involved when a small domestic hot water system or unitary solar equipment or single solar appliances such as solar stills, solar cookers, or solar dryers are to be installed. Not only do such appliances differ in engineering construction from region to region, there are also standardized commercially available units whose designs are already more or less optimized by the manufacturers, normally as a result of previous experimentation, both technical or otherwise. Such equipment is not described in this chapter.

The design concepts described in this chapter are applicable to domestic water heating, swimming pool heating, active space heating, industrial process heat, convective drying systems, and solar cooling systems.

6.4.1   Solar Collectors

Collector Types

A solar thermal collector is a heat exchanger that converts radiant solar energy into heat. In essence this consists of a receiver that absorbs the solar radiation and then transfers the thermal energy to a working

TABLE 6.4.1   Advantages and Disadvantages of Liquid and Air Systems

Characteristics	Liquid	Air
Efficiency	Collectors generally more efficient for a given temperature difference	Collectors generally operate at slightly lower efficiency
System configuration	Can be readily combined with service hot water and cooling systems	Space heat can be supplied directly but does not adapt easily to cooling. Can preheat hot water.
Freeze protection	May require antifreeze and heat exchangers that add cost and reduce efficiency	None needed
Maintenance	Precautions must be taken against leakage, corrosion and boiling	Low maintenance requirements. Leaks repaired readily with duct tape, but leaks may be difficult to find.
Space requirements	Insulated pipes take up nominal space and are more convenient to install in existing buildings	Duct work and rock storage units are bulky, but ducting is a standard HVAC installation technique.
Operation	Less energy required to pump liquids	More energy required by blowers to move air; noisier operation.
Cost	Collectors cost more.	Storage costs more.
State of the art	Has received considerable attention from solar industry	Has received less attention from solar industry

Source: From SERI, 1989.

TABLE 6.4.2   Types of Solar Thermal Collectors

Nontracking Collectors	Tracking Collectors
Basic flat-plate	Parabolic troughs
Flat-plate enhanced with side reflectors or V-troughs	Fresnel reflectors
Tubular collectors	Paraboloids
Compound parabolic concentrators (CPCs)	Heliostats with central receivers

fluid. Because of the nature of the radiant energy (its spectral characteristics, its diurnal and seasonal variability, changes in diffuse to global fraction, etc.) as well as the different types of applications for which solar thermal energy can be used, the analysis and design of solar collectors present unique and unconventional problems in heat transfer, optics, and material science. The classification of solar collectors can be made according to the type of working fluid (water, air, or oils) or the type of solar receiver used (nontracking or tracking).

Most commonly used working fluids are water (glycol being added for freeze protection) and air. Table 6.4.1 identifies the relative advantages and potential disadvantages of air and liquid collectors and associated systems. Because of the poorer heat transfer characteristics of air with the solar absorber, the air collector may operate at a higher temperature than a liquid-filled collector, resulting in greater thermal losses and, consequently, a lower efficiency. The choice of the working fluid is usually dictated by the application. For example, air collectors are suitable for space heating and convective drying applications, while liquid collectors are the obvious choice for domestic and industrial hot water applications. In certain high-temperature applications, special types of oils that provide better heat transfer characteristics are used.

The second criterion of collector classification is according to the presence of a mechanism to track the sun throughout the day and year in either a continuous or discreet fashion (see Table 6.4.2). The stationary flat-plate collectors are rigidly mounted, facing toward the equator with a tilt angle from the horizontal roughly equal to the latitude of the location for optimal year-round operation. The compound parabolic concentrators (CPCs) can be designed either as completely stationary devices or as devices that need seasonal adjustments only. On the other hand, Fresnel reflectors, paraboloids, and heliostats need two-axis tracking. Parabolic troughs have one axis tracking either along the east–west direction or the north–south direction. These collector types are described in other chapters.

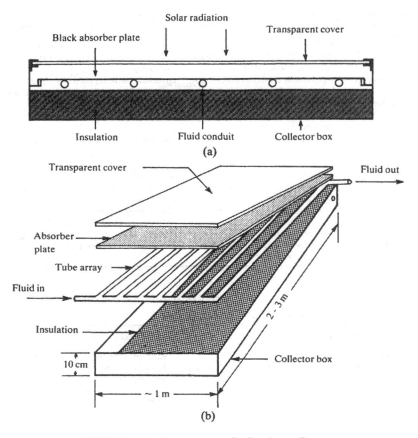

FIGURE 6.4.1 Cross-section of a flat-plate collector.

A third classification criterion is to distinguish between nonconcentrating and concentrating collectors. The main reason for using concentrating collectors is not that *more energy* can be collected but that the thermal energy is obtained at higher temperatures. This is done by decreasing the area from which heat losses occur (called the receiver area) with respect to the aperture area (i.e., the area that intercepts the solar radiation). The ratio of the aperture to receiver area is called the *concentration ratio*.

Flat-Plate Collectors

Description

The flat-plate collector is the most common conversion device in operation today, since it is most economical and appropriate for delivering energy at temperatures up to about 100°C. The construction of flat-plate collectors is relatively simple, and many commercial models are available.

Figure 6.4.1 shows the physical arrangements of the major components of a conventional flat-plate collector with a liquid working fluid. The blackened absorber is heated by radiation admitted via the transparent cover. Thermal losses to the surroundings from the absorber are contained by the cover, which acts as a black body to the infrared radiation (this effect is called the *greenhouse* effect), and by insulation provided under the absorber plate. Passages attached to the absorber are filled with a circulating fluid, which extracts energy from the hot absorber. The simplicity of the overall device makes for long service life.

The absorber is the most complex portion of the flat-plate collector, and a great variety of configurations are currently available for liquid and air collectors. Figure 6.4.2 illustrates some of these concepts

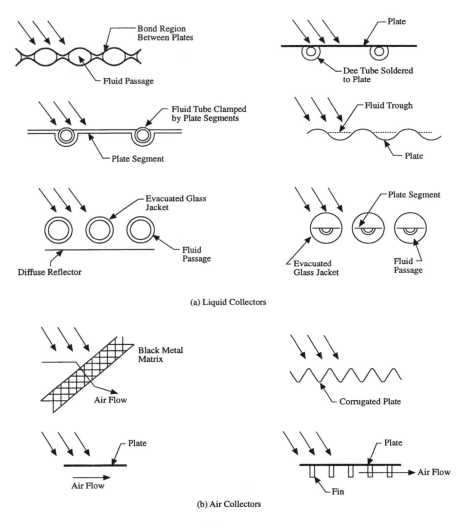

FIGURE 6.4.2   Typical flat-plate absorber configurations.

in absorber design for both liquid and air absorbers. Conventional materials are copper, aluminum, and steel. The absorber is either painted with a dull black paint or can be coated with a *selective surface* to improve performance (see "Improvements to Flat-Plate Collector Performance" for more details). Bonded plates having internal passageways perform well as absorber plates because the hydraulic passageways can be designed for optimal fluid and thermal performance. Such collectors are called *roll-bond* collectors. Another common absorber consists of tubes soldered or brazed to a single metal sheet, and mechanical attachments of the tubes to the plate have also been employed. This type of collector is called a *tube-and-sheet* collector. Heat pipe collectors have also been developed, though these are not as widespread as the previous two types. The so-called *trickle type* of flat-plate collector, with the fluid flowing directly over the corrugated absorber plate, dispenses entirely with fluid passageways. Tubular collectors have also been used because of the relative ease by which air can be evacuated from such collectors, thereby reducing convective heat losses from the absorber to the ambient air.

The absorber in an air collector normally requires a larger surface than in a liquid collector because of the poorer heat transfer coefficients of the flowing air stream. Roughness elements and producing turbulence by way of devices such as expanded metal foil, wool, and overlapping plates have been used as a means for increasing the heat transfer from the absorber to the working fluid. Another approach to enhance heat transfer is to use packed beds of expanded metal foils or matrices between the glazing and the bottom plate.

Modeling

A particular modeling approach and the corresponding degree of complexity in the model are dictated by the objective as well as by experience gained from past simulation work. For example, it has been found that transient collector behavior has insignificant influence when one is interested in determining the long-term performance of a solar thermal system. For complex systems or systems meant for non-standard applications, detailed modeling and careful simulation of system operation are a must initially, and simplifications in component models and system operation can subsequently be made. However, in the case of solar thermal systems, many of the possible applications have been studied to date and a backlog of experience is available not only concerning system configurations but also with reference to the degree of component model complexity.

Because of low collector time constants (about 5 min), heat capacity effects are usually small. Then the instantaneous (or hourly, because radiation data are normally available in hourly time increments only) steady-state useful energy q_C in Watts delivered by a solar flat-plate collector of surface area A_C is given by

$$q_c = A_c F' \left[I_T \eta_0 - U_L \left(T_{Cm} - T_a \right) \right]^+ \tag{6.4.1}$$

where F' is the plate efficiency factor, which is a measure of how good the heat transfer is between the fluid and the absorber plate; η_0 is the optical efficiency, or the product of the transmittance and absorptance of the cover and absorber of the collector; U_L is the overall heat loss coefficient of the collector, which is dependent on collector design only normally expressed in $\text{W/(m}^2\,^\circ\text{C)}$; T_{Cm} is the *mean* fluid temperature in the collector (in °C); and I_T is the radiation intensity on the plane of the collector (in W/m^2). The + sign denotes that only positive values are to be used, which physically implies that the collector should not be operated when q_C is negative (i.e., when the collector loses more heat than it can collect).

However, because T_{cm} is not a convenient quantity to use, it is more appropriate to express collector performance in terms of the fluid *inlet* temperature to the collector (T_{Ci}). This equation is known as the classical Hottel–Whillier–Bliss (HWB) equation and is most widely used to predict instantaneous collector performance:

$$q_c = A_c F_R \left[I_T \eta_0 - U_L \left(T_{Ci} - T_a \right) \right]^+ \tag{6.4.2}$$

where F_R is called the heat removal factor and is a measure of the solar collector performance as a heat exchanger, since it can be interpreted as the ratio of actual heat transfer to the maximum possible heat transfer. It is related to F' by

$$\frac{F_R}{F'} = \frac{\left(mc_p \right)_C}{A_C F' U_L} \left\{ 1 - \exp \left[-\frac{A_C U_L F'}{\left(mc_p \right)_C} \right] \right\} \tag{6.4.3}$$

where m_c is the total fluid flow rate through the collectors and c_{pc} is the specific heat of the fluid flowing through the collector. The variation of (F_R/F') with $[(mC_p)_c/A_C U_L F']$ is shown graphically in Figure 6.4.3. Note the asymptotic behavior of the plot, which suggests that increasing the fluid flow rate more than a certain amount results in little improvement in F_R (and hence in q_C) while causing a quadratic increase in the pressure drop.

Factors influencing solar collector performance are of three types: (1) constructional, that is related to collector design and materials used, (2) climatic, and (3) operational, that is, fluid temperature, flow rate, and so on. The plate efficiency factor F' is a factor that depends on the physical constructional

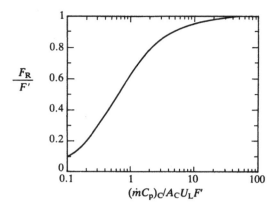

FIGURE 6.4.3 Variation of F_R/F' as a function of $[(mc_p)_c/(A_cU_LF')]$ (From Duffie, J.A. and Beckman, W.A., *Solar Engineering of Thermal Processes*, Wiley Interscience, New York. Copyright 1980. Reprinted with permission of John Wiley & Sons, Inc.)

features and is essentially a constant for a given liquid collector. (This is not true for air collectors, which require more careful analysis.) Operational features involve changes in m_C and T_{Ci}. While changes in m_C affect F_R as per Eq. (6.4.3), we note from Eq. (6.4.2) that to enhance q_C, T_{Ci} needs to be kept as low as possible. For solar collectors that are operated under more or less constant flow rates, specifying $F_R\eta_0$ and $F_R U_L$ is adequate to predict collector performance under varying climatic conditions.

There are a number of procedures by which collectors have been tested. The most common is a *steady-state procedure*, where transient effects due to collector heat capacity are minimized by performing tests only during periods when radiation and ambient temperature are steady. The procedure involves simultaneous and accurate measurements of the mass flow rate, the inlet and outlet temperatures of the collector fluid, and the ambient conditions (incident solar radiation, air temperature and wind speed). The most widely used test procedure is the ASHRAE Standard 93-77 (1978), whose test setup is shown in Figure 6.4.4. Though a solar simulator can be used to perform indoor testing, outdoor testing is always more realistic and less expensive. The procedure can be used for nonconcentrating collectors using air or liquid as the working fluid (but not two phase mixtures) that have a single inlet and a single outlet and contain no integral thermal storage.

Steady-state procedures have been in use for a relatively long period and though the basis is very simple the engineering setup is relatively expensive (see Figure 6.4.4). From an overall heat balance on the collector fluid and from Eq. (6.4.2), the expressions for the instantaneous collector efficiency under normal solar incidence are

$$\eta_C \equiv \frac{q_c}{A_C I_T} = \frac{\left(mc_p\right)_c\left(T_{Co} - T_{Ci}\right)}{A_C I_T} \tag{6.4.4}$$

$$= \left[F_R\eta_n - F_RU_L\left(\frac{T_{Ci} - T_a}{I_T}\right)\right] \tag{6.4.5}$$

where η_n is the optical efficiency at normal solar incidence.

From the test data, points of η_c against reduced temperature $[(T_{Ci} - T_a)/I_T]$ are plotted as shown in Figure 6.4.5. Then a linear fit is made to these data points by regression, from which the values of $F_R\eta_n$ and $F_R U_L$ are easily deduced. It will be noted that if the reduced term were to be taken as $[(T_{Cm} - T_a)/I_T]$, estimates of $F'\eta_n$ and $F'U_L$ would be correspondingly obtained.

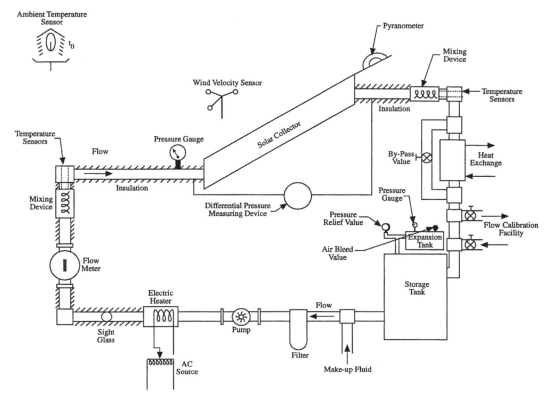

FIGURE 6.4.4 Set up for testing liquid collectors according to ASHRAE Standard 93-72.

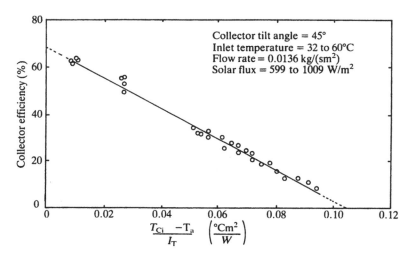

FIGURE 6.4.5 Thermal efficiency curve for a double glazed flat-plate liquid collector (ASHRAE 1978). Test conducted outdoors on a 1.2 m by 1.25 m panel with 10.2 cm of glass fiber back insulation and a flat copper absorber with black coating of emissivity of 0.97.

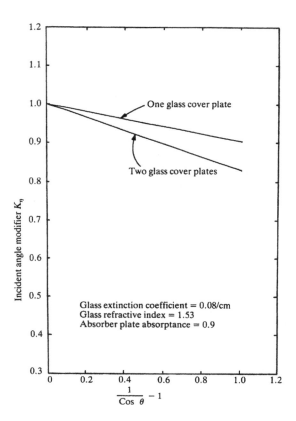

FIGURE 6.4.6 Incident angle modifiers for two flat-plate collectors with non-selective coating on the absorber. (Adapted from ASHRAE, 1978.)

Incidence Angle Modifier

The optical efficiency η_0 depends on the collector configuration and varies with the angle of incidence as well as with the relative values of diffuse and beam radiation. The incidence angle modifier is defined as $K_\eta \equiv (\eta_0/\eta_n)$. For flat plate collectors with 1 or 2 glass covers, K_η is almost unchanged up to incidence angles of 60°, after which it abruptly drops to zero.

A simple way to model the variation of K_η with incidence angle for flat plate collectors is to specify η_n, the optical efficiency of the collector at normal beam incidence, to assume the entire radiation to be beam, and to use the following expression for the angular dependence (ASHRAE, 1978)

$$K_\eta = 1 + b_0 \left(\frac{1}{\cos \theta} - 1 \right) \tag{6.4.6}$$

where θ is the solar angle of incidence on the collector plane (in degrees) and b_0 is a constant called the incidence angle modifier coefficient. Plotting K_η against $[(1/\cos \theta)-1]$ results in linear plots (see Figure 6.4.6), thus justifying the use of Eq. (6.4.6). We note that for one-glass and two-glass covers, approximate values of b_0 are –0.10 and –0.17, respectively.

In case the diffuse solar fraction is high, one needs to distinguish between beam, diffuse, and ground-reflected components. Diffuse radiation, by its very nature, has no single incidence angle. One simple way is to assume an equivalent incidence angle of 60° for diffuse and ground-reflected components. One would then use Eq. (6.4.6) for the beam component along with its corresponding value of θ and account for the contribution of diffuse and ground reflected components by assuming a value of $\theta = 60°$ in

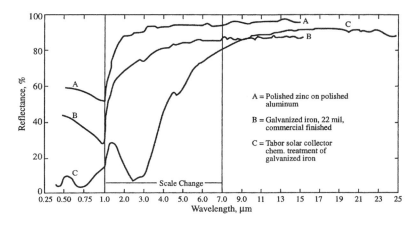

FIGURE 6.4.7   Spectral reflectance of several surfaces. (From Edwards, D.K. et al., 1960.)

Eq. (6.4.6). For more accurate estimation, one can use the relationship between the effective diffuse solar incidence angle versus collector tilt given in Duffie and Beckman (1980). It should be noted that the preceding equation gives misleading results with incidence angles close to 90°. An alternative functional form for the incidence angle modifier for both flat-plate and concentrating collectors has been proposed by Rabl (1981).

Improvements to Flat-Plate Collector Performance

There are a number of ways by which the performance of the basic flat-plate collectors can be improved. One way is to enhance optical efficiency by treatment of the glass cover thereby reducing reflection and enhancing performance. As much as a 4% increase has been reported (Anderson, 1977). Low-iron glass can also reduce solar absorption losses by a few percent.

These improvements are modest compared to possible improvements from reducing losses from the absorber plate. Essentially, the infrared upward reradiation losses from the heated absorber plate have to be decreased. One could use a second glass cover to reduce the losses, albeit at the expense of higher cost and lower optical efficiency. Usually for water heating applications, radiation accounts for about two-thirds of the losses from the absorber to the cover with convective losses making up the rest (conduction is less than about 5%). The most widely used manner of reducing these radiation losses is to use selective surfaces whose emissivity varies with wavelength (as against matte-black painted absorbers, which are essentially gray bodies). Note that 98% of the solar spectrum is at wavelengths less than 3.0 μm, whereas less than 1% of the black body radiation from a 200°C surface is at wavelengths less than 3.0 μm. Thus selective surfaces for solar collectors should have high solar absorptance (i.e., low reflectance in the solar spectrum) and low long-wave emittance (i.e., high reflectance in the long-wave spectrum). The spectral reflectance of some commonly used selective surfaces is shown in Figure 6.4.7. Several commercial collectors for water heating or low-pressure steam (for absorption cooling or process heat applications) that use selective surfaces are available.

Another technique to simultaneously reduce both convective and radiative losses between the absorber and the transparent cover is to use honeycomb material (Hollands, 1965). The honeycomb material can be reflective or transparent (the latter is more common) and should be sized properly. Glass honeycombs have had some success in reducing losses in high-temperature concentrating receivers, but plastics are usually recommended for use in flat-plate collectors. Because of the poor thermal aging properties, honeycomb flat-plate collectors have had little commercial success. Currently the most promising kind seems to be the simplest (both in terms of analysis and construction), namely collectors using horizontal rectangular slats (Meyer, 1978). Convection can be entirely suppressed provided the slats with the proper aspect ratio are used.

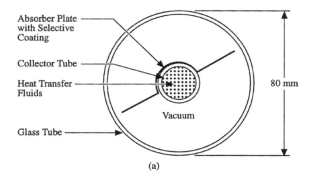

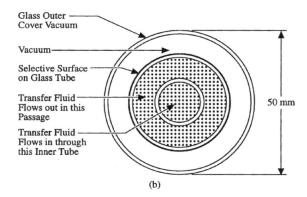

FIGURE 6.4.8   Evacuated tubular collectors. (From Charters, W.W.S. and Pryor, T.L., 1982.)

Finally, collector output can be enhanced by using side reflectors, for instance a sheet of anodized aluminum. The justification in using these is their low cost and simplicity. For instance, a reflector placed in front of a tilted collector cannot but increase collector performance because losses are unchanged and more solar radiation is intercepted by the collector. Reflectors in other geometries may cast a shadow on the collector and reduce performance. Note also that reflectors would produce rather nonuniform illumination over the day and during the year, which, though not a problem in thermal collectors, may drastically penalize the electric output of photovoltaic modules. Whether reflectors are cost-effective depends on the particular circumstances and practical questions such as aesthetics and space availability. The complexity involved in the analysis of collectors with planar reflectors can be reduced by assuming the reflector to be long compared to its width and treating the problem in two dimensions only. How optical performance of solar collectors are affected by side planar reflectors is discussed in several papers, for example Larson (1980) and Chiam (1981).

Other Collector Types

Evacuated Tubular Collectors

One method of obtaining temperatures between 100°C and 200°C is to use evacuated tubular collectors. The advantage in creating and being able to main a vacuum is that convection losses between glazing and absorber can be eliminated. There are different possible arrangements of configuring evacuated tubular collectors. Two designs are shown in Figure 6.4.8. The first is like a small flat-plate collector with the liquid to be heated making one pass through the collector tube. The second uses an all-glass construction with the glass absorber tube being coated selectively. The fluid being heated passes up the middle of the absorber tube and then back in contact with the hot absorber surface. Evacuated tubes can collect both direct and diffuse radiation and do not require tracking. Glass breakage and leaking

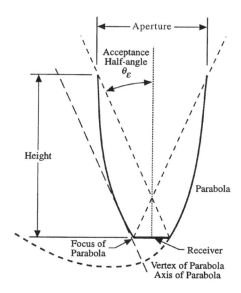

FIGURE 6.4.9 Cross-section of a symmetrical non-truncated CPC. (From Duffie, J.A. and Beckman, W.A., *Solar Engineering of Thermal Processes*, Wiley Interscience, New York. Copyright 1980. Reprinted with permission of John Wiley & Sons, Inc.)

joints due to thermal expansion are some of the problems which have been experienced with such collector types. Various reflector shapes (like flat-plate, V-groove, circular cylindrical, involute, etc.) placed behind the tubes are often used to usefully collect some of the solar energy, which may otherwise be lost, thus providing a small amount of concentration.

Compound Parabolic Concentrators (CPCs)

The CPC collector, discovered in 1966, consists of parabolic reflectors that funnel radiation from aperture to absorber rather than focusing it. The right and left halves belong to different parabolas (hence the name *compound*) with the edges of the receiver being the foci of the opposite parabola (see Figure 6.4.9). It has been proven that such collectors are *ideal* in that any solar ray, be it beam or diffuse, incident on the aperture within the acceptance angle will reach the absorber while all others will bounce back to and fro and re-emerge through the aperture. CPCs are also called *nonimaging* concentrators because they do not form clearly defined images of the solar disk on the absorber surface as achieved in classical concentrators. CPCs can be designed both as low-concentration devices with large acceptance angles or as high-concentration devices with small acceptance angles. CPCs with low concentration ratios (of about 2) and with east–west axes can be operated as stationary devices throughout the year or at most with seasonal adjustments only. CPCs, unlike other concentrators, are able to collect all the beam and a large portion of the diffuse radiation. Also they do not require highly specular surfaces and can thus better tolerate dust and degradation. A typical module made up of several CPCs is shown in Figure 6.4.10. The absorber surface is located at the bottom of the trough, and a glass cover may also be used to encase the entire module. CPCs show considerable promise for water heating close to the boiling point and for low-pressure steam applications. Further details about the different types of absorber and receiver shapes used, the effect of truncation of the receiver and the optics, can be found in Rabl (1985). In order to justify the significant investment in a solar heating system, the HVAC designer must weigh costs against energy production (Section 3.2). This section outlines how to calculate the energy production.

6.4.2 Long-Term Performance of Solar Collectors

Effect of Day-to-Day Changes in Insolation

Instantaneous or hourly performance of solar collectors has been discussed in "Flat-Plate Collectors." For example, one would be tempted to use the HWB equation (Eq. 6.4.2) to predict long-term collector

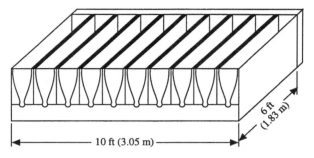

FIGURE 6.4.10   A CPC collector module. (From SERI, 1989.)

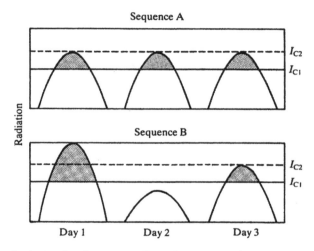

FIGURE 6.4.11   Effect of radiation distribution on collector long-term performance. (From Klein, S.A., Calculation of flat-plate collector utilizability, *Solar Energy*, 21, 393, 1978. With permission.)

performance at a prespecified and constant fluid inlet temperature T_{Ci} merely by assuming average hourly values of I_T and T_a. Such a procedure would be erroneous and lead to underestimation of collector output because of the presence of the control function, which implies that collectors are turned on only when $q_C > 0$, that is when radiation I_T exceeds a certain critical value I_C. This critical radiation value is found by setting q_C in Eq. (6.4.2) to zero:

$$I_C = U_L(T_{Ci} - T_a)/\eta_0 \qquad (6.4.7a)$$

To be more rigorous, a small increment δ to account for pumping power and stability of controls can also be included if needed by modifying the equation to

$$I_C = U_L(T_{Ci} + \delta - T_a)/\eta_0 \qquad (6.4.7b)$$

Then, Eq. (6.4.2) can be rewritten in terms of I_C as

$$q_C = A_C F_R \eta_0 \left[I_T - I_C\right]^+ \qquad (6.4.8)$$

Why one cannot simply assume a mean value of I_T in order to predict the mean value of q_C will be illustrated by the following simple concept (Klein, 1978). Consider the three identical day sequences shown in sequence A of Figure 6.4.11. If I_{C1} is the critical radiation intensity, and if it is constant over the whole day, the useful energy collected by the collector is represented by the sum of the shaded areas. If a higher critical radiation value shown as I_{C2} in Figure 6.4.11 is selected, we note that no useful energy is collected at all. Actual weather sequences would not look like that in sequence A but rather like that in sequence B, which is comprised of an excellent, a poor, and an average day. Even if both sequences have the same average radiation over 3 days, a collector subjected to sequence B will collect useful energy when the critical radiation is I_{C2}. Thus, neglecting the variation of radiation intensity from day to day over the long term and dealing with mean values would result in an underestimation of collector performance.

Loads are to a certain extent repetitive from day to day over a season or even the year. Consequently, one can also expect collectors to be subjected to a known diurnal repetitive pattern or mode of operation, that is, the collector inlet temperature T_{Ci} has a known repetitive pattern.

Individual Hourly Utilizability

In this mode, T_{Ci} is assumed to vary over the day but has the same variation for all the days over a period of N days (where $N = 30$ days for monthly and $N = 365$ for yearly periods). Then from Eq. (6.4.8), *total useful energy collected over N days during individual hour i of the day is*

$$q_{CN}(i) = A_C F_R \overline{\eta}_0 \overline{I}_{Ti} \sum_{i=1}^{N} \frac{\left[I_T - I_C \right]^+}{\overline{I}_{Ti}} \tag{6.4.9}$$

If we define the radiation ratio

$$X = I_{Ti} / \overline{I}_{Ti} \tag{6.4.10}$$

then the critical radiation ratio

$$X_C = I_C / \overline{I}_{Ti}$$

The HWB equation (Eq. 6.4.8) can be rewritten as

$$q_{CN}(i) = A_C F_R \overline{\eta}_0 \overline{I}_{Ti} \overline{N\phi}_i \tag{6.4.11}$$

where the individual hourly utilizability factor ϕ_i is identified as

$$\phi_i(X_C) = \frac{1}{N} \sum_{i=1}^{N} (X_i - X_C)^+ \tag{6.4.12}$$

Thus ϕ_i can be considered to be the fraction of the incident solar radiation that can be converted to useful heat by an ideal collector (i.e., whose $F_R \eta_0 = 1$. The utilizability factor is thus a *radiation statistic* in the sense that it depends solely on the radiation values at the specific location. As such, it is in no way dependent on the solar collector itself. Only after the radiation statistics have been applied is a collector dependent significance attached to X_C.

Hourly utilizability curves on a *monthly* basis that are independent of location were generated by Liu and Jordan (1963) more than 30 years ago for flat-plate collectors (see Figure 6.4.12). The key climatic parameter which permits generalization is the *monthly clearness index* $\overline{K}$ of the location defined as

$$\overline{K} = \overline{H} / \overline{H}_0 \tag{6.4.13}$$

where $\overline{H}$ is the monthly mean daily global radiation on the horizontal surface and $\overline{H}_0$ is the monthly mean daily extraterrestrial radiation on a horizontal surface.

Extensive tables giving monthly values of $\overline{K}$ for several different locations worldwide can be found in several books, for example, Duffie and Beckman (1980) or Reddy (1987). The curves apply to

(a)

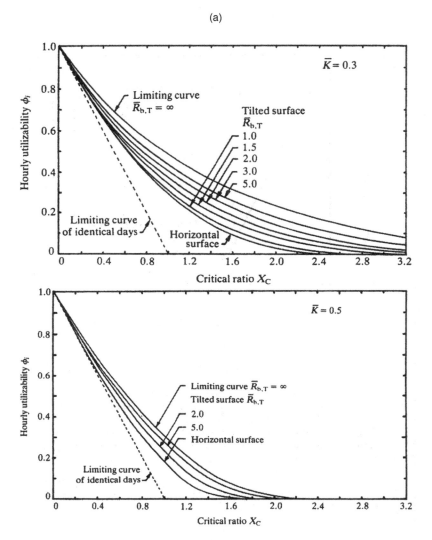

FIGURE 6.4.12(a) Generalized hourly utilizability curves of Liu and Jordan (1963) for three different monthly mean clearness indices $\overline{K}$.

equator-facing tilted collectors with the effect of collector tilt accounted for by the factor $\overline{R}_{b,T}$ which is the ratio of the monthly mean daily extraterrestrial radiation on the tilted collector to that on a horizontal surface. Monthly mean daily calculations can be made using the 15th of the month, though better accuracy is achieved using slightly different dates (Reddy, 1987). Clark et al. (1983), working from measured data from several U.S. cities, have proposed the following correlation for individual hourly utilizability over monthly time scales applicable to *flat-plate collectors only*:

$$\phi_i = 0 \qquad\qquad\qquad\qquad\qquad \text{for } X_C \geq X_{max} \qquad (6.4.14)$$

$$= \left(1 - X_C/X_{max}\right)^2 \qquad\qquad\qquad \text{for } X_{max} = 2$$

$$= \left| |a| - \left[a^2 + (1+2a)\left(1 - X_C/X_{max}\right)^2\right]^{1/2} \right| \quad \text{otherwise}$$

(b)

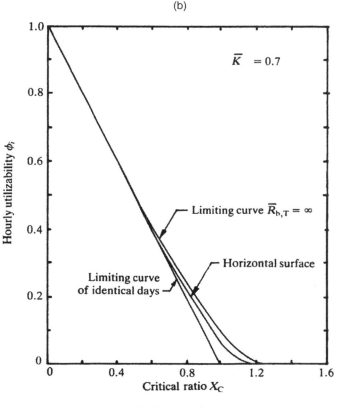

FIGURE 6.4.12(b)

where

$$a = \left(X_{max} - 1\right)\left(2 - X_{max}\right) \tag{6.4.15}$$

and

$$X_{max} = 1.85 + 0.169 \left(\bar{r}_T / \bar{k}^2\right) - 0.0696 \ \cos\beta / \bar{k}^2 - 0.981 \ \bar{k} / \left(\cos \delta\right)^2 \tag{6.4.16}$$

where $\bar{k}$ is the monthly mean *hourly* clearness index for the particular hour, δ is the solar declination, β is the tilt angle of the collector plane with respect to the horizontal, and $\bar{r}_T$ is the ratio of monthly average hourly global radiation on a tilted surface to that on a horizontal surface for that particular hour. For an isotropic sky assumption, $\bar{r}_T$ is given by

$$\bar{r}_T = \left(1 - \frac{\bar{I}_d}{\bar{I}}\right) r_{b,T} + \left(\frac{1 + \cos \beta}{2}\right) \frac{\bar{I}_d}{\bar{I}} + \left(\frac{1 - \cos \beta}{2}\right) \rho \tag{6.4.17}$$

where $\bar{I}_d$ and $\bar{I}$ are the hourly diffuse and global radiation on the horizontal surface, $r_{b,T}$ is the ratio of hourly beam radiation on the tilted surface to that on a horizontal surface (this is a purely astronomical quantity and can be calculated accurately from geometric considerations), and ρ is the ground albedo.

Daily Utilizability

Basis

In this mode, T_{Ci}, and hence the critical radiation level is assumed constant during all hours of the day. The *total* useful energy over N days that can be collected by solar collectors operated all day over n hours is given by

$$Q_{CN} = A_C F_R \bar{\eta}_0 \bar{H}_T N \bar{\phi} \tag{6.4.18}$$

where $\bar{H}_T$ is the average daily global radiation on the collector surface, and $\bar{\phi}$ (called "phibar") is the daily utilizability factor, defined as

$$\bar{\phi} = \sum^N \sum^n (I_T - I_C)^+ \Big/ \sum^N \sum^n I_T = \frac{1}{Nn} \sum^N \sum^n (X_i - \bar{X}_C)^+ \tag{6.4.19}$$

Generalized correlations have been developed both at monthly time scales and for annual time scales based on the parameter $\bar{K}$. Generalized (i.e., location and month independent) correlations for $\bar{\phi}$ on a *monthly* time scale have been proposed by Theilacker and Klein (1980). These are strictly applicable for flat-plate collectors only. Collares-Pereira and Rabl (1979) have also proposed generalized correlations for $\bar{\phi}$ on a monthly time scale which, though a little more tedious to use are applicable to concentrating collectors as well. The reader may refer to Rabl (1985) or Reddy (1987) for complete expressions.

Monthly Time Scales

The phibar method of determining the daily utilizability fraction proposed by Theilacker and Klein (1980) is based on correlating $\bar{\phi}$ to the following factors:

1. A geometry factor $\bar{R}_T / \bar{r}_{T,noon}$, which incorporates the effects of collector orientation, location, and time of year. $\bar{R}_T$ is the ratio of monthly average global radiation on the tilted surface to that on a horizontal surface. $\bar{r}_{T,noon}$ is the ratio of radiation at noon on the tilted surface to that on a horizontal surface for the average day of the month. Geometrically, $\bar{r}_{T,noon}$ is a measure of the maximum height of the radiation curve over the day, whereas $\bar{R}_T$ is a measure of the enclosed area. Generally the value $(\bar{R}_T / \bar{r}_{T,noon})$ is between 0.9 and 1.5.
2. A dimensionless critical radiation level $\bar{X}_{C,K}$ where

$$\bar{X}_{C,K} = I_C / \bar{I}_{T,noon} \tag{6.4.20}$$

with $\bar{I}_{T,noon}$, the radiation intensity on the tilted surface at noon, given by

$$\bar{I}_{T,noon} = \bar{r}_{noon} \bar{r}_{T,noon} \bar{H} \tag{6.4.21}$$

where $\bar{r}_{noon}$ is the ratio of radiation at noon to the daily global radiation on a horizontal surface during the mean day of the month which can be calculated from the following correlation proposed by Liu and Jordan (1960):

$$r(W) = \frac{I(W)}{H} = \frac{\pi}{24}(a + b \cos W) \frac{(\cos W - \cos W_S)}{\left(\sin W_S - \frac{\pi}{180} W_S \cos W_S\right)} \tag{6.4.22}$$

with

$$a = 0.409 + 0.5016 \sin(W_s - 60)$$

$$b = 0.6609 - 0.4767 \sin(W_s - 60)$$

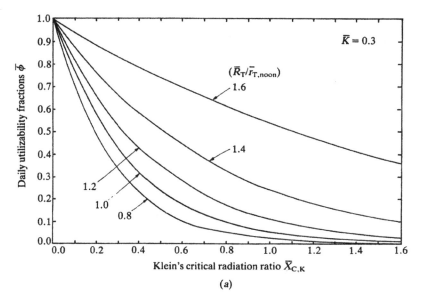

FIGURE 6.4.13 Generalized daily utilizability curves of Theilacker and Klein (1980) for three different K values.

where W is the hour angle corresponding to the midpoint of the hour (in degrees) and W_S is the sunset hour angle given by

$$\cos W_S = -\tan L \tan \delta \tag{6.4.23}$$

The fraction r is the ratio of hourly to daily global radiation on a horizontal surface. The factors $\bar{r}_{T,noon}$ and $\bar{r}_{noon}$ can be determined from Eqs. (6.4.17) and (6.4.22) respectively with $W = 0°$.

The Theilacker and Klein correlation for the daily utilizability for equator-facing flat-plate collectors is

$$\overline{\phi}(X_{C,K}) = \exp\left\{\left[a' + b'\left(\bar{r}_{T,noon}/\overline{R}_T\right)\right]\left[X_{C,K} + c' \; X_{C,K}^2\right]\right\} \tag{6.4.24}$$

where

$$a' = 7.476 - 20.0 \; \overline{K} + 11.188 \; \overline{K}^2 \tag{6.4.25}$$

$$b' = -8.562 + 18.679 \; \overline{K} - 9.948 \; \overline{K}^2$$

$$c' = -0.722 + 2.426 \; \overline{K} + 0.439 \; \overline{K}^2$$

How $\overline{\phi}$ varies with the critical radiation ratio $\overline{X}_{C,K}$ for three different values of $\overline{K}$ is shown in Figure 6.4.13.

Annual Time Scales

Generalized expressions for the *yearly* average energy delivered by the principal collector types with constant radiation threshold (i.e., when the fluid inlet temperature is constant for all hours during the day over the entire year) have been developed by Rabl (1981) based on data from several U.S. locations. The correlations are basically quadratic of the form

$$\frac{Q_{CY}}{A_C F_R \eta_n} = \tilde{a} + \tilde{b} I_C + \tilde{c} I_C^2 \tag{6.4.26}$$

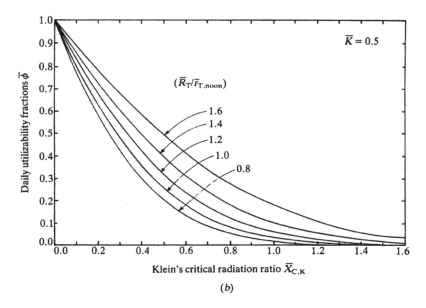

(b)

FIGURE 6.4.13(b)

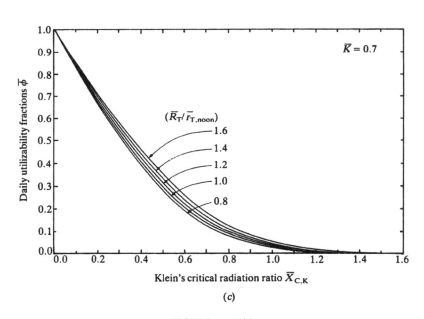

(c)

FIGURE 6.4.13(c)

where the coefficients $\bar{a}$, $\bar{b}$, and $\bar{c}$ are function of collector type and/or tracking mode, climate, and in some cases, latitude. The complete expressions as revised by Gordon and Rabl (1982) are given in Reddy (1987). Note that the yearly *daytime* average value of T_a should be used to determine I_C. If this is not available, the yearly mean *daily* average value can be used. Plots of Q_{CY} versus I_C for flat-plate collectors that face the equator with tilt equal to the latitude are shown in Figure 6.4.14. The solar radiation enters these expressions as $\bar{I}_{bn}$, the annual average beam radiation at normal incidence. This can be estimated from the following correlation

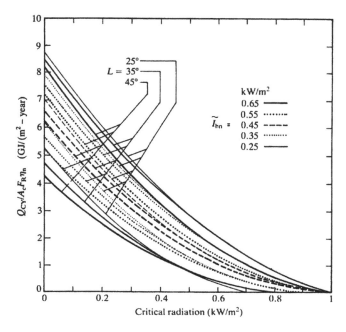

FIGURE 6.4.14 Yearly total energy delivered by flat-plate collectors with tilt equal to latitude. (From Gordon, J.M. and Rabl, A., Design, analysis, and optimization of solar industrial process heat plants without storage, *Solar Energy*, 28, 519, 1982. With permission.)

$$\tilde{I}_{bn} = 1.37 \ \tilde{K} - 0.34 \qquad (6.4.27)$$

where $\tilde{I}_{bn}$ is in kW/m² and $\tilde{K}$ is the annual average clearness index of the location. Values of $\tilde{K}$ for several locations worldwide are given in Reddy (1987).

This correlation is strictly valid for latitudes ranging from 25° to 48°. If used for lower latitudes, the correlation is said to lead to overprediction. Hence, it is recommended that for such lower latitudes a value of 25° should be used to compute Q_{CY}.

A direct comparison of the yearly performance of different collector types is given in Figure 6.4.15 (from Rabl, 1981). A latitude of 35°N is assumed and plots of Q_{cy} vs. $(T_{ci} - T_a)$ have been generated in a sunny climate with $\tilde{I}_{bn} = 0.6$ kW/m². Relevant collector performance data are given in Figure 6.4.15. The cross-over point between flat-plate and concentrating collectors is approximately 25°C above ambient temperature whether the climate is sunny or cloudy.

6.4.3 Solar Systems

Classification

Solar thermal systems used for HVAC can be divided into two categories: standalone or solar supplemented. They can be further classified by means of energy collection as active or passive, and by the type of storage they use into seasonal or daily systems.

Standalone and Solar Supplemented Systems
Standalone systems are systems in which solar energy is the only source of energy input used to meet the required load. Such systems are normally designed for applications where a certain amount of tolerance is permissible concerning the load requirement; in other words, where it is not absolutely imperative that the specified load be met each and every instant. This leniency is generally admissible in the case of certain residential and agricultural applications. The primary reasons for using such systems are their low cost and simplicity of operation.

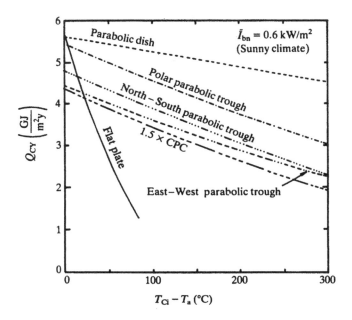

FIGURE 6.4.15 Figure illustrating the comparative performance (yearly collectible energy) of different collector types as a function of the difference between collector inlet temperature and ambient collector performance parameters $F'\eta_0$ and $F'U_L$ in W/(m² °C) are: flat plate (0.70 and 5.0), CPC (0.60 and 0.75), parabolic trough (0.65 and 0.67), and parabolic dish (0.61 and 0.27). (From Rabl, A., Yearly average performace of the principal solar collector types, *Solar Energy*, 27, 215, 1981. With permission.)

Solar-supplemented systems, widely used for both industrial and residential purposes, are those in which solar energy supplies part of the required heat load, the rest being met by an auxiliary source of heat input. Due to the daily variations in incident solar radiation, the portion of the required heat load supplied by the solar energy system may vary from day to day. However, the auxiliary source is so designed that at any instant it is capable of meeting the remainder of the required heat load. It is normal practice to incorporate an auxiliary heat source large enough to supply the entire heat load required. Thus, the benefit in the solar subsystem is not in its capacity credit (i.e., not that a smaller capacity conventional system can be used), but rather that a part of the conventional fuel consumption is displaced. The solar subsystem thus acts as a fuel economizer.

Solar-supplemented energy systems will be the primary focus of this chapter. Designing such systems has acquired a certain firm scientific rationale, and the underlying methodologies have reached a certain maturity and diversity, which may satisfy professionals from allied fields. On the other hand, unitary solar apparatus are not discussed here, since these are designed and sized based on local requirements, material availability, construction practices and practical experience. Simple rules of thumb based on prior experimentation are usually resorted to for designing such systems.

Active and Passive Systems

Active systems are those systems that need electric pumps or blowers to collect solar energy. It is evident that the amount of solar energy collected should be more than the electrical energy used. Active systems are invariably used for industrial applications and for most domestic and commercial applications as well. *Passive systems* are those systems that collect or use solar energy without direct recourse to any source of conventional power, such as electricity, to aid in the collection. Thus, either such systems operate by natural thermosyphon (for example domestic water heating systems) between collector, storage, and load or, in the case of space heating, the architecture of the building is such as to favor optimal use of solar energy. Use of a passive system for space heating applications, however, in no way precludes the use of a back-up auxiliary system. This chapter deals with active solar systems.

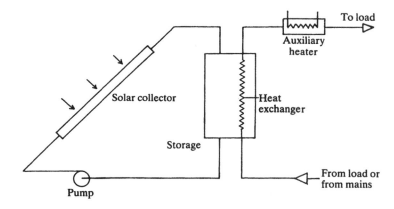

FIGURE 6.4.16 Schematic of a closed-loop solar system.

Daily and Seasonal Storage

By *daily storage* is meant systems having capacities equivalent to at most a few days of demand (i.e., just enough to tide over day-to-day climatic fluctuations). In *seasonal storage,* solar energy is stored during the summer for use in winter. The present-day economics of seasonal storage units do not usually make such systems an economical proposition except for community heating in cold climates.

Closed-Loop and Open-Loop Systems

The two possible configurations of solar thermal systems with daily storage are classified as closed-loop or open-loop systems. Though different authors define these differently, we shall define these as follows. A *closed-loop system* has been defined as a circuit in which the performance of the solar collector is directly dependent on the storage temperature. Figure 6.4.16 gives a schematic of a closed-loop system in which the fluid circulating in the collectors does not mix with the fluid supplying thermal energy to the load. Thus, these two subsystems are distinct in the sense that any combination of fluids (water or air) is theoretically feasible (a heat exchanger, as shown in the figure, is of course imperative when the fluids are different). However, in practice, only water–water or water–air combinations are used. From the point of system performance, the storage temperature normally varies over the day and, consequently, so does collector performance. Closed-loop system configurations have been widely used to date for domestic hot water and space heating applications. The flow rate per unit collector area is generally around 50 kg/(h m²) for liquid collectors. The storage volume makes about 5 to 10 passes through the collector during a typical sunny day, and this is why such systems are called *multipass* systems. The temperature rise for each pass is small, of the order of 2 to 5°C for systems with circulating pumps and about 10°C for thermosyphon systems. An expansion tank and a check valve to prevent reverse thermosyphoning at nights, although not shown in the figure, are essential for such system configurations.

Figure 6.4.17 illustrates one of the possible configurations of *open-loop or single-pass systems.* Open-loop systems are defined as systems in which the collector performance is independent of the storage temperature. The working fluid may be rejected (or a heat recuperator can be used) if contaminants are picked up during its passage through the load. Alternatively, the working fluid could be directly recalculated back to the entrance of the solar collector field. In all these open-loop configurations, the collector is subject to a given or known inlet temperature specified by the load requirements.

If the working fluid is water, instead of having a continuous flow rate (in which case the outlet temperature of the water will vary with insolation), a control valve can be placed just at the exit of the collector, set so as to open when the desired temperature level of the fluid in the collector is reached. The water is then discharged into storage, and fresh water is taken into the collector. The solar collector will thus operate in a discontinuous manner, but this will ensure that the temperature in the storage is always at the desired level. An alternative way of ensuring uniform collector outlet temperature is to vary

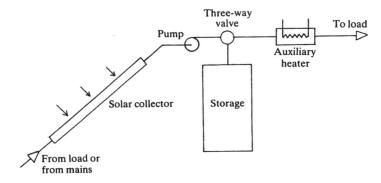

FIGURE 6.4.17   Schematic of an open-loop solar system.

the flow rate according to the incident radiation. One can collect a couple of percent more energy than with constant rate single-pass designs (Gordon and Zarmi, 1985). However, this entails changing the flow rate of the pump more or less continuously. Of all the three variants of the open-loop configuration, the first one, namely the single-pass open-loop solar thermal system configuration with constant flow rate and without a solenoid valve, is the most common. The single-pass design is not recommended for *variable* loads. The tank size is based on yearly daily load volumes, and efficient use of storage requires near-total depletion of the daily collected energy each day. If the load draw is markedly lower than its average value, the storage would get full relatively early the next day and solar collection would cease. It is because industrial loads tend to be more uniform, both during the day and over the year, than domestic applications that the single-pass open-loop configuration is recommended for such applications.

Description of a Typical Closed-Loop System

Figure 6.4.18 illustrates a typical closed-loop solar-supplemented liquid heating system. The useful energy is often (but not always) delivered to the storage tank via a collector–heat exchanger, which separates the collector fluid stream and the storage fluid. Such an arrangement is necessary either for antifreeze protection or to avoid corrosion of the collectors by untreated water containing gases and salts. A safety relief valve is provided because the system piping is normally nonpressurized, and any steam produced in the solar collectors will be let off from this valve. When this happens, energy dumping is said to take place. Fluid from storage is withdrawn and made to flow through the load–heat exchanger when the load calls for heat. Whenever possible, one should withdraw fluid directly from the storage and pass it through the load, and avoid incorporating the load–heat exchanger, since it introduces additional thermal penalties and involves extra equipment and additional parasitic power use. Heat is withdrawn from the storage tank at the top and reinjected at the bottom in order to derive maximum benefit from the thermal stratification that occurs in the storage tank. A bypass circuit is incorporated prior to the load heat exchanger and comes into play

1. When there is no heat in the storage tank (i.e., storage temperature T_S is less than the fluid temperature entering the load heat exchanger T_{Xi})
2. When T_S is such that the temperature of the fluid leaving the load heat exchanger is greater than that required by the load (i.e., $T_{Xo} > T_{Li}$, in which case the three-way valve bypasses part of the flow so that $T_{Xo} = T_{Li}$). The bypass arrangement is thus a differential control device which is said to modulate the flow such that the above condition is met. Another operational strategy for maintaining $T_{Xo} = T_{Li}$ is to operate pump in a "bang-bang" fashion (i.e., by short cycling the pump). Such an operation is not advisable, however, since it would lead to premature pump failure.

An auxiliary heater of the *topping-up type* supplies just enough heat to raise T_{Xo} to T_{Li}. After passing through the load, the fluid (which can be either water or air) can be recirculated or, in case of liquid contamination through the load, fresh liquid can be introduced. The auxiliary heater can also be placed

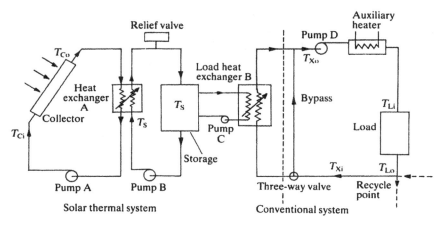

FIGURE 6.4.18 Schematic of a typical closed-loop system with auxiliary heater placed in series (also referred to as a topping-up type).

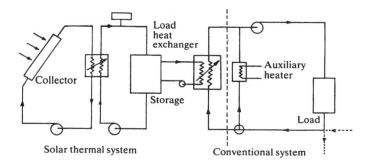

FIGURE 6.4.19 Schematic of a typical closed-loop system with auxiliary heater placed in parallel (also referred to as an all-or-nothing type).

in parallel with the load (see Figure 6.4.19) in which case it is called an *all-or-nothing type*. Although such an arrangement is thermally less efficient than the topping-up type, this type is widely used during the solar retrofit of heating systems because it involves little mechanical modifications or alterations to the auxiliary heater itself.

It is obvious that there could also be solar-supplemented energy systems that do not include a storage element in the system. Figure 6.4.20 shows such a system configuration with the auxiliary heater installed in series. The operation of such systems is not very different from that of systems with storage, the primary difference being that whenever instantaneous solar energy collection exceeds load requirements (i.e., $T_{Co} > T_{Li}$), energy dumping takes place. It is obvious that by definition there cannot be a closed-loop, no-storage solar thermal system. Solar thermal systems without storage are easier to construct and operate, and even though they may be effective for 8 to 10 hours a day, they are appropriate for applications such as process heat in industry.

Active closed-loop solar systems as described earlier are widely used for service hot water systems, that is for domestic hot water and process heat applications as well as for space heat. There are different variants to this generic configuration. A system without the collector–heat exchanger is referred to having collectors *directly coupled* to the storage tank (as against *indirect coupling* as in Figure 6.4.16). For domestic hot water systems, the system can be simplified by placing the auxiliary heater (which is simply an electric heater) directly inside the storage tank. One would like to maintain stratification in the tank so that the coolest fluid is at the bottom of the storage tank, thereby enhancing collection efficiency. Consequently, the electric heater is placed at about the upper third portion of the tank so as to assure good collection

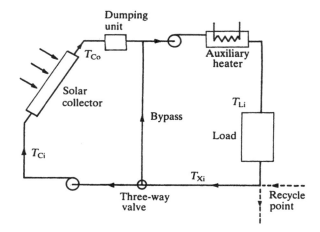

FIGURE 6.4.20   Simple solar thermal system without storage.

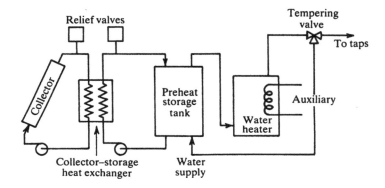

FIGURE 6.4.21   Schematic of a standard domestic hot-water system with double tank arrangement.

efficiency while assuring adequate hot water supply to the load. A more efficient but expensive option is widely used in the United States: the *double tank system*, shown in Figure 6.4.21. Here the functions of solar storage and auxiliary heating are separated, with the solar tank acting as a preheater for the conventional gas or electric unit. Note that a further system simplification can be achieved for domestic applications by placing the load heat exchanger directly inside the storage tank. In certain cases, one can even eliminate the heat exchanger completely.

Another system configuration is the *drain-back* (also called drain-out) system, where the collectors are emptied each time the solar system shuts off. Thus the system invariably loses collector fluid at least once, and often several times, each day. No collector–heat exchanger is needed, and freeze protection is inherent in such a configuration. However, careful piping design and installation, as well as a two-speed pump, are needed for the system to work properly (Newton and Gilman, 1977). The drain-back config-uration may be either open (vented to atmosphere) or closed (for better corrosion protection). Long-term experience in the United States with the drain-back system has shown it to be very reliable if engineered properly. A third type of system configuration is the *drain-down* system, where the fluid from the collector array is removed only when adverse conditions, such as freezing or boiling, occur. This design is used when freezing ambient temperatures are only infrequently encountered.

Active solar systems of the type described above are mostly used in countries such as the United States and Canada. Countries such as Australia, India, and Israel (where freezing is rare) usually prefer thermosyphon systems. No circulating pump is needed, the fluid circulation being driven by density difference between the cooler water in the inlet pipe and the storage tank and the hotter water in the outlet pipe of the collector and the storage tank. The low fluid flow in thermosyphon systems enhances thermal stratification in the storage tank. The system is usually fail-proof, and a study by Liu and Fanney (1980) reported that a thermosyphon system performed better than several pumped service hot water systems. If operated properly, thermosyphon and active solar systems are comparable in their thermal performance. A major constraint in installing thermosyphon systems in already existing residences is the requirement that the bottom of the storage tank be at least 20 cm higher than the top of the solar collector in order to avoid reverse thermosyphoning at night. To overcome this, spring-loaded one-way valves have been used, but with mixed success.

Thermal Storage Systems

Low-temperature solar thermal energy can be stored in liquids, solids, or phase change materials (PCMs). Water is the most frequently used liquid storage medium because of its low cost and high specific heat. The most widely used solid storage medium is rocks (usually of uniform circular size 25 to 40 mm in diameter). PCM storage is much less bulky because of the high latent heat of the PCM material, but this technology has yet to become economical and safe for widespread use.

Water storage would be the obvious choice when liquid collectors are used to supply hot water to a load. When hot air is required (for space heat or for convective drying), one has two options: an air collector with a pebble-bed storage or a system with liquid collectors, water storage, and a load heat exchanger to transfer heat from the hot water to the working air stream. Though a number of solar air systems have been designed and operated successfully (mainly for space heating), water storage is very often the medium selected. Water has twice the heat capacity of rock, so water storage tanks will be smaller than rock-bed containers. Moreover, rock storage systems require higher parasitic energy to operate, have higher installation costs and require more sophisticated controls. Water storage permits simultaneous charging and discharging while such an operation is not possible for rock storage systems. The various types of materials used as containers for water and rock-bed storage and the types of design, installation, and operation details one needs to take care of in such storage systems are described by Mueller Associates (1985) and SERI (1989).

Sensible storage systems, whether water or rock-bed, exhibit a certain amount of thermal stratification. Standard textbooks present relevant equations to model such effects. In the case of active closed-loop multipass hot water systems, storage stratification effects can be neglected for long-term system performance with little loss of accuracy. Moreover, this leads to conservative system design (i.e., solar contribution is underpredicted if stratification is neglected). A designer who wishes to account for the effect of stratification in the water storage can resort to a formulation by Phillips and Dave (1982), who showed that this effect can be fairly well modeled by introducing a *stratification coefficient* (which is a system constant that needs to be determined only once) and treating the storage subsystem as fully mixed. However, this approach is limited to the specific case of no (or very little) heat withdrawal from storage during the collection period. Even when water storage systems are highly stratified, simulation studies seem to indicate that modeling storage as a one-dimensional plug-flow three-node heat transfer problem yields satisfactory results of long-term solar system performance.

The thermal losses q_w from the storage tank can be modeled as

$$q_w = (UA_S)(T_S - T_{env}) \tag{6.4.28}$$

where (UA_S) is the storage overall heat loss per unit temperature difference and T_{env} is the temperature of the air surrounding the storage tank. Note that (UA_S) depends on the storage size, which is a parameter to be sized during system design, and on the configuration of the storage tank (i.e., on the length by diameter ratio in case of a cylindrical tank). For storage tanks, this ratio is normally in the range of 1.0 to 2.0.

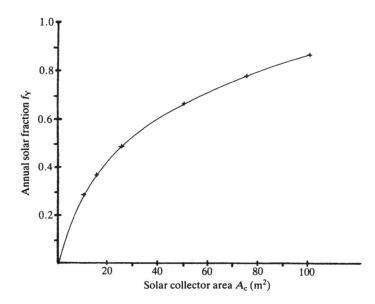

FIGURE 6.4.22   A typical solar system production function.

6.4.4   Solar System Sizing Methodology

Sizing of solar systems primarily involves determining the collector area and storage size that are most cost effective. We shall address standalone and solar-supplemented systems separately since the basic design problem is somewhat different.

Solar-Supplemented Systems

Production Functions

Because of the annual variation of incident solar radiation, it is not normally economical to size a solar subsystem such that it provides 100% of the heat demand. Most solar energy systems follow the *law of diminishing returns*. This implies that increasing the size of the solar collector subsystem results in a less than proportional increase in the annual fuel savings (or alternatively, in the annual solar fraction).

Any model has two types of variables: exogenous and endogenous. The *exogenous parameters* are also called the input variables, and these in turn may be of two kinds. *Variable* exogenous parameters are the collector area A_C, the collector performance parameters $F_R\eta_n$ and $F_R U_L$, the collector tilt, the thermal storage capacity $(Mc_p)_S$, the heat exchanger size, and the control strategies of the solar thermal system. On the other hand, the climatic data specified by radiation and the ambient temperature, as well as the end-use thermal demand characteristics, are called *constrained* exogenous parameters because they are imposed externally and cannot be changed. The *endogenous* parameters are the output parameters whose values are to be determined, the annual solar fraction being one of the parameters most often sought.

Figure 6.4.22 illustrates the law of diminishing results. The annual solar fraction f_Y is seen to increase with collector area but at a decreasing rate and at a certain point will reach saturation. Variation of any of the other exogenous parameters also exhibit a similar trend. The technical relationship between f_Y and one or several variable exogenous parameters for a given location is called the *yearly production function*.

It is only for certain simple types of solar thermal systems that an analytical expression for the production can be deduced directly from theoretical considerations. The most common approach is to carry out computer simulations of the particular system (solar plus auxiliary) over the complete year for several combinations of values of the exogenous parameters. The production function can subsequently be determined by an empirical curve fit to these discrete sets of points.

Economic Analysis

It is widely recognized that *discounted cash flow analysis* is most appropriate for applications such as sizing an energy system. This analysis takes into account both the initial cost incurred during the installation of the system and the annual running costs over its entire life span.

The economic objective function for optimal system selection can be expressed in terms of either the energy cost incurred or the energy savings. These two approaches are basically similar and differ in the sense that the objective function of the former has to be minimized while that of the latter has to be maximized. In our analysis, we shall consider the latter approach, which can further be subdivided into the following two methods:

1. Present worth or life cycle savings, wherein all running costs are discounted to the beginning of the first year of operation of the system
2. Annualized life cycle savings, wherein the initial expenditure incurred at the start as well as the running costs over the life of the installation are expressed as a yearly mean value.

Chapter 3.2 described the details. The optimization methods for Chapter 3.2 (see Figure 3.2.6) must be used. The HVAC designer must complete such a calculation for solar sizing since solar systems do not meet peak loads unlike fossil fuel based systems.

6.4.5 Solar System Design Methods

Classification

Design methods may be separated into three generic classes. The *simple* category, usually associated with the prefeasibility study phase (see the introduction) involves quick manual calculations of solar collector/system performance and rule-of-thumb engineering estimates. For example, the generalized yearly correlations proposed by Rabl (1981) and described in Section 6.4.2 could be conveniently used for year-round, more or less constant loads. The approach is directly valid for open-loop solar systems, while it could also be used for closed-loop systems if an *average* collector inlet temperature could be determined. A simple manner of selecting this temperature $\overline{T}_m$ for domestic closed-loop multipass systems is to assume the following empirical relation:

$$\overline{T}_m = T_{\text{mains}}/3 + (2/3)\, T_{\text{set}} \tag{6.4.29}$$

where T_{mains} is the average annual supply temperature and T_{set} is the required hot water temperature (about 60 to 80°C in most cases).

These manual methods often use general guidelines, graphs, and/or tables for sizing and performance evaluation. The designer should have a certain amount of knowledge and experience in solar system design in order to make pertinent assumptions and simplifications regarding the operation of the particular system.

Mid-level design methods are resorted to during the feasibility phase of a project. The main focus of this chapter has been toward this level, and a few of these design methods will be presented in this section. A personal computer is best suited to these design methods because they could be conveniently programmed to suit the designer's tastes and purpose (spreadsheet programs, or better still one of the numerous equation-solver software packages, are most convenient). Alternatively, commercially available software packages such as f-chart (Beckman et al., 1977) could also be used for certain specific system configurations.

Detailed design methods involve performing hourly simulations of the solar system over the entire year from which accurate optimization of solar collector and other equipment can be performed. Several simulation programs for active solar energy systems are available, TRNSYS (Klein et al., 1975, 1979) developed at the University of Wisconsin—Madison being perhaps the best known. This public-domain software has technical support and is being constantly upgraded. TRNSYS contains simulation models

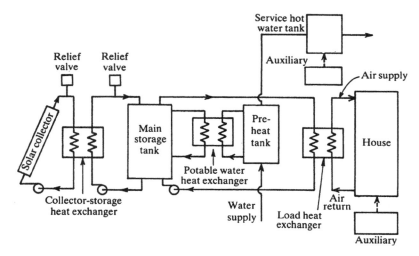

FIGURE 6.4.23　Schematic of the standard space heating liquid system configuration for the f-chart method. (From Duffie, J.A. and Beckman, W.A., *Solar Engineering of Thermal Processes*, Wiley Interscience, New York. Copyright 1980. Reprinted with permission of John Wiley & Sons, Inc.).

of numerous subsystem components (solar radiation, solar equipment, loads, mechanical equipment, controls, etc.) that comprise a solar energy system. A user can conveniently hook up components representative of a particular solar system to be analyzed and then simulate that system's performance at a level of detail that the user selects. Thus TRNSYS provides the design with large flexibility, diversity and convenience of usage.

As pointed out by Rabl (1985), the detailed computer simulations approach, though a valuable tool, has several problems. Judgment is needed both in the selection of the input and in the evaluation of the output. The very flexibility of big simulation programs has drawbacks. So many variables must be specified by the user that errors in interpretation or specification are common. Also, learning how to use the program is a time-consuming task. Because of the numerous system variables to be optimized, the program may have to be run for numerous sets of combinations, which adds to expense and time. The inexperienced user can be easily misled by the second-order details while missing first-order effects. For example, uncertainties in load, solar radiation, and economic variables are usually very large, and long-term performance simulation results are only accurate to within a certain degree. Nevertheless, detailed simulation programs, if properly used by experienced designers, can provide valuable information on system design and optimization aspects at the final stages of a project design.

There are basically three types of mid-level design approaches: the empirical correlation approach, the analytical approach, and the one-day repetitive methods approach (described fully in Reddy, 1987). We shall illustrate their use by means of specific applications.

Active Space Heating

The solar system configuration for this particular application has become more or less standardized. For example, for a liquid system, one would use the system shown in Figure 6.4.23. One of the most widely used design methods is the f-chart method (Beckman et al., 1977; Duffie and Beckman, 1980), which is applicable for standardized liquid and air heating systems as well as for standardized domestic hot water systems. The f-chart method basically involves using a simple algebraic correlation that has been deduced from numerous TRNSYS simulation runs of these standard solar systems subject to a wide range of climates and solar system parameters. Correlations were developed between monthly solar fractions and two easily calculated dimensionless variables X and Y, where

$$X = \left(A_C \, F'_R \, U_L \left(T_{\text{Ref}} - \overline{T}_a \right) \Delta t \right) \Big/ Q_{LM} \tag{6.4.30}$$

$$Y = A_C \, F_R' \, \overline{\eta}_0 \, \overline{H}_T N / Q_{LM} \qquad (6.4.31)$$

where A_C = collector area (m²)
  F_R' = collector-heat exchanger heat removal factor
  U_L = collector overall loss coefficient (W/(m² °C))
  Δt = total number of seconds in the month = $3{,}600 \times 24 \times N = 86{,}400 \times N$
  $\overline{T}_a$ = monthly average ambient temperature (°C)
  T_{Ref} = an empirically derived reference temperature, taken as 100°C
  Q_{LM} = monthly total heating load for space heating and/or hot water (J)
  $\overline{H}_T$ = monthly average daily radiation incident on the collector surface per unit area (J/m²)
  N = number of days in the month
  $\overline{\eta}_0$ = monthly average collector optical efficiency

The dimensionless variable X is the ratio of reference collector losses over the entire month to the monthly total heat load; the variable Y is the ratio of the monthly total solar energy absorbed by the collectors to the monthly total heat load. It will be noted that the collector area and its performance parameters are the predominant exogenous variables that appear in these expressions. For changes in secondary exogenous parameters, the following corrective terms X_C and Y_C should be applied for liquid systems:

1. for changes in storage capacity:

$$X_C / X = \left(\text{actual storage capacity}/\text{standard storage capacity}\right)^{-0.25} \qquad (6.4.32)$$

where the standard storage volume is 75 L/m² of collector area.

2. for changes in heat exchanger:

$$Y_C / Y = 0.39 + 0.65 \ \exp\left[-\left(0.139 \ (UA)_B \Big/ \left(E_L \left(mc_p\right)_{min}\right)\right)\right] \qquad (6.4.33)$$

The monthly solar fraction for liquid space heating can then be determined from the following empirical correlation:

$$f_M = 1.029 \ Y - 0.065 \ X - 0.245 \ Y^2 + 0.0018 \ X^2 + 0.0215 \ Y^3 \qquad (6.4.34)$$

subject to the conditions that $0 \le X \le 15$ and $0 \le Y \le 3$. This empirical correlation is shown graphically in Figure 6.4.24.

 A similar correlation has also been proposed for space heating systems using air collectors and pebble-bed storage. The procedure for exploiting the preceding empirical correlations is as follows. For a predetermined location, specified by its 12 monthly radiation and ambient temperature values, Eq. (6.4.34) is repeatedly used for each month of the year for a particular set of variable exogenous parameters. The monthly solar fraction f_M and the annual thermal energy delivered by the solar thermal system are easily deduced. Subsequently, the entire procedure is repeated for different values and combinations of variable exogenous parameters. Finally, an economic analysis is performed to determine optimal sizes of various solar system components. Care must be exercised that the exogenous parameters considered are not outside the range of validity of the f-chart empirical correlations.

Domestic Water Heating

The f-chart correlation can also be used to predict the monthly solar fraction for domestic hot water systems provided the water mains temperature T_{mains} is between 5 and 20°C and the minimum acceptable

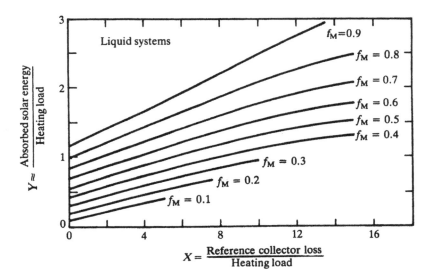

FIGURE 6.4.24 The f-chart correlation for liquid system configuration. (From Duffie, J.A. and Beckman, W.A., *Solar Engineering of Thermal Processes*, Wiley Interscience, New York. Copyright 1980. Reprinted with permission of John Wiley & Sons, Inc.)

hot water temperature drawn from the storage for end use (called the set water temperature T_w) is between 50 and 70°C. Further, the dimensionless parameter X must be corrected by the following ratio

$$X_w / X = \left(11.6 + 1.8\,T_w + 3.86\,T_{\text{mains}} - 2.32\,\overline{T}_a\right) / \left(100 - \overline{T}_a\right) \tag{6.4.35}$$

In case the domestic hot water load is much smaller than the space heat load, it is recommended that Eq. (6.4.34) be used without the above correction.

Industrial Process Heat

Open-Loop Single-Pass Systems

The advantages offered by open-loop single-pass systems over closed-loop multipass systems for meeting constant loads has been described in Section 6.4.3 under "Closed-Loop and Open-Loop Systems." Because industrial loads operate during the entire sun-up hours or even for 24 hours daily, the simplest solar thermal system is one with no heat storage. A sizable portion (between 25 and 70%) of the daytime thermal load can be supplied by such systems and consequently, the sizing of such systems will be described below (Gordon and Rabl, 1982). We shall assume that T_{Li} and T_{Xi} are constant for all hours during system operation. Because no storage is provided, excess solar energy collection (whenever $T_{Ci} > T_{Li}$) will have to be dumped out.

The maximum collector area $\hat{A}_C$ for which energy dumping does not occur at any time of the year can be found from the following instantaneous heat balance equation:

$$P_L = \hat{A}_C\,\hat{F}_R\left[I_{\max}\,\eta_n - U_L\left(T_{Ci} - T_a\right)\right] \tag{6.4.36}$$

where P_L the instantaneous thermal heat demand of the load (say, in kW) is given by

$$P_L = m_L\,c_p\left(T_{Li} - T_{Xi}\right) \tag{6.4.37}$$

and F_R is the heat removal factor of the collector field when its surface area is $\hat{A}_C$. Since $\hat{A}_C$ is as yet unknown, the value of $\hat{F}_R$ is also undetermined. (Note that though the *total* fluid flow rate is known, the flow rate per unit collector area is not known.) Recall that the plate efficiency factor F' for liquid collectors can be assumed constant and independent of fluid flow rate per unit collector area. Equation (6.4.36) can be expressed in terms of critical radiation level I_C:

$$P_L = \hat{A}_C \, \hat{F}_R \, \eta_n \left(I_{max} - I_C \right) \tag{6.4.38}$$

or

$$\hat{A}_C \, \hat{F}_R \, \eta_n = P_L \big/ \left(I_{max} - I_C \right) \tag{6.4.39}$$

Substituting for F_R and rearranging yields

$$\hat{A}_C = -\left(m_L c_p / F'U_L \right) \ln\left[1 - P_L U_L \big/ \eta_n \left(I_{max} - I_C \right) m_L c_p \right] \tag{6.4.40}$$

If the actual collector area A_C exceeds this value, dumping will occur as soon as the radiation intensity reaches a value I_D, whose value is determined from the following heat balance:

$$P_L = A_C \, F_R \, \eta_n \left(I_D - I_C \right) \tag{6.4.41}$$

Hence

$$I_D = I_C + P_L \big/ \left(A_C \, F_R \, \eta_n \right) \tag{6.4.42}$$

Note that the value of I_D decreases with increasing collector area A_C, thereby indicating that increasing amounts of solar energy will have to be dumped out.

Since the solar thermal system is operational during the entire sunshine hours of the year, the yearly total energy collected can be directly determined by the Rabl correlation given by Eq. (6.4.26). Similarly, the yearly total solar energy collected by the solar system which has got to be dumped out is

$$Q_{DY} = A_C \, F_R \, \eta_n \left(\tilde{a} + \tilde{b} I_D + \tilde{c} I_D^2 \right) \tag{6.4.43}$$

The yearly total solar energy delivered to the load is

$$Q_{UY} = Q_{CY} - Q_{DY}$$
$$= A_C \, F_R \, \eta_n \left[\tilde{b} \cdot \left(I_C - I_D \right) + \tilde{c} \cdot \left(I_C^2 - I_D^2 \right) \right] \tag{6.4.44}$$

$$= -\left(\tilde{b} + 2\tilde{c} I_C \right) P_L - \tilde{c} P_L^2 \big/ \left(A_C \, F_R \, \eta_n \right) \tag{6.4.45}$$

Replacing the value of F_R given by Eq. (6.4.3), the annual production function in terms of A_C is

$$Q_{UY} = -\left(\tilde{b} + 2\tilde{c} I_C \right) P_L - \frac{\tilde{c} P_L^2}{\left(\dfrac{F'\eta_n}{F'U_L} \right) \cdot \left(m_L c_p \right) \left[1 - \exp\left(-\dfrac{F'U_L A_C}{m_L c_p} \right) \right]} \tag{6.4.46}$$

subject to the condition that $A_C > \hat{A}_C$. If the thermal load is not needed during all days of the year due to holidays or maintenance shut-down, the production function can be reduced proportionally.

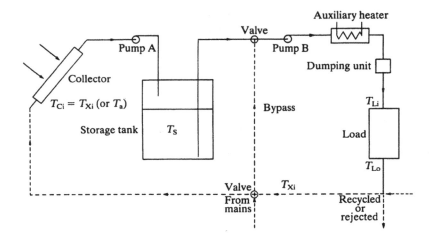

FIGURE 6.4.25　Open-loop solar industrial hot-water system with storage.

TABLE 6.4.3　Percentage of Total System Cost by Component

Cost Component	Percentage Range
Collectors	15–30
Collector installation	5–10
Collector support structure	5–20[a]
Storage tanks	5–7
Piping and specialties	10–30
Pumps	1–3
Heat exchangers	0–5[b]
Chiller	5–10
Miscellaneous	2–10
Instrumentation	1–3
Insulation	2–8
Control subsystem	4–9
Electrical	2–6

[a] For collectors mounted directly on a tilted roof.
[b] For systems without heat exchangers.
Source: From Mueller and Associates, 1985.

6.4.6　Design Recommendations and Costs

Design Recommendations

Design methods reduce computational effort compared to detailed computer simulations. Even with this decrease, the problem of optimal system design and sizing remains formidable due to:

1. The presence of several solar thermal system configuration alternatives.
2. The determination of optimal component sizes for a given system.
3. The presence of certain technical and economic constraints.
4. The choice of proper climatic, technical and economic input parameters.
5. The need to perform sensitivity analysis of both technical and economic parameters.

For most practical design work, a judicious mix of theoretical expertise and practical acumen is essential. Proper focus right from the start on the important input variables as well as the restriction of the normal range of variation would lead to a great decrease in design time and effort.

Several examples of successful case studies and system design recommendations are described in the published literature (see, for example, Kutcher et al., 1982).

Solar System Costs

How the individual components of the solar system contribute to the total cost can be gauged from Table 6.4.3. We note that collectors constitute the major fraction (from 15 to 30%), thus suggesting that collectors should be selected and sized with great care. Piping costs are next with other collector-related costs like installation and support structure being also important.

Costs vary by location. The HVAC designer should consult R.S. Means *Mechanical Cost Data* issued annually. Solar costs are in Division 13, "Period Construction," Section 13600.

References

Anderson, B. (1977). *Solar Energy: Fundamentals in Building Design*, McGraw-Hill, New York.

ASHRAE Standard 93-77 (1978). *Methods of Testing to Determine the Thermal Performance of Solar Collectors*, American Society of Heating, Refrigeration and Air Conditioning Engineers, New York.

ASHRAE (1985). *Fundamentals*, American Society of Heating, Refrigeration and Air Conditioning Engineers, New York.

Beckman, W.A., Klein, S.A. and Duffie, J.A. (1977). *Solar Heating Design by the f-Chart Method*, Wiley Interscience, New York.

Charters, W.W.S. and Pryor, T.L. (1982). *An Introduction to the Installation of Solar Energy Systems*, Victoria Solar Energy Council, Melbourne, Australia.

Chiam, H.F. (1981). Planar concentrators for flat-plate solar collectors, *Solar Energy*, 26, p. 503.

Clark, D.R., Klein, S.A. and Beckman, W.A. (1983). Algorithm for evaluating the hourly radiation utilizability function. *ASME Journal of Solar Energy Eng.*, 105, p. 281.

Collares-Pereira, M. and Rabl, A. (1979). Derivation of method for predicting the long-term average energy delivery of solar collectors, *Solar Energy*, 23, p. 223.

Collares-Pereira, M., Gordon, J.M., Rabl, A. and Zarmi, Y. (1984). Design and optimization of solar industrial hot water systems with storage, *Solar Energy*, 32, p. 121.

Connelly, M., Giellis, R., Jenson, G. and McMorchie, R. (1976). Solar heating and cooling computer analysis—A simplified method for non-thermal specialists, Proc. of the Int. Solar Energy Society Conf., Winnipeg, Canada.

de Winter (1975). Heat exchanger penalties in double loop solar water heating systems, *Solar Energy*, 17, p. 335.

Duffie, J.A. and Beckman, W.A. (1980). *Solar Engineering of Thermal Processes*, Wiley Interscience, New York.

Edwards, D.K., Nelson, K.E., Roddick, R.D. and Gier, J.T. (1960). Basic Studies on the Use of Solar Energy, Report no. 60-93, Dept. of Engineering, Univ. of California at Los Angeles, CA.

Emery, M. and Rogers, B.A. (1984) On a solar collector thermal performance test method for use in variable conditions, *Solar Energy*, 33, p. 117.

Feuermann, D., Gordon, J.M. and Zarmi, Y. (1985). A typical meteorological day (TMD) approach for predicting the long-term performance of solar energy systems, *Solar Energy*, 35, p. 63.

Gordon, J.M. and Zarmi, Y. (1981). Technical note: Thermosyphon systems: Single vs. multipass, *Solar Energy*, 27, p. 441.

Gordon, J.M. and Rabl, A. (1982). Design, analysis and optimization of solar industrial process heat plants without storage, *Solar Energy*, 28, p. 519.

Gordon, J.M. and Zarmi, Y. (1985). An analytic model for the long-term performance of solar thermal systems with well-mixed storage, *Solar Energy*, 35, p. 55.

Gordon, J.M. and Rabl, A. (1986). Design of solar industrial process heat systems, in *Reviews of Renewable Energy Sources*, M.S. Sodha, S.S. Mathur, M.A.S. Malik and T.C. Kandpal (eds.) Ch. 6, Wiley Easter, New Delhi.

Gordon, J.M. and Saltiel, C. (1986). Analysis and optimization of multistage solar collector systems, *ASME Journal of Solar Energy Eng.*, 108, p. 92.

Gordon, J.M. (1987). Optimal sizing of stand-alone photovoltaic systems, *Solar Cells*, 20, p. 295.

Grassie, S.L. and Sheridan, N.R. (1977). The use of planar reflectors for increasing the energy yield of flat-plate collectors, *Solar Energy*, 19, p. 663.

Hollands, K.G.T. (1965). Honeycomb devices in flat-plate solar collectors, *Solar Energy*, 9, p. 159.

Klein, S.A., Cooper, P.I., Freeman, T.L., Beekman, D.M., Beckman, W.A. and Duffie, J.A. (1975). A method of simulation of solar processes and its applications, *Solar Energy*, 17, p. 29.

Klein, S.A. (1978). Calculation of flat-plate collector utilizability, *Solar Energy*, 21, p. 393.

Klein, S.A. et al. (1979). TRNSYS-A Transient System Simulation User's Manual, University of Wisconsin-Madison Engineering Experiment Station Report 38-10.

Klein, S.A. and Beckman, W.A. (1979). A general design method for closed-loop solar energy systems, *Solar Energy*, 22, p. 269.

Kreider, J.F. (1979). *Medium and High Temperature Solar Processes*, Academic Press, New York.

Kutcher, C.F., Davenport, R.L., Dougherty, D.A., Gee, R.C., Masterson, P.M. and May, E.K. (1982). Design Approaches for Solar Industrial Process Heat Systems, SERI/TR-253-1356, Solar Energy Research Institute, Golden, CO.

Larson, D.C. (1980). Optimization of flat-plate collector flat mirror system, *Solar Energy*, 24, p. 203.

Larson, R.W. Vignola, F. and West, R. (1992). *Economics of Solar Energy Technologies*, American Solar Energy Society Report, Boulder, CO.

Liu, B.Y.H. and Jordan, R.C. (1960). The inter-relationship and characteristic distribution of direct, diffuse and total solar radiation, *Solar Energy*, 4, p. 1.

Liu, B.Y.H. and Jordan, R.C. (1963). A rational procedure for predicting the long-term average performance of flat-plate solar energy collectors, *Solar Energy*, 7, p. 53.

Liu, S.T. and Fanney, A.H. (1980). Comparing experimental and computer-predicted performance for solar hot water systems, *ASHRAE Journal*, 22, No. 5, p. 34.

Meyer, B.A. (1978). Natural convection heat transfer in small and moderate aspect ratio enclosures — An application to flat-plate collectors, in *Thermal Storage and Heat Transfer in Solar Energy Systems*, F. Kreith, R. Boehm, J. Mitchell and R. Bannerot (eds.), American Society of Mechanical Engineers, New York.

Mitchell, J.C., Theilacker, J.C. and Klein, S.A. (1981). Technical note: Calculation of monthly average collector operating time and parasitic energy requirements, *Solar Energy*, 26, p. 555.

Mueller Associates (1985). *Active Solar Thermal Design Manual*, funded by U.S. DOE (no. EG-77-C-01-4042), SERI(XY-2-02046-1) and ASHRAE (project no. 40), Baltimore, MD.

Newton, A.B. and Gilman, S.H. (1977). *Solar Collector Performance Manual*, funded by U.S. DOE (no. EG-77-C-01-4042), SERI(XH-9-8265-1) and ASHRAE (project no. 32, Task 3).

OTA (1991). Office of Technology Assessment, U.S. Congress, Washington, D.C.

Phillips, W.F. and Dave, R.N. (1982). Effect of stratification on the performance of liquid-based solar heating systems, *Solar Energy*, 29, p. 111.

Rabl, A. (1981). Yearly average performance of the principal solar collector types, *Solar Energy*, 27, p. 215.

Rabl, A. (1985). *Active Solar Collectors and Their Applications*, Oxford University Press, New York.

Reddy, T.A., Gordon, J.M. and de Silva, I.P.D. (1988). MIRA: A one-repetitive day method for predicting the long-term performance of solar energy systems, *Solar Energy*, 39, no. 2, p. 123.

Reddy, T.A. (1987). *The Design and Sizing of Active Solar Thermal Systems*, Oxford University Press, Oxford, U.K.

Saunier, G.Y. and Chungpaibulpatana, S. (1983). A new inexpensive dynamic method of testing to determine solar thermal collector performance, Int. Solar Energy Society World Congress, Perth, Australia.

SERI (1989). *Engineering Principles and Concepts for Active Solar Systems*, Hemisphere Publishing Company, New York.

Symons, J.G. (1976). *The Direct Measurement of Heat Loss from Flat-Plate Solar Collectors on an Indoor Testing Facility*, Technical Report TR7, Division of Mechanical Eng., Commonwealth Scientific and Industrial Research Organisation, Melbourne, Australia.

Theilacker, J.C. and Klein, S.A. (1980). Improvements in the utilizability relationships, American Section of the International Solar Energy Society Meeting Proceedings, P. 271, Phoenix, AZ.

7

Operation and Maintenance

James Braun
Purdue University

David E. Claridge
Texas A&M University

Srinivas Katipamula
*Pacific Northwest National
Laboratory*

Mingsheng Liu
University of Nebraska

Robert G. Pratt
*Pacific Northwest National
Laboratory*

7.1 HVAC System Commissioning

David E. Claridge and Mingsheng Liu

Commissioning was originally used by the Navy to ensure that battleships and submarines functioned properly before they were sent out to sea. It has been adopted and adapted within the building construction industry to apply to many different building systems. ASHRAE provides a definition of building commissioning in its Guideline 1-1996 (ASHRAE, 1996, p. 23):

> Commissioning is the process of ensuring systems are designed, installed, functionally tested, and operated in conformance with the design intent. Commissioning begins with planning and includes design, construction, start-up, acceptance, and training and can be applied throughout the life of the building. Furthermore, the commissioning process encompasses and coordinates the traditionally separate functions of systems documentation, equipment start-up, control system calibration, testing and balancing, and performance testing.

The ASHRAE commissioning efforts were restricted to new buildings, but it later became evident that while initial start-up problems were not an issue in older buildings, most of the other problems which commissioning tackled were even more prevalent in older systems. Commissioning of HVAC systems has been growing in popularity over the last decade; however, it is still not the norm in construction practice or building operation. One of the recommendations in the National Strategy for Building Commissioning (PECI, 1999) is "to develop a standard definition of commissioning."

The principal motivation for commissioning HVAC systems is to achieve HVAC systems that work properly to provide comfort to all the occupants of a building in an unobtrusive manner and at low cost, and to optimize HVAC system operation with minimal energy and operational costs. When HVAC systems are commissioned based on the design intents for a new building, the process is called *new building commissioning*. When HVAC systems are commissioned based on initial design intents for an existing building, the process is called *existing building commissioning*. When HVAC systems are commissioned based on actual use and the HVAC systems operation are optimized for different load conditions, the process is called *continuous commissioning*.

In principle, all building systems should be designed, installed, documented, tested, and staffed by personnel trained in their use. In practice, competitive pressures, fee structures, and financial pressures to occupy new buildings as quickly as possible result in buildings that are handed over to the owners with minimal contact between designers and operators, and are characterized by minimal functional testing of systems, documentation largely consisting of manufacturer system or component manuals, and little or no training for operators. This has lead to numerous problems including: mold growth in walls of new buildings, rooms that never cool properly, and air quality and comfort problems.

It has been estimated that new building commissioning will save 8% in energy cost alone compared with the average building that is not commissioned (PECI, 1999). This offers a payback for the cost of commissioning in just over 4 years from the energy savings alone and provides improved comfort and air quality. Traditional commissioning of existing buildings typically provides 12% in energy savings, with a payback of just over 1 year (PECI, 1999). The enhanced commissioning process, or continuous commissioning, significantly improves building comfort and typically decreases energy costs by 20% (Claridge et al., 1998) with payback of the project cost often less than 1.5 years.

Although commissioning provides higher quality buildings and results in fewer initial and subsequent operational problems, the direct and rapid payback of the commissioning expense from lowered operating costs is often the principal motivation for many owners. Documenting these lower operational costs is much easier if a specific plan is implemented to monitor and verify the results of the commissioning process. This is sometimes done with utility bill information, but is often more effective if measurement equipment is used on a temporary or permanent basis to record hourly or daily energy use data. The last section in this chapter addresses effective ways to monitor and verify savings from commissioning projects.

7.1.1 Commissioning New HVAC Systems

The goal of the commissioning process for a new HVAC system is to achieve a properly operating system that provides design comfort levels in every room in a building from the first day it is occupied. The motivation for commissioning a building is sometimes the desire to achieve this state as quickly, painlessly, and inexpensively as possible. In other cases, the primary motivation is to achieve operating savings and secondarily to minimize operating problems, while the motivation is more complex in other cases.

Disney Development Corporation has constructed over $10 billion in new facilities over the last decade and has concluded that commissioning is an essential element for their company. The corporation often uses innovative construction techniques and creative designs in highly utilized facilities where the occupants have very high expectations. Most of their facilities are expected to be aesthetically and operationally at the cutting edge of technology (Odom and Parsons, 1998). Other major private sector property owners who have adopted commissioning include Westin Hotels, Boeing, Chevron, Kaiser Permanente, and Target. The U.S. General Services Administration has begun to integrate commissioning into its design and construction program (PECI, 1999). State and local governments have also been leaders in the move toward commissioning, with significant programs at the state or local level in Florida, Idaho, Maryland, Montana, New York, Oregon, Tennessee, Texas, and Washington (Haasl and Wilkinson, 1998).

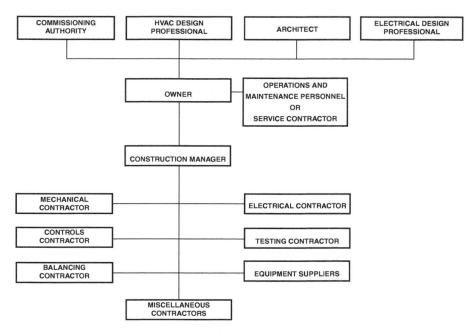

FIGURE 7.1.1   Commissioning organizational structure.

The Commissioning Process

Perhaps the major reason that commissioning is needed is that in many projects "commissioning" the project simply consists of turning everything on and verifying that all motors, chillers, and boilers run. The HVAC commissioning process ideally begins during the building programming phase and continues through the design phase, the construction phase, the acceptance phase, and into the post-acceptance phase. It requires the participation of the owner (or representative), the commissioning coordinator (or commissioning authority (CA)), design professionals, and the construction manager.

There is considerable agreement that a strong commissioning program requires a CA or a person or company who implements the overall commissioning process and coordinates commissioning related interactions between the other parties involved in the design, construction, and commissioning process.

The organizational structure of the commissioning process is shown in Figure 7.1.1. The CA reports to the owner and works with the other design professionals during the project. The construction manager then has primary responsibility of ensuring that the various contractors carry out the intent of the design developed, with the CA providing a detailed verification that the project, as built, does in fact meet the design intent.

The many facets of the commissioning process are shown schematically in Figure 7.1.2, specifically identifying the responsibilities of the owner, the CA, the design professional, and the constructions manager as they relate to the commissioning process in the programming, design, construction, acceptance, and post-acceptance phases of the project. In many projects, the commissioning process is implemented later in the design and construction process, decreasing the benefits of commissioning.

To maximize the benefits of commissioning, the owner selects the CA early in the programming phase so he or she can participate in the programming phase and develop a preliminary commissioning plan before the design phase begins. During the design phase, the principal responsibility of the CA is to review and comment on the design as it evolves and to update the commissioning plan as necessary. During the construction phase, the full commissioning team comes on board, and training of the building staff begins, while the CA continues to closely observe the construction process. The major commissioning activity occurs during the acceptance phase, with a multitude of checks and tests performed, further staff training, and finally, reporting and documentation of the process.

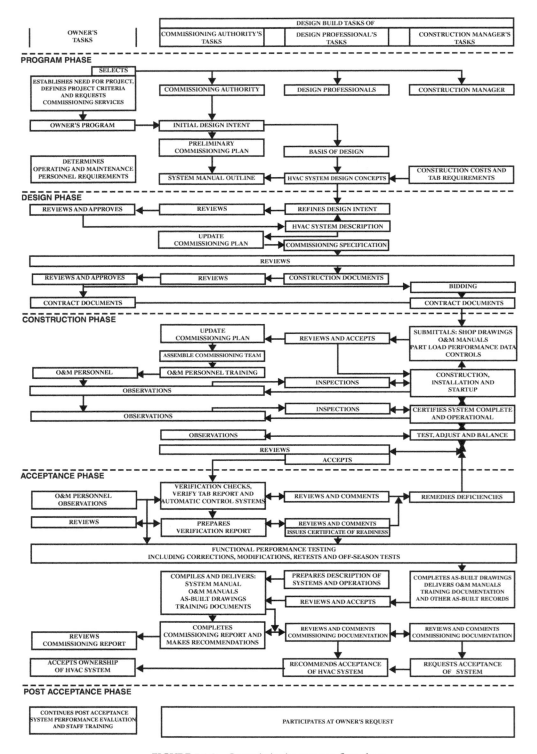

FIGURE 7.1.2 Commissioning process flow chart.

Commissioning Authority

The following is a more detailed listing of the duties of the commissioning authority (CA), as given by ASHRAE (1996):

1. Organize and lead the commissioning team.
2. Prepare the initial design intent document from the information contained in the owner's program.
3. Prepare a program-phase commissioning plan that describes in general the extent of the commissioning process to accomplish the owner's program and the initial design intent.
4. Develop the design-phase commissioning plan, which details the extent and activities of the commissioning process including commissioning team organization, schedule, training, and documentation requirements and all related testing, verification, and quality control procedures.
5. Review and comment on the impact of the design documents on the HVAC commissioning process for the mechanical, electrical, structural, plumbing, process, interior design, and other design professionals within the commissioning process, so that interfaces between systems are recognized and coordinated.
6. Prepare the construction-phase HVAC commissioning plan required as part of the commissioning specification. Include a list of all contractors for commissioning events by name, firm, and trade specialty.
7. Execute the HVAC commissioning process through organization of meetings, tests, demonstrations, training events, and performance verifications described in the contract documents and in the approved HVAC commissioning process. Organizational responsibilities include preparation of agendas, attendance lists, arrangements for facilities, and timely notification of participants for each commissioning event. The commissioning authority acts as chair at all commissioning events and ensures execution of all agenda items. The commissioning authority prepares minutes of every commissioning event and sends copies to all commissioning team members and attendees within five workdays of the event.
8. Review the plans and specifications with respect to their completeness in all areas relating to the HVAC commissioning process. This includes ensuring that the design phase commissioning plan is followed and adequate devices are included in the design in order to properly test, balance, and adjust the systems and to document the performance of each piece of equipment and each system. Any items required but not shown shall be brought to the attention of the construction manager prior to submittal of shop drawings.
9. Schedule the construction-phase coordination meeting within 90 days of the award of the contract at a convenient location and at a time suitable to the construction manager, the HVAC design professional, and the electrical design professional. This meeting is meant for reviewing the complete HVAC commissioning process and establishing tentative schedules for mechanical and electrical system orientation and inspections, O&M submittals, training sessions, system flushing and testing, job completion, testing, adjusting, and balancing (TAB) work, and functional performance testing.
10. Schedule the initial owner HVAC training session so that it is held immediately before the mechanical system orientation and inspection. This session is attended by the owner's O&M personnel, the HVAC design professional, the electrical design professional, the mechanical contractor, the electrical contractor, and the commissioning authority. The HVAC design professional will conduct this session with the assistance of the electrical design professional, giving an overview of the system, the system design goals, and the reasoning behind the selection of the equipment.
11. Coordinate the HVAC mechanical system orientation and inspection following the initial training session. The mechanical system orientation and inspection is conducted by the mechanical contractor. Its emphasis is on observation of the equipment's location with respect to accessibility. Prepare minutes of this meeting, with separate summaries of deficiency findings by the owner's staff and commissioning authority. Distribute to attendees and the owner.

12. Coordinate the HVAC electrical system orientation and inspection following the HVAC mechanical system orientation and inspection session. The electrical system orientation and inspection is conducted by the electrical contractor. Its emphasis is on observation of the equipment's location with respect to accessibility and function. Prepare minutes of this meeting, with separate summaries of deficiency findings by the owner and commissioning authority. Distribute to attendees and the owner.

13. Receive and review the operations and maintenance (O&M) manuals as submitted by the contractor. Ensure that they follow the specified outline and format. Insert the system's description as provided by the HVAC design professional in the Systems Manual.

14. Check the installation for adequate accessibility for maintenance and component replacement or repair.

15. Witness equipment, subsystem, and system start-up and testing. Ensure that the results are documented — including a summary of deficiencies — and incorporated in the O&M manuals.

16. Prior to initiating the TAB work, meet with the owner, mechanical contractor, HVAC design professional, and TAB contractor. The TAB contractor will outline TAB procedures and get concurrence from the HVAC design professional and commissioning authority. Ensure that the TAB contractor has all forms required for proper data collection and that he or she understands their importance and use.

17. Schedule the O&M training sessions. These training sessions are attended by the owner, the commissioning authority, the HVAC design professional, the electrical design professional, the construction manager, contractors, and equipment suppliers, as necessary. The format of these sessions follows the outline in the O&M manuals and includes hands-on training.

18. Upon receipt of notification from the construction manager that the HVAC system has been completed and is operational and the TAB report has been accepted by the HVAC design professional, proceed to verify the TAB report and the function of the control systems in accordance with the commissioning specification. Prepare a verification report, including all test data and identification of any deficiencies, and submit it to the owner and HVAC design professional for review.

19. Supervise the commissioning team members in the functional performance tests. The test data will be part of the commissioning report.

20. Review "as-built" drawings for accuracy with respect to the installed systems. Request revisions to achieve accuracy.

21. Ensure that the O&M manuals and all other "as-built" records have been updated to include all modifications made during the construction phase.

22. Prepare the Systems Manual.

23. Repeat functional performance tests to accommodate seasonal tests and/or correct any performance deficiencies. Revise and resubmit the commissioning report.

24. Assemble the final documentation, which will include the commissioning report, the Systems Manual, and all "as-built" records. Submit this documentation to the owner for review and acceptance.

25. Recommend acceptance of the HVAC system to the owner.

Commissioning Resources

Commissioning projects can be implemented at many levels of detail and a number of guidelines for implementing commissioning projects are available. *ASHRAE Guideline 1-1996: The HVAC Commissioning Process* (ASHRAE, 1996) is a product of the ASHRAE consensus process and as such has benefited from the input of all major stakeholders. It does not contain sets of forms and tables which are often helpful in scheduling and setting up the tests required in the process. The PECI guideline of Haasl and Sharp (1999) is helpful in this regard. They list 21 sources for commissioning guidelines, guide specifications, and sample functional performance tests. An abridged version of this list is provided in Table 7.1.1.

TABLE 7.1.1 Sources for Commissioning Guidelines, Guide Specifications, and Sample Functional Performance Tests

Source	Guidelines	Guide Specs	Sample Tests
Model Commissioning Plan and Guide Commissioning Specifications, USDOE/PECI, 1997. NTIS: # DE 97004564, 1-800-553-6847. PECI Web site: http://www.peci.org.	Some	Yes	Yes
The HVAC Commissioning Process, ASHRAE Guideline 1-1996, 1996. ASHRAE Publications Dept., 1791 Tullie Circle, NE, Atlanta, GA 30329.	Yes	Some	No
Engineering and Design Systems Commissioning Procedures, U.S. Army Corps of Engineers, 1995 (ER 1110-345-723). Department of the Army, U.S. Army Corps of Engineers, Washington, D.C. 20314-1000.	Some	Some	No
Commissioning Specifications, C-2000 Program, Canada, 1995. C-2000 Program, Energy Mines & Resources, Energy Efficiency Division, 7th Floor, 580 Booth St., Ottawa, Ontario, Canada K 1 A OE4.	No	Yes	No
Building Commissioning Guide, U.S. General Services Administration and USDOE, 1995. Prepared by Enviro-Management & Research, Inc., 703-642-5310.	Yes	No	No
Commissioning Guide Specification, Facility Management Office, University of Washington, 1993-6. http://weber.u.washington.edu/-fsesweb/	No	Yes	Some
Commissioning Guidelines, Instructions for Architects and Engineers, State of Washington, 1995. Dept. of General Administration, Div. of Engineering & Architectural Services, 360-902-7272.	Yes	No	No
Standard HVAC Control Systems Commissioning and Quality Verification User Guide, U.S. Army Const. Engineering Research Laboratories, 1994. Facilities Engineering Applications Program, U.S. Army Engineering and Housing Support Center, Ft. Belvoir, VA 22060-5516. FEAP-UG-GE-94/20.	No	No	Yes
Contractor Quality Control and Commissioning Program — Guidelines and Specification, Montgomery County Gov., State of Maryland, 1993. 301-217 6071.	Yes	Yes	Some
Procedural Standards for Building Systems Commissioning, National Environmental Balancing Bureau (NEBB), 1993. NEBB, 1385 Piccard Drive, Rockville, MD 20850. 301-977-3698	Yes	Some	Some
HVAC Systems Commissioning Manual, Sheet Metal and Air Conditioning Contractors' National Association (SMACNA), 1993. SMACNA, 4201 Lafayette Center Dr., Chantilly, VA 22021.	Yes	Some	Some
Guide Specification for Military Construction — Commissioning of HVAC Systems, Department of the Army, U.S. Army Corps of Engineers, January 1993. Department of the Army, U.S. Army Corps of Engineers, Washington, D.C. 20314-1000	No	Some	Yes
Commissioning Guide, Public Works Canada, Western Region, 1993. 403-497-3770.	Some	Yes	No
Building Commissioning Guidelines, Bonneville Power Administration/PECI, 1992. 503-230-7334.	Yes	Some	Some
The Building Commissioning Handbook, The Association of Higher Education Facilities Officers (APPA), written by John Heinz and Rick Casault, 1996. APPA, 1643 Prince Street, Alexandria, VA 22314.	Yes	Yes	No
HVAC Functional Inspection and Testing Guide, U.S. Dept. of Commerce and the General Services Administration, 1992. NTIS: 800-553-6847.	No	No	Yes
AABC Master Specification, Associated Air Balance Council (contains information on how the TAB fits into the commissioning process.) AABC National Headquarters, 202-737-0202.	No	Yes	No

Source: Abridged from Haasl and Sharp, 1999.

7.1.2 Case Study — Boeing Commercial Airplane Group Headquarters

Project Overview

The Boeing Company maintains and operates a large number of facilities in multiple locations. Over the years, Boeing has used many different methods to design, construct, and maintain its facilities. It currently uses an internal Facilities Asset Management Organization to handle the real estate, procurement, construction, maintenance, and asset accounting functions required to site, build, maintain, and manage all aspect of facilities infrastructure for the commercial airplane group.

The original case study description (Davenny, Doering, and McGuire, 1999), from which this description has been condensed and adapted, was written by two lead members of the Boeing project management team and the lead engineer for the commissioning agent.

The commissioning process used for this project is the result of many years of experience by Boeing's facilities personnel, and as such is not identical to the ASHRAE process. However, there are many common elements. The emphasis placed on the commissioning process is indicative of the direction in which the overall construction process at Boeing is headed. The company believes that the inherent benefits and efficiencies of the commissioning process increase staff effectiveness and help ensure success in the construction, operation, and maintenance of facilities.

The new commercial airplane group headquarters office building is a 309,000 square feet, 5-story office building located in Renton, Washington, and houses executive, administrative, and sales offices. The project was performed as a cost-plus-fixed-fee, design-build partnering effort. It began in October 1996 with initiation of the programming and preliminary design process. Ground was broken on May 14, 1997, and the building was occupied on October 2, 1998. (Costs below are for 1998.)

Commissioning Organizational Structure

The owner decided to expand the scope of the mechanical engineer's role to include the commissioning process. The mechanical engineer thus was assigned to act as the commissioning focal (CF), responsible for managing the commissioning work, in addition to being the owner's representative to the mechanical design and construction process, functioning as the liaison between the Boeing operating and maintenance staff and the design/construction team, reviewing design and equipment submittals, and resolving coordination problems and operational issues.

The owner then hired an independent commissioning agent who was assigned the responsibility of defining and executing the detailed quality assurance measures and system functional tests. Thus, the commissioning authority's (CA) responsibilities for the ASHRAE process were divided between the CF and the commissioning agent on this project. This allowed the CF to maintain an overview of the commissioning process, while still giving the required attention to other responsibilities. As the project developed, a commissioning team, which included representatives from the various contractor and facilities personnel, was formed. The coordination and communication role of the CA was identified, and the quality assurance and documentation duties which the company had traditionally viewed within the scope of commissioning were expanded to include organizing, scheduling, and reporting on the weekly commissioning team meetings, similar to the ASHRAE process recommendations. With a direct reporting line between the owner and the CA, the appropriate channel was available for decision making and problem resolution by the owner's staff.

Role and Responsibilities of a Commissioning Agent

The scope of work negotiated between the commissioning agent (CA) and the owner included the following specific responsibilities:

- Commissioning or Cx plan: This plan was prepared as a draft review document using input from all team members to establish respective roles, responsibilities, and communication pathways which were not articulated in the design/build contract documentation. The purpose was to clearly define the specifics of contractor relationships, reporting structures, and paper flow requirements relating to Cx. This plan became the focal point for the construction team to define, implement, and administer the Cx scope and process.
- Schedule: The CA assisted the general contractor with incorporating Cx into the master construction schedule. The Cx plan was translated into scheduled activities with specific milestones and scheduled time frames. These tasks were assigned work breakdown structure numbers to "nest" within the master schedule. Documentation requirements for each task were indicated on the MS project schedule document. This was to ensure that the Cx process enhanced project work flow as well as overall quality.

- Start-up documentation: The Cx team, led by the CA, reviewed and developed installation, start-up, and point-to-point checklists and appropriate follow-up documentation for subcontractor specialties. This step was incorporated into the Cx plan as the various specifications and responsibilities were reviewed.
- Test procedures and record sheets: Functional performance test procedures and record sheets for the various systems and components were written and executed. The systems included most of the mechanical and controls equipment within the building, including interface with other campus facilities. Electrical scope consisted of reviewing component test documentation by third party testing agents and witnessing emergency power system demonstrations.
- Quality assurance: Spot checking of test and balance scope for more than 400 VAV terminal boxes throughout the building was initially performed on 10% of the units. This QA scope expanded as a number of installation and operational irregularities were noted. Rather than having an adversarial role, the prefunctional testing allowed the team to resolve potential occupancy issues ahead of move-in.
- Cx meetings: The CA organized, scheduled, and conducted 38 regular meetings involving appropriate team members with a focus on the Cx process and related activities. These meetings supplemented other construction meetings as part of the Cx package and included the writing and distribution of meeting minutes, schedule generation and modifications, and task follow-up.
- Cx reports and Cx manuals: Reports were generated detailing site activities and items of importance during the construction and testing phases. These reports represented the summation of issues requiring resolution during construction and the performance of functional test procedures. Additionally, the final versions of all documentation relating to the scope outlined above were incorporated into appropriate format for a Cx manual.

Contract and Specification Issues

One of the biggest challenges presented to the commissioning effort was to revisit the client design criteria and review plans and specifications for enhanced compliance to Cx standards during construction. The design-build partnering aspects of the project facilitated the Cx in ways uncommon to bid-spec delivery, including such things as designated focals, time allotments, and extended cost mechanisms. The details, however, needed a fair amount of fleshing out. In some ways, the Cx team was playing a "catch-up" game by defining requirements as events occurred. Many times the Cx meeting forum identified technical issues during the job that were not addressed in the conventional construction meetings. The net effect was very positive.

Technical Issues

The building equipment, which was the focus of the HVAC commissioning work, included four custom air handling units (AHUs) with a total capacity of 350,000 cfm, associated fans with variable frequency drives (VFDs), pumps, coils and dampers, 400 VAV fan-powered terminal units, a building automation direct digital control system, and other miscellaneous systems and items.

One of the areas of greatest interest technically is that of VFDs. These devices control the air handler supply and return fan motors, and while they were provided by the electrical contractor, they are integral to the mechanical performance of the systems and are interfaced with the building automation system (BAS). Thus, there was great potential for conflict between various contractors as problems arose. Issues identified by the functional testing of these drives were added to a close-out punch list and resolved by the appropriate parties. These included:

- The need to provide additional wiring from the local unit disconnect back to the VFD to provide a shut down of the drive when the disconnect was open.
- Identification of proper VFD keypad operation to avoid conflict with the BAS.
- The need to repair fan intake damper closure so the drives did not fault due to inadvertent backward turning of the fans when disabled.

- Increase of minimum fan speed setting to avoid overvoltage occurrence at the drive when all flow was across a return fan in full recirculation mode.
- Identification of wiring problems at the motor controller when a drive was placed in bypass mode.

It was very beneficial to have these issue identified and resolved while all affected parties were still involved with project close-out.

The quality assurance scope included sampling the test and balance work performed at variable air volume terminal units throughout the space. This was particularly important to make sure that the staged building occupancy could proceed on a floor-by-floor basis with a minimum of disruption and comfort callbacks. This included checking the diffuser proportioning, primary air valve operation, and control of the air valves by the unit air flow sensors. Groups of terminal units were checked by changing control set points and trending zone responses in heating and cooling modes. Problems identified and corrected included:

- Improper installation of duct mounted air flow sensors
- Faulty proportioning of some terminal unit air flows
- Connection of a few zone temperature sensors to the wrong terminal units
- Improper setpoints at some terminal units

As a result of the problems identified, the scope was expanded to test more sample boxes. This process was repeated until all parties were comfortable with the performance of the equipment.

The commissioning process for this project also included the electrical systems and the fire alarms and smoke control systems.

Costs and Benefits of Commissioning on this Project

The services and deliverables described in the CA scope of work, as well as the ancillary support and testing work cost $0.58 per square foot. The overall cost picture should include the subcontractor costs associated with commissioning. Those figures were not available at the time of publication.

Some of the individual problems identified and corrected by the commissioning process have already been mentioned. During the construction process, the team initiated weekly commissioning meetings attended by designated representatives from the various contractors, subcontractors, and Boeing departments. These proved to be a valuable auxiliary forum for communication between partnering staff and the affected parties and facilitated the identification and resolution of technical and operational issues in a proactive fashion. The Cx team meetings complemented weekly foreman and owner's meetings and added depth and focus to many areas that are traditionally problematic. The meetings continued until well after the building was occupied and ensured post-construction continuity of design intent and owner satisfaction.

The Boeing project managers contend that:

The benefits of Cx work are easily recognizable to those involved with construction, operating, and maintaining buildings and related systems. That perspective is not always as easy to demonstrate to the business and financial entities within organizations. Placing a monetary value on items such as fewer change orders and contractor call-backs, expending less O&M staff time, and having fewer building occupant complaints, "sick" building scenarios, systems outages, and equipment warranty issued can be difficult. The best case for any owner can usually be made internally, when the total costs of projects performed without Cx are analyzed. This usually requires identifying and isolating costs encountered after the construction project is closed out. This can be a somewhat cumbersome and painful, but beneficial, undertaking for any company. In today's business climate, the value of avoided litigation should also be considered. When the total value of such avoided costs and realized benefits is truly accounted for, Cx is recognized as one of the best bargains in the construction marketplace today.

7.1.3   Commissioning Existing HVAC Systems: Continuous Commissioning

Commissioning of HVAC systems in existing buildings is intended to identify and correct any construction problems that have not been rectified, just as commissioning does in a new building. However, it is also intended to identify and correct other problems that develop during subsequent operation of the building.

Operators and maintenance personnel often increase utility consumption when dealing with an immediate problem. For example, the chilled water temperature might be decreased from 42°F to 39°F if one of the AHUs cannot provide 55°F cold air with 42°F chilled water since the control valve is stuck in a partially closed position. Or the static pressure of the VAV systems is set at a higher level than needed instead of locating the kink in the flex duct that limits flow and cooling in one zone. The efficiency issues associated with these solutions are ignored. During building operation, resolving comfort problems is a top priority. However, inefficient solutions such as those noted above tend to accumulate as time passes, and these solutions often lead to additional comfort problems. It is generally true that an older building has more comfort problems and opportunities to improve HVAC efficiency than a new building.

Commissioning of existing buildings is called by various names including: continuous commissioning (CC), retro-commissioning, and recommissioning. Common practice when commissioning an existing building emphasizes bringing the building operation into compliance with design intent. However, changes occur in most buildings as time passes so the operation of an existing building which is commissioned will generally differ from the original design intent. Some practitioners started using the term "recommissioning" to distinguish between the commissioning of new buildings and existing buildings, but this has met some resistance since it implies that the building was originally commissioned, which is seldom the case. The term "retro-commissioning" is used by many to indicate that commissioning was performed as a "retrofit" to an existing building.

"Continuous commissioning" (CC) is a term applied to the commissioning process developed by a group of researchers at the Energy Systems Laboratory at Texas A&M University. The continuous commissioning process assumes that building use and operation are sufficiently different from the original design intent; therefore, a new optimum operating strategy should be regularly identified and implemented. This process uses advanced approaches to optimize the HVAC system operation to meet the current needs of the building. An additional feature of the CC process is ongoing follow-up after the initial CC activities to maintain and continuously improve the facility operation. The CC process will be the commissioning process for existing buildings which is described in this chapter.

Benefits of Commissioning Existing Buildings

The specific benefits of commissioning existing buildings can be summarized as follows:

1. Identifies and solves system operating, control, and maintenance problems.
2. Provides cost savings that rapidly pay back the commissioning cost.
3. Normally provides a healthier, more comfortable, and productive working environment for occupants.
4. Optimizes the efficiency of the energy-using equipment subject to the comfort requirements of the building.
5. Reduces maintenance costs and premature equipment failure.
6. Provides better building documentation which expedites troubleshooting.
7. Provides training to operating staff, increasing skill levels.
8. Provides the basis for accurate retrofit recommendations to upgrade the facility.

Commissioning of existing buildings is very attractive economically, even if the only benefit considered is energy savings. Gregerson (1997) presented results from commissioning 44 existing buildings that showed simple paybacks which ranged from 0.1 years to 4.2 years, with 28 having a payback of less than one year, 12 between 1.0 and 2.0 years, and only 4 with a payback longer than 2.0 years. The buildings in this study were generally large buildings, with the smallest having 48,000 square feet, and only 12 were less than 100,000 square feet. Energy use in these buildings was reduced by 2% to 49% with an average reduction of 17.5%. The cost of commissioning was quite evenly distributed over the range from $0.03/square feet to $0.43/square feet with 11 buildings less than $0.10/square feet and 9 at more than $0.30/square feet.

Evaluating an Existing Building for Commissioning

The most effective way to evaluate the commissioning potential of an existing building is to conduct a commissioning screening survey of the building. The following characteristics typically indicate a building with high commissioning potential:

1. A significant level of comfort complaints. The systems in buildings that do not produce uniform comfort have generally been adjusted in ways that reduce efficiency in attempts to deal with the comfort complaints.
2. A high level of energy use for the building type. A building which uses more energy than similar buildings with comparable use patterns is generally a prime candidate to benefit from commissioning.
3. Indoor air quality problems. Buildings that experience complaints about indoor air quality often have HVAC systems adjusted in ways that may or may not resolve the IAQ problem, but that compromise effective and efficient operation.
4. Buildings with energy management and control systems (EMCS). The EMCSs installed in buildings are rarely used to the full extent of their capabilities. This may be due to one or more of the following: (a) failure of the operating staff to fully understand the system, (b) failure of the control contractor to adequately understand the HVAC system in the building, and/or (c) poor design specification from the mechanical engineers.

The presence of one or more of these characteristics, coupled with any other known operating problems, is normally good justification for performing a screening study for the commissioning potential of a building. A commissioning screening will generally cost approximately $0.01–$0.03/square foot for medium to large facilities.

The Process of Commissioning Existing Buildings

The process of commissioning an existing building can be viewed in terms of four phases: planning, investigation, implementation, and follow-up phases as shown in Table 7.1.2. The planning phase commissioning activities most closely parallel those during the conceptual or predesign phase for a new building. Some activities during the investigation phase overlap with the construction phase, while others overlap with the acceptance phase for a new building. Implementation phase activities generally parallel some of those in the acceptance phase for a new building, and the hand-off parallels the post-acceptance phase.

Planning Phase

The first step in planning the commissioning of an existing building is to evaluate the need for commissioning. The operating staff may be aware of problems in the building that have never been properly resolved due to time constraints or other factors. There may also be a strong sense that commissioning or tune-up of the building is likely to provide significant benefits. This should generally be followed by a screening visit by one or more experienced providers of commissioning services. This will typically require a few hours to a few days (depending on the size of the facility) to examine the systems and operating practices of a large building, examine selected EMCS settings, and make selective system measurements. Examination of available building documentation and analysis of historical utility data are normally part of the screening visit. After consultation with the facility staff, a commissioning proposal tailored to the needs of the building should be provided. The proposal includes a price, services to be provided, and specific benefits to be expected.

Investigation Phase

The investigation phase should begin with meetings with the facility manager and any members of the facility operating staff who have been assigned to be part of the commissioning team. They will review

TABLE 7.1.2 Process Comparison for Commissioning Existing Buildings and New Buildings

Existing Buildings	New Construction
1. Planning phase (a) Determine need for commissioning. (b) Review available documentation and obtain historical utility data. (c) Conduct commissioning screening study. (d) Hire commissioning provider. (e) Develop commissioning plan.	**1. Conception or predesign phase** (a) Develop commissioning objectives. (b) Hire commissioning provider. (c) Develop design phase commissioning requirements. (d) Choose the design team.
No design phase activities	**2. Design phase** (a) Do a commissioning review of design intent. (b) Write commissioning specifications for bid documents. (c) Award job to contractor. (d) Develop commissioning plan.
2. Investigation phase (a) Obtain and develop missing documentation. (b) Develop and implement M&V plan. (c) Develop and execute diagnostic monitoring and test plans. (d) Develop and execute functional test plans. (e) Analyze results. (f) Develop master list of deficiencies and improvements. (g) Develop optimized operating plan for implementation.	**3. Construction/installation phase** (a) Gather and review documentation. (b) Hold commissioning scoping meeting and finalize plan. (c) Develop pretest checklists. (d) Start up equipment or perform pretest checklists to ensure readiness for functional testing during acceptance.
3. Implementation phase (a) Implement repairs and improvements. (b) Retest and monitor for results. (c) Fine-tune improvements if needed. (d) Determine short-term energy savings.	**4. Acceptance phase** (a) Execute functional test and diagnostics. (b) Fix deficiencies. (c) Retest and monitor as needed. (d) Verify operator training. (e) Review O&M manuals. (f) Have building accepted by owner.
4. Project hand-off/integration phase (a) Prepare and submit final report. (b) Document savings. (c) Provide ongoing services.	**5. Post-acceptance phase** (a) Prepare and submit final report. (b) Perform deferred tests (if needed). (c) Develop recommissioning plan/schedule.

Source: Modified from Haasl and Sharp, 1999.

building operating practices, special client needs, and all known operating problems in the building. It may be necessary at this point to search for or develop additional documentation — for example, obtain manufacturer information specifications for chillers, AHUs, or other equipment. A request may be sent to the utility for 15-minute electrical data if it is recorded, but not routinely provided. A plan should be developed for verification of the results of the commissioning effort and additional instrumentation should be installed if needed. The commissioning provider must determine the diagnostic and functional tests needed, and then execute them with participation and assistance of the building operating staff. These tests will typically consist of some combination of setting up trend logs on the EMCS, a series of spot measurements on the building systems, and/or installation of temporary portable loggers to record critical system parameters for a day or more.

The results of these tests will be analyzed and a list of operating changes, equipment maintenance, and possibly equipment retrofit recommendations will be generated. This list should include the expected comfort improvements and/or operating savings that will result from these changes. The list may also include items that were evaluated, but that do not appear to be cost effective or offer significant comfort and other benefits.

Implementation Phase

The recommendations will be discussed with the operating staff or an owner's representative who will decide which recommendations will be implemented. Implementation may be handled by the building staff, the commissioning provider, or a third party, depending on the skills and preferences of the owner. It is desirable to use EMCS data, or other data collected on an hourly basis, to verify the impact of the changes in the first days following implementation. This often provides near-term positive feedback to the operating staff on the impact of the changes. It can also give rapid feedback to the provider if the changes are not as effective as anticipated, and provide the basis for further fine tuning. The short-term savings determined from monitored data can then be used for comparison with the original savings estimates, and revisions can be made as necessary.

Project Follow-up/Integration Phase

Report: At this point, a final report on the commissioning effort is prepared and delivered to the owner. This report should provide a clear explanation of the optimum operating strategy which has been implemented in a concise format useful for the operators.

Document savings: The savings should be documented with the measured hourly data or utility bills. The savings analysis will consider the impacts of weather variation, usage schedules, and occupancy changes. The savings should be documented as soon as possible after the procedures are implemented. Monthly or quarterly reports are desirable.

Provide ongoing services: After completing the initial commissioning process, the commission engineers should provide assistance whenever the building operating staff needs it. This assistance is often needed when there is a change in occupancy, equipment, or schedule. It is a good "rule of thumb" for the operating staff to seek input from the commissioning engineers any time they are ready to revert to earlier practices to resolve an occupant complaint or component malfunction.

7.1.4 Continuous Commissioning Guidelines

Continuous commissioning (CC) guidelines should define the objectives of the commissioning process and provide procedures a checklist to follow, and documentation requirements. The commissioning team should follow the guideline to provide quality services. An abbreviated set of example guidelines is provided next using air handlers as an example.

Sample CC Guidelines for AHU Commissioning

These guidelines include the following sections: objectives, common AHU problems, AHU information requirements, CC procedures, and CC documentation.

Objectives

Optimize the deck and static pressure reset schedules to maintain room comfort conditions; improve electrical and mechanical equipment operation; minimize the fan power, chilled water, and hot water consumption.

Common AHU Problems

1. Inefficient deck and static pressure reset schedules
2. Inability to maintain room comfort (temperature and/or RH)
3. VFD and valve hunting
4. Low differential temperature across the coils
5. Inability to maintain the deck setpoints
6. Too much cold and hot air leakage through dampers in the terminal boxes

AHU Information Requirements

1. Sketch a single line diagram for each AHU (fill in standard forms)
2. Fan: hp, type (VFD, inlet guide vane, eddy switch, or other)
3. VFD: type, hp, brand, working condition (% speed, hunting)
4. Automatic valve description: type (normally open or closed), range (3–8 psi or 0–13 psi), working condition, and control (by EMCS or stand alone controller, DDC or pneumatic)

5. Coil data: inlet and outlet temperature (design and measured) and differential pressure
6. Damper data: working condition (adjustable or not), actuator condition
7. Temperature sensors: EMCS readings and hand meter readings
8. Controller condition: working or disabled
9. Air flow: note setting for outside air flow, return air flow, maximum total flow, and minimum total flow
10. Condition of system air flows: measure temperature and CO_2 level for outside air, return air, and supply air
11. Control sequence: determine cold and hot deck setpoints, economizer control sequence, and static pressure control sequence

CC Procedures

Step 1: Commission temperature and pressure sensors.

Use a hand meter to verify accuracy of discharge air temperature sensors and differential pressure sensors. Make sure the readings from the EMCS or the control system agree with the field measurements. If the control system readings do not agree with the hand meter readings, repair or replacement should be performed. If a systematic bias exists, a software correction may be used but is not recommended.

Step 2: Determine the optimal static pressure for a VAV system.

Modulate the variable flow device, such as the VFD, eddy switch, or the inlet guide vane, to maintain the minimum static pressure level at preselected terminal boxes. Record the static pressure in the main duct as read by the control system. This pressure should be the setpoint for the current load condition.

Step 3: Test the optimal static pressure setpoint.

If the optimal setpoint is very different from the existing setpoint, reset the static pressure to the optimal level and wait for a while to see if any problems occur. If comfort problems occur in another area, correct the problem at the local level.

Step 4: Determine the cold and hot deck setpoints under the current conditions.

Field testing method: the optimal cold and hot deck setpoints can be determined by following an engineering procedure developed by the ESL.

Analytical method: both optimal hot and cold deck reset schedules can be determined by model simulation using AirModel.

Step 5: Determine the cold and hot deck reset schedules.

Step 6: Determine the outside air intake. Measure outside air, return air, and total supply air flow rates. Measure return air CO_2 levels. If the outside air intake is lower than the design value and the return air CO_2 level is lower than 800 ppm when the building is occupied, no minimum outside air increase is suggested. However, a spot check is suggested for the CO_2 levels in individual rooms. If the outside air intake is higher than the design value and the CO_2 levels are lower than 800 ppm, the minimum outside air flow should be adjusted based on the current standard. Make sure that the CO_2 level in the return air is not higher than 800 ppm when the building is occupied. Inspect damper actuators.

Step 7: Select a control sequence. Locate each sensor position and draw a schematic diagram. Draw a block diagram of the AHUs and control systems. Select a control sequence. This step is strongly system dependent. Commissioning engineers should be able to perform the task independently. Summarize the current control sequence and the proposed control sequence.

Step 8: Implement the optimal reset schedule. Inspect valve and VFD operation and trend data with a time interval of 10 sec. If any valve or VFD is hunting, PID fine tuning should be performed first. Change the control program and trending control parameters. Compare the setpoint and the measured data. If there are any problems, troubleshooting should be performed immediately.

CC Documentation

In addition to the physical characteristics and operational parameters noted in the Documentation Guidelines, the following information must be recorded when an AHU is commissioned:

1. Pre-CC and post-CC reset schedules
2. Repair list
3. Suggestions
4. Operational procedures

CC Guidelines for Water Loop Commissioning

These guidelines include the following sections: objectives, common waterside problems, water loop information requirements, CC procedures, and CC documentation.

Objectives

Identify optimal pump operating points or control schedules to (a) supply adequate water to each coil, (b) minimize pump energy consumption, and (c) maintain optimal differential temperature.

Common Waterside Problems

1. Coexistence of over-flow and under-flow
2. Low differential temperature for the whole building loop
3. Lack of flow in some coils
4. Poor automatic valve control performance due to high pressure
5. Over-pressurization of building loop
6. VFD hunting

Water Loop Information Requirements

1. Water loop riser diagram: differential pressure sensor position, temperature sensor position, automatic valve position, building bypass, coil bypass
2. Pump: single line diagram of pump and pipe line connections, hp, VFD, differential pressure across pump
3. VFD: operating conditions (working, manual, bypassed, damaged), % of speed or Hz, control logic
4. Automatic valves: operating condition (working, bypassed), type (normally open or closed), operating range, location, function, and position
5. Control: loop control logic, differential pressure reset schedule, return temperature reset schedule, automatic control valve control schedule
6. Water conditions: building supply and return temperature, coil supply and return temperature, differential pressure across building and each coil

CC Procedures

Step 1: Valve commissioning. Connect all valves to the control system or controllers. Troubleshoot malfunctioning controllers or control system. Fine tune PID gains to eliminate hunting.

Step 2: Verify valve working conditions. Measure air discharge temperature. If the discharge air temperature setpoint is maintained, the valves are working. If the discharge air temperature cannot be maintained in more than half of the coils, repeat step 1.

Step 3: Reset balance valves to adjust differential pressure to a correct level.

Step 4: Determine the minimum differential pressure under current conditions.

Step 5: Determine the reset schedules. Measure the building water return and supply temperatures and flow rate under the optimal differential pressure. The building energy consumption can be determined from the measured data. Determine the maximum load on the building and the differential temperature under the maximum load condition.

If a VFD is installed, the maximum differential pressure is then determined by the following formula:

$$\Delta P_{max} = \Delta P_{current}\left(\frac{Q_{max} \times \Delta T_{current}}{Q_{current} \times \Delta T_{max}}\right)^2$$

The ΔP_{max} is the maximum differential pressure setpoint at the maximum load, Q_{max} is the maximum load, ΔT_{max} is the differential temperature under the maximum load conditions, and $\Delta T_{current}$ is the differential temperature under the current load conditions. When there is no VFD in the loop, determine the minimum differential pressure; it is the same as the maximum pressure if a building bypass is used.

If there is no building bypass and VFD, the maximum differential pressure can be determined by the above equation.

Step 6: Implement the reset schedule. When a VFD exists, correlate the differential pressure with the outside air temperature. A linear equation is suggested. Program it into the controller. When neither VFD nor building bypass exist, modulate the balance valve in the main line to maintain the differential pressure at the maximum value. The impact of the main loop pressure variation can be considered by adding a possible drop to the maximum setpoint.

When a building bypass exists without VFD, the minimum differential pressure should be maintained. Note that the main loop impact on the building loop can be considered by adding a possible drop to the minimum setpoint.

CC Documentation

The following information should be documented:

1. Pre-CC control and post-CC control sequence
2. Valve and VFD performance
3. Energy performance
4. Operation procedures
5. Problem and repair lists
6. Other suggestions

CC Case Study — Texas Capitol Extension Building

The Texas Capitol Extension Building was built in 1992 as an energy-efficient building intended to surpass the performance of other buildings in the complex. It is located next to the State Capitol and is entirely below grade to preserve open space around the Capitol. The two upper floors are built around a covered atrium and House legislative offices and hearing rooms. Two lower floors are a parking garage. Total floor area is 55,100 m², while the conditioned area (the two upper floors) is 33,500 m².

The building receives both chilled water and steam from a central plant. Three secondary chilled water pumps (50 hp each) are used to circulate chilled water in the building. Heat exchangers are used to convert steam to hot water. Three hot water pumps circulate hot water to provide heating for the building.

Twenty-one dual duct VAV systems (DDVAV) are used to condition the office area. Eight single duct VAV systems (SDVAV) are used to condition 16 hearing rooms. Twelve single duct constant volume systems (SDCV) serve the central court area, one auditorium, and a pump room. Five constant volume units serve the kitchen and dining area. Outside air is pretreated by four variable volume units (OAHU-VFD) and supplied to each mechanical room. Four supply fans and four exhaust fans serve the two-story parking garage. A total of fifty AHUs and eight fans serve the building.

The modern DDC energy management and control system (EMCS) is used to control the operation of HVAC systems. When commissioning was begun, it was found that the EMCS was being used to implement:

1. Hot water supply temperature reset
2. Chilled water and hot water pump lead-lag sequence control
3. Static pressure control for AHUs
4. Cold deck reset for SDVAV
5. Cold deck reset and hot deck control for DDVAV systems

The measured energy consumption before CC was 8,798,275 kWh/yr ($306,444) for electricity; 54,007 MMBtu/yr ($175,524) for chilled water; and 14,931 MMBtu/yr ($57,340) for hot water. The energy cost index was $1.5/ft^2/yr based on conditioned floor area or $0.91/ft^2/yr based on gross area. The building was operated 24 hours a day and seven days a week.

The building was controlled at a satisfactory level except that the hearing rooms needed to be over-cooled before a meeting. Discomfort occurred when an unexpected meeting was scheduled at the last minute, leaving no time for the operating staff to react. To solve this problem, the room temperature was kept at 19°C to 21°C during unoccupied hours. However, when the hearing room was packed with people, the AHU could not cool the room satisfactorily. The problem persisted despite repeated attempts to deal with it.

CC Measures Implemented

The CC effort led to implementation of the following changes in the operation of this relatively new and efficient building:

1. *Set back VFD static pressure.* To maintain comfort conditions while minimizing energy consumption, the static pressure and minimum VFD speed were reduced to about half their normal values during the nominally unoccupied hours.
2. *Change control to maintain hearing room comfort.* Hearing rooms were being maintained at 66°F to 69°F during unoccupied hours, and the operators frequently changed the room temperature setpoint in an attempt to maintain room comfort conditions. Even so, room temperatures sometimes reached uncomfortably high levels when the rooms were packed with people. This was determined to be the result of inadequate cooling energy to the hearing rooms.

   After a detailed analysis, it was proposed that cooling energy to the room should be increased under full load conditions, while using the terminal box damper position to reset the static pressure and the cold deck temperature to maintain comfort and reduce energy use at part-load levels. The post-commissioning schedules provide more cooling energy to a room than the old schedules under maximum load conditions; they lower the static pressure and increase the cold deck temperature to reduce energy consumption as soon as the load decreases.

   The control schedule was first tested in one AHU and then implemented in all 8 SDVAVs. After these schedules were implemented, the complaints disappeared and room temperature was maintained in a range of 21°C to 22°C.
3. *Optimize the dual duct VAV system reset schedules.* Twenty-one dual duct VFD AHUs with VAV boxes condition all the offices which comprise 60% of the conditioned area in the building. The hot deck temperature set point was originally 27°C year-round. The cold deck temperature setpoint varied from 13°C to 18°C using a standard algorithm from the control company.

   After a field inspection, it was proposed that both cold and hot deck temperatures be reset based on the highest and lowest supply air temperatures required at the time in any zone. These schedules set the cold deck temperature to 13°C if the minimum supply temperature (T_{supmin}) needed by any zone is less than or equal to 13°C, to T_{supmin} if between 13°C and 18°C, and hold it at 18°C when T_{supmin} goes above 18°C. A similar schedule is used for the hot deck. The proposed schedules were tested first in one AHU, then were implemented in all 21 AHUs within a month.

   Since implementing these schedules, the hot deck temperatures have varied from 21°C to 24°C. There has been almost no heating consumption.

4. *Separate hot water control loops and reset supply temperature.* The Capitol building (CPB) and the Capitol extension (CPX) building have used the same hot water supply temperature control loop since the CPX Building was built. The CPB is an above-ground building with a lot of exterior surface, and the CPX is an underground building with very little surface exposed to ambient conditions. The outside air conditions play an important role in the CPB heating load but have little influence on loads in the CPX building. If hot water supply temperature satisfies the CPB, the consumption of steam for the CPX will increase when outside air temperature decreases.

   In order to satisfy the requirements of the CPB without increasing steam consumption for the CPX building, the following recommendations were implemented:
   a. Provide separate control of hot water supply temperature for the CPB and CPX.
   b. Lower hot water supply temperature from the range 27°C–38°C to the range 27°C–32°C.

5. *Shut down the AHUs that serve the central court area at night.* Eight single duct constant volume AHUs (SDCV) serve the central court area which is about 10% of the conditioned area of the building. Very few occupants are in the central court area at night, especially when the legislature is not in session. The eight SDCVs were shut down at night during nonsession periods. It was found that the space temperature in the central court area increased by about 2°F. The cooling and electricity consumption were reduced at night as expected.

6. *Implement delta-T control for the chilled water loop bypass valve.* Three constant volume chilled water pumps supply the chilled water to the AHUs. The control sequence keeps only one pump on-line most of the time. There is one bypass line with a bypass valve in parallel with the pumps. The chilled water flow typically ranged from 900 gpm to about 1150 gpm with a 2°C to 5.5°C building temperature differential. This caused some chilled water leakage through the coil valves and sometimes resulted in loss of control. ΔT control with a ΔT of 6.7°C was implemented instead of ΔP control. The chilled water flow was reduced to the range of 750 gpm to 1000 gpm. ΔT was maintained between 6.1°C and 6.7°C.

7. *Shut off steam during the summer.* Steam was provided from the central plant continuously. On June 25, 1996, the steam to the heat exchanger was shut off. The measured hot water consumption data showed that this measure reduced hot water consumption by up to 1 MMBtu/hr on days when the daily average temperature was above 24°C.

8. *Optimize outside air intake.* Four AHUs equipped with VFDs supply about 0.2 CFM/square foot of outside air to the building following the design specifications. The CO_2 levels in the building were measured in several rooms and ranged from 400 ppm to 550 ppm. This indicated that more outside air was being supplied to the building during nonsession periods than necessary, with a corresponding energy cost. On July 1, 1996, the speeds of the four AHUs were reduced by about 50% for nonsession periods. The CO_2 level generally increased to 550 ppm to 750 ppm with an average of 650 ppm. The maximum CO_2 level was found in a fully occupied hearing room where it was 950 ppm. The reduced outside air flows again reduced the electricity, the cooling, and the heating consumption.

Project Duration, Cost, and Savings

The project started in July 1995, and initial commissioning ended on July 1, 1996. During this 12-month period, the commissioning engineer's effort spanned over 5 months, and the measured savings were about $100,000.

The building had meters installed to measure hourly whole building electricity, chilled water consumption, and hot water consumption. The initial cost of these meter installations was approximately $15,000. The costs of the metering and savings analysis were about $8,000 for two years. The costs were paid back before the completion of the project.

Figure 7.1.3 compares the measured chilled water and hot water consumption for both the pre-CC and the post-CC period. The implementation of CC measures has significantly reduced the heating and cooling energy consumption.

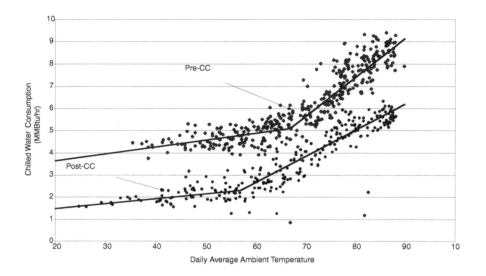

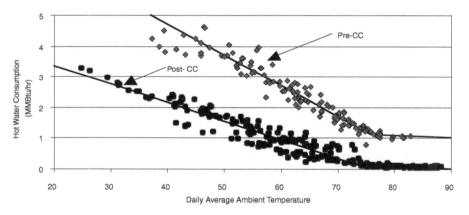

FIGURE 7.1.3   Comparison of chilled water and hot water consumption for the pre-CC and post-CC period.

The measured savings reached $102,700 during the first 8.5 months following completion of the commissioning. This corresponds to an annualized savings rate of 27%, or $144,700 relative to 1994.

7.1.5   Monitoring, Verification, and Commissioning

The need to analyze the energy performance of commercial buildings, to measure savings from energy efficiency retrofits, and to provide information for use in commissioning activities has increased dramatically in recent years. Energy Service Companies (ESCOs) are providing capital retrofits to save energy, with billions of dollars in contract volume. Many of these contracts guarantee a certain level of operating savings, with provision for rebates or penalties for savings not realized. This makes the determination of the savings resulting from these projects a very serious concern for ESCOs and building owners alike. The U.S. Department of Energy (DOE) began developing interim protocols in 1995 which led to the *North American Energy Measurement and Verification Protocol*, and the *International Performance Measurement and Verification Protocol* for measuring savings in contracts between ESCOs and building owners. ASHRAE is currently developing Guideline 14 for this purpose.

Importance of M&V for Commissioning

Monitoring and Verification (M&V) of energy savings for energy efficiency retrofits is growing rapidly. The major impetus for growth has been the tremendous increase in volume of energy service company business where large financial payments hinge on the determination of energy savings in specific buildings. However, a major factor in the willingness of many building owners to commission, particularly in existing buildings, is the expectation that the project will produce energy savings that will at least pay for the cost of the commissioning. A plot clearly showing that cooling costs for a building dropped by 30% following a commissioning project can quickly convince an owner of the value of the project. Likewise, plots which show a facility operator that changing a cold deck setpoint resulted in savings of $10/hour (nearly $90,000 per year in a continuously operated facility), can enlist his enthusiastic support for a commissioning program.

The second major application of measured energy use data is its use as a tool for diagnosis of building operating problems. Both of these applications are described below.

M&V Methods

What is needed to determine savings for a commissioning project? Early energy savings projects were typically evaluated by simple comparison of utility bills before and after measures were implemented. This works fine when the savings from the measures implemented result in obvious large savings. However, savings from commissioning efforts are probably most often in the range of 5%–20%, and at this level, there are many factors that can obscure the savings. This requires employing more sophisticated M&V methods that can normalize for changes in occupancy, schedule, and weather. Over the last five years, two major efforts have been implemented to develop standard methods for savings determination. The DOE initiated an effort which has since involved dozens of domestic and international organizations and resulted in publication of the *International Performance Measurement and Verification Protocol* (IPMVP, 1997). ASHRAE is currently developing a guideline for savings measurement through the ASHRAE consensus process (not completed at press time).

The process for determining savings as adopted in the IPMVP defines:

$$Energy\ Savings = Baseline\ Energy\ Use - Post\text{-}Installation\ Energy\ Use$$

where the *baseline energy use* is determined from a model of the building operation before the retrofit (or commissioning) which uses post-installation operating conditions (e.g., weather, occupancy, etc.). *Post-installation energy use* is simply the measured energy use, but it may be determined from a model, though we would seldom recommend this approach.

The IPMVP includes four different M&V techniques or options. These options may be summarized as Option A: stipulated savings, Option B: measurement at the system or device level, Option C: measurement at the whole building or facility level, and Option D: determination from calibrated simulation. Each option is described in more detail next.

Option A

This option focuses on a physical determination of equipment changes to ensure that the installation meets contract specifications. It determines savings by measuring the capacity or the efficiency of a system before and after retrofit, and then multiplies the difference by an agreed upon or "stipulated" factor such as the hours of operation, or the load on the system. Key performance factors (e.g., lighting wattage) are determined with spot or short-term measurements and operational factors (e.g., lighting operating hours) are stipulated based on historical data or spot measurement. Performance factors are measured or checked yearly. The accuracy of this method is generally inversely proportional to the complexity of the measures being evaluated. As such, it may be quite suitable for lighting retrofits, or replacement of motors operated at constant load with high efficiency motors. However, it is not suitable for the more complex changes typically implemented in the process of commissioning an existing building or applying the continuous commissioning process.

Option B

This option normally determines savings by continuous measurements taken throughout the project term at the device or system level. Individual loads or end uses are monitored continuously to determine performance and the long-term persistence of the measures installed. The data collected can be used to improve or optimize the system operation, and as such is particularly valuable for continuous commissioning projects. This option includes procedures for verifying that the proper equipment or systems were installed and that proper operating procedures have been implemented. Since measurements are taken throughout the project term, the savings determination is normally more accurate than with Option A, but cost is higher.

Option C

Option C determines savings by analysis of "whole building" or facility level data measured during the baseline period and the post-installation period. This option is required when it is desired to measure interaction effects, for example, the impact of a lighting retrofit on the cooling consumption as well as savings in lighting energy. The data used may be utility data, or sub-metered data, and may be recorded at monthly or shorter intervals.

Option C requires that installation of the proper systems/equipment and proper operating practices are confirmed. It determines savings from metered data taken throughout the project term. The major limitation in the use of Option C for savings determination is that the size of the savings must be larger than the error in the baseline model. The major challenge is accounting for changes other than those associated with the ECMs or commissioning changes implemented.

The following points should be carefully considered when using Option C, especially when using utility billing data. Many of these points are applicable to Option B as well.

1. All explanatory variables that affect energy consumption as well as possible interactive terms (i.e., combination of variables) need to be specified, whether or not they are accounted for in the model. Critical variables can include weather, occupancy patterns, setpoints, and operating schedules.
2. Independent variable data need to correspond to the time periods of the billing meter reading dates and intervals.
3. If the energy savings model discussed above incorporates weather in the form of heating degree-days and cooling degree-days, the following issues should be considered:
   (a) Use of the building "temperature balance point" for defining degree-days vs. an arbitrary degree-day temperature base.
   (b) The relationship between temperature and energy use tends to vary depending upon the time of year. For example, an ambient temperature of 55°F in January has a different implication for energy usage than the same temperature in August. Thus, season should be addressed in the model.
   (c) The nonlinear response to weather. For example, a 10°F change in temperature results in a very different energy use impact if that change is from 75°F to 85°F, rather than 35°F to 45°F.
   (d) Matching degree-day data with billing start and end dates.
4. The criteria used for identifying and eliminating outliers need to be documented. Outliers are data beyond the expected range of values (or 2–3 standard deviation away from the average of the data). Outliers should be defined using common sense as well as common statistical practice.
5. Statistical validity of the final regression model needs to be demonstrated. Validation checks make sure:
   (a) The model makes intuitive sense, e.g., the explanatory variables are reasonable and the coefficients have the expected sign (positive or negative) and are within an expected range (magnitude).
   (b) Modeled data is representative of the population.
   (c) Model form conforms to standard statistical practice.

(d) The number of coefficients is appropriate for the number of observations (approximately no more than one explanatory variable for every five data observations).

(e) All model data is thoroughly documented, and model limits (range of independent variables for which the model is valid) are specified. (IPMVP, 1997.)

Accurate determination of savings using Option C normally requires 12 months of continuous data before a retrofit and continuous data after a retrofit. However, for commissioning applications, a shorter period of data during which daily average ambient conditions cover a large fraction of normal yearly variation is generally adequate.

Option D

Savings are determined through simulation of the facility components and/or the whole facility. The most powerful application of this approach calibrates a simulation model to baseline consumption data. For commissioning applications, it is recommended that calibration be to daily or hourly data. This type of calibration may be done most rapidly if simulated data is compared to measured data as a function of ambient temperature.

Just as for the other options, the implementation of proper operating practices should be confirmed. It is particularly important that personnel experienced in the use of the particular simulation tool conduct the analysis. The simulation analysis needs to be well documented, with electronic and hard copies of the input and output decks preserved.

Measurement Channels

The minimum number of measurement channels recommended for evaluation of a commissioning project will be the number needed to separate heating, cooling, and other electric uses. The actual number of channels will vary, depending on whether pulses are taken from utility meters, or if two or three current transformers are installed to measure the three-phase power going into a chiller. Other channels may be added, depending on the specific measures being evaluated.

Use of M&V Data for Diagnostics

Most whole-building diagnostic procedures can be split into two major categories: examination of time series data, and use of physical or empirical models in the analysis of whole-building data streams.

Diagnostics with Time Series Data — the simplest form of diagnostics with whole-building data is manual or automated examination of the data to determine whether prescribed operational schedules are followed. The normal minimum set of whole-building data required for diagnostics are separate channels for heating, cooling, and other electrical uses. With these data streams, it is possible to identify probable opportunities for HVAC system shut-offs, excessive lighting operation, etc.

Shut-Off Opportunities — this is often the most intuitive of all diagnostic procedures. However, the use of whole-building data, even with heating and cooling removed, can cause some confusion since nighttime electric use in many buildings is 30%–70% of daytime use. If nighttime and weekend use seems high, then the connected load must be investigated to determine whether observed consumption patterns correspond to reasonable operating practices. Our experience indicates that while many, if not most, opportunities for equipment shut-off by an EMCS or other system-level action may have been already implemented, time series data analysis can still find opportunities in 10%–20% of buildings.

While these opportunities can be observed using plots that show several days of hourly data, it is often helpful to superimpose several days or weeks of hourly data on a single 24-hour plot to observe typical operating hours and the frequency of variations from the typical schedule.

Operating Anomalies — a slightly different category of opportunities can be identified using the same techniques. Mistakes in implementing changes in thermostat setup/setback schedules sometimes result in short-time simultaneous heating and cooling which show up as large spikes in consumption lasting only an hour or two. Time series plots of motor control centers often show that VAV systems seldom operate above their minimum box settings — and hence are essentially operating CAV systems. Comparisons between typical weather-independent operating profiles from one year to the next will often reveal "creep" in consumption which is often due to addition of new computers or other office equipment.

Blink Tests — a valuable way in which whole-building data can be used to identify the size of various equipment loads such as switchable connected lighting load, AHU consumption, etc., is the use of a short-term "blink test" such as that described in the example of the state office building discussed earlier.

Diagnostics with Models and Data

The description of the process used to diagnose opportunities for improved operation at the BSB building made heavy use of a physical simulation model. Calibration of simulation models has been regarded as time consuming making it appropriate only for research projects. However, this approach has been systematized by the authors using a series of energy "signatures" which have enabled the performance of calibrated simulation as a classroom assignment. Signatures have been developed for constant volume dual-duct AHUs, dual-duct VAV systems, single-duct CAV systems, and single-duct VAV systems.

These model-based approaches can readily be used in conjunction with limited field measurements to diagnose and determine the potential savings from correcting a large variety of systems problems which include:

- VAV behavior as CAV systems
- Simultaneous heating and cooling
- Excess supply air
- Excess OA
- Sub-optimal cold deck schedule
- Sub-optimal hot deck schedule
- High duct static pressure
- Others

Conclusions

Whole-building data for heating, cooling, and non-weather dependent electricity consumption can be used to identify a range of shut-off opportunities, scheduling changes, and operating anomalies due to improper control settings and other factors. It can also be used in conjunction with appropriate simulation tools and energy signatures to identify an entire range of nonoptimum system operating parameters. It is then very straightforward to reliably predict the energy savings which will be realized from correcting these problems.

It should be recognized that the systems diagnostics available from whole-building data and modeling are indications of probable cause. Additional field measurements are generally needed to confirm the probable cause.

References

ASHRAE, 1996, *ASHRAE Guideline 1-1996: The HVAC Commissioning Process*, American Society of Heating, Refrigerating and Air-Conditioning Engineers, Inc., Atlanta, GA.

Claridge, D.E., Liu, M., Turner, W.D., Zhu, Y., Abbas, M., and Haberl, J.S., Energy and Comfort Benefits of Continuous Commissioning in Buildings, *Proceedings of the International Conference Improving Electricity Efficiency in Commercial Buildings*, September 21–23, 1998, Amsterdam, The Netherlands, pp. 12.5.1–12.5.17.

Davenny, M., Doering, D., and McGuire, T., Case Study: Commissioning the Boeing Commercial Airplane Group Headquarters Office Building, *Proceedings of the 7th National Conference on Building Commissioning*, May 3–5, 1999, Portland, OR.

Haasl, T. and Sharp, T., *A Practical Guide for Commissioning Existing Buildings*, Portland Energy Conservation, Inc., and Oak Ridge National Laboratory for U.S. DOE, ORNL/TM-1999/34, 1999.

Haasl, T. and Wilkinson, R., Using Building Commissioning to Improve Performance in State Buildings, *Proceedings of the 11th Symposium on Improving Building Systems in Hot and Humid Climates*, June 1–2, 1998, Fort Worth, TX, pp. 166–175.

IPMVP, 1997, *International Performance Measurement and Verification Protocol*, U.S. Department of Energy, DOE/EE-0157.

Odom, J.D. and Parsons, S., The Evolution of Building Commissioning at Walt Disney World, 6th National Conference on Building Commissioning, Lake Buena Vista, FL, May 18–20, 1998.

PECI, 1999, *National Strategy for Building Commissioning*, PECI, Inc., Portland, OR.

7.2 Building Systems Diagnostics and Predictive Maintenance

Srinivas Katipamula, Robert G. Pratt, and James Braun

There has been an increasing interest in the development of methods and tools for automated fault detection and diagnostics (FDD) of building systems and components in the 1990s. This chapter will describe the status of these methods and methodologies as applied to heating, ventilation, air conditioning, and refrigeration (HVAC&R) and building systems and present illustrative case studies.

Building Systems Diagnostics

Operation problems associated with degraded equipment, failed sensors, improper installation, poor maintenance, and improperly implemented controls plague many commercial buildings. Today, most problems with building systems are detected as a result of occupant complaints or alarms provided by building automation systems (BASs). Building operators often respond by checking space temperatures or adjusting thermostat settings or other setpoints. The root cause of an operation problem often goes undiagnosed, so problems recur, and the operator responds again by making an adjustment. When the operator diagnoses problems more carefully by inspecting equipment, controls, or control algorithms, the process is time consuming and often based on rudimentary or incorrect physical reasoning and rules of thumb built on personal experience. Often a properly operating automatic control is overridden or turned off, when it *appears* to be the cause of a problem. Moreover, some "latent" problems do not manifest themselves in conditions that directly affect occupants in obvious ways and, as a result, go undetected — such as simultaneous heating and cooling. These undetected problems may affect energy costs or indoor air quality.

Operating problems lead to inefficient operation (energy costs), a loss in cooling/heating capacity (comfort), discomfort (loss of productivity and loss of tenants), and increased wear of components (reliability). However, too much maintenance leads to excessive maintenance costs. Automated diagnostics for building systems and equipment promise to help remedy these problems and improve building operation by automatically and continuously detecting performance problems and maintenance requirements and bringing them to the attention of building operators. In addition, early diagnosis of equipment problems using remote monitoring techniques can reduce the costs associated with repairs by improving scheduling and reducing on-site labor time. Furthermore, as performance contracting for services becomes more prevalent, the need for tools that ensure performance will increase.

Automation and data visualization are key elements of FDD systems. Because the building industry is cost sensitive and lacks a sufficient number of well-trained building operators, fully automated tools can help alleviate the problem. Data visualization is the key link between the building system and building operators in fully automated systems. Clear data presentation will help the building operator avoid scanning, sorting, and interpreting raw data, thus performing metrics, allowing time for correcting the problems identified by the FDD system, and improving equipment performance and efficiency.

Predictive Maintenance

Many buildings today use sophisticated BASs to manage a wide and varied range of building systems. Although the capabilities of BASs have increased, many buildings still are improperly operated and maintained. Lack of or improper commissioning, the inability of the building operators to grasp the complex controls, and lack of proper maintenance are some of the reasons for improper operations. A study of 60 commercial buildings found that more than half of them suffered from control problems. In addition, 40% had problems with the HVAC&R equipment, and a third had sensors that were not operating properly (PECI, 1997).

Effective maintenance extends equipment life, maintains comfort, improves equipment availability, and results in fewer complaints from building occupants; whereas, poorly maintained equipment will have a shorter life and will experience more frequent equipment failure, leading to low levels of equipment availability and occupant dissatisfaction. If regularly scheduled maintenance practices are adopted, they can be expensive. However, if there were a way to decide whether maintenance is required for a particular piece of equipment, it would certainly cut down on the cost of maintenance. The art of predicting when building systems need maintenance is generally referred to as predictive maintenance or condition-based maintenance.

There are many similarities between the FDD and the predictive maintenance methods because both require monitoring of building systems to detect abnormal conditions; therefore, a significant portion of this chapter is devoted to building systems diagnostics.

In the following section, we define the scope for the entire chapter, provide definitions of terms used, and present a generic overview of an FDD system.

7.2.1 Objectives and Scope

The primary objective of this chapter is to provide the HVAC&R engineer and researcher with a fundamental knowledge of (a) the methods and methodologies used in the detection and diagnosis of faults in building systems and components, and (b) predictive maintenance. The chapter contains

- A description of a generic FDD system
- The benefits of automated FDD and predictive maintenance applications
- Results of a detailed review of the literature to identify the methods and the methodologies used
- A discussion on cost vs. benefits, and how to select methods for FDD
- Detailed description of the FDD application on a few building systems
- A brief description of the FDD tools that are currently being used in the field, and application of the automated FDD methods to continuous commissioning of building systems
- Infrastructure requirements to deploy the automated FDD tools in the field
- The future of building systems diagnostics

Definition of Terms

Until recently most of the research and development in the areas of FDD have been limited to nuclear power plants, aircraft, process plants, and the automobile industry. A survey of the FDD literature indicates a lack of consistent terminology. Issermann and Ballé (1997) provide a set of definitions, used in this chapter with minor modifications, as given below.

Cause: A primary reason or explanation of the current fault or problem in the system.
Error: A deviation between a measured or computed value (of an output variable) and the true (actual) specified or theoretically correct value.
Disturbance: An uncontrolled (and possibly sometimes unknown) input acting on a system.
Failure: A permanent interruption of a system's ability to perform a required function under specified operating conditions.

Fault or problem: A deviation of at least one characteristic property or parameter of the system from the acceptable, usual, and/or standard state or condition.

Malfunction: An intermittent irregularity in a system's ability to perform a desired function.

Perturbation: Input acting on a system, which results in a temporary departure from the current state.

Residual or error: A fault indicator, based on a deviation between measurements and model- or equation-based computations.

Symptom: A deviation of an observable quantity from normal behavior.

Fault detection: Detection and time of detection of a fault or faults in the system.

Fault diagnosis: Determination of the kind, magnitude (size), location, time variant behavior, and time of detection of a fault. Follows fault detection. Includes fault isolation and identification.

Fault isolation: Determination of the kind, location, and time of detection of a fault. This usually follows fault detection.

Fault identification (evaluation): Determination of the magnitude (size) and time-variant behavior of a fault. Follows fault isolation.

Monitoring: A continuous real-time task of determining the conditions of a physical system by recording information, and recognizing and indicating anomalies in the behavior.

Supervision: Monitoring a physical system and taking appropriate actions to refine diagnoses and maintain the operation in case of faults.

Protective or proactive control: Means by which a potentially dangerous behavior of the system is suppressed or the consequences of a dangerous behavior are avoided or mitigated.

Commissioning: A systematic process by which proper installation and operation of building systems and equipment are checked and adjusted when necessary to improve performance.

Analytical redundancy: Analytical redundancy implies that values computed analytically can be compared with measured sensors, in contrast to physical redundancy where measurements from multiple sensors are compared to each other.

7.2.2 An Introduction to the FDD Process

There are several different ways to represent an FDD process depending on the methods used and the intended application. In this section, a generic FDD process that can be applied to many building systems is described. A similar process has been used widely in both critical and noncritical systems (Issermann, 1984). There are many similarities between the FDD system and a predictive maintenance system. In the next section, these similarities will be identified. The primary objective of an FDD system is early detection of the fault and diagnosis of the causes before the entire system fails. It is accomplished by continuously monitoring the operations of a system or process to detect, diagnose, evaluate, and respond to the faults arising from abnormal conditions. Therefore, a typical FDD system can be viewed as having four distinct functional processes, as shown in Figure 7.2.1. The first step in the FDD process is to monitor the building systems or subsystems and detect any abnormal (problem) conditions. This step is generally referred to as the fault detection phase. If an abnormal condition is detected, then the fault diagnosis process evaluates the fault and diagnoses the cause of the abnormal condition. Following diagnosis, fault evaluation assesses the size of the impact (energy and/or cost or availability of the plant) on system performance. Finally, a decision is made on how to react to the fault. In most cases, detection of faults is relatively easier than diagnosing the cause of the fault, or evaluating the size or impact arising from the fault.

Fault Detection

In the fault detection stage, the building system or component is continuously monitored and abnormal conditions are detected. There are several methods by which faults can be detected including comparison of the raw outputs that are directly measured from the components, or estimated characteristic quantities based on the available measurements with the expected values. A fault is indicated if the comparison residual (difference between actual value and expected value) exceeds a predefined threshold. The char-

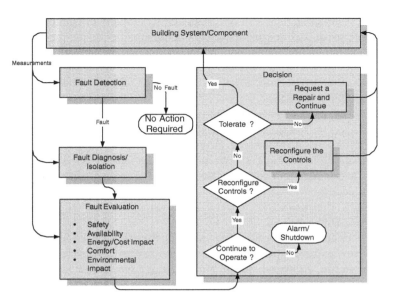

FIGURE 7.2.1   Generic fault detection and diagnostic system with proactive control/diagnostic test capability.

acteristic quantities are features that cannot be directly measured but can be computed from other measured quantities, for example, the outdoor-air fraction for the air-handling unit (AHU) or the coefficient of performance of an air conditioner. In addition to using the raw measured data and characteristic quantities, detailed mathematical models are also widely used in estimating the expected values (Gertler, 1988; Issermann, 1984) for comparison with the measured values accounting for data uncertainties.

In most cases, a model of some kind is essential to detect a fault because most building systems are dynamic in nature. For example, a characteristic quantity such as efficiency can be used to detect a fault in an air conditioner. In the absence of a model, the efficiency calculated from the measured values is compared to a fixed threshold. However, because the efficiency varies with the indoor and outdoor conditions, the threshold will have to be at the minimum efficiency value associated with the normal operation. With a model-based approach, the efficiency threshold can be dynamically calculated based on the other measured inputs.

Several different types of models are used for detection including detailed physical models, empirical models based on first principles, and black-box models. These models can be steady state, linear dynamic, or nonlinear dynamic. A brief discussion of different models is provided later in the chapter.

Fault Diagnosis

At the fault diagnosis stage, the residuals and other data are analyzed, and the cause of the fault is determined. Unlike fault detection, fault diagnosis is not a binary outcome (fault, no fault). A fault is diagnosed as soon as it is detected for FDD implementations at the subsystem or the component level with adequate measured data. Fault diagnosis is generally difficult when implemented at the system level, with multiple components, for example, air conditioner, chiller, and air handler, or at the component level with multiple subsystems. For example, if a fault with the air handler's air filter is detected because the pressured drop across the filter is excessive, the cause of the fault is a dirty or clogged filter. Therefore, additional diagnosis is not necessary in this case. However, if a deviation of the efficiency of the air conditioners is detected, a fault diagnosis is essential to isolate the actual cause because there is more than one possible cause for the deviation. In some cases, because of the lack of analytical redundancy, the fault diagnosis may yield more than one possible cause for a fault. Most buildings systems have limited sensors making the fault diagnosis step inevitable.

Most methods used for detection can also be used for diagnosis, but the criteria used are different. Generally, black-box approaches and statistical pattern recognition methods are well suited for the diagnostic step. A brief discussion of the different modeling techniques is provided later in the chapter.

Fault Evaluation

Following fault detection and diagnosis, the impact of the fault has to be evaluated. For most latent faults, the impacts have to be evaluated before a decision is made to stop, continue, or reconfigure the controls. The evaluation criteria depend on the application and severity of the fault. For critical processes, safety is the primary evaluation criterion. For FDD applied in a process industry, availability of the plant is important because it dictates the profit margin. Although for most building systems the cost of operations is the primary criterion, productivity impacts from lack of proper ventilation and comfort conditioning should not be neglected. Safety and environmental issues can also play an important role for building systems.

Fault evaluation is particularly important when the performance of a component is degrading slowly over time, such as heat exchanger fouling (Rossi and Braun, 1996). In this case, it is possible to detect and diagnose a fault well before it is severe enough to justify the service expense.

Decision on Course of Action

Finally, after the fault has been detected, diagnosed, and evaluated, a decision is needed on the course of action to be taken. The first step in the decision making process is to stop the system or send an alarm to shut it down if the fault is severe and the controls cannot be reconfigured to accommodate the fault. In some FDD applications, such as aeronautics and nuclear power plants where safety is critical, there is redundancy in controllers, actuators, and sensors. In such situations, corrective action can be taken to ensure continued safe operations using redundant fault-free components. For example, if a failure of one sensor of a redundant pair of sensors is diagnosed, then the supervisory system can reconfigure the controls such that the failed sensor is not used in making control decisions until it is replaced or fixed. This type of FDD system, which can enable corrective action to counteract the fault or make recommendations for altering the system operation, is referred to as fault-tolerant control system or an FDD supervisory system. In most cases, fault-tolerant control applications reconfigure the programmable parts of the control loop such that the system operates in a fault-free environment. In some cases, reconfiguration in the control loop may slightly degrade the reliability or the performance of the system. Operating the system in a degraded state, in some cases, is better than shutting it down. However, in other cases, the system can operate without any degradation in performance. For example, if the FDD system detects a sensor bias, it can reconfigure the controls to compensate for the bias.

If the fault is not severe (i.e., it is not a safety issue and will not damage the system or equipment) and the system controls cannot be reconfigured to accommodate it, the fault has to be tolerated or a request for repair needs to be made. Unlike critical systems, most building systems do not pose an immediate safety problem because of faulty operation; therefore, they lack redundancy and extensive instrumentation. For building systems, if the cost impact is small compared to the cost of correcting the fault, the fault can be tolerated. However, if the cost impact is large, the FDD system must provide a message leading to correction of the problem along with the impact it has on the operations.

Advanced supervisory systems can also have the capability to perform nonintrusive tests to refine a fault diagnosis. For example, if the FDD system detects a sensor failure but is able to pinpoint the failed sensor from among three sensors (e.g., return-, outdoor-, and mixed-air temperature) of an air-handling unit, the supervisory control system can perform additional nonintrusive tests during the unoccupied hours of the day to refine the diagnosis. For example, the air-handling unit can be operated at 100% outdoor-air and comparing the outdoor-air and mixed-air temperature signals, then operating the air-handling unit at 100% return-air and comparing the return-air and mixed-air temperature signals.

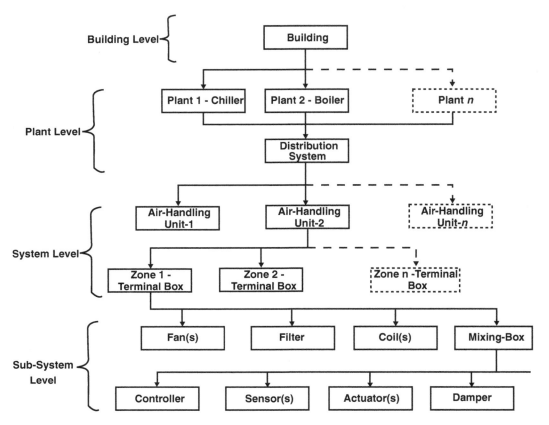

FIGURE 7.2.2   Hierarchical relationships of the various HVAC&R systems and subsystems in a building.

7.2.3 Hierarchical Relationships of the Various HVAC&R Systems and Subsystems in a Building

An FDD system can be deployed at several different levels in the building hierarchy, as shown in Figure 7.2.2. There are several different types of HVAC&R systems and subsystems in a building; some are independent while most are linked hierarchically to other systems in the building. An FDD system deployed at the building level can use the whole-building energy use (electric or thermal) to detect abnormal energy use (Dodier and Kreider, 1999). Although abnormal conditions can be detected at the whole-building level, their cause cannot be easily diagnosed because of insufficient resolution in the data.

Additional monitoring at lower levels is generally required for fault diagnosis. Almost any FDD method can be deployed at the building level. Regression and neural network models are probably a good choice for detection at this level. In contrast, for most implementations at the subsystem level, no fault diagnosis is needed. At that level, when a fault is detected, the cause of the fault is already known. FDD systems deployed at the intermediate plant and systems levels need methods for both detection and diagnosis.

7.2.4 Predictive Maintenance

Maintenance can be defined broadly as having three components: service, inspection, and repair (Patel and Kamrani, 1996). Service includes all steps taken to preserve the nominal state of the equipment and to prevent equipment failure. Inspection involves measuring and evaluating the current state of the equipment to detect the malfunction early and to prevent failure. Repair involves all steps taken to restore the nominal state of the equipment (Patel and Kamrani, 1996).

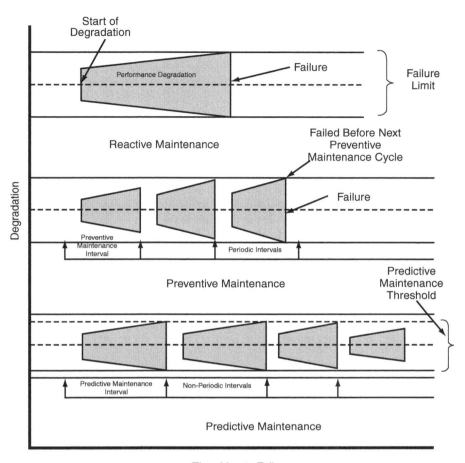

FIGURE 7.2.3   Comparison of the various maintenance practices.

Improper maintenance of building systems can lead to inefficiency, unreliable operations, and safety hazards. Maintenance costs are substantially determined by the chosen maintenance strategy. There are three commonly used maintenance practices: corrective or reactive maintenance, preventive maintenance, and predictive maintenance (Figure 7.2.3). In the corrective or reactive maintenance mode, the equipment is operated without maintenance until it breaks down. No attention is paid to ensuring that operating conditions are within the design envelope; consequently, the actual service performance and life span of the equipment may be substantially below the estimates of the manufacturer. While reactive maintenance may make sense in some instances (for example, replacing a light bulb), in the vast majority of building systems it is the most expensive option (Jarrell and Meador, 1997).

Corrective maintenance results in unscheduled downtime and may lead to unpredicted fatal failures. To avoid this problem, engineers started to maintain equipment at scheduled intervals throughout the life span of the equipment. This preventive maintenance method is the art of periodically checking the performance or condition of a piece of equipment to determine if the operating conditions and resulting degradation rate are within the expected limits. Statistics of past failures are used to define the periods for checking (for example, every 1000 working hours). If the periodic inspection reveals degradation in a part of the equipment, that part is replaced and no root cause analysis is usually undertaken. While good failure statistics allow the test interval to be optimized, catastrophic failures are still likely to occur. This method is labor intensive and, sometimes, parts are replaced unnecessarily and unjustifiably because no root cause analysis is performed. Preventive maintenance can be a cost-effective strategy when the life span of the equipment is consistent and well understood. Studies in the utility industry (power plants) have reported savings of 12 to 18% with preventive maintenance compared to reactive maintenance (Jarrell and Meador, 1997). Savings for the building systems have not been documented yet.

Although preventive maintenance leads to fewer unscheduled interruptions and extends the life span of the equipment, it increases the maintenance cost because it is conducted whether it is necessary or not. Instead of periodic maintenance, if the failure of a component can be predicted, the performance of the equipment can be optimized and enhanced while the maintenance expenditure is reduced (Patel and Kamrani, 1996). In the early 1990s, the concept of predictive maintenance was widely introduced. Predictive maintenance advocates measurements and procedures aimed at the detection of degradation mechanisms, thereby allowing the degradation to be understood, eliminated, or controlled prior to a complete failure (Jarrell and Meador, 1997). To predict the failure, the equipment must be constantly monitored for fault symptoms, symptoms must be analyzed for trends, and decisions must be made regarding the existence, location, cause, and severity of the fault.

Predictive maintenance strategy requires information about the state of wear and the remaining life span or rate of performance degradation of a system, and how long the system will be able to meet the design intent of the monitored device. Predictive maintenance can result in marked increase in equipment life, earlier corrective actions, decreased downtime, decreased maintenance, better quality product, decreased environmental impact, and reduced operational and energy costs. It is estimated that predictive maintenance can save between 8 and 12% over a good preventive maintenance program (Jarrell and Meador, 1997). However, there is an up-front cost in installation of additional sensors and the development of procedures to detect degradation.

Methods for Detecting Degradation

The basic requirements (Patel and Kamrani, 1996) for developing methods of predictive maintenance are

1. The system must identify abnormal conditions accurately.
2. The system must not give false alarms of abnormal conditions.
3. The system must report the level of confidence associated with each diagnosis.
4. The system must rank the conclusions.
5. The system must be able to handle insufficient data and uncertain situations.

These requirements are almost identical to the requirements of the FDD systems; therefore, many of the methods used in the FDD applications can also be used in detection of equipment degradation. However, most FDD applications detect faults at discrete time intervals and do not keep track of any discernable trends. Furthermore, most current FDD systems for building applications do not implement or discuss the fault evaluation and decision steps described previously (Figure 7.2.1). Fault evaluation is particularly important when the performance of a component is degrading slowly over time.

Predictive maintenance tools have been deployed in the telecommunications, automobile, aircraft, process, nuclear, and computer networks industries. In building applications, most research has focused on detection and diagnosis of faults (but not evaluating them) with the exception of Rossi and Braun (1996), Katipamula et al. (1999), and Dodier and Kreider (1999). Patel and Kamrani (1996) have tabulated more than 75 predictive maintenance research projects developed around the world. These projects are mostly related to manufacturing and process control and are based on some form of expert system.

Because many of the methods developed for FDD systems can also be used for predictive maintenance, the rest of the chapter will be devoted to FDD systems without discussion of their application to predictive maintenance.

7.2.5   Benefits from FDD Applications and Predictive Maintenance

Commercial buildings are increasingly using sophisticated BASs that have tremendous capabilities to monitor and control the building systems. Nonetheless, building systems routinely fail to perform as designed (Brambley et al., 1998; Katipamula et al., 1999). Although the BASs are sophisticated, they lack the tools necessary to detect and diagnose faults arising in the building systems. Furthermore, building operators generally overlook the symptoms because they lack a proper understanding of the control strategy and the failure symptoms. This leads to manual override of control strategies.

Faults that cause discomfort to the occupants are reported as complaints, while "latent" faults (such as simultaneous heating and cooling) go undetected. Such faults can have a significant impact on the operations of the building. Use of FDD applications has great potential to alleviate the problems associated with both the latent faults and the time needed to detect and correct conspicuous faults in building systems. FDD applications can improve energy efficiency, extend equipment life, reduce maintenance costs, reduce unscheduled equipment downtime, improve occupant comfort, health, and productivity, and reduce liability. Operating cost savings are the result of lower service and utility costs and extended equipment life. Productivity gains come from reduced equipment downtime and better overall comfort for the occupants.

Some of these benefits are tangible, i.e., the cost impact or the benefit from correcting existing problems can be quantified. On the other hand, the costs associated with poor indoor-air quality, lost productivity, and impact on equipment life are very difficult to quantify. Although difficult in most cases, energy and cost savings associated with the faults identified by the FDD applications in a building's equipment can be estimated. An example is given later in the chapter. One of the major barriers to widespread adoption of automated diagnostic tools is quantifying the impact of both tangible and intangible benefits.

Published reports indicate 3 to 50% of HVAC&R energy is wasted because of improper operations in existing buildings. The wide variation is primarily caused by the types of problems uncovered during the commissioning process and is, in part, the result of the various methods employed in the estimation of savings. Typical savings are expected to be between 5 and 15%. (Gregerson, 1997).

Published reports also indicate that many of the problems identified during the recommissioning process are related to controls. Unless the building is periodically recommissioned, these problems resurface. Fortunately, a number of problems related to improper controls can be detected and diagnosed in a continuous manner using automated tools.

7.2.6 Literature Review

While FDD was well established in the process, nuclear, aircraft, and automotive industries, it did not enter the building and HVAC&R industries until the mid 1990s (Braun, 1999). High reliability and safety were relatively less critical in building operations; therefore, FDD did not receive the same level of interest among building researchers, owners, and operators. The primary driver of building operations is still the operating cost and capital investment. Although FDD has been an active area of research among the buildings and HVAC&R community for several years, it is not widely used in the field. The primary reasons for slow adoption of FDD in buildings and HVAC&R areas include a relatively high cost-to-benefit ratio for an FDD implementation, partly caused by the lack of extensive instrumentation in the building and HVAC&R systems, and lack of data to quantify the benefits.

Because critical processes require high reliability and operational safety, the FDD system was an essential element of the plant operation. Early fault detection methods were generally limited to detecting values of measurable output when the signals had already exceeded the limit. Widespread use of microcomputers in the early 1980s led to advanced mathematical process models, which provided the ability to detect the fault earlier and to locate the fault by use of additional measurable signals (Isserman, 1984). Because reliability and safety are a primary concern, these plants have extensive and redundant sensors. Therefore, the FDD methods evolved around the data-rich environment. In the late 1970s, fault detection and diagnosis began to be applied to mass-produced consumer equipment such as automobiles and household appliances (Willsky, 1976).

Aeronautics, Nuclear, and Process Industry

Over the last 3 decades, several survey papers have summarized the FDD research in the aeronautics, nuclear, and process industries. The first major survey was written by Willsky (1976). Issermann (1984) surveyed various modeling and estimation methods for process fault detection. Gertler (1988) published a survey of model-based FDD in complex plants. Frank (1990) surveyed methods based on analytical and knowledge-based redundancy for fault diagnosis in dynamic systems. Issermann and Ballé (1997) published trends in applications of model-based FDD of technical processes. Frank (1997) published new developments using artificial intelligence in fault diagnosis. The developments in fault-detection methods up to the respective times are also summarized in books by Himmelblau (1978), Pau (1981), Patton et al. (1989), Mangoubi (1998), Gertler (1998), and Chen and Patton (1999).

The literature review shows a wide array of approaches used to detect and diagnose faults. The sequencing of the detection and diagnosis varies. In some cases, the detection system ran continuously, while the diagnostic system was triggered only upon the detection of a fault. In other applications, the detection and diagnostic systems ran in parallel and, in some instances, the detection and diagnostics were performed in a single step. The methods of detection and diagnosis can be broadly classified into two groups: model-based methods and model-free methods. In some cases, similar models were used for detection and diagnosis and, in others cases, different models were used.

Since the advent of computers in the process control industry, most practical FDD systems have used some form of fault detection and diagnosis (Gertler, 1988). Earlier deployments relied on simple limit checking for detection and diagnostic functions. Even the early fault detection systems for the space shuttles' main engines, while on the ground, primarily used limit checking with fixed thresholds on each measured variable (Cikanek, 1986). As the complexity of the control systems and use of computers increased, model-based FDD systems were developed. These systems rely on analytical redundancy by using an explicit mathematical model of the monitored plant to detect and diagnose faults. In contrast to physical redundancy (where measurements from multiple sensors are compared to each other), with analytical redundancy sensor measurements are compared to values computed analytically, with other measured variables serving as model inputs (Gertler, 1998).

The commonly used model-based methods for fault detection included observer, parity space, parameter estimation, frequency spectral analysis, and neural networks. The methods used for fault classification included neural nets, fuzzy logic, Bayes classification, and hypothesis testing (Issermann and Ballé, 1997). These methods were used mostly to detect and diagnose the faults with sensors, actuators, process, and control loop or controller. Details about the individual methods and how they are used in an FDD system can be found in the various survey papers and books written over the past two decades (Willsky, 1976; Issermann, 1984; Gertler, 1988; Frank, 1987; Frank, 1990; Issermann and Ballé, 1997; Frank, 1997; Himmelblau, 1978; Pau, 1981; Patton et al., 1989; Mangoubi, 1998; Gertler, 1998; and Chen and Patton, 1999).

Building Systems

Unlike process control systems, FDD research for building systems did not begin until the early 1990s. In the 1990s, several FDD applications for building systems were developed and tested in the laboratory, and were related to vapor compression equipment (refrigerators, air conditioners, heat pumps, and chillers) followed by the application of AHU. The methods used measured pressure and/or temperatures at various locations and the thermodynamic relationships to detect and diagnose common faults.

It is clear from the literature review that there is a lack of standard definitions of terms. For example, the term FDD is loosely used even when the described approach only detects faults. Furthermore, the words "fault" and "failure" are loosely used to mean the same thing — the fault. However, in the following literature review, the definitions provided earlier in the chapter are used in order to be consistent.

The available literature relating to building systems includes refrigerators, air conditioners and heat pumps, air-handling units, HVAC&R control systems, heating systems, pumps, thermal plant, several FDD applications for motors, and whole-building systems. In the following section, the FDD methods for refrigerators, air conditioners and heat pumps, chillers, and AHUs are summarized. For details on other building systems, refer to the relevant literature: HVAC&R plants (Pape et al., 1990; Dexter and Benouarets, 1996; Georgescu et al., 1993; Jiang et al., 1995; Han et al., 1999); HVAC&R control systems (Fasolo and Seborg, 1995); heating systems (Li et al., 1996; Li et al., 1997); pumps (Isserman and Nold,

1988; Dalton et al., 1995); thermal plant (Noura et al., 1993); several FDD applications for motors (Isserman and Ballé, 1997); and whole-building systems (Dodier and Kreider, 1999). In one of the earliest automated FDD systems used in industry, Kreider and Reinert (private communication, July 1999) deployed a diagnostic system using an expert system in a large computer manufacturing plant; the system detected and diagnosed the imminent failure of a large, centrifugal chiller.

In the early 1990s, the International Energy Agency (IEA) commissioned the Annex 25 collaborative research project on real-time simulation of HVAC&R systems for building optimization, fault detection, and diagnostics (Hyvärinen and Kärki, 1996). The Annex 25 study identified common faults for various types of HVAC&R systems, and investigated a wide variety of detection and diagnosis methods including physical models of HVAC&R systems and black-box models. The black-box models use classification techniques such as artificial neural networks, fuzzy models, and rule-based expert systems. The selected methods proved to be successful in detecting and diagnosing faults with simulated data; however, the effectiveness of the FDD systems in real building systems was not assessed.

Summary of Methods Used in Building Systems
The FDD literature related to the building systems is summarized in Table 7.2.1, and a more detailed review of the individual building system follows this section. In addition to the methods used to detect and diagnose faults, it summarizes whether or not the fault evaluation is addressed, whether the study included any discussion of the sensitivity of detection/diagnosis of faults vs. false alarms, and whether the FDD system was tested in the field.

In the 1990s, there was a significant contribution to FDD from a theoretical point of view; however, the practical aspects of implementing FDD systems in the field have not yet been thoroughly analyzed (sensitivity of diagnosis vs. false alarm and data gathering). Simplified physical models were mostly used for fault detection followed by rule-based methods and neural networks. Many studies did not address fault diagnosis, and some developed methods that combined fault detection and diagnosis into a single step. Most studies that addressed fault diagnosis used some type of classification approach, especially based on neural networks, rule-based knowledge systems, or fuzzy logic. Fault evaluation and sensitivity of the methods to detect and diagnose faults vs. false alarms were rarely addressed. With the exception of a couple of studies, detailed field tests were not conducted.

Review of Literature Related to Building Systems Applications
In this section, we briefly describe methods used in development of FDD applications for building systems. As the summary in Table 7.2.1 indicates, many researchers did not address the tradeoffs between the sensitivity of the methods to detect and diagnose faults vs. the false alarms and, with the exception of a couple of studies, none of the FDD systems were tested in the field. Because most of the studies did not discuss these issues unless otherwise mentioned, it should be assumed that the studies lacked such information.

Refrigerators — One of the early applications of FDD was to a vapor compression cycle based refrigerator (McKellar, 1987; Stallard, 1989). Although McKellar (1987) did not develop an FDD system, he identified common faults for a refrigerator based on the vapor compression cycle, and investigated the effects of the faults on the thermodynamic states of various points in the cycle. He concluded that the suction pressure (or temperature), discharge pressure (or temperature), and the discharge-to-suction pressure ratio were sufficient for developing an FDD system. The faults considered were compressor valve leakage, fan faults (condenser and evaporator), evaporator frosting, partially blocked capillary tube, and improper refrigerant charge (under- and over-charge).

Building upon McKellar's work, Stallard (1989) developed an automated FDD system for refrigerators. A rule-based expert system was used with simple limit checks for both detection and diagnosis. Condensing temperature, evaporating temperature, condenser inlet temperature, and the ratio of discharge-to-suction pressure were used directly as classification features. Faults were detected and diagnosed by comparing the change in the direction of the measured quantities with the expected values, and matching the changes to expected directional changes associated with each fault.

TABLE 7.2.1   Summary of FDD Literature Related to Building Systems

Reference	Method		Evaluation	Sensitivity/False Alarm	Field Testing
	Detection	Diagnosis			
Refrigerator					
McKellar (1987)	TM	None	No	No	No
Stallard (1989)	RB	PM	No	No	No
Air Conditioner					
Yoshimura and Ito (1989)	RB with FC[a]		No	No	No
Kumamaru et al. (1991)	RB[a]		No	No	No
Wagner and Shoureshi (1992)	Li/Tr and SPM[a]		No	Yes	No
Rossi and Braun (1996)	SPM	SRB	Yes	No	No[b]
Breuker and Braun (1999a, 1999b)	SPM	SRB	Yes	Yes	No
Chiller					
Grimmelius et al. (1995)	Empirical regression model with pattern matching[a]		No	No	No
Gordon and Ng (1995)	SPM	None	No	No	No
Stylianou and Nikanpour (1996)	SPM	PM	No	No	No
Tutsui and Kamimura (1996)	TCBM	None	No	Yes	No
Bailey et al. (2000)	NN[a]		No	Yes	Yes
Air Handling Unit					
Norford and Little (1993)	EM	Inferred	No	No	No
Glass et al. (1995)	QM	Inferred	No	No	No
Yoshida et al. (1996)	ARX & Kalman filter	None	No	No	No
Haves et al. (1996)	RBF/SPM	None	No	No	No
Lee et al. (1996a)	Li/ARX/ARMX[a]		No	No	No
Lee et al. (1996b)	NN[a]		No	No	No
Lee et al. (1997)	NN	NN	No	No	No
Peitsman and Soethout (1997)	ARX	ARX	No	No	No
Brambley et al. (1998), Katipamula et al. (1999)	SPM	KB and MI	Yes	Yes	Widely
Ngo and Dexter (1999)	FMB[a]		No	Yes	No
House et al. (1999)	NN, RB, Bayes, NNC, NPC	NN, RB, Bayes, NNC, NPC	No	—	No
Yoshida and Kumar (1999)	ARX/AFMM	None	No	No	No
Seem et al. (1999)	CQ	None	No	No	No
Kreider and Reinert (1997)	NN, TM	KB	No	No	Yes
HVAC&R Plants					
Pape et al. (1990)	SPM	SPM	No	No	No
Dexter and Benouarets (1996)	FMB[a]		No	No	No
Georgescu et al. (1993)	SPM	KB	No	No	No
Jiang et al. (1995)	CQ[a]		No	Yes	Yes
Han et al. (1999)	KB/RB[a]		No	No	No
HVAC&R Controls					
Fasolo and Seborg (1995)	CQ	None	No	Yes	No

TABLE 7.2.1 (continued)   Summary of FDD Literature Related to Building Systems

Reference	Method Detection	Method Diagnosis	Evaluation	Sensitivity/False Alarm	Field Testing
Heating Systems					
Li et al. (1996)	NN[a]		No	No	No
Li et al. (1997)	NN[a]		No	No	No
Pumps					
Issermann and Nold (1988)					
Dalton et al. (1995)					
Thermal Plants					
Noura et al. (1993)	EM	PM	No	No	No
Issermann and Ballé (1997)					
Whole Building					
Dodier and Kreider (1999)	NN	No	Yes	Yes	No

AFMM	Adaptive Forgetting through Multiple Models	NN	Neural Network
ARMX	Autoregressive Moving Average with Exogenous Input	NNC	Nearest Neighbor Classifier
		NPC	Nearest Prototype Classifier
ARX	Autoregressive Exogenous	PM	Pattern Matching
CQ	Characteristic Quantities	QM	Qualitative Model
EM	Empirical Model	RB	Rule Based Expert System
FC	Fuzzy Classification	RBF	Radial Basis Function
FMB	Fuzzy Model Based	SPM	Simplified Physical Models
KB	Knowledge Based	SRB	Statistical Rule Based System
Li	Limits	TCBM	Topological Case Based Modeling
MI	Mathematical Inference	TM	Thermodynamic Model
		Tr	Trends

[a] Fault detection and diagnosis was performed as a single step.
[b] A slightly modified version of the FDD system has been widely tested and is currently commercially available.

Air-Conditioners and Heat Pumps — There are many applications of FDD to air conditioners and heat pumps based on a vapor compression cycle; some of the studies are discussed below (Yoshimura and Ito, 1989; Kumamaru et al., 1991; Wagner and Shoureshi, 1992; Rossi, 1995; Rossi and Braun, 1996; Rossi and Braun, 1997; Breuker, 1997; Breuker and Braun, 1999a, 1999b).

Yoshimura and Ito (1989) used pressure and temperature measurements to detect problems with condenser, evaporator, compressor, expansion valve, and refrigerant charge on a packaged air conditioner. The difference between the measured values and the expected values was used to detect a fault. The expected value for comparison was estimated from the manufacturers' data, and the thresholds for fault detection were experimentally determined in the laboratory. The detection and diagnosis was conducted in a single step. No details were provided on how the thresholds for detection were selected.

Wagner and Shoureshi (1992) developed two different fault detection methods and compared their ability to detect five different faults in a small heat pump system in the laboratory. The five faults included condenser and evaporator fan fault, capillary tube blockage, compressor piston leakage, and seal system leakage. The first method was based on limit and trend checking, and the second method was a model-based approach in which the difference between the prediction from a simplified physical model and the monitored observations are transformed (or normalized) into useful statistical quantities. The transformed statistical quantities are then compared to predetermined thresholds to detect a fault.

The two fault detection strategies were operated in parallel on a heat pump in a psychrometric room. The model-free method was able to detect four of the five faults that were introduced abruptly, while the model-based method was successful in detecting only two faults. The selection of the thresholds for both methods is critical in avoiding false alarms and reduced sensitivity. Wagner and Shoureshi (1992) provide a brief discussion of how to trade off between sensitivity to diagnosis and false alarm. The implementation is only capable of detecting faults but lacks diagnosis, evaluation, and decision stages described in the previous section.

Rossi (1995) described the development of a statistical rule based fault detection and diagnostic method for air conditioning equipment with nine temperature measurements and one humidity measurement. The FDD method is capable of detecting and diagnosing condenser fouling, evaporator fouling, liquid-line restriction, compressor valve leakage, and refrigerant leakage. In addition to the detection and diagnosis, Rossi and Braun (1996) also described an implementation of the fault evaluation method. A detailed explanation of the fault evaluation method can be found in Rossi and Braun (1997). The methods were demonstrated in limited testing with a rooftop air conditioner in a laboratory.

Breuker (1997) performed a more detailed evaluation of the methods developed by Rossi (1995). The methods and results of the evaluation on a rooftop air conditioner in a laboratory environment (Breuker and Braun, 1999a, 1999b) are discussed in more detail later in the chapter.

Chillers — Several researchers have applied FDD methods to detect and diagnose faults in vapor compression based chillers; some of the studies are summarized below (Grimmelius et al, 1995; Gordon and Ng, 1995; Stylianou and Nikanpour, 1996; Tutsui and Kamimura, 1996; Peitsman and Bakker, 1996; Stylianou, 1997; Bailey et al., 2000; Gordon and Ng, 2000).

Grimmelius et al. (1995) developed a fault diagnostic system for a chiller. A reference model is used in parallel with the measured variables for fault detection and diagnosis. While the fault detection and diagnostics are carried out in a single step, their approach lacks the evaluation and decision steps. Twenty different measurements were used including temperature, pressure, power consumption, and compressor oil level. In addition to the measured variables, some derived variables were used, such as liquid subcooling, superheat, and pressure drop. The reference model is a multivariate linear regression model developed with the data from a properly operating chiller to estimate the process variables. These estimates were subsequently used to generate residuals by comparing the actual measured values with those estimated by the reference model. The inputs to the reference model included the environmental inputs and load conditions. The residuals were then used to score each fault symptom.

The chiller operation was classified into seven regions. Fault modes were associated with any component that was serviceable, which led to 58 different fault modes. The cause and effect study of the 58 fault modes helped establish the expected influence on the components and subsequent chiller behavior. The symptoms associated with the 58 fault modes on the measured and derived variables were generated. In the resulting symptom matrix, some fault modes were indistinguishable in terms of their respective symptoms because they either had identical or empty patterns. As a result, the symptom matrix was reduced from 58 to 37 fault modes and symptom patterns.

To diagnose a fault, scores are assigned to each known fault mode in the matrix. The score for a given symptom is not a constant, but is determined based on knowledge about the particular fault symptoms. A variable that shows a very distinct reaction to a fault mode becomes a higher score than a variable that shows only a limited reaction to the fault mode. For example, if there is increased resistance to flow of the evaporator on the chilled water side, the score associated with the decrease in suction pressure becomes higher than the decrease in the cooling water temperature difference. A symptom matrix for selected faults is shown in Table 7.2.2 (arrow pointing up, $\uparrow$, indicates increasing value as a result of the fault; likewise, arrow pointing down, $\downarrow$, indicates decreasing value as a result of the fault; a horizontal arrow, $\rightarrow$, indicates the fault has no effect on the variable).

TABLE 7.2.2   Symptom Patterns for Selected Faults in a Chiller

Fault	Compressor Suction Pressure	Compressor Suction Temperature	Compressor Discharge Pressure	Compressor Discharge Temperature	Compressor Pressure Ratio	Oil Pressure	Oil Temperature	Oil Level	Crankcase Pressure	Compressor Electric Power	Subcooling of Refrigerant	ΔT Refrigerant and Cooling Water	ΔT Cooling Water	Inlet Temperature at Expansion Valve	Filter Pressure Drop	Evaporator Outlet Pressure	Superheat	ΔT Chilled Water	Evaporator Outlet Temperature	Number of Active Cylinders
Compressor, suction side, increase in flow resistance	↓	→	→	→	→	↓	→	→	↓	↓	→	→	→	→	→	→	↑	↑	↓	→
Compressor, discharge side, increase in flow resistance	↑	→	↑	→	→	↑	→	→	↑	→	→	→	→	→	→	→	↑	↑	↓	→
Condenser, cooling water side, increase in flow resistance	→	→	↑	→	→	→	→	→	→	→	↑	↓	→	↑	→	→	→	→	→	→
Fluid line increase in flow resistance	→	→	→	→	→	→	→	→	→	↓	→	→	→	↓	→	→	↑	↑	→	→
Expansion valve, control unit, power element loose from pipe	↑	→	→	→	→	↑	→	→	↑	↑	→	→	→	→	→	→	↑	↓	↑	→
Evaporator, chilled water side, increase in flow resistance	↓	→	→	→	→	↓	→	→	↓	↓	→	→	→	→	→	→	↓	↑	↑	→

Source: Grimmelius et al., 1995.

Using the symptom matrix, a total score is generated by adding the individual scores of all expected symptoms that match the measured symptoms. A normalized score is calculated by dividing the total score by the total number of points per pattern. A normalized score of 0.9 or higher is used to indicate a probable fault, and a score between 0.5 and 0.9 is used to indicate a possible fault. Although the method proved effective in identifying faults in the systems before the system failed completely, faults with only a few symptoms got high scores more often. Because the reference model is a simple regression model developed with the data from the test chiller, the same model may not work on another chiller.

Gordon and Ng (1995, 2000) developed thermodynamic models for three commonly used chillers: reciprocating, centrifugal, and absorption. In addition, they also developed thermodynamic models for thermoacoustic and thermoelectric refrigerators. These models may not work to develop characteristic quantities for use within an FDD system. Although the models were used to demonstrate both the predictive and diagnostic capabilities, no full FDD system was developed.

Stylianou and Nikanpour (1996) used the reciprocating chiller model developed by Gordon and Ng (1995) and the pattern matching approach outlined by Grimmelius et al. (1995) as part of their FDD system. Like Grimmelius et al. (1995), they also perform the detection and diagnosis in a single step, and their approach lacks the evaluation and decision steps. The methods used in the FDD system included a thermodynamic model for fault detection, and pattern recognition from expert knowledge for diagnosis of selected faults. The diagnoses of the faults are performed by an approach similar to that outlined by Grimmelius et al. (1995). Seventeen different measurements were used, including pressures, temperatures, and flow rates, to detect four different faults: refrigerant leak, refrigerant line flow restriction, condenser water side flow resistance, and evaporator water side flow resistance.

The FDD system is subdivided into three parts: one used to detect problems when the chiller is off, one used during the start-up, and one used at the steady state condition. The off-cycle module is deployed when the chiller is turned off, and is primarily used to detect faults in the temperature sensors. The temperature sensor readings are compared to one another after the chiller is shut down. The differences are then compared to the values established during commissioning.

TABLE 7.2.3 Fault Patterns Used in the Diagnostic Module

Fault	Discharge Temperature	High Pressure Liquid Line Temperature	Discharge Pressure	Low Pressure Liquid Line Temperature	Suction Line Temperature	Suction Pressure	ΔT_{cond}	ΔT_{Evap}
Restriction in refrigerant line	↑	↓	↓	↓	↑	↓	↓	↑
Refrigerant leak	↑	↓	↓	↓	↑	↓	↓	↑
Restriction in cooling water	↑	↑	↑	↓	↓	↓	↑	↓
Restriction in chilled water	↑	↓	↓	↓	↓	↓	↓	↓

Source: Stylianou and Nikanpour, 1996.

The start-up module is deployed once the chiller is started and is left deployed for 15 minutes. The module used four measured inputs — discharge temperature, the crankcase oil temperature, and refrigerant temperature entering and leaving the evaporator — scanned at 5 sec intervals to detect refrigerant flow faults that are easier to detect before the system reaches steady state. To detect faults, the responses of the measured variables are compared to the baseline responses. For example, a shift (in time or magnitude) in the peak of the discharge temperature may indicate liquid refrigerant floodback, refrigerant loss, or refrigerant line restriction. Because the ambient conditions affect the baseline response, the response may have to be normalized before a comparison is made.

The steady state module is deployed after the chiller reaches steady state and stays deployed until the chiller is turned off. In this mode, it performs two functions: (1) to verify performance of the system, and (2) to detect and diagnose selected faults. Performance is verified using the thermodynamic models developed by Gordon and Ng (1995). For the fault diagnostics, linear regression models are used to generate estimates of pressure and temperature variables, similar to the approach outlined by Grimmelius et al. (1995). To diagnose faults, the estimated variables are compared to the measured values, and the residuals are matched using a rule-base to the patterns shown in Table 7.2.3.

Although Stylianou and Nikanpour (1996) extended the previous work of Gordon and Ng (1995) and Grimmelius et al. (1995), the evaluation of the FDD systems was not comprehensive and lacked several key elements including sensitivity and false alarm. In addition, it is not clear whether the start-up module can be easily generalized.

Tutsui and Kamimura (1996) developed a model based on a topological case based reasoning technique, and applied it to an absorption chiller. They showed that although the linear model had a better overall modeling error (mean error) than the topological case based model, the latter was better at identifying abnormal conditions.

Peitsman and Bakker (1996) used two types of black-box models, neural networks (NNs) and autoregressive with exogenous inputs (ARX), to detect faults at the system, as well as at the component level of a reciprocating chiller system. The inputs to the system models included condenser supply water temperature, evaporator supply glycol temperature, instantaneous power of compressor, and flow rate of cooling water entering the condenser (for NN only). The choice of the inputs was only limited to those that are commonly available in the field. Using the inputs with both the NN and ARX models, 14 outputs were estimated. For the NN models, inputs from the current and the previous time step and outputs from two previous time steps were used.

Peitsman and Bakker (1996) state that 14 system level models and 16 component level models were developed to detect faults in a chiller; however, only one example is described in their research. The intent was to use system level models to detect the fault at the system level and then use the component level models to isolate the fault. NN models appeared to have a slightly better performance than the ARX models in detecting faults at both the system and the component level. The evaluation and decision steps were not implemented.

Stylianou (1997) replaced the rule-based model used to match the pattern shown in Table 7.2.3 with a statistical pattern recognition algorithm. This algorithm uses the residuals generated from comparison of predicted (using linear regression) models, and measured pressures and temperatures to generate patterns that identify faults. Because this approach relies on the availability of training data for normal and faulty operation, it may be difficult to implement it in the field. There was only limited testing of the method.

Bailey, Kreider, and Curtiss (2000) also used the NN model to detect and diagnose faults in an air-cooled chiller with a screw compressor. The detection and diagnosis was carried out in a single step. The faults evaluated included: refrigerant under- and over-charge, oil under- and over-charge, condenser fan loss, and condenser fouling. The measured data included: superheat for circuits 1 and 2, subcooling from circuits 1 and 2, power consumption, suction pressure for circuits 1 and 2, discharge pressures for circuits 1 and 2, chilled water inlet and outlet temperatures from the evaporator, and chiller capacity.

Air-Handling Unit — There are several studies relating to the FDD method for the air-handling units (both the air side and the water side) and some of these are summarized in this section (Norford and Little, 1993; Glass et al., 1995; Yoshida et al., 1996; Haves et al., 1996, 1996a, 1996b, and 1997; Peitsman and Soethout, 1997; Brambley et al., 1998; Katipamula et al., 1999; House et al., 1999; Ngo and Dexter, 1999; Yoshida and Kumar, 1999; Seem et al., 1999).

Norford and Little (1993) classify faults in ventilating systems consisting of fans, ducts, dampers, heat exchangers, and controls. They review two forms of steady state parametric models for the electric power used by ventilation system fans and propose a third, that of correlating power with a variable speed drive control signal. The models are compared based on prediction accuracy, sensor requirements, and their ability to detect faults.

Using the three proposed models, four different types of faults associated with fan systems are detected: (1) failure to maintain supply-air temperature, (2) failure to maintain supply air pressure setpoint, (3) increased pressure drop, and (4) malfunction of fan motor coupling to fan and fan controls. Although Norford and Little's study lacked details on how the faults were evaluated, error analysis and associated model fits were discussed. The results indicate that all three models were able to identify at least three of the four faults. The diagnosis of the faults is inferred after the fault is detected.

Glass et al. (1995) used a qualitative model-based approach to detect faults in an air-handling unit. The method uses outdoor-, return-, and supply-air temperatures and control signals for the cooling coil, heating coil, and the damper system. Although Glass et al. (1995) mentioned that the diagnosis is inferred from the fault conditions, no clear explanation or examples were provided.

Detection starts by analyzing the measured variables to verify whether steady state conditions exist. Then, the controller values are converted to qualitative signal data and, using a model and the measured temperature data expected, qualitative signals are estimated. Faults are detected based on discrepancies between measured qualitative controller outputs and corresponding model predictions based on temperature measurements. Examples of qualitative states for the damper signal include: "maximum position," "minimum position," "closed," and "in between." When the quantitative value of the damper signal approaches maximum value, the corresponding qualitative value of "maximum" is assigned to the measured controller output.

The results of testing the method on a laboratory AHU were mixed because it requires steady state conditions to be achieved before fault detection is undertaken. Fault detection sensitivity and ability to deal with false alarms were not discussed.

Yoshida et al. (1996) used ARX and the extended Kalman filter approach to detect abrupt faults with simulated test data of an AHU. Although the fault diagnosis approach was clearly described, the authors noted that diagnosis is not feasible with the ARX method, but the Kalman filter approach could be used for diagnosis. Fault detection sensitivity and ability to deal with false alarms were not discussed.

Haves et al. (1996) used a combination of two models to detect coil fouling and valve leakage in the cooling coil of an AHU. The methodology was tested with data produced by the HVACSIM+ simulation

TABLE 7.2.4 Normalized Pattern for AHU Fault Diagnosis Used in NN Training

	Network Inputs — Residuals							Network Outputs								
Fault Diagnosis	Supply Pressure	Difference in Air Flow Supply and Return	Supply-Air Temperature	Control Signal to Cooling Coil	Supply Fan Speed	Return Fan Speed	Cooling Coil Valve Position									
Normal	0	0	0	0	0	0	0	1	0	0	0	0	0	0	0	0
Supply Fan	−1	−1	0	1	−1	0	0	0	1	0	0	0	0	0	0	0
Return Fan	0	1	0	0	0	−1	0	0	0	1	0	0	0	0	0	0
Pump	0	0	0	1	0	0	0	0	0	0	1	0	0	0	0	0
Cooling Coil Valve	0	0	0	0	0	0	1	0	0	0	0	1	0	0	0	0
Temperature Sensor	0	0	−1	−1	0	0	0	0	0	0	0	0	1	0	0	0
Pressure Transducer	−1	0	0	0	0	0	0	0	0	0	0	0	0	1	0	0
Supply Fan Flow Station	0	−1	0	0	0	0	0	0	0	0	0	0	0	0	1	0
Return Fan Flow Station	0	1	0	0	0	0	0	0	0	0	0	0	0	0	0	1

Source: From Lee et al., 1996b.

tool (Clark, 1985). A radial bias function (RBF) models the local behavior of the HVAC&R system and is updated using a recursive gradient-based estimator. The data generated by exercising the RBF over the operating range of the system are used in the estimation of the parameters of the physical model (UA and percent leakage) using a direct search method. Detection is accomplished by comparing estimated parameters to fault-free parameters.

Lee et al. (1996a) used two methods to detect eight different faults (mostly abrupt) in a laboratory test AHU. The first method used discrepancies between measured and expected variables (residuals) to detect the presence of a fault. The expected values were estimated at nominal operating conditions. The second method compared parameters that were estimated using autoregressive moving average with exogenous input (ARMX) and ARX models with the normal (or expected) parameters to detect faults. The faults evaluated included: complete failure of the supply and return fan, complete failure of the chilled water circulation pump, stuck cooling coil valve, complete failure of temperature sensor, complete failure of static pressure sensor, and failure of supply and return air fan flow station. Because each of the eight faults had a unique signature, no diagnosis was necessary.

Lee at al. (1996b) used an NN model to detect the same faults described previously (Lee et al., 1996a). NN was trained using the normal data and data representing each of the eight faults. Seven normalized residual values were used as inputs to the NN model and the nine output values consitute a pattern that represents normal operation or one of the eight fault modes. Instead of generating the training data with faults, idealized training patterns were specified by considering the dominant symptoms of each fault. For example, supply fan failure implies: supply fan speed of zero, supply-air pressure of zero, supply fan control signal of maximum, flow difference between the supply, and return ducts of zero.

Using similar reasoning, a pattern of dominant training residuals for each fault was generated and is shown in Table 7.2.4. The NN was trained using the pattern shown in Table 7.2.4. Normalized residuals were calculated for the faults that were artificially generated in the laboratory AHU. The normalized residuals vector at each time step was then used with the trained NN to identify the fault. Although the NN was successful in detecting the faults from laboratory data, it is not clear how successful this method will be, in general, because the faults generated in the laboratory setting were severe and without noise.

Lee et al. (1997) extended the previous work described in Lee et al. (1996b). In the 1997 analysis, two NN models were used to detect and diagnose the faults. The AHU was broken down into various subsystems such as: the pressure control subsystem, the flow control subsystem, the cooling coil subsystem, and the mixing damper subsystem. The first NN model is trained to identify the subsystem in which a fault occurs, while the second NN model is trained to diagnose the specific cause of a fault at the subsystem level. An approach similar to the one described previously (Lee et al., 1996b) is used to train both NN models.

Lee et al. (1997) noted that this two-stage approach simplifies generalization by replacing a single NN that encompasses all considered faults with a number of less complex NNs, each one dealing with a subset of the residuals and symptoms. Although 11 faults are identified for detection and diagnosis, fault detection and diagnosis for only one fault are presented in their study.

Peitsman and Soethout (1997) used several different ARX models to predict the performance of the AHU and compared the prediction to the measured values to detect faults in the AHU. The training data for the ARX models were generated using HVACSIM+. The AHU is modeled at two levels: (1) the system level where the complete AHU is modeled with one ARX model, and (2) the component level where the AHU is subdivided into several subsystems such as return fan, the mixing box, and the cooling coil. Each component is modeled with a separate ARX model. The first level ARX model is used to detect a problem, and the second level ARX models are used to diagnose the problem.

Most abrupt faults were correctly identified and diagnosed, while the slow evolving faults were not detected. In addition, there is a potential for conflict between the two levels; for example, the top level ARX model could detect a fault with the AHU, and the second level ARX models may not indicate any faults. Furthermore, there is a potential for multiple diagnoses at the second level. Peitsman and Soethout (1997) indicated that some of the multiple diagnoses could be discriminated by ranking them according to their improbability; however, no details were provided on how to implement such a scheme.

As part of its mission in commercial buildings research and development, the U.S. Department of Energy (DOE), in collaboration with industry, has developed a tool that automates detection and diagnosis of problems associated with outdoor-air ventilation and economizer operation. The tool, known as the outdoor-air/economizer (OAE) diagnostician, monitors the performance of AHUs and detects problems with outdoor-air control and economizer operation, using sensors that are commonly installed for control purposes (Brambley et al., 1998; Katipamula et al., 1999).

The tool diagnoses the operating conditions of AHUs using rules derived from engineering models of proper and improper air-handler performance. These rules are implemented in a decision tree structure in software. The diagnostician uses data collected periodically (e.g., from a BAS) to navigate the decision tree and reach conclusions regarding the operating state of the AHU. At each point in the tree, a rule is evaluated based on the data, and the result determines which branch the diagnosis follows. When the end of a branch is reached, a conclusion is reached regarding the current condition of the AHU. A detailed description of the methodology used is described later in the chapter.

House et al. (1999) compared several classification techniques for fault detection and diagnosis of seven different faults in an AHU. The data for the comparison were generated using a HVACSIM+ simulation model. Using the residuals, as defined in Lee et al. (1996a, 1996b), five different classification methods were evaluated and compared for their ability to detect and diagnose faults. The five classification methods include: NN classifier, nearest neighbor classifier, nearest prototype classifier, a rule-based classifier, and a Bayes classifier.

Based on the performance of classification methods, the Bayes classifier appeared to be a good choice for fault detection. For diagnosis, the rule-based method proved to be a better choice for the classification problems considered, where the various classes of faulty operations were well separated and could be distinguished by a single dominant symptom or feature.

Ngo and Dexter (1999) developed a semiqualitative analysis of the measured data using generic fuzzy reference models to diagnose faults with the cooling coil of an AHU. The method uses sets of training data with and without faults to develop generic fuzzy reference models for diagnosing faults in the cooling coil, including leaky valve, water side fouling, valve stuck closed, valve stuck midway, and valve stuck open. The fuzzy reference models describe in qualitative terms the steady state behavior of a particular class of equipment with no faults present and when each of the faults has occurred. The measured data are used to identify a partial fuzzy model that describes the steady state behavior of the equipment at a particular operating point. The partial fuzzy model is then compared to each of the reference models using a fuzzy matching scheme to determine the degree of similarity between the partial model and the reference models. The Ngo and Dexter (1999) study provides a detailed description of fault detection sensitive and false alarm rates.

Yoshida and Kumar (1999) evaluated two model-based methods to identify abrupt/sudden faults in an AHU. They reported that both ARX and adaptive forgetting through multiple models (AFMM) seem promising for use in on-line fault detection of the AHU. They report that ARX models require only a minimal knowledge of the system, and the potential limitation of the technique is that it requires long periods to stabilize its parameters. On the other hand, Yoshida and Kumar (1999) report that the AFMM method requires long moving averages to suppress false alarms. By doing so, faults of lesser magnitude cannot be easily detected. Implementation details were lacking, and only one example of fault detection was provided.

Seem et al. (1999) developed a method based on estimating performance indices that can be used for fault detection; however, no details were provided.

7.2.7 Costs and Benefits of Diagnostics and Predictive Maintenance

The cost of FDD implementation depends on several factors including the type of diagnostic method used, type of faults to be evaluated, number of sensors required (including any redundancy), and level of automation. The benefits from FDD can be classified into three categories: (1) improved health and safety, (2) improved reliability and availability, and (3) reduced cost of operations and maintenance.

Because safety is the overriding factor in the critical process, FDD applications with high cost can be easily justified. High availability of plant equipment is critical in the chemical or food process plants, where equipment failures and inefficiencies can have a significant impact on production costs. The economic impact of abnormal operations in the petrochemical process operations is about $20 billion per year in the U.S. (Mylaraswamy and Venkatsubramanian, 1997). Therefore, automated FDD systems are almost essential in reducing downtime and improving productivity. Most FDD research and applications development so far have been for critical and process industries because these industries can afford applications with a high cost, or because the benefits are so large that the cost of the FDD can be correspondingly high.

Cost vs. Benefits in Building Systems

In general, the health and safety benefits for building systems are lower than for critical or process plants and are generally limited to detection of problems relating to indoor-air quality, operations of fire systems, and elevator operations. Generally, FDD benefits must be derived entirely from reduction in operation and maintenance cost, and improved occupant comfort and health to offset the development and implementation costs. In comparison to critical or process plants, the cost savings are undoubtedly a smaller portion of the costs of operating the businesses that they serve. This means that FDD applications for noncritical building applications must have lower installed costs to achieve the same cost-to-benefit ratio (Braun, 1999).

Clearly, low installed costs are critical to wider adoption of FDD applications in building systems. Interest in FDD has grown as the costs of sensors and control hardware have gone down. In addition, there is increased emphasis on using information technology within the HVAC&R industry for scheduling, parts tracking, billing, and personnel management. This has provided an infrastructure and a higher expectation for the use of quantifiable information for better decision making. Finally, the structure of the industry that provides services for the operation and maintenance of buildings is changing. Companies are consolidating and offering whole-building operation and maintenance packages. In addition, utilities are in the process of being deregulated and are beginning to offer new services, which could ultimately include complete facility management. The cost-to-benefit ratio for FDD improves as the industry moves toward large organizations managing the operations and maintenance of many buildings. In particular, the cost of developing and managing the necessary software tools can be spread out over a larger revenue base.

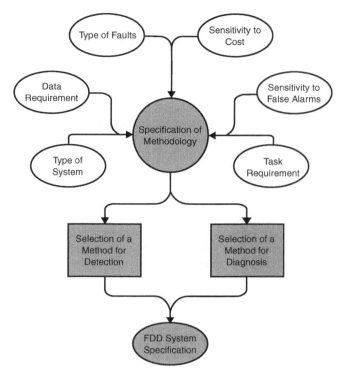

FIGURE 7.2.4   Schematic of a methodology specification.

7.2.8   Selection of Methods for FDD Applications

Selection of methods for FDD plays a critical part in the development of FDD systems. There is a wide range of methods available for FDD; most of them perform adequately in the laboratory or test setting, but many of them may not be suitable for field implementation. Some methods have fewer data requirements, while others require extensive data. This section provides a brief discussion on how to properly select a method for FDD.

There are several approaches to detecting and diagnosing faults in building systems. They differ widely depending on the type of system they are applied to, the necessary degree of knowledge about the diagnosed object, cost-to-benefit ratio (including monetary, as well as life safety related issues), the degree of automation, and the input data required. Most classical methods use alarm limits as fault criteria, whereas the advanced methods apply accurate mathematical models of the process. Between the two groups are various simplified empirical and heuristic knowledge-based methods of fault detection and diagnostics. Development of detailed physical models is expensive and impractical in most instances; therefore, either a simplified model based on first principles, or a heuristic knowledge base is widely used for FDD.

The success of the FDD system depends on proper selection of methods for both detection and diagnosis. Often methods are selected because of the interest of the developer or the availability of an existing tool. While this approach may yield satisfactory results for small-scale laboratory applications, it often leads to problems in full-scale real applications. For some FDD applications, fault diagnosis may not be needed because detection isolates the fault. On the other hand, fault diagnosis may not be possible because resolution of the data is not sufficient for diagnosis. Selection of the best method for detection and diagnosis depends on several factors, as shown in Figure 7.2.4.

The methods used for detection are often different from the methods used for diagnosis. During detection, the actual measurements (or estimated actual state/parameter) are compared to the expected measurements to identify an abnormal condition. Diagnosis is more involved and requires sophisticated methods to isolate the fault and the cause. From the survey of the literature both in critical processes and in the HVAC&R area, model-based approaches were widely used in detecting faults. Model-based approaches relied on mathematical models to predict the state or the output variables and compare them to the measured variables. For diagnosis, classification methods such as NNs, fuzzy clustering, and rule-based reasoning methods were widely used in the literature.

As mentioned earlier, almost any type of FDD method can be used at the building level (Figure 7.2.2); however, diagnosis at that level is limited. FDD systems deployed at the subsystems level, or component level, many not need a diagnosis method because when a fault is detected, the cause is already known. FDD systems deployed at the intermediate levels will most likely need both detection and diagnosis methods.

The amount of measured data plays a critical role in the selection of a method for both detection and diagnosis. A limited set of information will lead to selection of a detailed or moderately detailed physical model for detection. For diagnosis, it will then be necessary to have a set of fault models and a technique for selecting the fault models for a given set of inputs and outputs. In general, most building HVAC&R systems will have limited sensors — sensors that are required for controls purposes only. Additional sensor costs should be considered when selecting methods that require data beyond those that are normally provided for controls. On the other hand, methods that rely on a limited set of data may generate more false alarms.

Statistical pattern recognition techniques are often used to identifying the best matching model. If the system is extensively instrumented, classical limit checks and simplified empirical models are sufficient for detection, while rule-based or knowledge-based models are needed for diagnosing the cause.

Before selecting methods for detection and diagnosis, a good understanding of the anticipated faults is essential. Some faults influence the selection of the diagnostics method more than the detection. Examples of faults that make diagnosis difficult include faults that exhibit different symptoms at different times, faults that are intermittent, and multiple simultaneous faults. Not many methods can diagnose the fault that exhibits different symptoms at different times depending on the operational dynamics. For example, if the outdoor-air damper is stuck wide open and the outdoor-air conditions are favorable for economizing, there is no fault. However, if the outdoor-air conditions are unfavorable for economizing, it is a fault. In addition, multiple simultaneous faults make determining the cause of the fault difficult.

To a lesser extent, the cost of development and deployment of an FDD system influences the methods selected. Because the building industry is cost sensitive and safety is not an issue with the building systems, the methods used for detection and diagnosis have to rely on a limited set of measured data.

For noncritical applications, the methods used for detection and diagnosis should minimize the number of false positives (false alarms). If a number of false positive faults are detected and diagnosed, the operators may disable the FDD system completely. FDD methods applied to critical systems are tuned to be sensitive to fault detection; therefore, these applications may generate false alarms more often. On the other hand, FDD methods applied to noncritical systems (most building systems) are tuned to generate fewer false alarms.

The task requirement of the FDD system also plays a crucial role in the selection of the methods. If the FDD system is deployed in a decision support role, simple detection and diagnostic methods such as knowledge-based models are sufficient. On the other hand, if the FDD system is deployed as a fault-tolerant control system, more accurate and robust detection and diagnostics methods are required.

7.2.9 Detailed Descriptions of Three FDD Systems

In the next section, detailed descriptions of three FDD systems are presented: (1) a whole-building energy diagnostician, (2) an outdoor-air/economizer diagnostician, and (3) an automated FDD system for vapor compression systems. These three FDD applications were selected because they use different detection and diagnosis methods.

Whole-Building Energy Diagnostician

Energy consumption levels and patterns in buildings, when properly understood, can be indicators of building systems operation. Malfunctions of costly equipment can be identified by comparing "nominal" equipment behavior to that measured in real time during ongoing building operation. A statistically rigorous method was developed to detect problems in the whole-building energy consumption by organizing NNs into a higher-level model called a belief network, which can be viewed as a probabilistic database containing what is known about a system (Pearl, 1988). The whole-building energy (WBE) module described here is one module of a larger system for whole-building diagnostics developed by a team of private sector, national laboratory, and university researchers (Brambley et al., 1998). A summary of the WBE follows; for more information refer to Dodier and Kreider (1999).

Detection Variables in WBE

The WBE diagnostician determines the ratio of measured energy use to expected energy use accounting for the weather, time of day, day of week, and other features of building energy use that are time and day dependent. Specifically, WBE detections are based on the energy consumption index (ECI), which is defined as

$$\text{ECI} = \frac{(\text{Actual energy use})}{(\text{Expected energy use})}$$

A separate ECI is computed for each for the four major energy end uses: building total electric, building total thermal, HVAC&R electric other than chiller/packaged units, and chiller/packaged units. Therefore, the data required for the FDD systems include: outdoor-air temperature and humidity, whole-building electricity and thermal, electricity consumption of the chiller or packaged units, and other HVAC&R electricity consumption (less chiller or packaged units). If any of the consumption data are not available, detection for that end use is not performed.

The actual energy use is measured, while the expected energy use is computed as a function of time of day, day of year, day of week, outdoor-air dry-bulb temperature and relative humidity, and other optional weather and load predictors may be used as well (such as wind speed, production, historical or sales). NNs are used to predict each energy end use given the values of these weather and calendar variables. These predictor networks are calibrated by training them on data from the same building. The amount of training data depend on the end use; an end use that varies by outdoor conditions may need as much as 6 to 9 months, while others many need as little as 4 weeks.

The actual and expected energy use variables of interest are totals computed from hourly values recorded over a 24-hour period from midnight on one day to midnight the next day. Each of the four ECI values is calculated once a day. Installing the WBE in a new building requires that it be tuned (by training the neural network predictor models) especially for that building. Because energy use varies widely from one building to another, tuning the WBE for each building gives much more accurate energy use predictions than are possible with models that do not consider a building's particular characteristics.

Problem Detection Approach

The flow of data within WBE is shown schematically in Figure 7.2.5. In summary, the current belief network, current end use consumption data, current weather variables (dry-bulb temperature and relative humidity), and calendar variables (time of day, day of year, and a weekday/weekend flag) are the inputs. The time of day and day of year are represented as sine and cosine functions. The weekday/weekend flag is "1" if the day is a weekday, and "0" if it is a weekend. Detected problems and their costs are output. Essentially a belief network embodies, in a quantitative way, the relationships between known or measured influencing parameters (e.g., weather, schedule, occupancy in a building) and the energy end uses of interest. The WBE module compares the measured data with predicted data to detect problems. If a problem is detected, it estimates energy cost impacts.

Specifically, the WBE module uses probabilistic inference in the form of a belief network with continuous and discrete variables for problem detection. Problems to be detected are represented as variables

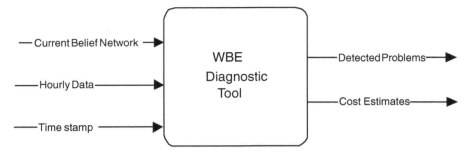

FIGURE 7.2.5   Data flow in WBE diagnostician (from Dodier and Kreider, 1999).

in the network. Metered data and other known values are also variables in the network. Some intermediate quantities (such as daily energy use totals and estimated values for missing sensor readings) are variables as well. Probability distributions are propagated through the belief network. The result of the propagation algorithm is the probability distribution over each variable conditional on observed values.

The probability distribution over each detection variable is used to compute the risk associated with each possible diagnostic message. A cost matrix for each detection variable is stored with the associated messages. The message with least risk is output from the WBE.

Belief Networks

A belief network is a probabilistic model composed of a number of submodels that compute the probability of a dependent variable x, given the values of the variables that have a cause or influence relation to x. These influential variables are called "parent" variables, and x is called a "child" variable. The model is called a "belief" network because it computes probabilities, representing degrees of belief. Although slightly dated from a technical point of view, Pearl (1988) remains the best general introduction to belief networks because of clarity and breadth; it discusses the interpretation of belief networks and how they represent the world.

Because the belief network summarizes what is known about a system, discussion of the fault detection system is centered on the belief network. A belief network allows heterogeneous data to be organized into a single structure; therefore, all relevant data is stored within the network.

Summary of WBE Diagnostician

The belief networks offer a workable approach to including both physical and statistical knowledge for detecting energy use problems. A strictly probabilistic approach supersedes the use of ad hoc "certainty factors" commonly used in expert systems. Neural networks can be used to predict whole-building energy use quickly and with sufficient accuracy to form the basis of WBE detection process.

Testing on field data indicates that the WBE approach is able to identify changes in HVAC&R systems (Dodier and Kreider, 1999) and to estimate the difference in energy use. Data from a large building have been analyzed using the WBE, with encouraging results. In practice, the most significant hurdle is to automatically train accurate prediction models for energy end uses, when only short data streams are available. The other important practical obstacle is baseline data. The WBE diagnostician automatically determines when there is sufficient data to make accurate predictions. Furthermore, the WBE automatically determines if it is not making sufficiently accurate predictions and if additional training data need to be collected.

Outdoor-Air/Economizer Diagnostician

The outdoor-air/economizer (OAE) diagnostician is part of a larger tool developed by the DOE (Brambley et al., 1998; Katipamula et al., 1999). It monitors the performance of AHUs and automatically detects problems with outdoor-air control and economizer operation using sensors that are commonly installed for control purposes. The OAE diagnostician can be used with most major types of economizer and

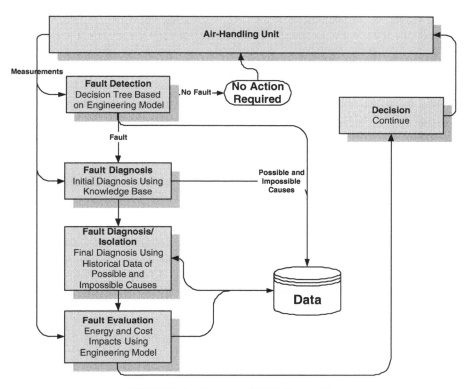

FIGURE 7.2.6 Overview of OAE diagnostician.

ventilation systems. It detects both over- and under-supply of outdoor air; thus, it can be used to ensure adequate outdoor-air supply for the occupants and eliminate excess heating or cooling.

Detection and Diagnostic Methods

As with any mechanical system, faults that diminish or eliminate an economizer's usefulness can occur. However, unlike the primary (mechanical) cooling system, a failure of the economizer may go completely unnoticed. Any failure, for example, that prevents outdoor air from being used for cooling when outdoor conditions are favorable may go unnoticed because the mechanical cooling system will pick up the load and maintain occupant comfort. Similarly, a failure that results in too much outdoor air may not be apparent in a reheat system. Reheating will ensure that the air supplied to the space is at a comfortable temperature. In both of these examples, however, the system would be using much more energy (and costing much more to operate) than necessary.

The OAE diagnostician is designed to monitor conditions of the system not normally observable by occupants, and to alert the building operator when there is evidence of an operational fault. The common types of outdoor-air ventilation and economizer problems handled by the DAE diagnostician include: stuck outdoor-air dampers, failures of temperature and humidity sensors, economizer and ventilation controller failures, supply-air controller problems, and air flow restrictions that cause unanticipated changes in overall system circulation. The diagnostician also performs some self-diagnosis to identify errors introduced by users in setup and configuration of the software tool.

An overview of the fault detection and diagnostic process is shown in Figure 7.2.6. The first step in the FDD process for the OAE diagnostician is fault detection. The diagnoses of the faults are carried out in two steps: (1) initial diagnosis of the fault is accomplished by using a knowledge base, and (2) the final diagnosis that refines the initial diagnosis is accomplished by reviewing the historical results. The initial and the final diagnoses are carried out for each time step. After the fault is detected and the cause of the fault is diagnosed, the fault is evaluated, and the energy and cost impact arising from the fault are estimated. Although the current version of the OAE diagnostician provides the user with information about the fault that is necessary to make a decision on whether to continue to run the system in the faulty mode or shut it down for repair, it does not take any corrective actions by itself. Details of the fault detection, diagnosis, and evaluation methods are described in the following subsections.

Fault Detection Approach

An overview of the logic tree used to identify operational states and to build the lists of possible failures is illustrated in Figure 7.2.7. The boxes represent major subprocesses necessary to determine the operating state of the air handler; diamonds represent tests (decisions), and ovals represent end states and contain brief descriptions of "OK" and "not OK" states. Only selected end states are shown in this overview.

The OAE diagnostician uses a logic tree to discern the operational "state" of outdoor-air ventilation and economizer systems at each point in time for which measured data are available. The tool uses rules derived from engineering models of proper and improper air-handler performance to diagnose operating conditions. The rules are implemented in a decision tree structure in the software. The diagnostician uses periodically measured conditions (temperature or enthalpy) of the various air flow streams, measured outdoor conditions, and status information to navigate the decision tree and reach conclusions regarding the operating state of the AHU. At each point in the tree, a rule is evaluated based on the data, and the result determines which branch the diagnosis follows. A conclusion is reached regarding the operational state of the AHU when the end of a branch is reached.

Many of the states that correspond to normal operation are dubbed "OK states." For example, one OK state is described as "ventilation and economizer OK; the economizer is correctly operating (fully open), and ventilation is more than adequate." Other states correspond to something operationally wrong with the system and are referred to as "problem states." An example problem state might be described as "economizer should not be off; cooling energy is being wasted because the economizer is not operating; it should be fully open to utilize cool outside air; ventilation is adequate." Other states may be tagged as incomplete diagnoses if the measured information is insufficient.

Fault Diagnosis Approach

The OAE diagnostician performs fault diagnosis in two steps. After a fault is detected, using a knowledge base, a list of possible and impossible causes is identified for the fault state. The knowledge base is populated *a priori* with possible causes and impossible causes for every problem state in the decision tree. In the example above, a bad or biased temperature sensor, stuck outdoor-air damper, an economizer controller failure, an actuator failure, a broken linkage, or perhaps an error in setting up the diagnostician could cause an economizer malfunction to be reported. Thus, at each measured time period, a list of possible and impossible causes is generated.

The list of possible causes can be rather long and often different at different time steps because the same fault can manifest itself in different problem states depending on the current operating conditions. For example, if the outdoor-air conditions are favorable for economizing and if the outdoor-air damper is stuck fully open, it is not a fault; but if the conditions are not favorable for economizing, then it is a fault. Thus, each set of observations leads to a different end branch in the decision tree. In the second stage diagnosis, the number of possible causes is reduced. The methodology uses a historical list of possible and impossible causes and reduces the list of possible causes. It does this by jointly considering the faults, possible causes, and impossible causes along with metrics of their statistical certainties over time to determine a reduced set (subset) of causes that are more likely to have caused the fault during that timespan.

Data Requirements

The OAE diagnostician uses two primary types of data — measured and setup. The measured data include information on mixed-air, return-air, and outdoor-air temperatures (and enthalpies for enthalpy-controlled economizers), supply fan on/off status, and heating/cooling on/off status. These data are typically available from BASs as trend logs or at requested intervals. Alternatively, measured data could be collected using custom metering and data collection systems, or the diagnostician could be used to process an

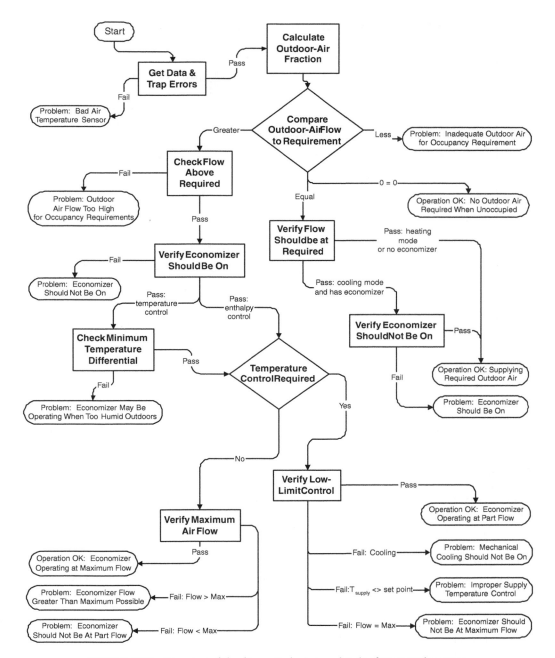

FIGURE 7.2.7 Overview of the diagnostic logic tree showing key operating states.

existing database containing the required data. The setup data, obtained by querying the user (building operator or installer), include information describing the type of economizer, its control strategies, setpoints, and building occupancy (and hence, ventilation) schedules.

Basic OAE Functionality

The OAE diagnostician detects about 25 different basic operational problems using the methodology described earlier. The results are presented using a color code to alert the building operator when a fault occurs and then provides assistance in identifying (diagnosing) the causes of the fault and correcting

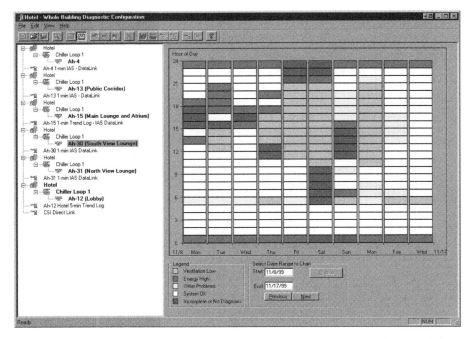

FIGURE 7.2.8   Diagnostic results showing proper and faulty operation with a data set having a faulty outdoor-air sensor.

them. Figure 7.2.8, for example, shows a representative OAE diagnostician window. Each cell in the diagram represents an hour. The color of the cell indicates the type of state. White cells identify "OK states," for which no faults were detected. Other colors represent problem states. "Clicking" the computer mouse on any colored cell brings up the specific detailed diagnostic results for that hour as shown in Figures 7.2.9 and 7.2.10.

Sensitivity vs. False Alarm

Adjustment of the sensitivity of the methods to detect and diagnose faults vs. generating false alarms is critical because the measured data in the field has both noise and bias. In the OAE, tolerances for each measured and static input variable are used to generate uncertainties that are propagated through all calculations and tests. For example, to test if the outdoor-air temperature is greater than the return-air temperature, not only should the outdoor-air temperature value be greater than the return-air temperature, the uncertainty of the test should also be less than a specified threshold. The uncertainty thresholds and tolerances on each variable are user specified. By specifying the tolerance and adjusting the uncertainty thresholds, false alarms can be reduced or sensitivity of detector increased.

Although field testing is ultimately required, simulations provide an effective way of generating data that would be more costly to generate in a laboratory or through field tests. The results are also valuable for illustrating the success of the diagnostician in detecting operation problems and their underlying causes. The general approach involves generating sets of data by simulation, where each set corresponds to an air handler with a specific underlying fault. These data sets are then processed by the OAE diagnostician to determine whether it detected problems and identified the correct cause (i.e., underlying problem). Although there are over 25 problem states defined in the OAE algorithm, only a few common fault states (problems) were tested with annual hourly simulated data sets. They include: bad sensors (outdoor-air sensor biased to read 10°F higher), outdoor-air damper stuck fully closed, outdoor-air damper stuck fully open, outdoor-air damper stuck at required ventilation position, outdoor-air damper stuck between fully closed and fully open.

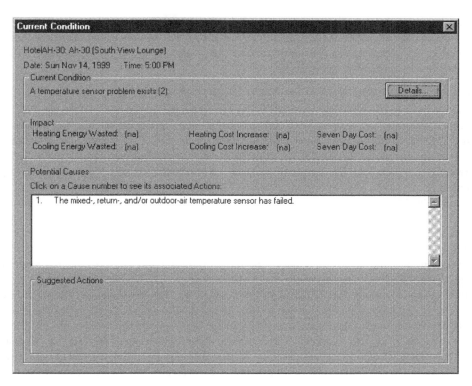

FIGURE 7.2.9 Pop-up windows providing a description of a problem, a list of reduced causes, and suggested actions to correct that cause.

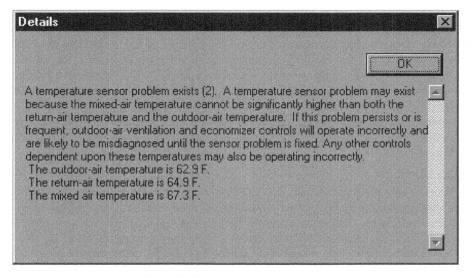

FIGURE 7.2.10 Details of diagnostic results.

Field Test Results

The OAE diagnostician was installed in three buildings for initial field testing. Field testing provides opportunities to investigate unanticipated practical problems and test usefulness in practice. The results obtained suggest that the OAE diagnostician will provide significant benefits.

Of the 18 air handlers monitored, more than half were found to have problems shortly after initial processing of data. The problems found included: sensor problems, return-air dampers not closing fully when outdoor-air dampers were fully open, and a chilled water controller problem. All problems have been confirmed by inspection of the AHUs.

Elements of an Automated FDD for Vapor Compression Systems

This section describes some of the methods being developed for automated FDD as applied to vapor compression equipment. In general, HVAC&R applications will not tolerate the use of expensive sensors. As a result, many of the methods being developed rely on the use of temperature, and in some cases, pressure measurements. As discussed earlier, contributions in the development of FDD methods for vapor compression equipment have been made by McKellar (1987), Stallard (1989), Yoshimura and Ito (1989), Kumamaru et al. (1991), Wagner and Shoureshi (1992), Inatsu et al. (1992), Grimmelius et al. (1995), Gordon and Ng (1995), Stylianou and Nikanpour (1996), Peitsman and Bakker (1996), Stylianou (1997), Rossi and Braun (1996, 1997), Breuker and Braun (1998a,b), and Bailey et al. (2000). The faults considered include: compressor valve leakage, heat exchanger fan failures, evaporator frosting, condenser fouling, evaporator air filter fouling, liquid line restrictions, and refrigerant leakage. The following subsections provide background and details on some of the more promising and well-documented methods. The presentation is organized according to the major elements of an FDD system.

Faults for vapor compression systems can be divided into two categories: (1) "hard" failures that occur abruptly and either cause the system to stop functioning or fail to meet comfort conditions, and (2) "soft" faults that cause a degradation in performance but allow continued operation of the system. Many of the most frequently occurring and expensive faults are associated with service in response to hard failures, such as compressor and electrical faults. Certainly, an automated FDD system should be able to diagnose "hard" faults. However, these faults are typically easy to detect and diagnose using inexpensive measurements. For instance, a compressor failure leads to a complete loss of refrigerant flow and can be easily diagnosed by monitoring the temperatures or pressures at the inlet and outlet of the compressor. Similarly, a fan motor failure could be diagnosed by measuring temperatures or pressures at the inlets and outlets of the heat exchangers (evaporator or condenser) that they serve. Other hard faults that should probably be included within an FDD system include common controls failures, blown fuses, and malfunctioning electrical components such as contactors. It would also be important to detect dangerous operating conditions, such as the possibility of a flooded start, which lead to "hard" failures. "Soft" faults, such as a slow loss of refrigerant or fouling of a heat exchanger, are more difficult to detect and diagnose. Furthermore, they often lead to premature failure of components, a loss in comfort, or excessive energy consumption.

The techniques developed for diagnosing "soft" faults in vapor compression cooling equipment can be described in terms of a series of steps, presented in Figure 7.2.11 (for discussion of the various steps refer to Section 7.2.2).

Fault Detection

Fault detection is accomplished by comparing measurements with some expectations for normal behavior, where the expectations are determined from a model. In the simplest system, the expectations could be that the measurements (e.g., suction and discharge pressure) should fall within acceptable ranges (low and high limits). Generally, the measurements vary with the operating conditions so the acceptable ranges need to be relatively large to avoid false alarms. In this case, only relatively large faults can be detected. Much better resolution can be obtained if an on-line model is utilized that relates expectations for measurements under normal operation to measurements of the operating conditions (e.g., ambient temperature). Because no model is perfect, the deviations of measurements from expected values need to be greater than some threshold that depends upon the uncertainty in the model and measurements.

COP as a Performance Expectation — If the only goal is to detect faults (without diagnosis), then one or two measurements are probably sufficient. In particular, cooling capacity and power consumption (or COP) are excellent performance indices, because it probably is not necessary to perform service unless

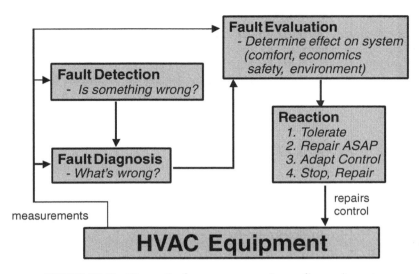

FIGURE 7.2.11 Diagnostics for vapor compression cooling equipment.

these indices change by a significant amount. Gordon and Ng (1995, 2000) presented a semiempirical model for predicting the COP of chillers during steady state operation that is useful for fault detection. Stylianou and Nikanpour (1996) used the model of Gordon and Ng for fault detection during steady state operation. This was one element of an overall FDD approach that was developed for a reciprocating chiller.

The model of Gordon and Ng (1995, 2000) was derived from a simple first and second law analysis using empirical relations for the irreversibilities associated with the heat exchangers. For a given chiller, COP is correlated using the following form.

$$\frac{1}{COP} = -1 + \frac{T_{c,i}}{T_{e,o}} + \frac{-a_0 + a_1 T_{c,i} - a_2 \dfrac{T_{c,i}}{T_{e,o}}}{\dot{Q}_e} \tag{7.2.1}$$

where $T_{c,i}$ is the temperature of the secondary working fluid (air or water) entering the condenser, $T_{e,o}$ is the temperature of the secondary working fluid leaving the evaporator (air, water, or water/glycol), $\dot{Q}_e$ is the rate of heat addition to the evaporator (cooling load), and a_0, a_1, and a_2 are empirical constants. The constants are determined using linear regression applied to a set of training data obtained from the equipment manufacturer, from laboratory tests, or from the field when the unit is operating normally. There are some advantages in using the model of Equation 7.2.1 as compared with polynomial correlations that are typically employed. In particular, less data are required to obtain an acceptable fit, and there is better confidence that the model extrapolates well to operating conditions outside of the range used to obtain the correlations.

It is necessary to establish thresholds for the identification of faults. Stylianou and Nikanpour (1996) did not directly address the issue of fault detection thresholds for their proposed method. However, it is not difficult to establish reasonable thresholds for deviations in COP. One criterion is that the thresholds should be significantly larger than the uncertainty of the models in predicting the expected values of the measurements to avoid false alarms. The semiempirical model of Gordon and Ng can predict cooling COP to within about 4%. Expert knowledge could be used to set larger thresholds that would guarantee that the detected faults are important and should be repaired. In this case, the fault evaluation step in Figure 7.2.11 could be skipped. For instance, a 10% loss in efficiency represents a significant fault and should probably be repaired as soon as possible.

Thermodynamic States as Expectations — Many diagnostic approaches utilize thermodynamic state measurements as inputs (see next section) for differentiating between faults. Because several measurements are necessary for diagnostics, these measurements can also be used for fault detection. Rossi and Braun (1997) and Breuker and Braun (1998a,b) developed and evaluated a complete FDD system for packaged air conditioning equipment that utilizes steady state models for both fault detection and diagnosis. All of the measurements required for fault diagnosis are used in the fault detection step (i.e., any measurement can trigger the detection of a fault). The output state measurements used by the technique are

1. Evaporating temperature (T_{evap})
2. Suction line superheat (T_{sh})
3. Condensing temperature (T_{cond}),
4. Liquid line subcooling (T_{sc})
5. Hot gas line or compressor outlet temperature (T_{hg})
6. Secondary fluid (air or water) temperature rise across the condenser (ΔT_c)
7. Secondary fluid (air or water) temperature drop across the evaporator (ΔT_e)

Seven steady state models are used to describe the relationship between the driving conditions and the expected output states in a normally operating system. In a normally operating, simple packaged air conditioning unit (on/off compressor control, fixed speed fans), all the output states ($\mathbf{Y}$) in the system are assumed to be functions of only three driving conditions ($\mathbf{U}$) that affect the operating states of the unit: the temperature of the ambient air into the condenser coil (T_{amb}), the temperature of the return air into the evaporator coil (T_{ra}), and the relative humidity (Φ_{ra}) or wet-bulb temperature (T_{wo}) of the return air into the evaporator coil.

Polynomial models were fit using steady state training data obtained in the laboratory and compared with a separate set of steady state test data. The form of the polynomial models are

$$
\begin{aligned}
y_i = a_1 + a_2 T_{wb} + a_3 T_{ra} + a_4 T_{amb} + a_5 T_{wb}^2 + a_6 T_{ra}^2 + a_7 T_{amb}^2 \\
+ a_8 T_{wb} T_{ra} + a_9 T_{ra} T_{amb} + a_{10} T_{wb} T_{amb} + a_{11} T_{wb}^3 + a_{12} T_{ra}^3 + a_{13} T_{amb}^3 \\
+ a_{14} T_{wb} T_{ra}^2 + a_{15} T_{wb} T_{amb}^2 + a_{16} T_{ra} T_{wb}^2 + a_{17} T_{ra} T_{amb}^2 + a_{18} T_{amb} T_{wb}^2 \\
+ a_{19} T_{amb} T_{ra}^2 + a_{20} T_{wb} T_{ra} T_{amb} + \ldots
\end{aligned}
\tag{7.2.2}
$$

where i_i is the ith output variable prediction and the As are coefficients determined using linear regression.

Table 7.2.5 gives the model orders used by Breuker and Braun (1998a,b) and model accuracy for the test data considered. Stylianou and Nikanpour (1996) used similar polynomial forms for thermodynamic states of a small water-cooled reciprocating chiller. In this case, the driving conditions were the temperatures of the secondary fluid for the condenser and evaporator.

TABLE 7.2.5 Example Model Evaluations

Variable	Best Model to Use	RMS Error (F)	Maximum Error (F)
T_{evap}	1st order	0.49	0.99
T_{sh}	3rd order with cross terms	1.39	3.03
T_{hg}	3rd order with cross terms	1.00	3.24
T_{cond}	1st order	0.31	0.61
T_{sc}	2nd order with cross terms	0.46	1.39
ΔT_c	1st order	0.18	0.48
ΔT_e	2nd order with cross terms	0.23	0.56

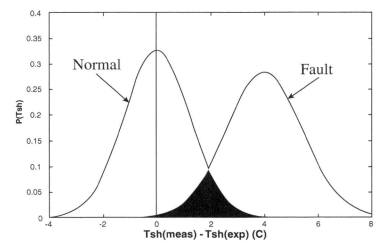

FIGURE 7.2.12   One-dimensional example of the fault detection classifier.

In the fault detection method described by Rossi and Braun (1997) and Breuker and Braun (1998a,b), a fault is identified whenever the current measurements are statistically different than the expected values. The detection algorithm uses the differences between the measurements and expected values (termed residuals) as features for a classifier. Figure 7.2.12 illustrates how this method works for a one-dimensional example. A probability distribution of the residual of the suction line superheat for both normal and faulty operation is shown. Under normal operation, there is a distribution of residuals that results from measurement noise and modeling errors. In the absence of modeling errors and with random noise, the distribution for normal operation would have zero mean. The introduction of a fault changes both the mean and/or standard deviation of the residuals. A fault is indicated whenever the overlap between the two distributions is less than a fixed threshold. The overlap is termed the fault detection error and the threshold is called the fault detection threshold. The overlap is related to the probability of erroneously classifying the current operation as faulty and decreases with the severity of the fault.

In the general case of m output measurements, the statistical fault detection method estimates the overlap between m-dimensional probability distributions of residuals for current and normal operation. The method assumes that residual distributions are Gaussian and can be characterized using a mean vector and covariance matrix and that the separation between the distributions for current and normal operation is dominated by mean vector differences as opposed to covariance matrix differences. The resulting classifier is termed an optimal linear classifier (Fukunaga, 1990). A fault is identified whenever the following inequality holds.

$$\left(\mathbf{M}_C - \mathbf{M}_N\right)^T \Sigma^{-1} Y - \frac{1}{2}\left(\mathbf{M}_N^T \Sigma^{-1} \mathbf{M}_N - \mathbf{M}_C^T \Sigma^{-1} \mathbf{M}_C\right) \geq 0 \tag{7.2.3}$$

where

$$\Sigma = s\Sigma_n + \left(1 - s\right)\Sigma_C \tag{7.2.4}$$

and where $\mathbf{Y}$ is a vector of current residuals, $\mathbf{M}_N$ is the mean vector and Σ_N is the covariance matrix that describes the distribution of residuals in the absence of any faults (i.e., normal operation), and $\mathbf{M}_C$ and Σ_C are the mean vector and covariance matrix that describe the current distribution of residuals determined using recent measurements. The average covariance matrix, Σ, is determined as the weighted average of Σ_N and Σ_C with Equation 7.2.3 where the weighting factor s is determined by minimizing the classification error (i.e., probability of making an erroneous decision). The classification error, ε, is determined by integrating the overlapping areas associated with the multidimensional normal and fault distributions using Fukunaga (1990).

$$\varepsilon = erfc\left(\frac{-\mathbf{V}^T\mathbf{M}_N - v_o}{\sqrt{2\sigma_N^2}}\right) + erfc\left(\frac{-\mathbf{V}^T\mathbf{M}_C - v_o}{\sqrt{2\sigma_C^2}}\right) \tag{7.2.5}$$

where

$$\mathbf{V} = \left(s\mathbf{\Sigma}_N + (1-s)\mathbf{\Sigma}_C\right)^{-1}\left(\mathbf{M}_C - \mathbf{M}_N\right)$$

$$v_o = -\frac{s\sigma_N^2\mathbf{V}^T\mathbf{M}_C + (1-s)\sigma_C^2\mathbf{V}^T\mathbf{M}_N}{s\sigma_N^2 + (1-s)\sigma_C^2}$$

$$\sigma_N^2 = \mathbf{V}^T\mathbf{\Sigma}_N\mathbf{V}$$

$$\sigma_C^2 = \mathbf{V}^T\mathbf{\Sigma}_C\mathbf{V}$$

The mean vector is determined by averaging differences between measured and model predictions of outputs over a specified measurement window. The uncertainty of the residuals characterized with the covariance matrix depends upon both measurement and modeling errors. Measurement errors impact output measurements directly and output model predictions indirectly through their effect on input measurements. Modeling errors can result from neglecting inputs that affect the output states, using a steady state model to characterize transient operation, and an imperfect mapping between the inputs and outputs.

The covariance matrix is determined if the modeling and measurement errors are independent and normally distributed. The measurement errors associated with the inputs are propagated through the steady state model using a first-order Taylor series about the known operating point, so that the elements of the covariance matrix for the model form of Equation 7.2.2 are

$$\Sigma_{ij} \approx \sigma_T^2 + \sigma_{M,i}^2 + \left(\frac{\partial y_i}{\partial T_{amb}}\right)^2\sigma_T^2 + \left(\frac{\partial y_i}{\partial T_{ra}}\right)^2\sigma_T^2 + \left(\frac{\partial y_i}{\partial T_{wb}}\right)^2\sigma_{wb}^2, \qquad i = j \tag{7.2.6}$$

$$\Sigma_{ij} \approx \left(\frac{\partial y_i}{\partial T_{amb}}\right)\left(\frac{\partial y_j}{\partial T_{amb}}\right)\sigma_T^2 + \left(\frac{\partial y_i}{\partial T_{ra}}\right)\left(\frac{\partial y_j}{\partial T_{ra}}\right)\sigma_T^2 + \left(\frac{\partial y_i}{\partial T_{wb}}\right)\left(\frac{\partial y_j}{\partial T_{wb}}\right)\sigma_{wb}^2, \qquad i \neq j \tag{7.2.7}$$

where

Σ_{ij} is the element in the ith row and jth column of the covariance matrix

y_i is the steady state model prediction for output i

σ_T^2 $E(w_T^2)$, where w_T is zero mean noise added to the dry-bulb temperature measurements (uncertainty in temperature measurement)

$\sigma_{M,i}^2$ $E(w_{M,i}^2)$, where $w_{M,i}$ is zero mean noise added to the model predictions (modeling uncertainty) for output i

σ_{wb}^2 $E(w_{wb}^2)$, where w_{wb} is zero mean noise added to the wet-bulb temperature measurements (uncertainty in wet-bulb measurement)

$E()$ is the expected value operator

The modeling uncertainty for the ith output model can be estimated as the variance associated with the fit to the training data (approximated by the sum of the squares of the errors). In general, measurement uncertainty is caused by both random and systematic errors. Random errors are associated with noise in the instrumentation system and can be characterized using specifications from the sensor manufacturer. Systematic errors refer to measurements that are biased in one direction (i.e., higher or lower than the actual value). Systematic measurement errors may be caused by miscalibration or drift in sensors. If the models are trained using the installed sensors, miscalibration is not an issue. However, if systematic errors are not considered as part of the measurement uncertainty, the fault detection method will identify a fault condition when sensors drift. Generally, it is prudent to assign a measurement uncertainty that allows for some sensor drift. Reasonable values for the standard deviations of the temperature and wet-bulb measurements are $\sigma_T = 0.5$ C and $\sigma_{wb} = 1.0$ C.

Method Comparisons — There are some advantages and disadvantages associated with each of the two approaches for fault detection presented in this section. The use of COP as a performance index is very straightforward to implement, but requires costly measurements of cooling capacity and power. The use of temperature measurements as performance indices has lower sensor cost but is more complicated to implement. Either of these methods could be utilized for packaged air conditioning or chiller equipment. However, the model forms and driving conditions are different for the two applications and may depend upon the method used for capacity control.

Fault Diagnosis

Once a fault has been detected, it is necessary to identify its cause. This may involve sending a technician to the site to perform additional testing and analysis. However, a fully automated FDD would perform some diagnoses using the available measurements. Several investigators have proposed the use of thermodynamic impact to diagnose faults which will be illustrated using the following example.

Consider a packaged air conditioner with a fixed orifice as the expansion device, a reciprocating compressor with on/off control, fixed condenser, and evaporator air flows, with R22 as the refrigerant. Figure 7.2.13 shows a P-h diagram for three cases of steady state operation at a given set of secondary fluid inlet conditions to the evaporator and condenser: normal, fouled condenser, and low refrigerant charge. Condenser fouling is equivalent to having a smaller condenser and leads to higher condensing temperatures and pressures than for the normal (no fault) case. For a system with a fixed orifice, the higher condensing pressures lead to a greater condenser-to-evaporator pressure differential that tends to increase the refrigerant flow rate. Furthermore, the increased flow rate tends to reduce the amount of condenser subcooling and evaporator superheat and increase the evaporating temperature. In contrast, the loss of refrigerant tends to lower the pressure throughout the system leading to reductions in both evaporating and condensing temperatures. The lower evaporating pressure and corresponding vapor density leads to a lower refrigerant flow rate, which results in higher evaporator superheat and a higher refrigerant discharge temperature from the compressor. This example illustrates that condenser fouling and low refrigerant can be distinguished by their unique effects on thermodynamic measurements.

Rule-Based Classifiers — Some of the proposed diagnostic methods for vapor compression cooling equipment use differences between measurements and normal expectations of thermodynamic states at steady state for diagnoses of faults. Fault diagnosis is then performed using a set of rules that relate each fault to the direction that each measurement changes when the fault occurs. Table 7.2.6 gives the diagnostic rules for the five faults and seven output measurements developed by Breuker and Braun (1998a,b) for a rooftop air conditioner. The arrows in Table 7.2.6 indicate whether a particular measurement increases (↑) or decreases (↓) in response to a particular fault at steady state conditions. For instance, as previously shown, the loss of refrigerant generally causes the superheat of the refrigerant entering the compressor to increase above its "normal" value at any steady state condition. Each of the faults results in a different combination of increasing or decreasing measurements with respect to their normal values. The rules of Table 7.2.6 are effectively fault models that are generic for this type of air conditioner and

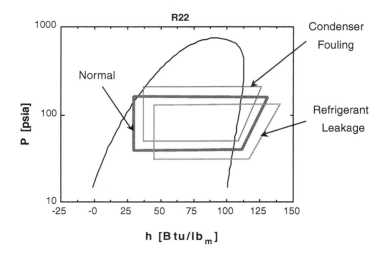

FIGURE 7.2.13 Effect of faults on thermodynamic states.

TABLE 7.2.6   Rules for the Diagnostic Classifier

Fault	T_{evap}	T_{sh}	T_{cond}	T_{sc}	T_{hg}	ΔT_{ca}	ΔT_{ea}
Refrigerant Leak	↓	↑	↓	↓	↑	↓	↓
Compressor Valve Leakage	↑	↓	↓	↓	↑	↓	↓
Liquid-Line Restriction	↓	↑	↓	↑	↑	↓	↓
Condenser Fouling	↑	↓	↑	↓	↑	↑	↓
Evaporator Fouling	↓	↓	↓	↓	↓	↓	↑

do not require any on-line learning. Similar rules were developed by Grimmelius et al. (1995) and Stylianou and Nikanpour (1996) for chillers (see Section 7.2.6).

Similar to the fault detection problem, it is necessary to have thresholds for diagnostics. The diagnostic classifier should evaluate the probability that each fault applies to the current operation, and the evidence should be high for a particular fault before any recommendations are made. Rossi and Braun (1997) addressed the issue of diagnostic thresholds in the development of their statistical rule-based FDD method. The diagnostic classifier evaluates the probability that each fault applies to the current operation. It estimates the degree to which the probability distribution characterizing the current residuals overlaps the region of the m-dimensional space defined by the set of rules corresponding to that fault.

Figure 7.2.14 illustrates the fault diagnostic classification method for two possible faults (refrigerant leakage and liquid-line restriction) with two input features (superheat and subcooling residuals). The progression of changes in the contours of two-dimensional probability distributions are shown as the two different faults are slowly introduced. Normal operation is shown as the distribution centered at the zero point. As a fault develops, the contour moves along a curve. When the overlap between the normal performance distribution and the current distribution (as indicated by the classification error, ε), is small enough for the false alarm rate to be acceptable (e.g., $\varepsilon < 0.001$), a fault is signaled by the fault detector. The different diagnostic classes are separated by the axis. The overlap of the current distribution with each of the modeled classes is calculated and represents the probability that the fault class is the correct diagnosis. A diagnosis is indicated when the probability (overlap) of the most likely class is larger than the second most likely class by a specified threshold (e.g., factor of 2). As the fault becomes more severe, confidence in the fault detection and diagnosis increases as the current distribution moves further from the normal distribution, and from the axis separating the classes. The choice of a diagnostic threshold results from a tradeoff between diagnostic sensitivity and the rate of false diagnoses.

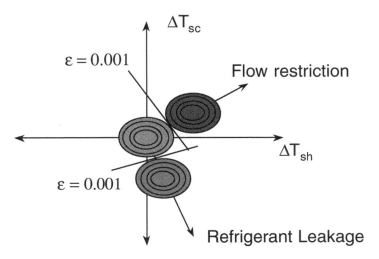

FIGURE 7.2.14   Fault diagnostic classifier (two-dimensional example).

To perform the classification for diagnostics, the probability that each rule applies to the current operation is evaluated. The probability of each hypothesis is determined by the degree to which the distribution characterizing the current residuals overlaps each class. The overlap is evaluated by integrating the area under the m-dimensional Gaussian probability distribution that falls within each class's region of the domain. Assuming that each dimension is independent, then the probabilities in each dimension can be "ANDed" together such that:

$$w_j = \prod_{k=1}^{m} \frac{1}{2}\left[1 + C_{jk}\, erf\left(\frac{\mathbf{M}_C(k) - \mathbf{M}_N(k)}{\sqrt{2\Sigma_C(k,k)}}\right)\right] \tag{7.2.8}$$

where C_{jk} = +1 if $(\mathbf{M}_c(k) - \mathbf{M}_N(k))$ falls within the domain for the jth fault (i.e., $(\mathbf{M}_c(k) - \mathbf{M}_N(k))$ has the same sign as defined in Table 7.2.6 for the appropriate fault) and C_{jk} = −1, otherwise. For diagnoses, the current distribution has been shifted to give zero mean for normal operation. A nonzero residual mean could occur for normal operation with an imperfect model.

Fault Evaluation

It is possible to design an FDD system that can detect and diagnose faults well before there would be a need to repair the unit. In general, an FDD system should evaluate the impact of the fault before recommending a course of action. These recommendations should be based upon the severity of the fault with respect to four criteria:

1. Impact on equipment safety
2. Environmental impact
3. Loss of comfort
4. Economics

Equipment safety primarily relates to the compressor and motor. The compressor/motor should not operate under conditions that will lead it to fail prematurely. These conditions include liquid entering the compressor, high compressor superheat, high pressure ratio, high discharge pressure, high motor temperatures, low oil, etc. Existing controllers generally have safeties that will shut down the unit in case of operation at adverse conditions. Under these circumstances, the FDD system could add an explanation regarding the probable fault that led to the shutdown. In addition, lower level warning limits should be established for these variables. When these limits are exceeded, the evaluator might recommend that service be performed when convenient.

The environmental criterion primarily relates to refrigerant leakage. Refrigerant leakage is an environmental hazard and should be repaired quickly. This is particularly true if the refrigerant is toxic (ammonia). However, when a refrigerant leak is detected and diagnosed, the actual output of the evaluator might depend on the rate of refrigerant leakage and type of refrigerant. For a small leak, it may be acceptable to keep the unit running and schedule repairs for the near future. Conversely, for a large leak, it may be appropriate to shut down the unit and call for immediate repairs.

Ideally, the evaluator should be able to identify if the current "health" of the equipment is such that it will not have sufficient cooling capacity to maintain comfort in the future. Once a fault has been identified, this feature would allow scheduling of service to address this need rather than requiring immediate service in response to a loss of comfort (i.e., complaints). This could involve the use of on-line models for predicting cooling capacity and cooling needs.

If a fault has been identified, but the current operation is not adversely affecting the equipment life or the environment and the system can maintain comfort both now and in the future, then service should be performed only if it is economical to do so. In this case, the best decision results from a tradeoff between service and energy costs. Service costs money but reduces energy costs. Rossi and Braun (1996) developed a simple method for optimal maintenance scheduling for cleaning heat exchangers and replacing air-side filters. The method relies on measurements of power consumption, estimates of cost per service, and utility rates, but does not require any forecasting. At any time, t, a decision to recommend service is based upon evaluation of the following inequality.

$$\int_0^t \gamma_{on} h(\tau) d\tau + \frac{C_s}{C_e} > t \cdot h(t) \tag{7.2.9}$$

where C_s is the cost for performing the service (\$), C_e is the cost per unit energy (\$/kW), γ_{on} is an on/off indicator (one if the unit is on and zero otherwise), and $h(\tau)$ is the extra power required to provide the necessary cooling caused by the performance degradation. At any time,

$$h(t) = P(t) - P^*(t) \tag{7.2.10}$$

where $P(t)$ is a measurement of the current power and $P^*(t)$ is a prediction of the power at the current operating conditions if the unit was operating normally (no fouling).

Equation (7.2.9) was derived by applying optimization theory with simplifying assumptions to a cost function that combines energy and service costs. This simplified method gave nearly identical results as a detailed optimization when tested through simulation for a range of situations. The combined energy and cost savings were found to between 5 and 15% for optimal vs. regular maintenance scheduling. The savings primarily depend upon the ratio of service to energy costs, the rate of fouling, and the baseline regular service interval.

Steady State Detectors

Many vapor compression cooling systems utilize "on/off" control and spend a significant amount of time in a transient condition. When a steady state model is used to predict normal operating states, a steady state detector must be used to distinguish between transient and steady state operation. The FDD system should only indicate a fault and provide a diagnosis when the system is in steady state.

Steady state detection can be implemented using the time rate of change in measurements during a moving window. Steady state is indicated whenever the "smoothed" time derivatives are less than a fixed threshold (e.g., 0.1 F/h for temperature measurements). Another approach is the exponentially weighted variance method of Glass et al. (1995). This algorithm estimates the sample variance about the mean of output measurements over a moving exponentially weighted window. In general, the variance decreases

TABLE 7.2.7 Performance of FDD Prototype (3 Input, 10 Output Temperatures)

Performance Index	Refrigerant Leakage (% Leakage)		Liquid-Line Restriction (% ΔP)		Compressor Valve Leak (% $\Delta\eta_v$)		Condenser Fouling (% lost area)		Evaporator Fouling (% lost flow)	
	1st	All	1st	All	1st	All	1st	All	1st	All
Fault Level (%)	5.4	Max	2.1	4.1	3.6	7.0	11.2	17.4	9.7	20.3
% Loss Capacity	3.4	>8	1.8	3.4	3.7	7.3	2.5	3.5	5.4	11.5
% Loss COP	2.8	>4.6	1.3	2.5	3.9	7.9	3.4	5.1	4.9	10.3
ΔT_{sh}	5.4	>11	2.3	4.8	−1.8	−3.6	−0.6	−1.6	−1.7	−2.7
ΔT_{hg}	4.8	>10	2.4	4.8	0.0	0.0	1.8	2.3	−1.2	−2.7

as the system approaches steady state. Two parameters of the steady state detector that impact FDD performance are the forgetting factor, and the threshold for steady state detection. The forgetting factor varies between zero and one and dictates the weighting of previous measurements (one for equal weighting, and zero for zero weighting of all previous measurements). The steady state detector threshold is the output variance below which an output is considered to be at steady state. As the threshold is decreased, the residuals of the steady state operating points should decrease, and FDD sensitivity should be improved. However, fewer operating points are available for FDD. In general, all of the output measurements could be used in testing for steady state behavior. However, it is also possible to select measurements having slower transients, such as compressor shell temperatures.

FDD System Performance

Only limited testing has been performed on complete FDD systems for vapor compression cooling equipment. To adequately test an FDD system in the field, it would be necessary to install several systems and collect data for several years before enough faults could develop and experience could be collected to make general assessments. Laboratory testing allows a more thorough evaluation of FDD performance in a shorter time frame, but may not include some important effects that occur in the field. Grimmelius et al. (1995), Stylianou and Nikanpour (1996), and Bailey et al. (2000) performed laboratory evaluations of FDD systems for chillers. In these tests, a limited number of faults were simulated in the laboratory and the performance of the methods was evaluated in terms of whether the method could correctly identify the fault. In some cases, the misclassification (or false alarm) rate was estimated.

Breuker and Braun (1998a,b) performed an extensive evaluation of the FDD technique developed by Rossi and Braun (1997). Steady state and transient tests were performed on a simple rooftop air conditioner in a laboratory over a range of conditions and fault levels. The data without faults were used to train the models for normal operation and determine statistical thresholds for fault detection, while the transient data with faults were used to evaluate FDD performance.

Table 7.2.7 shows results that characterize the sensitivity of the FDD method for detecting and diagnosing faults. The levels at which each fault could be detected at one point ("First Detected") and at all steady state points ("All Detected") from the database of transient test results are presented for the five faults along with the corresponding percent loss in capacity and COP, and the change in superheat and subcooling at these detectable levels. These results show that the faults can generally be detected and diagnosed before a decrease in capacity or efficiency of 5% is reached. In terms of the effect on performance, the technique is less sensitive to compressor valve leakage and evaporator fouling. At these levels, the changes in compressor superheat and hot gas temperature were probably not large enough to have an impact on the life of the compressor.

7.2.10 Application of Diagnostics Methods and Tools for Continuous Commissioning of Building Systems

In most cases, FDD systems installed on-line can also be used to continuously commission a building system. Commissioning is a systematic process by which proper installation and operation of building systems and equipment are checked and adjusted as necessary to improve performance (adjustment is virtually always required; only the degree of correction and performance impact differ among buildings). Proper commissioning begins during design, continues through construction (remodeling or retrofit), and includes establishment of a good preventive maintenance program (PECI, 1997; US DOE/PECI, 1997). Although a distinction is made between commissioning of new buildings (Cx) and commissioning of existing buildings (retro-commissioning or Rx), in this chapter we will refer to both generically as Cx. Cx is active, i.e., test and analyze, while Rx is passive, i.e., observe and analyze.

Despite the benefits, the commercial buildings market has been slow to widely adopt Cx. One reason is the first cost associated with performing Cx. Although Cx has been shown in many cases to be cost-effective on a life-cycle basis, the importance of first cost still dominates many decisions in the buildings industry.

Automation is already used for some commissioning tasks. Spreadsheets, for example, are used for processing, tabulating, and graphing input data and results. Handheld personal computers are used for such tasks as inputting data directly into a database in the field. Cx and Rx are discussed in detail in Chapter 7.1.

Prospects for Improvement

Automation provides several opportunities for improving the process of commissioning. Generic improvements that automated tools can provide include:

- Speeding up the process of preparing a commissioning plan
- Ensuring compliance with standards/guidelines, and providing consistency across projects
- Speeding up the process of detecting and diagnosing problems with operation of heating, ventilating, and air conditioning equipment and systems
- Eliminating errors that occur during manual data entry
- Disseminating expert knowledge by embedding it in software tools
- Ensuring consistency in fault detection and diagnosis across buildings, projects, and different commissioning agents through the use of that embedded knowledge
- Archiving data electronically for future reference or use

The promise of automation is higher quality commissioning at lower cost. Higher quality results from better quality control (in data management and analysis) and from making expert knowledge readily available in an easy-to-use form. Lower costs are the result of reducing use of expensive labor for mundane tasks such as recording data.

Tools and methodologies that implement some of these generic capabilities are available today. In some cases, tools have been implemented by individual commissioning agents or companies in spreadsheets or specialty programs. In other cases, software that assists with data management, analysis, or diagnosis is available commercially, from professional organizations, or from government agencies. Some examples are identified later in this chapter.

A Tool for Commissioning Outdoor-Air Handling

This section describes in some detail how the OAE software tool developed for detecting and diagnosing problems with outdoor-air control can be used to facilitate and potentially improve commissioning of air-handling units.

As mentioned earlier, the OAE diagnostician monitors the performance of the air-handling units and can detect more than 25 different basic operation problems with outdoor-air control and economizer operation.

OAE Application in Commissioning

The OAE diagnostician can be used to commission AHUs. Data can be collected over a short term, and batch processed or continuously collected and processed on-line. The first of these is easier and, therefore, less costly to implement. It requires no installation of the OAE diagnostician onsite, no direct connection to data sources, such as a BAS, and no operator training. Data are typically collected by establishing trend logs in a BAS to collect the necessary data. Alternatively, temporarily installed data loggers can be used to collect some or all of the required data. Data should be collected for approximately 2 weeks. If the data are already available from historical trend logs, these logs can be used in place of collecting new data.

When available, data are processed by the OAE diagnostician, which identifies problems, possible causes, and corresponding corrective actions, and estimates energy and cost impacts caused by improper operations during the observation period. The energy and cost impacts can be used to prioritize actions to address the problems found. When problems are found, further investigation may be required and actions should be taken to correct them. After implementing corrective actions, the commissioning team should collect metered data for another week or two to confirm correct operation. In some cases, the OAE diagnostician may detect new and previously undetected problems during this follow-up period. These problems should be corrected and proper operation verified by an additional week of metering and processing by the OAE. Problems not ordinarily found during commissioning may be detected this way.

As with commissioning in general, there is a set of operational problems that cannot be detected easily during normal operation unless commissioning tests are performed over the full range of weather and occupancy conditions. Tests must be performed for these during other seasons, or systems artificially tested in these modes. Because the OAE diagnostician is a passive diagnostician, it can only examine system performance for conditions that exist during data collection. To commission the outdoor-air handling system completely, testing by the OAE diagnostician must be conducted during other times of the year. Ideally, a complete check would require testing under occupied and unoccupied conditions for each of the following operation modes:

- Heating
- Economizer cooling with throttling: outdoor-air temperature (or enthalpy) lower than supply-air temperature (or enthalpy)
- Economizer with outdoor-air damper fully open: outdoor conditions lower than return-air conditions, but higher than the supply-air conditions
- Mechanical cooling with economizer locked out (closed to the minimum required ventilation position) because outdoor-air conditions are warm and/or humid

The commissioning plan should specify the full range of tests to be conducted, for how long, and at what times of the year. Partial testing is far better than no testing at all; however, retesting at various times of the year (and operating modes) is necessary to be reasonably assured that all problems have been identified. Having set up the necessary BAS trend logs for initial use of the OAE, collection of data during various times of year and processing by the OAE should be relatively simple. Furthermore, periodic testing in the long term (i.e., periodic recommissioning) can help ensure that good performance persists.

The OAE also provides a continuous record of attempts to meet the standard of best practice to provide adequate outdoor-air ventilation. This record can be used to help establish due diligence on the part of the building owner or operators in the event of a lawsuit related to indoor-air quality (IAQ). The importance of maintaining proper outdoor-air ventilation is emphasized by a recent EPA study (Daisey and Angell, 1998) which found that the leading cause of IAQ problems in schools is simply inadequate outdoor-air supply.

Significant energy and money savings, associated productivity improvements, and carbon emission reductions are possible from proper Cx followed by steps that help ensure the persistence of the Cx improvements. Productivity savings are the most significant of these. However, the penetration of commissioning in the building stock is very low. Steps are underway to promote greater use of commissioning, including demonstration projects, publication of case studies, documentation of savings, and government programs that encourage Cx. Improving and reducing the cost of the Cx process is also an important component of a multifaceted approach to bringing the benefits of Cx to the entire building stock. Automation shows promise as a tool for improving the process. Automating parts of the commissioning process will reduce cost, improve effectiveness, and ensure persistence of the benefits of Cx.

7.2.11   Infrastructure Requirements for Deploying FDD Systems in Buildings

Networked software applications, which can harness the vast potential of integrating control networks with the Internet, require access to data from control panels or sensing devices that may be distributed across buildings. Being able to exchange data and information between field devices and software applications is the key to successful implementation of the networked software applications (Bayne, 1999). Although software applications are independent from the process of gathering data, the capability to gather data is dependent on the functions provided by the BAS and the type of interface (gateway) and communications protocols it uses.

An infrastructure supporting the next generation of software tools that owners and operators will use to manage distributed facilities requires

- A control network with a BAS or network of intelligent devices (in each building)
- A mechanism or a transport layer that ties field panels and other devices on the control networks to the Internet
- The "killer software applications" that enhance facility management

Networking Developments

BASs have evolved over the past two decades from pneumatic and mechanical devices to direct digital controls (DDC). Today's BASs consist of electronic devices with microprocessors and communication capabilities. Widespread use of powerful, low-cost microprocessors, use of standard cabling, and adoption of standard protocols (such as BACnet, LonWorks) have led to today's improved BASs. Most modern BASs have powerful microprocessors in the field panels and controllers, and the prevalence of microprocessors embedded in the sensors is growing as well. Therefore, in addition to providing better functionality at a lower cost, these BASs also allow for distributing the processing and control functions including FDD within the field panels and controllers without having to rely on a central supervisory controller.

Many BAS manufacturers support either BACnet or LonWorks protocols; some support both (EUN, 1999). Recently, ASHRAE has approved a BACnet/IP addendum that makes it easier to monitor and control building systems from remote locations over the Internet. LonWorks is also heading in the same direction.

The manufacturers of BASs are developing gateways to connect modern proprietary control networks to the Internet, making it easy for distributed software applications to share information. However, there are many legacy BASs in the field for which gateways are needed but do not exist or will never be developed. In such situations, there are three options to connect these systems to the Internet: (1) DDE (dynamic data exchange), (2) OLE (object link and embedding), and (3) developing a custom interface between the BAS and the Internet for legacy systems that do not support either DDE or OLE.

Data-Gathering Tools

Without easy access to data from meters, controllers, and equipment that are distributed throughout the facility, it would be difficult to realize all the benefits of distributed facilities management. Although the details of data gathering depend on the type of BAS and the protocols it supports, integrated networks provide some standard methods to access data from geographically distributed facilities.

As part of a larger U.S. Department of Energy project to develop an automated diagnostician (whole building diagnostician (WBD)), prototype tools were developed to collect data from BASs locally or over the Internet. These tools allow building-generated data to be collected at any frequency and stored in a database.

Many BAS manufacturers provide DDE/OLE servers to facilitate data exchange between controllers/devices and software application programs. The WBD data collection tools, running in the background, initiate a DDE "conversation" between the manufacturer's DDE server (provided by the BAS manufacturer) and the WBD database, and collect data at time intervals set by the operator. Relationships defined during setup of the software for the building are used to map data from the sensors for each of the AHUs and building end-use meters into the WBD database. The data-gathering tools are independent of the WBD diagnostic modules; therefore, any application can use the data-gathering infrastructure. By querying the database, raw data can be retrieved for use by other software applications, such as programs used to reconcile metered data with utility bills and charge tenants for energy use.

7.2.12 Estimating Cost and Energy Impacts from Use of Diagnostic and Predictive Maintenance Tools

It is very important that FDD systems evaluate the cost impacts of faults detected in building systems for two basic reasons: (1) justifying the expense of developing and/or purchasing the FDD system by quantifying its benefits, and (2) providing perspective on the magnitude of the fault to prompt the user to fix the faults with high cost impacts, prioritize correction of faults with moderate impact, and neglect faults with low impacts. This section will discuss how *energy* cost impacts of faults can be estimated. Other impacts, such as comfort, health, and damage or shortened lifetimes for equipment are also very important, but are difficult to quantify and often very specific to the buildings and systems involved. We will not attempt to provide guidance on estimating these impacts here, although their importance and the added value to an FDD system of providing this information cannot be overemphasized.

We express the energy cost at any instant of time (t) as the product of the demand for power, its price, and the interval being analyzed (Δt)

$$\text{Cost}(t) = \text{Demand}(t)\,\text{Price}(t)\,\Delta t \tag{7.2.11}$$

Denoting quantities under faulted conditions with a prime ($'$), the cost increment of a fault is

$$\Delta\text{Cost}(t) = \text{Cost}'(t) - \text{Cost}(t) = \text{Demand}'(t)\,\text{Price}'(t)\Delta t - \text{Demand}(t)\,\text{Price}(t)\Delta t$$

$$= \big[\Delta\text{Demand}(t)\,\text{Price}(t) + \text{Demand}'(t)\Delta\text{Price}(t)\big]\Delta t \tag{7.2.12}$$

where, for any general property, X

$$\Delta X(t) = X'(t) - X(t) \tag{7.2.13}$$

Equation 7.2.12 includes the general case where the price of power, usually electricity, is dependent upon time (real-time pricing or time-of-day rates), or the demand itself (either at the current time or during some time period used by a utility to define peak demand). If the price is not a function of demand, as is often the case (at least for the fault's impact), then the second term drops out and Equation 7.2.12 reduces to

$$\Delta\text{Cost}(t) = \Delta\text{Demand}(t)\,\text{Price}(t)\Delta t \tag{7.2.14}$$

Often a building system uses more than one fuel to supply the services it is designed to provide. An example is an AHU supplying both heating and cooling services in a gas or steam heated building. In such cases Equations 7.2.11 and 7.2.13 must be applied twice, once for each fuel affected by a given fault.

Direct Estimation of Impacts Based on Measured Consumption

In cases where the FDD technique directly measures the demand for energy and compares it against some expected value (see the whole-building energy diagnostician, described earlier), the change in demand caused by the fault at any given time can be directly estimated as the difference between the actual and expected consumption. Estimating the impact of the fault then reduces to integrating ΔCost over time (more on this later).

In such methods, the expected consumption is typically based on some type of model of *average* consumption for the time-of-day, and/or time-of-week, weather, and other conditions. These models are either empirical (regression, neural network, bin method, etc.), or engineering-based (presumably calibrated to fit historical consumption patterns). At best, random deviations of the actual consumption from the average for the time and conditions are to be expected, with the size of the deviations proportional to the accuracy of the model. Nonrandom deviations can also be expected when, as is usually the case, the input data or the mathematical form of the model imperfectly captures important effects, including nonlinearity.

Even for very accurate models, these errors tend to become larger as the time interval for the FDD cost analysis (Δt) gets smaller, i.e., weekly models are more stable than daily models, and daily models are more stable than hourly models. Subhourly models exhibit even more "noise," including that from such tangible effects as the cycle time of equipment in a building. Cost impacts, therefore, will generally be more accurate when integrated over a number of time intervals (Katipamula et al., 1996).

The advantage of this approach to estimating cost impacts is its simplicity and directness. This advantage is strong enough that it is worthwhile considering directly measuring and modeling the demand of the system or subsystem, which is the focus of the FDD system. This will become clearer when an alternative approach to estimating energy cost impacts is described below. However, there are two primary limitations to this approach related to the need to measure and model demand. The first limitation is cost; if it is not used as an inherent part of the FDD method, extra instrumentation and analysis capability is added to it solely for the purpose of estimating energy cost impacts.

The second limitation is complexity. While obvious at the whole building and boiler/chiller plant levels, the demand caused by other systems is often indirect and hard to quantify. A good example of this is an AHU in a built-up HVAC system; it consumes energy in the form of hot and chilled water from the plant, some electricity for fan power, and its operation impacts the subsequent need for terminal reheat in the zones it serves. In principal, it might be possible to measure and model each of these three modes of consumption separately. This may be less expensive if some proxy measurements are used. Examples are valve positions or temperature differences across coils instead of Btu meters, for estimating the energy in constant volume flows. The complexities of such an approach become evident, however, when considering how to handle faults that may lie in other systems (the terminal boxes, or the hot and chilled water reset schedule, for example) but that reflect themselves in the consumption patterns of the AHU.

Estimation of Impacts from First Principles

An alternative means of estimating energy cost impacts of faults in buildings is to base them on a first principles analysis. This approach is useful when

- Impacts are expected to be a small fraction of a measured consumption total
- Attribution of impacts among multiple faults is desired
- The expected impacts are about the same as the expected accuracy (i.e., the "noise") of an empirical model of the measured consumption.

An example is failure of a lighting occupancy sensor in one office of a zone encompassing many offices. Here the expected impacts may be much less than 5%, but are virtually certain to exist. In such cases, it may be preferable to estimate the impact based on first principles, for accuracy, simplicity or both. If the lights involved do not impact heating or cooling loads, because they are not in a conditioned space, for example, then the impact could be estimated simply as the product of the lighting power density and the floor area per occupancy sensor.

The principal complicating factor in using this approach is when the fault impacts the heating or cooling loads, directly through the energy conversion efficiency or the outdoor-air ventilation of the system, or indirectly because it changes the internal heat gains of the space. This discussion will focus on cases where there are direct effects on heating and cooling loads, or there are enough indirect effects that it is necessary or desirable to consider them.

To make such estimates, we consider the demand to be comprised of two components, one related to the thermal (heating and cooling) loads, and the other related to all other (non-HVAC) demand. This approach implicitly assumes that the building system being analyzed is heated or cooled at a given time, but not both. For buildings where simultaneous heating of some zones and cooling of other zones can legitimately occur, this implies that the impact analysis must be computed at a lower level in the building hierarchy where this assumption is valid or is a more reasonable approximation.

The demand for energy is expressed as the sum of demands for cooling (Cool), heating (Heat), and other loads

$$\text{Demand}(t) = \text{Cool}(t) + \text{Heat}(t) + \text{Dist}_c(t) + \text{Lights}(t) + \text{Plugs}(t) + \text{Ext}(t) \tag{7.2.15}$$

where we explicitly account for other loads as the sum of fan and pump loads in constant volume systems (Dist_c), lights (Lights), plug loads (Plugs), and loads external to the building envelope (Ext), such as exterior lighting.

The zone heat balance or thermal load, Load(t), of a zone at any time is comprised of

- Internal heat gained from lights, Lights(t); plug loads, Plugs(t); occupants, Occ(t); and solar radiation (through glazed and opaque surfaces), Solar(t)
- Heat lost by conduction through the zone envelope to the outdoors, Cond(t)
- Heat conducted into the zone's internal thermal mass, Mass(t)
- Heat carried by the air to the outdoors by the ventilation air required for the zone's occupants, OA(t)

$$\text{Load}(t) = f_L \text{Lights}(t) + \text{Plugs}(t) + \text{Occ}(t) + \text{Solar}(t) - \text{Cond}(t) - \text{Mass}(t) - \text{OA}(t) \tag{7.2.16}$$

The fraction of the lighting energy that ends up as internal heat gain to the zone, f_L, is usually close to 100%. Heat from lights that enters the return air stream directly is included in f_L because it will return to the zone after passing through the AHU. The effect of exhausting some or all the return air with the outdoor-air/economizer system is accounted for by the OA(t) term. However, f_L will be less than 100% if some of the heat from lighting fixtures is conducted through the roof from top-floor ceiling plenums, for example.

The heating and cooling demands can be expressed as the ratio of the thermal load from the space that is seen by the HVAC system (positive for heating, negative for cooling) divided by the overall *system* energy conversion efficiency of the heating, cooling, or reheat system (COP). Because, in general, heating and cooling loads are served by systems with different thermal efficiencies or COP, the associated demands must be accounted for as separate terms. So the demand can be expressed as

$$\text{Demand}(t) = \frac{\left(\text{Load}(t) - \text{Econ}(t) + f_d \text{Dist}_c(t)\right)^+}{\text{COP}_{cool}(t)} - \frac{\left(\text{Load}(t) + f_d \text{Dist}_c(t)\right)^-}{\text{COP}_{heat}(t)} \tag{7.2.17}$$
$$+ \text{Dist}_c(t) + \text{Lights}(t) + \text{Plugs}(t) + \text{Ext}(t)$$

Econ(t) is the heat exhausted from the return air stream by an economizer introducing extra outside air beyond that required for the occupants when conditions are suitable for free cooling. $\text{Dist}_c(t)$ is the consumption of the distribution system (fans and pumps) that is relatively constant with respect to the amount of heating or cooling supplied, and f_d is the fraction of that energy entering the air and water flows as heat. (We will also include a term that is convenient for variable-volume distribution systems in the system COPs in the following discussion of specific types of faults.)

The terms enclosed in ( )$^+$ and ( )$^-$ in Equation 7.2.17 are the *net* cooling and heating loads seen by the HVAC system, respectively. That is, this is the amount of heating or cooling that must be delivered by the system to the space, less free cooling delivered by the economizer and heat gained from fans and pumps. By this notation, we define these terms as nonzero only when the enclosed term (the net thermal load) is positive and negative, respectively. These terms imply that normally there is no simultaneous heating and cooling for a zone (except for reheat). That is, there is a cooling demand only when the net load on the system is positive, and a heating demand only when it is negative.

The system COP includes the primary energy conversion efficiency of the heating or cooling sources at the current conditions (temperature, humidity, and part load). It also must include all duct losses, and may be defined (optionally, see later discussion) to include the energy consumption by fans and pumps for distribution, auxiliary loads, and any reheat energy required by the HVAC system for proper temperature control.

To proceed, we now make a series of assumptions and approximations. The phenomena involved in the conduction and mass terms are diverse and highly building- and even zone-specific. Without a detailed thermal simulation of the building it may be impossible to come up with good estimates of the resulting heat transfer rates at any given time. So, it is desirable to eliminate their effect from the impact estimate. It is convenient to assume that it is not necessary to estimate energy cost impacts for faults that result in appreciable changes to the zone temperature. Because this would presumably be reported by occupants or basic BAS alarms, it is not a primary target fault for FDD systems. Further, regulating it is the primary function of the control system and loss of comfort control presumably has impacts that far exceed associated energy impacts.

If the fault does not appreciably impact zone temperature, it can be assumed that the conduction and thermal mass terms are nearly the same in both the faulted and unfaulted conditions. The same assumption will be made for the plug loads and other external loads, because these are not generally the subject of either control or FDD systems.

Because we have assumed that the conduction, mass, plug loads, and occupancy terms are not appreciably affected by the fault, we can express the change in the zone thermal load as

$$\Delta \mathrm{Load}(t) = f_L \Delta \mathrm{Lights}(t) + \Delta \mathrm{Solar}(t) - \Delta \mathrm{OA}(t) \tag{7.2.18}$$

where f_L is the fraction of the lighting impact energy that is dissipated within the conditioned space. The estimated change in demand resulting from correction of a fault is

$$\cong \mathrm{Demand}(t) = \left\{ \mathrm{Cool}'(t) - \mathrm{Cool}(t) \right\} + \left\{ \mathrm{Heat}'(t) - \mathrm{Heat}(t) \right\} + \left\{ \mathrm{Dist}'_c(t) - \mathrm{Dist}_c(t) \right\}$$

$$+ \left\{ \mathrm{Lights}'(t) - \mathrm{Lights}(t) \right\} \tag{7.2.19}$$

where, again, we denote condition X under normal operation as X′ when a fault exists. Equation 7.2.19 must be applied once for each fuel impacted by the fault, dropping terms for demands not served by the fuel.

Assume that the normal operating heating and cooling system COPs can be estimated or computed, and that measured heating and cooling demands are available for the system being diagnosed, (i.e., those actually occurring at time t, *including the impact of any faults*). Alternatively, normal heating and cooling demands may be known from some type of theoretical model or an empirical model of past performance.

The cooling, heating, and reheat demands that would occur if the fault was fixed are

$$\mathrm{Cool}(t) = \frac{\left(\left\{ \mathrm{Load}'(t) - \Delta \mathrm{Load}(t) \right\} - \left\{ \mathrm{Econ}'(t) - \Delta \mathrm{Econ}(t) \right\} + f_d \left\{ \mathrm{Dist}'_c(t) - \Delta \mathrm{Dist}_c(t) \right\} \right)^+}{\mathrm{COP}_{\mathrm{cool}}(t)} \tag{7.2.20}$$

$$\mathrm{Heat}(t) = -\frac{\left(\left\{ \mathrm{Load}'(t) - \Delta \mathrm{Load}(t) \right\} + f_d \left\{ \mathrm{Dist}'_c(t) - \Delta \mathrm{Dist}_c(t) \right\} \right)^-}{\mathrm{COP}_{\mathrm{heat}}(t)} \tag{7.2.21}$$

The zone thermal load can be expressed as the difference in the net heating and cooling loads (estimated as the products of the cooling and heating COPs and demands), plus the economizer free cooling, less the heat gain from fans and pumps

$$\begin{aligned}
\text{Load}'(t) &= \text{Cool}'(t)\text{COP}'_{\text{cool}}(t) - \text{Heat}'(t)\text{COP}'_{\text{heat}}(t) + \text{Econ}'(t) - f_d\text{Dist}'_c(t) \\
&= \text{Cool}'(t)\{\text{COP}_{\text{cool}}(t) + \Delta\text{COP}_{\text{cool}}(t)\} \\
&\quad - \text{Heat}'(t)\{\text{COP}_{\text{heat}}(t) + \Delta\text{COP}_{\text{heat}}(t)\} + \text{Econ}'(t) - f_d\text{Dist}'_c(t)
\end{aligned} \tag{7.2.22}$$

For faults that degrade HVAC system performance, ΔCOP is normally negative. Note that Equation 7.2.22 expresses the net load in terms of both cooling and heating, and both must be included even if different fuels are used to supply them. So, Equation (7.2.22) is also valid when normal or faulted operation results in simultaneous heating and cooling.

The estimated change in demand caused by a fault can be expressed as

$$\begin{aligned}
\Delta\text{Demand}(t) &= \text{Cool}'(t) + \text{Heat}'(t) + \Delta\text{Lights}(t) + \Delta\text{Dist}_c(t) \\
&\quad - \frac{\left(\begin{array}{l}\text{Cool}'(t)\{\text{COP}_{\text{cool}}(t) + \Delta\text{COP}_{\text{cool}}(t)\} - \text{Heat}'(t)\{\text{COP}_{\text{heat}}(t) + \Delta\text{COP}_{\text{heat}}(t)\} \\ +\Delta\text{Econ}(t) - f_d\Delta\text{Dist}_c(t) - f_L\Delta\text{Lights}(t) - \Delta\text{Solar}(t) + \Delta\text{OA}(t)\end{array}\right)^+}{\text{COP}_{\text{cool}}(t)} \\
&\quad + \frac{\left(\begin{array}{l}\text{Cool}'(t)\{\text{COP}_{\text{cool}}(t) + \Delta\text{COP}_{\text{cool}}(t)\} - \text{Heat}'(t)\{\text{COP}_{\text{heat}}(t) + \Delta\text{COP}_{\text{heat}}(t)\} \\ +\Delta\text{Econ}'(t) - f_d\Delta\text{Dist}_c(t) - f_L\Delta\text{Lights}(t) - \Delta\text{Solar}(t) + \Delta\text{OA}(t)\end{array}\right)^-}{\text{COP}_{\text{heat}}(t)}
\end{aligned} \tag{7.2.23}$$

Note that Equation 7.2.23 must be applied separately for each fuel impacted by the fault. Entire terms for demands not served by the fuel are dropped (either both of the cooling demand terms, or both of the heating demand terms). All the subterms of the expressions for the net heating and cooling loads and the terms inside the ()$^+$ and ()$^-$ must be retained in their entirety to accurately characterize the net thermal load seen by the HVAC system. For example, if the impact on electrical demand is desired for a gas-heated building, the second and last terms are dropped and

$$\begin{aligned}
\Delta\text{Demand}(t) &= \text{Cool}'(t) + \Delta\text{Lights}(t) + \Delta\text{Reheat1}(t) + \Delta\text{Dist}_c(t) + \Delta\text{Aux1}(t) \\
&\quad - \frac{\left(\begin{array}{l}\text{Cool}'(t)\{\text{COP}_{\text{cool}}(t) + \Delta\text{COP}_{\text{cool}}(t)\} - \text{Heat}'(t)\{\text{COP}_{\text{heat}}(t) + \Delta\text{COP}_{\text{heat}}(t)\} \\ +\Delta\text{Econ}(t) - f_D\Delta\text{Dist}_c(t) - f_L\Delta\text{Lights}(t) - \Delta\text{Solar}(t) + \Delta\text{OA}(t)\end{array}\right)^+}{\text{COP}_{\text{cool}}(t)}
\end{aligned} \tag{7.2.24}$$

(7.2.24)
[electrical demand impact, nonelectric heating]

Application of First Principles Method

It is useful to consider Equation 7.2.23, which is generally applicable to multiple faults, as applied to faults associated with three specific energy using subsystems building systems: outdoor-air ventilation/economizers, lighting, and heating/cooling. These three specific applications of Equation 7.2.23 are considered in this subsection.

Ventilation Fault Impacts

Ventilation fault impacts involving outdoor-air supply generally occur at the AHU level of the building system hierarchy, and affect the net load of all the zones served by the AHU. These faults may involve supply of excess ventilation air or failure of the economizer operation to supply free cooling. In either case, we will assume that the fault manifests itself as a change in the volumetric flow rate of outside air (ΔF), and that the magnitude of this flow rate error is available from or can be estimated as a byproduct of the FDD algorithm. Further, to apply Equation 7.2.23 in its full form implies that we also know the heating and/or cooling delivered to the zones served by the AHU. This may involve metering of energy consumption in unitary packaged equipment or flow rates and enthalpies or temperatures in air handlers served by a boiler/chiller plant.

The impacts on outdoor air (OA) and economizer (Econ) are combined into a single fault. Assume (for this example) that there is no impact on the system COP, and no impact on the fan and pump consumption in constant volume systems. So, $\Delta COP(t)$ and $\Delta Dist_c(t)$ are zero. Then, the change in heat loss via the outdoor ventilation air can be expressed in terms of the air density (ρ) and the difference of the return and outdoor-air enthalpies ($h_r - h_o$) to estimate the demand impacts of ventilation system faults as

$$\Delta Demand(t) = Cool'(t) + Heat'(t)$$

$$- \frac{\left(Cool'(t)COP_{cool}(t) - Heat'(t)COP_{heat}(t) + (h_r - h_o)\Delta F(t)\right)^+}{COP_{cool}(t)}$$

$$+ \frac{\left(Cool'(t)COP_{cool}(t) - Heat'(t)COP_{heat}(t) + Econ'(t) + (h_r - h_o)\Delta F(t)\right)^-}{COP_{heat}(t)}$$

(7.2.25)
[ventilation
faults with no
system COP
impacts]

As before, terms in Equation 7.2.25 for demands not served by a given fuel are dropped. For example, if the impact on electrical demand is desired for a nonelectrically heated building, one would not include the second and last terms.

It may not be possible to measure or estimate the heating and cooling demands at the air handler level, and the assumption must be made that the mode of the air handler (heating or cooling) would be unchanged if the fault was corrected. Then, Equation 7.2.25 reduces to

$$\Delta Demand(t) = \frac{\bar{n}(h_o - h_r)\Delta F(t)}{COP_{cool}(t)}$$

(7.2.26)
[cooling impact, no change in
cooling mode caused by fault]

for impacts when cooling and

$$\Delta Demand(t) = \frac{\bar{n}(h_r - h_o)\Delta F(t)}{COP_{heat}(t)}$$

(7.2.27)
[heating impact, no change in
heating mode caused by fault]

for heating, because Econ'(t) in a heating mode is zero if the entire fault is included in ΔF. The OAE diagnostician used Equation 7.2.26 and Equation 7.2.27 to estimate the energy impacts from improper operation of the outdoor-air controls or economizer.

Note that the impact of faults in ventilation systems can increase or decrease the demand for cooling, depending upon the difference in the indoor and outdoor temperatures and humidities, and whether the fault results in too much or too little flow at any given time. For example, cooling loads are increased by excess flow when it is hotter or more humid outside than inside. They decrease when either the relative indoor and outdoor conditions or the sign of the flow rate error are reversed, but not both.

When the system is in the heating mode, the outdoor-air temperature is almost always less than the indoor-air temperature, otherwise heating would not be required. One notable, but rare exception to this generalization is during warmups on relative mild mornings. Therefore, the outdoor-air enthalpy is almost always less than the indoor-air enthalpy, and excess outdoor ventilation air flow will not result in a change in mode from heating to cooling, and Equation 7.2.27 usually applies for faults involving excess ventilation air. Note that negative flow rate errors during heating modes suggest that the outdoor-air volumes are less than needed for the occupants, with consequent negative effects on indoor air quality.

Whether during heating or cooling, faults that reduce ventilation air flow to levels below those required by the occupancy have negative air-quality impacts that should be considered far more valuable than the positive energy benefit. This suggests that positive cost benefits for such faults probably should not be displayed to users.

Lighting Fault Impacts

Impacts of faults resulting in excess gains from solar radiation, such as errors in day lighting sensors controlling active shading devices, are entirely analogous to the impacts of lighting faults. So, the discussion and the equations developed here can be used to estimate impacts for them by simply substituting the relevant load terms for the lighting terms.

Assume, for this example, there is no impact of the lighting fault on the system COP or the fan and pump consumption in constant volume systems. So, $\Delta COP(t)$ and $\Delta Dist_c(t)$ are equal to zero. If the lighting fault affects only one zone served by a constant-volume multizone AHU and if that zone is reheating, the change in lighting demand will simply be offset by an equivalent change in the need for reheat and the following analysis will not apply. In the case of electric resistance heating (COP = 1.0) and if f_L is 1.0, then net impact on the total demand will be zero. However, generally, lighting faults are likely to impact all zones served by an AHU. In such cases, the effect of changed reheat requirements is minimal and will be neglected as a second order effect. There is only a similar effect on reheat for variable air-volume systems for zones where air flow is at minimum, so this will also be neglected here.

From Equation 7.2.23, the estimated reduction in demand that would result from correcting a fault with lighting control causing excess consumption $\Delta Lights$, but not affecting the system COPs or the fan/pump power in constant volume system, is

$$\Delta Demand(t) = Cool'(t) + Heat'(t) + \Delta Lights(t)$$

$$- \frac{\left(Cool'(t)COP_{cool}(t) - Heat'(t)COP_{heat}(t) + \Delta Econ(t) - f_L \Delta Lights(t)\right)^+}{COP_{cool}(t)}$$

$$+ \frac{\left(Cool'(t)COP_{cool}(t) - Heat'(t)COP_{heat}(t) + Econ'(t) - f_L \Delta Lights(t)\right)^-}{COP_{heat}(t)} \qquad (7.2.28)$$

But, any change in lighting load may simply be absorbed by a corresponding change in the economizer operation, if it is not already operating at full flow. Because there is no fault in the economizer operation in this example, this implies that outdoor conditions are suitable for free cooling and there is no cooling demand, i.e., $Cool'(t)$ is zero. Therefore, the contribution of a normally operating economizer toward meeting the cooling loads in faulted and unfaulted conditions must be estimated.

Let the volumetric flow rate required for the occupants at any time (of day and week) be $F_{req}(t)$. If the flow rate under faulted conditions, $F'(t)$, is known or can be estimated, then

$$Econ'(t) = \tilde{n}\left\{F'(t) - F_{req}(t)\right\}\left(h_r - h_o\right) \qquad (7.2.29)$$

Let the maximum achievable flow rate be F_{max} when the economizer should fully open the outside-air dampers. An economizer normally operates to minimize cooling whenever possible. The maximum cooling load displaced by the economizer ($Econ'_{max}$) is the product of the air density, the volumetric flow rate, the return- and outdoor-air enthalpy difference, and two control functions

$$Econ'_{max}(t) = \left\{F_{max} - F_{req}(t)\right\}(h_r - h_o) \overbrace{\frac{(y_c - y_o)^+}{y_c - y_o}}^{1^{st} \text{ function}} \overbrace{\frac{T_r - max[T_o, T_{min}]}{T_r - T_o}}^{2^{nd} \text{ function}} \tag{7.2.30}$$

where the first control function defines whether the economizer is operating or not (values of zero or one), and the second control function defines the fraction of full flow at which it operates (values between zero and one). The variable y in the first control function is either temperature or enthalpy corresponding to the basis for the economizer control. The subscript c indicates the controlling variable, either return air for differential control or a high limit (usually temperature) for high-limit control.

Then, the maximum *increase* in the heat exhausted by the economizer, $\Delta Econ_{max}(t)$, is

$$\Delta Econ_{max}(t) = Econ_{max}(t) - Econ'(t) \tag{7.2.31}$$

and the maximum *decrease* in the heat exhausted by the economizer is equal to $Econ'(t)$.

For faults that *increase* the lighting load, i.e., $\Delta Lights(t)$ greater than zero, increased flow in the economizer will absorb the increased cooling load resulting from the higher lighting level until the economizer reaches maximum flow. Thus

$$\Delta Demand(t) = Cool'(t) + Heat'(t) + \Delta Lights(t)$$

$$- \frac{\left(Cool'(t)COP_{cool}(t) - Heat'(t)COP_{heat}(t) + \left(\Delta Econ_{max}(t) - f_L \Delta Lights(t)\right)^-\right)^+}{COP_{cool}(t)}$$

$$+ \frac{\left(Cool'(t)COP_{cool}(t) - Heat'(t)COP_{heat}(t) + Econ'(t) - f_L \Delta Lights(t)\right)^-}{COP_{heat}(t)} \tag{7.2.32}$$

[fault which increases lighting demand]

For faults that *decrease* the lighting load, i.e., $\Delta Lights(t)$ less than zero, the lower cooling load resulting from decreased lighting is offset by decreased flow in the economizer until it reaches the minimum required flow. Thus $\Delta Econ(t)$ is equal to $-Econ_{max}(t)$ and

$$\Delta Demand(t) = Cool'(t) + Heat'(t) + \Delta Lights(t)$$

$$- \frac{\left(Cool'(t)COP_{cool}(t) - Heat'(t)COP_{heat}(t) + \left(-Econ_{max}(t) - f_L \Delta Lights(t)\right)^+\right)^+}{COP_{cool}(t)}$$

$$+ \frac{\left(Cool'(t)COP_{cool}(t) - Heat'(t)COP_{heat}(t) + Econ'(t) - f_L \Delta Lights(t)\right)^-}{COP_{heat}(t)} \tag{7.2.33}$$

[fault which decreases lighting demand]

When the system is cooling in the faulted condition and correcting the lighting fault would not result in a change in mode from cooling to heating, then the economizer is already at maximum flow during free cooling conditions and Equation 7.2.32 and Equation 7.2.33 reduce to

$$\Delta Demand(t) = \Delta Lights(t) + \frac{f_L \Delta Lights(t)}{COP_{cool}(t)} \tag{7.2.34}$$

[cooling mode in faulted and unfaulted conditions]

where the lighting fault increases the impact of the fault.

For the simple case where the system is heating in the faulted condition and correcting the lighting fault would not result in a change in mode from heating to cooling, then Econ'(t) is zero and both Equation 7.2.32 and Equation 7.2.33 reduce to

$$\Delta \text{Demand}(t) = \Delta \text{Lights}(t) - \frac{f_L \Delta \text{Lights}(t)}{\text{COP}_{\text{heat}}(t)}$$

(7.2.35)
[electric heat, heating mode in faulted and unfaulted conditions]

where the change in the lights is partially offset by the increased requirement for heat, as expected. For nonelectric heating, Equation 7.2.35 must be applied twice, once for the heating fuel impact with the lighting term, $\Delta \text{Lights}(t)$, equal to zero, and once for the electricity impact when the second term is dropped.

HVAC Equipment Fault Impacts

For a fault whose impact is confined to the system COPs, the impact on the total demand is (from Equation 7.2.23)

$$\Delta \text{Demand}(t) = \text{Cool}'(t) + \text{Heat}'(t)$$

$$- \frac{\left(\text{Cool}'(t) \left\{ \text{COP}_{\text{cool}}(t) - \Delta \text{COP}_{\text{cool}}(t) \right\} + \text{Heat}'(t) \left\{ \text{COP}_{\text{heat}}(t) + \Delta \text{COP}_{\text{heat}}(t) \right\} + \Delta \text{Econ}(t) \right)^{+}}{\text{COP}_{\text{cool}}(t)}$$

(7.2.36)

$$+ \frac{\left(\text{Cool}'(t) \left\{ \text{COP}_{\text{cool}}(t) - \Delta \text{COP}_{\text{cool}}(t) \right\} + \text{Heat}'(t) \left\{ \text{COP}_{\text{heat}}(t) + \Delta \text{COP}_{\text{heat}}(t) \right\} + \text{Econ}'(t) \right)^{-}}{\text{COP}_{\text{heat}}(t)}$$

The system COPs must be estimated over the range of operating conditions, under both normal and faulted operations. For systems without preheat or dehumidification loads, the heating and cooling delivered by the system (exclusive of the economizer and constant distribution loads to be consistent with Equation 7.2.23 is

$$\text{Delivered}(t) = \text{Cool}(t)\text{COP}_{\text{cp}}(t) - \text{Heat}(t)\text{COP}_{\text{hp}} - \text{Reheat}(t)\text{COP}_{\text{rh}}(t) - f_d \text{Dist}_v(t) + f_D \text{Loss}(t)$$

(7.2.37)

where Reheat(t) is the heat demand required for proper temperature control in multizone systems; $\text{Dist}_v(t)$ is the fan and pump energy that varies with the load; Loss(t) is heat loss from the ducts and pipes caused by conduction and air leakage; f_D is the fraction of the duct loss that is retained in the conditioned space; and $\text{COP}_{\text{cp}}(t)$, $\text{COP}_{\text{hp}}(t)$, and $\text{COP}_{\text{rh}}(t)$ are the COPs of the primary cooling, heating, and reheating energy conversion equipment at the current load and temperature conditions.

For systems without preheat or dehumidification loads, the demand of the system (exclusive of the constant distribution loads to be consistent with Equation 7.2.23 is

$$\text{Demand}(t) = \text{Cool}(t) - \text{Heat}(t) - \text{Reheat}(t)\text{COP}_{\text{rh}}(t) - \text{Dist}_v(t) + \text{Aux}(t)$$

(7.2.38)

where Aux(t) is the energy consumption of auxiliary HVAC equipment such as condenser and cooling towers fans.

The system COPs are defined as the ratio of the delivered energy to the demand. The cooling system COP is

$$\text{COP}_{\text{cool}}(t) = \frac{\text{COP}_{\text{cp}} + \text{COP}_{\text{rh}} \dfrac{\text{Reheat}(t)}{\text{Cool}(t)} + f_d \dfrac{\text{Distrib}(t)}{\text{Cool}(t)} + f_D \dfrac{\text{Loss}(t)}{\text{Cool}(t)}}{1 + \dfrac{\text{Reheat}(t)}{\text{Cool}(t)} + \dfrac{\text{Distrib}(t)}{\text{Cool}(t)} + \dfrac{\text{Aux}(t)}{\text{Cool}(t)}}$$

(7.2.39)

and the heating system COP is

$$
COP_{heat}(t) = \frac{COP_{hp} + f_d \dfrac{Distrib(t)}{Cool(t)} - f_D \dfrac{Loss(t)}{Cool(t)}}{1 + \dfrac{Distrib(t)}{Cool(t)} + \dfrac{Aux(t)}{Cool(t)}}
\tag{7.2.40}
$$

Under unfaulted operation, the primary COPs can be estimated from manufacturer's data. These formulations of the system COPs are convenient because they express the effect of reheat, distribution, auxiliary consumption, and duct losses as ratios that tend to remain somewhat constant in many situations and that can be readily approximated for many systems. Alternatively, they can be estimated based on design calculations. Estimation of the system COPs under faulted conditions typically requires an estimate of the impact of the fault on only one of the terms.

In general (for an air conditioner, for example) the primary COP is a function of the latent cooling fraction, the supply and outdoor temperatures, and the part load ratio. These can be estimated from the manufacturer's data for various conditions under normal operation. The effect of the faults must then be estimated also. The primary COP is often discontinuous, such as for staged cooling devices. At rated conditions, the primary COP for combustion equipment is often between 0.8 to 0.95, for absorption cooling equipment between 0.45 to 0.7, for cooling equipment between 3 to 5, for electric heat pumps between 3 to 4, and for electric resistance heating equipment it is 1.0.

At full load conditions, fan power for air distribution systems is typically in the range of 5 to 10% of the delivered energy. For water distribution systems, this is typically in the range of 2 to 5%. These fractions are building and system specific, and are primarily dependent on flow rates, supply temperatures, and pipe and duct sizes and lengths. In constant volume systems, the distribution power is essentially constant, whereas in variable volume systems it varies approximately with the square of the flow rate (and may have a lower limit corresponding to a minimum flow rate).

The distribution power also generally results in heat being added to the flow. This displaces some need for thermal energy from the primary heating source and adds to that from the primary cooling source. Because of friction in the fan and duct or pump and pipe, virtually all the mechanical power input into the fan or pump is converted to heat. The total (thermal and fluidic) power input is the product of the volumetric flow rate (F) and total pressure increase across the fan or pump (Δp), with proper unit conversion factor (k)

$$
Distrib(t) = \frac{F\Delta pk}{\eta}
\tag{7.2.41}
$$

where (η) is the efficiency of a fan motor outside an air stream whose waste heat is not added to the flow; otherwise η is 1.0.

Reheat occurs in single-duct multizone HVAC systems in the cooling mode whenever air volumes are constant. Even in variable-air volume systems designed to reduce the need for reheat, once air volumes are reduced to a minimum (determined by the need for outdoor air), some reheat is necessary.

Reheat is necessary because each zone has its flow rate set based on design conditions. Differences in balance temperatures (defined below) among zones served by a single AHU will cause their cooling loads to drop at different rates as the outdoor temperature drops below design conditions. Their cooling loads do not decrease in proportion to their design flow rates, so some zones will receive more cooling than they require to satisfy the zone with the highest relative cooling load.

A simple analysis can be developed that expresses the reheat required as a ratio to the zone's cooling load as follows:

$$\frac{\text{Reheat}(t)}{\text{Load}(t)} = \left(\frac{T_o^* - T_b}{T_o - T_b}\right)\left(\frac{T_o - T_b''}{T_o^* - T_b''}\right) - 1 \qquad (7.2.42)$$

where the balance temperature for the zone, T_b, is expressed as a function of the zone temperature (T_z), the internal heat gains to the zone, and the heat loss coefficient (UA).

$$T_b = T_z - \frac{f_L \text{Lights}(t) + \text{Plugs}(t) + \text{Occ}(t) + \text{Solar}(t) + f_D \text{Dist}(t)}{\text{UA}} \qquad (7.2.43)$$

So, if the balance temperatures for the zones served by an AHU are known or estimated, the ratio of their reheat to their load can be estimated from Equation 7.2.42.

Integrating of Impacts Over Time

The primary purpose of providing energy impact information to users of FDD systems is to give a sense of scale for problems detected so that operators can prioritize corrective actions among their other work duties. Problems with small impacts can be ignored or fixing them can be put off, while problems with large impacts may justify immediate action. Clearly, FDD systems will not have any beneficial effect on building operation if significant problems are not corrected. Users and potential users of FDD systems have consistently placed high value on such feedback.

The time scale over which a problem's impact is presented fundamentally determines its value. It also seems likely that users will compare this value or cost to fix a problem, perhaps an hourly labor rate or the cost of a service call. With 168 hours in a week, presenting the impact in terms of a weekly cost impact magnifies the user's *perception* of the value of fixing the problem by two orders of magnitude compared to presenting an hourly impact. Small problems can result in large cost impacts if they affect every hour of every day. Since most problems targeted by FDD systems go undetected for months, if not years, presenting impact estimates over even longer intervals is probably desirable.

However, this raises a set of issues about how to construct such estimates. First, the impact of problems is not steady, either from hour-to-hour within a day, from one day to another, or from one season to another. An economizer that fails to operate when it should provides a simple illustration of this. At night or on weekends, the impact may be zero if the cooling system is shut off. The impact may also be zero in the winter and the summer when the economizer cannot operate anyway. Providing an annual estimate of the impact may be the fairest way to present FDD impacts, but is not useful in setting immediate priorities for action. In this example, fixing the economizer in the winter is not an *immediate* priority, while getting it fixed by spring is. The value of the corrective action and the benefit of the FDD system are best represented by the annual number.

There are also procedural difficulties in computing impacts over time. When hourly (for example) time-series impacts are computed by the FDD algorithm, it may be quite simple and useful to display the sum of the hourly impacts for the last week or month. The issue here is that a new problem may have a large impact, but this will not be apparent if these impacts are just starting to accumulate. This suggests that either the onset of the problem is taken into account, adding another layer of complexity to the FDD system, or future impacts are forecast based on the nature of the problem. The latter requires some type of model to project impacts. This model could be theoretical, based on assumptions about loads as a function of weather and a typical weather year, for example. On the other hand, the model could be empirical, based on the conditions seen over the last year, if such data has been stored by the FDD system. In either case, the effect of the problem must be superimposed on the model. Examples of how this can be done were presented in the previous section, but making such estimates reasonably accurate over widely varying conditions is complex.

In summary, it is important to keep in mind the purpose for the impact estimates. For the reasons cited above, it may be appropriate to provide impact estimates over more than one time interval, and perhaps targeted at various uses and users. For this to be effective, multiple impact results must be presented by the FDD system to clearly distinguish their basis and their use or target audience. It is also important to keep in mind that making some kind of impact estimate, even one with significant uncertainty, is probably always better than providing none at all.

Since increased energy costs are not the only impact of many problems in buildings, it is very important to place energy cost impacts in their proper context. For example, faults that cause inadequate outdoor air to be supplied to building spaces (i.e., below the amount specified by code) actually *save* energy and lower costs substantially in most conditions. Nevertheless, they have adverse impacts on occupant health and productivity that overshadow their energy cost benefits and open up the owner/operator to potential liability. Similarly, faults in a chiller may result in a lack of capacity that limits consumption, but any associated energy savings are likely to be overshadowed by the failure to provide comfort conditions and potential damage to the chiller that may result from continued operation.

Therefore, FDD systems should be careful about displaying energy cost *benefits* from faults (perhaps avoiding doing so altogether) that may distract the user from the real issues involved. Even if nonenergy impacts of faults are not quantified by an FDD tool, they should be presented to the user in qualitative fashion to prompt their consideration as the operators decide on a response to a detected fault.

7.2.13   The Future of Diagnostics in Buildings

In the 1990s, there was significant growth in the development of fault detection and diagnostic methods and methodologies for building systems. However, very few commercial products exist today, and the ones that exist are very specialized or not fully automated. There are several reasons for lack of widespread availability and deployment of FDD systems: lack of sensors on building systems, unavailability of low cost reliable sensors, high cost-to-benefit ratio of deploying FDD systems with current sensor technologies, lack of acceptable benchmarks to quantify the potential benefits from deploying FDD systems, lack of easy access to real-time data, and lack of infrastructure to gather data from existing BASs.

The functionality/benefits and costs of a fully automated FDD system differ significantly from those of a service tool. With the development of low-cost reliable sensor technology, FDD systems would soon be integrated into individual equipment controllers and would provide continuous monitoring, fault detection and diagnostic outputs, and recommendations for when service should be performed. Ultimately, as networking infrastructure matures, the use of automated FDD systems could allow a small support staff to operate, monitor, and maintain a large number of different systems from a remote, centralized location. Local FDD systems would communicate across a network to provide a status report on the "health" of the equipment that they monitor. Failures that lead to loss of comfort could be identified quickly before there is a significant impact on comfort. In many cases, degradation faults could be identified well before they lead to loss of comfort or uneconomical operation, allowing more efficient scheduling (lower cost) of service.

As the cost of sensors and control hardware continues to drop, chillers will probably be the first application of automated FDD within the HVAC&R industry because of a low cost-to-benefit ratio. Once fully developed, the technology could be integrated into all controllers associated with vapor compression cooling equipment. When fully mature, the costs associated with implementing the technology should be primarily the result of the addition of low-cost temperature sensors. These costs should be a relatively small fraction of the controller costs. The same technology would also be applicable to refrigeration and residential space cooling. Furthermore, the technology could be implemented in add-on systems to existing cooling equipment, which would increase the rate of market penetration.

Open communication standards for building automation systems are catching on as well, and use of Internet and Intranet technologies is pervasive. These developments enable FDD systems to be deployed more readily. In addition, the structure of the industry that provides services for the operations and maintenance of buildings is changing; companies are consolidating and offering whole-building operations and maintenance packages. Furthermore, as utilities are deregulated they will begin to offer new services, including complete facility management. With complete and distributed facility management, the cost-to-benefit of deploying FDD systems will improve because the cost can be spread over a large number of buildings (Katipamula et al., 1999). To benefit from these changes, facility managers, owners, operators, and energy service providers are challenged to acquire or develop new capabilities and resources to better manage this information and, in the end, their buildings and facilities.

Although the technology and incentives for application of FDD systems for vapor compression cooling equipment have never been greater, there still are several obstacles to their development and deployment. First, there is a need to quantify the potential benefits to establish benchmarks for acceptable costs and to provide marketing information. Specific research issues related to FDD methods include development of methods for detection and diagnosis of sensor faults and multiple simultaneous faults, identification of appropriate models and training approaches, and evaluation of the tradeoffs between sensors (type and quality) and FDD performance. The testing of FDD methods should be performed first in the laboratory and then in the field.

Acknowledgments

The work described in this chapter was partially funded by the Office of Building Technology, State and Community Systems of the U.S. Department of Energy as part of the Building Systems Program at Pacific Northwest National Laboratory. The laboratory is operated for the U.S. Department of Energy by Battelle Memorial Institute under contract DE-AC06-76RLO 1830. The authors would also like to thank their colleagues Peter Armstrong, at Pacific Northwest National Laboratory, and Agami Reddy, at Drexel University, for their valuable suggestions and review of the chapter.

References

Agrusa, R. and Singers, R.R. 1999. Control Your World in a Glance. *Building Systems Innovation, Heating, Piping and Air-Conditioning Supplement to the January Issue of HPAC Engineering*, pp. 21–32, Penton Media, Cleveland, Ohio.

Anderson, D., Graves, L., Reinert, W., and Kreider, J.F. 1989. A Quasi-Real-Time Expert System for Commercial Building HVAC Diagnostics. *ASHRAE Transactions*, Vol. 95, Part 2, pp. 954–960.

Bailey, M.B., Kreider, J.F., and Curtiss, P.S. 2000. Results of a Probabilistic Fault Detection and Diagnosis Method for Vapor Compression Cycle Equipment. *Proceedings of the International Conference of Chartered Institution of Building Services Engineers and ASHRAE*, College of Surgeons, Dublin, Ireland. September 21–23, 2000; see also by the first author, *The Design and Viability of a Probabilistic Fault Detection and Diagnosis Method for Vapor Compression Cycle Equipment*. Ph.D. Thesis, JCEM, College of Engineering, University of Colorado, Boulder, Colorado, 1998.

Bayne, J.S. 1999. Unleashing the Power of Networks. *Supplement to the January Issue of HPAC Engineering*, Penton Media, Cleveland, Ohio.

Brambley, M.R., Pratt, R.G., and Katipamula, S. 1999. Use of Automated Tools for Building Commissioning. Presented at the 7th National Conference on Building Commissioning, May 1999.

Brambley, M.R., Pratt, R.G., Chassin, D.P., and Katipamula, S. 1998. Automated Diagnostics for Outdoor Air Ventilation and Economizers. *ASHRAE Journal*, Vol. 40, No. 10, pp. 49–55.

Braun, J.E. 1999. Automated Fault Detection and Diagnostics for the HVAC&R Industry. *HVAC&R Research*, Vol. 5, No. 2, pp. 85–86.

Breuker, M.S. 1997. *Evaluation of a Statistical, Rule-Based Fault Detection and Diagnostics Method for Vapor Compression Air Conditioners*. Master's thesis, School of Mechanical Engineering, Purdue University, Purdue, Indiana.

Breuker, M.S. and Braun, J.E. 1998a. Common Faults and Their Impacts for Rooftop Air Conditioners. *International Journal of Heating, Ventilating, and Air Conditioning and Refrigerating Research*, Vol. 4, No. 2, pp. 303–318.

Breuker, M.S. and Braun, J.E. 1998b. Evaluating the Performance of a Fault Detection and Diagnostic System for Vapor Compression Equipment. *International Journal of Heating, Ventilating, and Air Conditioning and Refrigerating Research*, Vol. 4, No. 4, pp. 401–425.

Cikanek, H.A. 1986. Space Shuttle Main Engine Failure Detection. *IEEE Transactions on Automatic Control*, Vol. 6, pp. 13–18.

Clark, D.R. 1985. *HVACSIM+ Program Reference Manual*. NBSIR 84-2996, National Institute of Standards and Testing, Gaithersburg, Maryland.

Chen, J. and Patton, R.J. 1999. *Robust Model-Based Fault-Diagnosis for Dynamic Systems*. Kluwer Academic Publishers, Norwell, Massachusetts.

Daisey, J.M. and Angell, W.J. 1998. *A Survey and Critical Review of the Literature on Indoor Air Quality, Ventilation and Health Symptoms in Schools*. Report No. LBNL-41517, Lawrence Berkeley National Laboratory, Berkeley, California.

Dalton T., Patton, R.J., and Miller, P.J.H. 1995. Methods of Fault Detection for a Centrifugal Pump System. *On-Line Fault Detection and Supervision in the Chemical Process Industries, IFAC Workshop*, Newcastle Upon Tyne, U.K., Pergamon Press, New York.

Dexter, A.L. and Benouarets, M. 1996. Generic Approach to Identifying Faults in HVAC Plants. *ASHRAE Transactions*, Vol. 102, No. 1, pp. 550–556.

Dodier, R.H. and Kreider, J.F. 1999. Detecting Whole Building Energy Problems. *ASHRAE Transactions*, Vol. 105, No. 1.

EUN, *Energy User News*, Vol. 24, No. 3, pp. 43–48, March 1999.

Fasolo, P.S. and Seborg, D.E. 1995. Monitoring and Fault Detection for an HVAC Control System. *International Journal of Heating, Ventilation, and Air-Conditioning and Refrigeration Research*, Vol. 99, Part 1, pp. 3–13.

Frank, P.M. 1987. Fault Diagnosis in Dynamic Systems via State Estimation: A Survey. *System Fault Diagnostics, Reliability and Related Knowledge-Based Approaches*, Vol. 1, pp. 35–98, D. Reidel Publishing Company, Dordrecht, Holland.

Frank, P.M. 1990. Fault Diagnosis in Dynamic Systems Using Analytical and Knowledge-Based Redundancy — A Survey and Some New Results. *Automatica*, Vol. 26, pp. 459–474.

Frank, P.M. 1997. New Developments Using AI in Fault Diagnosis. *Engng. Applic. Artif. Intell.* Vol. 10, No. 1, pp. 3–14.

Fukunaga, K. 1990. *Introduction to Statistical Pattern Recognition*. Academic Press, Purdue University, W. Lafayette, Indiana.

Georgescu, C., Afshari, A., and Bornard, G. 1993. A Model-Based Adaptive Predictor Fault Detection Method Applied to Building Heating, Ventilating, and Air-Conditioning Process. *TOOLDIAG '93*, Organized by Département d'Études et de Recherches en Automatique, Toulouse, Cedex, France.

Gertler, J. 1988. Survey of Model-Based Failure Detection and Isolation in Complex Plants. *IEEE Control Systems Magazine*, Vol. 8, No. 6, pp. 3–11.

Gertler, J. 1998. *Fault Detection and Diagnosis in Engineering Systems*. Marcel Dekker, New York.

Glass, A.S., Gruber, P., Roos, M., and Todtli, J. 1995. Qualitative Model-Based Fault Detection in Air-Handling Units. *IEEE Control Systems Magazine*, Vol. 15, No. 4, pp. 11–22.

Gordon, J.M. and Ng, K.C. 1995. Predictive and Diagnostic Aspects of a Universal Thermodynamic Model for Chillers. *International Journal of Heat and Mass Transfer*, Vol. 38, No. 5, pp. 807–818.

Gordon, J.M. and Ng, K.C. 2000. *Cool Thermodynamics*, Cambridge International Scientific Publishers, Cambridge, U.K.

Gregerson, J. 1997. *Commissioning Existing Buildings, E-Source Tech Update*. TU–97–3, E-Source Inc., Boulder, Colorado.

Grimmelius, H.T., Woud, J.K., and Been, G. 1995. On-line Failure Diagnosis for Compression Refrigerant Plants. *International Journal of Refrigeration*, Vol. 18, No. 1, pp. 31–41.

Han, C.Y., Xiao, Y., and Ruther, C.J. 1999. Fault Detection and Diagnosis of HVAC Systems. *ASHRAE Transactions*, Vol. 105, Part 1.

Haves, P., Salsbury, T., and Wright, J.A. 1996. Condition Monitoring in HVAC Subsystems Using First Principles. *ASHRAE Transactions*, Vol. 102, Part 1, pp. 519–527.

Himmelblau, D.M. 1978. *Fault Detection and Diagnosis in Chemical and Petrochemical Processes*. Elsevier Scientific Publishing Company, New York.

House, J.M., Lee, W.Y., and Shin, D.R. 1999. Classification Techniques for Fault Detection and Diagnosis of an Air-Handling Unit. *ASHRAE Transactions*, Vol. 105, Part 1.

Hyvärinen, J. and Kärki, S., Eds. 1996. *International Energy Agency Building Optimisation and Fault Diagnosis Source Book*. Published by Technical Research Centre of Finland, Laboratory of Heating and Ventilation, Espoo, Finland.

Inatsu, H., Matsuo, H., Fujiwara, K., Yamada, K., and Nishizawa, K. 1992. Development of Refrigerant Monitoring Systems for Automotive Air-Conditioning Systems. *Society of Automotive Engineers*, SAE Paper No. 920212.

Issermann, R. 1984. Process Fault Detection Based on Modeling and Estimation Methods — A Survey. *Automatica*, Vol. 20, No. 4, pp. 387–404.

Issermann, R. and Nold, S. 1988. Model Based Fault Detection for Centrifugal Pumps and AC Drives. In *11th IMEKO World Congress*. Houston, Texas, U.S.A., pp. 16–21.

Issermann, R. and Ballé, P. 1997. Trends in the Application of Model-Based Fault Detection and Diagnosis of Technical Process. *Control Engineering Practice*, Vol. 5. No. 5, pp. 709–719.

Jarrell, D.B. and Meador, R.J. 1997. *Twenty-nine Palms Final Report*, Volume I, *PNNL-11582*, Pacific Northwest National Laboratory, Richland, Washington.

Jiang, Y., Li, J., and Yang, X. 1995. Fault Direction Space Method for On-Line Fault Detection. *ASHRAE Transactions*, Vol. 101, Part 2, pp. 219–228.

Katipamula, S.T., Reddy, A., and Claridge, D.E., Effect of time resolution on Statistical modeling of cooling energy use in large commericial buildings, *ASHRAE Transactions*, Vol. 101, Part 2, pp. 172–185, 1995.

Katipamula, S., Pratt, R.G., Chassin, D.P., Taylor, Z.T., Gowri, K., and Brambley, M.R. 1999. Automated Fault Detection and Diagnostics for Outdoor-Air Ventilation Systems and Economizers: Methodology and Results from Field Testing. *ASHRAE Transactions*, Vol. 105, Part 1.

Kumamaru, T., Utsunomiya, T., Iwasaki, Y., Shoda, I., and Obayashi, M. 1991. A Fault Diagnosis Systems for District Heating and Cooling Facilities. *Proceedings of the International Conference on Industrial Electronics, Control, and Instrumentation*, Kobe, Japan (IECON 91), pp. 131–136.

Lee, W.Y., Park, C., and Kelly, G.E. 1996a. Fault Detection of an Air-Handling Unit Using Residual and Recursive Parameter Identification Methods. *ASHRAE Transactions*, Vol. 102, Part 1, pp. 528–539.

Lee, W.Y., House, J.M., Park, C., and Kelly, G.E. 1996b. Fault Diagnosis of an Air-Handling Unit Using Artificial Neural Networks. *ASHRAE Transactions*, Vol. 102, Part 1, pp. 540–549.

Lee, W.Y., House, J.M., and Shin, D.R. 1997. Fault Detection of an Air-Handling Unit Using Residual and Recursive Parameter Identification Methods. *ASHRAE Transactions*, Vol. 102, Part 1, pp. 528–539.

Li, X., Hossein, V., and Visier, J. 1996. Development of a Fault Diagnosis Method for Heating Systems Using Neural Networks. *ASHRAE Transactions*, Vol. 102, Part 1, pp. 607–614.

Li, X., Visier, J., and Vaezi-Nejad, H. 1997. A Neural Network Prototype for Fault Detection and Diagnosis of Heating Systems. *ASHRAE Transactions*, Vol. 103, Part 1, pp. 634–644.

Mangoubi, R.S. 1998. *Robust Estimation and Failure Detection*. Springer-Verlag, New York.

McKellar, M. G. 1987. *Failure Diagnosis for a Household Refrigerator*. Master's thesis, School of Mechanical Engineering, Purdue University, Purdue, Indiana.

Mylaraswamy, D. and Venkatsubramanian, V. 1997. A Hybrid Framework for Large Scale Process Fault Diagnosis. *Computers Chem. Engng*. Vol. 21, pp. S935–S940.

Ngo, D. and Dexter, A.L. 1999. A Robust Model-Based Approach to Diagnosing Faults in Air-Handling Units. *ASHRAE Transactions*, Vol. 105, Part 1.

Norford, L.K. and Little, R.D. 1993. Fault Detection and Monitoring in Ventilation Systems. *ASHRAE Transactions*, Vol. 99, Part 1, pp. 590–602.

Noura, H., Aubrun, C., Sauter, D., and Robert, M. 1993. A Fault Diagnosis and Reconfiguration Method Applied to Thermal Plant. *TOOLDIAG '93*, Organized by Département d'Études et de Recherches en Automatique, Toulouse, Cedex, France.

Pape, F.L.F., Mitchell, J.W., and Beckman, W.A. 1990. Optimal Control and Fault Detection in Heating, Ventilating, and Air-Conditioning Systems. *ASHRAE Transactions*, Vol. 97, Part 1, pp. 729–736.

Patel, S.A. and Kamrani, A.K. 1996. Intelligent Decision Support System for Diagnosis and Maintenance of Automated Systems. *Computers and Industrial Engineering*, Vol. 30, No. 2, pp. 297–319.

Patton, R., Frank, P., and Clark, R. 1989. *Fault Diagnosis in Dynamic Systems: Theory and Application.* Prentice Hall, Englewood Cliffs, NJ.

Pau, L.F. 1981. *Failure Diagnosis and Performance Monitoring*, Marcel Dekker, New York.

Peitsman, H.C. and Bakker, V. 1996. Application of Black-Box Models to HVAC Systems for Fault Detection. *ASHRAE Transactions*, Vol. 102, Part 1, pp. 628–640.

Peitsman, H.C and Soethout, L.L. 1997. ARX Models and Real-Time Model-Based Diagnosis. *ASHRAE Transactions*, Vol. 103, Part 1, pp. 657–671.

Pearl, J. 1988. *Probabilistic Reasoning in Intelligent Systems.* Morgan Kaufmann, San Mateo, California.

Portland Energy Conservation Inc. (PECI). 1997. *Commissioning for Better Buildings in Oregon*, Oregon Office of Energy, Salem, Oregon.

Rossi, T.M. 1995. *Detection, Diagnosis, and Evaluation of Faults in Vapor Compression Cycle Equipment.* Ph.D. thesis, School of Mechanical Engineering, Purdue University, Purdue, Indiana.

Rossi, T.M. and Braun, J.E. 1996. Minimizing Operating Costs of Vapor Compression Equipment With Optimal Service Scheduling. *International Journal of Heating, Ventilating, and Air Conditioning and Refrigerating Research*, Vol. 2, No. 1, pp. 3–26.

Rossi, T.M. and Braun, J.E. 1997. A Statistical, Rule-Based Fault Detection and Diagnostic Method for Vapor Compression Air Conditioners. *International Journal of Heating, Ventilating, and Air Conditioning and Refrigerating Research*, Vol. 3, No. 1 pp. 19–37.

Seem, J., House, J.M., and Monroe, R.H. 1999. On-Line Monitoring and Fault Detection, *ASHRAE Journal*, Vol. 41, No. 7, pp. 21–26.

Stallard, L.A. 1989. *Model Based Expert System for Failure Detection and Identification of Household Refrigerators.* Master's thesis, School of Mechanical Engineering, Purdue University, Purdue, Indiana.

Stylianou, M. and Nikanpour, D. 1996. Performance Monitoring, Fault Detection, and Diagnosis of Reciprocating Chillers. *ASHRAE Transactions*, Vol. 102, Part 1, pp. 615–627.

Stylianou, M. 1997. Classification Functions to Chiller Fault Detection and Diagnosis. *ASHRAE Transactions*. Vol. 103, Part 1, pp. 645–648.

Tutsui, H. and Kamimura, K. 1996. Chiller Condition Monitoring Using Topological Case-Based Modeling. *ASHRAE Transactions*, Vol. 102, Part 1, pp. 641–648.

U.S. DOE/PECI. 1998. *Model Commissioning Plan and Guide Commissioning Specifications, Version 2.05*, PECI, Portland, OR, February.

Wagner, J. and Shoureshi, R. 1992. Failure Detection Diagnostics for Thermofluid Systems. *Journal of Dynamic Systems, Measurement, and Control*, Vol. 114, No. 4, pp. 699–706.

Willsky, A.S. 1976. A Survey of Design Methods for Failure Detection in Dynamic Systems. *Automatica*, Vol. 29, pp. 601–611.

Yoshida, H., Iwami, T., Yuzawa, H., and Suzuki, M. 1996. Typical Faults of Air-Conditioning Systems, and Fault Detection by ARX Model and Extended Kalman Filter. *ASHRAE Transactions*, Vol. 102, Part 1, pp. 557–564.

Yoshida, H. and Kumar, S. 1999. ARX and AFMM Model-Based On-Line Real-Time Data Base Diagnosis of Sudden Fault in AHU of VAV System. *Energy Conversion and Management*, Vol. 40, pp. 1191–1206.

Yoshimura, M. and Ito, N. 1989. Effective Diagnosis Methods for Air-Conditioning Equipment in Telecommunications Buildings. *INTELEC 89: The Eleventh International Telecommunications Energy Conference*, October 15–18, Centro dei, Firenze, Vol. 21, pp. 1–7.

Bibliography

Bagby, D.G. and Cormier, R.A. 1989. A Heat Exchanger Expert System. *ASHRAE Transactions*, Vol. 95, No. 2, pp. 927–933.

Chassin, D. P. 1999. Computer Software Architecture to Support Automated Diagnostics, In *Proceedings of CIB W78 Workshop on Information Technologies in Construction*, Vancouver B.C., Canada, June.

Clark, D.R. and May, W.B. 1985. *HVACSIM+ Users' Guide.* NBSIR 85-3243, National Institute of Standards and Testing, Gaithersburg, Maryland.

Clark D.R., Hurley, C.W., and Hill, C.R. 1985. Dynamic Models for HVAC System Components. *ASHRAE Transactions,* Vol. 91, No. 1B, pp. 737–751.

Culp, C.H. 1989. Expert Systems in Preventive Maintenance and Diagnosis. *ASHRAE Journal,* Vol. 31, No. 8, pp. 13–18.

Culp, C.H., Haberl, J.S., Norford, L., Brothers, P., and Hall, J.D. 1990. The Impact of AI Technology Within the HVAC Industry. *ASHRAE Journal,* Vol. 31. No. 12, pp. 12–22.

De Kleer, J. and Williams, B. 1987. Diagnosing Multiple Faults. *Artificial Intelligence,* Vol. 32, No. 1, pp. 97–130.

Dexter, A.L. 1993. Fault Detection in Air-Conditioning Systems Using Fuzzy Models. *IEEE Colloquium 'Two Decades of Fuzzy Control—Part 2'.* Digest No. 1993/118.

Dexter, A.L., Fargus, R.S., and Haves, P. 1994. Fault Detection in Air-Conditioning Systems Using A.I. Techniques. In *Proceedings of the Second BEPAC Conference BEP 94,* York, U.K.

Dexter, A.L. and Hepworth, S.J. 1993. *A Comparison of Fuzzy and Neural Methods of Detecting Faults in an Air-Handling Unit.* University of Oxford, Dept. of Eng. Science, Report No. OUEL 1981/93, Oxford University, Oxford, England.

Dexter, A.L. and Mok, B.K.K. 1993. *Fault Detection in HVAC Systems Using Fuzzy Models.* University of Oxford, Dept. of Eng. Science, Report No. OUEL 1977/93, Oxford University, Oxford, England.

Dexter, A.L. and Benouarets, M. 1996. A Generic Approach to Modeling of HVAC Plants for Fault Diagnosis, *Proceeding of 4th IBPSA International Conference:* Building Simulation 95, Madison, Wisconsin, pp. 339–345.

Dexter, A.L. and Ngo, D. 1997. Fault Diagnosis in Large-Scale Air-Conditioning Systems. *Proceeding of IFAC Symposium SAFEPROCESS 97,* Vol. 2, pp. 737–741.

Dialynas, E.N., Machias, A.V., and Souflis, J.L. 1987. Reliability and Fault Diagnosis Methods of Power System Components. *System Fault Diagnostics, Reliability and Related Knowledge-Based Approaches,* Vol. 1, pp. 327–341, D. Reidel Publishing Company, Dordrecht, Holland.

Dodier, R.H., Curtiss, P.S., and Kreider, J.F. 1997. *Small Scale, On-Line Diagnostics for an HVAC System.* RP-883, JCEM Technical Report TR/96/30, University of Colorado, Colorado.

Dodier, R.H., Curtiss, P.S., and Kreider, J.F. 1998. Small-Scale On-Line Diagnostics for an HVAC System. *ASHRAE Transactions,* Vol. 104, No. 1, pp. 530–539.

Duyar, A. and Merill, W. 1992. Fault Diagnosis for the Space Shuttle Main Engine. *AIAA, Journal of Guidance, Control, and Dynamics,* Vol. 15, No. 2, pp. 384–389.

Fasolo, P.S. 1993. *On-Line Statistical Methods for Fault Detection in an HVAC Process.* Master's thesis, University of California, Santa Barbara, California.

Fasolo, P.S. and Seborg, D.E. 1994. An SQC Approach to Monitoring and Fault Detection in HVAC Control Systems. In *Proceedings of the 1994 American Control Conference,* pp. 3055–3059, IEEE, New York.

Formera, L., Glass, A.S., Gruber, P., and Todtli, J. 1994. Qualitative Fault Detection based on Logical Programming Applied to a VAV Air Handling Unit. *Second IFAC Workshop on Computer Software Structures Integrating AI/KBS Systems in Process Control,* Lund, August 10–12.

Grimmelius, H.T., Woud, J.K., and Been, G. 1995. On-line Failure Diagnosis for Compression Refrigeration Plants. *International Journal of Refrigeration,* Vol. 18, No. 1, pp. 31–41.

Haberl, J.S. and Claridge, D.E. 1987. An Expert System for Building Energy Consumption Analysis. *ASHRAE Transactions,* Vol. 93, No. 1, pp. 979–998.

Haberl, J.S., Norford, L.K., and Spadaro, J.S. 1989. Expert Systems for Diagnosing Operation Problems in HVAC Systems. *ASHRAE Journal,* Vol. 31, No. 6.

Haves, P., Salsbury, T.I., and Wright, J.A. 1996. Condition Monitoring in HVAC Subsystems Using First Principles Models. *ASHRAE Transactions,* Vol. 102, Part 1, pp. 519–527.

Haves, P., Norford, L.K., DeSimone, M.A. 1998. Standard Simulation Test Bed for the Evaluation of Control Algorithms and Strategies. *ASHRAE Transactions,* Vol. 104, Part 1A, pp. 460–473.

Himmelblau, D.M. 1992. Fault Detection in Heat Exchangers. In *Proceedings of the 1992 American Control Conference*, pp. 2369–2372. IEEE, New York.

Hiroshi, I.H., Matsuo, K., Fujiwara, Yamada, K., and Nishizawa, K. 1992. Development of Refrigerant Monitoring Systems for Automotive Air-Conditioning Systems. *Society of Automotive Engineers*, SAE Paper No. 920212.

Hyvarinen, J. and Kohonen, R., Eds. 1993. *Building Optimisation and Fault Diagnosis System Concept*. Tech. Research Centre of Finland (VTT), Laboratory of Heating and Ventilation. (ISBN 952-9601-16-6). VTT, Espoo, Finland.

Jardine, A.K.S. 1973. *Maintenance, Replacement, and Reliability*. Halsted Press, John Wiley & Sons Inc., New York.

Jones, A.H. and Burge, S.E. 1987. An Expert System Design Using Cause-Effect Representations and Simulation for Fault Detection. *System Fault Diagnostics, Reliability and Related Knowledge-Based Approaches*, Vol. 2, pp. 71–80, D. Reidel Publishing Company, Dordrecht, Holland.

Johnson, D.M. 1996. A Review of Fault Management Techniques Used in Safety-Critical Avionic Systems. *Prog. Aerospace Sci*. Vol. 32, pp. 415–431.

Kaler, G.M. 1988. Expert System Predicts Service. *Heating, Piping, and Air Conditioning*, Vol. 11, pp. 99–101.

Kaler, G.M. 1990. Embedded Expert System Development for Monitoring Packaged HVAC Equipment. *ASHRAE Transactions*, Vol. 96, Part 2., p. 733.

Kitamura, M. 1980. Detection of Sensor Failures in Nuclear Plant Using Analytic Redundancy. *Transactions of the American Nuclear Society*, Vol. 34, pp. 581–583.

Koscielny, J.M. 1995. Fault Isolation in Industrial Process by the Dynamic Table of States Method. *Automatica*, Vol. 31, No. 5, pp. 747–753.

Liu, S.T and Kelly, G.E. 1988. Knowledge-based front-end input generating program for building system simulation. *ASHRAE Transactions*, Vol. 94, Part 1, pp. 1074–1084.

Liu, S.T. and Kelly, G.E. 1989. Rule-Based Diagnostic Method for HVAC Fault Detection. In *Proceedings of IBPSA Building Simulation 93*, pp. 319–324, Vancouver, Canada.

Loparo, K.A., Buchner, M.R., and Vasudeva, K.S. 1991. Leak Detection in an Experimental Heat Exchanger Process: A Multiple Model Approach. *IEEE Trans. Auto. Control*, Vol. 36, Part 2, pp. 167–177.

Ngo, D. and Dexter, A.L. 1998. Fault Diagnosis in Air-Conditioning Systems Using Generic Models of HVAC Plants. *Proceedings of the International Conference on System Simulation in Buildings*, SSB 98, Liege, Belgium.

Park, C., Clark, D.R., and Kelly, G.E. 1986. *HVACSIM+ Building Loads Calculation*. NBSIR 86-3331, National Institutes of Standards, Gaithersburg, Maryland.

Park, C. and Bushby, S.T. 1989. Simulation of a Large Office Building System Using the HVACSIM+ program. *ASHRAE Transactions*, Vol. 95, Part 1, pp. 642–651.

Rossi, T. and Braun, J. 1995. Thermodynamic Impact of Detecting Refrigerant Leaks in Vapor Compression Equipment. *1995 American Control Conference*, IEEE Control Systems Society, Seattle, Washington.

Salsbury, T.I., Haves, P., and Wright, J.A. 1995. A Fault Detection and Diagnosis Method Based on First Principles Models and Expert Rules. *Proceedings of HVAC 95 2nd International Symposium on Heating Ventilating and Air Conditioning*, Beijing, China.

Sami, S.M., Zhou, Y., and Tulej, P.J. 1992. Development of a Diagnostic Expert System for Heat Pumps. *Proceedings of the 2nd Annual Conference on Heat Pumps in Cold Climates*, Moncton, New Brunswick, August 16–17, pp. 475–489.

Tzafestas, S.G. 1987. A Look at the Knowledge-Based Approach to System Fault Diagnosis and Supervisory Control. *System Fault Diagnostics, Reliability and Related Knowledge-Based Approaches*, Vol. 2 pp. 3–15, D. Reidel Publishing Company, Dordrecht, Holland.

Usoro, P.B. and Schick, I.C. 1985. A Hierarchical Approach to HVAC System Fault Detection and Identification. In *Proceedings of Dynamic Systems: Modeling and Control, Winter Annual Meeting of ASME*, pp. 285–291, Miami, Florida.

Usoro, P.B., Schick, I.C., and Negahdaripour. 1985. An Innovation-Based Methodology for HVAC System Fault Detection, *ASME Journal of Dynamic Systems, Measurements, and Control*, Vol. 107, pp. 284–289.

Visier, J.C., Vaezi-Nejad, H., and Corrales, P. 1999. A Fault Detection Tool for School Buildings. *ASHRAE Transactions*, Vol. 105, Part 1.

Watanabe, K., Hirota, S., Hou, L., and Himmelblau, D.M. 1994. Diagnosis of Multiple Simultaneous Faults via Hierarchical Artificial Neural Networks. *AIChE Journal*, Vol. 40, No. 5, pp. 839–848.

Xiao, Y. and Han, C.Y. 1998. An OOM-KRES Approach for Fault Detection and Diagnosis. *Springer Verlag Lecture Notes on AI*, Vol. 1415, pp. 831–839.

Yang, C.H.Y. and Jiang, Y. 1995. Sensor Fault Detection of HVAC System — System Constraint and Voting. *Proceedings of Pan Pacific Symposium on Building and Urban Environmental Conditioning in Asia*, Nagoya, Japan.

Yu, C.C. and Lee, C. 1991. Fault Diagnosis Based on Qualitative/Quantitative Process Knowledge, *AIChE Journal*, Vol. 4., p. 617.

8

Appendices

Paul Norton

National Renewable Energy Laboratory

Appendix A Properties of Gases and Vapors

TABLE A.1 Properties of Dry Air at Atmospheric Pressure

Symbols and Units:

K = absolute temperature, degrees Kelvin

deg C = temperature, degrees Celsius

deg F = temperature, degrees Fahrenheit

ρ = density, kg/m³ (sea level)

c_p = specific heat capacity, kJ/kg·K

c_p/c_v = specific heat capacity ratio, dimensionless

μ = viscosity, N·s/m² × 10⁶ (For N·s/m² (= kg/m·s) multiply tabulated values by 10⁻⁶)

k = thermal conductivity, W/m·k × 10³ (For W/m·K multiply tabulated values by 10⁻³)

Pr = Prandtl number, dimensionless

h = enthalpy, kJ/kg

V_s = sound velocity, m/s

Temperature			Properties							
K	deg C	deg F	ρ	c_p	c_p/c_v	μ	k	Pr	h	V_s
100	−173.15	−280	3.598	1.028		6.929	9.248	.770	98.42	198.4
110	−163.15	−262	3.256	1.022	1.420 2	7.633	10.15	.768	108.7	208.7
120	−153.15	−244	2.975	1.017	1.416 6	8.319	11.05	.766	118.8	218.4
130	−143.15	−226	2.740	1.014	1.413 9	8.990	11.94	.763	129.0	227.6
140	−133.15	−208	2.540	1.012	1.411 9	9.646	12.84	.761	139.1	236.4
150	−123.15	−190	2.367	1.010	1.410 2	10.28	13.73	.758	149.2	245.0
160	−113.15	−172	2.217	1.009	1.408 9	10.91	14.61	.754	159.4	253.2
170	−103.15	−154	2.085	1.008	1.407 9	11.52	15.49	.750	169.4	261.0
180	−93.15	−136	1.968	1.007	1.407 1	12.12	16.37	.746	179.5	268.7
190	−83.15	−118	1.863	1.007	1.406 4	12.71	17.23	.743	189.6	276.2
200	−73.15	−100	1.769	1.006	1.405 7	13.28	18.09	.739	199.7	283.4
205	−68.15	−91	1.726	1.006	1.405 5	13.56	18.52	.738	204.7	286.9
210	−63.15	−82	1.684	1.006	1.405 3	13.85	18.94	.736	209.7	290.5
215	−58.15	−73	1.646	1.006	1.405 0	14.12	19.36	.734	214.8	293.9
220	−53.15	−64	1.607	1.006	1.404 8	14.40	19.78	.732	219.8	297.4
225	−48.15	−55	1.572	1.006	1.404 6	14.67	20.20	.731	224.8	300.8
230	−43.15	−46	1.537	1.006	1.404 4	14.94	20.62	.729	229.8	304.1
235	−38.15	−37	1.505	1.006	1.404 2	15.20	21.04	.727	234.9	307.4
240	−33.15	−28	1.473	1.005	1.404 0	15.47	21.45	.725	239.9	310.6
245	−28.15	−19	1.443	1.005	1.403 8	15.73	21.86	.724	244.9	313.8
250	−23.15	−10	1.413	1.005	1.403 6	15.99	22.27	.722	250.0	317.1
255	−18.15	−1	1.386	1.005	1.403 4	16.25	22.68	.721	255.0	320.2
260	−13.15	8	1.359	1.005	1.403 2	16.50	23.08	.719	260.0	323.4
265	−8.15	17	1.333	1.005	1.403 0	16.75	23.48	.717	265.0	326.5
270	−3.15	26	1.308	1.006	1.402 9	17.00	23.88	.716	270.1	329.6
275	+1.85	35	1.285	1.006	1.402 6	17.26	24.28	.715	275.1	332.6
280	6.85	44	1.261	1.006	1.402 4	17.50	24.67	.713	280.1	335.6
285	11.85	53	1.240	1.006	1.402 2	17.74	25.06	.711	285.1	338.5
290	16.85	62	1.218	1.006	1.402 0	17.98	25.47	.710	290.2	341.5
295	21.85	71	1.197	1.006	1.401 8	18.22	25.85	.709	295.2	344.4
300	26.85	80	1.177	1.006	1.401 7	18.46	26.24	.708	300.2	347.3
305	31.85	89	1.158	1.006	1.401 5	18.70	26.63	.707	305.3	350.2
310	36.85	98	1.139	1.007	1.401 3	18.93	27.01	.705	310.3	353.1
315	41.85	107	1.121	1.007	1.401 0	19.15	27.40	.704	315.3	355.8
320	46.85	116	1.103	1.007	1.400 8	19.39	27.78	.703	320.4	358.7

Source: Condensed and computed from "Tables of Thermal Properties of Gases", National Bureau of Standards Circular 564, U.S. Government Printing Office, November 1955.

TABLE A.1 (continued) **Properties of Dry Air at Atmospheric Pressure**

Temperature			Properties							
K	deg C	deg F	ρ	c_p	c_p/c_v	μ	k	Pr	h	V_s
325	51.85	125	1.086	1.008	1.400 6	19.63	28.15	.702	325.4	361.4
330	56.85	134	1.070	1.008	1.400 4	19.85	28.53	.701	330.4	364.2
335	61.85	143	1.054	1.008	1.400 1	20.08	28.90	.700	335.5	366.9
340	66.85	152	1.038	1.008	1.399 9	20.30	29.28	.699	340.5	369.6
345	71.85	161	1.023	1.009	1.399 6	20.52	29.64	.698	345.6	372.3
350	76.85	170	1.008	1.009	1.399 3	20.75	30.03	.697	350.6	375.0
355	81.85	179	0.994 5	1.010	1.399 0	20.97	30.39	.696	355.7	377.6
360	86.85	188	0.980 5	1.010	1.398 7	21.18	30.78	.695	360.7	380.2
365	91.85	197	0.967 2	1.010	1.398 4	21.38	31.14	.694	365.8	382.8
370	96.85	206	0.953 9	1.011	1.398 1	21.60	31.50	.693	370.8	385.4
375	101.85	215	0.941 3	1.011	1.397 8	21.81	31.86	.692	375.9	388.0
380	106.85	224	0.928 8	1.012	1.397 5	22.02	32.23	.691	380.9	390.5
385	111.85	233	0.916 9	1.012	1.397 1	22.24	32.59	.690	386.0	393.0
390	116.85	242	0.905 0	1.013	1.396 8	22.44	32.95	.690	391.0	395.5
395	121.85	251	0.893 6	1.014	1.396 4	22.65	33.31	.689	396.1	398.0
400	126.85	260	0.882 2	1.014	1.396 1	22.86	33.65	.689	401.2	400.4
410	136.85	278	0.860 8	1.015	1.395 3	23.27	34.35	.688	411.3	405.3
420	146.85	296	0.840 2	1.017	1.394 6	23.66	35.05	.687	421.5	410.2
430	156.85	314	0.820 7	1.018	1.393 8	24.06	35.75	.686	431.7	414.9
440	166.85	332	0.802 1	1.020	1.392 9	24.45	36.43	.684	441.9	419.6
450	176.85	350	0.784 2	1.021	1.392 0	24.85	37.10	.684	452.1	424.2
460	186.85	368	0.767 7	1.023	1.391 1	25.22	37.78	.683	462.3	428.7
470	196.85	386	0.750 9	1.024	1.390 1	25.58	38.46	.682	472.5	433.2
480	206.85	404	0.735 1	1.026	1.389 2	25.96	39.11	.681	482.8	437.6
490	216.85	422	0.720 1	1.028	1.388 1	26.32	39.76	.680	493.0	442.0
500	226.85	440	0.705 7	1.030	1.387 1	26.70	40.41	.680	503.3	446.4
510	236.85	458	0.691 9	1.032	1.386 1	27.06	41.06	.680	513.6	450.6
520	246.85	476	0.678 6	1.034	1.385 1	27.42	41.69	.680	524.0	454.9
530	256.85	494	0.665 8	1.036	1.384 0	27.78	42.32	.680	534.3	459.0
540	266.85	512	0.653 5	1.038	1.382 9	28.14	42.94	.680	544.7	463.2
550	276.85	530	0.641 6	1.040	1.381 8	28.48	43.57	.680	555.1	467.3
560	286.85	548	0.630 1	1.042	1.380 6	28.83	44.20	.680	565.5	471.3
570	296.85	566	0.619 0	1.044	1.379 5	29.17	44.80	.680	575.9	475.3
580	306.85	584	0.608 4	1.047	1.378 3	29.52	45.41	.680	586.4	479.2
590	316.85	602	0.598 0	1.049	1.377 2	29.84	46.01	.680	596.9	483.2
600	326.85	620	0.588 1	1.051	1.376 0	30.17	46.61	.680	607.4	486.9
620	346.85	656	0.569 1	1.056	1.373 7	30.82	47.80	.681	628.4	494.5
640	366.85	692	0.551 4	1.061	1.371 4	31.47	48.96	.682	649.6	502.1
660	386.85	728	0.534 7	1.065	1.369 1	32.09	50.12	.682	670.9	509.4
680	406.85	764	0.518 9	1.070	1.366 8	32.71	51.25	.683	692.2	516.7
700	426.85	800	0.504 0	1.075	1.364 6	33.32	52.36	.684	713.7	523.7
720	446.85	836	0.490 1	1.080	1.362 3	33.92	53.45	.685	735.2	531.0
740	466.85	872	0.476 9	1.085	1.360 1	34.52	54.53	.686	756.9	537.6
760	486.85	908	0.464 3	1.089	1.358 0	35.11	55.62	.687	778.6	544.6
780	506.85	944	0.452 4	1.094	1.355 9	35.69	56.68	.688	800.5	551.2
800	526.85	980	0.441 0	1.099	1.354	36.24	57.74	.689	822.4	557.8
850	576.85	1 070	0.415 2	1.110	1.349	37.63	60.30	.693	877.5	574.1
900	626.85	1 160	0.392 0	1.121	1.345	38.97	62.76	.696	933.4	589.6
950	676.85	1 250	0.371 4	1.132	1.340	40.26	65.20	.699	989.7	604.9
1 000	726.85	1 340	0.352 9	1.142	1.336	41.53	67.54	.702	1 046	619.5
1 100	826.85	1 520	0.320 8	1.161	1.329	43.96			1 162	648.0
1 200	926.85	1 700	0.294 1	1.179	1.322	46.26			1 279	675.2
1 300	1 026.85	1 880	0.271 4	1.197	1.316	48.46			1 398	701.0
1 400	1 126.85	2 060	0.252 1	1.214	1.310	50.57			1 518	725.9
1 500	1 220.85	2 240	0.235 3	1.231	1.304	52.61			1 640	749.4
1 600	1 326.85	2 420	0.220 6	1.249	1.299	54.57			1 764	772.6
1 800	1 526.85	2 780	0.196 0	1.288	1.288	58.29			2 018	815.7
2 000	1 726.85	3 140	0.176 4	1.338	1.274				2 280	855.5
2 400	2 126.85	3 860	0.146 7	1.574	1.238				2 853	924.4
2 800	2 526.85	4 580	0.124 5	2.259	1.196				3 599	983.1

TABLE A.2 Psychrometric Table: Properties of Moist Air at 101 325 N/m²

Symbols and Units:

P_s = pressure of water vapor at saturation, N/m²

W_s = humidity ratio at saturation, mass of water vapor associated with unit mass of dry air

V_a = specific volume of dry air, m³/kg

V_s = specific volume of saturated mixture, m³/kg dry air

h_a^a = specific enthalpy of dry air, kJ/kg

h_s = specific enthalpy of saturated mixture, kJ/kg dry air

s_s = specific entropy of saturated mixture, J/K·kg dry air

Temperature			Properties						
C	K	F	P_s	W_s	V_a	V_s	h_a^a	h_s	s_s
−40	233.15	−40	12.838	0.000 079 25	0.659 61	0.659 68	−22.35	−22.16	−90.659
−30	243.15	−22	37.992	0.000 234 4	0.688 08	0.688 33	−12.29	−11.72	−46.732
−25	248.15	−13	63.248	0.000 390 3	0.702 32	0.702 75	−7.265	−6.306	−24.706
−20	253.15	−4	103.19	0.000 637 1	0.716 49	0.717 24	−2.236	−0.6653	−2.2194
−15	258.15	+5	165.18	0.001 020	0.730 72	0.731 91	+2.794	5.318	21.189
−10	263.15	14	259.72	0.001 606	0.744 95	0.746 83	7.823	11.81	46.104
−5	268.15	23	401.49	0.002 485	0.759 12	0.762 18	12.85	19.04	73.365
0	273.15	32	610.80	0.003 788	0.773 36	0.778 04	17.88	27.35	104.14
5	278.15	41	871.93	0.005 421	0.787 59	0.794 40	22.91	36.52	137.39
10	283.15	50	1 227.2	0.007 658	0.801 76	0.811 63	27.94	47.23	175.54
15	288.15	59	1 704.4	0.010 69	0.816 00	0.829 98	32.97	59.97	220.22
20	293.15	68	2 337.2	0.014 75	0.830 17	0.849 83	38.00	75.42	273.32
25	298.15	77	3 167.0	0.020 16	0.844 34	0.871 62	43.03	94.38	337.39
30	303.15	86	4 242.8	0.027 31	0.858 51	0.896 09	48.07	117.8	415.65
35	308.15	95	5 623.4	0.036 73	0.872 74	0.924 06	53.10	147.3	512.17
40	313.15	104	7 377.6	0.049 11	0.886 92	0.956 65	58.14	184.5	532.31
45	318.15	113	9 584.8	0.065 36	0.901 15	0.995 35	63.17	232.0	783.06
50	323.15	122	12 339	0.086 78	0.915 32	1.042 3	68.21	293.1	975.27
55	328.15	131	15 745	0.115 2	0.929 49	1.100 7	73.25	372.9	1 221.5
60	333.15	140	19 925	0.153 4	0.943 72	1.174 8	78.29	478.5	1 543.5
65	338.15	149	25 014	0.205 5	0.957 90	1.272 1	83.33	621.4	1 973.6
70	343.15	158	31 167	0.278 8	0.972 07	1.404 2	88.38	820.5	2 564.8
75	348.15	167	38 554	0.385 8	0.986 30	1.592 4	93.42	1 110	3 412.8
80	353.15	176	47 365	0.551 9	1.000 5	1.879 1	98.47	1 557	4 710.9
85	358.15	185	57 809	0.836 3	1.014 6	2.363 2	103.5	2 321	6 892.6
90	363.15	194	70 112	1.416	1.028 8	3.340 9	108.6	3 876	11 281

Note: The P_s column in this table gives the vapor pressure of pure water at temperature intervals of five degrees Celsius. For the latest data on vapor pressure at intervals of 0.1 deg C, from 0–100 deg C, see "Vapor Pressure Equation for Water", A. Wexler and L. Greenspan, *J. Res. Nat. Bur. Stand.*, 75A(3):213–229, May–June 1971.

[a] For very low barometric pressures and high wet-bulb temperatures, the values of h_a in this table are somewhat low; for corrections see *ASHRAE Handbook of Fundamentals*, 2001.

Source: Computed from Psychrometric Tables, in *ASHRAE Handbook of Fundamentals*, American Society of Heating, Refrigerating, and Air-Conditioning Engineers, 2001.

TABLE A.3 Water Vapor at Low Pressures: Perfect Gas Behavior pv/T = R = 0.461 51 kJ/kg·K

Symbols and Units:

t = thermodynamic temperature, deg C

T = thermodynamic temperature, K

pv = RT, kJ/kg

u_o = specific internal energy at zero pressure, kJ/kg

h_o = specific enthalpy at zero pressure, kJ/kg

s_l = specific entropy of semiperfect vapor at 0.1 MN/m², kJ/kg·K

ψ_l = specific Helmholtz free energy of semiperfect vapor at 0.1 MN/m², kJ/kg

ψ_l = specific Helmholtz free energy of semiperfect vapor at 0.1 MN/m², kJ/kg

ζ_l = specific Gibbs free energy of semiperfect vapor at 0.1 MN/m², kJ/kg

p_r = relative pressure, pressure of semiperfect vapor at zero entropy, TN/m²

v_r = relative specific volume, specific volume of semiperfect vapor at zero entropy, mm³/kg

c_{po} = specific heat capacity at constant pressure for zero pressure, kJ/kg·K

c_{vo} = specific heat capacity at constant volume for zero pressure, kJ/kg·K

k = c_{po}/c_{vo} = isentropic exponent, $-(\partial \log p/\partial \log v)_s$

t	T	pv	u_o	h_o	s_l	ψ_l	ζ_l	p_r	v_r	c_{po}	c_{vo}	k
0	273.15	126.06	2 375.5	2 501.5	6.804 2	516.9	643.0	.252 9	498.4	1.858 4	1.396 9	1.330 4
10	283.15	130.68	2 389.4	2 520.1	6.871 1	443.9	574.6	.292 3	447.0	1.860 1	1.398 6	1.330 0
20	293.15	135.29	2 403.4	2 538.7	6.935 7	370.2	505.5	.336 3	402.4	1.862 2	1.400 7	1.329 5
30	303.15	139.91	2 417.5	2 557.4	6.998 2	296.0	435.9	.385 0	363.4	1.864 7	1.403 1	1.328 9
40	313.15	144.52	2 431.5	2 576.0	7.058 7	221.1	365.6	.439 0	329.2	1.867 4	1.405 9	1.328 3
50	323.15	149.14	2 445.6	2 594.7	7.117 5	145.6	294.7	.498 6	299.1	1.870 5	1.409 0	1.327 5
60	333.15	153.75	2 459.7	2 613.4	7.174 5	69.5	223.2	.564 2	272.5	1.873 8	1.412 3	1.326 8
70	343.15	158.37	2 473.8	2 632.2	7.230 0	−7.2	151.2	.636 3	248.9	1.877 4	1.415 9	1.325 9
80	353.15	162.98	2 488.0	2 651.0	7.284 0	−84.3	78.6	.715 2	227.9	1.881 2	1.419 7	1.325 1
90	363.15	167.60	2 502.2	2 669.8	7.336 6	−162.1	5.5	.801 5	209.1	1.885 2	1.423 7	1.324 2
100	373.15	172.21	2 516.5	2 688.7	7.387 8	−240.3	−68.1	.895 7	192.26	1.889 4	1.427 9	1.323 2
120	393.15	181.44	2 545.1	2 726.6	7.486 7	−398.3	−216.8	1.109 7	163.50	1.898 3	1.436 7	1.321 2
140	413.15	190.67	2 573.9	2 764.6	7.581 1	−558.2	−367.5	1.361 7	140.03	1.907 7	1.446 2	1.319 1
160	433.15	199.90	2 603.0	2 802.9	7.671 5	−720.0	−520.1	1.656 4	120.69	1.917 7	1.456 2	1.316 9
180	453.15	209.13	2 632.2	2 841.3	7.758 3	−883.5	−674.4	1.999 1	104.61	1.928 1	1.466 6	1.314 7
200	473.15	218.4	2 661.6	2 880.0	7.841 8	−1 048.7	−830.4	2.396	91.15	1.938 9	1.477 4	1.312 4
300	573.15	264.5	2 812.3	3 076.8	8.218 9	−1 898.4	−1 633.9	5.423	48.77	1.997 5	1.536 0	1.300 5
400	673.15	310.7	2 969.0	3 279.7	8.545 1	−2 783.1	−2 472.5	10.996	28.25	2.061 4	1.599 9	1.288 5
500	773.15	356.8	3 132.4	3 489.2	8.835 2	−3 699	−3 342	20.61	17.310	2.128 7	1.667 2	1.276 8
600	873.15	403.0	3 302.5	3 705.5	9.098 2	−4 642	−4 239	36.45	11.056	2.198 0	1.736 5	1.265 8
700	973.15	449.1	3 479.7	3 928.8	9.340 3	−5 610	−5 161	61.58	7.293	2.268 3	1.806 8	1.255 4
800	1 073.15	495.3	3 663.9	4 159.2	9.565 5	−6 601	−6 106	100.34	4.936	2.338 7	1.877 1	1.245 9
900	1 173.15	541.4	3 855.1	4 396.5	9.776 9	−7 615	−7 073	158.63	3.413	2.407 8	1.946 2	1.237 1
1 000	1 273.15	587.6	4 053.1	4 640.6	9.976 6	−8 649	−8 061	244.5	2.403	2.474 4	2.012 8	1.299 3
1 100	1 373.15	633.7	4 257.5	4 891.2	10.166 1	−9 702	−9 068	368.6	1.719	2.536 9	2.075 4	1.222 4
1 200	1 473.15	679.9	4 467.9	5 147.8	10.346 4	−10 774	−10 094	544.9	1.248	2.593 8	2.132 3	1.216 4

Source: Adapted from Steam Tables, J.H. Keenan, F.G. Keyes, P.G. Hill, and J.G. Moore, John Wiley & Sons, Inc., New York, 1969 (International Edition — Metric Units).

REFERENCE

For other steam tables in metric units, see *Steam Tables in SI Units*, Ministry of Technology, London, 1970.

TABLE A.4 Properties of Saturated Water and Steam

Part a. Temperature Table

Temp. °C	Press. bars	Specific Volume m³/kg		Internal Energy kJ/kg		Enthalpy kJ/kg			Entropy kJ/kg · K		Temp. °C
		Sat. Liquid $v_f \times 10^3$	Sat. Vapor v_g	Sat. Liquid u_f	Sat. Vapor u_g	Sat. Liquid h_f	Evap. h_{fg}	Sat. Vapor h_g	Sat. Liquid s_f	Sat. Vapor s_g	
.01	0.00611	1.0002	206.136	0.00	2375.3	0.01	2501.3	2501.4	0.0000	9.1562	.01
4	0.00813	1.0001	157.232	16.77	2380.9	16.78	2491.9	2508.7	0.0610	9.0514	4
5	0.00872	1.0001	147.120	20.97	2382.3	20.98	2489.6	2510.6	0.0761	9.0257	5
6	0.00935	1.0001	137.734	25.19	2383.6	25.20	2487.2	2512.4	0.0912	9.0003	6
8	0.01072	1.0002	120.917	33.59	2386.4	33.60	2482.5	2516.1	0.1212	8.9501	8
10	0.01228	1.0004	106.379	42.00	2389.2	42.01	2477.7	2519.8	0.1510	8.9008	10
11	0.01312	1.0004	99.857	46.20	2390.5	46.20	2475.4	2521.6	0.1658	8.8765	11
12	0.01402	1.0005	93.784	50.41	2391.9	50.41	2473.0	2523.4	0.1806	8.8524	12
13	0.01497	1.0007	88.124	54.60	2393.3	54.60	2470.7	2525.3	0.1953	8.8285	13
14	0.01598	1.0008	82.848	58.79	2394.7	58.80	2468.3	2527.1	0.2099	8.8048	14
15	0.01705	1.0009	77.926	62.99	2396.1	62.99	2465.9	2528.9	0.2245	8.7814	15
16	0.01818	1.0011	73.333	67.18	2397.4	67.19	2463.6	2530.8	0.2390	8.7582	16
17	0.01938	1.0012	69.044	71.38	2398.8	71.38	2461.2	2532.6	0.2535	8.7351	17
18	0.02064	1.0014	65.038	75.57	2400.2	75.58	2458.8	2534.4	0.2679	8.7123	18
19	0.02198	1.0016	61.293	79.76	2401.6	79.77	2456.5	2536.2	0.2823	8.6897	19
20	0.02339	1.0018	57.791	83.95	2402.9	83.96	2454.1	2538.1	0.2966	8.6672	20
21	0.02487	1.0020	54.514	88.14	2404.3	88.14	2451.8	2539.9	0.3109	8.6450	21
22	0.02645	1.0022	51.447	92.32	2405.7	92.33	2449.4	2541.7	0.3251	8.6229	22
23	0.02810	1.0024	48.574	96.51	2407.0	96.52	2447.0	2543.5	0.3393	8.6011	23
24	0.02985	1.0027	45.883	100.70	2408.4	100.70	2444.7	2545.4	0.3534	8.5794	24
25	0.03169	1.0029	43.360	104.88	2409.8	104.89	2442.3	2547.2	0.3674	8.5580	25
26	0.03363	1.0032	40.994	109.06	2411.1	109.07	2439.9	2549.0	0.3814	8.5367	26
27	0.03567	1.0035	38.774	113.25	2412.5	113.25	2437.6	2550.8	0.3954	8.5156	27
28	0.03782	1.0037	36.690	117.42	2413.9	117.43	2435.2	2552.6	0.4093	8.4946	28
29	0.04008	1.0040	34.733	121.60	2415.2	121.61	2432.8	2554.5	0.4231	8.4739	29
30	0.04246	1.0043	32.894	125.78	2416.6	125.79	2430.5	2556.3	0.4369	8.4533	30
31	0.04496	1.0046	31.165	129.96	2418.0	129.97	2428.1	2558.1	0.4507	8.4329	31
32	0.04759	1.0050	29.540	134.14	2419.3	134.15	2425.7	2559.9	0.4644	8.4127	32
33	0.05034	1.0053	28.011	138.32	2420.7	138.33	2423.4	2561.7	0.4781	8.3927	33·
34	0.05324	1.0056	26.571	142.50	2422.0	142.50	2421.0	2563.5	0.4917	8.3728	34
35	0.05628	1.0060	25.216	146.67	2423.4	146.68	2418.6	2565.3	0.5053	8.3531	35
36	0.05947	1.0063	23.940	150.85	2424.7	150.86	2416.2	2567.1	0.5188	8.3336	36
38	0.06632	1.0071	21.602	159.20	2427.4	159.21	2411.5	2570.7	0.5458	8.2950	38
40	0.07384	1.0078	19.523	167.56	2430.1	167.57	2406.7	2574.3	0.5725	8.2570	40
45	0.09593	1.0099	15.258	188.44	2436.8	188.45	2394.8	2583.2	0.6387	8.1648	45

TABLE A.4 (continued) **Properties of Saturated Water and Steam**

Temp. °C	Press. bars	Specific Volume m³/kg		Internal Energy kJ/kg		Enthalpy kJ/kg			Entropy kJ/kg · K		Temp. °C
		Sat. Liquid $v_f \times 10^3$	Sat. Vapor v_g	Sat. Liquid u_f	Sat. Vapor u_g	Sat. Liquid h_f	Evap. h_{fg}	Sat. Vapor h_g	Sat. Liquid s_f	Sat. Vapor s_g	
50	.1235	1.0121	12.032	209.32	2443.5	209.33	2382.7	2592.1	.7038	8.0763	50
55	.1576	1.0146	9.568	230.21	2450.1	230.23	2370.7	2600.9	.7679	7.9913	55
60	.1994	1.0172	7.671	251.11	2456.6	251.13	2358.5	2609.6	.8312	7.9096	60
65	.2503	1.0199	6.197	272.02	2463.1	272.06	2346.2	2618.3	.8935	7.8310	65
70	.3119	1.0228	5.042	292.95	2469.6	292.98	2333.8	2626.8	.9549	7.7553	70
75	.3858	1.0259	4.131	313.90	2475.9	313.93	2321.4	2635.3	1.0155	7.6824	75
80	.4739	1.0291	3.407	334.86	2482.2	334.91	2308.8	2643.7	1.0753	7.6122	80
85	.5783	1.0325	2.828	355.84	2488.4	355.90	2296.0	2651.9	1.1343	7.5445	85
90	.7014	1.0360	2.361	376.85	2494.5	376.92	2283.2	2660.1	1.1925	7.4791	90
95	.8455	1.0397	1.982	397.88	2500.6	397.96	2270.2	2668.1	1.2500	7.4159	95
100	1.014	1.0435	1.673	418.94	2506.5	419.04	2257.0	2676.1	1.3069	7.3549	100
110	1.433	1.0516	1.210	461.14	2518.1	461.30	2230.2	2691.5	1.4185	7.2387	110
120	1.985	1.0603	0.8919	503.50	2529.3	503.71	2202.6	2706.3	1.5276	7.1296	120
130	2.701	1.0697	0.6685	546.02	2539.9	546.31	2174.2	2720.5	1.6344	7.0269	130
140	3.613	1.0797	0.5089	588.74	2550.0	589.13	2144.7	2733.9	1.7391	6.9299	140
150	4.758	1.0905	0.3928	631.68	2559.5	632.20	2114.3	2746.5	1.8418	6.8379	150
160	6.178	1.1020	0.3071	674.86	2568.4	675.55	2082.6	2758.1	1.9427	6.7502	160
170	7.917	1.1143	0.2428	718.33	2576.5	719.21	2049.5	2768.7	2.0419	6.6663	170
180	10.02	1.1274	0.1941	762.09	2583.7	763.22	2015.0	2778.2	2.1396	6.5857	180
190	12.54	1.1414	0.1565	806.19	2590.0	807.62	1978.8	2786.4	2.2359	6.5079	190
200	15.54	1.1565	0.1274	850.65	2595.3	852.45	1940.7	2793.2	2.3309	6.4323	200
210	19.06	1.1726	0.1044	895.53	2599.5	897.76	1900.7	2798.5	2.4248	6.3585	210
220	23.18	1.1900	0.08619	940.87	2602.4	943.62	1858.5	2802.1	2.5178	6.2861	220
230	27.95	1.2088	0.07158	986.74	2603.9	990.12	1813.8	2804.0	2.6099	6.2146	230
240	33.44	1.2291	0.05976	1033.2	2604.0	1037.3	1766.5	2803.8	2.7015	6.1437	240
250	39.73	1.2512	0.05013	1080.4	2602.4	1085.4	1716.2	2801.5	2.7927	6.0730	250
260	46.88	1.2755	0.04221	1128.4	2599.0	1134.4	1662.5	2796.6	2.8838	6.0019	260
270	54.99	1.3023	0.03564	1177.4	2593.7	1184.5	1605.2	2789.7	2.9751	5.9301	270
280	64.12	1.3321	0.03017	1227.5	2586.1	1236.0	1543.6	2779.6	3.0668	5.8571	280
290	74.36	1.3656	0.02557	1278.9	2576.0	1289.1	1477.1	2766.2	3.1594	5.7821	290
300	85.81	1.4036	0.02167	1332.0	2563.0	1344.0	1404.9	2749.0	3.2534	5.7045	300
320	112.7	1.4988	0.01549	1444.6	2525.5	1461.5	1238.6	2700.1	3.4480	5.5362	320
340	145.9	1.6379	0.01080	1570.3	2464.6	1594.2	1027.9	2622.0	3.6594	5.3357	340
360	186.5	1.8925	0.006945	1725.2	2351.5	1760.5	720.5	2481.0	3.9147	5.0526	360
374.14	220.9	3.155	0.003155	2029.6	2029.6	2099.3	0	2099.3	4.4298	4.4298	374.14

TABLE A.4 (continued)   **Properties of Saturated Water and Steam**

Part b. Pressure Table

Press. bars	Temp. °C	Specific Volume m³/kg		Internal Energy kJ/kg		Enthalpy kJ/kg			Entropy kJ/kg · K		Press. bars
		Sat. Liquid $v_f \times 10^3$	Sat. Vapor v_g	Sat. Liquid u_f	Sat. Vapor u_g	Sat. Liquid h_f	Evap. h_{fg}	Sat. Vapor h_g	Sat. Liquid s_f	Sat. Vapor s_g	
0.04	28.96	1.0040	34.800	121.45	2415.2	121.46	2432.9	2554.4	0.4226	8.4746	0.04
0.06	36.16	1.0064	23.739	151.53	2425.0	151.53	2415.9	2567.4	0.5210	8.3304	0.06
0.08	41.51	1.0084	18.103	173.87	2432.2	173.88	2403.1	2577.0	0.5926	8.2287	0.08
0.10	45.81	1.0102	14.674	191.82	2437.9	191.83	2392.8	2584.7	0.6493	8.1502	0.10
0.20	60.06	1.0172	7.649	251.38	2456.7	251.40	2358.3	2609.7	0.8320	7.9085	0.20
0.30	69.10	1.0223	·5.229	289.20	2468.4	289.23	2336.1	2625.3	0.9439	7.7686	0.30
0.40	75.87	1.0265	3.993	317.53	2477.0	317.58	2319.2	2636.8	1.0259	7.6700	0.40
0.50	81.33	1.0300	3.240	340.44	2483.9	340.49	2305.4	2645.9	1.0910	7.5939	0.50
0.60	85.94	1.0331	2.732	359.79	2489.6	359.86	2293.6	2653.5	1.1453	7.5320	0.60
0.70	89.95	1.0360	2.365	376.63	2494.5	376.70	2283.3	2660.0	1.1919	7.4797	0.70
0.80	93.50	1.0380	2.087	391.58	2498.8	391.66	2274.1	2665.8	1.2329	7.4346	0.80
0.90	96.71	1.0410	1.869	405.06	2502.6	405.15	2265.7	2670.9	1.2695	7.3949	0.90
1.00	99.63	1.0432	1.694	417.36	2506.1	417.46	2258.0	2675.5	1.3026	7.3594	1.00
1.50	111.4	1.0528	1.159	466.94	2519.7	467.11	2226.5	2693.6	1.4336	7.2233	1.50
2.00	120.2	1.0605	0.8857	504.49	2529.5	504.70	2201.9	2706.7	1.5301	7.1271	2.00
2.50	127.4	1.0672	0.7187	535.10	2537.2	535.37	2181.5	2716.9	1.6072	7.0527	2.50
3.00	133.6	1.0732	0.6058	561.15	2543.6	561.47	2163.8	2725.3	1.6718	6.9919	3.00
3.50	138.9	1.0786	0.5243	583.95	2546.9	584.33	2148.1	2732.4	1.7275	6.9405	3.50
4.00	143.6	1.0836	0.4625	604.31	2553.6	604.74	2133.8	2738.6	1.7766	6.8959	4.00
4.50	147.9	1.0882	0.4140	622.25	2557.6	623.25	2120.7	2743.9	1.8207	6.8565	4.50
5.00	151.9	1.0926	0.3749	639.68	2561.2	640.23	2108.5	2748.7	1.8607	6.8212	5.00
6.00	158.9	1.1006	0.3157	669.90	2567.4	670.56	2086.3	2756.8	1.9312	6.7600	6.00
7.00	165.0	1.1080	0.2729	696.44	2572.5	697.22	2066.3	2763.5	1.9922	6.7080	7.00
8.00	170.4	1.1148	0.2404	720.22	2576.8	721.11	2048.0	2769.1	2.0462	6.6628	8.00
9.00	175.4	1.1212	0.2150	741.83	2580.5	742.83	2031.1	2773.9	2.0946	6.6226	9.00
10.0	179.9	1.1273	0.1944	761.68	2583.6	762.81	2015.3	2778.1	2.1387	6.5863	10.0
15.0	198.3	1.1539	0.1318	843.16	2594.5	844.84	1947.3	2792.2	2.3150	6.4448	15.0
20.0	212.4	1.1767	0.09963	906.44	2600.3	908.79	1890.7	2799.5	2.4474	6.3409	20.0
25.0	224.0	1.1973	0.07998	959.11	2603.1	962.11	1841.0	2803.1	2.5547	6.2575	25.0
30.0	233.9	1.2165	0.06668	1004.8	2604.1	1008.4	1795.7	2804.2	2.6457	6.1869	30.0
35.0	242.6	1.2347	0.05707	1045.4	2603.7	1049.8	1753.7	2803.4	2.7253	6.1253	35.0
40.0	250.4	1.2522	0.04978	1082.3	2602.3	1087.3	1714.1	2801.4	2.7964	6.0701	40.0
45.0	257.5	1.2692	0.04406	1116.2	2600.1	1121.9	1676.4	2798.3	2.8610	6.0199	45.0
50.0	264.0	1.2859	0.03944	1147.8	2597.1	1154.2	1640.1	2794.3	2.9202	5.9734	50.0
60.0	275.6	1.3187	0.03244	1205.4	2589.7	1213.4	1571.0	2784.3	3.0267	5.8892	60.0
70.0	285.9	1.3513	0.02737	1257.6	2580.5	1267.0	1505.1	2772.1	3.1211	5.8133	70.0
80.0	295.1	1.3842	0.02352	1305.6	2569.8	1316.6	1441.3	2758.0	3.2068	5.7432	80.0
90.0	303.4	1.4178	0.02048	1350.5	2557.8	1363.3	1378.9	2742.1	3.2858	5.6772	90.0
100.0	311.1	1.4524	0.01803	1393.0	2544.4	1407.6	1317.1	2724.7	3.3596	5.6141	100.0
110.0	318.2	1.4886	0.01599	1433.7	2529.8	1450.1	1255.5	2705.6	3.4295	5.5527	110.0
120.0	324.8	1.5267	·0.01426	1473.0	2513.7	1491.3	1193.6	2684.9	3.4962	5.4924	120.0
130.0	330.9	1.5671	0.01278	1511.1	2496.1	1531.5	1130.7	2662.2	3.5606	5.4323	130.0
140.0	336.8	1.6107	0.01149	1548.6	2476.8	1571.1	1066.5	2637.6	3.6232	5.3717	140.0
150.0	342.2	1.6581	0.01034	1585.6	2455.5	1610.5	1000.0	2610.5	3.6848	5.3098	150.0
160.0	347.4	1.7107	0.009306	1622.7	2431.7	1650.1	930.6	2580.6	3.7461	5.2455	160.0
170.0	352.4	1.7702	0.008364	1660.2	2405.0	1690.3	856.9	2547.2	3.8079	5.1777	170.0
180.0	357.1	1.8397	0.007489	1698.9	2374.3	1732.0	777.1	2509.1	3.8715	5.1044	180.0
190.0	361.5	1.9243	0.006657	1739.9	2338.1	1776.5	688.0	2464.5	3.9388	5.0228	190.0
200.0	365.8	2.036	0.005834	1785.6	2293.0	1826.3	583.4	2409.7	4.0139	4.9269	200.0
220.9	374.1	3.155	0.003155	2029.6	2029.6	2099.3	0	2099.3	4.4298	4.4298	220.9

TABLE A.5 **Properties of Superheated Steam**

Symbols and Units:

T = temperature, °C

T_{sat} = saturation temperature, °C

v = specific volume, m³/kg

u = internal energy, kJ/kg

h = enthalpy, kJ/kg

S = entropy, kJ/kg·K

p = pressure, bar and μPa

T °C	v m³/kg	u kJ/kg	h kJ/kg	s kJ/kg · K	v m³/kg	u kJ/kg	h kJ/kg	s kJ/kg · K
	p = 0.06 bar = 0.006 MPa (T_{sat} = 36.16°C)				p = 0.35 bar = 0.035 MPa (T_{sat} = 72.69°C)			
Sat.	23.739	2425.0	2567.4	8.3304	4.526	2473.0	2631.4	7.7158
80	27.132	2487.3	2650.1	8.5804	4.625	2483.7	2645.6	7.7564
120	30.219	2544.7	2726.0	8.7840	5.163	2542.4	2723.1	7.9644
160	33.302	2602.7	2802.5	8.9693	5.696	2601.2	2800.6	8.1519
200	36.383	2661.4	2879.7	9.1398	6.228	2660.4	2878.4	8.3237
240	39.462	2721.0	2957.8	9.2982	6.758	2720.3	2956.8	8.4828
280	42.540	2781.5	3036.8	9.4464	7.287	2780.9	3036.0	8.6314
320	45.618	2843.0	3116.7	9.5859	7.815	2842.5	3116.1	8.7712
360	48.696	2905.5	3197.7	9.7180	8.344	2905.1	3197.1	8.9034
400	51.774	2969.0	3279.6	9.8435	8.872	2968.6	3279.2	9.0291
440	54.851	3033.5	3362.6	9.9633	9.400	3033.2	3362.2	9.1490
500	59.467	3132.3	3489.1	10.1336	10.192	3132.1	3488.8	9.3194
	p = 0.70 bar = 0.07 MPa (T_{sat} = 89.95°C)				p = 1.0 bar = 0.10 MPa (T_{sat} = 99.63°C)			
Sat.	2.365	2494.5	2660.0	7.4797	1.694	2506.1	2675.5	7.3594
100	2.434	2509.7	2680.0	7.5341	1.696	2506.7	2676.2	7.3614
120	2.571	2539.7	2719.6	7.6375	1.793	2537.3	2716.6	7.4668
160	2.841	2599.4	2798.2	7.8279	1.984	2597.8	2796.2	7.6597
200	3.108	2659.1	2876.7	8.0012	2.172	2658.1	2875.3	7.8343
240	3.374	2719.3	2955.5	8.1611	2.359	2718.5	2954.5	7.9949
280	3.640	2780.2	3035.0	8.3162	2.546	2779.6	3034.2	8.1445
320	3.905	2842.0	3115.3	8.4504	2.732	2841.5	3114.6	8.2849
360	4.170	2904.6	3196.5	8.5828	2.917	2904.2	3195.9	8.4175
400	4.434	2968.2	3278.6	8.7086	3.103	2967.9	3278.2	8.5435
440	4.698	3032.9	3361.8	8.8286	3.288	3032.6	3361.4	8.6636
500	5.095	3131.8	3488.5	8.9991	3.565	3131.6	3488.1	8.8342
	p = 1.5 bars = 0.15 MPa (T_{sat} = 111.37°C)				p = 3.0 bars = 0.30 MPa (T_{sat} = 133.55°C)			
Sat.	1.159	2519.7	2693.6	7.2233	0.606	2543.6	2725.3	6.9919
120	1.188	2533.3	2711.4	7.2693				
160	1.317	2595.2	2792.8	7.4665	0.651	2587.1	2782.3	7.1276
200	1.444	2656.2	2872.9	7.6433	0.716	2650.7	2865.5	7.3115
240	1.570	2717.2	2952.7	7.8052	0.781	2713.1	2947.3	7.4774
280	1.695	2778.6	3032.8	7.9555	0.844	2775.4	3028.6	7.6299
320	1.819	2840.6	3113.5	8.0964	0.907	2838.1	3110.1	7.7722
360	1.943	2903.5	3195.0	8.2293	0.969	2901.4	3192.2	7.9061
400	2.067	2967.3	3277.4	8.3555	1.032	2965.6	3275.0	8.0330
440	2.191	3032.1	3360.7	8.4757	1.094	3030.6	3358.7	8.1538
500	2.376	3131.2	3487.6	8.6466	1.187	3130.0	3486.0	8.3251
600	2.685	3301.7	3704.3	8.9101	1.341	3300.8	3703.2	8.5892

TABLE A.5 (continued)   **Properties of Superheated Steam**

Symbols and Units:

T = temperature, °C

T_{sat} = saturation temperature, °C

v = specific volume, m³/kg

u = internal energy, kJ/kg

h = enthalpy, kJ/kg

S = entropy, kJ/kg·K

p = pressure, bar and μPa

T °C	v m³/kg	u kJ/kg	h kJ/kg	s kJ/kg · K	v m³/kg	u kJ/kg	h kJ/kg	s kJ/kg . k
	p = 5.0 bars = 0.50 MPa (T_{sat} = 151.86°C)				p = 7.0 bars = 0.70 MPa (T_{sat} = 164.97°C)			
Sat.	0.3749	2561.2	2748.7	6.8213	0.2729	2572.5	2763.5	6.7080
180	0.4045	2609.7	2812.0	6.9656	0.2847	2599.8	2799.1	6.7880
200	0.4249	2642.9	2855.4	7.0592	0.2999	2634.8	2844.8	6.8865
240	0.4646	2707.6	2939.9	7.2307	0.3292	2701.8	2932.2	7.0641
280	0.5034	2771.2	3022.9	7.3865	0.3574	2766.9	3017.1	7.2233
320	0.5416	2834.7	3105.6	7.5308	0.3852	2831.3	3100.9	7.3697
360	0.5796	2898.7	3188.4	7.6660	0.4126	2895.8	3184.7	7.5063
400	0.6173	2963.2	3271.9	7.7938	0.4397	2960.9	3268.7	7.6350
440	0.6548	3028.6	3356.0	7.9152	0.4667	3026.6	3353.3	7.7571
500	0.7109	3128.4	3483.9	8.0873	0.5070	3126.8	3481.7	7.9299
600	0.8041	3299.6	3701.7	8.3522	0.5738	3298.5	3700.2	8.1956
700	0.8969	3477.5	3925.9	8.5952	0.6403	3476.6	3924.8	8.4391

T °C	v m³/kg	u kJ/kg	h kJ/kg	s kJ/kg · K	v m³/kg	u kJ/kg	h kJ/kg	s kJ/kg . k
	p = 10.0 bars = 1.0 MPa (T_{sat} = 179.91°C)				p = 15.0 bars = 1.5 MPa (T_{sat} = 198.32°C)			
Sat.	0.1944	2583.6	2778.1	6.5865	0.1318	2594.5	2792.2	6.4448
200	0.2060	2621.9	2827.9	6.6940	0.1325	2598.1	2796.8	6.4546
240	0.2275	2692.9	2920.4	6.8817	0.1483	2676.9	2899.3	6.6628
280	0.2480	2760.2	3008.2	7.0465	0.1627	2748.6	2992.7	6.8381
320	0.2678	2826.1	3093.9	7.1962	0.1765	2817.1	3081.9	6.9938
360	0.2873	2891.6	3178.9	7.3349	0.1899	2884.4	3169.2	7.1363
400	0.3066	2957.3	3263.9	7.4651	0.2030	2951.3	3255.8	7.2690
440	0.3257	3023.6	3349.3	7.5883	0.2160	3018.5	3342.5	7.3940
500	0.3541	3124.4	3478.5	7.7622	0.2352	3120.3	3473.1	7.5698
540	0.3729	3192.6	3565.6	7.8720	0.2478	3189.1	3560.9	7.6805
600	0.4011	3296.8	3697.9	8.0290	0.2668	3293.9	3694.0	7.8385
640	0.4198	3367.4	3787.2	8.1290	0.2793	3364.8	3783.8	7.9391

T °C	v m³/kg	u kJ/kg	h kJ/kg	s kJ/kg · K	v m³/kg	u kJ/kg	h kJ/kg	s kJ/kg . k
	p = 20.0 bars = 2.0 MPa (T_{sat} = 212.42°C)				p = 30.0 bars = 3.0 MPa (T_{sat} = 233.90°C)			
Sat.	0.0996	2600.3	2799.5	6.3409	0.0667	2604.1	2804.2	6.1869
240	0.1085	2659.6	2876.5	6.4952	0.0682	2619.7	2824.3	6.2265
280	0.1200	2736.4	2976.4	6.6828	0.0771	2709.9	2941.3	6.4462
320	0.1308	2807.9	3069.5	6.8452	0.0850	2788.4	3043.4	6.6245
360	0.1411	2877.0	3159.3	6.9917	0.0923	2861.7	3138.7	6.7801
400	0.1512	2945.2	3247.6	7.1271	0.0994	2932.8	3230.9	6.9212
440	0.1611	3013.4	3335.5	7.2540	0.1062	3002.9	3321.5	7.0520
500	0.1757	3116.2	3467.6	7.4317	0.1162	3108.0	3456.5	7.2338
540	0.1853	3185.6	3556.1	7.5434	0.1227	3178.4	3546.6	7.3474
600	0.1996	3290.9	3690.1	7.7024	0.1324	3285.0	3682.3	7.5085
640	0.2091	3362.2	3780.4	7.8035	0.1388	3357.0	3773.5	7.6106
700	0.2232	3470.9	3917.4	7.9487	0.1484	3466.5	3911.7	7.7571

TABLE A.5 (continued) **Properties of Superheated Steam**

Symbols and Units:

T = temperature, °C　　　　　　　　　　h = enthalpy, kJ/kg
T_{sat} = saturation temperature, °C　　　　S = entropy, kJ/kg·K
v = specific volume, m³/kg　　　　　　p = pressure, bar and μPa
u = internal energy, kJ/kg

T °C	v m³/kg	u kJ/kg	h kJ/kg	s kJ/kg · K	v m³/kg	u kJ/kg	h kJ/kg	s kJ/kg · K
	p = 40 bars = 4.0 MPa (T_{sat} = 250.4°C)				p = 60 bars = 6.0 MPa (T_{sat} = 275.64°C)			
Sat.	0.04978	2602.3	2801.4	6.0701	0.03244	2589.7	2784.3	5.8892
280	0.05546	2680.0	2901.8	6.2568	0.03317	2605.2	2804.2	5.9252
320	0.06199	2767.4	3015.4	6.4553	0.03876	2720.0	2952.6	6.1846
360	0.06788	2845.7	3117.2	6.6215	0.04331	2811.2	3071.1	6.3782
400	0.07341	2919.9	3213.6	6.7690	0.04739	2892.9	3177.2	6.5408
440	0.07872	2992.2	3307.1	6.9041	0.05122	2970.0	3277.3	6.6853
500	0.08643	3099.5	3445.3	7.0901	0.05665	3082.2	3422.2	6.8803
540	0.09145	3171.1	3536.9	7.2056	0.06015	3156.1	3517.0	6.9999
600	0.09885	3279.1	3674.4	7.3688	0.06525	3266.9	3658.4	7.1677
640	0.1037	3351.8	3766.6	7.4720	0.06859	3341.0	3752.6	7.2731
700	0.1110	3462.1	3905.9	7.6198	0.07352	3453.1	3894.1	7.4234
740	0.1157	3536.6	3999.6	7.7141	0.07677	3528.3	3989.2	7.5190

T °C	v m³/kg	u kJ/kg	h kJ/kg	s kJ/kg · K	v m³/kg	u kJ/kg	h kJ/kg	s kJ/kg · K
	p = 80 bars = 8.0 MPa (T_{sat} = 295.06°C)				p = 100 bars = 10.0 MPa (T_{sat} = 311.06°C)			
Sat.	0.02352	2569.8	2758.0	5.7432	0.01803	2544.4	2724.7	5.6141
320	0.02682	2662.7	2877.2	5.9489	0.01925	2588.8	2781.3	5.7103
360	0.03089	2772.7	3019.8	6.1819	0.02331	2729.1	2962.1	6.0060
400	0.03432	2863.8	3138.3	6.3634	0.02641	2832.4	3096.5	6.2120
440	0.03742	2946.7	3246.1	6.5190	0.02911	2922.1	3213.2	6.3805
480	0.04034	3025.7	3348.4	6.6586	0.03160	3005.4	3321.4	6.5282
520	0.04313	3102.7	3447.7	6.7871	0.03394	3085.6	3425.1	6.6622
560	0.04582	3178.7	3545.3	6.9072	0.03619	3164.1	3526.0	6.7864
600	0.04845	3254.4	3642.0	7.0206	0.03837	3241.7	3625.3	6.9029
640	0.05102	3330.1	3738.3	7.1283	0.04048	3318.9	3723.7	7.0131
700	0.05481	3443.9	3882.4	7.2812	0.04358	3434.7	3870.5	7.1687
740	0.05729	3520.4	3978.7	7.3782	0.04560	3512.1	3968.1	7.2670

T °C	v m³/kg	u kJ/kg	h kJ/kg	s kJ/kg · K	v m³/kg	u kJ/kg	h kJ/kg	s kJ/kg · K
	p = 120 bars = 12.0 MPa (T_{sat} = 324.75°C)				p = 140 bars = 14.0 MPa (T_{sat} = 336.75°C)			
Sat.	0.01426	2513.7	2684.9	5.4924	0.01149	2476.8	2637.6	5.3717
360	0.01811	2678.4	2895.7	5.8361	0.01422	2617.4	2816.5	5.6602
400	0.02108	2798.3	3051.3	6.0747	0.01722	2760.9	3001.9	5.9448
440	0.02355	2896.1	3178.7	6.2586	0.01954	2868.6	3142.2	6.1474
480	0.02576	2984.4	3293.5	6.4154	0.02157	2962.5	3264.5	6.3143
520	0.02781	3068.0	3401.8	6.5555	0.02343	3049.8	3377.8	6.4610
560	0.02977	3149.0	3506.2	6.6840	0.02517	3133.6	3486.0	6.5941
600	0.03164	3228.7	3608.3	6.8037	0.02683	3215.4	3591.1	6.7172
640	0.03345	3307.5	3709.0	6.9164	0.02843	3296.0	3694.1	6.8326
700	0.03610	3425.2	3858.4	7.0749	0.03075	3415.7	3846.2	6.9939
740	0.03781	3503.7	3957.4	7.1746	0.03225	3495.2	3946.7	7.0952

TABLE A.6   Chemical, Physical, and Thermal Properties of Gases: Gases and Vapors, Including Fuels and Refrigerants, English and SI Units

Common name(s)	Air [mixture]	Hydrogen	Methane	Nitrogen	Oxygen
Chemical formula		H_2	CH_4	N_2	O_2
Refrigerant number	729	702	50	728	732
CHEMICAL AND PHYSICAL PROPERTIES					
Molecular weight	28.966	2.016	16.044	28.013 4	31.998 8
Specific gravity, air = 1	1.00	0.070	0.554	0.967	1.105
Specific volume, ft³/lb	13.5	194.	24.2	13.98	12.24
Specific volume, m³/kg	0.842	12.1	1.51	0.872	0.764
Density of liquid (at atm bp), lb/ft³	54.6	4.43	26.3	50.46	71.27
Density of liquid (at atm bp), kg/m³	879.	71.0	421.	808.4	1 142.
Vapor pressure at 25 deg C, psia					
Vapor pressure at 25 deg C, MN/m²					
Viscosity (abs), lbm/ft·sec	12.1×10^{-6}	6.05×10^{-6}	7.39×10^{-6}	12.1×10^{-6}	13.4×10^{-6}
Viscosity (abs), centipoises[a]	0.018	0.009	0.011	0.018	0.020
Sound velocity in gas, m/sec	346	1 315.	446.	353.	329.
THERMAL AND THERMODYNAMIC PROPERTIES					
Specific heat, c_p, Btu/lb·deg F or cal/g·deg C	0.240 3	3.42	0.54	0.249	0.220
Specific heat, c_p, J/kg·K	1 005.	14 310.	2 260.	1 040.	920.
Specific heat ratio, c_p/c_v	1.40	1.405	1.31	1.40	1.40
Gas constant R, ft-lb/lb·deg R	53.3	767.	96.	55.2	48.3
Gas constant R, J/kg·deg C	286.8	4 126.	518.	297.	260.
Thermal conductivity, Btu/hr·ft·deg F	0.015 1	0.105	0.02	0.015	0.015
Thermal conductivity, W/m·deg C	0.026	0.018 2	0.035	0.026	0.026
Boiling point (sat 14.7 psia), deg F	−320	−423.	−259.	−320.4	−297.3
Boiling point (sat 760 mm), deg C	−195	20.4 K	−434.2	−195.8	−182.97
Latent heat of evap (at bp), Btu/lb	88.2	192.	219.2	85.5	91.7
Latent heat of evap (at bp), J/kg	205 000.	447 000.	510 000.	199 000.	213 000.
Freezing (melting) point, deg F (1 atm)	−357.2	−434.6	−296.6	−346.	−361.1
Freezing (melting) point, deg C (1 atm)	−216.2	−259.1	−182.6	−210.	−218.4
Latent heat of fusion, Btu/lb	10.0	25.0	14.	11.1	5.9
Latent heat of fusion, J/kg	23 200	58 000.	32 600.	25 800.	13 700.
Critical temperature, deg F	−220.5	−399.8	−116.	−232.6	−181.5
Critical temperature, deg C	−140.3	−240.0	−82.3	−147.	−118.6
Critical pressure, psia	550.	189.	673.	493.	726.
Critical pressure, MN/m²	3.8	1.30	4.64	3.40	5.01
Critical volume, ft³/lb	0.050	0.53	0.099	0.051	0.040
Critical volume, m³/kg	0.003	0.033	0.006 2	0.003 18	0.002 5
Flammable (yes or no)	No	Yes	Yes	No	No
Heat of combustion, Btu/ft³	—	320.	985.	—	—
Heat of combustion, Btu/lb	—	62 050.	2 290.	—	—
Heat of combustion, kJ/kg	—	144 000.	—	—	—

[a]For N·sec/m² divide by 1 000.

Note: The properties of pure gases are given at 25°C (77°F, 298 K) and atmospheric pressure (except as stated).

TABLE A.7 Ideal Gas Properties of Dry Air

Part a. SI Units

T(K), h and u(kJ/kg), s^o(kJ/kg·K)

T	h	p_r	u	v_r	s^o	T	h	p_r	u	v_r	s^o
200	199.97	0.3363	142.56	1707.0	1.29559	450	451.80	5.775	322.62	223.6	2.11161
210	209.97	0.3987	149.69	1512.0	1.34444	460	462.02	6.245	329.97	211.4	2.13407
220	219.97	0.4690	156.82	1346.0	1.39105	470	472.24	6.742	337.32	200.1	2.15604
230	230.02	0.5477	164.00	1205.0	1.43557	480	482.49	7.268	344.70	189.5	2.17760
240	240.02	0.6355	171.13	1084.0	1.47824	490	492.74	7.824	352.08	179.7	2.19876
250	250.05	0.7329	178.28	979.0	1.51917	500	503.02	8.411	359.49	170.6	2.21952
260	260.09	0.8405	185.45	887.8	1.55848	510	513.32	9.031	366.92	162.1	2.23993
270	270.11	0.9590	192.60	808.0	1.59634	520	523.63	9.684	374.36	154.1	2.25997
280	280.13	1.0889	199.75	738.0	1.63279	530	533.98	10.37	381.84	146.7	2.27967
285	285.14	1.1584	203.33	706.1	1.65055	540	544.35	11.10	389.34	139.7	2.29906
290	290.16	1.2311	206.91	676.1	1.66802	550	554.74	11.86	396.86	133.1	2.31809
295	295.17	1.3068	210.49	647.9	1.68515	560	565.17	12.66	404.42	127.0	2.33685
300	300.19	1.3860	214.07	621.2	1.70203	570	575.59	13.50	411.97	121.2	2.35531
305	305.22	1.4686	217.67	596.0	1.71865	580	586.04	14.38	419.55	115.7	2.37348
310	310.24	1.5546	221.25	572.3	1.73498	590	596.52	15.31	427.15	110.6	2.39140
315	315.27	1.6442	224.85	549.8	1.75106	600	607.02	16.28	434.78	105.8	2.40902
320	320.29	1.7375	228.42	528.6	1.76690	610	617.53	17.30	442.42	101.2	2.42644
325	325.31	1.8345	232.02	508.4	1.78249	620	628.07	18.36	450.09	96.92	2.44356
330	330.34	1.9352	235.61	489.4	1.79783	630	638.63	19.84	457.78	92.84	2.46048
340	340.42	2.149	242.82	454.1	1.82790	640	649.22	20.64	465.50	88.99	2.47716
350	350.49	2.379	250.02	422.2	1.85708	650	659.84	21.86	473.25	85.34	2.49364
360	360.58	2.626	257.24	393.4	1.88543	660	670.47	23.13	481.01	81.89	2.50985
370	370.67	2.892	264.46	367.2	1.91313	670	681.14	24.46	488.81	78.61	2.52589
380	380.77	3.176	271.69	343.4	1.94001	680	691.82	25.85	496.62	75.50	2.54175
390	390.88	3.481	278.93	321.5	1.96633	690	702.52	27.29	504.45	72.56	2.55731
400	400.98	3.806	286.16	301.6	1.99194	700	713.27	28.80	512.33	69.76	2.57277
410	411.12	4.153	293.43	283.3	2.01699	710	724.04	30.38	520.23	67.07	2.58810
420	421.26	4.522	300.69	266.6	2.04142	720	734.82	32.02	528.14	64.53	2.60319
430	431.43	4.915	307.99	251.1	2.06533	730	745.62	33.72	536.07	62.13	2.61803
440	441.61	5.332	315.30	236.8	2.08870	740	756.44	35.50	544.02	59.82	2.63280
750	767.29	37.35	551.99	57.63	2.64737	1300	1395.97	330.9	1022.82	11.275	3.27345
760	778.18	39.27	560.01	55.54	2.66176	1320	1419.76	352.5	1040.88	10.747	3.29160
770	789.11	41.31	568.07	53.39	2.67595	1340	1443.60	375.3	1058.94	10.247	3.30959
780	800.03	43.35	576.12	51.64	2.69013	1360	1467.49	399.1	1077.10	9.780	3.32724
790	810.99	45.55	584.21	49.86	2.70400	1380	1491.44	424.2	1095.26	9.337	3.34474
800	821.95	47.75	592.30	48.08	2.71787	1400	1515.42	450.5	1113.52	8.919	3.36200
820	843.98	52.59	608.59	44.84	2.74504	1420	1539.44	478.0	1131.77	8.526	3.37901
840	866.08	57.60	624.95	41.85	2.77170	1440	1563.51	506.9	1150.13	8.153	3.39586
860	888.27	63.09	641.40	39.12	2.79783	1460	1587.63	537.1	1168.49	7.801	3.41247
880	910.56	68.98	657.95	36.61	2.82344	1480	1611.79	568.8	1186.95	7.468	3.42892
900	932.93	75.29	674.58	34.31	2.84856	1500	1635.97	601.9	1205.41	7.152	3.44516
920	955.38	82.05	691.28	32.18	2.87324	1520	1660.23	636.5	1223.87	6.854	3.46120
940	977.92	89.28	708.08	30.22	2.89748	1540	1684.51	672.8	1242.43	6.569	3.47712
960	1000.55	97.00	725.02	28.40	2.92128	1560	1708.82	710.5	1260.99	6.301	3.49276
980	1023.25	105.2	741.98	26.73	2.94468	1580	1733.17	750.0	1279.65	6.046	3.50829
1000	1046.04	114.0	758.94	25.17	2.96770	1600	1757.57	791.2	1298.30	5.804	3.52364
1020	1068.89	123.4	776.10	23.72	2.99034	1620	1782.00	834.1	1316.96	5.574	3.53879
1040	1091.85	133.3	793.36	22.39	3.01260	1640	1806.46	878.9	1335.72	5.355	3.55381
1060	1114.86	143.9	810.62	21.14	3.03449	1660	1830.96	925.6	1354.48	5.147	3.56867
1080	1137.89	155.2	827.88	19.98	3.05608	1680	1855.50	974.2	1373.24	4.949	3.58335
1100	1161.07	167.1	845.33	18.896	3.07732	1700	1880.1	1025	1392.7	4.761	3.5979
1120	1184.28	179.7	862.79	17.886	3.09825	1750	1941.6	1161	1439.8	4.328	3.6336
1140	1207.57	193.1	880.35	16.946	3.11883	1800	2003.3	1310	1487.2	3.944	3.6684
1160	1230.92	207.2	897.91	16.064	3.13916	1850	2065.3	1475	1534.9	3.601	3.7023
1180	1254.34	222.2	915.57	15.241	3.15916	1900	2127.4	1655	1582.6	3.295	3.7354
1200	1277.79	238.0	933.33	14.470	3.17888	1950	2189.7	1852	1630.6	3.022	3.7677
1220	1301.31	254.7	951.09	13.747	3.19834	2000	2252.1	2068	1678.7	2.776	3.7994
1240	1324.93	272.3	968.95	13.069	3.21751	2050	2314.6	2303	1726.8	2.555	3.8303
1260	1348.55	290.8	986.90	12.435	3.23638	2100	2377.4	2559	1775.3	2.356	3.8605
1280	1372.24	310.4	1004.76	11.835	3.25510	2150	2440.3	2837	1823.8	2.175	3.8901
						2200	2503.2	3138	1872.4	2.012	3.9191
						2250	2566.4	3464	1921.3	1.864	3.9474

TABLE A.7 (continued) Ideal Gas Properties of Dry Air

Part b. English Units

$T(°R)$, h and u (Btu/lb), $s°$ (Btu/lb · °R)

T	h	p_r	u	v_r	$s°$	T	h	p_r	u	v_r	$s°$
360	85.97	0.3363	61.29	396.6	0.50369	940	226.11	9.834	161.68	35.41	0.73509
380	90.75	0.4061	64.70	346.6	0.51663	960	231.06	10.61	165.26	33.52	0.74030
400	95.53	0.4858	68.11	305.0	0.52890	980	236.02	11.43	168.83	31.76	0.74540
420	100.32	0.5760	71.52	270.1	0.54058	1000	240.98	12.30	172.43	30.12	0.75042
440	105.11	0.6776	74.93	240.6	0.55172	1040	250.95	14.18	179.66	27.17	0.76019
460	109.90	0.7913	78.36	215.33	0.56235	1080	260.97	16.28	186.93	24.58	0.76964
480	114.69	0.9182	81.77	193.65	0.57255	1120	271.03	18.60	194.25	22.30	0.77880
500	119.48	1.0590	85.20	174.90	0.58233	1160	281.14	21.18	201.63	20.29	0.78767
520	124.27	1.2147	88.62	158.58	0.59172	1200	291.30	24.01	209.05	18.51	0.79628
537	128.34	1.3593	91.53	146.34	0.59945	1240	301.52	27.13	216.53	16.93	0.80466
540	129.06	1.3860	92.04	144.32	0.60078	1280	311.79	30.55	224.05	15.52	0.81280
560	133.86	1.5742	95.47	131.78	0.60950	1320	322.11	34.31	231.63	14.25	0.82075
580	138.66	1.7800	98.90	120.70	0.61793	1360	332.48	38.41	239.25	13.12	0.82848
600	143.47	2.005	102.34	110.88	0.62607	1400	342.90	42.88	246.93	12.10	0.83604
620	148.28	2.249	105.78	102.12	0.63395	1440	353.37	47.75	254.66	11.17	0.84341
640	153.09	2.514	109.21	94.30	0.64159	1480	363.89	53.04	262.44	10.34	0.85062
660	157.92	2.801	112.67	87.27	0.64902	1520	374.47	58.78	270.26	9.578	0.85767
680	162.73	3.111	116.12	80.96	0.65621	1560	385.08	65.00	278.13	8.890	0.86456
700	167.56	3.446	119.58	75.25	0.66321	1600	395.74	71.73	286.06	8.263	0.87130
720	172.39	3.806	123.04	70.07	0.67002	1650	409.13	80.89	296.03	7.556	0.87954
740	177.23	4.193	126.51	65.38	0.67665	1700	422.59	90.95	306.06	6.924	0.88758
760	182.08	4.607	129.99	61.10	0.68312	1750	436.12	101.98	316.16	6.357	0.89542
780	186.94	5.051	133.47	57.20	0.68942	1800	449.71	114.0	326.32	5.847	0.90308
800	191.81	5.526	136.97	53.63	0.69558	1850	463.37	127.2	336.55	5.388	0.91056
820	196.69	6.033	140.47	50.35	0.70160	1900	477.09	141.5	346.85	4.974	0.91788
840	201.56	6.573	143.98	47.34	0.70747	1950	490.88	157.1	357.20	4.598	0.92504
860	206.46	7.149	147.50	44.57	0.71323	2000	504.71	174.0	367.61	4.258	0.93205
880	211.35	7.761	151.02	42.01	0.71886	2050	518.61	192.3	378.08	3.949	0.93891
900	216.26	8.411	154.57	39.64	0.72438	2100	532.55	212.1	388.60	3.667	0.94564
920	221.18	9.102	158.12	37.44	0.72979	2150	546.54	233.5	399.17	3.410	0.95222

TABLE A.7 (continued) **Ideal Gas Properties of Dry Air**

$T(°R)$, h and u(Btu/lb), $s°$(Btu/lb · °R)

T	h	p_r	u	v_r	$s°$	T	h	p_r	u	v_r	$s°$
2200	560.59	256.6	409.78	3.176	0.95868	3700	998.11	2330	744.48	.5882	1.10991
2250	574.69	281.4	420.46	2.961	0.96501	3750	1013.1	2471	756.04	.5621	1.11393
2300	588.82	308.1	431.16	2.765	0.97123	3800	1028.1	2618	767.60	.5376	1.11791
2350	603.00	336.8	441.91	2.585	0.97732	3850	1043.1	2773	779.19	.5143	1.12183
2400	617.22	367.6	452.70	2.419	0.98331	3900	1058.1	2934	790.80	.4923	1.12571
2450	631.48	400.5	463.54	2.266	0.98919	3950	1073.2	3103	802.43	.4715	1.12955
2500	645.78	435.7	474.40	2.125	0.99497	4000	1088.3	3280	814.06	.4518	1.13334
2550	660.12	473.3	485.31	1.996	1.00064	4050	1103.4	3464	825.72	.4331	1.13709
2600	674.49	513.5	496.26	1.876	1.00623	4100	1118.5	3656	837.40	.4154	1.14079
2650	688.90	556.3	507.25	1.765	1.01172	4150	1133.6	3858	849.09	.3985	1.14446
2700	703.35	601.9	518.26	1.662	1.01712	4200	1148.7	4067	860.81	.3826	1.14809
2750	717.83	650.4	529.31	1.566	1.02244	4300	1179.0	4513	884.28	.3529	1.15522
2800	732.33	702.0	540.40	1.478	1.02767	4400	1209.4	4997	907.81	.3262	1.16221
2850	746.88	756.7	551.52	1.395	1.03282	4500	1239.9	5521	931.39	.3019	1.16905
2900	761.45	814.8	562.66	1.318	1.03788	4600	1270.4	6089	955.04	.2799	1.17575
2950	776.05	876.4	573.84	1.247	1.04288	4700	1300.9	6701	978.73	.2598	1.18232
3000	790.68	941.4	585.04	1.180	1.04779	4800	1331.5	7362	1002.5	.2415	1.18876
3050	805.34	1011	596.28	1.118	1.05264	4900	1362.2	8073	1026.3	.2248	1.19508
3100	820.03	1083	607.53	1.060	1.05741	5000	1392.9	8837	1050.1	.2096	1.20129
3150	834.75	1161	618.82	1.006	1.06212	5100	1423.6	9658	1074.0	.1956	1.20738
3200	849.48	1242	630.12	.9546	1.06676	5200	1454.4	10539	1098.0	.1828	1.21336
3250	864.24	1328	641.46	.9069	1.07134	5300	1485.3	11481	1122.0	.1710	1.21923
3300	879.02	1418	652.81	.8621	1.07585						
3350	893.83	1513	664.20	.8202	1.08031						
3400	908.66	1613	675.60	.7807	1.08470						
3450	923.52	1719	687.04	.7436	1.08904						
3500	938.40	1829	698.48	.7087	1.09332						
3550	953.30	1946	709.95	.6759	1.09755						
3600	968.21	2068	721.44	.6449	1.10172						
3650	983.15	2196	732.95	.6157	1.10584						

Source: Adapted from M.J. Moran and H.N. Shapiro, *Fundamentals of Engineering Thermodynamics*, 3rd. ed., Wiley & Sons, New York, 1995, as based on J.H. Keenan and J. Kaye, *Gas Tables*, John Wiley and Sons, New York, 1945. With permission.

Appendix B Properties of Liquids

TABLE B.1 Properties of Liquid Water

Symbols and Units:

ρ = density, lbm/ft^3. For g/cm^3 multiply by 0.016018. For kg/m^3 multiply by 16.018.

c_p = specific heat, Btu/lbm·deg R = cal/g·K. For J/kg·K multiply by 4186.8

μ = viscosity. For lbf·sec/ft^2 = slugs/sec·ft, multiply by 10^{-7}. For lbm·sec·ft multiply by 10^{-7} and by 32.174. For g/sec·cm (poises) multiply by 10^{-7} and by 478.80. For N·sec/m^2 multiply by 10^{-7} and by 478.880.

k = thermal conductivity, Btu/hr·ft·deg R. For W/m·K multiply by 1.7307.

Temp, °F	At 1 atm or 14.7 psia				At 1,000 psia				At 10,000 psia			
	ρ	c_p	μ	k	ρ	c_p	μ	k	ρ	c_p	μ	k^a
32	62.42	1.007	366	0.3286	62.62	0.999	365	0.3319	64.5	0.937	357	0.3508
40	62.42	1.004	323	0.334	62.62	0.997	323	0.337	64.5	0.945	315	0.356
50	62.42	1.002	272	0.3392	62.62	0.995	272	0.3425	64.5	0.951	267	0.3610
60	62.38	1.000	235	0.345	62.58	0.994	235	0.348	64.1	0.956	233	0.366
70	62.31	0.999	204	0.350	62.50	0.994	204	0.353	64.1	0.960	203	0.371
80	62.23	0.998	177	0.354	62.42	0.994	177	0.358	64.1	0.962	176	0.376
90	62.11	0.998	160	0.359	62.31	0.994	160	0.362	63.7	0.964	159	0.380
100	62.00	0.998	142	0.3633	62.19	0.994	142	0.3666	63.7	0.965	142	0.3841
110	61.88	0.999	126	0.367	62.03	0.994	126	0.371	63.7	0.966	126	0.388
120	61.73	0.999	114	0.371	61.88	0.995	114	0.374	63.3	0.967	114	0.391
130	61.54	0.999	105	0.374	61.73	0.995	105	0.378	63.3	0.968	105	0.395
140	61.39	0.999	96	0.378	61.58	0.996	96	0.381	63.3	0.969	98	0.398
150	61.20	1.000	89	0.3806	61.39	0.996	89	0.3837	63.0	0.970	91	0.4003
160	61.01	1.001	83	0.383	61.20	0.997	83	0.386	62.9	0.971	85	0.403
170	60.79	1.002	77	0.386	60.98	0.998	77	0.389	62.5	0.972	79	0.405
180	60.57	1.003	72	0.388	60.75	0.999	72	0.391	62.5	0.973	74	0.407
190	60.35	1.004	68	0.390	60.53	1.001	68	0.393	62.1	0.974	70	0.409
200	60.10	1.005	62.5	0.3916	60.31	1.002	62.9	0.3944	62.1	0.975	65.4	0.4106
250	boiling point 212°F				59.03	1.001	47.8	0.3994	60.6	0.981	50.6	0.4158
300					57.54	1.024	38.4	0.3993	59.5	0.988	41.3	0.4164
350					55.83	1.044	32.1	0.3944	58.1	0.999	35.1	0.4132
400					53.91	1.072	27.6	0.3849	56.5	1.011	30.6	0.4064
500					49.11	1.181	21.6	0.3508	52.9	1.051	24.8	0.3836
600					boiling point 544.58°F				48.3	1.118	21.0	0.3493

^{a} At 7,500 psia.

Source: "1967 ASME Steam Tables", American Society of Mechanical Engineers, Tables 9, 10, and 11 and Figures 6, 7, 8, and 9.

Note: The ASME compilation is a 330-page book of tables and charts, including a 2½ × 3½-ft Mollier chart. All values have been computed in accordance with the 1967 specifications of the International Formulation Committee (IFC) and are in conformity with the 1963 International Skeleton Tables. This standardization of tables began in 1921 and was extended through the (1963) and Glasgow (1966). Based on these worldwide standard data, the 1967 ASME volume represents detailed computer output in both tabular and graphic form. Included are density and volume, enthalpy, entropy, specific heat, viscosity, thermal conductivity, Prandtl number, isentropic exponent, choking velocity, p-v product, etc., over the entire range (to 1500 psia 1500°F). English units are used, but all conversion factors are given.

TABLE B.2 Physical and Thermal Properties of Common Liquids

Part a. SI Units

(At 1.0 Atm Pressure (0.101 325 MN/m²), 300 K, except as noted.)

Common name	Density, kg/m³	Specific heat, kJ/kg·K	Viscosity, N·s/m²	Thermal conductivity, W/m·K	Freezing point, K	Latent heat of fusion, kJ/kg	Boiling point, K	Latent heat of evaporation, kJ/kg	Coefficient of cubical expansion per K
Acetic acid	1 049	2.18	.001 155	0.171	290	181	391	402	0.001 1
Acetone	784.6	2.15	.000 316	0.161	179.0	98.3	329	518	0.001 5
Alcohol, ethyl	785.1	2.44	.001 095	0.171	158.6	108	351.46	846	0.001 1
Alcohol, methyl	786.5	2.54	.000 56	0.202	175.5	98.8	337.8	1 100	0.001 4
Alcohol, propyl	800.0	2.37	.001 92	0.161	146	86.5	371	779	
Ammonia (aqua)	823.5	4.38		0.353					
Benzene	873.8	1.73	.000 601	0.144	278.68	126	353.3	390	0.001 3
Bromine		.473	.000 95		245.84	66.7	331.6	193	0.001 2
Carbon disulfide	1 261	.992	.000 36	0.161	161.2	57.6	319.40	351	0.001 3
Carbon tetrachloride	1 584	.866	.000 91	0.104	250.35	174	349.6	194	0.001 3
Castor oil	956.1	1.97	.650	0.180	263.2				
Chloroform	1 465	1.05	.000 53	0.118	209.6	77.0	334.4	247	0.001 3
Decane	726.3	2.21	.000 859	0.147	243.5	201	447.2	263	
Dodecane	754.6	2.21	.001 374	0.140	247.18	216	489.4	256	
Ether	713.5	2.21	.000 223	0.130	157	96.2	307.7	372	0.001 6
Ethylene glycol	1 097	2.36	.016 2	0.258	260.2	181	470	800	
Fluorine refrigerant R-11	1 476	.870ᵃ	.000 42	0.093ᵃ	162		297.0	180ᵇ	
Fluorine refrigerant R-12	1 311	.971ᵃ		0.071ᵃ	115	34.4	243.4	165ᵇ	
Fluorine refrigerant R-22	1 194	1.26ᵃ		0.086ᵃ	113	183	232.4	232ᵇ	
Glycerine	1 259	2.62	.950	0.287	264.8	200	563.4	974	0.000 54
Heptane	679.5	2.24	.000 376	0.128	182.54	140	371.5	318	
Hexane	654.8	2.26	.000 297	0.124	178.0	152	341.84	365	
Iodine		2.15			386.6	62.2	457.5	164	
Kerosene	820.1	2.09	.001 64	0.145				251	
Linseed oil	929.1	1.84	.033 1		253		560		
Mercury		.139	.001 53		234.3	11.6	630	295	0.000 18
Octane	698.6	2.15	.000 51	0.131	216.4	181	398	298	0.000 72
Phenol	1 072	1.43	.008 0	0.190	316.2	121	455		0.000 90
Propane	493.5	2.41ᵃ	.000 11		85.5	79.9	231.08	428ᵇ	
Propylene	514.4	2.85	.000 09		87.9	71.4	225.45	342	
Propylene glycol	965.3	2.50	.042		213		460	914	
Sea water	1 025	3.76–4.10			270.6				
Toluene	862.3	1.72	.000 550	0.133	178	71.8	383.6	363	
Turpentine	868.2	1.78	.001 375	0.121	214		433	293	0.000 99
Water	997.1	4.18	.000 89	0.609	273	333	373	2 260	0.000 20

ᵃAt 297 K, liquid.
ᵇAt .101 325 meganewtons, saturation temperature.

TABLE B.2 (continued) Physical and Thermal Properties of Common Liquids

Part b. English Units

(At 1.0 Atm Pressure 77°F (25°C), except as noted.)

For viscosity in N·s/m² (=kg m·s), multiply values in centipoises by 0.001. For surface tension in N/m, multiply values in dyne/cm by 0.001.

Common name	Density, $\frac{lb}{ft^3}$	Specific gravity	Viscosity $lb_m/ft\ sec$ $\times 10^4$	Viscosity cp	Sound velocity, $\frac{meters}{sec}$	Dielectric constant	Refractive index
Acetic acid	65.493	1.049	7.76	1.155	1584[50]	6.15	1.37
Acetone	48.98	.787	2.12	0.316	1174	20.7	1.36
Alcohol, ethyl	49.01	.787	7.36	1.095	1144	24.3	1.36
Alcohol, methyl	49.10	.789	3.76	0.56	1103	32.6	1.33
Alcohol, propyl	49.94	.802	12.9	1.92	1205	20.1	1.38
Ammonia (aqua)	51.411	.826	—	—	—	16.9	—
Benzene	54.55	.876	4.04	0.601	1298	2.2	1.50
Bromine	—	—	6.38	0.95	—	3.20	—
Carbon disulfide	78.72	1.265	2.42	0.36	1149	2.64	1.63
Carbon tetrachloride	98.91	1.59	6.11	0.91	924	2.23	1.46
Castor oil	59.69	0.960	—	650	1474	4.7	—
Chloroform	91.44	1.47	3.56	0.53	995	4.8	1.44
Decane	45.34	.728	5.77	0.859	—	2.0	1.41
Dodecane	47.11	—	9.23	1.374	—	—	1.41
Ether	44.54	0.715	1.50	0.223	985	4.3	1.35
Ethylene glycol	68.47	1.100	109	16.2	1644	37.7	1.43
Fluorine refrigerant R–11	92.14	1.480	2.82	0.42	—	2.0	1.37
Fluorine refrigerant R–12	81.84	1.315	—	—	—	2.0	1.29
Fluorine refrigerant R–22	74.53	1.197	—	—	—	2.0	1.26
Glycerine	78.62	1.263	6380	950	1909	40	1.47
Heptane	42.42	.681	2.53	0.376	1138	1.92	1.38
Hexane	40.88	.657	2.00	0.297	1203	—	1.37
Iodine	—	—	—	—	—	11	—
Kerosene	51.2	0.823	11.0	1.64	1320	—	—
Linseed oil	58.0	0.93	222	33.1	—	3.3	—
Mercury	—	13.633	10.3	1.53	1450	—	—
Octane	43.61	.701	3.43	0.51	1171	—	1.40
Phenol	66.94	1.071	54	8.0	1274[100]	9.8	—
Propane	30.81	.495	0.74	0.11	—	1.27	1.34
Propylene	32.11	.516	0.60	0.09	—	—	1.36
Propylene glycol	60.26	.968	—	42	—	—	1.43
Sea water	64.0	1.03	—	—	1535	—	—
Toluene	53.83	0.865	3.70	0.550	1275[30]	2.4	1.49
Turpentine	54.2	0.87	9.24	1.375	1240	—	1.47
Water	62.247	1.00	6.0	0.89	1498	78.54[a]	1.33

[a]The dielectric constant of water near the freezing point is 87.8; it decreases with increase in temperature to about 55.6 near the boiling point.

Appendix C Properties of Solids

TABLE C.1 Properties of Common Solids

Material	Specific gravity	Specific heat		Thermal conductivity	
		$\dfrac{Btu}{lbm \cdot deg\ R}$	$\dfrac{kJ}{kg \cdot K}$	$\dfrac{Btu}{hr \cdot ft \cdot deg\ F}$	$\dfrac{W}{m \cdot K}$
Asbestos cement board	1.4	0.2	.837	0.35	0.607
Asbestos millboard	1.0	0.2	.837	0.08	0.14
Asphalt	1.1	0.4	1.67		
Beeswax	0.95	0.82	3.43		
Brick, common	1.75	0.22	.920	0.42	0.71
Brick, hard	2.0	0.24	1.00	0.75	1.3
Chalk	2.0	0.215	.900	0.48	0.84
Charcoal, wood	0.4	0.24	1.00	0.05	0.088
Coal, anthracite	1.5	0.3	1.26		
Coal, bituminous	1.2	0.33	1.38		
Concrete, light	1.4	0.23	.962	0.25	0.42
Concrete, stone	2.2	0.18	.753	1.0	1.7
Corkboard	0.2	0.45	1.88	0.025	0.04
Earth, dry	1.4	0.3	1.26	0.85	1.5
Fiberboard, light	0.24	0.6	2.51	0.035	0.058
Fiber hardboard	1.1	0.5	2.09	0.12	0.2
Firebrick	2.1	0.25	1.05	0.8	1.4
Glass, window	2.5	0.2	.837	0.55	0.96
Gypsum board	0.8	0.26	1.09	0.1	0.17
Hairfelt	0.1	0.5	2.09	0.03	0.050
Ice (32°)	0.9	0.5	2.09	1.25	2.2
Leather, dry	0.9	0.36	1.51	0.09	0.2
Limestone	2.5	0.217	.908	1.1	1.9
Magnesia (85%)	0.25	0.2	.837	0.04	0.071
Marble	2.6	0.21	.879	1.5	2.6
Mica	2.7	0.12	.502	0.4	0.71
Mineral wool blanket	0.1	0.2	.837	0.025	0.04
Paper	0.9	0.33	1.38	0.07	0.1
Paraffin wax	0.9	0.69	2.89	0.15	0.2
Plaster, light	0.7	0.24	1.00	0.15	0.2
Plaster, sand	1.8	0.22	.920	0.42	0.71
Plastics, foamed	0.2	0.3	1.26	0.02	0.03
Plastics, solid	1.2	0.4	1.67	0.11	0.19
Porcelain	2.5	0.22	.920	0.9	1.5
Sandstone	2.3	0.22	.920	1.0	1.7
Sawdust	0.15	0.21	.879	0.05	0.08
Silica aerogel	0.11	0.2	.837	0.015	0.02
Vermiculite	0.13	0.2	.837	0.035	0.058
Wood, balsa	0.16	0.7	2.93	0.03	0.050
Wood, oak	0.7	0.5	2.09	0.10	0.17
Wood, white pine	0.5	0.6	2.51	0.07	0.12
Wool, felt	0.3	0.33	1.38	0.04	0.071
Wool, loose	0.1	0.3	1.26	0.02	0.3

Source: Compiled from several sources.

TABLE C.2 Density of Various Solids: Approximate Density of Solids at Ordinary Atmospheric Temperature

Substance	Grams per cu cm	Pounds per cu ft	Substance	Grams per cu cm	Pounds per cu ft	Substance	Grams per cu cm	Pounds per cu ft
Agate	2.5–2.7	156–168	Glass			Tallow		
Alabaster			Common	2.4–2.8	150–175	Beef	0.94	59
Carbonate	2.69–2.78	168–173	Flint	2.9–5.9	180–370	Mutton	0.94	59
Sulfate	2.26–2.32	141–145	Glue	1.27	79	Tar	1.02	66
Albite	2.62–2.65	163–165	Granite	2.64–2.76	165–172	Topaz	3.5–3.6	219–223
Amber	1.06–1.11	66–69	Graphite[a]	2.30–2.72	144–170	Tourmaline	3.0–3.2	190–200
Amphiboles	2.9–3.2	180–200	Gum arabic	1.3–1.4	81–87	Wax, sealing	1.8	112
Anorthite	2.74–2.76	171–172	Gypsum	2.31–2.33	144–145	Wood (seasoned)		
Asbestos	2.0–2.8	125–175	Hematite	4.9–5.3	306–330	Alder	0.42–0.68	26–42
Asbestos slate	1.8	112	Hornblende	3.0	187	Apple	0.66–0.84	41–52
Asphalt	1.1–1.5	69–94	Ice	0.917	57.2	Ash	0.65–0.85	40–53
Basalt	2.4–3.1	150–190	Ivory	1.83–1.92	114–120	Balsa	0.11–0.14	7–9
Beeswax	0.96–0.97	60–61	Leather, dry	0.86	54	Bamboo	0.31–0.40	19–25
Beryl	2.69–2.7	168–169	Lime, slaked	1.3–1.4	81–87	Basswood	0.32–0.59	20–37
Biotite	2.7–3.1	170–190	Limestone	2.68–2.76	167–171	Beech	0.70–0.90	32–56
Bone	1.7–2.0	106–125	Linoleum	1.18	74	Birch	0.51–0.77	32–48
Brick	1.4–2.2	87–137	Magnetite	4.9–5.2	306–324	Blue gum	1.00	62
Butter	0.86–0.87	53–54	Malachite	3.7–4.1	231–256	Box	0.95–1.16	59–72
Calamine	4.1–4.5	255–280	Marble	2.6–2.84	160–177	Butternut	0.38	24
Calcspar	2.6–2.8	162–175	Meerschaum	0.99–1.28	62–80	Cedar	0.49–0.57	30–35
Camphor	0.99	62	Mica	2.6–3.2	165–200	Cherry	0.70–0.90	43–56
Caoutchouc	0.92–0.99	57–62	Muscovite	2.76–3.00	172–187	Dogwood	0.76	47
Cardboard	0.69	43	Ochre	3.5	218	Ebony	1.11–1.33	69–83
Celluloid	1.4	87	Opal	2.2	137	Elm	0.54–0.60	34–37
Cement, set	2.7–3.0	170–190	Paper	0.7–1.15	44–72	Hickory	0.60–0.93	37–58
Chalk	1.9–2.8	118–175	Paraffin	0.87–0.91	54–57	Holly	0.76	47
Charcoal			Peat blocks	0.84	52	Juniper	0.56	35
Oak	0.57	35	Pitch	1.07	67	Larch	0.50–0.56	31–35
Pine	0.28–0.44	18–28	Porcelain	2.3–2.5	143–156	Lignum vitae	1.17–1.33	73–83
Cinnabar	8.12	507	Porphyry	2.6–2.9	162–181	Locust	0.67–0.71	42–44
Clay	1.8–2.6	112–162	Pressed wood			Logwood	0.91	57
Coal			pulp board	0.19	12	Mahogany		
Anthracite	1.4–1.8	87–112	Pyrite	4.95–5.1	309–318	Honduras	0.66	41
Bituminous	1.2–1.5	75–94	Quartz	2.65	165	Spanish	0.85	53
Cocoa butter	0.89–0.91	56–57	Resin	1.07	67	Maple	0.62–0.75	39–47
Coke	1.0–1.7	62–105	Rock salt	2.18	136	Oak	0.60–0.90	37–56
Copal	1.04–1.14	65–71	Rubber, hard	1.19	74	Pear	0.61–0.73	38–45
Cork	0.22–0.26	14–16	Rubber, soft			Pine		
Cork linoleum	0.54	34	Commercial	1.1	69	Pitch	0.83–0.85	52–53
Corundum	3.9–4.0	245–250	Pure gum	0.91–0.93	57–58	White	0.35–0.50	22–31
Diamond	3.01–3.52	188–220	Sandstone	2.14–2.36	134–147	Yellow	0.37–0.60	23–37
Dolomite	2.84	177	Serpentine	2.50–2.65	156–165	Plum	0.66–0.78	41–49
Ebonite	1.15	72	Silica			Poplar	0.35–0.5	22–31
Emery	4.0	250	Fused trans-			Satinwood	0.95	59
Epidote	3.25–3.50	203–218	parent	2.21	138	Spruce	0.48–0.70	30–44
Feldspar	2.55–2.75	159–172	Translucent	2.07	129	Sycamore	0.40–0.60	24–37
Flint	2.63	164	Slag	2.0–3.9	125–240	Teak		
Fluorite	3.18	198	Slate	2.6–3.3	162–205	Indian	0.66–0.88	41–55
Galena	7.3–7.6	460–470	Soapstone	2.6–2.8	162–175	African	0.98	61
Gamboge	1.2	75	Spermaceti	0.95	59	Walnut	0.64–0.70	40–43
Garnet	3.15–4.3	197–268	Starch	1.53	95	Water gum	1.00	62
Gas carbon	1.88	117	Sugar	1.59	99	Willow	0.40–0.60	24–37
Gelatin	1.27	79	Talc	2.7–2.8	168–174			

[a] Some values reported as low as 1.6

Source: Based largely on "Smithsonian Physical Tables", 9th rev. ed., W.E. Forsythe, Ed., The Smithsonian Institute, 1956, p. 292.

Note: In the case of substances with voids, such as paper or leather, the bulk density is indicated rather than the density of the solid portion. For density in kg/m³, multiply values in g/cm³ by 1,000.

TABLE C.3 Thermal Properties of Pure Metals—Metric Units

Metal	Melting point, °C	Boiling point, °C	Latent heat of fusion, cal/g**	AT ATMOSPHERIC PRESSURE — At 100°K Thermal conductivity, watts/cm °C	At 100°K Specific heat, cal/g °C**	At 25°C (77°F) Specific heat, cal/g °C**	At 25°C Coeff. of linear expansion (×10^6)(°C)^{-1}	At 25°C Thermal conductivity, watts/cm °C	Specific heat (liquid) at 2000°K, cal/g °C**	LIQUID METAL Vapor pressure 10^{-3} atm (Boiling point temperatures, °K)	10^{-6} atm	10^{-9} atm
Aluminum	660.0	2441.0	95	3.00*	0.115	0.215	25	2.37	0.26	1,782	1,333	1,063
Antimony	630.0	1440.0	38.5	—	0.040	0.050	9	0.185	0.062	1,007	741	612
Beryllium	1285.0	2475.0	324.	—	0.049	0.436	12	2.18	0.78	1,793	1,347	1,085
Bismuth	271.4	1660.0	12.4	—	0.026	0.030	13	0.084	0.036	1,155	851	677
Cadmium	321.0	767.0	13.2	1.03	0.047	0.055	30	0.93	0.063	655	486	388
Chromium	1860.0	2670.0	79	1.58	0.046	0.110	6	0.91	0.224	1,992	1,530	1,247
Cobalt	1495.0	2925.0	66	—	0.057	0.10	12	0.69	0.164	2,167	1,652	1,345
Copper	1084.0	2575.0	49	4.83*	0.061	0.092	16.6	3.98	0.118	1,862	1,391	1,120
Gold	1063.0	2800.0	15	3.45*	0.026	0.031	14.2	3.15	0.0355	2,023	1,510	1,211
Iridium	2450.0	4390.0	33	—	0.022	0.031	6	1.47	0.0434	3,253	2,515	2,062
Iron	1536.0	2870.0	65	1.32*	0.052	0.108	12	0.803	0.197	2,093	1,594	1,297
Lead	327.5	1750.0	5.5	0.396	0.028	0.031	29	0.346	0.033	1,230	889	698
Magnesium	650.0	1090.0	88.0	1.69	0.016	0.243	25	1.59	0.32	857	638	509
Manganese	1244.0	2060.0	64	—	0.064	0.114	22	—	0.20	1,495	1,131	913
Mercury	−38.86	356.55	2.7	—	—	0.033	—	0.0839	—	393	287	227
Molybdenum	2620.0	4651.0	69	1.79	0.029	0.060	5	1.4	0.089	3,344	2,558	2,079
Nickel	1453.0	2800.0	71	1.58	0.033	0.106	13	0.899	0.175	2,156	1,646	1,343
Niobium (Columbium)	2470.0	4740.0	68	0.552	0.055	0.064	7	0.52	0.083	3,523	2,721	2,232
Osmium	3025.0	4225.0	34	—	0	0.031	5	0.61	0.039	—	—	—
Platinum	1770.0	3825.0	24	0.79*	0.024	0.032	9	0.73	0.043	2,817	2,155	1,757
Plutonium	640.0	3230.0	3	—	0.019	0.032	54	0.08	0.041	2,200	1,596	1,252
Potassium	63.3	760.0	14.5	—	0.150	0.180	83	0.99	—	606	430	335
Rhodium	1965.0	3700.0	50	—	—	0.058	8	1.50	0.092	—	—	—
Selenium	217.0	700.0	16	—	—	0.077	37	0.005	—	—	—	—
Silicon	1411.0	3280.0	430	—	0.062	0.17	3	0.835	0.217	2,340	1,749	1,427
Silver	961.0	2212.0	26.5	4.50*	0.045	0.057	19	4.27	0.068	1,582	1,179	952
Sodium	97.83	884.0	27	—	0.234	0.293	70	1.34	—	701	504	394
Tantalum	2980.0	5365.0	41	0.592	0.026	0.034	6.5	0.54	0.040	3,959	3,052	2,495
Thorium	1750.0	4800.0	17	—	0.024	0.03	12	0.41	0.047	3,251	2,407	1,919
Tin	232.0	2600.0	14.1	0.85	0.039	0.054	20	0.64	0.058	1,857	1,366	1,080
Titanium	1670.0	3290.0	100	0.312	0.072	0.125	8.5	0.2	0.188	2,405	1,827	1,484
Tungsten	3400.0	5550.0	46	2.35*	0.021	0.032	4.5	1.78	0.040	4,139	3,228	2,656
Uranium	1132.0	4140.0	12	—	0.022	0.028	13.4	0.25	0.048	2,861	2,128	1,699
Vanadium	1900.0	3400.0	98	—	0.061	0.116	8	0.60	0.207	2,525	1,948	1,591
Zinc	419.5	910.0	27	1.32	0.063	0.093	35	1.15	—	752	559	449

* Temperatures of maximum thermal conductivity (conductivity values in watts/cm °C): Aluminum 13°K, cond. = 71.5; copper 10°K, cond. = 196; gold 10°K, cond. = 28.2; iron 20°K, cond. = 9.97; platinum 8°K, cond. = 12.9; silver 7°K, cond. = 193; tungsten 8°K, cond. = 85.3.

** To convert to SI units note that 1 cal = 4.186 J.

TABLE C.4   Miscellaneous Properties of Metals and Alloys

Part a. Pure Metals

At Room Temperature

Common name	PROPERTIES (TYPICAL ONLY)						
	Thermal conductivity, Btu/hr ft °F	Specific gravity	Coeff. of linear expansion, μ in./ in. °F	Electrical resistivity, microhm-cm	Poisson's ratio	Modulus of elasticity, millions of psi	Approximate melting point, °F
Aluminum	137	2.70	14	2.655	0.33	10.0	1220
Antimony	10.7	6.69	5	41.8		11.3	1170
Beryllium	126	1.85	6.7	4.0	0.024–.030	42	2345
Bismuth	4.9	9.75	7.2	115		4.6	521
Cadmium	54	8.65	17	7.4		8	610
Chromium	52	7.2	3.3	13		36	3380
Cobalt	40	8.9	6.7	9		30	2723
Copper	230	8.96	9.2	1.673	0.36	17	1983
Gold	182	19.32	7.9	2.35	0.42	10.8	1945
Iridium	85.0	22.42	3.3	5.3		75	4440
Iron	46.4	7.87	6.7	9.7		28.5	2797
Lead	20.0	11.35	16	20.6	0.40–.45	2.0	621
Magnesium	91.9	1.74	14	4.45	0.35	6.4	1200
Manganese		7.21–7.44	12	185		23	2271
Mercury	4.85	13.546		98.4			−38
Molybdenum	81	10.22	3.0	5.2	0.32	40	4750
Nickel	52.0	8.90	7.4	6.85	0.31	31	2647
Niobium (Columbium)	30	8.57	3.9	13		15	4473
Osmium	35	22.57	2.8	9		80	5477
Platinum	42	21.45	5	10.5	0.39	21.3	3220
Plutonium	4.6	19.84	30	141.4	0.15–.21	14	1180
Potassium	57.8	0.86	46	7.01			146
Rhodium	86.7	12.41	4.4	4.6		42	3569
Selenium	0.3	4.8	21	12.0		8.4	423
Silicon	48.3	2.33	2.8	1×10^5		16	2572
Silver	247	10.50	11	1.59	0.37	10.5	1760
Sodium	77.5	0.97	39	4.2			208
Tantalum	31	16.6	3.6	12.4	0.35	27	5400
Thorium	24	11.7	6.7	18	0.27	8.5	3180
Tin	37	7.31	11	11.0	0.33	6	450
Titanium	12	4.54	4.7	43	0.3	16	3040
Tungsten	103	19.3	2.5	5.65	0.28	50	6150
Uranium	14	18.8	7.4	30	0.21	24	2070
Vanadium	35	6.1	4.4	25		19	3450
Zinc	66.5	7	19	5.92	0.25	12	787

Appendix D Gases and Vapors

TABLE D.1 SI Units — Definitions, Abbreviations and Prefixes

BASIC UNITS—MKS

Length	meter	m	Electric current	ampere	A
Mass	kilogram	kg	Thermodynamic temperature	kelvin	K
Time	second	s	Luminous intensity	candela	cd

DERIVED UNITS

Property	Units†	Abbreviations and dimensions	
Acceleration	meter per second squared	m/s^2	
Activity (of radioactive source)	1 per second	s^{-1}	
Angular acceleration	radian per second squared	rad/s^{-1}	
Angular velocity	radian per second	rad/s	
Area	square meter	m^2	
Density	kilogram per cubic meter	kg/m^3	
Dynamic viscosity	newton-second per sq meter	$N{\cdot}s/m^2$	
Electric capacitance	farad	F	$(A{\cdot}s/V)$
Electric charge	coulomb	C	$(A{\cdot}s)$
Electric field strength	volt per meter	V/m	
Electric resistance	ohm		(V/A)
Entropy	joule per kelvin	J/K	
Force	newton	N	$(kg{\cdot}m/s^2)$
Frequency	hertz	hz	(s^{-1})
Illumination	lux	lx	(lm/m^2)
Inductance	henry	H	$(V{\cdot}s/A)$
Kinematic viscosity	sq meter per second	m^2/s	
Luminance	candela per sq meter	cd/m^2	
Luminous flux	lumen	lm	$(cd{\cdot}sr)$
Magnetomotive force	ampere	A	
Magnetic field strength	ampere per meter	A/m	
Magnetic flux	weber	Wb	$(V{\cdot}s)$
Magnetic flux density	tesla	T	(Wb/m^2)
Power	watt	W	(J/s)
Pressure	newton per square meter	N/m^2	
Radiant intensity	watt per steradian	W/sr	
Specific heat	joule per kilogram kelvin	$J/kg\ K$	
Thermal conductivity	watt per meter kelvin	$W/m\ K$	
Velocity	meter per second	m/s	
Volume	cubic meter	m^3	
Voltage, potential difference, electromotive force	volt	V	(W/A)
Wave number	1 per meter	m^{-1}	
Work, energy, quantity of heat	joule	J	$(N{\cdot}m)$

PREFIX NAMES OF MULTIPLES AND SUBMULTIPLES OF UNITS

Decimal equivalent	Prefix	Pronun-ciation	Symbol	Exponential expression
1,000,000,000,000	tera	tĕr′å	T	10^{+12}
1,000,000,000	giga	jĭ′gå	G	10^{+9}
1,000,000	mega	mĕg′å	M	10^{+6}
1,000	kilo	kĭl′ō	k	10^{+3}
100	hecto	hĕk′tō	h	10^{+2}
10	deka	dĕk′å	da	10
0.1	deci	dĕs′ĭ	d	10^{-1}
0.01	centi	sĕn′tĭ	c	10^{-2}
0.001	milli	mĭl′ĭ	m	10^{-3}
0.000 001	micro	mī′krō	μ	10^{-6}
0.000 000 001	nano	năn′ō	n	10^{-9}
0.000 000 000 001	pico	pē′kō	p	10^{-12}
0.000 000 000 000 001	femto	fĕm′tō	f	10^{-15}
0.000 000 000 000 000 001	atto	ăt′tō	a	10^{-18}

Appendix E Composition and Heating Values of Common Fuels

TABLE E.1 Properties of Typical Gaseous and Liquid Commercial Fuels

Gaseous fuels	Composition, percent by volume								Mol wt of fuel	Theor. air/fuel ratio by wt	Higher heating value, Btu/lb$_m$	Density, lb$_m$/ft^3
	H_2	N_2	O_2	CH_4	CO	CO_2	C_2H_4	C_6H_6				
Blast furnace gas	1.0	60.0	—	—	27.5	11.5	—	—	29.6	0.667	1,170	.075 5 [a]
Blue water gas	47.3	8.3	0.7	1.3	37.0	5.4	—	—	16.4	3.759	6,550	.042 2 [a]
Carb. water gas	40.5	2.9	0.5	10.2	34.0	3.0	6.1	2.8	18.3	7.299	11,350	.046 6 [a]
Coal gas	54.5	4.4	0.2	24.2	10.9	3.0	1.5	1.3	12.1	10.87	16,500	.031 1 [a]
Coke-oven gas	46.5	8.1	0.8	32.1	6.3	2.2	3.5	0.5	13.7	17.24	17,000	.032 6 [a]
Natural gas (15.8% C_2H_6)	—	0.8	—	83.4	—	—	—	—	18.3	17.24	24,100	.045 1 [a]
Producer gas	14.0	50.9	0.6	3.0	27.0	4.5	—	—	24.7	14.29	2,470	.063 6 [a]

Liquid commercial fuels	Vapor		Gravity, API, 60°F	Distillation			Flash point, °F	Viscosity, centi- stokes, 100°F	Mol wt of fuel	Theor. air/fuel ratio by wt	Higher heating value, Btu/lb$_m$	Density, lb$_m$/ft^3
	c_p, 60°F	c_p/c_v, 60°F		10%, °F	90%, °F	End point, °F						
	(approximately)											
Gasoline	0.4	1.05	63	121	320	397	0	—	113	14.93	20,460	43.8 [b]
Gasoline	0.4	1.05	63	118	330	410	0	—	126 [c]	14.97	20,260	46.1 [b]
Kerosene	0.4	1.05	41.9	370	510	546	130	—	154 [c]	14.99	19,750	51.5 [b]
Diesel oil (1-D)	0.4	1.05	42	—	550	—	100	1.4–2.5	170	15.02	19,240	54.6 [b]
Diesel oil (2-D)	0.4	1.05	36	—	—	—	125	2.0–5.8	184	15.06	19,110	57.4 [b]
Diesel oil (4-D)	0.4	1.05	—	—	540–576	—	130	5.8–26.4	198	14.93	18,830	59.9 [b]

[a] Based on dry air at 25°C and 760 mm Hg.

[b] Based on H_2O at 60°F, 1 atm (ρ = 62.367 lb$_m$/ft^3).

[c] Estimated.

Source: Abridged from *Engineering Experimentation*, G.L. Tuve and L.C. Domholdt, McGraw-Hill Book Company, 1966; and *The Internal Combustion Engine*, 2nd ed., C.F. Taylor and E.S. Taylor, Textbook Co., 1961. With permission.

Note: For heating value in J/kg, multiply the value in Btu/lb$_m$ by 2324. For density in kg/m^3, multiply the value in lb/ft^3 by 16.02.

TABLE E.2 Combustion Data for Hydrocarbons

Hydrocarbon	Formula	Higher heating value (vapor), Btu/lb_m	Theor. air/fuel ratio, by mass	Max flame speed, ft/sec	Adiabatic flame temp (in air), °F	Ignition temp (in air), °F	Flash point, °F	Flammability limits (in air), % by volume	
PARAFFINS OR ALKANES									
Methane	CH_4	23875	17.195	1.1	3484	1301	gas	5.0	15.0
Ethane	C_2H_6	22323	15.899	1.3	3540	968–1166	gas	3.0	12.5
Propane	C_3H_8	21669	15.246	1.3	3573	871	gas	2.1	10.1
n-Butane	C_4H_{10}	21321	14.984	1.2	3583	761	−76	1.86	8.41
iso-Butane	C_4H_{10}	21271	14.984	1.2	3583	864	−117	1.80	8.44
n-Pentane	C_5H_{12}	21095	15.323	1.3	4050	588	< −40	1.40	7.80
iso-Pentane	C_5H_{12}	21047	15.323	1.2	4055	788	< −60	1.32	9.16
Neopentane	C_5H_{12}	20978	15.323	1.1	4060	842	gas	1.38	7.22
n-Hexane	C_6H_{14}	20966	15.238	1.3	4030	478	−7	1.25	7.0
Neohexane	C_6H_{14}	20931	15.238	1.2	4055	797	−54	1.19	7.58
n-Heptane	C_7H_{16}	20854	15.141	1.3	3985	433	25	1.00	6.00
Triptane	C_7H_{16}	20824	15.141	1.2	4035	849	—	1.08	6.69
n-Octane	C_8H_{18}	20796	15.093	—	—	428	56	0.95	3.20
iso-Octane	C_8H_{18}	20770	15.093	1.1	—	837	10	0.79	5.94
OLEFINS OR ALKENES									
Ethylene	C_2H_4	21636	14.807	2.2	4250	914	gas	2.75	28.6
Propylene	C_3H_6	21048	14.807	1.4	4090	856	gas	2.00	11.1
Butylene	C_4H_8	20854	14.807	1.4	4030	829	gas	1.98	9.65
iso-Butene	C_4H_8	20737	14.807	1.2	—	869	gas	1.8	9.0
n-Pentene	C_5H_{10}	20720	14.807	1.4	4165	569	—	1.65	7.70
AROMATICS									
Benzene	C_6H_6	18184	13.297	1.3	4110	1044	12	1.35	6.65
Toluene	C_7H_8	18501	13.503	1.2	4050	997	40	1.27	6.75
p-Xylene	C_8H_{10}	18663	13.663	—	4010	867	63	1.00	6.00
OTHER HYDROCARBONS									
Acetylene	C_2H_2	21502	13.297	4.6	4770	763–824	gas	2.50	81.0
Naphthalene	$C_{10}H_8$	17303	12.932	—	4100	959	174	0.90	5.9

Source: Based largely on *Gas Engineers' Handbook* , American Gas Association, Inc., Industrial Press, 1967.

Notes: For heating value in J/kg, multiply the value in Btu/lb_m by 2324. For flame speed in m/s, multiply the value in ft/s by 0.3048.

The higher heating value is obtained when all of the water formed by combustion is condensed to a liquid. The lower heating value is obtained when all of the water formed by combustion is a vapor. Table E.3 shows some example values. For other fuels, subtract from the HHV the heat of vaporization of water at standard conditions (e.g., ~1050 Btu/lb@ 77°F) multiplied by the ratio of the number of pounds (kg) of water produced per pound of methane burned. Therefore, the difference between higher and lower HVs if 2.25 x 1050 = 2363 Btu/lb or 5486 $\frac{KJ}{Kg}$.

REFERENCES

American Institute of Physics Handbook, 2nd ed., D.E. Gray, Ed., McGraw-Hill Book Company, 1963.

Chemical Engineers' Handbook, 4th ed., R.H. Perry, C.H. Chilton, and S.D. Kirkpatrick, Eds., McGraw-Hill Book Company, 1963.

Handbook of Chemistry and Physics, 53rd ed., R.C. Weast, Ed., The Chemical Rubber Company, 1972; gives the heat of combustion of 500 organic compounds.

Handbook of Laboratory Safety, 2nd ed., N.V. Steere, Ed., The Chemical Rubber Company, 1971.

Physical Measurements in Gas Dynamics and Combustion, Princeton University Press, 1954.

Table E.3 Heating Values in kJ/kg of Selected Hydrocarbons at 25°C

Hydrocarbon	Formula	Higher Value[a]		Lower Value[b]	
		Liquid Fuel	Gas. Fuel	Liquid Fuel	Gas. Fuel
Methane	CH_4	—	55,496	—	50,010
Ethane	C_2H_6	—	51,875	—	47,484
Propane	C_3H_8	49,973	50,343	45,982	46,352
n-Butane	C_4H_{10}	49,130	49,500	45,344	45,714
n-Octane	C_8H_{18}	47,893	48,256	44,425	44,788
n-Dodecane	$C_{12}H_{26}$	47,470	47,828	44,109	44,467
Methanol	CH_3OH	22,657	23,840	19,910	21,093
Ethanol	C_3H_5OH	29,676	30,596	26,811	27,731

a H_2O liquid in the products.
b H_2O vapor in the products.

Index

D

I